ASTRONOMICAL TABLES OF THE SUN, MOON AND PLANETS

SECOND EDITION

JEAN MEEUS

FOREWORD BY ROGER SINNOTT

Published by:

Willmann-Bell, Inc.

**Publishers and Booksellers Serving
Astronomers Worldwide Since 1972**

P.O. Box 35025 • Richmond, Virginia 23235

Published by Willmann–Bell, Inc.
P.O. Box 35025, Richmond, Virginia 23235

Printed in the United States of America

Library of Congress Cataloging-in-Publication Data.
Meeus, Jean.
    Astronomical tables of the sun, moon and planets / Jean Meeus. –
2nd ed.
       p.   cm.
    Includes bibliographical references and index.
    ISBN 0-943396-45-X
    1. Ephemerides.   I. Title.
    QB12.M44   1995
    528–dc20   95-3657
                                             CIP

95 96 97 98 00 01 9 8 7 6 5 4 3 2 1

Foreword

To find out the Moon's phase or which bright planets are up, most people depend on TV weather forecasters or the night-sky columnist of the local newspaper. Such commentators, in turn, rely on astronomy magazines and almanacs. But these publications tend to focus on the here and now—what's going to happen tonight, next month, or later in the current year—with little regard for long-term trends.

To understand the sky better, amateur astronomers often expand their horizons with computer programs that plot sky objects for any date and time in the past or future. The better software packages are reasonably trustworthy over a span of many centuries. But there's a catch. To look for a specific type of planetary grouping, a transit of Venus across the Sun's disk, or the best possible observing conditions for Mars, you need to know the approximate date of the event.

In fact, some of the most obvious astronomical questions are not about what happened (or will happen) on a certain date, but just the reverse. When was the last time Venus appeared so close to Jupiter? In what year will Saturn's rings be edgewise again? When can I see the Moon occult Antares from my backyard? Planetarium-type programs are a fun way to attack such puzzles, but you could find yourself staring at the video monitor while the machine cycles through the days, months, and years. To discover an event's date more efficiently calls for astronomical insight, special programming tricks, and a thorough grasp of the clockwork of the heavens—or, better yet, this book!

For example, I get frequent phone calls from calendar makers. They need to know the exact dates of lunar phases and the start of spring, summer, fall, and winter several years in advance. No doubt someone has said, "Just call *Sky & Telescope*." What these people don't know is where I always turn for the information they seek. The first edition of Jean Meeus's *Astronomical Tables* appeared in 1983, and it is one of the few books I've kept within arm's length at my desk ever since. In return for supplying this arcane information to the callers, I've also acquired lots of free calendars.

Early in 1994 Meeus wrote me that his first project upon retirement would be to recalculate everything for an update of this book. Months later, when I saw the page proofs, I grabbed my dog-eared 1983 edition to compare the calendrical data I had trusted so long. The newly calculated times of the lunar phases are indeed different—by typically 1 to 3 seconds! The starts of the seasons starts have been refined too, by less than a half minute for recent times but by 10 minutes or so during the era of the Roman empire. Thus I gained a feel, not only of the subtle improvements made recently in astronomers' knowledge of solar-system dynamics, but also of the great attention to detail that has earned Meeus worldwide renown in calculations of this nature. At the same time, he has extended most tables of the book at least 15 years further into the future.

Who, then, will use this book? To many skywatchers it is an observing guide for the better part of a lifetime. All the most spectacular planetary conjunctions are listed here. The dates of greatest elongation of Mercury and Venus from the Sun tell when these two inner planets are most readily viewed with a telescope. Down through years, observers have confirmed a curious finding by Johann Schröter in the late 18th century, that Venus and Mercury appear just half illuminated (at "dichotomy"') several days before an evening elongation and after a morning one, rather than at the elongation itself. For an outer planet like Mars, the opposition date is very close to the time it presents the largest possible disk for telescopic study. Oppositions are also good times for spotting the brighter asteroids in binoculars. Those who watch Jupiter's moons dip in and out of eclipse will enjoy page 58 of this book, listing the intervals when Callisto avoids the giant planet's shadow entirely. At the endpoints of these intervals, rare partial eclipses may be seen.

A major section (pages 217–346) deals with lunar occultations, those spectacular moments when the Moon hides and later uncovers a bright star or planet. No ordinary almanac is much help with the exact times to look for these events because they depend so much on the observer's geographic location. The star may wink out in Montreal almost an hour after it does in Denver, or the Moon may miss the star completely. Commercial software can be a poor guide too, as I discovered a few days before a recent occultation of Spica. A flurry of messages on the CompuServe computer network warned that various programs differed in their predictions by three to five minutes, even for a specific latitude and longitude. Later, when some actual timings were posted, I confirmed that my secret source—the Besselian elements in this book's first edition—yielded times that were correct to within 2 seconds. Meeus has now extended the tables of occultation elements 20 years further, including on page 234 a Basic program to aid with the calculations.

Even armchair astronomers will find much of interest here. The 11-year sunspot cycle, tabulated in such detail on pages 347–382, is more than a numerical curiosity. Years with high sunspot counts are frequently the years when spectacular aurorae burst upon the night sky and shortwave communications undergo brief disruptions. Years with the fewest sunspots produce solar eclipses with the longest coronal streamers. Many people have tried to find correlations between sunspots and the weather. Some have even wondered if earthquakes and volcanic eruptions are more likely during perigees of the Moon, when the tidal stresses within our planet are largest. The data in this book can play a vital but impartial role in queries of this sort, providing a sound basis for rejecting the wilder speculations.

Anyone curious how this book's calculations were done should consult the same author's *Astronomical Algorithms* (Willmann-Bell, 1991). Computer programmers are having a bonanza with that impressive work, but for countless other astronomy buffs it is equally rewarding to ponder the outcome of such calculations, compiled by a supreme master of the art, in the book you now hold.

Roger W. Sinnott, *Sky & Telescope* magazine

Preface to the Second Edition

The present work is a completely rewritten version of the first edition of our *Astronomical Tables* (Willmann-Bell, 1983). We have now made use of the modern analytical theories constructed at the Bureau des Longitudes, Paris: the ELP-2000 lunar theory by Chapront, and the VSOP87 planetary theory by Bretagnon.

The tables which were given to the year 2005 in the first edition have now been extended to A.D. 2020.

The elements for the occultations of the planets and the first-magnitude stars, given for the years 1980 − 2000 in the first edition, are now provided for the years 1990 − 2020.

Some new tables are given, such as the oppositions of bright minor planets, or data about the Jewish and Moslem calendars. And, of course, the part on the solar activity has been brought up to date.

Jean Meeus

Table of Contents

Note on Time Reckoning

Calendar Dates

For the calendar dates, the Julian Calendar is used till 1582 October 4, while the Gregorian Calendar is used from 1582 October 15 onwards.

In general, we have written the dates as is usual in Astronomy, namely in the order year – month – day of month. For the names of the months, three-letter abbreviations are used. For instance, 1994 Aug 6 means the 6th day of August, 1994. This 'scientific' form for calendar dates, which reads from the largest to the smallest unit of time, contrasts with the common 'American' form (August 6, 1994), and with the 'European' form (6 August 1994). Anyway, it is recommended to spell out the month, because one person's 8/6/94 is another's 6/8/94.

The dates are based either on the Dynamical Time, or on the Universal Time (see below). In Part 1, Universal Time is used. For example, on page 18 we read that on 1993 March 9 Mercury was in inferior conjunction with the Sun; this means that this conjunction took place between 0^h00^m and 24^h00^m Universal Time on 1993 March 9.

Dynamical Time and Universal Time

The Universal Time (UT), or Greenwich Civil Time, is based on the rotation of the Earth. The UT is necessary for civil life and for the astronomical calculations where local hour angles are involved.

However, the Earth's rotation is generally slowing down and, moreover, this occurs with unpredictable irregularities. For this reason, the UT is not a uniform time.

But the astronomers need a uniform time scale for their accurate calculations (celestial mechanics, orbits, ephemerides). From 1960 to 1983, in the great astronomical almanacs such as the *Astronomical Ephemeris*, use was made of the uniform time scale called the *Ephemeris Time* (ET) which was defined by the laws of dynamics and was based on the planetary motions. In 1984, the ET was replaced by the *Dynamical Time*, which is defined by atomic clocks. In practice, the Dynamical Time is a prolongation of the Ephemeris Time.

ASTRONOMICAL TABLES

TABLE A

$\Delta T = \text{TD} - \text{UT}$ (in seconds) for the beginning of some years

Subtract the value given by this Table in order to convert TD to UT

Year	ΔT	Year	ΔT	Year	ΔT	Year	ΔT	Year	ΔT
1620	+124	1690	+10	1800	+13.7	1870	+ 1.6	1940	+24.3
1622	115	1692	9	1802	13.1	1872	− 1.0	1942	25.3
1624	106	1696	9	1804	12.7	1874	− 2.7	1944	26.2
1626	98	1700	9	1806	12.5	1876	− 3.6	1946	27.3
1628	91	1706	9	1808	12.5	1878	− 4.7	1948	28.2
1630	+85	1708	+10	1810	+12.5	1880	− 5.4	1950	+29.1
1632	79	1712	10	1812	12.5	1882	− 5.2	1952	30.0
1634	74	1716	10	1814	12.5	1884	− 5.5	1954	30.7
1636	70	1718	11	1816	12.5	1886	− 5.6	1956	31.4
1638	65	1722	11	1818	12.3	1888	− 5.8	1958	32.2
1640	+62	1727	+11	1820	+12.0	1890	− 5.9	1960	+33.1
1642	58	1732	11	1822	11.4	1892	− 6.2	1962	34.0
1644	55	1734	12	1824	10.6	1894	− 6.4	1964	35.0
1646	53	1738	12	1826	9.6	1896	− 6.1	1966	36.5
1648	50	1742	12	1828	8.6	1898	− 4.7	1968	38.3
1650	+48	1744	+13	1830	+ 7.5	1900	− 2.7	1970	+40.2
1652	46	1750	13	1832	6.6	1902	− 0.0	1972	42.2
1654	44	1752	14	1834	6.0	1904	+ 2.6	1974	44.5
1656	42	1756	14	1836	5.7	1906	5.4	1976	46.5
1658	40	1758	15	1838	5.6	1908	7.7	1978	48.5
1660	+37	1764	+15	1840	+ 5.7	1910	+10.5	1980	+50.5
1662	35	1766	16	1842	5.9	1912	13.4	1982	52.2
1664	33	1770	16	1844	6.2	1914	16.0	1984	53.8
1666	31	1774	16	1846	6.5	1916	18.2	1986	54.9
1668	28	1776	17	1848	6.8	1918	20.2	1988	55.8
1670	+26	1780	+17	1850	+ 7.1	1920	+21.2	1990	+56.9
1672	24	1782	17	1852	7.3	1922	22.4	1992	58.3
1674	22	1784	17	1854	7.5	1924	23.5	1994	60.0
1676	20	1786	17	1856	7.7	1926	23.9		
1678	18	1788	17	1858	7.8	1928	24.3		
1680	+16	1790	+17	1860	+ 7.9	1930	+24.0		
1682	14	1792	16	1862	7.5	1932	23.9		
1684	13	1794	16	1864	6.4	1934	23.9		
1686	12	1796	15	1866	5.4	1936	23.7		
1688	11	1798	14	1868	2.9	1938	24.0		

TABLE B

Approximate values of ΔT

Year	Minutes	Year	Minutes	Year	Minutes
0	177	1000	35	2075	4
100	158	1100	27	2200	8
200	140	1200	20	2300	13
300	123	1300	14	2400	19
400	107	1400	9	2500	26
500	93	1500	5	2600	34
600	79	1600	2	2700	43
700	66	1700	0	2800	53
800	55	1800	0	2900	64
900	45	1980	1	3000	76

One distinguishes a Barycentric Dynamical Time (TDB) and a Terrestrial Dynamical Time (TDT). These times differ by at most 0.0017 second, the difference being related to the motion of the Earth on its elliptical orbit around the Sun (relativistic effect). Because this very small difference can be neglected for most practical purposes, we will not make the distinction between TDB and TDT, and we will name both simply TD (Dynamical Time).

The exact value of the difference $\Delta T = \text{TD} - \text{UT}$ can be deduced only from observations. Table A gives the value of ΔT for the *beginning* of some years. Except for the four last values, they are taken from the *Astronomical Almanac* for 1988 (Washington, D.C.), pages K8 and K9.

For the year 1996, the value $\Delta T = +62$ seconds can be used. For the next years, increase ΔT by $+1$ second per year, in order to obtain an extrapolated provisional value.

For epochs outside the years $1620 - 2020$, an *approximate* value of ΔT, in seconds, can be deduced from the following relation due to Morrison and Stephenson:

$$\Delta T = -15 + 0.00325 \, (\text{year} - 1810)^2$$

This leads to Table B. It should be noted that for the very past and future, the tabulated values of ΔT may be in error by many minutes, as the fluctuations due to the variable rotation of the Earth are unknown for those remote epochs.

Tabulated times in TD may be converted into Universal Time by *subtracting* from them the quantity ΔT, since we have $\text{UT} = \text{TD} - \Delta T$. In Part 1 of this book, all times have already been converted from TD to UT.

Example 1. — On page 184 we read that First Quarter took place on 1980 May 21, at $19^h16^m48^s$ Dynamical Time. In 1980, ΔT was equal to $+51$ seconds. Consequently, that lunar phase occurred at $19^h15^m57^s$ Universal Time.

Example 2. — On page 150, we find that in A.D. 1999 the March equinox will take place on March 21 at $1^h46^m53^s$ TD. According to the rule given above, we expect that in 1999 the quantity ΔT will be equal to $+65$ seconds. Consequently, in 1999 the spring equinox will occur on March 21 at $1^h46^m53^s - 65^s = 1^h45^m48^s$ UT, which may be rounded to 1^h46^m UT. For other time zones, this reduces for example to:

> March 21 at 2^h46^m Central European Time,
> March 20 at 20^h46^m Eastern Standard Time,
> March 20 at 17^h46^m Pacific Standard Time.

Example 3. — On page 65, we find that in A.D. 133 the opposition of Mars took place on January 17 at 0^h18^m TD.

In A.D. 133, the quantity ΔT was equal to approximately $+152$ minutes, or 2 hours and 32 minutes. Consequently, in the year 133 Mars' opposition occurred on January 16 at 21^h46^m Universal Time.

1

Planetary Phenomena

1990 − 2020

Contents

General Remarks

Notwithstanding its title, the present part contains some data for years outside the period 1990 − 2020.

Most data in the present part have been deduced by the author from accurate ephemerides calculated using the VSOP87 planetary theory by P. Bretagnon, of the Bureau des Longitudes, Paris. The data for Pluto are deduced from a numerical integration, based on the osculating elements by P. K. Seidelmann *et al.*, in *Icarus*, Vol. 44, page 20 (1980).

All times are expressed in *Universal Time*, or Greenwich Civil Time. All dates, too, are based on the UT, even for years before 1925 when the Greenwich *Mean* Time (that is, measured from Greenwich Mean Noon) was in use in the astronomical ephemerides. Thus, when we write that Neptune reached the ascending node of its orbit on 1920 June 3, this means that the passage through that node took place between 0^h00^m and 24^h00^m Universal Time on 1920 June 3.

Contents

Tables I and II give the dates of all inferior and superior conjunctions of Mercury with the Sun taking place from A.D. 1990 to 2020.

Tables III lists, from 1990 to 2020, the greatest elongations of Mercury, that is when the angular distance between Mercury and the center of the Sun's disk, as seen from the Earth, is a maximum.

During 1990 − 2020, the extreme values of Mercury's greatest elongations are:

17°52′ on 1990 September 24, 2003 September 26, and 2010 September 19;
27°50′ on 2000 March 28, and 2013 March 31.

There is a periodicity of 13 years, a periodicity of 33 years and a much better periodicity of 46 years (which is a combination of those of 13 and 33 years). Examples are given in the table below.

13 years		*33 years*		*46 years*	
1953 Aug 13	18°50′ W	1939 Aug 28	18°16′ W	1913 Aug 22	18°26′ W
1966 Aug 16	18°43′ W	1972 Aug 25	18°20′ W	1959 Aug 23	18°25′ W
1979 Aug 19	18°36′ W	2005 Aug 23	18°24′ W	2005 Aug 23	18°24′ W
1992 Aug 21	18°30′ W				
2005 Aug 23	18°24′ W				

Table IV gives the dates of all inferior conjunctions of Venus with the Sun during the period 1975 − 2020, together with the planet's least distances to the Earth.

During the 20th century, the greatest geocentric latitudes of Venus took place at the inferior conjunctions of 1919 September 13 and 1927 September 10, when in both cases it was −8°43′. Venus will reach no larger geocentric latitude (at inferior conjunction) until the second half of the 21st century (for instance, +8°50′ at the inferior conjunctions of 2081 March 6 and 2089 March 3).

Table V gives the dates of the superior conjunctions of Venus with the Sun which take place during the period 1975 – 2020.

Table VI provides, from 1990 to 2020, the greatest elongations of Venus, that is when the angular distance between Venus and the center of the Sun's disk, as seen from the Earth, is a maximum. During the 20th century, the greatest and least of Venus' greatest elongations are the following ones :

$$47°19' \quad \text{on} \quad 1901 \text{ December } 5$$
$$45°23' \quad \text{on} \quad 1999 \text{ June } 11$$

Table VII contains informations about the oppositions of the planet Mars with the Sun which take place during the years 1960 to 2020: the date and the Universal Time of the geometric opposition in celestial longitude; the common heliocentric longitude of the planet and the Earth at the time of opposition, referred to the mean equinox of the date; the planet's visual magnitude at the time of opposition; and the planetocentric declination D_E of the Earth at the time of opposition. — More data on Mars oppositions can be found in Part 2.

Tables VIII to XII contain informations about the oppositions of the planets Jupiter to Pluto during the years 1979 to 2020. In each table, the columns give successively :
— the calendar date and the Universal Time of the geometric opposition in celestial longitude. 'Geometric' means that the effects of light-time and of aberration have not been taken into account;
— the common heliocentric longitude of the planet and the Earth at the time of opposition, referred to the mean equinox of the date;
— the apparent geocentric declination of the planet at the time of opposition. The effects of light-time, aberration and nutation have been taken into account here;
— the planet's visual magnitude at the time of opposition (except for Pluto);
— the instant (calendat date and UT) when the planet is nearest to the Earth;
— the value of this least distance, respectively in astronomical units (a.u.) and in millions of kilometers (m.k.);
— the value of the apparent diameter of the planet at the time of least distance to the Earth. For Jupiter and Saturn, the equatorial diameter is given. This information is not given for Pluto.

The stellar magnitudes of Mars, Jupiter and Saturn have been obtained from the formulae of G. Müller (1893). Those of Uranus and Neptune are calculated from

$$\text{Uranus :} \qquad -6.85 \; + \; 5 \log r\Delta$$
$$\text{Neptune :} \qquad -7.05 \; + \; 5 \log r\Delta$$

where r and Δ are the planet's distances to the Sun and to the Earth, respectively, expressed in astronomical units. The logarithms are to the base 10. See also my *Astronomical Algorithms*, Chapter 40 (Willmann-Bell, ed.; 1991).

The following values have been used for the diameters at unit distance :

Jupiter 196″.94   Saturn 166″.66   Uranus 68″.56   Neptune 73″.12

During the period 1960 – 2020, a *transit of the Earth* over the Sun's disk occurs at the following dates (see also my *Transits*; Wilmann-Bell, ed.; 1989) :

for Mars :	1984 May 11		*for Uranus :*	1981 May 19
for Jupiter :	1972 Jun 24			1982 May 24
	1977 Dec 23			1983 May 29
	1984 Jun 29			1984 Jun 2
	1989 Dec 27			1985 Jun 6
	1996 Jul 4			1986 Jun 11
	2002 Jan 1			1987 Jun 16
	2008 Jul 9			1988 Jun 20
	2014 Jan 5		*for Neptune :*	2000 Jul 28
for Saturn :	1961 Jul 19			2001 Jul 30
	1990 Jul 14			2002 Aug 2
	2005 Jan 13			2003 Aug 4
	2020 Jul 20			2004 Aug 6
				2005 Aug 8
				2006 Aug 11

During the 20th century, that is the years 1901 to 2000, the least distances of the planets Venus through Pluto to the Earth are as follows:

Venus	1906 Nov 30	0.26506 *a.u.*	40 *millions of km*
Mars	1924 Aug 22	0.37285	56
Jupiter	1951 Oct 2	3.94871	591
Saturn	1914 Dec 21	8.03042	1201
Uranus	1966 Mar 9	17.29212	2587
Neptune	1901 Dec 22	28.90119	4324
Pluto	1990 May 7	28.68720	4292

Table XIII gives the dates of the conjunctions of the superior planets with the Sun taking place from A.D. 1979 to 2020.

Tables XIV and XV contain the times of the perihelion and aphelion passages of Earth and Mars, respectively, and the corresponding values of the radius vector expressed in astronomical units (1 a.u. = 149 597 870 km).

Table XVI contains the dates of the perihelion and aphelion passages of the outer planets, and the corresponding values of the radius vector: Jupiter from 1950 to 2020, Saturn and Uranus from 1900 to 2050, Neptune from 1850 to 2100. Pluto is closer to the Sun than Neptune from 1979 February 7 to 1999 February 10.

Table XVII contains the times of the passages of Mars through the nodes of its orbit, from 1960 to 2020.

Table XVIII contains the dates on which the outer planets reach the nodes of their orbits and their greatest heliocentric latitudes: Jupiter from 1970 to 2020, Saturn and Uranus from 1940 to 2020, Neptune and Pluto from 1900 to 2020. All these data refer to the ecliptic of the date.

Table XIX provides some informations about the extreme northern and southern declinations of Venus and Mars. Furthermore, the complete list is given of the dates when the outer planets cross the celestial equator and when they reach their extreme northern or southern declinations: Jupiter and Saturn for the years 1970 to 2020, Uranus from 1950 to 2020, Neptune and Pluto from 1920 to 2020.

All these data refer to the apparent declinations of the planets, that is, the corrections for light-time, aberration and nutation have been taken into account.

In Table XX the extreme northern and southern apparent geocentric declinations of the center of the solar disk are given for the years 1985 to 2020. Note the gradual decrease of these values due to the decrease of the obliquity of the ecliptic, and the oscillation with a period of 18.6 years due to the nutation: a maximum occurs in 1987, a minimum around 1997, another maximum around 2005, and so on.

Table XXI gives the complete list of all planetary conjunctions in *right ascension* (*not* in celestial longitude) during the period 1990 − 2020. Conjunctions with Pluto are not considered here. The conjunctions are given separately for each couple of planets. The columns give the date and time (UT) of the conjunction in right ascension, the difference in declination between the two bodies at that instant, and the elongation or angular distance to the Sun's center (E = east from the Sun = visible in the evening sky; W = west from the Sun = visible in the morning sky).

The difference in declination is positive (negative) if the first planet passes to the North (South) of the second planet. For instance, on 1990 September 14 at 15^h UT Mercury was 3°21′ south of Venus.

The tabulated elongation is that of the planet which is closest to the Sun.

Multiple conjunctions are indicated with a bracket. It appears that *quintuple* conjunctions are possible between Mercury and Venus (as in 2004 − 2005), and between Mercury and Mars (as in 1995 − 1996).

The instant of conjunction in right ascension is not the same as that of the least apparent separation. The difference can be important in some cases. A striking example took place in August 1991, when there was a triple conjunction (in right ascension) between Mercury and Venus, but only one least separation between the two planets, namely 2°05′ on August 8.

It should further be noted that a conjunction in right ascension is not the same as a conjunction in celestial (ecliptic) longitude. In limiting cases, there may be a conjunction in right ascension but not in longitude, or inversely.

Conjunctions which occur close to the Sun are not observable. An accurate limit cannot be given, since the visibility depends on the magnitudes of the planets, and on the inclination of the ecliptic on the local horizon near the time of sunrise or sunset. But a general rule is that a conjunction certainly is not visible if the two planets are less than 10 degrees from the Sun.

Some interesting facts can be seen by examination of the lists. For example, for the planets Mercury - Venus the difference of the declinations at the time of conjunction in right ascension can exceed 10 degrees, as for instance on 1999 August 26.

For the planets Mercury - Mars, single conjunctions are much rarer than multiple ones. During the period 1990 − 2020, there is only one single conjunction in right ascension, namely in 2002, against 9 triple ones and 4 quintuple ones.

For the Mercury - Saturn conjunctions, the dates show a remarkable pattern. During four or five consecutive years, the conjunctions occur at nearly the same date of the year, for example :

1998 May 12   1999 May 13   2000 May 10   2001 May 7

and then there is a jump of approximately fifty days (2002 July 2), corresponding to the beginning of a new series (2002 − 2005), etc. Triple conjunctions occur when one series ends and another begins.

While triple conjunctions between Venus and Mars are frequent (six cases during the period 1990 − 2020), during the same period there are only two triple conjunctions for the couple Venus-Jupiter, two for Venus-Saturn, none for Venus-Uranus, and one for Venus-Neptune.

During the period 1990 − 2020, there is no triple conjunction of Mars with either Jupiter, Saturn, Uranus, or Neptune. The last triple conjunction (in right ascension) between Mars and Jupiter took place in 1979 − 1980, between Mars and Saturn in 1945 − 1946, between Mars and Uranus in 1964 − 1965, between Mars and Neptune in 1932 − 1933.

During the period 1990 − 2017, Mars always passes south of Neptune. On 2018 December 7, Mars will pass north of Neptune, for the first time since 1952 January 2. On 2013 March 22, Mars will pass north of Uranus, for the first time since 1988 February 22.

From 1990 to 2020, all Mars - Saturn conjunctions take place every second year. In fact, this happens since 1954, and will last till A.D. 2136. In other words, from 1954 to 2136, there is just one geocentric Mars - Saturn conjunction in even years (1954, 1956, etc.), and none in the intermediate years !

The Mars - Uranus conjunctions, from 1986 to 2001, occur at intervals of 2 years minus one or two weeks, while the elongation from the Sun gradually changes from West to East. Then there is a discontinuity, and in 2003 another series begins. Similar series exist for the conjunctions of Mars with Jupiter, Saturn, and Neptune.

In 1981 there was a triple conjunction between Jupiter and Saturn. The previous triple conjunction between these planets took place in 1940 − 1941, while the next one will not occur before 2238 − 2239.

Triple conjunctions between Jupiter and Uranus occurred in 1927 − 1928, 1954 − 1955, and 1968 − 1969 ; between Jupiter and Neptune in 1919 − 1920 and 1971 ; between Saturn and Uranus in 1896 − 1897 ; between Saturn and Neptune in 1952 − 1953 when, moreover, both Saturn and Neptune came in triple conjunction with the star Spica !

Those conjunctions in right ascension of the period 1990 − 2020 for which the separation in declination is less than 0°06′00″, are given with more details in Table XXII. The times of the conjunction in right ascension and of least separation are given to the nearest minute of time, while the difference of the declinations (at the time of conjunction in right ascension) and the least distance between the centers of the disks of the two planets are given to the nearest second of a degree. All these values are geocentric.

During the period 1990 − 2020, the least geocentric separation between two planets will be 39″ (Mars-Uranus, on 2013 March 22), and there will be no occultation of a planet by another planet.

Sometimes, two planets approach each other within five degrees, although there is no conjunction in right ascension between these bodies. Such approaches-without-conjunction have been called *quasi-conjunctions* by the author [*L'Astronomie*, Vol. 88, 141 (April 1974)]. Table XXIII lists all quasi-conjunctions between the naked-eye planets from 1990 to 2020. The instants and the values of the least separations are given.

Table XXIV gives the *heliocentric* conjunctions between the outer planets, generally from 1950 to 2020, except for the couples Jupiter-Pluto and Saturn-Pluto. After the date, the common heliocentric longitude, referred to the mean equinox of the date, is given.

Table XXV gives, for the ten brightest minor planets, the oppositions and the least distances to the Earth during the period 1990 – 2020. For 433 *Eros*, there is a periodicity of 81 years and for this minor planet the list extends from the previous very favorable opposition (1975) to the next (2056). — The tabulated times are those of the opposition in celestial longitude, *not* in right ascension. The declination is given for the time of opposition and refers to the equinox of the date. The least distances to Earth are given in astronomical units. The last column gives the brightest visual magnitude at each apparition; for the calculation of these magnitudes, use has been made of the photometric parameters given in the *Minor Planet Circulars* 17257 and 17258.

For the ten minor planets mentioned in Table XXV, the closest approaches to the Earth in the period 1980 – 2060 are as follows:

1	Ceres	1.5832	2009 Feb 25
2	Pallas	1.2309	2014 Feb 23
3	Juno	1.0359	2018 Nov 16
4	Vesta	1.1379	2047 Jun 14
6	Hebe	0.9741	2044 Sep 21
7	Iris	0.8483	2006 Nov 12
8	Flora	0.8760	2020 Oct 31
15	Eunomia	1.1903	2028 Oct 22
18	Melpomene	0.8142	2002 Sep 22
433	Eros	0.1498	2056 Jan 24

Table XXVI mentions eclipse series of Callisto, the fourth satellite of Jupiter. In the calculations, both the Sun and the satellite have been considered as being points.

Table XXVII gives the dates of the passages of the Earth and the Sun through the ring-plane of Saturn from 1600 to 2030. For the calculations, we used for the ring the elements given by G. Dourneau [*Astronomy & Astrophysics*, Vol. 267, p. 297 (1993)]. N→S means a passage from North to South, S→N a passage from South to North.

I
INFERIOR CONJUNCTIONS OF MERCURY

1990 Jan 9	1997 Dec 17	2005 Nov 24	2013 Nov 1
1990 May 3	1998 Apr 6	2006 Mar 12	2014 Feb 15
1990 Sep 8	1998 Aug 13	2006 Jul 18	2014 Jun 19
1990 Dec 24	1998 Dec 1	2006 Nov 8 *T*	2014 Oct 16
1991 Apr 14	1999 Mar 19	2007 Feb 23	2015 Jan 30
1991 Aug 21	1999 Jul 26	2007 Jun 28	2015 May 30
1991 Dec 8	1999 Nov 15 *T*	2007 Oct 23	2015 Sep 30
1992 Mar 26	2000 Mar 1	2008 Feb 6	2016 Jan 14
1992 Aug 2	2000 Jul 6	2008 Jun 7	2016 May 9 *T*
1992 Nov 21	2000 Oct 30	2008 Oct 6	2016 Sep 12
1993 Mar 9	2001 Feb 13	2009 Jan 20	2016 Dec 28
1993 Jul 15	2001 Jun 16	2009 May 18	2017 Apr 20
1993 Nov 6 *T*	2001 Oct 14	2009 Sep 20	2017 Aug 26
1994 Feb 20	2002 Jan 27	2010 Jan 4	2017 Dec 13
1994 Jun 25	2002 May 27	2010 Apr 28	2018 Apr 1
1994 Oct 21	2002 Sep 27	2010 Sep 3	2018 Aug 9
1995 Feb 3	2003 Jan 11	2010 Dec 20	2018 Nov 27
1995 Jun 5	2003 May 7 *T*	2011 Apr 9	2019 Mar 15
1995 Oct 5	2003 Sep 11	2011 Aug 17	2019 Jul 21
1996 Jan 18	2003 Dec 27	2011 Dec 4	2019 Nov 11 *T*
1996 May 15	2004 Apr 17	2012 Mar 21	2020 Feb 26
1996 Sep 17	2004 Aug 23	2012 Jul 28	2020 Jul 1
1997 Jan 2	2004 Dec 10	2012 Nov 17	2020 Oct 25
1997 Apr 25	2005 Mar 29	2013 Mar 4	
1997 Aug 31	2005 Aug 5	2013 Jul 9	

T means that there is a transit of Mercury over the Sun's disk

II
SUPERIOR CONJUNCTIONS OF MERCURY

1990 Mar 19	1998 Feb 22	2006 Jan 26	2013 Dec 29
1990 Jul 2	1998 Jun 10	2006 May 18	2014 Apr 26
1990 Oct 22	1998 Sep 25	2006 Sep 1	2014 Aug 8
1991 Mar 2	1999 Feb 4	2007 Jan 7	2014 Dec 8
1991 Jun 17	1999 May 25	2007 May 3 O	2015 Apr 10
1991 Oct 3	1999 Sep 8	2007 Aug 15	2015 Jul 23
1992 Feb 12	2000 Jan 16	2007 Dec 17	2015 Nov 17 O
1992 May 31	2000 May 9 O	2008 Apr 16	2016 Mar 23
1992 Sep 15	2000 Aug 22	2008 Jul 29	2016 Jul 7
1993 Jan 23	2000 Dec 25	2008 Nov 25	2016 Oct 27
1993 May 16 O	2001 Apr 23	2009 Mar 31	2017 Mar 7
1993 Aug 29	2001 Aug 5	2009 Jul 14	2017 Jun 21
1994 Jan 3	2001 Dec 4	2009 Nov 5 O	2017 Oct 8
1994 Apr 30 O	2002 Apr 7	2010 Mar 14	2018 Feb 17
1994 Aug 13	2002 Jul 21	2010 Jun 28	2018 Jun 6
1994 Dec 14	2002 Nov 14 O	2010 Oct 17	2018 Sep 21
1995 Apr 14	2003 Mar 21	2011 Feb 25	2019 Jan 30
1995 Jul 28	2003 Jul 5	2011 Jun 12	2019 May 21
1995 Nov 23	2003 Oct 25	2011 Sep 28	2019 Sep 4
1996 Mar 28	2004 Mar 4	2012 Feb 7	2020 Jan 10
1996 Jul 11	2004 Jun 18	2012 May 27	2020 May 4 O
1996 Nov 1	2004 Oct 5	2012 Sep 10	2020 Aug 17
1997 Mar 11	2005 Feb 14	2013 Jan 18	2020 Dec 20
1997 Jun 25	2005 Jun 3	2013 May 11 O	
1997 Oct 13	2005 Sep 18	2013 Aug 24	

O means that Mercury is occulted by the Sun's disk

III
GREATEST ELONGATIONS OF MERCURY

E = greatest eastern elongation   *W = greatest western elongation*

1990	Feb 1	25°09′ W	1996	Jan 2	19°28′ E	2002	Jan 11	19°01′ E
	Apr 13	19 37 E		Feb 11	25 55 W		Feb 21	26 35 W
	May 31	24 41 W		Apr 23	20 14 E		May 4	20 58 E
	Aug 11	27 25 E		Jun 10	23 42 W		Jun 21	22 44 W
	Sep 24	17 52 W		Aug 21	27 24 E		Sep 1	27 13 E
	Dec 6	21 07 E		Oct 3	17 55 W		Oct 13	18 04 W
				Dec 15	20 27 E		Dec 26	19 52 E
1991	Jan 14	23 41 W						
	Mar 27	18 47 E	1997	Jan 24	24 32 W	2003	Feb 4	25 21 W
	May 12	26 12 W		Apr 6	19 13 E		Apr 16	19 46 E
	Jul 25	27 02 E		May 22	25 22 W		Jun 3	24 26 W
	Sep 7	18 01 W		Aug 4	27 19 E		Aug 14	27 26 E
	Nov 19	22 23 E		Sep 16	17 53 W		Sep 26	17 52 W
	Dec 27	22 14 W		Nov 28	21 38 E		Dec 9	20 56 E
1992	Mar 9	18 17 E	1998	Jan 6	23 04 W	2004	Jan 17	23 55 W
	Apr 23	27 17 W		Mar 20	18 32 E		Mar 29	18 53 E
	Jul 6	26 05 E		May 4	26 44 W		May 14	26 00 W
	Aug 21	18 30 W		Jul 17	26 41 E		Jul 27	27 07 E
	Oct 31	23 43 E		Aug 31	18 11 W		Sep 9	17 58 W
	Dec 9	20 53 W		Nov 11	22 57 E		Nov 21	22 11 E
				Dec 20	21 38 W		Dec 29	22 27 W
1993	Feb 21	18 07 E						
	Apr 5	27 48 W	1999	Mar 3	18 11 E	2005	Mar 12	18 20 E
	Jun 17	24 43 E		Apr 16	27 35 W		Apr 26	27 10 W
	Aug 4	19 20 W		Jun 28	25 33 E		Jul 9	26 15 E
	Oct 14	25 00 E		Aug 14	18 48 W		Aug 23	18 24 W
	Nov 22	19 45 W		Oct 24	24 17 E		Nov 3	23 31 E
				Dec 3	20 23 W		Dec 12	21 05 W
1994	Feb 4	18 16 E						
	Mar 19	27 41 W	2000	Feb 15	18 09 E	2006	Feb 24	18 08 E
	May 30	23 07 E		Mar 28	27 50 W		Apr 8	27 46 W
	Jul 17	20 31 W		Jun 9	24 03 E		Jun 20	24 56 E
	Sep 26	26 08 E		Jul 27	19 48 W		Aug 7	19 11 W
	Nov 6	18 52 W		Oct 6	25 31 E		Oct 17	24 49 E
				Nov 15	19 20 W		Nov 25	19 54 W
1995	Jan 19	18 44 E						
	Mar 1	27 01 W	2001	Jan 28	18 26 E	2007	Feb 7	18 14 E
	May 12	21 34 E		Mar 11	27 28 W		Mar 22	27 44 W
	Jun 29	22 01 W		May 22	22 27 E		Jun 2	23 22 E
	Sep 9	26 58 E		Jul 9	21 08 W		Jul 20	20 19 W
	Oct 20	18 15 W		Sep 18	26 32 E		Sep 29	25 59 E
				Oct 29	18 34 W		Nov 8	18 59 W

III (cont.)
Greatest Elongations of Mercury

E = greatest eastern elongation   *W = greatest western elongation*

2008	Jan 22	18°39' E	2012	Mar 5	18°13' E	2017	Jan 19	24°08' W
	Mar 3	27 09 W		Apr 18	27 30 W		Apr 1	19 00 E
	May 14	21 48 E		Jul 1	25 45 E		May 17	25 47 W
	Jul 1	21 47 W		Aug 16	18 42 W		Jul 30	27 12 E
	Sep 11	26 52 E		Oct 26	24 05 E		Sep 12	17 56 W
	Oct 22	18 19 W		Dec 4	20 33 W		Nov 24	22 00 E
			2013	Feb 16	18 08 E			
2009	Jan 4	19 21 E		Mar 31	27 50 W	2018	Jan 1	22 40 W
	Feb 13	26 06 W		Jun 12	24 17 E		Mar 15	18 24 E
	Apr 26	20 25 E		Jul 30	19 38 W		Apr 29	27 01 W
	Jul 13	23 27 W		Oct 9	25 20 E		Jul 12	26 25 E
	Aug 24	27 22 E		Nov 18	19 29 W		Aug 26	18 19 W
	Oct 6	17 57 W					Nov 6	23 19 E
	Dec 18	20 18 E	2014	Jan 31	18 22 E		Dec 15	21 16 W
				Mar 14	27 33 W			
				May 25	22 41 E			
2010	Jan 27	24 45 W		Jul 12	20 55 W	2019	Feb 27	18 08 E
	Apr 8	19 21 E		Sep 21	26 24 E		Apr 11	27 43 W
	May 26	25 08 W		Nov 1	18 40 W		Jun 23	25 09 E
	Aug 7	27 22 E					Aug 9	19 03 W
	Sep 19	17 52 W	2015	Jan 14	18 54 E		Oct 20	24 38 E
	Dec 1	21 27 E		Feb 24	26 45 W		Nov 28	20 04 W
				May 7	21 11 E			
				Jun 24	22 29 W			
2011	Jan 9	23 17 W		Sep 4	27 08 E	2020	Feb 10	18 12 E
	Mar 23	18 37 E		Oct 16	18 07 W		Mar 24	27 47 W
	May 7	26 33 W		Dec 29	19 43 E		Jun 4	23 36 E
	Jul 20	26 49 E					Jul 22	20 08 W
	Sep 3	18 07 W	2016	Feb 7	25 33 W		Oct 1	25 49 E
	Nov 14	22 45 E		Apr 18	19 56 E		Nov 10	19 06 W
	Dec 23	21 51 W		Jun 5	24 11 W			
				Aug 16	27 26 E			
				Sep 28	17 53 W			
				Dec 11	20 46 E			

IV
INFERIOR CONJUNCTIONS OF VENUS

The geocentric latitude of Venus at the time of inferior conjunction is given. The sign +
(or −) indicates that Venus is north (or south) of the ecliptic. At the inferior conjunctions
of 2004 and 2012, there is a transit of Venus over the Sun's disk.

Inferior Conjunction				*Least Distance to Earth*			
1961 Apr	10	24h	+7°08′	1961 Apr	10	20h	0.28371 *a.u.*
1962 Nov	12	20	−4 04	1962 Nov	13	12	0.26792
1964 Jun	19	23	−1 49	1964 Jun	19	21	0.28949
1966 Jan	26	9	+6 55	1966 Jan	25	12	0.26881
1967 Aug	29	22	−8 30	1967 Aug	30	8	0.28636
1969 Apr	8	15	+7 20	1969 Apr	8	11	0.28334
1970 Nov	10	9	−4 25	1970 Nov	11	2	0.26842
1972 Jun	17	15	−1 29	1972 Jun	17	14	0.28942
1974 Jan	23	21	+6 40	1974 Jan	23	1	0.26830
1975 Aug	27	13	−8 26	1975 Aug	27	22	0.28675
1977 Apr	6	6	+7 31	1977 Apr	6	2	0.28296
1978 Nov	7	22	−4 45	1978 Nov	8	16	0.26903
1980 Jun	15	7	−1 10	1980 Jun	15	6	0.28932
1982 Jan	21	10	+6 24	1982 Jan	20	14	0.26783
1983 Aug	25	5	−8 20	1983 Aug	25	13	0.28712
1985 Apr	3	22	+7 42	1985 Apr	3	17	0.28260
1986 Nov	5	10	−5 05	1986 Nov	6	6	0.26965
1988 Jun	12	24	−0 50	1988 Jun	12	22	0.28918
1990 Jan	18	23	+6 07	1990 Jan	18	3	0.26734
1991 Aug	22	20	−8 14	1991 Aug	23	4	0.28745
1993 Apr	1	13	+7 52	1993 Apr	1	8	0.28224
1994 Nov	2	23	−5 24	1994 Nov	3	19	0.27023
1996 Jun	10	16	−0 30	1996 Jun	10	14	0.28905
1998 Jan	16	11	+5 49	1998 Jan	15	16	0.26688
1999 Aug	20	12	−8 07	1999 Aug	20	19	0.28782
2001 Mar	30	4	+8 01	2001 Mar	29	22	0.28187
2002 Oct	31	12	−5 42	2002 Nov	1	8	0.27088
2004 Jun	8	9	−0 11 *T*	2004 Jun	8	7	0.28888
2006 Jan	13	24	+5 31	2006 Jan	13	6	0.26649
2007 Aug	18	4	−7 59	2007 Aug	18	10	0.28816
2009 Mar	27	19	+8 10	2009 Mar	27	12	0.28147
2010 Oct	29	1	−5 59	2010 Oct	29	22	0.27150
2012 Jun	6	1	+0 09 *T*	2012 Jun	5	24	0.28870
2014 Jan	11	12	+5 11	2014 Jan	10	20	0.26612
2015 Aug	15	19	−7 50	2015 Aug	16	0	0.28844
2017 Mar	25	10	+8 18	2017 Mar	25	2	0.28105
2018 Oct	26	14	−6 15	2018 Oct	27	11	0.27212
2020 Jun	3	18	+0 29	2020 Jun	3	17	0.28858

V
SUPERIOR CONJUNCTIONS OF VENUS

1976 Jun 18 O	1987 Aug 23	1998 Oct 30	2010 Jan 11
1978 Jan 22	1989 Apr 4	2000 Jun 11 O	2011 Aug 16
1979 Aug 25	1990 Nov 1	2002 Jan 14	2013 Mar 28
1981 Apr 7	1992 Jun 13 O	2003 Aug 18	2014 Oct 25
1982 Nov 4	1994 Jan 17	2005 Mar 31	2016 Jun 6 O
1984 Jun 15 O	1995 Aug 20	2006 Oct 27	2018 Jan 9
1986 Jan 19	1997 Apr 2	2008 Jun 9 O	2019 Aug 14

O means that Venus is occulted by the Sun's disk

VI
GREATEST ELONGATIONS OF VENUS

1990 Mar 30	46°29′ W	2001 Jan 17	47°06′ E	2011 Jan 8	46°57′ W
1991 Jun 13	45 23 E	2001 Jun 8	45 50 W	2012 Mar 27	46 02 E
1991 Nov 2	46 31 W	2002 Aug 22	46 00 E	2012 Aug 15	45 48 W
1993 Jan 19	47 04 E	2003 Jan 11	46 58 W	2013 Nov 1	47 04 E
1993 Jun 10	45 49 W	2004 Mar 29	46 00 E	2014 Mar 22	46 33 W
1994 Aug 24	46 03 E	2004 Aug 17	45 49 W	2015 Jun 6	45 24 E
1995 Jan 13	46 58 W	2005 Nov 3	47 06 E	2015 Oct 26	46 26 W
1996 Apr 1	45 58 E	2006 Mar 25	46 32 W	2017 Jan 12	47 09 E
1996 Aug 20	45 50 W	2007 Jun 9	45 23 E	2017 Jun 3	45 52 W
1997 Nov 6	47 08 E	2007 Oct 28	46 28 W	2018 Aug 17	45 56 E
1998 Mar 27	46 30 W	2009 Jan 14	47 07 E	2019 Jan 6	46 57 W
1999 Jun 11	45 23 E	2009 Jun 5	45 51 W	2020 Mar 24	46 05 E
1999 Oct 30	46 29 W	2010 Aug 20	45 58 E	2020 Aug 13	45 47 W

VII
OPPOSITIONS OF MARS

Date	UT	Helioc. long.	Magn.	D_E
1960 Dec 30	10h	98°45′	−1.3	+ 4°
1963 Feb 4	12	134 57	−1.0	+17
1965 Mar 9	12	168 44	−1.0	+23
1967 Apr 15	11	204 49	−1.3	+20
1969 May 31	16	250 00	−2.0	+ 7
1971 Aug 10	7	317 00	−2.6	−15
1973 Oct 25	3	31 34	−2.3	−19
1975 Dec 15	14	82 58	−1.5	− 2
1978 Jan 22	0	121 37	−1.1	+12
1980 Feb 25	6	155 47	−1.0	+21
1982 Mar 31	10	190 23	−1.2	+22
1984 May 11	9	230 51	−1.7	+14
1986 Jul 10	5	287 41	−2.4	− 6
1988 Sep 28	3	5 14	−2.5	−22
1990 Nov 27	20	65 20	−1.8	− 9
1993 Jan 7	23	107 40	−1.2	+ 7
1995 Feb 12	2	142 54	−1.0	+19
1997 Mar 17	8	176 46	−1.1	+23
1999 Apr 24	18	214 06	−1.5	+19
2001 Jun 13	18	262 46	−2.1	+ 3
2003 Aug 28	18	335 01	−2.7	−19
2005 Nov 7	8	45 01	−2.1	−16
2007 Dec 24	20	92 37	−1.4	+ 1
2010 Jan 29	20	129 48	−1.1	+15
2012 Mar 3	20	163 39	−1.0	+22
2014 Apr 8	21	198 57	−1.3	+21
2016 May 22	11	241 48	−1.8	+10
2018 Jul 27	5	304 09	−2.6	−11
2020 Oct 13	23	21 05	−2.4	−20

See more information in Part 2.

VIII

OPPOSITIONS OF JUPITER

Date		UT	Helioc. long.	Declin.	Magn.	Date		UT	a.u.	m.k.	diam.
		h	° ′	° ′				h			″
1979 Jan	24	15	124 02	+19 53	−2.2	1979 Jan	23	22	4.29222	642	45.88
1980 Feb	24	18	155 17	+10 47	−2.1	1980 Feb	24	17	4.40255	659	44.73
1981 Mar	26	6	185 29	− 0 43	−2.0	1981 Mar	26	24	4.45245	666	44.23
1982 Apr	26	0	215 28	−11 58	−2.0	1982 Apr	27	7	4.42999	663	44.46
1983 May	27	22	246 04	−20 26	−2.1	1983 May	29	11	4.34016	649	45.38
1984 Jun	29	16	278 05	−23 08	−2.2	1984 Jun	30	23	4.20568	629	46.83
1985 Aug	4	11	312 02	−18 01	−2.4	1985 Aug	5	3	4.06677	608	48.43
1986 Sep	10	21	347 52	− 6 11	−2.4	1986 Sep	10	13	3.97189	594	49.58
1987 Oct	18	14	24 40	+ 8 05	−2.5	1987 Oct	17	12	3.95915	592	49.74
1988 Nov	23	3	61 03	+19 22	−2.4	1988 Nov	21	16	4.03397	603	48.82
1989 Dec	27	14	95 48	+23 11	−2.3	1989 Dec	26	9	4.16549	623	47.28
1991 Jan	29	0	128 35	+18 51	−2.1	1991 Jan	28	8	4.30610	644	45.74
1992 Feb	29	0	159 40	+ 9 13-	−2.0	1992 Feb	29	3	4.41180	660	44.64
1993 Mar	30	12	189 47	− 2 25	−2.0	1993 Mar	31	7	4.45423	666	44.21
1994 Apr	30	9	219 47	−13 25	−2.0	1994 May	1	17	4.42272	662	44.53
1995 Jun	1	11	250 31	−21 14	−2.1.	1995 Jun	2	23	4.32470	647	45.54
1996 Jul	4	11	282 46	−22 53	−2.2	1996 Jul	5	18	4.18608	626	47.05
1997 Aug	9	13	317 00	−16 40	−2.4	1997 Aug	10	1	4.04908	606	48.64
1998 Sep	16	3	353 03	− 4 11	−2.5	1998 Sep	15	17	3.96289	593	49.70
1999 Oct	23	19	29 56	+10 00	−2.5	1999 Oct	22	14	3.96288	593	49.70
2000 Nov	28	2	66 09	+20 26	−2.4	2000 Nov	26	15	4.04936	606	48.63
2002 Jan	1	6	100 38	+23 01	−2.3	2001 Dec	31	1	4.18747	626	47.03
2003 Feb	2	9	133 06	+17 43	−2.1	2003 Feb	1	19	4.32715	647	45.51
2004 Mar	4	5	163 58	+ 7 38	−2.0	2004 Mar	4	9	4.42566	662	44.50
2005 Apr	3	15	193 58	− 4 03	−2.0	2005 Apr	4	14	4.45664	667	44.19
2006 May	4	14	224 00	−14 46	−2.0	2006 May	5	24	4.41270	660	44.63
2007 Jun	5	23	254 55	−21 54	−2.1	2007 Jun	7	12	4.30437	644	45.75
2008 Jul	9	7	287 28	−22 29	−2.3	2008 Jul	10	11	4.16101	622	47.33
2009 Aug	14	18	322 04	−15 10	−2.4	2009 Aug	15	3	4.02782	603	48.89
2010 Sep	21	11	358 23	− 2 06	−2.5	2010 Sep	20	21	3.95393	591	49.81
2011 Oct	29	2	35 17	+11 53	−2.5	2011 Oct	27	19	3.96976	594	49.61
2012 Dec	3	2	71 18	+21 21	−2.4	2012 Dec	1	15	4.06854	609	48.41
2014 Jan	5	21	105 27	+22 40	−2.2	2014 Jan	4	18	4.21044	630	46.77
2015 Feb	6	18	137 38	+16 28	−2.1	2015 Feb	6	7	4.34621	650	45.31
2016 Mar	8	11	168 18	+ 5 59	−2.0	2016 Mar	8	18	4.43535	664	44.40
2017 Apr	7	21	198 15	− 5 42	−2.0	2017 Apr	8	21	4.45490	666	44.21
2018 May	9	0	228 22	−16 04	−2.0	2018 May	10	12	4.39982	658	44.76
2019 Jun	10	15	259 29	−22 26	−2.1	2019 Jun	12	3	4.28390	641	45.97
2020 Jul	14	8	292 20	−21 55	−2.3	2020 Jul	15	10	4.13932	619	47.58

IX
OPPOSITIONS OF SATURN

| | | At opposition | | | | | Nearest to Earth | | |
	Date	UT	Helioc. long.	Declin.	Magn.	Date	UT	a.u.	m.k.	diam.
		h	° ′	° ′			h			″
1979 Mar	1	17	160 33	+ 9 30	+0.5	1979 Mar	1	14	8.33814	1247	19.99
1980 Mar	14	2	173 39	+ 4 44	+0.8	1980 Mar	13	24	8.44780	1264	19.73
1981 Mar	27	5	186 26	− 0 07	+0.6	1981 Mar	27	4	8.55942	1280	19.47
1982 Apr	9	2	198 55	− 4 51	+0.5	1982 Apr	9	2	8.66692	1297	19.23
1983 Apr	21	19	211 07	− 9 17	+0.4	1983 Apr	21	20	8.76574	1311	19.01
1984 May	3	8	223 05	−13 17	+0.3	1984 May	3	10	8.85252	1324	18.83
1985 May	15	18	234 50	−16 41	+0.2	1985 May	15	21	8.92436	1335	18.67
1986 May	28	0	246 25	−19 23	+0.2	1986 May	28	5	8.97866	1343	18.56
1987 Jun	9	5	257 52	−21 17	+0.2	1987 Jun	9	9	9.01350	1348	18.49
1988 Jun	20	9	269 15	−22 18	+0.2	1988 Jun	20	13	9.02779	1351	18.46
1989 Jul	2	13	280 37	−22 24	+0.2	1989 Jul	2	17	9.02116	1350	18.47
1990 Jul	14	18	292 01	−21 34	+0.3	1990 Jul	14	22	8.99384	1345	18.53
1991 Jul	27	0	303 30	−19 50	+0.3	1991 Jul	27	4	8.94664	1338	18.63
1992 Aug	7	10	315 08	−17 16	+0.4	1992 Aug	7	13	8.88119	1329	18.77
1993 Aug	19	23	326 57	−13 58	+0.5	1993 Aug	20	2	8.79997	1316	18.94
1994 Sep	1	16	339 00	−10 01	+0.7	1994 Sep	1	19	8.70620	1302	19.14
1995 Sep	14	15	351 20	− 5 35	+0.9	1995 Sep	14	17	8.60386	1287	19.37
1996 Sep	26	19	3 58	− 0 50	+0.7	1996 Sep	26	19	8.49761	1271	19.61
1997 Oct	10	4	16 54	+ 4 05	+0.4	1997 Oct	10	3	8.39239	1255	19.86
1998 Oct	23	19	30 10	+ 8 56	+0.2	1998 Oct	23	17	8.29343	1241	20.10
1999 Nov	6	14	43 42	+13 27	−0.0	1999 Nov	6	12	8.20571	1228	20.31
2000 Nov	19	12	57 30	+17 21	−0.2	2000 Nov	19	11	8.13334	1217	20.49
2001 Dec	3	14	71 29	+20 18	−0.3	2001 Dec	3	12	8.08059	1209	20.62
2002 Dec	17	17	85 36	+22 03	−0.3	2002 Dec	17	14	8.05194	1205	20.70
2003 Dec	31	21	99 46	+22 25	−0.3	2003 Dec	31	17	8.05013	1204	20.70
2005 Jan	13	23	113 53	+21 20	−0.2	2005 Jan	13	19	8.07562	1208	20.64
2006 Jan	27	23	127 52	+18 58	−0.0	2006 Jan	27	19	8.12684	1216	20.51
2007 Feb	10	18	141 38	+15 32	+0.2	2007 Feb	10	15	8.20037	1227	20.32
2008 Feb	24	10	155 09	+11 20	+0.4	2008 Feb	24	7	8.29141	1240	20.10
2009 Mar	8	20	168 22	+ 6 41	+0.7	2009 Mar	8	17	8.39446	1256	19.85
2010 Mar	22	0	181 17	+ 1 51	+0.7	2010 Mar	21	22	8.50382	1272	19.60
2011 Apr	3	24	193 52	− 2 57	+0.6	2011 Apr	3	23	8.61394	1289	19.35
2012 Apr	15	18	206 10	− 7 31	+0.4	2012 Apr	15	19	8.71962	1304	19.11
2013 Apr	28	8	218 12	−11 42	+0.3	2013 Apr	28	9	8.81620	1319	18.90
2014 May	10	18	230 01	−15 21	+0.3	2014 May	10	20	8.89968	1331	18.73
2015 May	23	1	241 38	−18 21	+0.2	2015 May	23	3	8.96670	1341	18.59
2016 Jun	3	6	253 07	−20 35	+0.2	2016 Jun	3	10	9.01490	1349	18.49
2017 Jun	15	10	264 30	−21 58	+0.2	2017 Jun	15	14	9.04268	1353	18.43
2018 Jun	27	13	275 51	−22 27	+0.2	2018 Jun	27	17	9.04882	1354	18.42
2019 Jul	9	17	287 13	−22 01	+0.3	2019 Jul	9	21	9.03279	1351	18.45
2020 Jul	20	22	298 39	−20 39	+0.3	2020 Jul	21	3	8.99470	1346	18.53

X
OPPOSITIONS OF URANUS

	At opposition					Nearest to Earth				
Date	UT	Helioc. long.	Declin.	Magn.		Date	UT	a.u.	m.k.	diam.
	h	° ′	° ′				h			″
1979 May 10	5	228 59	−17 08	5.74		1979 May 10	11	17.67660	2644	3.88
1980 May 14	5	233 34	−18 23	5.76		1980 May 14	9	17.73795	2654	3.87
1981 May 19	4	238 06	−19 31	5.77		1981 May 19	6	17.80170	2663	3.85
1982 May 24	3	242 37	−20 32	5.79		1982 May 24	3	17.86691	2673	3.84
1983 May 29	1	247 07	−21 24	5.81		1983 May 28	24	17.93305	2683	3.82
1984 Jun 1	22	251 34	−22 08	5.82		1984 Jun 1	19	17.99935	2693	3.81
1985 Jun 6	19	255 59	−22 44	5.84		1985 Jun 6	15	18.06572	2703	3.80
1986 Jun 11	14	260 23	−23 11	5.85		1986 Jun 11	8	18.13176	2712	3.78
1987 Jun 16	10	264 44	−23 29	5.87		1987 Jun 16	3	18.19768	2722	3.77
1988 Jun 20	4	269 03	−23 39	5.88		1988 Jun 19	19	18.26355	2732	3.75
1989 Jun 24	22	273 20	−23 40	5.90		1989 Jun 24	12	18.32957	2742	3.74
1990 Jun 29	14	277 36	−23 33	5.91		1990 Jun 29	2	18.39547	2752	3.73
1991 Jul 4	7	281 49	−23 18	5.93		1991 Jul 3	17	18.46091	2762	3.71
1992 Jul 7	22	286 01	−22 54	5.94		1992 Jul 7	6	18.52529	2771	3.70
1993 Jul 12	13	290 12	−22 24	5.96		1993 Jul 11	21	18.58808	2781	3.69
1994 Jul 17	4	294 21	−21 46	5.97		1994 Jul 16	9	18.64852	2790	3.68
1995 Jul 21	17	298 29	−21 01	5.98		1995 Jul 20	23	18.70593	2798	3.67
1996 Jul 25	7	302 36	−20 10	6.00		1996 Jul 24	11	18.75980	2806	3.65
1997 Jul 29	19	306 42	−19 13	6.01		1997 Jul 28	24	18.80946	2814	3.64
1998 Aug 3	7	310 46	−18 11	6.02		1998 Aug 2	10	18.85461	2821	3.64
1999 Aug 7	18	314 49	−17 04	6.03		1999 Aug 6	22	18.89519	2827	3.63
2000 Aug 11	5	318 51	−15 52	6.04		2000 Aug 10	8	18.93163	2832	3.62
2001 Aug 15	15	322 52	−14 37	6.04		2001 Aug 14	18	18.96423	2837	3.62
2002 Aug 20	1	326 51	−13 18	6.05		2002 Aug 19	2	18.99340	2841	3.61
2003 Aug 24	10	330 50	−11 55	6.05		2003 Aug 23	12	19.01913	2845	3.60
2004 Aug 27	18	334 49	−10 30	6.06		2004 Aug 26	20	19.04162	2849	3.60
2005 Sep 1	3	338 47	− 9 02	6.06		2005 Aug 31	5	19.06042	2851	3.60
2006 Sep 5	11	342 45	− 7 32	6.07		2006 Sep 4	12	19.07541	2854	3.59
2007 Sep 9	19	346 42	− 6 00	6.07		2007 Sep 8	21	19.08609	2855	3.59
2008 Sep 13	2	350 40	− 4 26	6.07		2008 Sep 12	4	19.09218	2856	3.59
2009 Sep 17	9	354 38	− 2 52	6.07		2009 Sep 16	13	19.09297	2856	3.59
2010 Sep 21	17	358 36	− 1 17	6.07		2010 Sep 20	20	19.08821	2856	3.59
2011 Sep 26	0	2 34	+ 0 19	6.07		2011 Sep 25	5	19.07750	2854	3.59
2012 Sep 29	7	6 33	+ 1 55	6.06		2012 Sep 28	13	19.06133	2852	3.60
2013 Oct 3	14	10 31	+ 3 30	6.06		2013 Oct 2	21	19.03992	2848	3.60
2014 Oct 7	21	14 30	+ 5 04	6.05		2014 Oct 7	4	19.01411	2844	3.61
2015 Oct 12	4	18 30	+ 6 38	6.05		2015 Oct 11	12	18.98432	2840	3.61
2016 Oct 15	10	22 30	+ 8 10	6.04		2016 Oct 14	20	18.95114	2835	3.62
2017 Oct 19	17	26 31	+ 9 40	6.03		2017 Oct 19	4	18.91463	2830	3.62
2018 Oct 24	1	30 33	+11 08	6.02		2018 Oct 23	12	18.87524	2824	3.63
2019 Oct 28	8	34 37	+12 34	6.01		2019 Oct 27	21	18.83285	2817	3.64
2020 Oct 31	16	38 41	+13 57	6.00		2020 Oct 31	5	18.78762	2811	3.65

ASTRONOMICAL TABLES

XI

OPPOSITIONS OF NEPTUNE

At opposition					Nearest to Earth				
Date	UT	Helioc. long.	Declin.	Magn.	Date	UT	a.u.	m.k.	diam.
	h	° ′	° ′			h			″
1979 Jun 10	14	259 07	−21 33	7.69	1979 Jun 11	0	29.2672	4378	2.50
1980 Jun 12	3	261 18	−21 45	7.69	1980 Jun 12	12	29.2639	4378	2.50
1981 Jun 14	16	263 29	−21 55	7.69	1981 Jun 14	24	29.2606	4377	2.50
1982 Jun 17	4	265 40	−22 03	7.69	1982 Jun 17	13	29.2565	4377	2.50
1983 Jun 19	17	267 51	−22 09	7.69	1983 Jun 19	24	29.2515	4376	2.50
1984 Jun 21	6	270 03	−22 14	7.68	1984 Jun 21	13	29.2451	4375	2.50
1985 Jun 23	19	272 14	−22 16	7.68	1985 Jun 24	0	29.2372	4374	2.50
1986 Jun 26	8	274 26	−22 16	7.68	1986 Jun 26	14	29.2280	4372	2.50
1987 Jun 28	21	276 37	−22 14	7.68	1987 Jun 29	1	29.2179	4371	2.50
1988 Jun 30	10	278 48	−22 09	7.68	1988 Jun 30	14	29.2079	4369	2.50
1989 Jul 2	22	281 00	−22 03	7.68	1989 Jul 3	1	29.1983	4368	2.50
1990 Jul 5	11	283 11	−21 55	7.68	1990 Jul 5	13	29.1896	4367	2.51
1991 Jul 8	0	285 22	−21 45	7.68	1991 Jul 8	1	29.1817	4366	2.51
1992 Jul 9	13	287 34	−21 33	7.67	1992 Jul 9	12	29.1746	4364	2.51
1993 Jul 12	2	289 45	−21 18	7.67	1993 Jul 12	1	29.1677	4363	2.51
1994 Jul 14	15	291 57	−21 02	7.67	1994 Jul 14	13	29.1606	4362	2.51
1995 Jul 17	4	294 09	−20 44	7.67	1995 Jul 17	2	29.1529	4361	2.51
1996 Jul 18	18	296 22	−20 24	7.67	1996 Jul 18	14	29.1442	4360	2.51
1997 Jul 21	7	298 34	−20 02	7.67	1997 Jul 21	3	29.1342	4358	2.51
1998 Jul 23	20	300 46	−19 39	7.67	1998 Jul 23	16	29.1228	4357	2.51
1999 Jul 26	9	302 58	−19 13	7.66	1999 Jul 26	4	29.1105	4355	2.51
2000 Jul 27	23	305 11	−18 46	7.66	2000 Jul 27	17	29.0978	4353	2.51
2001 Jul 30	12	307 23	−18 17	7.66	2001 Jul 30	4	29.0856	4351	2.51
2002 Aug 2	1	309 35	−17 47	7.66	2002 Aug 1	17	29.0743	4349	2.51
2003 Aug 4	14	311 47	−17 15	7.66	2003 Aug 4	4	29.0642	4348	2.52
2004 Aug 6	3	314 00	−16 42	7.66	2004 Aug 5	17	29.0553	4347	2.52
2005 Aug 8	16	316 12	−16 07	7.66	2005 Aug 8	5	29.0472	4345	2.52
2006 Aug 11	5	318 25	−15 30	7.65	2006 Aug 10	17	29.0398	4344	2.52
2007 Aug 13	18	320 38	−14 53	7.65	2007 Aug 13	6	29.0323	4343	2.52
2008 Aug 15	7	322 51	−14 14	7.65	2008 Aug 14	18	29.0245	4342	2.52
2009 Aug 17	21	325 04	−13 33	7.65	2009 Aug 17	8	29.0158	4341	2.52
2010 Aug 20	10	327 17	−12 52	7.65	2010 Aug 19	20	29.0060	4339	2.52
2011 Aug 22	23	329 31	−12 09	7.65	2011 Aug 22	9	28.9951	4338	2.52
2012 Aug 24	12	331 44	−11 26	7.65	2012 Aug 23	21	28.9838	4336	2.52
2013 Aug 27	1	333 57	−10 42	7.64	2013 Aug 26	10	28.9728	4334	2.52
2014 Aug 29	14	336 11	− 9 56	7.64	2014 Aug 28	22	28.9624	4333	2.52
2015 Sep 1	3	338 24	− 9 10	7.64	2015 Aug 31	11	28.9533	4331	2.53
2016 Sep 2	16	340 37	− 8 24	7.64	2016 Sep 1	22	28.9453	4330	2.53
2017 Sep 5	5	342 51	− 7 36	7.64	2017 Sep 4	11	28.9387	4329	2.53
2018 Sep 7	18	345 05	− 6 48	7.64	2018 Sep 6	23	28.9328	4328	2.53
2019 Sep 10	7	347 19	− 5 59	7.64	2019 Sep 9	11	28.9276	4328	2.53
2020 Sep 11	20	349 33	− 5 10	7.64	2020 Sep 11	1	28.9223	4327	2.53

XII
OPPOSITIONS OF PLUTO

At opposition				Nearest to Earth			
Date	UT	Helioc. long.	Declin.	Date	UT	a.u.	m.k.
	h	° ′	° ′		h		
1979 Apr 8	7	197 51	+ 9 22	1979 Apr 10	6	29.3093	4385
1980 Apr 10	3	200 24	+ 8 28	1980 Apr 12	1	29.2028	4369
1981 Apr 12	23	202 58	+ 7 32	1981 Apr 14	19	29.1054	4354
1982 Apr 15	21	205 33	+ 6 35	1982 Apr 17	14	29.0172	4341
1983 Apr 18	18	208 09	+ 5 37	1983 Apr 20	8	28.9382	4329
1984 Apr 20	16	210 45	+ 4 38	1984 Apr 22	5	28.8687	4319
1985 Apr 23	14	213 22	+ 3 39	1985 Apr 24	23	28.8095	4310
1986 Apr 26	12	215 58	+ 2 39	1986 Apr 27	19	28.7612	4303
1987 Apr 29	10	218 35	+ 1 39	1987 Apr 30	14	28.7243	4297
1988 May 1	9	221 11	+ 0 39	1988 May 2	9	28.6996	4293
1989 May 4	7	223 48	− 0 22	1989 May 5	4	28.6872	4292
1990 May 7	5	226 23	− 1 22	1990 May 7	23	28.6872	4292
1991 May 10	3	228 59	− 2 23	1991 May 10	18	28.6988	4293
1992 May 12	1	231 34	− 3 23	1992 May 12	12	28.7217	4297
1993 May 14	23	234 08	− 4 23	1993 May 15	8	28.7554	4302
1994 May 17	20	236 41	− 5 22	1994 May 18	1	28.7994	4308
1995 May 20	17	239 13	− 6 21	1995 May 20	20	28.8534	4316
1996 May 22	14	241 44	− 7 18	1996 May 22	13	28.9172	4326
1997 May 25	10	244 14	− 8 15	1997 May 25	7	28.9910	4337
1998 May 28	5	246 42	− 9 10	1998 May 27	22	29.0747	4350
1999 May 30	24	249 08	−10 04	1999 May 30	14	29.1691	4364
2000 Jun 1	18	251 32	−10 57	2000 Jun 1	5	29.2742	4379
2001 Jun 4	12	253 55	−11 49	2001 Jun 3	19	29.3906	4397
2002 Jun 7	4	256 17	−12 39	2002 Jun 6	9	29.5177	4416
2003 Jun 9	20	258 36	−13 27	2003 Jun 8	21	29.6551	4436
2004 Jun 11	12	260 54	−14 14	2004 Jun 10	11	29.8020	4458
2005 Jun 14	3	263 11	−14 59	2005 Jun 12	22	29.9576	4482
2006 Jun 16	17	265 25	−15 42	2006 Jun 15	11	30.1213	4506
2007 Jun 19	7	267 38	−16 23	2007 Jun 17	21	30.2920	4532
2008 Jun 20	19	269 49	−17 03	2008 Jun 19	8	30.4694	4558
2009 Jun 23	7	271 59	−17 40	2009 Jun 21	18	30.6528	4586
2010 Jun 25	19	274 06	−18 15	2010 Jun 24	3	30.8424	4614
2011 Jun 28	5	276 12	−18 48	2011 Jun 26	12	31.0381	4643
2012 Jun 29	15	278 15	−19 20	2012 Jun 27	18	31.2405	4674
2013 Jul 1	24	280 17	−19 49	2013 Jun 30	2	31.4496	4705
2014 Jul 4	8	282 17	−20 16	2014 Jul 2	7	31.6653	4737
2015 Jul 6	15	284 15	−20 42	2015 Jul 4	13	31.8873	4770
2016 Jul 7	22	286 12	−21 05	2016 Jul 5	17	32.1147	4804
2017 Jul 10	4	288 07	−21 27	2017 Jul 7	22	32.3469	4839
2018 Jul 12	10	290 01	−21 47	2018 Jul 10	2	32.5829	4874
2019 Jul 14	15	291 53	−22 04	2019 Jul 12	5	32.8219	4910
2020 Jul 15	19	293 44	−22 20	2020 Jul 13	9	33.0633	4946

XIII
CONJUNCTIONS OF SUPERIOR PLANETS WITH THE SUN

Year	Mars	Jupiter	Saturn	Uranus	Neptune	Pluto
1979	Jan 20	Aug 13	Sep 10	Nov 14	Dec 12	Oct 13
1980	—	Sep 13	Sep 23	Nov 18 *O*	Dec 14	Oct 14
1981	Apr 2	Oct 14	Oct 6	Nov 22 *O*	Dec 16	Oct 17
1982	—	Nov 13	Oct 18	Nov 27 *O*	Dec 19	Oct 20
1983	Jun 3	Dec 14	Oct 31	Dec 2 *O*	Dec 21	Oct 23
1984	—	—	Nov 11	Dec 5 *O*	Dec 22	Oct 25
1985	Jul 18	Jan 14 *O*	Nov 23	Dec 10 *O*	Dec 25	Oct 28
1986	—	Feb 18	Dec 4	Dec 14 *O*	Dec 27	Oct 31
1987	Aug 25	Mar 27	Dec 16	Dec 19 *O*	Dec 29	Nov 2
1988	—	May 2	Dec 26	Dec 22 *O*	Dec 31	Nov 4
1989	Sep 29	Jun 9	—	Dec 27 *O*	—	Nov 7
1990	—	Jul 15 *O*	Jan 6	Dec 31	Jan 2	Nov 10
1991	Nov 8 *O*	Aug 17	Jan 18 *O*	—	Jan 5	Nov 13
1992	—	Sep 17	Jan 29	Jan 5	Jan 7	Nov 14
1993	Dec 27	Oct 18	Feb 9	Jan 8	Jan 8	Nov 17
1994	—	Nov 17	Feb 21	Jan 12	Jan 11	Nov 20
1995	—	Dec 18 *O*	Mar 6	Jan 17	Jan 13	Nov 23
1996	Mar 4	—	Mar 17	Jan 21	Jan 16	Nov 24
1997	—	Jan 19	Mar 30	Jan 24	Jan 17	Nov 27
1998	May 12 *O*	Feb 23	Apr 13	Jan 28	Jan 19	Nov 30
1999	—	Apr 1	Apr 27	Feb 2	Jan 22	Dec 2
2000	Jul 1	May 8	May 10	Feb 6	Jan 24 *O*	Dec 4
2001	—	Jun 14	May 25	Feb 9	Jan 26 *O*	Dec 7
2002	Aug 10	Jul 20	Jun 9	Feb 13	Jan 28 *O*	Dec 9
2003	—	Aug 22	Jun 24	Feb 17	Jan 30 *O*	Dec 12
2004	Sep 15	Sep 21	Jul 8	Feb 22	Feb 2 *O*	Dec 13
2005	—	Oct 22	Jul 23	Feb 25	Feb 3 *O*	Dec 16
2006	Oct 23	Nov 21	Aug 7	Mar 1	Feb 6 *O*	Dec 18
2007	—	Dec 23 *O*	Aug 21	Mar 5	Feb 8 *O*	Dec 21
2008	Dec 5	—	Sep 4	Mar 8	Feb 11	Dec 22
2009	—	Jan 24	Sep 17	Mar 13	Feb 12	Dec 24
2010	—	Feb 28	Oct 1	Mar 17	Feb 14	Dec 27
2011	Feb 4	Apr 6	Oct 13	Mar 21	Feb 17	Dec 29
2012	—	May 13	Oct 25	Mar 24	Feb 19	Dec 30
2013	Apr 18	Jun 19 *O*	Nov 6	Mar 29	Feb 21	—
2014	—	Jul 24	Nov 18	Apr 2	Feb 23	Jan 1
2015	Jun 14	Aug 26	Nov 30	Apr 6	Feb 26	Jan 3
2016	—	Sep 26	Dec 10	Apr 9	Feb 28	Jan 6
2017	Jul 27	Oct 26	Dec 21	Apr 14	Mar 2	Jan 7
2018	—	Nov 26	—	Apr 18	Mar 4	Jan 9
2019	Sep 2	Dec 27 *O*	Jan 2	Apr 22	Mar 7	Jan 11 *O*
2020	—	—	Jan 13 *O*	Apr 26	Mar 8	Jan 13

O means that the planet is occulted by the Sun's disk

XIV

EARTH IN PERIHELION

1980 Jan	3	15 h	0.983255 a.u.
1981 Jan	2	2	319
1982 Jan	4	11	295
1983 Jan	2	15	263
1984 Jan	3	22	279
1985 Jan	3	20	0.983221
1986 Jan	2	5	279
1987 Jan	4	23	301
1988 Jan	3	24	252
1989 Jan	1	22	308
1990 Jan	4	17	0.983303
1991 Jan	3	3	281
1992 Jan	3	15	324
1993 Jan	4	3	283
1994 Jan	2	6	301
1995 Jan	4	11	0.983302
1996 Jan	4	7	223
1997 Jan	1	23	267
1998 Jan	4	21	300
1999 Jan	3	13	281
2000 Jan	3	5	0.983321
2001 Jan	4	9	286
2002 Jan	2	14	290
2003 Jan	4	5	320
2004 Jan	4	18	265
2005 Jan	2	1	0.983297
2006 Jan	4	16	327
2007 Jan	3	20	260
2008 Jan	2	24	280
2009 Jan	4	15	273
2010 Jan	3	0	0.983290
2011 Jan	3	19	341
2012 Jan	5	1	284
2013 Jan	2	5	290
2014 Jan	4	12	335
2015 Jan	4	7	0.983277
2016 Jan	2	23	304
2017 Jan	4	14	309
2018 Jan	3	6	284
2019 Jan	3	5	301
2020 Jan	5	8	244

EARTH IN APHELION

1980 Jul	5	18 h	1.016725 a.u.
1981 Jul	3	23	749
1982 Jul	4	14	683
1983 Jul	6	10	744
1984 Jul	3	6	759
1985 Jul	5	10	1.016689
1986 Jul	5	10	745
1987 Jul	4	1	737
1988 Jul	5	24	717
1989 Jul	4	12	723
1990 Jul	4	5	1.016652
1991 Jul	6	15	703
1992 Jul	3	12	740
1993 Jul	4	22	666
1994 Jul	5	19	724
1995 Jul	4	2	1.016742
1996 Jul	5	19	717
1997 Jul	4	19	754
1998 Jul	3	24	696
1999 Jul	6	23	718
2000 Jul	3	24	1.016741
2001 Jul	4	14	643
2002 Jul	6	4	688
2003 Jul	4	6	728
2004 Jul	5	11	694
2005 Jul	5	5	1.016742
2006 Jul	3	23	697
2007 Jul	6	24	706
2008 Jul	4	8	754
2009 Jul	4	2	666
2010 Jul	6	11	1.016702
2011 Jul	4	15	740
2012 Jul	5	4	675
2013 Jul	5	15	708
2014 Jul	4	0	682
2015 Jul	6	20	1.016682
2016 Jul	4	16	751
2017 Jul	3	20	676
2018 Jul	6	17	696
2019 Jul	4	22	754
2020 Jul	4	12	694

XV

MARS IN PERIHELION	MARS IN APHELION

1960 May 26	14 *h*	1.38156 *a.u.*	1961 May 5	4 *h*	1.66586 *a.u.*
1962 Apr 13	13	1.38138	1963 Mar 23	1	1.66586
1964 Feb 29	16	1.38151	1965 Feb 6	23	1.66594
1966 Jan 16	10	1.38149	1966 Dec 26	3	1.66600
1967 Dec 4	12	1.38123	1968 Nov 12	0	1.66601
1969 Oct 21	16	1.38135	1970 Sep 30	1	1.66603
1971 Sep 8	14	1.38152	1972 Aug 17	4	1.66598
1973 Jul 26	12	1.38139	1974 Jul 5	2	1.66589
1975 Jun 13	16	1.38141	1976 May 21	23	1.66591
1977 Apr 30	11	1.38154	1978 Apr 9	2	1.66603
1979 Mar 18	10	1.38125	1980 Feb 25	1	1.66605
1981 Feb 2	16	1.38126	1982 Jan 12	1	1.66602
1982 Dec 21	15	1.38151	1983 Nov 30	3	1.66592
1984 Nov 7	10	1.38148	1985 Oct 17	1	1.66591
1986 Sep 25	14	1.38133	1987 Sep 3	22	1.66599
1988 Aug 12	13	1.38154	1989 Jul 22	1	1.66604
1990 Jun 30	9	1.38133	1991 Jun 9	2	1.66602
1992 May 17	14	1.38120	1993 Apr 25	23	1.66604
1994 Apr 4	16	1.38141	1995 Mar 14	2	1.66599
1996 Feb 20	11	1.38149	1997 Jan 29	2	1.66592
1998 Jan 7	11	1.38133	1998 Dec 16	22	1.66590
1999 Nov 25	13	1.38150	2000 Nov 2	21	1.66599
2001 Oct 12	7	1.38141	2002 Sep 21	1	1.66613
2003 Aug 30	11	1.38115	2004 Aug 7	24	1.66614
2005 Jul 17	16	1.38130	2006 Jun 26	1	1.66603
2007 Jun 4	13	1.38148	2008 May 13	2	1.66594
2009 Apr 21	10	1.38133	2010 Mar 30	23	1.66594
2011 Mar 9	14	1.38138	2012 Feb 15	21	1.66598
2013 Jan 24	9	1.38149	2014 Jan 3	0	1.66606
2014 Dec 12	8	1.38121	2015 Nov 20	23	1.66606
2016 Oct 29	13	1.38124	2017 Oct 7	22	1.66609
2018 Sep 16	13	1.38144	2019 Aug 26	1	1.66606
2020 Aug 3	9	1.38138			

XVI
THE OUTER PLANETS IN PERIHELION AND APHELION

Jupiter	perihelion	1951 Nov 21	4.94841 *a.u.*
	aphelion	1957 Oct 23	5.45641
	perihelion	1963 Sep 26	4.95133
	aphelion	1969 Sep 3	5.45374
	perihelion	1975 Aug 12	4.95264
	aphelion	1981 Jul 28	5.45351
	perihelion	1987 Jul 10	4.95242
	aphelion	1993 Jun 14	5.45428
	perihelion	1999 May 20	4.95047
	aphelion	2005 Apr 14	5.45652
	perihelion	2011 Mar 17	4.94839
	aphelion	2017 Feb 17	5.45652
Saturn	aphelion	1900 Jul 9	10.06987 *a.u.*
	perihelion	1915 Feb 21	9.01353
	aphelion	1929 Nov 11	10.04668
	perihelion	1944 Sep 8	9.02885
	aphelion	1959 May 29	10.06641
	perihelion	1974 Jan 8	9.01531
	aphelion	1988 Sep 11	10.04440
	perihelion	2003 Jul 26	9.03090
	aphelion	2018 Apr 17	10.06565
	perihelion	2032 Nov 28	9.01492
	aphelion	2047 Jul 15	10.04615
Uranus	aphelion	1925 Apr 1	20.09732 *a.u.*
	perihelion	1966 May 21	18.28479
	aphelion	2009 Feb 27	20.09888
	perihelion	2050 Aug 17	18.28302
Neptune	perihelion	1876 Aug 28	29.81484 *a.u.*
	(a)	1881 Dec 12	29.82133
	(b)	1886 Jul 11	29.81741
	aphelion	1959 Jul 13	30.33174
	(c)	1965 Oct 6	30.32272
	(d)	1968 Nov 21	30.32407
	perihelion	2042 Sep 5	29.80642
	(a)	2049 Oct 24	29.816711
	(b)	2050 Jun 25	29.816696
Pluto	perihelion	1989 Sep 4	29.6558 *a.u.*
	aphelion	2114 Feb	49.32

(a) Maximum of the radius vector, which does not correspond to an aphelion.
(b) Second minimum of the radius vector.
(c) Minimum of the radius vector, which does not correspond to a perihelion.
(d) Second minimum of the radius vector.

XVII

MARS IN ASCENDING NODE			
1960	Sep	25	11 *h*
1962	Aug	13	9
1964	Jun	30	9
1966	May	18	8
1968	Apr	4	6
1970	Feb	20	5
1972	Jan	8	5
1973	Nov	25	4
1975	Oct	13	3
1977	Aug	30	2
1979	Jul	18	0
1981	Jun	3	23
1983	Apr	21	23
1985	Mar	8	22
1987	Jan	24	20
1988	Dec	11	20
1990	Oct	29	19
1992	Sep	15	17
1994	Aug	3	17
1996	Jun	20	16
1998	May	8	15
2000	Mar	25	14
2002	Feb	10	13
2003	Dec	29	11
2005	Nov	15	11
2007	Oct	3	11
2009	Aug	20	9
2011	Jul	8	8
2013	May	25	8
2015	Apr	12	6
2017	Feb	27	5
2019	Jan	15	5
2020	Dec	2	3

MARS IN DESCENDING NODE			
1961	Oct	13	2 *h*
1963	Aug	31	1
1965	Jul	18	0
1967	Jun	4	24
1969	Apr	21	22
1971	Mar	9	21
1973	Jan	24	21
1974	Dec	12	19
1976	Oct	29	18
1978	Sep	16	18
1980	Aug	3	16
1982	Jun	21	15
1984	May	8	14
1986	Mar	26	13
1988	Feb	11	12
1989	Dec	29	11
1991	Nov	16	10
1993	Oct	3	9
1995	Aug	21	7
1997	Jul	8	6
1999	May	26	5
2001	Apr	12	5
2003	Feb	28	4
2005	Jan	15	3
2006	Dec	3	1
2008	Oct	20	0
2010	Sep	6	23
2012	Jul	24	22
2014	Jun	11	21
2016	Apr	28	20
2018	Mar	16	19
2020	Feb	1	18

XVIII

ZERO AND EXTREME LATITUDES OF THE OUTER PLANETS

Jupiter	1972 Sep 14	Descending node	
	1975 Jul 2	Greatest heliocentric latitude South :	$-1°18'20''$
	1978 Apr 10	Ascending node	
	1981 May 28	Greatest heliocentric latitude North :	$+1°18'21''$
	1984 Jul 27	Descending node	
	1987 May 15	Greatest heliocentric latitude South :	$-1°18'19''$
	1990 Feb 19	Ascending node	
	1993 Apr 7	Greatest heliocentric latitude North :	$+1°18'18''$
	1996 Jun 6	Descending node	
	1999 Mar 25	Greatest heliocentric latitude South :	$-1°18'17''$
	2001 Dec 30	Ascending node	
	2005 Feb 17	Greatest heliocentric latitude North :	$+1°18'13''$
	2008 Apr 17	Descending node	
	2011 Feb 1	Greatest heliocentric latitude South :	$-1°18'12''$
	2013 Nov 9	Ascending node	
	2016 Dec 28	Greatest heliocentric latitude North :	$+1°18'11''$
	2020 Feb 26	Descending node	
Saturn	1946 Feb 28	Ascending node	
	1953 Mar 22	Greatest heliocentric latitude North :	$+2°29'20''$
	1961 Apr 4	Descending node	
	1968 Nov 23	Greatest heliocentric latitude South :	$-2°29'19''$
	1975 Aug 10	Ascending node	
	1982 Aug 27	Greatest heliocentric latitude North :	$+2°29'09''$
	1990 Sep 4	Descending node	
	1998 Apr 20	Greatest heliocentric latitude South :	$-2°29'07''$
	2005 Jan 8	Ascending node	
	2012 Jan 31	Greatest heliocentric latitude North :	$+2°29'13''$
	2020 Feb 13	Descending node	
Uranus	1945 Jul 19	Ascending node	
	1965 May 25	Greatest heliocentric latitude North :	$+0°46'21''$
	1984 Dec 21	Descending node	
	2007 Jan 3	Greatest heliocentric latitude South :	$-0°46'20''$
Neptune	1920 Jun 3	Ascending node	
	1961 Dec 19	Greatest heliocentric latitude North :	$+1°46'24''$
	2003 Aug 11	Descending node	
Pluto	1930 Sep 9	Ascending node *(a)*	
	1980 Feb 4	Greatest heliocentric latitude North :	$+17°08'15''$
	2018 Oct 24	Descending node *(b)*	

(a) Passage through the node on the ecliptic of 2000.0 on 1930 September 1

(b) Passage through the node on the ecliptic of 2000.0 on 2018 October 25

XIX
ZERO AND EXTREME VALUES
OF THE DECLINATIONS OF THE PLANETS

Venus

Each year Venus reaches its greatest northern declination. Sometimes this extreme value is less than $+22°$, as in 1993. During the period $1960 - 2020$, Venus' declination reaches a maximum greater than $+27°$ on the following dates:

1964	May	6	$+27°33'$
1972	May	6	$+27°37'$
1980	May	6	$+27°41'$
1988	May	6	$+27°44'$
1996	May	5	$+27°47'$ (greatest northern declination of Venus
2004	May	5	$+27°49'$ during the 20th century)
2012	May	4	$+27°49'$
2020	May	4	$+27°49'$

During the same period $1960 - 2020$, Venus reaches a maximum southern declination south of $\delta = -26°$ on the following dates:

1965	Nov	7	$-26°45'$
1973	Nov	7	$-26°48'$
1981	Nov	7	$-26°52'$
1989	Nov	7	$-26°57'$
1997	Nov	6	$-27°01'$
2005	Nov	6	$-27°05'$
2013	Nov	6	$-27°10'$

Venus' southernmost declination during the 20th century was $-27°51'$, on 1906 November 3. On 1874 November 6, Venus reached an extreme southern declination of $-28°05'$. Venus' southernmost declination during the 21st century will be $-28°00'$, on 2093 November 6. On 2125 November 6, the planet will reach an extreme southern declination of $-28°11'$.

Mars

During the period $1900 - 2020$, Mars reaches a maximum northern declination north of $+26°$, or a maximum southern declination south of $-28°$, on the following dates:

1914	Jan	26	$+27°11'$		1907	Jul	29	$-28°54'$
1929	Jan	1	$+26°47'$		1954	Jul	19	$-28°22'$
1946	Feb	7	$+26°35'$		1986	Aug	2	$-28°43'$
1961	Jan	17	$+27°13'$					
1975	Dec	20	$+26°04'$					
1993	Jan	30	$+27°01'$					
2008	Jan	7	$+26°59'$					

Jupiter	1972	Oct	10	−23°28′51″	
	1975	Mar	28	0°	S → N
	1978	Apr	21	+23°27′06″	
	1980	Nov	10	0°	N → S ⎤
	1981	Apr	9	0°	S → N ⎟
	1981	Jul	8	0°	N → S ⎦
	1984	Sep	17	−23°29′45″	
	1987	Mar	13	0°	S → N
	1990	Mar	27	+23°29′27″	
	1992	Oct	22	0°	N → S
	1996	Sep	11	−23°23′58″	
	1999	Feb	24	0°	S → N
	2002	Mar	12	+23°27′24″	
	2004	Oct	6	0°	N → S
	2007	Dec	22	−23°15′15″	
	2010	Jul	8	0°	S → N ⎤
	2010	Jul	31	0°	N → S ⎟
	2011	Feb	5	0°	S → N ⎦
	2014	Mar	11	+23°16′15″	
	2016	Sep	21	0°	N → S
	2019	Dec	7	−23°18′15″	
Saturn	1974	May	14	+22°45′05″	
	1980	Nov	1	0°	N → S ⎤
	1981	Mar	30	0°	S → N ⎟
	1981	Jul	29	0°	N → S ⎦
	1989	Oct	2	−22°46′51″	
	1996	May	25	0°	S → N ⎤
	1996	Aug	30	0°	N → S ⎟
	1997	Feb	16	0°	S → N ⎦
	2004	Apr	2	+22°48′46″	
	2010	Sep	8	0°	N → S
	2018	Oct	21	−22°46′17″	
Uranus	1950	Feb	27	+23°42′51″	
	1968	Oct	25	0°	N → S ⎤
	1969	Mar	26	0°	S → N ⎟
	1969	Aug	11	0°	N → S ⎦
	1989	Sep	1	−23°42′27″	
	2011	Apr	9	0°	S → N ⎤
	2011	Oct	16	0°	N → S ⎟
	2012	Jan	28	0°	S → N ⎦
Neptune	1943	Nov	1	0°	N → S ⎤
	1944	Mar	7	0°	S → N ⎟
	1944	Sep	4	0°	N → S ⎦
	1986	Oct	29	−22°21′37″	
Pluto	1946	Apr	9	+24°06′43″	
	1987	Nov	12	0°	N → S ⎤
	1988	Feb	27	0°	S → N ⎟
	1988	Sep	5	0°	N → S ⎦

S → N means that the planet crosses the celestial equator from South to North.

N → S means that the planet crosses the celestial equator from North to South.

Triple crossings are indicated with a bracket.

XX
EXTREME DECLINATIONS OF THE SUN

Year	June Solstice	December Solstice	Year	June Solstice	December Solstice
1985	+23°26′35″	−23°26′35″	2003	+23°26′24″	−23°26′25″
1986	35	35	2004	26	27
1987	36	36	2005	26	27
1988	35	34	2006	27	27
1989	33	33	2007	26	24
1990	32	30	2008	24	24
1991	28	26	2009	22	19
1992	25	23	2010	18	16
1993	20	19	2011	15	12
1994	18	17	2012	10	09
1995	15	14	2013	08	08
1996	13	14	2014	06	05
1997	14	13	2015	04	05
1998	13	14	2016	05	04
1999	15	16	2017	04	05
2000	16	17	2018	07	07
2001	19	21	2019	08	10
2002	22	22	2020	11	14

XXI
PLANETARY CONJUNCTIONS

(a)   MERCURY − VENUS

1990 Feb 4	6 h	−7°06′	25°W		2004 Jun 13	1 h	+1°21′	7°W
1990 Sep 14	15	−3 21	11 W		2004 Dec 29	5	+1 12	22 W
1990 Oct 16	5	+0 02	4 W		2005 Jan 14	1	−0 21	19 W
1990 Dec 18	23	+1 27	12 E		2005 Mar 28	23	+4 49	1 W
1991 Aug 7	6	+2 08	22 E		2005 Jun 27	20	−0 05	23 E
1991 Aug 20	19	+3 43	5 E		2005 Jul 7	8	−1 38	26 E
1991 Aug 29	5	+6 04	12 W		2006 Jan 17	2	−7 53	6 W
1992 Apr 5	22	+2 20	17 W		2006 Nov 7	14	−1 14	3 E
1992 May 28	2	+0 30	4 W		2007 Aug 15	23	+10 04	2 E
1992 Jul 25	15	−6 03	12 E		2008 Feb 26	3	+1 20	26 W
1993 Apr 16	11	−8 25	23 W		2008 Mar 23	10	−1 03	20 W
1993 Nov 14	13	+0 45	15 W		2008 Jun 8	1	−2 59	0 W
1993 Dec 25	8	−0 59	5 W		2008 Aug 23	5	−1 15	21 E
1994 Feb 15	19	+5 04	7 E		2008 Sep 11	5	−3 34	25 E
1994 Nov 12	18	+5 26	16 W		2009 Mar 27	12	−10 37	4 W
1995 Jun 19	7	−3 36	17 W		2010 Jan 5	8	+3 26	2 W
1995 Jul 20	14	+0 23	9 W		2010 Oct 24	11	+7 11	5 E
1995 Sep 28	21	−5 01	11 E		2011 Aug 15	23	−6 21	1 W
1996 Jun 23	12	+1 36	19 W		2012 Jun 1	21	+0 12	7 E
1997 Jan 12	13	+2 43	19 W		2013 Mar 6	7	+5 20	5 W
1997 Mar 2	0	−0 51	8 W		2013 May 25	4	+1 22	15 E
1997 Apr 21	18	+3 04	5 E		2013 Jun 20	18	−1 57	22 E
1998 Jan 26	17	−8 11	17 W		2014 Jan 7	10	−6 27	6 E
1998 Aug 25	23	−2 45	16 W		2014 Oct 17	8	−2 44	2 W
1998 Sep 11	0	+0 21	13 W		2015 Aug 5	9	+8 11	13 E
1998 Nov 28	12	+0 15	7 E		2016 May 13	21	−0 26	7 W
1999 Aug 26	12	+10 10	13 W		2016 Jul 16	18	+0 32	11 E
2000 Mar 15	0	+2 30	22 W		2016 Aug 27	5	−5 16	22 E
2000 Apr 28	9	−0 21	12 W		2017 Mar 16	23	−9 33	10 E
2000 Jul 2	9	−4 57	6 E		2017 Dec 15	16	+2 13	6 W
2001 Apr 6	21	−10 00	14 W		2018 Mar 5	18	+1 24	13 E
2002 Jan 26	1	+4 28	3 E		2018 Mar 18	1	+3 53	16 E
2002 Nov 3	8	+6 34	7 W		2018 Oct 14	15	+6 49	16 E
2003 May 28	0	−2 21	22 W		2019 Jul 24	11	−5 43	6 W
2003 Jun 21	2	−0 25	16 W		2019 Sep 13	22	−0 20	8 E
2003 Sep 7	3	−6 07	5 E		2019 Oct 30	8	−2 43	20 E
					2020 May 22	8	−0 54	19 E

(b) MERCURY - MARS

1991 Oct 14	19 h	−0°15'	8°E
1991 Dec 13	15	+2 56	11 W
1992 Jan 10	20	+0 39	19 W
1993 Oct 6	17	−2 20	23 E
1993 Oct 28	6	−2 29	17 E
1994 Jan 1	10	−0 50	2 W
1994 Feb 27	1	+4 26	14 W
1994 Apr 4	2	−1 28	22 W
1995 Dec 23	9	−1 08	16 E
1996 Jan 13	0	+2 48	12 E
1996 Mar 23	20	−0 54	4 W
1996 May 31	5	−3 42	19 W
1996 Jun 14	14	−3 05	22 W
1998 Mar 11	14	+1 11	15 E
1998 Mar 30	5	+4 06	10 E
1998 Jun 5	11	−0 16	6 W
2000 May 19	15	+1 07	12 E
2000 Jul 7	3	−5 40	2 W
2000 Aug 10	13	−0 05	12 W
2002 Jul 25	12	+0 40	5 E
2004 Jul 10	23	+0 10	22 E
2004 Aug 17	3	−6 14	10 E
2004 Sep 29	12	+0 51	5 W
2006 Sep 15	21	−0 10	12 E
2006 Nov 11	18	+0 39	6 W
2006 Dec 9	20	+1 02	15 W
2008 Sep 12	21	−3 26	25 E
2008 Sep 19	5	−4 08	23 E
2008 Nov 29	6	−0 34	2 E
2009 Jan 26	18	+4 25	13 W
2009 Mar 1	20	−0 36	22 W
2010 Nov 21	1	−1 41	19 E
2010 Dec 14	4	+1 02	13 E
2011 Feb 20	14	−1 04	4 W
2011 Apr 19	8	+0 47	15 W
2011 May 20	1	−2 21	22 W
2013 Feb 8	21	+0 18	15 E
2013 Feb 24	23	+4 15	12 E
2013 May 7	22	−0 26	5 W
2015 Apr 23	7	+1 23	14 E
2015 May 27	15	−1 41	5 E
2015 Jul 16	4	−0 08	9 W
2017 Jun 28	18	+0 47	9 E
2017 Sep 2	0	−4 06	11 W
2017 Sep 16	18	+0 03	17 W

(c) MERCURY - JUPITER

1990 Jul 7	8 h	+1°30'	6°E
1991 Jul 15	8	−0 05	25 E
1991 Aug 22	22	−5 20	4 W
1991 Sep 10	10	−0 04	18 W
1992 Sep 16	6	+0 29	2 E
1993 Sep 24	12	−2 01	18 E
1994 Nov 28	18	−0 24	9 W
1995 Dec 8	10	−2 08	8 E
1997 Feb 12	14	−1 02	19 W
1998 Feb 22	14	−1 05	1 E
1999 May 1	10	−1 45	22 W
2000 May 8	22	+0 52	0 W
2001 May 16	17	+2 47	21 E
2001 Jun 18	15	−3 39	3 W
2001 Jul 12	22	−1 56	21 W
2002 Jul 20	13	+1 15	0 W
2003 Jul 26	1	+0 23	20 E
2004 Sep 28	21	+0 40	5 W
2005 Oct 6	7	−1 28	13 E
2006 Oct 25	22	−3 56	21 E
2006 Oct 28	17	−3 43	19 E
2006 Dec 10	17	+0 08	15 W
2007 Dec 20	22	−1 48	2 E
2008 Dec 31	6	−1 17	19 E
2009 Jan 18	6	+3 15	5 E
2009 Feb 24	3	−0 37	24 W
2010 Mar 7	19	−1 11	6 W
2011 Mar 16	17	+2 20	16 E
2011 Apr 10	20	+3 31	3 W
2011 May 10	23	−2 12	25 W
2012 May 22	7	+0 24	6 W
2013 May 27	10	+2 22	17 E
2014 Aug 2	17	+0 58	6 W
2015 Aug 7	4	+0 35	15 E
2016 Oct 11	4	+0 52	12 W
2017 Oct 18	15	−1 01	6 E
2018 Oct 30	4	−3 16	21 E
2018 Nov 27	24	+0 28	2 W
2018 Dec 21	15	+0 52	20 W
2020 Jan 2	15	−1 30	5 W
2019 Jun 18	15 h	+0°14'	24°E
2019 Jul 7	14	−3 50	18 E
2019 Sep 3	11	+0 42	1 W

XXI (cont.)

(d)   MERCURY - SATURN

1989 Dec 16	22 h	−2°30′	19°E
1990 Jan 10	5	+2 49	3 W
1990 Feb 3	15	+0 13	25 W
1991 Feb 5	16	−1 14	16 W
1992 Feb 4	16	−1 28	5 W
1993 Feb 1	22	−0 55	7 E
1994 Feb 2	4	+1 21	17 E
1994 Feb 17	16	+5 19	4 E
1994 Mar 24	8	−0 15	27 W
1995 Mar 26	0	−0 35	18 W
1996 Mar 23	11	+0 20	5 W
1997 Mar 21	3	+2 06	9 E
1998 May 12	16	−0 49	25 W
1999 May 13	18	+0 40	14 W
2000 May 10	3	+2 16	1 E
2001 May 7	16	+3 40	15 E
2002 Jul 2	11	−0 14	19 W
2003 Jul 1	1	+1 33	5 W
2004 Jun 26	21	+2 06	10 E
2005 Jun 26	6	+1 25	22 E
2006 Aug 20	23	+0 31	11 W
2007 Aug 18	11	+0 30	3 E
2008 Aug 16	0	−0 42	16 E
2009 Aug 18	21	−3 27	25 E
2009 Sep 20	12	−5 24	3 W
2009 Oct 8	9	−0 19	18 W
2010 Oct 8	15	−0 35	7 W
2011 Oct 7	9	−1 52	6 E
2012 Oct 6	7	−3 29	17 E
2013 Oct 10	19	−5 24	24 E
2013 Oct 28	21	−4 06	8 E
2013 Nov 26	4	−0 20	17 W
2014 Nov 26	9	−1 43	7 W
2015 Nov 25	13	−2 46	4 E
2016 Nov 24	1	−3 28	15 E
2017 Nov 28	9	−3 03	21 E
2017 Dec 6	12	−1 21	14 E
2018 Jan 13	7	−0 39	20 W
2019 Jan 13	11	−1 43	10 W
2020 Jan 12	5	−2 03	1 E

(e)   MERCURY - URANUS

1990 Dec 10	7 h	−1°20′	20°E
1990 Dec 18	6	+0 37	13 E
1991 Jan 23	17	+0 27	22 W
1992 Jan 20	3	−0 38	15 W
1993 Jan 14	8	−1 16	6 W
1994 Jan 9	2	−1 37	4 E
1995 Jan 4	1	−1 39	12 E
1996 Jan 1	1	−0 53	19 E
1996 Jan 16	15	+3 17	5 E
1996 Feb 16	22	+0 15	25 W
1997 Feb 13	0	−0 54	19 W
1998 Feb 8	5	−1 23	10 W
1999 Feb 2	18	−1 31	1 W
2000 Jan 28	4	−1 19	9 E
2001 Jan 22	17	−0 24	17 E
2001 Feb 14	13	+4 38	5 W
2001 Mar 10	11	+0 08	27 W
2002 Mar 9	3	−1 13	22 W
2003 Mar 4	13	−1 32	14 W
2004 Feb 26	23	−1 27	5 W
2005 Feb 20	1	−1 00	5 E
2006 Feb 14	16	+0 02	14 E
2007 Apr 1	7	−1 37	25 W
2008 Mar 27	10	−1 45	17 W
2009 Mar 21	22	−1 24	8 W
2010 Mar 15	18	−0 44	2 E
2011 Mar 9	18	+0 22	11 E
2012 Mar 6	24	+3 06	17 E
2012 Mar 16	2	+4 36	8 E
2012 Apr 22	2	−2 08	26 W
2013 Apr 19	21	−2 02	20 W
2014 Apr 14	16	−1 23	11 W
2015 Apr 8	10	−0 31	2 W
2016 Mar 31	24	+0 38	9 E
2017 Mar 27	6	+2 25	17 E
2017 Apr 28	18	−0 09	13 W
2017 May 7	23	−2 14	22 W
2018 May 12	21	−2 24	22 W
2019 May 8	8	−1 23	14 W
2020 May 1	2	−0 19	4 W

XXI (cont.)

(f) MERCURY – NEPTUNE				
1989 Dec 15	4 h	−3°07′	18°E	
1990 Jan 13	20	+2 33	11 W	
1990 Jan 27	23	+0 46	25 W	
1991 Jan 26	14	−1 04	21 W	
1992 Jan 21	11	−1 53	14 W	
1993 Jan 14	11	−2 22	5 W	
1994 Jan 8	5	−2 40	3 E	
1995 Jan 2	2	−2 45	11 E	
1995 Dec 28	2	−2 25	19 E	
1996 Jan 20	13	+2 57	4 W	
1996 Feb 11	14	+0 04	26 W	
1997 Feb 7	20	−1 21	21 W	
1998 Feb 2	11	−1 59	13 W	
1999 Jan 27	8	−2 18	5 W	
2000 Jan 20	22	−2 23	4 E	
2001 Jan 13	17	−2 13	12 E	
2002 Jan 9	5	−1 20	19 E	
2002 Jan 26	13	+3 13	2 E	
2002 Feb 24	13	−0 30	26 W	
2003 Feb 20	24	−1 34	20 W	
2004 Feb 15	9	−1 58	13 W	
2005 Feb 8	1	−2 04	4 W	
2006 Feb 1	12	−1 57	4 E	
2007 Jan 26	7	−1 28	13 E	
2008 Jan 23	4	+0 20	18 E	
2008 Feb 1	20	+3 11	9 E	
2008 Mar 9	3	−0 56	26 W	
2009 Mar 5	1	−1 39	20 W	
2010 Feb 27	5	−1 48	12 W	
2011 Feb 20	17	−1 41	3 W	
2012 Feb 14	1	−1 18	6 E	
2013 Feb 6	21	−0 28	14 E	
2014 Mar 22	12	−1 15	26 W	
2015 Mar 17	24	−1 36	19 W	
2016 Mar 10	22	−1 30	11 W	
2017 Mar 4	6	−1 08	2 W	
2018 Feb 25	10	−0 29	7 E	
2019 Feb 19	11	+0 46	15 E	
2019 Mar 22	7	+3 24	13 W	
2019 Apr 2	19	+0 23	25 W	
2020 Apr 3	15	−1 25	25 W	

(g) VENUS – MARS				
1991 Jun 23	11 h	+0°16′	45°E	
1991 Jul 22	6	−3 36	35 E	
1992 Feb 19	22	+0 51	29 W	
1994 Jan 6	6	+0 18	3 W	
1995 Nov 22	22	−0 11	24 E	
1996 Jun 30	4	−4 12	26 W	
1996 Sep 4	15	−2 51	45 W	
1997 Oct 26	23	−2 07	47 E	
1997 Dec 22	12	+1 07	33 E	
1998 Aug 5	3	−0 51	23 W	
2000 Jun 21	20	−0 18	3 E	
2002 May 10	21	+0 18	28 E	
2004 Dec 5	7	+1 15	27 W	
2006 Oct 24	20	+0 43	1 W	
2008 Sep 11	21	+0 20	26 E	
2009 Apr 18	16	+5 36	30 W	
2009 Jun 19	14	−2 02	45 W	
2010 Aug 23	21	−2 27	45 E	
2010 Sep 29	6	−6 30	34 E	
2011 May 22	15	−1 03	23 W	
2013 Apr 6	16	−0 42	3 E	
2015 Feb 21	20	−0 28	28 E	
2015 Aug 29	5	−9 25	21 W	
2015 Nov 3	16	−0 42	46 W	
2017 Oct 5	13	+0 13	23 W	
2019 Aug 24	13	+0 19	3 E	

XXI (cont.)

(h)   VENUS - JUPITER

1990 Aug 12	23 h	+0°03′	21°W
1991 Jun 17	23	+1 14	45 E
1991 Aug 23	7	−9 36	4 W
1991 Oct 17	4	−2 28	45 W
1992 Aug 23	3	+0 17	19 E
1993 Nov 8	17	+0 23	17 W
1995 Jan 14	9	+2 49	46 W
1995 Nov 19	12	−1 16	23 E
1997 Feb 5	24	−0 20	14 W
1998 Apr 23	2	+0 18	45 W
1999 Feb 23	21	+0 09	28 E
2000 May 17	11	+0 01	7 W
2001 Aug 5	24	−1 12	38 W
2002 Jun 3	18	+1 39	34 E
2003 Aug 21	6	+0 34	1 E
2004 Nov 4	21	+0 36	34 W
2005 Sep 2	12	−1 22	39 E
2006 Nov 15	23	−0 27	5 E
2008 Feb 1	13	+0 35	32 W
2008 Dec 1	1	−2 02	43 E
2010 Feb 16	21	−0 35	9 E
2011 May 11	9	−0 37	26 W
2012 Mar 15	11	+3 16	45 E
2013 May 28	21	+1 00	16 E
2014 Aug 18	4	+0 12	18 W
2015 Jul 1	14	−0 24	42 E
2015 Jul 31	20	−6 26	19 E
2015 Oct 26	8	−1 04	46 W
2016 Aug 27	22	+0 04	22 E
2017 Nov 13	6	+0 17	14 W
2019 Jan 22	6	+2 26	46 W
2019 Nov 24	14	−1 24	26 E

(i)   VENUS - SATURN

1989 Nov 15	19 h	−3°54′	47°E
1990 Feb 7	4	+7 06	28 W
1990 Feb 14	18	+6 32	35 W
1991 Jan 1	15	−1 11	15 E
1992 Feb 29	2	+0 08	27 W
1992 Dec 21	16	−1 04	45 E
1994 Feb 14	3	−0 02	7 E
1995 Apr 13	17	+0 34	34 W
1996 Feb 3	2	+1 16	39 E
1997 Mar 31	21	+0 57	1 W
1998 May 29	2	+0 16	39 W
1999 Mar 20	21	+2 34	32 E
2000 May 18	20	+1 14	6 W
2001 Jul 15	5	−0 44	42 W
2002 May 7	18	+2 25	27 E
2003 Jul 8	8	+0 49	11 W
2004 Sep 1	1	−1 57	45 W
2005 Jun 25	21	+1 18	23 E
2006 Aug 26	23	+0 04	16 W
2007 Jul 2	1	−0 46	43 E
2007 Aug 9	9	−8 29	11 E
2007 Oct 15	14	−2 55	46 W
2008 Aug 13	19	−0 14	18 E
2009 Oct 13	16	−0 34	22 W
2010 Aug 10	2	−3 08	44 E
2011 Sep 30	11	−1 24	12 E
2012 Nov 27	5	−0 34	29 W
2013 Sep 20	0	−3 45	42 E
2014 Nov 13	9	−1 36	5 E
2016 Jan 9	4	+0 05	36 W
2016 Oct 30	8	−3 02	37 E
2017 Dec 25	18	−1 08	3 W
2019 Feb 18	14	+1 05	43 W
2019 Dec 11	5	−1 49	30 E

XXI (cont.)

(j)	VENUS	–	URANUS			(k)	VENUS	–	NEPTUNE	
1990 Dec 19	10 h	−0°36′	12°E		1990 Dec 23	3 h	−1°51′	13°E
1992 Feb 7	7	+0 54	32 W		1992 Feb 8	15	−0 17	31 W
1992 Nov 26	11	−1 54	41 E		1992 Nov 27	13	−3 01	41 E
1994 Jan 13	13	−0 22	1 W		1994 Jan 12	6	−1 24	1 W
1995 Mar 2	5	+1 30	42 W		1995 Feb 26	10	+0 41	43 W
1995 Dec 20	13	−1 16	31 E		1995 Dec 16	17	−2 16	30 E
1997 Feb 7	12	−0 12	13 W		1997 Feb 1	11	−0 59	15 W
1998 Mar 19	7	+3 20	46 W		⌈ 1997 Dec 7	20	−2 46	42 E
1999 Jan 13	19	−0 56	18 E		1998 Jan 9	17	+4 00	10 E
2000 Mar 4	0	−0 04	25 W		⌊ 1998 Mar 7	10	+3 46	45 W
2000 Dec 23	21	−1 19	46 E		1999 Jan 5	8	−1 45	16 E
2002 Feb 7	10	−0 45	6 E		2000 Feb 22	6	−0 30	28 W
2003 Mar 28	13	+0 03	37 W		2000 Dec 11	21	−2 32	44 E
2004 Jan 15	1	−0 56	36 E		2002 Jan 25	12	−1 20	3 E
2005 Mar 4	4	−0 41	7 W		2003 Mar 12	20	+0 11	40 W
2006 Apr 18	12	+0 19	45 W		2003 Dec 30	7	−1 53	33 E
2007 Feb 7	13	−0 44	25 E		2005 Feb 14	19	−0 58	11 W
2008 Mar 28	17	−0 45	19 W		2006 Mar 26	21	+1 52	47 W
2009 Jan 23	16	+1 24	46 E		2007 Jan 18	18	−1 25	20 E
2010 Mar 3	23	−0 40	12 E		2008 Mar 6	20	−0 36	24 W
2011 Apr 22	19	−0 55	30 W		2008 Dec 27	2	−1 30	46 E
2012 Feb 10	5	+0 20	41 E		2010 Feb 7	23	−1 04	7 E
2013 Mar 28	17	−0 43	1 E		2011 Mar 27	1	−0 09	36 W
2014 May 15	13	−1 16	40 W		2012 Jan 13	7	−1 10	36 E
2015 Mar 4	20	+0 06	31 E		2013 Feb 28	8	−0 46	7 W
2016 Apr 22	14	−0 53	12 W		2014 Apr 12	8	+0 42	45 W
2017 Jun 2	15	−1 47	45 W		2015 Feb 1	11	−0 50	24 E
2018 Mar 29	0	−0 04	19 E		2016 Mar 20	14	−0 32	20 W
2019 May 18	8	−1 09	23 W		2017 Jan 13	2	+0 25	47 E
2020 Mar 9	15	+2 25	45 E		2018 Feb 21	14	−0 35	10 E
					2019 Apr 10	4	−0 18	33 W
					2020 Jan 27	19	−0 05	40 E

XXI (cont.)

(l) MARS - JUPITER

1991 Jun 14	5 h	+0°38′	48°E
1993 Sep 7	0	−0 54	32 E
1995 Nov 16	8	−1 11	26 E
1998 Jan 21	1	−0 12	26 E
2000 Apr 6	23	+1 06	23 E
2002 Jul 3	6	+0 49	12 E
2004 Sep 27	5	−0 12	4 W
2006 Dec 11	24	−0 49	16 W
2009 Feb 17	10	−0 35	19 W
2011 May 1	11	+0 24	18 W
2013 Jul 22	6	+0 47	24 W
2015 Oct 17	14	+0 25	40 W
2018 Jan 7	4	−0 13	59 W
2020 Mar 20	6	−0 43	67 W

(n) MARS - URANUS

1990 Feb 9	14 h	−0°13′	43°W
1992 Jan 29	21	−0 23	24 W
1994 Jan 18	13	−0 29	6 W
1996 Jan 7	24	−0 34	13 E
1997 Dec 26	20	−0 36	32 E
1999 Dec 14	5	−0 39	52 E
2001 Nov 26	10	−0 48	77 E
2003 Jun 20	23	−3 11	116 W
2005 May 14	20	−1 11	74 W
2007 Apr 28	19	−0 45	51 W
2009 Apr 15	4	−0 28	31 W
2011 Apr 3	18	−0 14	12 W
2013 Mar 22	18	+0 01	6 E
2015 Mar 11	20	+0 17	24 E
2017 Feb 27	8	+0 37	43 E
2019 Feb 13	20	+1 03	64 E

(m) MARS - SATURN

1990 Feb 28	17 h	−1°00′	48°W
1992 Mar 6	13	−0 26	33 W
1994 Mar 14	10	+0 22	18 W
1996 Mar 22	20	+1 16	4 W
1998 Apr 3	12	+2 01	9 E
2000 Apr 16	23	+2 22	20 E
2002 May 4	17	+2 13	30 E
2004 May 24	23	+1 35	37 E
2006 Jun 17	23	+0 35	42 E
2008 Jul 11	6	−0 42	46 E
2010 Aug 1	20	−1 56	52 E
2012 Aug 17	9	−2 54	60 E
2014 Aug 27	13	−3 34	74 E
2016 Aug 25	18	−4 23	97 E
2018 Apr 2	12	−1 16	94 W
2020 Mar 31	11	−0 55	71 W

(o) MARS - NEPTUNE

1990 Feb 17	6 h	−1°27′	45°W
1992 Feb 1	8	−1 31	24 W
1994 Jan 16	7	−1 33	5 W
1996 Jan 1	7	−1 35	14 E
1997 Dec 15	19	−1 38	34 E
1999 Nov 28	14	−1 45	56 E
2001 Nov 4	18	−2 11	83 E
2003 May 13	14	−2 02	99 W
2005 Apr 13	0	−1 15	66 W
2007 Mar 25	7	−1 00	43 W
2009 Mar 8	4	−0 48	23 W
2011 Feb 20	21	−0 38	3 W
2013 Feb 4	16	−0 26	16 E
2015 Jan 19	21	−0 14	36 E
2017 Jan 1	7	−0 01	59 E
2018 Dec 7	15	+0 02	88 E
2020 Jun 12	12	−1 45	91 W

XXI (cont.)

(p)   JUPITER – SATURN

1981 Jan	14	8 h	−1°09′	104°W
1981 Feb	19	7	−1 09	141 W
1981 Jul	30	22	−1 12	58 E
2000 May	31	10	+1 11	17 W
2020 Dec	21	14	−0 06	30 E

(q)   JUPITER – URANUS

1983 Feb	17	14 h	+0°46′	80°W
1983 May	16	13	+0 50	167 W
1983 Sep	24	22	+0 27	64 E
1997 Feb	16	8	+0 10	22 W
2010 Jun	6	19	−0 28	76 W
2010 Sep	22	19	−0 53	178 E
2011 Jan	2	14	−0 34	75 E

(r)   JUPITER – NEPTUNE

1984 Jan	19	17 h	−0°52′	29°W
1997 Jan	8	17	−0 47	8 E
2009 May	25	13	−0 24	98 W
2009 Jul	13	19	−0 37	146 W
2009 Dec	20	5	−0 34	56 E

(s)   SATURN – URANUS

1988 Feb	13	1 h	+1°17′	54°W
1988 Jun	27	2	+1 21	173 E
1988 Oct	18	3	+1 04	63 E

Next :   2032 June 28

(t)   SATURN – NEPTUNE

1989 Mar	3	2 h	−0°14′	61°W
1989 Jun	24	17	−0 18	172 W
1989 Nov	12	21	−0 30	50 E

Next :   triple conjunction in 2025 − 2026

(u)   URANUS – NEPTUNE

1993 Jan	25	20 h	−1°06′	17°W
1993 Sep	22	23	−1 08	108 E

*Notes about the Uranus-
Neptune conjunctions
of 1993 :*

The previous case was a triple conjunction in 1821.

The next case will be a triple conjunction in 2164.

The case of 1993 September 22 was an exceptional, limiting event. There was no actual conjunction (in right ascension) between the *centers* of the two planetary disks. But as the difference between the right ascensions reached a minimum which was almost zero, there was a conjunction between a *part* of the two tiny planetary disks — though in declination the two planets were separated by 1°08′.

XXII
CLOSE PLANETARY CONJUNCTIONS

Date	Planets	Conjunction in right ascension		Least Distance	
		UT	Δδ	UT	Dist.
		h m	"	h m	"
1990 Aug 12	Venus - Jupiter	23 24	+153	23 34	151
1990 Oct 16	Mercury - Venus	5 18	+103	6 02	88
1991 Jul 15	Mercury - Jupiter	7 53	−282	7 12	256
1991 Sep 10	Mercury - Jupiter	10 25	−221	10 11	217
1994 Feb 14	Venus - Saturn	2 40	− 92	2 51	86
1996 Feb 11	Mercury - Neptune	14 26	+230	14 24	230
2000 Mar 4	Venus - Uranus	0 14	−244	0 37	234
2000 May 17	Venus - Jupiter	10 35	+ 44	10 30	42
2000 Aug 10	Mercury - Mars	12 59	−292	12 47	289
2003 Mar 28	Venus - Uranus	13 04	+165	12 46	157
2005 Jun 27	Mercury - Venus	20 19	−295	16 01	233
2006 Feb 14	Mercury - Uranus	15 42	+ 95	15 31	83
2006 Aug 26	Venus - Saturn	23 09	+267	23 36	257
2013 Mar 22	Mars - Uranus	18 26	+ 43	18 17	39
2015 Mar 4	Venus - Uranus	19 31	+349	18 41	317
2016 Jan 9	Venus - Saturn	3 56	+310	4 14	306
2016 Aug 27	Venus - Jupiter	21 47	+265	22 31	240
2017 Jan 1	Mars - Neptune	6 38	− 74	6 54	68
2017 Sep 16	Mercury - Mars	18 23	+203	18 43	199
2018 Mar 29	Venus - Uranus	0 15	−264	0 50	243
2018 Dec 7	Mars - Neptune	14 46	+142	14 08	128
2020 Jan 27	Venus - Neptune	19 24	−270	20 04	244

XXIII
QUASI-CONJUNCTIONS

*Approaches of two bright planets within 5 degrees of each other,
without a conjunction in right ascension.*
In the last column, the least angular separation is given.

1996 Dec 22	15 *h*	Mercury - Jupiter	3°56'	
1997 Jul 27	14	Mercury - Venus	4°27'	
1999 Mar 6	8	Mercury - Jupiter	3°59'	
2001 Oct 29	22	Mercury - Venus	0°35' ⎫	(*)
2001 Nov 4	0	Mercury - Venus	0°39' ⎭	
2002 Oct 10	13	Mercury - Mars	2°50'	
2002 Dec 6	12	Venus - Mars	1°33'	
2006 Aug 10	21	Mercury - Venus	2°11'	
2010 Apr 4	8	Mercury - Venus	3°00'	
2011 May 8	6	Mercury - Venus	1°26' ⎫	(**)
2011 May 18	7	Mercury - Venus	1°21' ⎭	
2011 Nov 1	23	Mercury - Venus	1°58'	
2011 Nov 13	8	Mercury - Venus	1°58'	
2012 Jul 1	8	Venus - Jupiter	4°48'	
2015 Jan 11	1	Mercury - Venus	0°39'	
2016 Feb 13	3	Mercury - Venus	4°00'	
2016 Aug 20	6	Mercury - Jupiter	3°47'	
2019 Apr 16	19	Mercury - Venus	4°17'	
2020 May 18	5	Jupiter - Saturn	4°42'	

(*) The separation between Mercury and Venus will be less than 1° during eleven days. Although there is no conjunction in right ascension, there are two conjunctions in *celestial longitude*, namely on 2001 October 30 at 19[h] and on November 3 at 7[h].

(**) Although there is no conjunction in right ascension, there are two conjunctions in *celestial longitude*, namely on 2001 May 9 at 16[h] and on May 16 at 9[h].

XXIV
HELIOCENTRIC CONJUNCTIONS
BETWEEN THE OUTER PLANETS

Jupiter - Saturn	1961	Apr	16	293°41'
	1981	Apr	16	187 08
	2000	Jun	22	52 01
	2020	Nov	2	301 50
Jupiter - Uranus	1955	Jan	26	115°47'
	1969	Apr	3	182 08
	1983	Jun	12	247 17
	1997	Mar	28	305 20
	2010	Sep	24	358 37
Jupiter - Neptune	1958	Jul	20	213°56'
	1971	May	24	241 42
	1984	Mar	16	269 28
	1996	Dec	24	297 19
	2009	Sep	19	325 16
Saturn - Uranus	1988	Jun	9	268°56'
	2032	Jul	20	87 47
Saturn - Neptune	1953	Feb	18	202°12'
	1989	Jul	18	281 05
	2025	Dec	11	1 14

Uranus - Neptune   1993 Apr 20   289°16'
(Previous one in September 1821)
(Next one in December 2164)

Uranus - Pluto   1966 Jan 7   166°49'
(Previous one in Februari 1851)
(Next one in A.D. 2104)

Neptune - Pluto   1892 Jan 29   68°01'

XXV
OPPOSITIONS OF BRIGHT MINOR PLANETS

1 Ceres

Opposition				*Least*		*distance*	
Date	*UT*	*Declin.*		*Date*	*UT*	*Dist.*	*Magn.*
1991 Apr 17	16 *h*	+ 2°		1991 Apr 16	15 *h*	1.6489	7.0
1992 Jul 25	22	−30		1992 Jul 23	22	1.9467	7.5
1993 Oct 22	19	− 1		1993 Oct 24	11	1.8915	7.4
1995 Feb 3	1	+30		1995 Feb 3	24	1.5972	6.9
1996 May 29	24	−18		1996 May 27	21	1.7616	7.0
1997 Aug 30	3	−23		1997 Aug 29	8	1.9964	7.7
1998 Nov 28	22	+17		1998 Dec 1	6	1.7563	7.0
2000 Mar 22	9	+15		2000 Mar 22	4	1.6005	6.9
2001 Jul 7	18	−29		2001 Jul 5	6	1.8892	7.3
2002 Oct 4	7	− 9		2002 Oct 5	6	1.9590	7.6
2004 Jan 9	14	+30		2004 Jan 11	10	1.6258	6.8
2005 May 8	18	− 9		2005 May 7	1	1.6867	7.0
2006 Aug 12	15	−28		2006 Aug 11	2	1.9851	7.6
2007 Nov 9	15	+ 8		2007 Nov 11	19	1.8321	7.2
2009 Feb 25	14	+24		2009 Feb 25	24	1.5832	6.9
2010 Jun 18	22	−25		2010 Jun 16	10	1.8250	7.0
2011 Sep 16	16	−17		2011 Sep 16	20	1.9883	7.6
2012 Dec 18	9	+25		2012 Dec 20	10	1.6843	6.7
2014 Apr 15	5	+ 3		2014 Apr 14	8	1.6434	7.0
2015 Jul 25	8	−30		2015 Jul 23	5	1.9394	7.5
2016 Oct 21	5	− 1		2016 Oct 22	20	1.8997	7.4
2018 Jan 31	13	+30		2018 Feb 1	15	1.6019	6.9
2019 May 28	22	−18		2019 May 26	21	1.7513	7.0
2020 Aug 28	12	−24		2020 Aug 27	15	1.9963	7.7

2 Pallas

Opposition				Least distance			
Date	UT	Declin.		Date	UT	Dist.	Magn.
1991 Mar 6	8 h	− 3°		1991 Mar 3	18 h	1.2369	6.7
1992 Jun 18	23	+25		1992 Jun 12	5	2.4623	9.5
1993 Aug 25	2	+ 8		1993 Aug 29	12	2.3563	9.1
1994 Nov 8	20	−27		1994 Nov 13	14	1.6421	8.1
1996 Apr 18	21	+21		1996 Apr 5	22	1.7274	8.2
1997 Jul 19	10	+20		1997 Jul 19	8	2.5704	9.6
1998 Sep 16	5	− 3		1998 Sep 21	5	2.0764	8.2
2000 Jan 27	3	−26		2000 Feb 5	16	1.3251	7.4
2001 May 27	18	+26		2001 May 16	8	2.2504	9.2
2002 Aug 12	11	+13		2002 Aug 15	16	2.4740	9.4
2003 Oct 13	15	−17		2003 Oct 18	4	1.7941	8.2
2005 Mar 23	7	+ 9		2005 Mar 15	14	1.3677	7.1
2006 Jul 1	19	+23		2006 Jun 28	1	2.5227	9.5
2007 Sep 2	24	+ 4		2007 Sep 7	17	2.2520	8.8
2008 Dec 4	17	−32		2008 Dec 13	12	1.5620	8.0
2010 May 4	4	+24		2010 Apr 20	12	1.9474	8.6
2011 Jul 29	14	+18		2011 Jul 31	2	2.5483	9.5
2012 Sep 25	2	− 8		2012 Sep 30	2	1.9559	8.3
2014 Feb 22	9	−12		2014 Feb 23	17	1.2309	7.0
2015 Jun 12	1	+26		2015 Jun 3	15	2.3984	9.4
2016 Aug 20	12	+10		2016 Aug 24	13	2.3984	9.2
2017 Oct 28	23	−24		2017 Nov 2	7	1.6953	8.2
2019 Apr 10	1	+18		2019 Mar 29	14	1.5899	7.9
2020 Jul 13	2	+21		2020 Jul 11	22	2.5591	9.6

3 Juno

Opposition				Least distance			
Date	UT	Declin.		Date	UT	Dist.	Magn.
1990 May 8	3 h	− 2°		1990 May 7	2 h	2.3583	10.1
1991 Jul 16	23	− 5		1991 Jul 25	19	1.9124	9.4
1992 Dec 28	2	+ 0		1992 Dec 21	2	1.1320	7.7
1994 Apr 15	1	+ 0		1994 Apr 10	16	2.1971	9.8
1995 Jun 18	15	− 5		1995 Jun 23	21	2.2368	10.0
1996 Oct 5	0	− 4		1996 Oct 11	1	1.1137	7.5
1998 Mar 19	9	+ 3		1998 Mar 12	1	1.8661	9.1
1999 May 25	19	− 3		1999 May 27	8	2.3714	10.2
2000 Aug 12	8	− 5		2000 Aug 22	15	1.5775	8.7
2002 Feb 11	1	+ 4		2002 Feb 2	5	1.4280	8.3

3 Juno (cont.)

Opposition				*Least distance*				
Date		*UT*	*Declin.*	*Date*		*UT*	*Dist.*	*Magn.*
2003 May	3	11 *h*	− 1°	2003 May	1	12 *h*	2.3336	10.1
2004 Jul	9	4	− 5	2004 Jul	17	5	2.0083	9.6
2005 Dec	9	8	− 2	2005 Dec	4	11	1.0629	7.5
2007 Apr	10	2	+ 1	2007 Apr	4	23	2.1341	9.7
2008 Jun	12	16	− 5	2008 Jun	17	3	2.2784	10.0
2009 Sep	21	8	− 4	2009 Sep	29	4	1.1991	7.6
2011 Mar	12	10	+ 4	2011 Mar	4	17	1.7747	8.8
2012 May	19	24	− 3	2012 May	20	17	2.3729	10.2
2013 Aug	4	1	− 5	2013 Aug	13	22	1.6899	9.0
2015 Jan	29	23	+ 4	2015 Jan	21	10	1.3242	8.1
2016 Apr	27	3	− 1	2016 Apr	24	8	2.2957	10.0
2017 Jul	2	13	− 5	2017 Jul	9	17	2.0901	9.7
2018 Nov	17	21	− 3	2018 Nov	16	8	1.0359	7.4
2020 Apr	2	20	+ 2	2020 Mar	27	23	2.0571	9.5

4 Vesta

Opposition				*Least distance*				
Date		*UT*	*Declin.*	*Date*		*UT*	*Dist.*	*Magn.*
1990 Nov	15	19 *h*	+10°	1990 Nov	13	23 *h*	1.5756	6.5
1992 Mar	9	4	+15	1992 Mar	13	17	1.3390	6.0
1993 Aug	28	4	−19	1993 Aug	23	2	1.3056	5.9
1994 Dec	25	2	+21	1994 Dec	26	2	1.5765	6.3
1996 May	8	20	− 7	1996 May	11	23	1.1662	5.6
1997 Oct	17	5	− 2	1997 Oct	13	9	1.4997	6.4
1999 Feb	4	8	+22	1999 Feb	7	20	1.4630	6.2
2000 Jul	16	18	−24	2000 Jul	13	17	1.1765	5.4
2001 Nov	27	17	+14	2001 Nov	26	18	1.5882	6.5
2003 Mar	26	23	+ 9	2003 Mar	31	14	1.2813	5.9
2004 Sep	13	7	−14	2004 Sep	8	8	1.3699	6.1
2006 Jan	5	23	+23	2006 Jan	7	17	1.5521	6.2
2007 May	30	14	−14	2007 May	31	23	1.1428	5.4
2008 Oct	30	2	+ 4	2008 Oct	27	3	1.5383	6.4
2010 Feb	18	5	+20	2010 Feb	22	5	1.4117	6.1

4 Vesta (cont.)

Opposition				*Least distance*			
Date	*UT*	*Declin.*		*Date*	*UT*	*Dist.*	*Magn.*
2011 Aug 5	10 *h*	−23°		2011 Aug 1	5 *h*	1.2272	5.6
2012 Dec 9	8	+18		2012 Dec 9	5	1.5885	6.4
2014 Apr 13	12	+ 3		2014 Apr 17	20	1.2301	5.7
2015 Sep 29	3	− 9		2015 Sep 24	11	1.4267	6.2
2017 Jan 18	0	+23		2017 Jan 20	13	1.5217	6.2
2018 Jun 19	20	−20		2018 Jun 19	8	1.1416	5.3
2019 Nov 12	9	+ 8		2019 Nov 10	5	1.5649	6.5

6 Hebe

Opposition				*Least distance*			
Date	*UT*	*Declin.*		*Date*	*UT*	*Dist.*	*Magn.*
1990 Apr 22	15 *h*	+ 8°		1990 Apr 25	11 *h*	1.9288	9.9
1991 Aug 27	6	−17		1991 Aug 31	22	1.0201	7.6
1993 Feb 16	16	+16		1993 Feb 11	12	1.6897	9.2
1994 May 12	18	+ 4		1994 May 18	9	1.8313	9.8
1995 Oct 19	4	−15		1995 Oct 14	21	0.9981	7.8
1997 Mar 7	24	+15		1997 Mar 4	15	1.8387	9.6
1998 Jun 4	10	− 1		1998 Jun 12	20	1.6523	9.4
1999 Dec 9	6	− 1		1999 Dec 1	21	1.1574	8.3
2001 Mar 26	6	+13		2001 Mar 25	3	1.9295	9.9
2002 Jul 1	20	− 7		2002 Jul 11	17	1.3970	8.8
2004 Jan 12	23	+10		2004 Jan 5	20	1.3769	8.6
2005 Apr 13	10	+10		2005 Apr 14	19	1.9489	9.9
2006 Aug 5	0	−14		2006 Aug 13	0	1.1292	7.8
2008 Feb 6	24	+15		2008 Jan 31	23	1.5887	8.8
2009 May 2	13	+ 6		2009 May 6	18	1.8918	9.9
2010 Sep 21	6	−18		2010 Sep 21	9	0.9761	7.7
2012 Feb 27	17	+16		2012 Feb 23	8	1.7715	9.4
2013 May 23	19	+ 1		2013 May 30	22	1.7518	9.6
2014 Nov 15	18	− 9		2014 Nov 9	2	1.0653	8.0
2016 Mar 17	6	+14		2016 Mar 14	21	1.8941	9.8
2017 Jun 17	2	− 4		2017 Jun 26	11	1.5389	9.2
2018 Dec 28	2	+ 5		2018 Dec 20	15	1.2574	8.4
2020 Apr 4	7	+11		2020 Apr 4	9	1.9509	9.9

7 Iris

Opposition				*Least*	*distance*		
Date	*UT*	*Declin.*		*Date*	*UT*	*Dist.*	*Magn.*
1990 May 13	21 *h*	−22°		1990 May 15	23 *h*	1.9152	9.4
1991 Sep 8	22	+ 4		1991 Sep 17	11	1.0157	7.5
1993 Mar 17	8	− 7		1993 Mar 11	13	1.6772	9.1
1994 Jun 9	0	−23		1994 Jun 14	11	1.7816	9.0
1995 Nov 30	3	+24		1995 Nov 25	7	0.8739	6.8
1997 Apr 14	3	−16		1997 Apr 11	21	1.8787	9.4
1998 Jul 10	23	−18		1998 Jul 19	15	1.4923	8.6
2000 Feb 2	6	+ 9		2000 Jan 23	21	1.2476	8.1
2001 May 9	4	−22		2001 May 10	14	1.9259	9.5
2002 Aug 27	11	− 1		2002 Sep 5	21	1.0905	7.7
2004 Mar 11	14	− 5		2004 Mar 5	1	1.6228	9.0
2005 Jun 3	19	−23		2005 Jun 8	16	1.8190	9.2
2006 Nov 14	20	+24		2006 Nov 12	18	0.8483	6.8
2008 Apr 9	9	−15		2008 Apr 6	13	1.8563	9.4
2009 Jul 4	8	−19		2009 Jul 12	13	1.5537	8.7
2011 Jan 24	8	+12		2011 Jan 14	17	1.1721	7.9
2012 May 4	12	−21		2012 May 5	7	1.9305	9.5
2013 Aug 16	17	− 5		2013 Aug 26	12	1.1651	7.9
2015 Mar 6	13	− 3		2015 Feb 27	9	1.5662	8.9
2016 May 29	18	−23		2016 Jun 3	1	1.8491	9.2
2017 Oct 29	23	+21		2017 Oct 30	21	0.8489	6.9
2019 Apr 5	9	−13		2019 Apr 1	22	1.8276	9.4
2020 Jun 28	2	−21		2020 Jul 5	17	1.6116	8.8

8 Flora

Opposition				*Least*	*distance*		
Date	*UT*	*Declin.*		*Date*	*UT*	*Dist.*	*Magn.*
1990 Jun 29	14 *h*	−21°		1990 Jul 5	24 *h*	1.2987	9.0
1992 Jan 26	8	+22		1992 Jan 19	20	1.1566	8.6
1993 May 26	6	−14		1993 May 30	6	1.4789	9.6
1994 Dec 5	2	+15		1994 Nov 30	7	0.9221	8.0
1996 Apr 26	5	− 5		1996 Apr 27	5	1.5481	9.8
1997 Sep 27	6	− 9		1997 Sep 30	3	0.9030	8.1
1999 Mar 28	13	+ 6		1999 Mar 26	8	1.4988	9.6
2000 Jul 30	23	−21		2000 Aug 6	19	1.1165	8.6
2002 Feb 24	1	+16		2002 Feb 18	22	1.3363	9.2
2003 Jun 20	19	−20		2003 Jun 26	18	1.3520	9.2

8 Flora (cont.)

Opposition				Least distance			
Date	UT	Declin.		Date	UT	Dist.	Magn.
2005 Jan 14	14 h	+22°		2005 Jan 7	24 h	1.0930	8.3
2006 May 18	24	−12		2006 May 22	8	1.5063	9.6
2007 Nov 19	1	+ 9		2007 Nov 15	19	0.8896	8.0
2009 Apr 19	9	− 2		2009 Apr 19	14	1.5464	9.8
2010 Sep 11	3	−14		2010 Sep 15	16	0.9419	8.2
2012 Mar 20	9	+ 8		2012 Mar 17	11	1.4688	9.6
2013 Jul 20	0	−22		2013 Jul 26	22	1.1794	8.7
2015 Feb 15	10	+18		2015 Feb 9	18	1.2802	9.0
2016 Jun 11	12	−18		2016 Jun 16	21	1.4014	9.4
2018 Jan 2	18	+21		2017 Dec 27	6	1.0288	8.2
2019 May 11	22	−10		2019 May 14	12	1.5274	9.7
2020 Nov 1	7	+ 3		2020 Oct 31	0	0.8760	8.0

15 Eunomia

Opposition				Least distance			
Date	UT	Declin.		Date	UT	Dist.	Magn.
1991 Feb 1	0 h	+12°		1991 Jan 25	13 h	1.6590	8.7
1992 Apr 22	11	−28		1992 Apr 23	16	2.1607	9.8
1993 Jul 17	14	−21		1993 Jul 24	16	1.5771	8.4
1994 Dec 28	11	+29		1994 Dec 22	5	1.3799	8.1
1996 Mar 31	13	−20		1996 Mar 30	1	2.1265	9.8
1997 Jun 17	11	−32		1997 Jun 24	1	1.8407	9.2
1998 Nov 10	15	+36		1998 Nov 7	23	1.2061	7.9
2000 Mar 8	12	− 9		2000 Mar 4	12	1.9900	9.5
2001 May 23	6	−34		2001 May 27	21	2.0412	9.6
2002 Sep 14	5	+13		2002 Sep 17	13	1.2512	8.0
2004 Feb 11	12	+ 6		2004 Feb 5	11	1.7583	9.0
2005 Apr 30	7	−30		2005 May 2	10	2.1505	9.8
2006 Jul 30	0	−14		2006 Aug 5	20	1.4844	8.3
2008 Jan 10	17	+23		2008 Jan 4	1	1.4695	8.2
2009 Apr 8	10	−23		2009 Apr 7	19	2.1516	9.8

15 Eunomia (cont.)

	Opposition					Least	distance	
	Date	UT	Declin.			Date	UT	Dist.	Magn.

	Date	UT	Declin.		Date	UT	Dist.	Magn.
2010	Jun 27	2 h	−29°	2010	Jul 4	1 h	1.7603	9.0
2011	Nov 28	21	+37	2011	Nov 24	13	1.2386	7.9
2013	Mar 17	1	−13	2013	Mar 13	19	2.0491	9.6
2014	May 31	18	−34	2014	Jun 6	3	1.9858	9.5
2015	Oct 3	11	+23	2015	Oct 4	23	1.2060	7.9
2017	Feb 20	11	+ 0	2017	Feb 14	23	1.8475	9.2
2018	May 8	12	−32	2018	May 11	24	2.1264	9.8
2019	Aug 13	6	− 6	2019	Aug 19	12	1.3962	8.2

18 Melpomene

	Opposition				Least	distance	

	Date	UT	Declin.		Date	UT	Dist.	Magn.
1990	Mar 5	20 h	+10°	1990	Mar 1	21 h	1.6845	10.0
1991	Jun 8	1	− 6	1991	Jun 17	5	1.4655	9.7
1993	Jan 10	10	+10	1993	Jan 1	14	1.2193	9.1
1994	Apr 20	4	+ 1	1994	Apr 22	17	1.7918	10.4
1995	Sep 15	7	−11	1995	Sep 18	7	0.8276	7.8
1997	Mar 9	12	+ 9	1997	Mar 6	2	1.7084	10.0
1998	Jun 13	8	− 7	1998	Jun 22	23	1.4183	9.6
2000	Jan 17	9	+11	2000	Jan 8	21	1.2685	9.1
2001	Apr 23	19	+ 1	2001	Apr 26	22	1.7791	10.3
2002	Sep 28	1	−10	2002	Sep 28	6	0.8142	7.8
2004	Mar 13	3	+ 8	2004	Mar 10	5	1.7324	10.1
2005	Jun 18	17	− 8	2005	Jun 28	19	1.3687	9.5
2007	Jan 23	1	+11	2007	Jan 14	23	1.3171	9.2
2008	Apr 27	11	− 0	2008	May 1	5	1.7659	10.3
2009	Oct 10	22	− 8	2009	Oct 8	9	0.8143	7.9
2011	Mar 17	18	+ 8	2011	Mar 15	8	1.7533	10.2
2012	Jun 24	16	− 8	2012	Jul 5	4	1.3120	9.4
2014	Jan 28	8	+12	2014	Jan 20	15	1.3673	9.3
2015	May 2	3	− 1	2015	May 6	11	1.7478	10.3
2016	Oct 23	23	− 6	2016	Oct 18	22	0.8295	8.0
2018	Mar 21	8	+ 7	2018	Mar 19	10	1.7709	10.2
2019	Jul 2	1	− 9	2019	Jul 12	20	1.2525	9.2

433 Eros

Opposition				Least	distance		
Date		UT	Declin.	Date		UT	Dist.	Magn.
1975 Jan 13	22 h	+37°		1975 Jan 23	8 h	0.1511	7.7
1977 Jun 23	2	−38		1977 Jun 16	22	0.7028	11.9
1979 Aug 17	6	− 5		1979 Aug 20	17	0.7394	12.0
1981 Nov 13	21	+55		1981 Dec 29	8	0.3085	10.1
1984 Jun 11	6	−42		1984 Jun 2	15	0.6563	11.8
1986 Aug 7	5	−12		1986 Aug 8	22	0.7595	12.0
1988 Oct 16	17	+39		1988 Nov 6	15	0.4676	11.0
1991 May 29	5	−46		1991 May 17	3	0.5916	11.6
1993 Jul 28	7	−19		1993 Jul 28	7	0.7675	11.9
1995 Sep 28	19	+26		1995 Oct 11	15	0.5660	11.5
1998 May 11	15	−47		1998 Apr 23	4	0.5024	11.2
2000 Jul 18	8	−25		2000 Jul 16	17	0.7636	12.0
2002 Sep 14	0	+15		2002 Sep 22	23	0.6370	11.7
2005 Apr 16	21	−45		2005 Mar 8	22	0.3656	10.5
2007 Jul 9	7	−30		2007 Jul 5	24	0.7478	12.0
2009 Sep 1	17	+ 6		2009 Sep 8	2	0.6889	11.9
2012 Mar 1	9	−26		2012 Jan 31	11	0.1787	8.5
2014 Jun 29	1	−35		2014 Jun 23	22	0.7203	12.0
2016 Aug 22	0	− 1		2016 Aug 26	10	0.7250	12.0
2018 Dec 7	17	+62		2019 Jan 15	6	0.2086	9.1
2021 Jun 18	3	−40		2021 Jun 10	20	0.6819	11.9
2023 Aug 12	21	− 8		2023 Aug 15	11	0.7506	12.0
2025 Oct 29	21	+47		2025 Nov 30	2	0.3976	10.7
2028 Jun 5	7	−44		2028 May 26	5	0.6268	11.7
2030 Aug 2	22	−15		2030 Aug 3	19	0.7648	12.0
2032 Oct 7	9	+33		2032 Oct 23	11	0.5203	11.3
2035 May 21	19	−47		2035 May 7	7	0.5517	11.4
2037 Jul 24	1	−21		2037 Jul 23	6	0.7671	11.9
2039 Sep 21	21	+21		2039 Oct 2	14	0.6032	11.6
2042 May 1	14	−47		2042 Apr 6	19	0.4461	11.0
2044 Jul 14	3	−27		2044 Jul 11	16	0.7575	12.0
2046 Sep 8	11	+11		2046 Sep 16	2	0.6639	11.8
2049 Mar 31	7	−41		2049 Feb 12	6	0.2726	9.9
2051 Jul 5	1	−32		2051 Jun 30	21	0.7358	12.0
2053 Aug 28	1	+ 3		2053 Sep 2	10	0.7076	11.9
2056 Jan 17	20	+31		2056 Jan 24	11	0.1498	7.6

XXVI
ECLIPSE SERIES OF CALLISTO

Callisto, satellite IV of Jupiter, is the only one of the four Galilean satellites which is not eclipsed during the course of each of its revolutions. During each revolution of Jupiter around the Sun, there are two series of eclipses of Callisto, so that the series occur at intervals of six years. The table mentions the eclipse series of Callisto taking place from A.D. 1900 to 2075. For instance, a series began on 1989 November 26; it ended on 1992 November 2, and included 65 eclipses.

The calculations are based on the theory E-2 of J.H. Lieske, improved by the corrections by the same author as given in *Astronomy and Astrophysics*, Vol. 176, pp. 146-158 (1987).

Date of the first eclipse of the series	Date of the last eclipse of the series	Number of eclipses in the series	Date of the first eclipse of the series	Date of the last eclipse of the series	Number of eclipses in the series
1901 May 1	1904 Mar 21	64	1989 Nov 26	1992 Nov 2	65
1906 Dec 22	1910 Feb 3	69	1996 Mar 9	1998 Nov 23	60
1913 Mar 18	1916 Jan 21	63	2001 Sep 27	2004 Sep 20	66
1918 Nov 8	1921 Dec 4	68	2008 Jan 9	2010 Sep 23	60
1925 Feb 2	1927 Nov 21	62	2013 Jul 28	2016 Jul 21	66
1930 Sep 9	1933 Sep 18	67	2019 Nov 9	2022 Aug 10	61
1936 Dec 20	1939 Sep 22	61	2025 May 29	2028 Jun 7	67
1942 Jul 27	1945 Jul 20	66	2031 Sep 9	2034 Jun 28	62
1948 Nov 7	1951 Jul 23	60	2037 Mar 29	2040 Apr 8	67
1954 May 27	1957 May 20	66	2043 Jul 11	2046 May 15	63
1960 Sep 7	1963 May 24	60	2049 Jan 27	2052 Feb 24	68
1966 Mar 28	1969 Mar 4	65	2055 May 11	2058 Mar 15	63
1972 Jul 8	1975 Mar 24	60	2060 Nov 28	2064 Jan 11	69
1978 Jan 26	1981 Jan 2	65	2067 Mar 28	2070 Jan 30	63
1984 May 9	1987 Jan 22	60	2072 Oct 15	2075 Nov 11	68

XXVII
PASSAGES OF EARTH AND SUN
THROUGH THE RING-PLANE OF SATURN

Passages of the Earth				*Passages of the Sun*			
1613	Feb	16	N→S	1612	Dec	28	N→S
1626	Sep	3	S→N	1626	Aug	24	S→N
1642	Mar	28	N→S				
1642	Oct	21	S→N	1642	Jun	13	N→S
1642	Dec	15	N→S				
1655	Oct	19	S→N				
1656	Mar	11	N→S	1656	Feb	18	S→N
1656	Jul	15	S→N				
1671	May	27	N→S				
1671	Jul	25	S→N	1671	Dec	3	N→S
1672	Feb	11	N→S				
1685	Aug	30	S→N	1685	Aug	3	S→N
1701	Mar	23	N→S	1701	May	17	N→S
1714	Oct	14	S→N				
1715	Mar	23	N→S	1715	Jan	28	S→N
1715	Jul	9	S→N				
1730	May	13	N→S				
1730	Aug	12	S→N	1730	Nov	6	N→S
1731	Feb	5	N→S				
1744	Aug	25	S→N	1744	Jul	12	S→N
1760	Mar	15	N→S	1760	Apr	16	N→S
1773	Oct	7	S→N				
1774	Apr	6	N→S	1774	Jan	1	S→N
1774	Jun	28	S→N				
1789	May	1	N→S				
1789	Aug	28	S→N	1789	Oct	4	N→S
1790	Jan	28	N→S				
1802	Dec	17	S→N				
1803	Jan	5	N→S	1803	Jun	17	S→N
1803	Aug	20	S→N				

XXVII (cont.)

Passages of the Earth				*Passages of the Sun*			
1819	Mar	9	N→S	1819	Mar	14	N→S
1832	Sep	30	S→N				
1833	Apr	30	N→S	1832	Dec	2	S→N
1833	Jun	9	S→N				
1848	Apr	21	N→S				
1848	Sep	14	S→N	1848	Aug	31	N→S
1849	Jan	18	N→S				
1861	Nov	22	S→N				
1862	Feb	2	N→S	1862	May	16	S→N
1862	Aug	12	S→N				
1878	Mar	1	N→S	1878	Feb	4	N→S
1891	Sep	22	S→N	1891	Oct	29	S→N
1907	Apr	12	N→S				
1907	Oct	4	S→N	1907	Jul	26	N→S
1908	Jan	7	N→S				
1920	Nov	7	S→N				
1921	Feb	22	N→S	1921	Apr	11	S→N
1921	Aug	3	S→N				
1936	Jun	26	N→S				
1936	Jul	1	S→N	1936	Dec	28	N→S
1937	Feb	20	N→S				
1950	Sep	14	S→N	1950	Sep	20	S→N
1966	Apr	2	N→S				
1966	Oct	29	S→N	1966	Jun	15	N→S
1966	Dec	17	N→S				
1979	Oct	27	S→N				
1980	Mar	12	N→S	1980	Mar	2	S→N
1980	Jul	23	S→N				
1995	May	21	N→S				
1995	Aug	11	S→N	1995	Nov	18	N→S
1996	Feb	11	N→S				
2009	Sep	4	S→N	2009	Aug	10	S→N
2025	Mar	23	N→S	2025	May	6	N→S

2

Oppositions of Mars

0 − 3000

Oppositions of Mars

The list which follows gives all the oppositions of Mars with the Sun taking place during the period A.D. 0 − 3000.

The two first columns give the calendar date (year, month, day of the month) and the Dynamical Time (TD) of the instant of opposition, this being the instant when the true heliocentric longitudes of the Earth and Mars, referred to the mean equinox of the date, are equal. Hence, the times are those of the *geometric* oppositions of Mars. The *apparent* opposition occurs approximately 7 minutes later than the geometric opposition, by reason of the aberration of light and the effect of light-time. To convert Dynamical Time to Universal Time, subtract the quantity ΔT, as explained on pages 6−8.

The next column gives the geocentric apparent declination of Mars at the time of the opposition.

The fourth and fifth columns give the calendar date and the Dynamical Time (TD) of the instant of Mars' least distance to the Earth. The next two columns contain the value of this least distance, in astronomical units (a.u.) and in millions of kilometers (m.k.), respectively.

The next column gives the planet's apparent diameter in seconds of arc at the instant of minimum distance, the adopted diameter at unit distance being $9''.36$.

For the computation of the heliocentric positions of the Earth and Mars, use was made of Bretagnon's VSOP87 theories for these planets.

During the period 0 to 3000, the extreme values of the shortest distance between Marsand Earth are 0.37200 a.u. (55.65 millions of km) at the perihelic opposition of 2729, and 0.67850 a.u. (101.50 millions of km) at the aphelic opposition of 2832.

The time interval between Mars' opposition and that of its least distance to the Earth can be as large as 8 ½ days, as this happened for instance at the opposition of 1969. On the other hand, that time interval is less than ten minutes at the oppositions of 2208 and 2232.

In the years 69, 274, 558, 763, 968, 1047, 1252, 1536, 1630, 1709, 1835, 2040, 2119, 2245, 2450, 2529, 2655, 2734, 2860, and 2939, the opposition occurs in early January, while Mars' least distance takes place at the end of December of the *preceding* year.

In the years 656 and 1728 the opposition of Mars took place on the bissextile day, February 29.

Opposition

Date		TD	Decl.
		h m	° '
0 Sep	18	16 15	− 6 15
2 Nov	9	14 58	+17 34
4 Dec	17	6 55	+27 10
7 Jan	20	14 47	+25 07
9 Feb	23	19 00	+14 17
11 Apr	6	14 40	− 3 45
13 Jun	3	22 09	−25 57
15 Aug	23	11 29	−18 20
17 Oct	22	17 47	+10 08
19 Dec	4	3 08	+25 01
22 Jan	7	10 41	+27 05
24 Feb	10	17 30	+19 19
26 Mar	20	1 53	+ 3 59
28 May	8	16 47	−17 43
30 Jul	22	23 08	−27 44
32 Oct	2	1 29	+ 0 24
34 Nov	18	23 25	+20 58
36 Dec	25	3 08	+27 37
39 Jan	28	5 38	+23 17
41 Mar	4	6 54	+10 42
43 Apr	17	16 50	− 8 47
45 Jun	20	17 20	−28 57
47 Sep	8	9 21	−11 21
49 Nov	2	4 45	+14 48
51 Dec	12	8 36	+26 31
54 Jan	15	0 43	+26 02
56 Feb	18	16 45	+16 24
58 Mar	29	19 06	− 0 29
60 May	22	21 32	−22 44
62 Aug	10	2 06	−23 00
64 Oct	14	5 03	+ 6 23
66 Nov	27	19 06	+23 37
69 Jan	1	19 35	+27 28
71 Feb	4	22 37	+21 00
73 Mar	13	3 00	+ 6 48
75 Apr	29	16 11	−13 58
77 Jul	8	14 51	−29 23
79 Sep	22	21 04	− 4 20
81 Nov	12	0 20	+18 43
83 Dec	20	8 31	+27 22
86 Jan	22	14 40	+24 30
88 Feb	26	22 40	+13 05
90 Apr	9	5 48	− 5 17

Nearest to Earth

Date		TD	Least distance		diam.
		h m	a.u.	m.k.	"
0 Sep	10	20 15	0.43427	64.97	21.55
2 Nov	3	0 53	0.56179	84.04	16.66
4 Dec	14	8 11	0.65005	97.25	14.40
7 Jan	21	12 55	0.67583	101.10	13.85
9 Feb	28	14 06	0.63588	95.13	14.72
11 Apr	14	16 06	0.53403	79.89	17.53
13 Jun	10	6 09	0.40636	60.79	23.03
15 Aug	17	16 09	0.39317	58.82	23.81
17 Oct	15	1 30	0.51362	76.84	18.22
19 Nov	29	16 42	0.62245	93.12	15.04
22 Jan	6	20 22	0.67361	100.77	13.90
24 Feb	14	0 37	0.65938	98.64	14.20
26 Mar	27	1 15	0.58024	86.80	16.13
28 May	16	22 29	0.45313	67.79	20.66
30 Jul	21	19 52	0.37458	56.04	24.99
32 Sep	23	22 31	0.46218	69.14	20.25
34 Nov	13	2 51	0.58582	87.64	15.98
36 Dec	23	0 36	0.66116	98.91	14.16
39 Jan	30	0 14	0.67279	100.65	13.91
41 Mar	9	21 33	0.61860	92.54	15.13
43 Apr	26	2 27	0.50548	75.62	18.52
45 Jun	24	23 26	0.38709	57.91	24.18
47 Sep	1	4 23	0.41377	61.90	22.62
49 Oct	26	1 51	0.54087	80.91	17.31
51 Dec	8	18 18	0.63890	95.58	14.65
54 Jan	15	7 02	0.67606	101.14	13.84
56 Feb	22	20 21	0.64761	96.88	14.45
58 Apr	6	9 55	0.55563	83.12	16.85
60 May	30	9 04	0.42594	63.72	21.98
62 Aug	6	2 02	0.38123	57.03	24.55
64 Oct	6	4 38	0.49066	73.40	19.08
66 Nov	22	17 35	0.60711	90.82	15.42
68 Dec	31	13 26	0.66933	100.13	13.98
71 Feb	7	13 50	0.66671	99.74	14.04
73 Mar	19	12 23	0.59852	89.54	15.64
75 May	8	3 12	0.47656	71.29	19.64
77 Jul	9	23 58	0.37637	56.30	24.87
79 Sep	14	22 01	0.43957	65.76	21.29
81 Nov	5	13 58	0.56672	84.78	16.52
83 Dec	17	14 10	0.65246	97.61	14.35
86 Jan	23	17 15	0.67553	101.06	13.86
88 Mar	2	21 55	0.63278	94.66	14.79
90 Apr	17	8 13	0.52847	79.06	17.71

Opposition

Date		TD	Decl.
		h m	° ′
92 Jun	7	13 25	−26 54
94 Aug	27	3 29	−16 31
96 Oct	25	8 05	+11 37
98 Dec	6	6 19	+25 34
101 Jan	9	10 27	+26 45
103 Feb	12	18 21	+18 20
105 Mar	22	9 25	+ 2 36
107 May	12	19 11	−19 08
109 Jul	27	0 14	−26 43
111 Oct	5	23 54	+ 2 11
113 Nov	21	5 02	+21 52
115 Dec	28	3 17	+27 36
118 Jan	30	5 46	+22 30
120 Mar	6	11 30	+ 9 26
122 Apr	20	10 22	−10 18
124 Jun	24	13 44	−29 17
126 Sep	11	19 29	− 9 24
128 Nov	4	16 46	+16 06
130 Dec	14	10 55	+26 51
133 Jan	17	0 18	+25 31
135 Feb	20	19 18	+15 17
137 Apr	1	6 53	− 1 58
139 May	27	7 12	−23 57
141 Aug	13	23 00	−21 26
143 Oct	17	21 59	+ 8 00
145 Nov	29	23 14	+24 19
148 Jan	4	19 39	+27 15
150 Feb	6	22 51	+20 06
152 Mar	15	8 48	+ 5 28
154 May	2	14 33	−15 27
156 Jul	12	15 30	−28 57
158 Sep	26	0 41	− 2 27
160 Nov	14	8 11	+19 47
162 Dec	22	9 00	+27 29
165 Jan	24	14 19	+23 49
167 Mar	1	2 05	+11 52
169 Apr	11	19 30	− 6 46
171 Jun	12	4 35	−27 41
173 Aug	30	18 27	−14 39
175 Oct	28	22 09	+13 03
177 Dec	8	9 33	+26 03
180 Jan	12	10 05	+26 21
182 Feb	14	19 46	+17 18

Nearest to Earth

Date		TD	Least distance		diam.
		h m	a.u.	m.k.	″
92 Jun	13	14 48	0.40190	60.12	23.29
94 Aug	21	1 38	0.39628	59.28	23.62
96 Oct	17	18 02	0.51881	77.61	18.04
98 Dec	2	0 34	0.62581	93.62	14.96
101 Jan	9	0 44	0.67429	100.87	13.88
103 Feb	16	5 52	0.65739	98.34	14.24
105 Mar	29	11 36	0.57590	86.15	16.25
107 May	20	22 52	0.44798	67.02	20.89
109 Jul	25	6 29	0.37509	56.11	24.95
111 Sep	27	19 39	0.46742	69.92	20.02
113 Nov	15	12 56	0.58997	88.26	15.87
115 Dec	26	5 07	0.66303	99.19	14.12
118 Feb	1	4 46	0.67197	100.53	13.93
120 Mar	12	5 54	0.61503	92.01	15.22
122 Apr	28	19 51	0.50017	74.82	18.71
124 Jun	28	7 23	0.38451	57.52	24.34
126 Sep	4	9 30	0.41852	62.61	22.36
128 Oct	28	17 37	0.54619	81.71	17.14
130 Dec	11	0 43	0.64178	96.01	14.58
133 Jan	17	10 38	0.67624	101.16	13.84
135 Feb	25	2 29	0.64505	96.50	14.51
137 Apr	8	23 04	0.55049	82.35	17.00
139 Jun	3	15 00	0.42078	62.95	22.24
141 Aug	9	13 17	0.38321	57.33	24.43
143 Oct	9	23 30	0.49589	74.18	18.88
145 Nov	25	2 05	0.61083	91.38	15.32
148 Jan	3	17 35	0.67051	100.31	13.96
150 Feb	9	17 55	0.66512	99.50	14.07
152 Mar	21	20 51	0.59453	88.94	15.74
154 May	11	0 59	0.47128	70.50	19.86
156 Jul	13	9 46	0.37537	56.16	24.94
158 Sep	17	22 41	0.44460	66.51	21.05
160 Nov	8	1 57	0.57125	85.46	16.39
162 Dec	19	18 36	0.65474	97.95	14.30
165 Jan	25	20 50	0.67520	101.01	13.86
167 Mar	6	4 37	0.62971	94.20	14.86
169 Apr	19	22 57	0.52333	78.29	17.89
171 Jun	17	20 39	0.39820	59.57	23.51
173 Aug	24	8 57	0.40024	59.87	23.39
175 Oct	21	11 45	0.52437	78.44	17.85
177 Dec	4	7 23	0.62914	94.12	14.88
180 Jan	12	3 51	0.67496	100.97	13.87
182 Feb	18	10 31	0.65531	98.03	14.28

Opposition   *Nearest to Earth*

Date	TD	Decl.	Date	TD	Least distance		diam.
	h m	° '		*h m*	*a.u.*	*m.k.*	"
184 Mar 24	18 55	+ 1 09	184 Mar 31	22 48	0.57115	85.44	16.39
186 May 16	0 29	−20 31	186 May 24	2 35	0.44239	66.18	21.16
188 Jul 31	0 34	−25 32	188 Jul 28	18 52	0.37563	56.19	24.92
190 Oct 8	21 10	+ 3 56	190 Sep 30	17 52	0.47259	70.70	19.81
192 Nov 23	10 47	+22 44	192 Nov 17	22 25	0.59411	88.88	15.75
194 Dec 30	3 41	+27 31	194 Dec 28	9 00	0.66469	99.44	14.08
197 Feb 1	5 38	+21 41	197 Feb 3	8 02	0.67083	100.35	13.95
199 Mar 9	15 42	+ 8 09	199 Mar 15	12 37	0.61146	91.47	15.31
201 Apr 23	4 02	−11 47	201 May 1	14 40	0.49493	74.04	18.91
203 Jun 29	11 09	−29 27	203 Jul 2	15 10	0.38211	57.16	24.50
205 Sep 15	4 06	− 7 27	205 Sep 7	12 36	0.42316	63.30	22.12
207 Nov 8	2 25	+17 19	207 Nov 1	6 46	0.55098	82.43	16.99
209 Dec 16	11 59	+27 07	209 Dec 13	5 07	0.64446	96.41	14.52
212 Jan 19	23 39	+24 57	212 Jan 20	13 21	0.67643	101.19	13.84
214 Feb 22	21 30	+14 08	214 Feb 27	7 42	0.64244	96.11	14.57
216 Apr 3	17 57	− 3 26	216 Apr 11	12 16	0.54554	81.61	17.16
218 May 30	17 24	−25 03	218 Jun 6	19 06	0.41626	62.27	22.49
220 Aug 17	18 41	−19 44	220 Aug 12	22 49	0.38595	57.74	24.25
222 Oct 20	15 32	+ 9 35	222 Oct 12	19 58	0.50159	75.04	18.66
224 Dec 2	3 57	+24 57	224 Nov 27	9 56	0.61470	91.96	15.23
227 Jan 6	19 34	+26 58	227 Jan 5	20 34	0.67166	100.48	13.94
229 Feb 8	23 34	+19 08	229 Feb 11	21 39	0.66353	99.26	14.11
231 Mar 18	16 09	+ 4 05	231 Mar 25	6 26	0.59028	88.30	15.86
233 May 5	14 50	−16 54	233 May 14	1 50	0.46548	69.63	20.11
235 Jul 17	16 30	−28 19	235 Jul 17	21 51	0.37436	56.00	25.00
237 Sep 29	2 10	− 0 35	237 Sep 20	23 26	0.44959	67.26	20.82
239 Nov 17	15 10	+20 47	239 Nov 11	11 57	0.57567	86.12	16.26
241 Dec 24	9 50	+27 32	241 Dec 21	22 26	0.65686	98.26	14.25
244 Jan 27	13 52	+23 06	244 Jan 28	23 26	0.67458	100.92	13.88
246 Mar 3	4 49	+10 39	246 Mar 8	10 02	0.62652	93.73	14.94
248 Apr 14	9 44	− 8 15	248 Apr 22	15 36	0.51820	77.52	18.06
250 Jun 15	21 39	−28 20	250 Jun 21	2 33	0.39468	59.04	23.72
252 Sep 3	8 45	−12 43	252 Aug 27	14 44	0.40411	60.45	23.16
254 Oct 31	11 02	+14 26	254 Oct 24	3 13	0.52939	79.20	17.68
256 Dec 10	11 44	+26 28	256 Dec 6	12 31	0.63228	94.59	14.80
259 Jan 14	9 27	+25 53	259 Jan 14	6 18	0.67561	101.07	13.85
261 Feb 16	21 17	+16 13	261 Feb 20	15 01	0.65316	97.71	14.33
263 Mar 28	3 40	− 0 17	263 Apr 4	10 33	0.56656	84.76	16.52
265 May 19	4 58	−21 49	265 May 27	4 05	0.43729	65.42	21.40
267 Aug 4	23 31	−24 12	267 Aug 2	5 47	0.37692	56.39	24.83
269 Oct 11	17 46	+ 5 39	269 Oct 3	16 31	0.47830	71.55	19.57
271 Nov 26	16 36	+23 31	271 Nov 21	6 52	0.59841	89.52	15.64
274 Jan 1	4 01	+27 21	273 Dec 30	12 16	0.66632	99.68	14.05

Opposition   Nearest to Earth

Date	TD	Decl.	Date	TD	Least distance		diam.
	h m	° '		h m	a.u.	m.k.	"
276 Feb 4	5 34	+20 49	276 Feb 6	11 00	0.66978	100.20	13.97
278 Mar 11	21 07	+ 6 49	278 Mar 17	20 53	0.60768	90.91	15.40
280 Apr 26	0 25	−13 17	280 May 4	12 47	0.48911	73.17	19.14
282 Jul 3	10 15	−29 24	282 Jul 6	2 15	0.37968	56.80	24.65
284 Sep 18	11 05	− 5 31	284 Sep 10	16 35	0.42769	63.98	21.89
286 Nov 10	11 56	+18 28	286 Nov 3	18 38	0.55568	83.13	16.84
288 Dec 18	13 36	+27 18	288 Dec 15	9 42	0.64708	96.80	14.47
291 Jan 21	23 20	+24 19	291 Jan 22	16 11	0.67630	101.17	13.84
293 Feb 24	23 34	+12 58	293 Mar 1	12 56	0.63963	95.69	14.63
295 Apr 7	5 07	− 4 53	295 Apr 15	2 37	0.54069	80.89	17.31
297 Jun 3	4 44	−26 03	297 Jun 9	22 19	0.41186	61.61	22.73
299 Aug 22	14 03	−17 56	299 Aug 17	6 33	0.38874	58.16	24.08
301 Oct 23	7 08	+11 06	301 Oct 15	13 25	0.50678	75.81	18.47
303 Dec 5	6 56	+25 30	303 Nov 30	15 33	0.61821	92.48	15.14
306 Jan 8	18 59	+26 37	306 Jan 7	23 06	0.67276	100.64	13.91
308 Feb 12	0 03	+18 09	308 Feb 15	1 25	0.66195	99.03	14.14
310 Mar 20	22 56	+ 2 42	310 Mar 27	16 46	0.58605	87.67	15.97
312 May 8	14 44	−18 18	312 May 17	0 37	0.46006	68.82	20.35
314 Jul 21	16 21	−27 29	314 Jul 21	9 14	0.37416	55.97	25.02
316 Oct 2	3 03	+ 1 15	316 Sep 24	1 13	0.45511	68.08	20.57
318 Nov 19	23 09	+21 44	318 Nov 13	22 15	0.58041	86.83	16.13
320 Dec 26	10 53	+27 31	320 Dec 24	2 44	0.65902	98.59	14.20
323 Jan 29	13 39	+22 19	323 Jan 31	2 37	0.67403	100.83	13.89
325 Mar 5	9 08	+ 9 22	325 Mar 10	17 52	0.62321	93.23	15.02
327 Apr 18	2 24	− 9 45	327 Apr 26	10 50	0.51263	76.69	18.26
329 Jun 19	16 37	−28 49	329 Jun 24	11 32	0.39100	58.49	23.94
331 Sep 7	21 34	−10 46	331 Aug 31	21 20	0.40793	61.03	22.95
333 Nov 2	22 32	+15 44	333 Oct 26	16 29	0.53431	79.93	17.52
335 Dec 13	13 49	+26 47	335 Dec 9	17 45	0.63533	95.04	14.73
338 Jan 16	9 10	+25 22	338 Jan 16	9 33	0.67599	101.13	13.85
340 Feb 19	22 13	+15 08	340 Feb 23	19 47	0.65084	97.36	14.38
342 Mar 30	12 16	− 1 42	342 Apr 6	22 57	0.56198	84.07	16.66
344 May 22	11 07	−23 02	344 May 30	4 34	0.43233	64.68	21.65
346 Aug 8	22 21	−22 42	346 Aug 5	14 35	0.37844	56.61	24.73
348 Oct 14	13 18	+ 7 19	348 Oct 6	12 55	0.48359	72.34	19.36
350 Nov 28	21 29	+24 14	350 Nov 23	14 24	0.60235	90.11	15.54
353 Jan 3	3 44	+27 08	353 Jan 1	15 33	0.66790	99.92	14.01
355 Feb 6	5 44	+19 54	355 Feb 8	14 56	0.66871	100.04	14.00
357 Mar 14	2 50	+ 5 29	357 Mar 20	6 37	0.60384	90.33	15.50
359 Apr 29	19 57	−14 44	359 May 8	8 50	0.48365	72.35	19.35
361 Jul 7	8 23	−29 09	361 Jul 9	13 08	0.37799	56.55	24.76
363 Sep 22	16 46	− 3 36	363 Sep 14	21 09	0.43282	64.75	21.63
365 Nov 12	21 26	+19 34	365 Nov 6	6 13	0.56079	83.89	16.69

Opposition				*Nearest to Earth*				
Date		*TD*	*Decl.*	*Date*		*TD*	*Least distance*	*diam.*
		h m	*° ′*			*h m*	*a.u.*	*m.k.*	*″*
367 Dec	21	15 06	+27 25	367 Dec	18	14 53	0.64975	97.20	14.41
370 Jan	23	22 56	+23 38	370 Jan	24	19 39	0.67625	101.17	13.84
372 Feb	28	2 11	+11 46	372 Mar	3	19 38	0.63686	95.27	14.70
374 Apr	9	18 36	− 6 22	374 Apr	17	19 16	0.53542	80.10	17.48
376 Jun	6	19 06	−26 56	376 Jun	13	4 50	0.40723	60.92	22.98
378 Aug	26	8 10	−16 05	378 Aug	20	14 47	0.39161	58.58	23.90
380 Oct	25	21 49	+12 34	380 Oct	18	5 02	0.51188	76.58	18.29
382 Dec	7	10 26	+25 59	382 Dec	2	22 37	0.62171	93.01	15.06
385 Jan	10	18 54	+26 12	385 Jan	10	3 13	0.67365	100.78	13.89
387 Feb	14	0 42	+17 07	387 Feb	17	6 30	0.66008	98.75	14.18
389 Mar	23	6 03	+ 1 18	389 Mar	30	4 09	0.58178	87.03	16.09
391 May	12	15 28	−19 39	391 May	20	22 16	0.45485	68.05	20.58
393 Jul	25	16 48	−26 27	393 Jul	24	18 47	0.37421	55.98	25.01
395 Oct	6	3 06	+ 3 03	395 Sep	28	0 27	0.46036	68.87	20.33
397 Nov	22	5 25	+22 36	397 Nov	16	7 21	0.58470	87.47	16.01
399 Dec	29	10 42	+27 25	399 Dec	27	6 36	0.66101	98.89	14.16
402 Jan	31	13 20	+21 29	402 Feb	2	6 31	0.67344	100.75	13.90
404 Mar	7	13 00	+ 8 06	404 Mar	13	2 13	0.61984	92.73	15.10
406 Apr	20	18 16	−11 13	406 Apr	29	4 11	0.50724	75.88	18.45
408 Jun	23	11 11	−29 07	408 Jun	27	20 59	0.38801	58.05	24.12
410 Sep	11	7 59	− 8 48	410 Sep	4	4 11	0.41241	61.70	22.70
412 Nov	5	10 29	+17 00	412 Oct	29	6 21	0.53973	80.74	17.34
414 Dec	15	16 36	+27 03	414 Dec	12	0 38	0.63852	95.52	14.66
417 Jan	18	8 50	+24 47	417 Jan	18	13 34	0.67640	101.19	13.84
419 Feb	22	0 03	+13 59	419 Feb	26	2 03	0.64854	97.02	14.43
421 Apr	1	23 36	− 3 09	421 Apr	9	13 36	0.55706	83.34	16.80
423 May	26	19 53	−24 11	423 Jun	3	8 15	0.42710	63.89	21.92
425 Aug	12	21 07	−21 05	425 Aug	9	0 18	0.38002	56.85	24.63
427 Oct	18	7 48	+ 8 56	427 Oct	10	7 04	0.48879	73.12	19.15
429 Dec	1	1 51	+24 52	429 Nov	25	22 46	0.60620	90.69	15.44
432 Jan	6	3 38	+26 50	432 Jan	4	20 08	0.66924	100.12	13.99
434 Feb	8	5 55	+18 57	434 Feb	10	19 51	0.66729	99.83	14.03
436 Mar	16	7 44	+ 4 09	436 Mar	22	16 05	0.59990	89.74	15.60
438 May	2	15 56	−16 08	438 May	11	3 24	0.47835	71.56	19.57
440 Jul	11	8 02	−28 42	440 Jul	12	22 28	0.37657	56.33	24.86
442 Sep	25	21 18	− 1 42	442 Sep	17	23 13	0.43778	65.49	21.38
444 Nov	15	5 56	+20 36	444 Nov	8	17 52	0.56543	84.59	16.55
446 Dec	23	15 49	+27 27	446 Dec	20	20 01	0.65216	97.56	14.35
449 Jan	25	22 20	+22 54	449 Jan	26	23 38	0.67609	101.14	13.84
451 Mar	2	5 07	+10 32	451 Mar	7	3 13	0.63393	94.83	14.77
453 Apr	12	8 02	− 7 50	453 Apr	20	10 27	0.53020	79.32	17.65
455 Jun	11	8 13	−27 40	455 Jun	17	12 06	0.40320	60.32	23.21
457 Aug	29	23 56	−14 11	457 Aug	24	0 12	0.39510	59.11	23.69

Opposition			Nearest to Earth				
Date	TD	Decl.	Date	TD	Least distance		diam.
	h m	° ′		h m	a.u.	m.k.	″
459 Oct 29	12 36	+13 59	459 Oct 21	21 27	0.51754	77.42	18.09
461 Dec 9	13 59	+26 24	461 Dec 5	6 34	0.62533	93.55	14.97
464 Jan 13	18 32	+25 44	464 Jan 13	7 25	0.67449	100.90	13.88
466 Feb 16	1 35	+16 03	466 Feb 19	11 48	0.65824	98.47	14.22
468 Mar 25	14 27	− 0 06	468 Apr 1	15 49	0.57727	86.36	16.21
470 May 15	18 54	−20 58	470 May 23	22 37	0.44928	67.21	20.83
472 Jul 29	18 10	−25 15	472 Jul 28	4 38	0.37434	56.00	25.00
474 Oct 9	1 30	+ 4 48	474 Sep 30	21 19	0.46544	69.63	20.11
476 Nov 24	11 23	+23 24	476 Nov 18	17 35	0.58887	88.09	15.89
478 Dec 31	11 13	+27 15	478 Dec 29	11 53	0.66282	99.16	14.12
481 Feb 2	13 19	+20 37	481 Feb 4	11 16	0.67244	100.60	13.92
483 Mar 10	16 49	+ 6 48	483 Mar 16	10 24	0.61623	92.19	15.19
485 Apr 23	10 54	−12 40	485 May 1	20 28	0.50199	75.10	18.65
487 Jun 28	7 07	−29 14	487 Jul 2	5 33	0.38524	57.63	24.30
489 Sep 14	18 04	− 6 50	489 Sep 7	9 45	0.41682	62.36	22.46
491 Nov 8	21 33	+18 11	491 Nov 1	20 40	0.54473	81.49	17.18
493 Dec 17	17 55	+27 14	493 Dec 14	6 25	0.64130	95.94	14.60
496 Jan 21	7 56	+24 09	496 Jan 21	17 15	0.67665	101.22	13.83
498 Feb 24	1 51	+12 50	498 Feb 28	8 15	0.64610	96.66	14.49
500 Apr 4	9 56	− 4 36	500 Apr 12	1 58	0.55212	82.60	16.95
502 May 30	3 15	−25 13	502 Jun 6	12 19	0.42231	63.18	22.16
504 Aug 16	17 10	−19 23	504 Aug 12	10 33	0.38230	57.19	24.48
506 Oct 21	1 18	+10 30	506 Oct 13	1 48	0.49446	73.97	18.93
508 Dec 3	6 41	+25 25	508 Nov 28	8 00	0.61022	91.29	15.34
511 Jan 8	3 45	+26 29	511 Jan 7	0 34	0.67058	100.32	13.96
513 Feb 10	6 16	+17 57	513 Feb 13	0 23	0.66588	99.61	14.06
515 Mar 19	14 24	+ 2 46	515 Mar 26	1 46	0.59580	89.13	15.71
517 May 5	15 18	−17 33	517 May 14	1 23	0.47271	70.72	19.80
519 Jul 16	9 23	−28 03	519 Jul 17	8 18	0.37514	56.12	24.95
521 Sep 29	1 00	+ 0 12	521 Sep 20	23 30	0.44259	66.21	21.15
523 Nov 18	13 47	+21 33	523 Nov 12	5 56	0.56995	85.26	16.42
525 Dec 25	16 40	+27 25	525 Dec 23	1 17	0.65440	97.90	14.30
528 Jan 28	22 05	+22 07	528 Jan 30	3 45	0.67554	101.06	13.86
530 Mar 4	7 42	+ 9 18	530 Mar 9	9 40	0.63076	94.36	14.84
532 Apr 14	20 50	− 9 16	532 Apr 23	0 06	0.52511	78.56	17.82
534 Jun 14	22 40	−28 15	534 Jun 20	18 27	0.39935	59.74	23.44
536 Sep 2	14 56	−12 14	536 Aug 27	8 01	0.39871	59.65	23.48
538 Nov 1	2 08	+15 20	538 Oct 24	13 58	0.52273	78.20	17.91
540 Dec 11	16 24	+26 44	540 Dec 7	13 11	0.62847	94.02	14.89
543 Jan 15	17 40	+25 12	543 Jan 15	10 40	0.67520	101.01	13.86
545 Feb 18	2 39	+14 57	545 Feb 21	16 49	0.65625	98.17	14.26
547 Mar 28	22 58	− 1 31	547 Apr 5	2 37	0.57265	85.67	16.35
549 May 18	21 46	−22 11	549 May 27	0 26	0.44408	66.43	21.08

Opposition   *Nearest to Earth*

Date		TD	Decl.	Date		TD	Least distance		diam.
		h m	° '			h m	a.u.	m.k.	"
551 Aug	3	17 14	−23 55	551 Aug	1	15 36	0.37516	56.12	24.95
553 Oct	11	23 06	+ 6 31	553 Oct	3	18 59	0.47106	70.47	19.87
555 Nov	27	17 37	+24 07	555 Nov	22	4 05	0.59336	88.76	15.77
558 Jan	2	11 40	+27 01	557 Dec	31	16 10	0.66465	99.43	14.08
560 Feb	5	13 14	+19 42	560 Feb	7	14 53	0.67149	100.45	13.94
562 Mar	12	21 32	+ 5 29	562 Mar	18	17 52	0.61268	91.65	15.28
564 Apr	26	5 42	−14 06	564 May	4	15 28	0.49646	74.27	18.85
566 Jul	2	5 35	−29 10	566 Jul	5	13 54	0.38239	57.20	24.48
568 Sep	18	2 44	− 4 52	568 Sep	10	12 32	0.42116	63.01	22.22
570 Nov	11	7 15	+19 18	570 Nov	4	10 10	0.54947	82.20	17.03
572 Dec	19	19 27	+27 21	572 Dec	16	11 49	0.64394	96.33	14.54
575 Jan	23	7 33	+23 27	575 Jan	23	20 38	0.67663	101.22	13.83
577 Feb	26	3 42	+11 38	577 Mar	2	13 25	0.64332	96.24	14.55
579 Apr	7	20 27	− 6 01	579 Apr	15	14 07	0.54721	81.86	17.11
581 Jun	2	12 33	−26 08	581 Jun	9	16 58	0.41776	62.50	22.41
583 Aug	21	13 25	−17 35	583 Aug	16	21 13	0.38480	57.57	24.32
585 Oct	23	18 38	+12 01	585 Oct	15	21 28	0.49986	74.78	18.73
587 Dec	6	10 21	+25 55	587 Dec	1	15 30	0.61383	91.83	15.25
590 Jan	10	2 57	+26 04	590 Jan	9	3 22	0.67175	100.49	13.93
592 Feb	13	6 41	+16 56	592 Feb	16	4 12	0.66439	99.39	14.09
594 Mar	21	20 46	+ 1 24	594 Mar	28	10 32	0.59169	88.52	15.82
596 May	8	13 16	−18 53	596 May	16	23 59	0.46729	69.91	20.03
598 Jul	20	8 19	−27 13	598 Jul	20	17 59	0.37440	56.01	25.00
600 Oct	2	2 16	+ 2 02	600 Sep	23	23 17	0.44794	67.01	20.90
602 Nov	20	21 15	+22 26	602 Nov	14	17 12	0.57475	85.98	16.29
604 Dec	27	17 44	+27 19	604 Dec	25	5 40	0.65670	98.24	14.25
607 Jan	30	21 36	+21 17	607 Feb	1	6 34	0.67509	100.99	13.86
609 Mar	6	11 05	+ 8 02	609 Mar	11	15 40	0.62765	93.89	14.91
611 Apr	18	12 36	−10 44	611 Apr	26	17 03	0.51975	77.75	18.01
613 Jun	18	17 10	−28 41	613 Jun	24	1 49	0.39539	59.15	23.67
615 Sep	7	5 55	−10 14	615 Aug	31	14 25	0.40232	60.19	23.27
617 Nov	3	15 05	+16 37	617 Oct	27	6 10	0.52773	78.95	17.74
619 Dec	14	18 45	+26 59	619 Dec	10	18 48	0.63159	94.48	14.82
622 Jan	17	17 15	+24 37	622 Jan	17	13 31	0.67571	101.08	13.85
624 Feb	21	3 57	+13 50	624 Feb	24	21 05	0.65395	97.83	14.31
626 Mar	31	6 49	− 2 55	626 Apr	7	12 39	0.56811	84.99	16.48
628 May	22	1 12	−23 19	628 May	30	1 48	0.43905	65.68	21.32
630 Aug	7	16 32	−22 25	630 Aug	5	3 02	0.37626	56.29	24.88
632 Oct	14	19 37	+ 8 11	632 Oct	6	17 06	0.47647	71.28	19.64
634 Nov	29	22 29	+24 46	634 Nov	24	12 01	0.59734	89.36	15.67
637 Jan	4	11 11	+26 43	637 Jan	2	18 46	0.66624	99.67	14.05
639 Feb	7	12 49	+18 45	639 Feb	9	17 33	0.67050	100.30	13.96
641 Mar	15	2 23	+ 4 09	641 Mar	21	1 16	0.60898	91.10	15.37

Opposition Date			TD h m	Decl. ° ′	Nearest to Earth Date			TD h m	Least distance a.u.	m.k.	diam. ″
643	Apr	30	0 23	−15 30	643	May	8	11 55	0.49093	73.44	19.07
645	Jul	6	2 20	−28 54	645	Jul	8	22 29	0.38021	56.88	24.62
647	Sep	22	9 22	− 2 55	647	Sep	14	15 31	0.42606	63.74	21.97
649	Nov	13	17 12	+20 22	649	Nov	6	23 21	0.55463	82.97	16.88
651	Dec	22	21 14	+27 23	651	Dec	19	16 34	0.64680	96.76	14.47
654	Jan	25	6 59	+22 43	654	Jan	25	23 06	0.67671	101.23	13.83
656	Feb	29	6 04	+10 25	656	Mar	4	18 31	0.64071	95.85	14.61
658	Apr	10	8 36	− 7 28	658	Apr	18	4 25	0.54224	81.12	17.26
660	Jun	6	1 19	−26 57	660	Jun	12	21 37	0.41296	61.78	22.67
662	Aug	25	9 42	−15 41	662	Aug	20	5 42	0.38730	57.94	24.17
664	Oct	26	10 12	+13 28	664	Oct	18	15 37	0.50496	75.54	18.54
666	Dec	8	13 26	+26 19	666	Dec	3	21 16	0.61732	92.35	15.16
669	Jan	12	2 36	+25 35	669	Jan	11	5 57	0.67273	100.64	13.91
671	Feb	15	7 05	+15 52	671	Feb	18	7 35	0.66256	99.12	14.13
673	Mar	24	2 49	+ 0 02	673	Mar	30	19 22	0.58744	87.88	15.93
675	May	12	12 26	−20 09	675	May	20	23 02	0.46194	69.11	20.26
677	Jul	24	8 25	−26 11	677	Jul	24	5 51	0.37403	55.95	25.02
679	Oct	6	3 38	+ 3 51	679	Sep	28	0 58	0.45325	67.81	20.65
681	Nov	23	4 32	+23 15	681	Nov	17	2 51	0.57918	86.64	16.16
683	Dec	30	17 47	+27 09	683	Dec	28	8 39	0.65875	98.55	14.21
686	Feb	1	20 46	+20 25	686	Feb	3	8 45	0.67462	100.92	13.87
688	Mar	8	14 45	+ 6 45	688	Mar	13	22 12	0.62446	93.42	14.99
690	Apr	21	3 25	−12 08	690	Apr	29	10 47	0.51440	76.95	18.20
692	Jun	22	9 18	−28 57	692	Jun	27	7 40	0.39197	58.64	23.88
694	Sep	10	17 43	− 8 16	694	Sep	3	19 10	0.40643	60.80	23.03
696	Nov	6	2 46	+17 50	696	Oct	29	20 15	0.53307	79.75	17.56
698	Dec	16	21 05	+27 10	698	Dec	13	0 05	0.63490	94.98	14.74
701	Jan	19	16 42	+23 58	701	Jan	19	16 06	0.67627	101.17	13.84
703	Feb	23	4 56	+12 40	703	Feb	27	1 12	0.65180	97.51	14.36
705	Apr	2	16 19	− 4 20	705	Apr	10	1 17	0.56348	84.30	16.61
707	May	26	9 04	−24 24	707	Jun	3	4 28	0.43371	64.88	21.58
709	Aug	11	16 55	−20 46	709	Aug	8	13 42	0.37742	56.46	24.80
711	Oct	18	15 29	+ 9 49	711	Oct	10	14 23	0.48163	72.05	19.43
713	Dec	2	3 28	+25 20	713	Nov	26	19 25	0.60126	89.95	15.57
716	Jan	7	11 09	+26 21	716	Jan	5	21 55	0.66772	99.89	14.02
718	Feb	9	12 53	+17 46	718	Feb	11	21 00	0.66920	100.11	13.99
720	Mar	17	7 13	+ 2 49	720	Mar	23	9 35	0.60511	90.52	15.47
722	May	2	18 48	−16 50	722	May	11	7 53	0.48556	72.64	19.28
724	Jul	9	23 53	−28 26	724	Jul	12	8 47	0.37839	56.61	24.74
726	Sep	25	15 34	− 0 59	726	Sep	17	19 43	0.43101	64.48	21.72
728	Nov	16	2 10	+21 20	728	Nov	9	10 04	0.55938	83.68	16.73
730	Dec	24	21 46	+27 20	730	Dec	21	20 15	0.64926	97.13	14.42
733	Jan	27	5 51	+21 56	733	Jan	28	1 17	0.67669	101.23	13.83

Opposition				*Nearest to Earth*				
Date	*TD*	*Decl.*		*Date*	*TD*	*Least distance*		*diam.*
	h m	° ′			h m	a.u.	m.k.	″
735 Mar 3	8 08	+ 9 12		735 Mar 8	0 01	0.63800	95.44	14.67
737 Apr 12	20 36	− 8 53		737 Apr 20	20 06	0.53711	80.35	17.43
739 Jun 10	12 59	−27 38		739 Jun 17	1 32	0.40857	61.12	22.91
741 Aug 29	2 22	−13 46		741 Aug 23	12 15	0.39032	58.39	23.98
743 Oct 30	1 02	+14 51		743 Oct 22	7 57	0.51043	76.36	18.34
745 Dec 10	17 19	+26 39		745 Dec 6	4 17	0.62114	92.92	15.07
748 Jan 15	2 26	+25 02		748 Jan 14	9 25	0.67381	100.80	13.89
750 Feb 17	7 31	+14 46		750 Feb 20	11 48	0.66091	98.87	14.16
752 Mar 26	10 28	− 1 22		752 Apr 2	6 50	0.58325	87.25	16.05
754 May 15	14 42	−21 24		754 May 23	22 33	0.45641	68.28	20.51
756 Jul 28	10 34	−24 58		756 Jul 27	17 23	0.37369	55.90	25.05
758 Oct 9	4 00	+ 5 38		758 Oct 1	1 10	0.45835	68.57	20.42
760 Nov 25	10 50	+23 59		760 Nov 19	11 34	0.58343	87.28	16.04
763 Jan 1	17 59	+26 55		762 Dec 30	12 32	0.66068	98.84	14.17
765 Feb 3	20 29	+19 30		765 Feb 5	12 24	0.67383	100.80	13.89
767 Mar 11	18 01	+ 5 27		767 Mar 17	5 44	0.62096	92.89	15.07
769 Apr 23	18 16	−13 31		769 May 2	3 59	0.50909	76.16	18.39
771 Jun 27	3 00	−29 01		771 Jul 1	16 20	0.38893	58.18	24.07
773 Sep 14	4 59	− 6 16		773 Sep 7	1 59	0.41073	61.44	22.79
775 Nov 9	14 40	+19 00		775 Nov 2	9 34	0.53814	80.51	17.39
777 Dec 18	23 05	+27 16		777 Dec 15	5 32	0.63786	95.42	14.67
780 Jan 22	15 43	+23 16		780 Jan 22	18 57	0.67670	101.23	13.83
782 Feb 25	6 18	+11 30		782 Mar 1	6 40	0.64956	97.17	14.41
784 Apr 5	2 24	− 5 45		784 Apr 12	15 13	0.55870	83.58	16.75
786 May 29	15 23	−25 22		786 Jun 6	5 32	0.42869	64.13	21.83
788 Aug 15	13 39	−19 03		788 Aug 11	21 15	0.37910	56.71	24.69
790 Oct 21	9 33	+11 22		790 Oct 13	9 01	0.48718	72.88	19.21
792 Dec 4	8 15	+25 49		792 Nov 29	3 40	0.60550	90.58	15.46
795 Jan 9	11 15	+25 55		795 Jan 8	2 10	0.66928	100.12	13.99
797 Feb 11	12 46	+16 44		797 Feb 14	1 05	0.66804	99.94	14.01
799 Mar 20	12 39	+ 1 28		799 Mar 26	19 19	0.60133	89.96	15.57
801 May 5	16 23	−18 10		801 May 14	4 16	0.48001	71.81	19.50
803 Jul 15	1 28	−27 46		803 Jul 16	20 49	0.37662	56.34	24.85
805 Sep 28	20 58	+ 0 56		805 Sep 20	23 21	0.43580	65.19	21.48
807 Nov 19	10 56	+22 15		807 Nov 12	21 20	0.56396	84.37	16.60
809 Dec 26	22 57	+27 14		809 Dec 24	1 38	0.65168	97.49	14.36
812 Jan 30	5 40	+21 05		812 Jan 31	5 36	0.67638	101.19	13.84
814 Mar 5	10 51	+ 7 56		814 Mar 10	7 31	0.63489	94.98	14.74
816 Apr 15	9 03	−10 17		816 Apr 23	11 15	0.53196	79.58	17.60
818 Jun 14	1 22	−28 09		818 Jun 20	7 27	0.40454	60.52	23.14
820 Sep 1	19 03	−11 47		820 Aug 26	20 58	0.39369	58.90	23.77
822 Nov 1	15 48	+16 11		822 Oct 24	23 45	0.51583	77.17	18.15
824 Dec 12	20 14	+26 55		824 Dec 8	11 05	0.62453	93.43	14.99

Opposition   *Nearest to Earth*

Date	TD (h m)	Decl. (° ′)	Date	TD (h m)	Least distance (a.u.)	(m.k.)	diam. (″)
827 Jan 17	1 30	+24 27	827 Jan 16	12 50	0.67464	100.93	13.87
829 Feb 19	7 57	+13 39	829 Feb 22	16 41	0.65919	98.61	14.20
831 Mar 29	18 06	− 2 45	831 Apr 5	18 24	0.57884	86.59	16.17
833 May 18	16 08	−22 33	833 May 26	20 50	0.45101	67.47	20.75
835 Aug 2	9 24	−23 38	835 Aug 1	1 34	0.37391	55.94	25.03
837 Oct 12	1 48	+ 7 20	837 Oct 3	22 17	0.46373	69.37	20.18
839 Nov 28	17 22	+24 38	839 Nov 22	21 49	0.58802	87.97	15.92
842 Jan 3	18 51	+26 36	842 Jan 1	17 58	0.66276	99.15	14.12
844 Feb 6	20 24	+18 32	844 Feb 8	16 51	0.67309	100.69	13.91
846 Mar 13	22 21	+ 4 08	846 Mar 19	14 27	0.61755	92.38	15.16
848 Apr 26	12 19	−14 54	848 May 4	22 03	0.50368	75.35	18.58
850 Jul 1	1 20	−28 55	850 Jul 5	3 40	0.38583	57.72	24.26
852 Sep 17	16 31	− 4 14	852 Sep 10	9 07	0.41500	62.08	22.55
854 Nov 12	1 50	+20 05	854 Nov 4	23 11	0.54311	81.25	17.23
856 Dec 21	0 47	+27 18	856 Dec 17	11 47	0.64070	95.85	14.61
859 Jan 24	15 27	+22 31	859 Jan 24	23 27	0.67685	101.26	13.83
861 Feb 27	8 00	+10 17	861 Mar 3	13 13	0.64695	96.78	14.47
863 Apr 8	12 00	− 7 08	863 Apr 16	3 49	0.55376	82.84	16.90
865 Jun 1	22 05	−26 13	865 Jun 9	8 22	0.42392	63.42	22.08
867 Aug 20	10 13	−17 15	867 Aug 16	6 46	0.38127	57.04	24.55
869 Oct 24	3 31	+12 52	869 Oct 16	3 09	0.49267	73.70	19.00
871 Dec 7	12 48	+26 15	871 Dec 2	12 25	0.60928	91.15	15.36
874 Jan 11	10 45	+25 26	874 Jan 10	6 16	0.67056	100.31	13.96
876 Feb 14	12 51	+15 40	876 Feb 17	5 46	0.66671	99.74	14.04
878 Mar 22	18 58	+ 0 06	878 Mar 29	5 28	0.59726	89.35	15.67
880 May 8	13 58	−19 27	880 May 17	0 30	0.47447	70.98	19.73
882 Jul 19	0 25	−26 55	882 Jul 20	5 13	0.37526	56.14	24.94
884 Oct 1	23 49	+ 2 47	884 Sep 23	23 28	0.44085	65.95	21.23
886 Nov 21	18 58	+23 04	886 Nov 15	9 22	0.56892	85.11	16.45
888 Dec 29	0 08	+27 03	888 Dec 26	7 24	0.65424	97.87	14.31
891 Feb 1	5 25	+20 12	891 Feb 2	9 49	0.67610	101.14	13.84
893 Mar 7	13 31	+ 6 40	893 Mar 12	14 21	0.63198	94.54	14.81
895 Apr 18	23 08	−11 41	895 Apr 27	2 08	0.52681	78.81	17.77
897 Jun 17	18 15	−28 32	897 Jun 23	16 44	0.40038	59.90	23.38
899 Sep 6	11 57	− 9 45	899 Aug 31	6 50	0.39712	59.41	23.57
901 Nov 4	5 45	+17 27	901 Oct 27	15 42	0.52098	77.94	17.97
903 Dec 15	23 03	+27 05	903 Dec 11	18 27	0.62772	93.91	14.91
906 Jan 19	1 14	+23 47	906 Jan 18	17 13	0.67527	101.02	13.86
908 Feb 22	9 15	+12 30	908 Feb 25	22 29	0.65697	98.28	14.25
910 Apr 1	2 12	− 4 08	910 Apr 8	5 32	0.57411	85.89	16.30
912 May 21	18 00	−23 37	912 May 29	21 05	0.44582	66.69	21.00
914 Aug 6	8 46	−22 07	914 Aug 4	11 20	0.37459	56.04	24.99
916 Oct 15	0 11	+ 9 01	916 Oct 6	19 27	0.46921	70.19	19.95

		Opposition					*Nearest*	*to*	*Earth*	
Date		*TD*	*Decl.*		*Date*		*TD*	*Least distance*		*diam.*
		h m	° ′				h m	a.u.	m.k.	″
918 Nov	30	23 18	+25 13		918 Nov	25	8 09	0.59224	88.60	15.80
921 Jan	5	18 33	+26 13		921 Jan	3	22 02	0.66446	99.40	14.09
923 Feb	8	19 58	+17 33		923 Feb	10	20 41	0.67221	100.56	13.92
925 Mar	16	2 35	+ 2 49		925 Mar	21	22 15	0.61402	91.86	15.24
927 Apr	30	5 39	− 16 15		927 May	8	15 09	0.49821	74.53	18.79
929 Jul	4	21 19	− 28 38		929 Jul	8	11 08	0.38302	57.30	24.44
931 Sep	21	23 50	− 2 17		931 Sep	14	11 45	0.41945	62.75	22.31
933 Nov	14	11 38	+21 05		933 Nov	7	13 00	0.54825	82.02	17.07
935 Dec	24	2 44	+27 15		935 Dec	20	18 01	0.64367	96.29	14.54
938 Jan	26	15 04	+21 43		938 Jan	27	3 14	0.67703	101.28	13.83
940 Mar	1	9 52	+ 9 04		940 Mar	5	18 47	0.64442	96.40	14.52
942 Apr	10	23 39	− 8 33		942 Apr	18	16 40	0.54886	82.11	17.05
944 Jun	5	9 18	− 26 58		944 Jun	12	15 13	0.41908	62.69	22.33
946 Aug	24	8 55	− 15 18		946 Aug	19	19 15	0.38355	57.38	24.40
948 Oct	26	21 31	+14 19		948 Oct	18	22 37	0.49800	74.50	18.80
950 Dec	9	16 37	+26 35		950 Dec	4	20 26	0.61288	91.69	15.27
953 Jan	13	10 23	+24 53		953 Jan	12	10 00	0.67167	100.48	13.94
955 Feb	16	13 36	+14 35		955 Feb	19	10 23	0.66501	99.48	14.08
957 Mar	25	0 49	− 1 15		957 Mar	31	14 09	0.59297	88.71	15.78
959 May	12	10 52	− 20 39		959 May	20	21 14	0.46909	70.17	19.95
961 Jul	22	23 14	− 25 54		961 Jul	23	13 44	0.37439	56.01	25.00
963 Oct	6	1 56	+ 4 36		963 Sep	27	22 59	0.44608	66.73	20.98
965 Nov	24	2 30	+23 50		965 Nov	17	21 05	0.57348	85.79	16.32
968 Jan	1	0 32	+26 48		967 Dec	29	11 43	0.65634	98.19	14.26
970 Feb	3	4 28	+19 17		970 Feb	4	12 43	0.67564	101.07	13.85
972 Mar	9	16 35	+ 5 23		972 Mar	14	20 37	0.62889	94.08	14.88
974 Apr	21	13 52	− 13 03		974 Apr	29	17 33	0.52148	78.01	17.95
976 Jun	21	9 56	− 28 45		976 Jun	26	23 05	0.39644	59.31	23.61
978 Sep	10	1 10	− 7 45		978 Sep	3	12 43	0.40074	59.95	23.36
980 Nov	6	18 31	+18 38		980 Oct	30	8 18	0.52633	78.74	17.78
982 Dec	18	1 40	+27 12		982 Dec	14	0 54	0.63115	94.42	14.83
985 Jan	21	0 48	+23 05		985 Jan	20	20 24	0.67597	101.12	13.85
987 Feb	24	10 21	+11 20		987 Feb	28	2 52	0.65494	97.98	14.29
989 Apr	3	10 44	− 5 31		989 Apr	10	15 40	0.56968	85.22	16.43
991 May	25	23 12	− 24 37		991 Jun	3	0 17	0.44061	65.91	21.24
993 Aug	10	10 55	− 20 28		993 Aug	8	0 40	0.37551	56.18	24.93
995 Oct	18	21 43	+10 39		995 Oct	10	17 40	0.47456	70.99	19.72
997 Dec	3	4 26	+25 44		997 Nov	27	16 53	0.59620	89.19	15.70
1000 Jan	8	18 32	+25 47		1000 Jan	7	1 28	0.66602	99.64	14.05
1002 Feb	10	20 04	+16 31		1002 Feb	13	0 09	0.67099	100.38	13.95
1004 Mar	18	7 16	+ 1 29		1004 Mar	24	5 35	0.61011	91.27	15.34
1006 May	2	23 29	− 17 32		1006 May	11	10 17	0.49274	73.71	19.00
1008 Jul	8	17 30	− 28 09		1008 Jul	11	18 21	0.38073	56.96	24.58

Opposition   Nearest to Earth

Date	TD	Decl.		Date	TD	Least distance		diam.
	h m	° ′			h m	a.u.	m.k.	″
1010 Sep 25	7 21	− 0 20		1010 Sep 17	14 18	0.42423	63.46	22.06
1012 Nov 16	21 53	+22 01		1012 Nov 10	2 51	0.55322	82.76	16.92
1014 Dec 26	3 56	+27 08		1014 Dec 22	22 38	0.64626	96.68	14.48
1017 Jan 28	13 54	+20 53		1017 Jan 29	5 25	0.67706	101.29	13.82
1019 Mar 4	11 54	+ 7 50		1019 Mar 8	23 38	0.64186	96.02	14.58
1021 Apr 13	11 05	− 9 56		1021 Apr 21	5 50	0.54390	81.37	17.21
1023 Jun 9	19 38	−27 35		1023 Jun 16	19 20	0.41434	61.98	22.59
1025 Aug 28	3 15	−13 22		1025 Aug 23	3 31	0.38607	57.75	24.24
1027 Oct 30	12 43	+15 41		1027 Oct 22	17 00	0.50341	75.31	18.59
1029 Dec 11	20 04	+26 50		1029 Dec 7	3 19	0.61671	92.26	15.18
1032 Jan 16	10 14	+24 16		1032 Jan 15	13 03	0.67289	100.66	13.91
1034 Feb 18	13 55	+13 28		1034 Feb 21	13 46	0.66344	99.25	14.11
1036 Mar 27	7 28	− 2 38		1036 Apr 2	22 56	0.58893	88.10	15.89
1038 May 15	11 47	−21 49		1038 May 23	22 14	0.46370	69.37	20.19
1040 Jul 27	2 18	−24 40		1040 Jul 27	3 13	0.37382	55.92	25.04
1042 Oct 9	4 48	+ 6 25		1042 Oct 1	0 48	0.45133	67.52	20.74
1044 Nov 26	10 05	+24 30		1044 Nov 20	7 31	0.57788	86.45	16.20
1047 Jan 3	1 03	+26 29		1046 Dec 31	15 13	0.65839	98.49	14.22
1049 Feb 5	4 15	+18 20		1049 Feb 6	15 34	0.67496	100.97	13.87
1051 Mar 12	20 20	+ 4 05		1051 Mar 18	2 57	0.62546	93.57	14.96
1053 Apr 24	4 05	−14 23		1053 May 2	10 19	0.51613	77.21	18.14
1055 Jun 26	1 36	−28 48		1055 Jul 1	4 06	0.39294	58.78	23.82
1057 Sep 13	13 48	− 5 44		1057 Sep 6	17 28	0.40481	60.56	23.12
1059 Nov 10	6 52	+19 45		1059 Nov 2	23 29	0.53156	79.52	17.61
1061 Dec 20	3 50	+27 13		1061 Dec 16	6 02	0.63419	94.87	14.76
1064 Jan 23	23 43	+22 20		1064 Jan 23	22 23	0.67647	101.20	13.84
1066 Feb 26	11 13	+10 08		1066 Mar 2	6 33	0.65284	97.66	14.34
1068 Apr 5	19 54	− 6 54		1068 Apr 13	3 26	0.56507	84.53	16.56
1070 May 29	5 11	−25 31		1070 Jun 6	2 55	0.43534	65.13	21.50
1072 Aug 14	9 19	−18 44		1072 Aug 11	11 10	0.37664	56.34	24.85
1074 Oct 21	17 01	+12 11		1074 Oct 13	15 06	0.47995	71.80	19.50
1076 Dec 5	9 38	+26 09		1076 Nov 30	1 03	0.60050	89.83	15.59
1079 Jan 10	18 50	+25 17		1079 Jan 9	4 53	0.66777	99.90	14.02
1081 Feb 12	20 04	+15 28		1081 Feb 15	3 20	0.66990	100.22	13.97
1083 Mar 21	12 21	+ 0 08		1083 Mar 27	13 26	0.60651	90.73	15.43
1085 May 5	19 32	−18 48		1085 May 14	7 49	0.48737	72.91	19.21
1087 Jul 13	18 12	−27 28		1087 Jul 16	6 11	0.37871	56.65	24.72
1089 Sep 28	15 31	+ 1 38		1089 Sep 20	19 11	0.42920	64.21	21.81
1091 Nov 20	7 22	+22 53		1091 Nov 13	14 33	0.55796	83.47	16.78
1093 Dec 28	5 05	+26 57		1093 Dec 25	2 33	0.64874	97.05	14.43
1096 Jan 31	13 28	+20 00		1096 Feb 1	8 02	0.67693	101.27	13.83
1098 Mar 6	14 26	+ 6 34		1098 Mar 11	5 15	0.63888	95.57	14.65
1100 Apr 15	22 43	−11 17		1100 Apr 23	20 46	0.53869	80.59	17.38

Opposition

Date	TD h m	Decl. ° '
1102 Jun 13	6 46	− 28 03
1104 Aug 31	20 46	− 11 22
1106 Nov 2	4 29	+ 16 59
1108 Dec 13	23 54	+ 27 01
1111 Jan 18	9 32	+ 23 37
1113 Feb 20	14 09	+ 12 19
1115 Mar 30	14 41	− 4 00
1117 May 18	12 40	− 22 54
1119 Aug 1	2 23	− 23 19
1121 Oct 12	4 08	+ 8 08
1123 Nov 29	16 27	+ 25 06
1126 Jan 5	1 45	+ 26 06
1128 Feb 8	3 55	+ 17 20
1130 Mar 14	23 42	+ 2 47
1132 Apr 26	20 17	− 15 43
1134 Jun 29	22 07	− 28 39
1136 Sep 17	3 45	− 3 39
1138 Nov 12	19 31	+ 20 48
1140 Dec 22	6 11	+ 27 10
1143 Jan 25	23 25	+ 21 31
1145 Feb 28	13 08	+ 8 55
1147 Apr 9	5 25	− 8 16
1149 Jun 1	10 49	− 26 18
1151 Aug 19	6 31	− 16 54
1153 Oct 24	11 48	+ 13 41
1155 Dec 8	14 32	+ 26 29
1158 Jan 12	18 32	+ 24 43
1160 Feb 15	19 32	+ 14 23
1162 Mar 23	17 30	− 1 12
1164 May 8	16 09	− 20 01
1166 Jul 17	17 15	− 26 37
1168 Oct 1	19 47	+ 3 30
1170 Nov 22	16 00	+ 23 39
1172 Dec 30	6 32	+ 26 41
1175 Feb 2	13 16	+ 19 04
1177 Mar 8	16 59	+ 5 17
1179 Apr 19	11 42	− 12 38
1181 Jun 16	21 30	− 28 23
1183 Sep 5	16 24	− 9 17
1185 Nov 4	19 55	+ 18 14
1187 Dec 17	3 03	+ 27 07
1190 Jan 20	9 17	+ 22 54
1192 Feb 23	15 05	+ 11 09

Nearest to Earth

Date	TD h m	Least distance a.u.	Least distance m.k.	diam. "
1102 Jun 19	22 41	0.40991	61.32	22.83
1104 Aug 26	10 19	0.38902	58.20	24.06
1106 Oct 25	10 50	0.50882	76.12	18.40
1108 Dec 9	9 51	0.62029	92.79	15.09
1111 Jan 17	15 29	0.67384	100.80	13.89
1113 Feb 23	17 09	0.66178	99.00	14.14
1115 Apr 6	9 24	0.58475	87.48	16.01
1117 May 26	21 46	0.45814	68.54	20.43
1119 Jul 31	14 28	0.37340	55.86	25.07
1121 Oct 4	1 06	0.45659	68.30	20.50
1123 Nov 23	16 29	0.58249	87.14	16.07
1126 Jan 2	19 14	0.66060	98.82	14.17
1128 Feb 9	18 41	0.67441	100.89	13.88
1130 Mar 20	9 51	0.62226	93.09	15.04
1132 May 5	4 58	0.51086	76.42	18.32
1134 Jul 4	13 58	0.38973	58.30	24.02
1136 Sep 10	1 05	0.40911	61.20	22.88
1138 Nov 5	13 40	0.53661	80.28	17.44
1140 Dec 18	11 24	0.63718	95.32	14.69
1143 Jan 26	1 31	0.67681	101.25	13.83
1145 Mar 4	12 09	0.65031	97.28	14.39
1147 Apr 16	16 52	0.56013	83.79	16.71
1149 Jun 9	3 21	0.43022	64.36	21.76
1151 Aug 15	18 58	0.37817	56.57	24.75
1153 Oct 16	11 09	0.48545	72.62	19.28
1155 Dec 3	8 37	0.60450	90.43	15.48
1158 Jan 11	8 04	0.66915	100.10	13.99
1160 Feb 18	6 24	0.66875	100.04	14.00
1162 Mar 29	22 21	0.60274	90.17	15.53
1164 May 17	4 40	0.48173	72.07	19.43
1166 Jul 19	17 41	0.37684	56.37	24.84
1168 Sep 23	22 28	0.43404	64.93	21.56
1170 Nov 16	1 24	0.56285	84.20	16.63
1172 Dec 27	7 53	0.65148	97.46	14.37
1175 Feb 3	11 50	0.67685	101.26	13.83
1177 Mar 13	11 56	0.63606	95.15	14.72
1179 Apr 27	13 00	0.53370	79.84	17.54
1181 Jun 23	5 14	0.40571	60.69	23.07
1183 Aug 30	19 40	0.39233	58.69	23.86
1185 Oct 28	3 02	0.51414	76.91	18.20
1187 Dec 12	16 21	0.62369	93.30	15.01
1190 Jan 19	19 13	0.67465	100.93	13.87
1192 Feb 26	22 24	0.65982	98.71	14.19

Opposition				*Nearest*	*to*	*Earth*		
Date	*TD*	*Decl.*		*Date*	*TD*	*Least distance*		*diam.*
	h m	*° ′*			*h m*	*a.u.*	*m.k.*	*″*
1194 Apr 1	22 07	− 5 21		1194 Apr 8	21 11	0.58013	86.79	16.13
1196 May 21	13 33	−23 54		1196 May 29	19 55	0.45262	67.71	20.68
1198 Aug 5	1 16	−21 49		1198 Aug 3	23 17	0.37346	55.87	25.06
1200 Oct 15	2 55	+ 9 48		1200 Oct 6	23 39	0.46195	69.11	20.26
1202 Dec 1	23 23	+25 37		1202 Nov 26	2 16	0.58692	87.80	15.95
1205 Jan 7	2 10	+25 39		1205 Jan 4	23 42	0.66247	99.10	14.13
1207 Feb 10	3 18	+16 19		1207 Feb 11	22 16	0.67367	100.78	13.89
1209 Mar 17	3 46	+ 1 28		1209 Mar 22	18 12	0.61889	92.59	15.12
1211 Apr 30	13 13	−17 00		1211 May 8	23 14	0.50541	75.61	18.52
1213 Jul 3	17 43	−28 20		1213 Jul 8	0 08	0.38652	57.82	24.22
1215 Sep 21	13 45	− 1 40		1215 Sep 14	7 20	0.41331	61.83	22.65
1217 Nov 15	6 02	+21 45		1217 Nov 8	2 09	0.54175	81.04	17.28
1219 Dec 25	8 03	+27 03		1219 Dec 21	17 31	0.64035	95.79	14.62
1222 Jan 27	23 14	+20 41		1222 Jan 28	5 45	0.67723	101.31	13.82
1224 Mar 2	14 31	+ 7 41		1224 Mar 6	18 07	0.64799	96.94	14.44
1226 Apr 11	15 30	− 9 38		1226 Apr 19	6 28	0.55544	83.09	16.85
1228 Jun 4	19 56	−26 59		1228 Jun 12	7 03	0.42539	63.64	22.00
1230 Aug 23	6 23	−14 56		1230 Aug 19	5 36	0.38028	56.89	24.61
1232 Oct 27	7 05	+15 07		1232 Oct 19	6 00	0.49094	73.44	19.07
1234 Dec 10	19 29	+26 45		1234 Dec 5	17 22	0.60834	91.01	15.39
1237 Jan 14	18 36	+24 06		1237 Jan 13	12 39	0.67045	100.30	13.96
1239 Feb 17	20 16	+13 15		1239 Feb 20	11 50	0.66727	99.82	14.03
1241 Mar 25	23 59	− 2 33		1241 Apr 1	9 25	0.59846	89.53	15.64
1243 May 12	12 55	−21 09		1243 May 21	0 21	0.47609	71.22	19.66
1245 Jul 21	15 47	−25 35		1245 Jul 23	2 35	0.37533	56.15	24.94
1247 Oct 5	23 28	+ 5 20		1247 Sep 28	0 05	0.43908	65.69	21.32
1249 Nov 25	0 31	+24 21		1249 Nov 18	13 10	0.56766	84.92	16.49
1252 Jan 2	7 26	+26 21		1251 Dec 30	13 13	0.65380	97.81	14.32
1254 Feb 4	12 32	+18 07		1254 Feb 5	15 36	0.67654	101.21	13.84
1256 Mar 10	19 27	+ 4 01		1256 Mar 15	18 54	0.63319	94.72	14.78
1258 Apr 22	1 22	−13 58		1258 Apr 30	4 14	0.52850	79.06	17.71
1260 Jun 20	12 34	−28 33		1260 Jun 26	13 50	0.40149	60.06	23.31
1262 Sep 9	7 49	− 7 15		1262 Sep 3	4 37	0.39566	59.19	23.66
1264 Nov 7	9 20	+19 23		1264 Oct 30	18 07	0.51947	77.71	18.02
1266 Dec 19	6 09	+27 08		1266 Dec 15	0 02	0.62727	93.84	14.92
1269 Jan 22	9 18	+22 08		1269 Jan 21	23 56	0.67555	101.06	13.86
1271 Feb 25	16 13	+ 9 57		1271 Mar 1	4 09	0.65792	98.42	14.23
1273 Apr 4	6 36	− 6 43		1273 Apr 11	9 05	0.57574	86.13	16.26
1275 May 25	17 14	−24 50		1275 Jun 2	20 18	0.44747	66.94	20.92
1277 Aug 9	3 53	−20 09		1277 Aug 7	10 06	0.37409	55.96	25.02
1279 Oct 19	2 57	+11 27		1279 Oct 10	21 57	0.46749	69.94	20.02
1281 Dec 4	5 48	+26 03		1281 Nov 28	12 48	0.59117	88.44	15.83
1284 Jan 10	2 24	+25 08		1284 Jan 8	4 38	0.66422	99.37	14.09

ASTRONOMICAL TABLES

Opposition			Nearest to Earth				
Date	TD	Decl.	Date	TD	Least distance		diam.
	h m	° '		h m	a.u.	m.k.	''
1286 Feb 12	3 40	+ 15 15	1286 Feb 14	3 17	0.67268	100.63	13.91
1288 Mar 19	8 32	+ 0 08	1288 Mar 25	3 18	0.61506	92.01	15.22
1290 May 3	6 21	− 18 14	1290 May 11	16 12	0.49977	74.76	18.73
1292 Jul 7	13 43	− 27 50	1292 Jul 11	9 01	0.38363	57.39	24.40
1294 Sep 24	22 09	+ 0 18	1294 Sep 17	11 54	0.41780	62.50	22.40
1296 Nov 17	16 53	+ 22 38	1296 Nov 10	16 31	0.54690	81.82	17.11
1298 Dec 27	10 11	+ 26 51	1298 Dec 24	0 12	0.64315	96.21	14.55
1301 Jan 29	22 35	+ 19 47	1301 Jan 30	9 41	0.67734	101.33	13.82
1303 Mar 5	16 17	+ 6 26	1303 Mar 10	0 14	0.64553	96.57	14.50
1305 Apr 14	3 05	− 10 59	1305 Apr 21	19 44	0.55053	82.36	17.00
1307 Jun 9	5 37	− 27 32	1307 Jun 16	12 48	0.42047	62.90	22.26
1309 Aug 27	3 18	− 12 58	1309 Aug 22	16 23	0.38241	57.21	24.48
1311 Oct 31	0 14	+ 16 28	1311 Oct 23	0 17	0.49638	74.26	18.86
1313 Dec 12	23 25	+ 26 56	1313 Dec 8	1 53	0.61227	91.59	15.29
1316 Jan 17	18 34	+ 23 26	1316 Jan 16	17 11	0.67182	100.50	13.93
1318 Feb 19	20 57	+ 12 07	1318 Feb 22	16 50	0.66583	99.61	14.06
1320 Mar 28	6 05	− 3 54	1320 Apr 3	18 40	0.59446	88.93	15.75
1322 May 15	11 32	− 22 15	1322 May 23	21 30	0.47080	70.43	19.88
1324 Jul 25	18 23	− 24 21	1324 Jul 26	13 11	0.37444	56.02	25.00
1326 Oct 9	4 00	+ 7 10	1326 Oct 1	1 14	0.44444	66.49	21.06
1328 Nov 27	9 03	+ 24 58	1328 Nov 21	1 48	0.57230	85.61	16.36
1331 Jan 4	8 30	+ 25 58	1331 Jan 1	18 39	0.65600	98.14	14.27
1333 Feb 6	12 34	+ 17 07	1333 Feb 7	19 57	0.67601	101.13	13.85
1335 Mar 13	23 19	+ 2 42	1335 Mar 19	2 39	0.62981	94.22	14.86
1337 Apr 24	16 07	− 15 16	1337 May 2	19 46	0.52296	78.23	17.90
1339 Jun 25	3 51	− 28 33	1339 Jun 30	21 09	0.39744	59.46	23.55
1341 Sep 12	21 52	− 5 12	1341 Sep 6	12 02	0.39925	59.73	23.44
1343 Nov 10	23 07	+ 20 27	1343 Nov 3	11 12	0.52489	78.52	17.83
1345 Dec 21	9 05	+ 27 05	1345 Dec 17	7 18	0.63051	94.32	14.85
1348 Jan 25	8 34	+ 21 19	1348 Jan 25	3 27	0.67608	101.14	13.84
1350 Feb 27	17 03	+ 8 45	1350 Mar 3	8 54	0.65590	98.12	14.27
1352 Apr 6	15 08	− 8 03	1352 Apr 13	19 37	0.57123	85.46	16.39
1354 May 28	21 23	− 25 40	1354 Jun 5	22 55	0.44213	66.14	21.17
1356 Aug 13	4 15	− 18 23	1356 Aug 10	21 47	0.37482	56.07	24.97
1358 Oct 21	23 37	+ 12 59	1358 Oct 13	18 36	0.47285	70.74	19.79
1360 Dec 6	11 04	+ 26 23	1360 Nov 30	22 24	0.59541	89.07	15.72
1363 Jan 12	2 49	+ 24 33	1363 Jan 10	9 04	0.66607	99.64	14.05
1365 Feb 14	3 54	+ 14 09	1365 Feb 16	7 24	0.67168	100.48	13.94
1367 Mar 22	13 14	− 1 12	1367 Mar 28	10 58	0.61144	91.47	15.31
1369 May 6	1 17	− 19 26	1369 May 14	11 15	0.49450	73.98	18.93
1371 Jul 12	13 10	− 27 08	1371 Jul 15	17 59	0.38128	57.04	24.55
1373 Sep 28	8 22	+ 2 18	1373 Sep 20	16 06	0.42269	63.23	22.14
1375 Nov 21	4 04	+ 23 27	1375 Nov 14	7 25	0.55195	82.57	16.96

Opposition			Nearest to Earth			
Date	TD	Decl.	Date	TD	Least distance	diam.
	h m	° '		h m	a.u.	m.k.	"
1377 Dec 29	11 49	+26 35	1377 Dec 26	5 40	0.64578	96.61	14.49
1380 Feb 1	22 12	+18 51	1380 Feb 2	13 04	0.67727	101.32	13.82
1382 Mar 7	18 59	+ 5 10	1382 Mar 12	6 12	0.64267	96.14	14.56
1384 Apr 16	14 36	−12 18	1384 Apr 24	8 53	0.54525	81.57	17.17
1386 Jun 12	15 12	−27 56	1386 Jun 19	18 01	0.41557	62.17	22.52
1388 Aug 30	22 04	−10 57	1388 Aug 26	2 10	0.38485	57.57	24.32
1390 Nov 2	16 32	+17 45	1390 Oct 25	19 19	0.50186	75.08	18.65
1392 Dec 15	3 30	+27 02	1392 Dec 10	9 55	0.61592	92.14	15.20
1395 Jan 19	18 12	+22 42	1395 Jan 18	20 31	0.67283	100.65	13.91
1397 Feb 21	21 00	+10 57	1397 Feb 24	20 21	0.66422	99.37	14.09
1399 Mar 31	12 40	− 5 14	1399 Apr 7	3 31	0.59036	88.32	15.85
1401 May 18	11 31	−23 16	1401 May 26	21 36	0.46533	69.61	20.11
1403 Jul 30	19 23	−22 59	1403 Jul 31	0 13	0.37362	55.89	25.05
1405 Oct 12	5 41	+ 8 54	1405 Oct 4	1 05	0.44957	67.25	20.82
1407 Nov 30	16 07	+25 29	1407 Nov 24	12 46	0.57690	86.30	16.22
1410 Jan 6	9 10	+25 30	1410 Jan 3	22 51	0.65830	98.48	14.22
1412 Feb 9	12 25	+16 05	1412 Feb 10	23 17	0.67550	101.05	13.86
1414 Mar 16	2 34	+ 1 24	1414 Mar 21	8 42	0.62667	93.75	14.94
1416 Apr 27	6 48	−16 31	1416 May 5	11 47	0.51787	77.47	18.07
1418 Jun 28	22 01	−28 22	1418 Jul 4	4 09	0.39390	58.93	23.76
1420 Sep 16	13 09	− 3 07	1420 Sep 9	18 52	0.40342	60.35	23.20
1422 Nov 13	12 36	+21 28	1422 Nov 6	3 55	0.53019	79.32	17.65
1424 Dec 23	11 44	+26 57	1424 Dec 19	13 08	0.63356	94.78	14.77
1427 Jan 27	8 08	+20 28	1427 Jan 27	6 13	0.67654	101.21	13.84
1429 Mar 1	18 47	+ 7 31	1429 Mar 5	13 33	0.65352	97.77	14.32
1431 Apr 10	0 40	− 9 24	1431 Apr 17	7 18	0.56626	84.71	16.53
1433 Jun 1	2 30	−26 23	1433 Jun 9	2 06	0.43673	65.33	21.43
1435 Aug 18	2 51	−16 33	1435 Aug 15	9 06	0.37582	56.22	24.91
1437 Oct 24	19 53	+14 28	1437 Oct 16	16 50	0.47832	71.56	19.57
1439 Dec 9	16 41	+26 39	1439 Dec 4	7 26	0.59957	89.70	15.61
1442 Jan 14	2 52	+23 56	1442 Jan 12	12 23	0.66757	99.87	14.02
1444 Feb 17	3 30	+13 03	1444 Feb 19	10 08	0.67052	100.31	13.96
1446 Mar 24	17 53	− 2 31	1446 Mar 30	18 05	0.60785	90.93	15.40
1448 May 8	20 33	−20 34	1448 May 17	7 56	0.48906	73.16	19.14
1450 Jul 16	11 46	−26 16	1450 Jul 19	3 43	0.37898	56.69	24.70
1452 Oct 1	14 46	+ 4 12	1452 Sep 23	18 42	0.42742	63.94	21.90
1454 Nov 23	12 45	+24 10	1454 Nov 16	19 24	0.55678	83.29	16.81
1456 Dec 31	13 05	+26 14	1456 Dec 28	10 03	0.64850	97.01	14.43
1459 Feb 3	21 52	+17 53	1459 Feb 4	15 56	0.67733	101.33	13.82
1461 Mar 9	21 14	+ 3 53	1461 Mar 14	11 22	0.63994	95.73	14.63
1463 Apr 20	2 35	−13 36	1463 Apr 27	23 09	0.54034	80.83	17.32
1465 Jun 16	4 18	−28 13	1465 Jun 22	23 00	0.41121	61.52	22.76
1467 Sep 4	18 52	− 8 51	1467 Aug 30	11 26	0.38795	58.04	24.13

Opposition   Nearest to Earth

Date		TD	Decl.		Date		TD	Least distance		diam.
		h m	° ′				h m	a.u.	m.k.	″
1469 Nov	5	9 47	+ 18 58		1469 Oct	28	15 08	0.50743	75.91	18.45
1471 Dec	18	7 35	+ 27 03		1471 Dec	13	16 43	0.61952	92.68	15.11
1474 Jan	21	17 58	+ 21 56		1474 Jan	20	23 11	0.67381	100.80	13.89
1476 Feb	24	22 09	+ 9 45		1476 Feb	28	0 21	0.66238	99.09	14.13
1478 Apr	2	20 12	− 6 35		1478 Apr	9	13 43	0.58586	87.64	15.98
1480 May	21	11 41	− 24 12		1480 May	29	21 51	0.45962	68.76	20.36
1482 Aug	3	19 15	− 21 29		1482 Aug	3	12 14	0.37306	55.81	25.09
1484 Oct	15	5 35	+ 10 34		1484 Oct	7	1 53	0.45486	68.05	20.58
1486 Dec	2	23 03	+ 25 56		1486 Nov	26	22 26	0.58143	86.98	16.10
1489 Jan	8	9 56	+ 24 59		1489 Jan	6	2 37	0.66029	98.78	14.18
1491 Feb	11	11 40	+ 15 01		1491 Feb	13	1 38	0.67483	100.95	13.87
1493 Mar	18	5 51	+ 0 06		1493 Mar	23	14 47	0.62347	93.27	15.01
1495 Apr	30	22 54	− 17 44		1495 May	9	6 26	0.51253	76.67	18.26
1497 Jul	2	17 06	− 28 01		1497 Jul	7	12 08	0.39043	58.41	23.97
1499 Sep	21	1 47	− 1 05		1499 Sep	14	0 17	0.40749	60.96	22.97
1501 Nov	16	0 25	+ 22 23		1501 Nov	8	18 07	0.53530	80.08	17.49
1503 Dec	26	13 57	+ 26 45		1503 Dec	22	18 27	0.63679	95.26	14.70
1506 Jan	29	7 57	+ 19 34		1506 Jan	29	9 18	0.67714	101.30	13.82
1508 Mar	3	20 32	+ 6 16		1508 Mar	7	18 32	0.65127	97.43	14.37
1510 Apr	12	9 58	− 10 43		1510 Apr	19	19 53	0.56172	84.03	16.66
1512 Jun	4	9 35	− 26 59		1512 Jun	12	4 14	0.43179	64.59	21.68
1514 Aug	22	3 33	− 14 35		1514 Aug	18	20 11	0.37753	56.48	24.79
1516 Oct	27	16 23	+ 15 53		1516 Oct	19	14 54	0.48401	72.41	19.34
1518 Dec	11	22 03	+ 26 51		1518 Dec	6	15 10	0.60362	90.30	15.51
1521 Jan	16	3 03	+ 23 15		1521 Jan	14	15 34	0.66902	100.08	13.99
1523 Feb	19	3 55	+ 11 54		1523 Feb	21	13 46	0.66922	100.11	13.99
1525 Mar	26	23 55	− 3 51		1525 Apr	2	3 25	0.60374	90.32	15.50
1527 May	12	17 19	− 21 40		1527 May	21	6 15	0.48317	72.28	19.37
1529 Jul	20	10 37	− 25 14		1529 Jul	22	15 21	0.37695	56.39	24.83
1531 Oct	5	20 04	+ 6 02		1531 Sep	27	22 34	0.43233	64.68	21.65
1533 Nov	25	22 16	+ 24 48		1533 Nov	19	7 01	0.56171	84.03	16.66
1536 Jan	3	14 46	+ 25 50		1535 Dec	31	14 58	0.65105	97.40	14.38
1538 Feb	5	21 17	+ 16 53		1538 Feb	6	18 44	0.67714	101.30	13.82
1540 Mar	11	23 41	+ 2 37		1540 Mar	16	17 08	0.63718	95.32	14.69
1542 Apr	22	15 17	− 14 52		1542 Apr	30	15 17	0.53534	80.09	17.48
1544 Jun	19	18 03	− 28 19		1544 Jun	26	4 19	0.40678	60.85	23.01
1546 Sep	8	13 07	− 6 46		1546 Sep	2	19 00	0.39097	58.49	23.94
1548 Nov	8	0 08	+ 20 05		1548 Oct	31	6 52	0.51268	76.70	18.26
1550 Dec	20	10 30	+ 27 00		1550 Dec	15	22 48	0.62313	93.22	15.02
1553 Jan	23	17 55	+ 21 07		1553 Jan	23	2 46	0.67488	100.96	13.87
1555 Feb	26	23 00	+ 8 33		1555 Mar	2	5 01	0.66062	98.83	14.17
1557 Apr	5	3 28	− 7 54		1557 Apr	12	1 00	0.58160	87.01	16.09
1559 May	25	14 03	− 25 03		1559 Jun	2	21 36	0.45430	67.96	20.60

Opposition *Nearest to Earth*

Date	TD	Decl.	Date	TD	Least distance		diam.
	h m	° ′		h m	a.u.	m.k.	″
1561 Aug 7	21 48	− 19 48	1561 Aug 7	0 11	0.37325	55.84	25.08
1563 Oct 19	6 52	+ 12 13	1563 Oct 11	3 12	0.46052	68.89	20.32
1565 Dec 5	6 51	+ 26 17	1565 Nov 29	8 36	0.58596	87.66	15.97
1568 Jan 11	10 45	+ 24 24	1568 Jan 9	6 57	0.66224	99.07	14.13
1570 Feb 13	11 52	+ 13 55	1570 Feb 15	5 34	0.67406	100.84	13.89
1572 Mar 20	10 51	− 1 13	1572 Mar 25	23 46	0.61984	92.73	15.10
1574 May 3	15 52	− 18 56	1574 May 12	1 44	0.50678	75.81	18.47
1576 Jul 6	12 02	− 27 30	1576 Jul 10	22 03	0.38705	57.90	24.18
1578 Sep 24	12 30	+ 0 55	1578 Sep 17	6 52	0.41170	61.59	22.73
1580 Nov 18	11 43	+ 23 13	1580 Nov 11	7 08	0.54047	80.85	17.32
1583 Jan 7	16 11	+ 26 28	1583 Jan 4	0 15	0.63980	95.71	14.63
1585 Feb 10	7 27	+ 18 38	1585 Feb 10	12 38	0.67739	101.34	13.82
1587 Mar 16	21 38	+ 5 02	1587 Mar 20	23 37	0.64893	97.08	14.42
1589 Apr 24	19 55	− 12 01	1589 May 2	9 35	0.55700	83.33	16.80
1591 Jun 18	18 21	− 27 28	1591 Jun 26	6 59	0.42671	63.83	21.94
1593 Sep 5	2 11	− 12 34	1593 Sep 1	4 56	0.37925	56.73	24.68
1595 Nov 10	10 40	+ 17 13	1595 Nov 2	9 31	0.48933	73.20	19.13
1597 Dec 24	2 41	+ 26 57	1597 Dec 18	23 14	0.60763	90.90	15.40
1600 Jan 29	3 14	+ 22 31	1600 Jan 27	19 58	0.67057	100.32	13.96
1602 Mar 3	4 28	+ 10 44	1602 Mar 5	18 38	0.66796	99.93	14.01
1604 Apr 8	5 53	− 5 11	1604 Apr 14	13 49	0.59985	89.74	15.60
1606 May 25	14 33	− 22 40	1606 Jun 3	2 31	0.47781	71.48	19.59
1608 Aug 3	11 49	− 24 00	1608 Aug 5	3 10	0.37560	56.19	24.92
1610 Oct 19	2 13	+ 7 53	1610 Oct 11	3 06	0.43771	65.48	21.38
1612 Dec 8	7 42	+ 25 21	1612 Dec 1	18 55	0.56662	84.77	16.52
1615 Jan 15	15 56	+ 25 22	1615 Jan 12	20 13	0.65344	97.75	14.32
1617 Feb 17	21 07	+ 15 51	1617 Feb 18	22 51	0.67685	101.25	13.83
1619 Mar 25	2 59	+ 1 19	1619 Mar 30	0 58	0.63404	94.85	14.76
1621 May 5	5 11	− 16 07	1621 May 13	7 51	0.52977	79.25	17.67
1623 Jul 4	8 26	− 28 17	1623 Jul 10	12 03	0.40235	60.19	23.26
1625 Sep 22	4 59	− 4 42	1625 Sep 16	3 20	0.39419	58.97	23.74
1627 Nov 21	14 23	+ 21 07	1627 Nov 13	22 29	0.51802	77.49	18.07
1630 Jan 1	14 10	+ 26 52	1629 Dec 28	6 29	0.62662	93.74	14.94
1632 Feb 5	17 42	+ 20 15	1632 Feb 5	6 58	0.67559	101.07	13.85
1634 Mar 10	23 34	+ 7 20	1634 Mar 14	10 07	0.65871	98.54	14.21
1636 Apr 17	11 34	− 9 13	1636 Apr 24	12 53	0.57726	86.36	16.21
1638 Jun 7	17 07	− 25 48	1638 Jun 15	20 55	0.44896	67.16	20.85
1640 Aug 21	22 46	− 18 02	1640 Aug 20	9 40	0.37347	55.87	25.06
1642 Nov 1	5 23	+ 13 45	1642 Oct 24	0 43	0.46581	69.68	20.09
1644 Dec 17	12 32	+ 26 34	1644 Dec 11	17 55	0.59030	88.31	15.86
1647 Jan 23	10 54	+ 23 46	1647 Jan 21	11 46	0.66421	99.36	14.09
1649 Feb 25	11 57	+ 12 49	1649 Feb 27	10 17	0.67329	100.72	13.90
1651 Apr 2	14 55	− 2 32	1651 Apr 8	8 25	0.61634	92.20	15.19

Opposition   Nearest to Earth

Date	TD	Decl.	Date	TD	Least distance		diam.
	h m	° ′		h m	a.u.	m.k.	″
1653 May 16	8 53	−20 03	1653 May 24	19 06	0.50148	75.02	18.66
1655 Jul 21	9 45	−26 47	1655 Jul 25	8 57	0.38437	57.50	24.35
1657 Oct 7	23 31	+ 2 55	1657 Sep 30	14 00	0.41650	62.31	22.47
1659 Dec 1	23 50	+23 58	1659 Nov 24	21 45	0.54575	81.64	17.15
1662 Jan 9	18 38	+26 08	1662 Jan 6	7 10	0.64268	96.14	14.56
1664 Feb 13	7 15	+17 39	1664 Feb 13	17 06	0.67755	101.36	13.81
1666 Mar 19	0 08	+ 3 45	1666 Mar 23	6 53	0.64628	96.68	14.48
1668 Apr 27	7 49	−13 18	1668 May 5	0 16	0.55176	82.54	16.96
1670 Jun 22	3 17	−27 49	1670 Jun 29	11 49	0.42156	63.07	22.20
1672 Sep 8	22 51	−10 32	1672 Sep 4	14 46	0.38121	57.03	24.55
1674 Nov 13	4 10	+18 28	1674 Nov 5	3 33	0.49481	74.02	18.92
1676 Dec 26	7 08	+26 58	1676 Dec 21	8 07	0.61153	91.48	15.31
1679 Jan 31	3 09	+21 44	1679 Jan 30	0 32	0.67173	100.49	13.93
1681 Mar 5	4 26	+ 9 33	1681 Mar 7	23 14	0.66649	99.71	14.04
1683 Apr 11	11 34	− 6 29	1683 Apr 17	23 17	0.59590	89.15	15.71
1685 May 28	12 40	−23 37	1685 Jun 5	22 56	0.47240	70.67	19.81
1687 Aug 8	12 52	−22 38	1687 Aug 9	12 46	0.37434	56.00	25.00
1689 Oct 22	5 16	+ 9 37	1689 Oct 14	3 14	0.44272	66.23	21.14
1691 Dec 11	15 27	+25 48	1691 Dec 5	6 31	0.57125	85.46	16.39
1694 Jan 17	16 52	+24 50	1694 Jan 15	1 51	0.65586	98.12	14.27
1696 Feb 20	21 06	+14 46	1696 Feb 22	3 27	0.67653	101.21	13.84
1698 Mar 27	6 17	+ 0 01	1698 Apr 1	8 41	0.63093	94.39	14.84
1700 May 8	19 28	−17 19	1700 May 16	23 12	0.52466	78.49	17.84
1702 Jul 9	0 51	−28 04	1702 Jul 14	21 06	0.39858	59.63	23.48
1704 Sep 26	21 43	− 2 35	1704 Sep 20	13 30	0.39814	59.56	23.51
1706 Nov 25	5 25	+22 05	1706 Nov 17	15 42	0.52370	78.34	17.87
1709 Jan 4	17 27	+26 39	1708 Dec 31	14 23	0.62997	94.24	14.86
1711 Feb 8	17 25	+19 21	1711 Feb 8	11 15	0.67620	101.16	13.84
1713 Mar 14	1 06	+ 6 06	1713 Mar 17	16 00	0.65659	98.22	14.26
1715 Apr 21	20 48	−10 31	1715 Apr 29	1 00	0.57241	85.63	16.35
1717 Jun 11	20 55	−26 27	1717 Jun 19	22 58	0.44336	66.33	21.11
1719 Aug 27	22 28	−16 11	1719 Aug 25	19 55	0.37401	55.95	25.03
1721 Nov 5	2 27	+15 12	1721 Oct 27	21 02	0.47118	70.49	19.86
1723 Dec 21	18 37	+26 45	1723 Dec 16	4 38	0.59455	88.94	15.74
1726 Jan 26	11 32	+23 05	1726 Jan 24	16 57	0.66587	99.61	14.06
1728 Feb 29	11 50	+11 40	1728 Mar 2	14 36	0.67219	100.56	13.92
1730 Apr 5	19 24	− 3 50	1730 Apr 11	16 32	0.61273	91.66	15.28
1732 May 20	3 32	−21 08	1732 May 28	13 11	0.49614	74.22	18.87
1734 Jul 26	8 30	−25 53	1734 Jul 29	18 10	0.38168	57.10	24.52
1736 Oct 12	8 29	+ 4 51	1736 Oct 4	17 48	0.42105	62.99	22.23
1738 Dec 5	9 54	+24 37	1738 Nov 28	11 43	0.55075	82.39	16.99
1741 Jan 12	20 00	+25 43	1741 Jan 9	12 57	0.64552	96.57	14.50
1743 Feb 16	7 03	+16 39	1743 Feb 16	21 10	0.67771	101.38	13.81

Opposition				*Nearest to Earth*				
Date	*TD*	*Decl.*		*Date*	*TD*	*Least distance*		*diam.*
	h m	° ′			h m	a.u.	m.k.	″
1745 Mar 22	2 23	+ 2 29		1745 Mar 26	13 01	0.64367	96.29	14.54
1747 May 1	18 51	− 14 33		1747 May 9	12 49	0.54688	81.81	17.12
1749 Jun 26	13 38	− 28 02		1749 Jul 3	18 12	0.41699	62.38	22.45
1751 Sep 14	20 11	− 8 26		1751 Sep 10	2 44	0.38402	57.45	24.37
1753 Nov 16	22 13	+ 19 39		1753 Nov 8	23 25	0.50063	74.89	18.70
1755 Dec 30	11 58	+ 26 54		1755 Dec 25	17 12	0.61530	92.05	15.21
1758 Feb 3	3 13	+ 20 54		1758 Feb 2	4 45	0.67284	100.66	13.91
1760 Mar 8	5 28	+ 8 21		1760 Mar 11	4 09	0.66481	99.45	14.08
1762 Apr 14	19 20	− 7 48		1762 Apr 21	9 44	0.59141	88.47	15.83
1764 Jun 1	12 43	− 24 29		1764 Jun 9	22 31	0.46659	69.80	20.06
1766 Aug 13	13 37	− 21 07		1766 Aug 13	22 57	0.37326	55.84	25.08
1768 Oct 26	7 25	+ 11 18		1768 Oct 18	2 58	0.44785	67.00	20.90
1770 Dec 14	23 12	+ 26 10		1770 Dec 8	18 47	0.57592	86.16	16.25
1773 Jan 20	17 55	+ 24 15		1773 Jan 18	7 00	0.65802	98.44	14.22
1775 Feb 23	20 49	+ 13 41		1775 Feb 25	7 09	0.67588	101.11	13.85
1777 Mar 30	9 10	− 1 16		1777 Apr 4	14 52	0.62780	93.92	14.91
1779 May 12	10 02	− 18 29		1779 May 20	14 15	0.51951	77.72	18.02
1781 Jul 12	18 42	− 27 41		1781 Jul 18	4 47	0.39473	59.05	23.71
1783 Oct 1	11 53	− 0 32		1783 Sep 24	19 59	0.40191	60.13	23.29
1785 Nov 27	17 55	+ 22 57		1785 Nov 20	7 55	0.52886	79.12	17.70
1788 Jan 7	19 48	+ 26 22		1788 Jan 3	20 32	0.63315	94.72	14.78
1790 Feb 10	17 07	+ 18 24		1790 Feb 10	14 43	0.67684	101.25	13.83
1792 Mar 16	2 44	+ 4 51		1792 Mar 19	21 07	0.65439	97.90	14.30
1794 Apr 24	5 56	− 11 48		1794 May 1	11 58	0.56775	84.93	16.49
1796 Jun 15	2 08	− 26 59		1796 Jun 23	2 31	0.43830	65.57	21.36
1798 Aug 31	23 41	− 14 13		1798 Aug 29	9 05	0.37535	56.15	24.94
1800 Nov 9	0 48	+ 16 37		1800 Oct 31	20 22	0.47707	71.37	19.62
1802 Dec 25	0 56	+ 26 52		1802 Dec 19	14 42	0.59884	89.59	15.63
1805 Jan 29	11 51	+ 22 20		1805 Jan 27	20 43	0.66747	99.85	14.02
1807 Mar 4	12 18	+ 10 30		1807 Mar 6	18 18	0.67102	100.38	13.95
1809 Apr 9	1 03	− 5 09		1809 Apr 15	0 39	0.60881	91.08	15.37
1811 May 24	23 08	− 22 09		1811 Jun 2	9 51	0.49033	73.35	19.09
1813 Jul 31	6 50	− 24 51		1813 Aug 3	3 04	0.37907	56.71	24.69
1815 Oct 17	15 04	+ 6 42		1815 Oct 9	19 40	0.42566	63.68	21.99
1817 Dec 8	19 15	+ 25 11		1817 Dec 2	1 01	0.55564	83.12	16.85
1820 Jan 16	21 47	+ 25 14		1820 Jan 13	18 18	0.64811	96.96	14.44
1822 Feb 19	6 34	+ 15 37		1822 Feb 20	0 11	0.67753	101.36	13.81
1824 Mar 25	4 17	+ 1 13		1824 Mar 29	17 53	0.64091	95.88	14.60
1826 May 5	6 46	− 15 46		1826 May 13	2 19	0.54194	81.07	17.27
1828 Jul 1	2 08	− 28 05		1828 Jul 7	23 47	0.41234	61.69	22.70
1830 Sep 19	16 13	− 6 19		1830 Sep 14	12 08	0.38667	57.85	24.21
1832 Nov 20	14 26	+ 20 44		1832 Nov 12	18 34	0.50597	75.69	18.50
1835 Jan 2	15 15	+ 26 46		1834 Dec 28	23 50	0.61891	92.59	15.12

Date		TD	Decl.	Date		TD	Least distance		diam.
		h m	° ′			*h m*	*a.u.*	*m.k.*	″
1837 Feb	6	2 48	+20 02	1837 Feb	5	7 37	0.67398	100.83	13.89
1839 Mar	12	6 27	+ 7 08	1839 Mar	15	8 12	0.66314	99.20	14.11
1841 Apr	18	1 58	− 9 05	1841 Apr	24	18 42	0.58722	87.85	15.94
1843 Jun	6	12 19	−25 15	1843 Jun	14	22 42	0.46130	69.01	20.29
1845 Aug	18	15 10	−19 27	1845 Aug	18	11 37	0.37302	55.80	25.09
1847 Oct	31	9 17	+12 55	1847 Oct	23	4 21	0.45355	67.85	20.64
1849 Dec	18	6 56	+26 27	1849 Dec	12	5 32	0.58059	86.86	16.12
1852 Jan	24	18 49	+23 36	1852 Jan	22	10 51	0.66007	98.75	14.18
1854 Feb	26	20 35	+12 34	1854 Feb	28	9 51	0.67521	101.01	13.86
1856 Apr	2	13 23	− 2 34	1856 Apr	7	21 29	0.62436	93.40	14.99
1858 May	16	2 44	−19 36	1858 May	24	9 09	0.51381	76.86	18.22
1860 Jul	17	13 11	−27 07	1860 Jul	22	11 57	0.39088	58.48	23.95
1862 Oct	6	0 23	+ 1 30	1862 Sep	29	0 41	0.40576	60.70	23.07
1864 Dec	1	6 07	+23 43	1864 Nov	23	23 10	0.53399	79.88	17.53
1867 Jan	10	22 18	+26 01	1867 Jan	7	2 12	0.63625	95.18	14.71
1869 Feb	13	16 38	+17 25	1869 Feb	13	17 25	0.67720	101.31	13.82
1871 Mar	20	3 ·50	+ 3 36	1871 Mar	24	1 05	0.65210	97.55	14.35
1873 Apr	27	14 38	−13 03	1873 May	4	23 12	0.56324	84.26	16.62
1875 Jun	20	8 26	−27 24	1875 Jun	28	5 12	0.43319	64.80	21.61
1877 Sep	5	23 36	−12 12	1877 Sep	2	20 24	0.37665	56.35	24.85
1879 Nov	12	20 04	+17 55	1879 Nov	4	17 32	0.48242	72.17	19.40
1881 Dec	27	5 12	+26 53	1881 Dec	21	21 49	0.60280	90.18	15.53
1884 Feb	1	11 36	+21 32	1884 Jan	30	23 33	0.66907	100.09	13.99
1886 Mar	6	12 23	+ 9 20	1886 Mar	8	21 33	0.66987	100.21	13.97
1888 Apr	11	6 09	− 6 26	1888 Apr	17	8 38	0.60500	90.51	15.47
1890 May	27	18 58	−23 05	1890 Jun	5	7 30	0.48494	72.55	19.30
1892 Aug	4	6 07	−23 38	1892 Aug	6	14 09	0.37736	56.45	24.80
1894 Oct	20	22 09	+ 8 33	1894 Oct	13	0 10	0.43099	64.48	21.72
1896 Dec	11	5 35	+25 40	1896 Dec	4	13 44	0.56076	83.89	16.69
1899 Jan	18	23 25	+24 42	1899 Jan	15	22 42	0.65071	97.34	14.38
1901 Feb	22	6 04	+14 32	1901 Feb	23	2 38	0.67741	101.34	13.82
1903 Mar	29	7 23	− 0 05	1903 Apr	2	23 45	0.63802	95.45	14.67
1905 May	8	20 00	−16 57	1905 May	16	18 28	0.53659	80.27	17.44
1907 Jul	6	15 21	−27 59	1907 Jul	13	4 47	0.40757	60.97	22.97
1909 Sep	24	10 02	− 4 13	1909 Sep	18	19 03	0.38947	58.26	24.03
1911 Nov	25	4 52	+21 43	1911 Nov	17	11 14	0.51121	76.48	18.31
1914 Jan	5	18 28	+26 34	1914 Jan	1	6 00	0.62245	93.12	15.04
1916 Feb	10	2 32	+19 08	1916 Feb	9	10 31	0.67485	100.96	13.87
1918 Mar	15	6 37	+ 5 55	1918 Mar	18	11 33	0.66130	98.93	14.15
1920 Apr	21	8 36	−10 21	1920 Apr	28	4 34	0.58306	87.22	16.05
1922 Jun	10	14 03	−25 56	1922 Jun	18	22 42	0.45594	68.21	20.53
1924 Aug	23	16 55 ·	−17 41	1924 Aug	22	23 50	0.37285	55.78	25.10
1926 Nov	4	9 24	+14 27	1926 Oct	27	5 11	0.45883	68.64	20.40

Opposition

Date	TD	Decl.
	h m	° '
1928 Dec 21	13 28	+26 39
1931 Jan 27	18 59	+22 54
1933 Mar 1	20 21	+11 26
1935 Apr 6	17 27	− 3 52
1937 May 19	18 30	−20 40
1939 Jul 23	7 56	−26 24
1941 Oct 10	12 40	+ 3 30
1943 Dec 5	18 24	+24 24
1946 Jan 14	0 45	+25 36
1948 Feb 17	16 08	+16 25
1950 Mar 23	5 37	+ 2 20
1952 May 1	1 25	−14 17
1954 Jun 24	17 14	−27 41
1956 Sep 10	21 51	−10 08
1958 Nov 16	14 26	+19 08
1960 Dec 30	10 14	+26 49
1963 Feb 4	11 50	+20 42
1965 Mar 9	12 22	+ 8 08
1967 Apr 15	11 24	− 7 43
1969 May 31	15 44	−23 57
1971 Aug 10	6 46	−22 15
1973 Oct 25	3 20	+10 18
1975 Dec 15	13 51	+26 03
1978 Jan 22	0 04	+24 06
1980 Feb 25	5 36	+13 27
1982 Mar 31	10 06	− 1 21
1984 May 11	8 46	−18 06
1986 Jul 10	5 21	−27 44
1988 Sep 28	3 25	− 2 07
1990 Nov 27	20 27	+22 38
1993 Jan 7	22 36	+26 16
1995 Feb 12	2 25	+18 10
1997 Mar 17	7 48	+ 4 40
1999 Apr 24	17 31	−11 37
2001 Jun 13	17 40	−26 30
2003 Aug 28	17 53	−15 49
2005 Nov 7	7 51	+15 54
2007 Dec 24	19 40	+26 46
2010 Jan 29	19 37	+22 09
2012 Mar 3	20 04	+10 17
2014 Apr 8	20 57	− 5 08
2016 May 22	11 11	−21 39
2018 Jul 27	5 07	−25 30

Nearest to Earth

Date	TD	Least distance		diam.
	h m	a.u.	m.k.	"
1928 Dec 15	14 34	0.58498	87.51	16.00
1931 Jan 25	14 18	0.66214	99.05	14.14
1933 Mar 3	13 05	0.67460	100.92	13.87
1935 Apr 12	5 04	0.62101	92.90	15.07
1937 May 28	3 38	0.50853	76.08	18.41
1939 Jul 27	20 52	0.38788	58.03	24.13
1941 Oct 3	7 32	0.41047	61.40	22.80
1943 Nov 28	13 19	0.53946	80.70	17.35
1946 Jan 10	7 40	0.63934	95.64	14.64
1948 Feb 17	20 13	0.67758	101.36	13.81
1950 Mar 27	6 15	0.64971	97.20	14.41
1952 May 8	13 31	0.55824	83.51	16.77
1954 Jul 2	8 01	0.42779	64.00	21.88
1956 Sep 7	4 54	0.37809	56.56	24.76
1958 Nov 8	13 15	0.48770	72.96	19.19
1960 Dec 25	5 46	0.60682	90.78	15.42
1963 Feb 3	3 23	0.67045	100.30	13.96
1965 Mar 12	1 13	0.66847	100.00	14.00
1967 Apr 21	17 39	0.60120	89.94	15.57
1969 Jun 9	4 15	0.47955	71.74	19.52
1971 Aug 12	2 32	0.37569	56.20	24.91
1973 Oct 17	4 11	0.43604	65.23	21.47
1975 Dec 9	0 09	0.56549	84.60	16.55
1978 Jan 19	3 07	0.65319	97.72	14.33
1980 Feb 26	6 06	0.67731	101.32	13.82
1982 Apr 5	6 36	0.63511	95.01	14.74
1984 May 19	10 44	0.53146	79.51	17.61
1986 Jul 16	10 59	0.40357	60.37	23.19
1988 Sep 22	3 19	0.39315	58.81	23.81
1990 Nov 20	3 59	0.51692	77.33	18.11
1993 Jan 3	13 33	0.62609	93.66	14.95
1995 Feb 11	14 20	0.67569	101.08	13.85
1997 Mar 20	16 51	0.65938	98.64	14.20
1999 May 1	17 28	0.57846	86.54	16.18
2001 Jun 21	22 57	0.45017	67.34	20.79
2003 Aug 27	9 52	0.37272	55.76	25.11
2005 Oct 30	3 26	0.46406	69.42	20.17
2007 Dec 18	23 47	0.58935	88.17	15.88
2010 Jan 27	19 02	0.66398	99.33	14.10
2012 Mar 5	17 01	0.67368	100.78	13.89
2014 Apr 14	12 54	0.61756	92.39	15.16
2016 May 30	21 36	0.50321	75.28	18.60
2018 Jul 31	7 51	0.38496	57.59	24.31

Opposition				Nearest to Earth					
Date		TD	Decl.	Date		TD	Least distance		diam.
		h m	° ′			h m	a.u.	m.k.	″
2020 Oct	13	23 20	+ 5 27	2020 Oct	6	14 19	0.41492	62.07	22.56
2022 Dec	8	5 36	+25 00	2022 Dec	1	2 18	0.54447	81.45	17.19
2025 Jan	16	2 32	+25 07	2025 Jan	12	13 38	0.64228	96.08	14.57
2027 Feb	19	15 45	+15 22	2027 Feb	20	0 14	0.67792	101.42	13.81
2029 Mar	25	7 43	+ 1 04	2029 Mar	29	12 56	0.64722	96.82	14.46
2031 May	4	11 57	−15 29	2031 May	12	3 50	0.55336	82.78	16.91
2033 Jun	28	1 24	−27 49	2033 Jul	5	11 19	0.42302	63.28	22.13
2035 Sep	15	19 33	− 8 02	2035 Sep	11	14 21	0.38041	56.91	24.61
2037 Nov	19	9 04	+20 16	2037 Nov	11	8 00	0.49358	73.84	18.96
2040 Jan	2	15 21	+26 41	2039 Dec	28	14 47	0.61092	91.39	15.32
2042 Feb	6	11 59	+19 50	2042 Feb	5	7 57	0.67174	100.49	13.93
2044 Mar	11	12 44	+ 6 56	2044 Mar	14	6 07	0.66708	99.79	14.03
2046 Apr	17	18 01	− 9 00	2046 Apr	24	4 33	0.59704	89.32	15.68
2048 Jun	3	14 45	−24 45	2048 Jun	12	1 41	0.47366	70.86	19.76
2050 Aug	14	7 46	−20 44	2050 Aug	15	12 55	0.37405	55.96	25.02
2052 Oct	28	6 28	+11 58	2052 Oct	20	5 12	0.44091	65.96	21.23
2054 Dec	17	22 09	+26 20	2054 Dec	11	11 44	0.57015	85.29	16.42
2057 Jan	24	1 26	+23 27	2057 Jan	21	9 03	0.65552	98.06	14.28
2059 Feb	27	5 25	+12 20	2059 Feb	28	10 32	0.67681	101.25	13.83
2061 Apr	2	12 47	− 2 38	2061 Apr	7	13 54	0.63199	94.54	14.81
2063 May	14	22 15	−19 11	2063 May	23	1 56	0.52637	78.74	17.78
2065 Jul	13	20 57	−27 19	2065 Jul	19	19 51	0.39959	59.78	23.42
2067 Oct	2	19 50	− 0 01	2067 Sep	26	12 58	0.39670	59.34	23.59
2069 Nov	30	10 14	+23 26	2069 Nov	22	19 16	0.52221	78.12	17.92
2072 Jan	11	0 59	+25 55	2072 Jan	6	20 23	0.62939	94.16	14.87
2074 Feb	14	1 53	+17 12	2074 Feb	13	18 25	0.67646	101.20	13.84
2076 Mar	19	8 50	+ 3 26	2076 Mar	22	22 30	0.65743	98.35	14.24
2078 Apr	27	1 33	−12 51	2078 May	4	5 09	0.57391	85.86	16.31
2080 Jun	16	20 22	−26 58	2080 Jun	24	22 53	0.44497	66.57	21.04
2082 Sep	1	17 34	−13 53	2082 Aug	30	19 01	0.37356	55.88	25.06
2084 Nov	10	6 02	+17 16	2084 Nov	2	0 35	0.46986	70.29	19.92
2086 Dec	27	2 32	+26 47	2086 Dec	21	10 54	0.59387	88.84	15.76
2089 Jan	31	20 17	+21 21	2089 Jan	30	0 26	0.66578	99.60	14.06
2091 Mar	6	20 09	+ 9 06	2091 Mar	8	21 44	0.67272	100.64	13.91
2093 Apr	11	2 15	− 6 25	2093 Apr	16	22 22	0.61382	91.83	15.25
2095 May	26	6 21	−22 36	2095 Jun	3	16 13	0.49747	74.42	18.82
2097 Jul	31	3 33	−24 26	2097 Aug	3	18 19	0.38187	57.13	24.51
2099 Oct	18	8 02	+ 7 20	2099 Oct	10	18 50	0.41922	62.71	22.33
2101 Dec	11	15 37	+25 29	2101 Dec	4	15 51	0.54942	82.19	17.04
2104 Jan	20	4 10	+24 34	2104 Jan	16	19 58	0.64504	96.50	14.51
2106 Feb	22	15 28	+14 18	2106 Feb	23	4 34	0.67788	101.41	13.81
2108 Mar	28	9 07	− 0 11	2108 Apr	1	18 51	0.64456	96.42	14.52
2110 May	7	22 12	−16 37	2110 May	15	15 57	0.54854	82.06	17.06

Opposition   *Nearest   to   Earth*

Date		TD	Decl.	Date		TD	Least distance		diam.
		h m	° ′			h m	a.u.	m.k.	″
2112 Jul	2	11 01	−27 49	2112 Jul	9	16 49	0.41835	62.58	22.37
2114 Sep	20	16 28	− 5 55	2114 Sep	16	1 10	0.38284	57.27	24.45
2116 Nov	23	2 03	+21 19	2116 Nov	15	2 02	0.49902	74.65	18.76
2119 Jan	5	19 07	+26 28	2118 Dec	31	22 59	0.61458	91.94	15.23
2121 Feb	9	11 28	+18 55	2121 Feb	8	11 58	0.67296	100.67	13.91
2123 Mar	15	13 18	+ 5 42	2123 Mar	18	11 02	0.66558	99.57	14.06
2125 Apr	21	0 43	−10 15	2125 Apr	27	14 33	0.59283	88.69	15.79
2127 Jun	8	12 52	−25 26	2127 Jun	16	22 47	0.46831	70.06	19.99
2129 Aug	19	7 33	−19 06	2129 Aug	19	21 52	0.37328	55.84	25.08
2131 Nov	2	9 12	+13 34	2131 Oct	25	5 16	0.44645	66.79	20.97
2133 Dec	21	6 23	+26 33	2133 Dec	15	0 20	0.57509	86.03	16.28
2136 Jan	28	2 24	+22 44	2136 Jan	25	14 29	0.65780	98.41	14.23
2138 Mar	2	5 08	+11 12	2138 Mar	3	14 29	0.67632	101.18	13.84
2140 Apr	5	16 13	− 3 55	2140 Apr	10	21 03	0.62880	94.07	14.89
2142 May	18	13 40	−20 15	2142 May	26	17 42	0.52088	77.92	17.97
2144 Jul	18	15 01	−26 43	2144 Jul	24	4 53	0.39536	59.15	23.67
2146 Oct	7	9 49	+ 2 01	2146 Sep	30	20 02	0.40021	59.87	23.39
2148 Dec	3	22 44	+24 09	2148 Nov	26	11 06	0.52736	78.89	17.75
2151 Jan	14	3 37	+25 29	2151 Jan	10	3 30	0.63257	94.63	14.80
2153 Feb	17	1 35	+16 11	2153 Feb	16	22 28	0.67691	101.26	13.83
2155 Mar	23	9 47	+ 2 11	2155 Mar	27	3 33	0.65516	98.01	14.29
2157 Apr	30	9 58	−14 03	2157 May	7	15 39	0.56935	85.17	16.44
2159 Jun	22	0 26	−27 18	2159 Jun	30	1 19	0.43991	65.81	21.28
2161 Sep	6	18 09	−11 51	2161 Sep	4	6 52	0.37459	56.04	24.99
2163 Nov	15	3 23	+18 33	2163 Nov	6	21 49	0.47536	71.11	19.69
2165 Dec	30	7 30	+26 44	2165 Dec	24	20 05	0.59793	89.45	15.65
2168 Feb	4	19 46	+20 31	2168 Feb	3	3 55	0.66742	99.85	14.02
2170 Mar	9	20 11	+ 7 55	2170 Mar	12	1 32	0.67170	100.49	13.93
2172 Apr	14	6 58	− 7 41	2172 Apr	20	6 02	0.61008	91.27	15.34
2174 May	30	0 15	−23 28	2174 Jun	7	10 33	0.49209	73.62	19.02
2176 Aug	5	1 02	−23 14	2176 Aug	8	1 58	0.37961	56.79	24.66
2178 Oct	22	15 07	+ 9 09	2178 Oct	14	20 54	0.42427	63.47	22.06
2180 Dec	14	1 47	+25 54	2180 Dec	7	6 08	0.55471	82.98	16.87
2183 Jan	22	6 08	+23 57	2183 Jan	19	1 52	0.64782	96.91	14.45
2185 Feb	24	14 59	+13 12	2185 Feb	25	7 49	0.67787	101.41	13.81
2187 Mar	31	11 36	− 1 27	2187 Apr	5	0 30	0.64184	96.02	14.58
2189 May	10	11 14	−17 46	2189 May	18	6 06	0.54335	81.28	17.23
2191 Jul	6	23 48	−27 41	2191 Jul	14	0 10	0.41332	61.83	22.65
2193 Sep	24	12 29	− 3 47	2193 Sep	19	11 32	0.38522	57.63	24.30
2195 Nov	26	18 12	+22 16	2195 Nov	18	20 48	0.50432	75.45	18.56
2198 Jan	7	22 34	+26 11	2198 Jan	3	6 30	0.61817	92.48	15.14
2200 Feb	12	11 09	+17 57	2200 Feb	11	15 29	0.67392	100.82	13.89
2202 Mar	18	13 47	+ 4 29	2202 Mar	21	15 04	0.66376	99.30	14.10

Opposition   Nearest to Earth

Date	TD	Decl.	Date	TD	Least distance		diam.
	h m	° ′		h m	a.u.	m.k.	″
2204 Apr 24	6 38	− 11 28	2204 Apr 30	22 48	0.58865	88.06	15.90
2206 Jun 12	11 42	− 26 01	2206 Jun 20	21 47	0.46305	69.27	20.21
2208 Aug 24	8 56	− 17 20	2208 Aug 24	9 01	0.37279	55.77	25.11
2210 Nov 6	10 44	+ 15 06	2210 Oct 29	4 57	0.45181	67.59	20.72
2212 Dec 24	13 08	+ 26 40	2212 Dec 18	10 52	0.57950	86.69	16.15
2215 Jan 31	2 31	+ 21 59	2215 Jan 28	18 01	0.65988	98.72	14.18
2217 Mar 5	4 38	+ 10 02	2217 Mar 6	17 21	0.67577	101.09	13.85
2219 Apr 9	19 51	− 5 10	2219 Apr 15	3 27	0.62551	93.58	14.96
2221 May 22	4 53	− 21 14	2221 May 30	10 20	0.51558	77.13	18.15
2223 Jul 24	8 09	− 25 59	2223 Jul 29	11 15	0.39186	58.62	23.89
2225 Oct 11	22 29	+ 4 01	2225 Oct 5	1 09	0.40444	60.50	23.14
2227 Dec 8	11 45	+ 24 47	2227 Dec 1	3 46	0.53296	79.73	17.56
2230 Jan 17	6 24	+ 25 00	2230 Jan 13	9 38	0.63585	95.12	14.72
2232 Feb 21	0 59	+ 15 08	2232 Feb 21	1 05	0.67739	101.34	13.82
2234 Mar 26	11 18	+ 0 56	2234 Mar 30	7 48	0.65294	97.68	14.34
2236 May 3	19 47	− 15 14	2236 May 11	3 11	0.56459	84.46	16.58
2238 Jun 26	7 32	− 27 31	2238 Jul 4	5 46	0.43439	64.98	21.55
2240 Sep 11	18 40	− 9 46	2240 Sep 8	19 10	0.37559	56.19	24.92
2242 Nov 18	22 53	+ 19 44	2242 Nov 10	19 04	0.48064	71.90	19.47
2245 Jan 2	12 12	+ 26 36	2244 Dec 28	4 16	0.60187	90.04	15.55
2247 Feb 7	19 52	+ 19 38	2247 Feb 6	7 20	0.66889	100.06	13.99
2249 Mar 12	20 08	+ 6 43	2249 Mar 15	4 41	0.67034	100.28	13.96
2251 Apr 18	11 27	− 8 55	2251 Apr 24	13 08	0.60626	90.70	15.44
2253 Jun 2	19 24	− 24 15	2253 Jun 11	7 05	0.48679	72.82	19.23
2255 Aug 10	23 59	− 21 52	2255 Aug 13	11 48	0.37765	56.50	24.78
2257 Oct 26	22 10	+ 10 54	2257 Oct 19	0 02	0.42925	64.21	21.81
2259 Dec 18	11 10	+ 26 13	2259 Dec 11	18 42	0.55952	83.70	16.73
2262 Jan 25	6 50	+ 23 18	2262 Jan 22	5 35	0.65033	97.29	14.39
2264 Feb 28	14 02	+ 12 06	2264 Feb 29	9 59	0.67781	101.40	13.81
2266 Apr 3	14 05	− 2 43	2266 Apr 8	5 43	0.63907	95.60	14.65
2268 May 13	23 00	− 18 50	2268 May 21	20 11	0.53827	80.52	17.39
2270 Jul 11	11 14	− 27 23	2270 Jul 18	4 04	0.40887	61.17	22.89
2272 Sep 29	5 58	− 1 41	2272 Sep 23	18 34	0.38835	58.10	24.10
2274 Nov 30	9 29	+ 23 06	2274 Nov 22	15 10	0.51002	76.30	18.35
2277 Jan 11	2 24	+ 25 49	2277 Jan 6	13 13	0.62190	93.04	15.05
2279 Feb 15	10 50	+ 16 58	2279 Feb 14	18 03	0.67491	100.97	13.87
2281 Mar 20	14 18	+ 3 15	2281 Mar 23	18 16	0.66202	99.04	14.14
2283 Apr 27	14 18	− 12 41	2283 May 4	8 49	0.58432	87.41	16.02
2285 Jun 15	14 23	− 26 32	2285 Jun 23	23 46	0.45733	68.42	20.47
2287 Aug 29	11 21	− 15 28	2287 Aug 28	22 26	0.37225	55.69	25.14
2289 Nov 9	11 04	+ 16 32	2289 Nov 1	5 49	0.45695	68.36	20.48
2291 Dec 27	19 47	+ 26 42	2291 Dec 21	20 17	0.58386	87.34	16.03
2294 Feb 2	2 59	+ 21 10	2294 Jan 30	21 38	0.66183	99.01	14.14

Opposition   Nearest to Earth

Date	TD	Decl.	Date	TD	Least distance		diam.
	h m	° ′		h m	a.u.	m.k.	″
2296 Mar 7	4 15	+ 8 52	2296 Mar 8	20 08	0.67491	100.96	13.87
2298 Apr 11	23 00	− 6 25	2298 Apr 17	9 29	0.62213	93.07	15.05
2300 May 25	19 41	−22 09	2300 Jun 3	3 48	0.51039	76.35	18.34
2302 Jul 29	2 08	−25 05	2302 Aug 2	18 17	0.38865	58.14	24.08
2304 Oct 16	10 34	+ 5 58	2304 Oct 9	6 15	0.40879	61.15	22.90
2306 Dec 11	23 11	+25 18	2306 Dec 4	17 33	0.53803	80.49	17.40
2309 Jan 20	7 56	+24 27	2309 Jan 16	14 03	0.63875	95.56	14.65
2311 Feb 23	23 55	+14 04	2311 Feb 24	3 10	0.67783	101.40	13.81
2313 Mar 29	12 35	− 0 19	2313 Apr 2	12 11	0.65065	97.34	14.39
2315 May 8	5 15	−16 22	2315 May 15	15 56	0.55981	83.75	16.72
2317 Jun 30	14 09	−27 36	2317 Jul 8	7 41	0.42936	64.23	21.80
2319 Sep 17	16 02	− 7 41	2319 Sep 14	3 58	0.37731	56.44	24.81
2321 Nov 22	17 45	+20 50	2321 Nov 14	16 06	0.48638	72.76	19.24
2324 Jan 6	17 35	+26 23	2324 Jan 1	12 16	0.60616	90.68	15.44
2326 Feb 10	19 57	+18 42	2326 Feb 9	10 28	0.67040	100.29	13.96
2328 Mar 15	20 03	+ 5 30	2328 Mar 18	7 44	0.66909	100.09	13.99
2330 Apr 21	17 08	−10 10	2330 Apr 27	21 47	0.60244	90.12	15.54
2332 Jun 6	16 59	−24 58	2332 Jun 15	5 41	0.48108	71.97	19.46
2334 Aug 16	0 54	−20 21	2334 Aug 18	0 29	0.37560	56.19	24.92
2336 Oct 31	3 11	+12 35	2336 Oct 23	3 25	0.43408	64.94	21.56
2338 Dec 21	19 18	+26 26	2338 Dec 15	4 55	0.56416	84.40	16.59
2341 Jan 28	7 49	+22 35	2341 Jan 25	9 47	0.65272	97.65	14.34
2343 Mar 3	13 27	+10 58	2343 Mar 4	12 49	0.67750	101.35	13.82
2345 Apr 6	16 04	− 3 58	2345 Apr 11	11 09	0.63607	95.15	14.72
2347 May 18	10 52	−19 51	2347 May 26	11 39	0.53324	79.77	17.55
2349 Jul 16	0 18	−26 56	2349 Jul 22	8 45	0.40478	60.55	23.12
2351 Oct 4	23 22	+ 0 24	2351 Sep 29	1 36	0.39171	58.60	23.90
2353 Dec 4	0 20	+23 52	2353 Nov 26	7 21	0.51536	77.10	18.16
2356 Jan 15	5 17	+25 23	2356 Jan 10	19 10	0.62533	93.55	14.97
2358 Feb 18	9 53	+15 57	2358 Feb 17	20 38	0.67582	101.10	13.85
2360 Mar 23	14 54	+ 2 01	2360 Mar 26	22 31	0.66024	98.77	14.18
2362 Apr 30	21 42	−13 52	2362 May 7	20 10	0.57997	86.76	16.14
2364 Jun 19	15 34	−26 55	2364 Jun 27	22 45	0.45194	67.61	20.71
2366 Sep 3	10 34	−13 32	2366 Sep 2	8 06	0.37239	55.71	25.14
2368 Nov 13	9 14	+17 52	2368 Nov 5	4 48	0.46257	69.20	20.23
2370 Dec 31	2 15	+26 39	2370 Dec 25	5 25	0.58854	88.05	15.90
2373 Feb 5	3 36	+20 19	2373 Feb 3	1 41	0.66383	99.31	14.10
2375 Mar 11	3 41	+ 7 42	2375 Mar 12	23 12	0.67418	100.86	13.88
2377 Apr 15	2 59	− 7 40	2377 Apr 20	17 12	0.61873	92.56	15.13
2379 May 29	13 35	−23 02	2379 Jun 6	23 35	0.50479	75.52	18.54
2381 Aug 1	23 53	−24 01	2381 Aug 6	5 51	0.38536	57.65	24.29
2383 Oct 20	21 37	+ 7 53	2383 Oct 13	12 50	0.41305	61.79	22.66
2385 Dec 14	10 14	+25 44	2385 Dec 7	6 08	0.54298	81.23	17.24

Opposition				Nearest to Earth				
Date	TD	Decl.		Date	TD	Least distance		diam.
	h m	° ′			h m	a.u.	m.k.	″
2388 Jan 23	9 50	+23 50		2388 Jan 19	19 35	0.64168	95.99	14.59
2390 Feb 25	23 30	+12 58		2390 Feb 26	6 36	0.67803	101.43	13.80
2392 Mar 31	14 05	− 1 34		2392 Apr 4	17 40	0.64806	96.95	14.44
2394 May 10	14 32	−17 27		2394 May 18	5 12	0.55507	83.04	16.86
2396 Jul 3	21 16	−27 33		2396 Jul 11	9 07	0.42459	63.52	22.04
2398 Sep 21	13 37	− 5 34		2398 Sep 17	12 03	0.37936	56.75	24.67
2400 Nov 25	11 37	+21 50		2400 Nov 17	10 19	0.49186	73.58	19.03
2403 Jan 8	21 31	+26 06		2403 Jan 3	19 33	0.60998	91.25	15.34
2405 Feb 12	19 16	+17 45		2405 Feb 11	13 47	0.67173	100.49	13.93
2407 Mar 18	19 51	+ 4 17		2407 Mar 21	11 38	0.66782	99.90	14.02
2409 Apr 23	22 48	−11 22		2409 Apr 30	7 48	0.59845	89.53	15.64
2411 Jun 10	14 11	−25 35		2411 Jun 19	2 06	0.47548	71.13	19.69
2413 Aug 19	23 59	−18 44		2413 Aug 21	10 47	0.37422	55.98	25.01
2415 Nov 4	6 10	+14 10		2415 Oct 27	5 35	0.43937	65.73	21.30
2417 Dec 24	4 04	+26 34		2417 Dec 17	16 27	0.56925	85.16	16.44
2420 Jan 31	9 16	+21 49		2420 Jan 28	15 23	0.65529	98.03	14.28
2422 Mar 5	13 00	+ 9 48		2422 Mar 6	16 36	0.67724	101.31	13.82
2424 Apr 8	19 09	− 5 13		2424 Apr 13	18 30	0.63311	94.71	14.78
2426 May 21	1 17	−20 51		2426 May 29	4 22	0.52796	78.98	17.73
2428 Jul 19	16 50	−26 19		2428 Jul 25	17 45	0.40043	59.90	23.37
2430 Oct 8	16 30	+ 2 29		2430 Oct 2	10 40	0.39504	59.10	23.69
2432 Dec 6	13 52	+24 31		2432 Nov 28	22 01	0.52052	77.87	17.98
2435 Jan 17	7 49	+24 54		2435 Jan 13	1 40	0.62864	94.04	14.89
2437 Feb 20	9 36	+14 54		2437 Feb 20	0 41	0.67647	101.20	13.84
2439 Mar 26	15 34	+ 0 47		2439 Mar 30	3 42	0.65810	98.45	14.22
2441 May 3	4 53	−15 00		2441 May 10	7 22	0.57546	86.09	16.27
2443 Jun 23	17 49	−27 10		2443 Jul 1	21 23	0.44673	66.83	20.95
2445 Sep 7	10 31	−11 32		2445 Sep 5	16 44	0.37296	55.79	25.10
2447 Nov 17	7 29	+19 07		2447 Nov 9	2 10	0.46807	70.02	20.00
2450 Jan 2	8 23	+26 31		2449 Dec 27	15 03	0.59281	88.68	15.79
2452 Feb 8	3 28	+19 26		2452 Feb 6	6 03	0.66560	99.57	14.06
2454 Mar 13	3 18	+ 6 30		2454 Mar 15	3 19	0.67336	100.73	13.90
2456 Apr 17	7 39	− 8 54		2456 Apr 23	2 18	0.61515	92.03	15.22
2458 Jun 1	6 52	−23 49		2458 Jun 9	17 19	0.49928	74.69	18.75
2460 Aug 5	19 56	−22 48		2460 Aug 9	15 37	0.38254	57.23	24.47
2462 Oct 24	5 51	+ 9 41		2462 Oct 16	17 55	0.41772	62.49	22.41
2464 Dec 16	20 40	+26 04		2464 Dec 9	19 26	0.54836	82.03	17.07
2467 Jan 25	11 46	+23 10		2467 Jan 22	2 05	0.64470	96.45	14.52
2469 Feb 27	23 05	+11 52		2469 Feb 28	10 46	0.67822	101.46	13.80
2471 Apr 3	15 44	− 2 49		2471 Apr 7	23 56	0.64556	96.58	14.50
2473 May 13	1 54	−18 32		2473 May 20	19 06	0.55009	82.29	17.02
2475 Jul 8	8 36	−27 21		2475 Jul 15	15 21	0.41951	62.76	22.31
2477 Sep 25	11 54	− 3 24		2477 Sep 20	22 37	0.38148	57.07	24.54

Opposition

Date	TD	Decl.
	h m	° '
2479 Nov 29	4 54	+22 44
2482 Jan 11	1 39	+25 44
2484 Feb 15	19 13	+16 45
2486 Mar 20	20 21	+ 3 04
2488 Apr 26	4 48	−12 33
2490 Jun 13	11 29	−26 06
2492 Aug 23	23 43	−17 00
2494 Nov 7	9 12	+15 40
2496 Dec 26	11 43	+26 36
2499 Feb 2	9 22	+21 00
2501 Mar 8	12 13	+ 8 39
2503 Apr 12	22 04	− 6 27
2505 May 24	15 11	−21 46
2507 Jul 25	8 25	−25 33
2509 Oct 13	5 46	+ 4 29
2511 Dec 11	2 54	+25 05
2514 Jan 20	11 00	+24 20
2516 Feb 24	9 20	+13 50
2518 Mar 29	16 36	− 0 28
2520 May 6	14 15	−16 08
2522 Jun 27	23 22	−27 20
2524 Sep 12	12 19	− 9 27
2526 Nov 21	5 05	+20 17
2529 Jan 5	13 30	+26 18
2531 Feb 11	3 20	+18 30
2533 Mar 16	3 20	+ 5 17
2535 Apr 21	11 41	−10 06
2537 Jun 4	23 51	−24 31
2539 Aug 11	17 01	−21 28
2541 Oct 28	13 26	+11 26
2543 Dec 21	6 38	+26 18
2546 Jan 28	12 55	+22 26
2548 Mar 2	22 04	+10 44
2550 Apr 6	17 46	− 4 04
2552 May 16	13 40	−19 32
2554 Jul 12	18 33	−27 00
2556 Sep 30	6 21	− 1 18
2558 Dec 2	21 00	+23 31
2561 Jan 14	5 29	+25 18
2563 Feb 18	18 52	+15 44
2565 Mar 23	20 34	+ 1 50
2567 Apr 30	11 24	−13 43
2569 Jun 17	11 40	−26 31

Nearest to Earth

Date	TD	Least distance		diam.
	h m	a.u.	m.k.	"
2479 Nov 21	3 53	0.49717	74.38	18.83
2482 Jan 6	3 57	0.61368	91.81	15.25
2484 Feb 14	18 23	0.67285	100.66	13.91
2486 Mar 23	16 50	0.66613	99.65	14.05
2488 May 2	17 45	0.59424	88.90	15.75
2490 Jun 21	21 55	0.47015	70.33	19.91
2492 Aug 24	19 25	0.37322	55.83	25.08
2494 Oct 30	5 48	0.44466	66.52	21.05
2496 Dec 20	3 50	0.57387	85.85	16.31
2499 Jan 30	20 04	0.65745	98.35	14.24
2501 Mar 9	20 16	0.67685	101.26	13.83
2503 Apr 18	1 39	0.63001	94.25	14.86
2505 Jun 1	19 27	0.52263	78.18	17.91
2507 Jul 31	2 09	0.39648	59.31	23.61
2509 Oct 6	18 06	0.39881	59.66	23.47
2511 Dec 3	13 40	0.52610	78.70	17.79
2514 Jan 16	9 37	0.63212	94.56	14.81
2516 Feb 24	5 01	0.67712	101.30	13.82
2518 Apr 2	9 08	0.65605	98.14	14.27
2520 May 13	19 22	0.57089	85.40	16.40
2522 Jul 6	0 22	0.44132	66.02	21.21
2524 Sep 10	4 12	0.37364	55.90	25.05
2526 Nov 12	22 38	0.47342	70.82	19.77
2528 Dec 31	0 39	0.59685	89.29	15.68
2531 Feb 9	10 30	0.66717	99.81	14.03
2533 Mar 18	7 46	0.67215	100.55	13.93
2535 Apr 27	10 05	0.61133	91.45	15.31
2537 Jun 13	10 06	0.49396	73.89	18.95
2539 Aug 14	23 25	0.38013	56.87	24.62
2541 Oct 20	20 38	0.42251	63.21	22.15
2543 Dec 14	9 12	0.55331	82.77	16.92
2546 Jan 25	7 33	0.64729	96.83	14.46
2548 Mar 3	13 56	0.67825	101.46	13.80
2550 Apr 11	5 45	0.64293	96.18	14.56
2552 May 24	8 16	0.54505	81.54	17.17
2554 Jul 19	21 30	0.41478	62.05	22.57
2556 Sep 25	8 33	0.38406	57.46	24.37
2558 Nov 24	22 12	0.50291	75.23	18.61
2561 Jan 9	12 20	0.61761	92.39	15.16
2563 Feb 17	22 17	0.67399	100.83	13.89
2565 Mar 26	20 55	0.66457	99.42	14.08
2567 May 7	2 51	0.59013	88.28	15.86
2569 Jun 25	21 07	0.46464	69.51	20.14

Opposition   *Nearest to Earth*

Date	TD	Decl.	Date	TD	Least distance		diam.
	h m	° ′		h m	a.u.	m.k.	″
2571 Aug 30	2 15	−15 07	2571 Aug 30	6 19	0.37238	55.71	25.14
2573 Nov 11	11 07	+17 06	2573 Nov 3	5 00	0.44982	67.29	20.81
2575 Dec 30	18 39	+26 34	2575 Dec 24	15 06	0.57823	86.50	16.19
2578 Feb 5	9 52	+20 09	2578 Feb 3	0 42	0.65948	98.66	14.19
2580 Mar 10	11 57	+ 7 28	2580 Mar 12	0 02	0.67610	101.14	13.84
2582 Apr 15	1 12	− 7 41	2582 Apr 20	8 17	0.62657	93.73	14.94
2584 May 27	5 26	−22 37	2584 Jun 4	10 20	0.51742	77.40	18.09
2586 Jul 29	0 50	−24 39	2586 Aug 3	8 42	0.39290	58.78	23.82
2588 Oct 16	18 52	+ 6 26	2588 Oct 9	23 58	0.40285	60.26	23.23
2590 Dec 13	15 39	+25 32	2590 Dec 6	6 07	0.53144	79.50	17.61
2593 Jan 22	12 52	+23 43	2593 Jan 18	15 17	0.63516	95.02	14.74
2595 Feb 26	8 01	+12 45	2595 Feb 26	7 23	0.67760	101.37	13.81
2597 Mar 31	17 30	− 1 42	2597 Apr 4	13 19	0.65394	97.83	14.31
2599 May 9	23 03	−17 13	2599 May 17	5 48	0.56623	84.71	16.53
2601 Jul 2	3 42	−27 21	2601 Jul 10	3 15	0.43609	65.24	21.46
2603 Sep 18	10 25	− 7 22	2603 Sep 15	15 01	0.37485	56.08	24.97
2605 Nov 25	0 12	+21 20	2605 Nov 16	19 18	0.47909	71.67	19.54
2608 Jan 9	18 40	+26 01	2608 Jan 4	9 52	0.60116	89.93	15.57
2610 Feb 14	3 32	+17 32	2610 Feb 12	14 19	0.66884	100.06	13.99
2612 Mar 19	3 07	+ 4 05	2612 Mar 21	10 55	0.67102	100.38	13.95
2614 Apr 24	16 46	−11 18	2614 Apr 30	17 34	0.60761	90.90	15.40
2616 Jun 8	20 18	−25 09	2616 Jun 17	6 57	0.48849	73.08	19.16
2618 Aug 16	17 47	−19 57	2618 Aug 19	9 35	0.37780	56.52	24.77
2620 Nov 1	21 14	+13 08	2620 Oct 24	23 17	0.42731	63.92	21.90
2622 Dec 24	15 54	+26 27	2622 Dec 17	22 18	0.55810	83.49	16.77
2625 Jan 31	13 52	+21 40	2625 Jan 28	12 06	0.64980	97.21	14.40
2627 Mar 6	21 31	+ 9 35	2627 Mar 7	16 53	0.67801	101.43	13.81
2629 Apr 9	19 49	− 5 17	2629 Apr 14	10 52	0.63998	95.74	14.63
2631 May 21	0 31	−20 29	2631 May 28	20 52	0.54004	80.79	17.33
2633 Jul 17	5 08	−26 31	2633 Jul 24	1 34	0.41031	61.38	22.81
2635 Oct 6	0 07	+ 0 47	2635 Sep 30	16 19	0.38704	57.90	24.18
2637 Dec 6	12 25	+24 13	2637 Nov 28	16 46	0.50840	76.06	18.41
2640 Jan 18	8 42	+24 48	2640 Jan 13	18 50	0.62106	92.91	15.07
2642 Feb 21	17 48	+14 41	2642 Feb 21	0 26	0.67493	100.97	13.87
2644 Mar 26	20 45	+ 0 36	2644 Mar 30	0 05	0.66289	99.17	14.12
2646 May 3	18 30	−14 51	2646 May 10	12 03	0.58582	87.64	15.98
2648 Jun 21	12 01	−26 50	2648 Jun 29	21 57	0.45914	68.69	20.39
2650 Sep 4	2 18	−13 12	2650 Sep 3	17 58	0.37201	55.65	25.16
2652 Nov 15	10 57	+18 24	2652 Nov 7	4 47	0.45531	68.11	20.56
2655 Jan 3	1 29	+26 26	2654 Dec 28	1 22	0.58300	87.22	16.05
2657 Feb 8	10 31	+19 15	2657 Feb 6	4 29	0.66168	98.99	14.15
2659 Mar 14	11 29	+ 6 16	2659 Mar 16	2 31	0.67545	101.05	13.86
2661 Apr 18	4 34	− 8 54	2661 Apr 23	13 59	0.62337	93.25	15.02

Opposition

Date	TD h m	Decl. ° '
2663 May 31	21 30	−23 24
2665 Aug 2	21 08	−23 34
2667 Oct 22	8 03	+ 8 22
2669 Dec 17	3 15	+25 54
2672 Jan 26	14 48	+23 02
2674 Mar 1	7 26	+11 38
2676 Apr 3	18 53	− 2 56
2678 May 13	7 58	−18 15
2680 Jul 5	9 24	−27 14
2682 Sep 22	8 39	− 5 15
2684 Nov 27	19 46	+22 17
2687 Jan 11	23 30	+25 39
2689 Feb 16	2 50	+16 33
2691 Mar 22	2 46	+ 2 52
2693 Apr 26	21 51	−12 28
2695 Jun 12	16 03	−25 42
2697 Aug 20	16 06	−18 21
2699 Nov 6	1 20	+14 42
2701 Dec 28	0 13	+26 30
2704 Feb 4	15 16	+20 50
2706 Mar 9	20 51	+ 8 25
2708 Apr 12	22 02	− 6 31
2710 May 24	13 46	−21 25
2712 Jul 21	20 29	−25 53
2714 Oct 10	19 33	+ 2 54
2716 Dec 10	3 47	+24 50
2719 Jan 21	11 47	+24 14
2721 Feb 24	17 21	+13 36
2723 Mar 30	21 40	− 0 37
2725 May 7	1 12	−15 56
2727 Jun 26	12 15	−27 02
2729 Sep 9	2 07	−11 14
2731 Nov 20	9 58	+19 38
2734 Jan 6	7 44	+26 14
2736 Feb 12	10 27	+18 19
2738 Mar 17	10 30	+ 5 05
2740 Apr 21	8 12	−10 05
2742 Jun 4	14 10	−24 07
2744 Aug 7	15 56	−22 22
2746 Oct 26	17 58	+10 11
2748 Dec 20	14 19	+26 10
2751 Jan 29	17 01	+22 18
2753 Mar 4	6 57	+10 29

Nearest to Earth

Date	TD h m	Least distance a.u.	m.k.	diam. "
2663 Jun 9	4 15	0.51214	76.62	18.28
2665 Aug 7	16 38	0.38930	58.24	24.04
2667 Oct 15	4 53	0.40698	60.88	23.00
2669 Dec 9	20 40	0.53645	80.25	17.45
2672 Jan 22	20 17	0.63807	95.45	14.67
2674 Mar 1	10 00	0.67788	101.41	13.81
2676 Apr 7	17 38	0.65140	97.45	14.37
2678 May 20	17 25	0.56138	83.98	16.67
2680 Jul 13	5 39	0.43103	64.48	21.72
2682 Sep 19	1 18	0.37641	56.31	24.87
2684 Nov 19	16 54	0.48469	72.51	19.31
2687 Jan 6	17 34	0.60514	90.53	15.47
2689 Feb 14	16 37	0.67025	100.27	13.96
2691 Mar 24	13 30	0.66982	100.20	13.97
2693 May 3	1 17	0.60383	90.33	15.50
2695 Jun 21	4 36	0.48293	72.24	19.38
2697 Aug 22	19 55	0.37587	56.23	24.90
2699 Oct 29	1 16	0.43237	64.68	21.65
2701 Dec 21	9 25	0.56312	84.24	16.62
2704 Feb 1	16 23	0.65247	97.61	14.35
2706 Mar 10	19 13	0.67787	101.41	13.81
2708 Apr 17	15 48	0.63719	95.32	14.69
2710 Jun 1	12 46	0.53495	80.03	17.50
2712 Jul 28	7 48	0.40580	60.71	23.07
2714 Oct 5	0 18	0.39015	58.37	23.99
2716 Dec 2	10 02	0.51366	76.84	18.22
2719 Jan 17	0 46	0.62445	93.42	14.99
2721 Feb 24	3 09	0.67574	101.09	13.85
2723 Apr 3	4 05	0.66086	98.86	14.16
2725 May 13	22 09	0.58137	86.97	16.10
2727 Jul 4	21 06	0.45376	67.88	20.63
2729 Sep 8	4 47	0.37200	55.65	25.16
2731 Nov 12	4 47	0.46080	68.94	20.31
2733 Dec 31	10 07	0.58737	87.87	15.94
2736 Feb 10	7 32	0.66352	99.26	14.11
2738 Mar 19	4 50	0.67473	100.94	13.87
2740 Apr 26	20 53	0.62002	92.75	15.10
2742 Jun 12	23 29	0.50661	75.79	18.48
2744 Aug 12	1 37	0.38607	57.75	24.24
2746 Oct 19	9 55	0.41137	61.54	22.75
2748 Dec 13	9 53	0.54176	81.05	17.28
2751 Jan 26	1 38	0.64127	95.93	14.60
2753 Mar 4	12 51	0.67828	101.47	13.80

Opposition   Nearest to Earth

Date	TD	Decl.	Date	TD	Least distance		diam.
	h m	° ′		h m	a.u.	m.k.	″
2755 Apr 7	20 22	− 4 10	2755 Apr 11	22 25	0.64908	97.10	14.42
2757 May 16	18 02	−19 15	2757 May 24	6 51	0.55672	83.28	16.81
2759 Jul 10	18 40	−26 59	2759 Jul 18	8 30	0.42596	63.72	21.97
2761 Sep 27	8 38	− 3 04	2761 Sep 23	10 41	0.37820	56.58	24.75
2763 Dec 2	14 13	+23 08	2763 Nov 24	12 25	0.49005	73.31	19.10
2766 Jan 15	3 36	+25 13	2766 Jan 10	0 23	0.60893	91.09	15.37
2768 Feb 20	2 38	+15 31	2768 Feb 18	19 52	0.67154	100.46	13.94
2770 Mar 25	2 55	+ 1 39	2770 Mar 27	17 13	0.66831	99.98	14.01
2772 Apr 30	3 04	−13 36	2772 May 6	10 22	0.59971	89.72	15.61
2774 Jun 16	12 26	−26 08	2774 Jun 25	1 20	0.47737	71.41	19.61
2776 Aug 25	14 58	−16 38	2776 Aug 27	7 02	0.37436	56.00	25.00
2778 Nov 10	5 21	+16 11	2778 Nov 2	4 37	0.43758	65.46	21.39
2780 Dec 30	9 02	+26 29	2780 Dec 23	20 23	0.56793	84.96	16.48
2783 Feb 6	15 59	+19 59	2783 Feb 3	20 43	0.65480	97.96	14.29
2785 Mar 11	19 46	+ 7 14	2785 Mar 12	21 55	0.67764	101.37	13.81
2787 Apr 16	0 43	− 7 43	2787 Apr 20	22 19	0.63433	94.89	14.76
2789 May 27	2 58	−22 16	2789 Jun 4	5 16	0.52973	79.25	17.67
2791 Jul 26	10 10	−25 06	2791 Aug 1	13 47	0.40157	60.07	23.31
2793 Oct 14	10 58	+ 4 54	2793 Oct 8	6 56	0.39357	58.88	23.78
2795 Dec 13	16 54	+25 19	2795 Dec 6	0 39	0.51912	77.66	18.03
2798 Jan 23	14 35	+23 36	2798 Jan 19	7 08	0.62810	93.96	14.90
2800 Feb 27	17 02	+12 31	2800 Feb 27	6 42	0.67665	101.23	13.83
2802 Apr 1	22 06	− 1 51	2802 Apr 5	8 32	0.65901	98.59	14.20
2804 May 9	8 57	−17 01	2804 May 16	9 46	0.57707	86.33	16.22
2806 Jun 29	16 33	−27 07	2806 Jul 7	21 25	0.44837	67.08	20.88
2808 Sep 13	4 42	−. 9 07	2808 Sep 11	15 34	0.37230	55.70	25.14
2810 Nov 23	9 07	+20 46	2810 Nov 15	3 33	0.46619	69.74	20.08
2813 Jan 8	14 02	+25 56	2813 Jan 2	19 12	0.59161	88.50	15.82
2815 Feb 14	10 41	+17 20	2815 Feb 12	11 43	0.66529	99.53	14.07
2817 Mar 19	10 27	+ 3 52	2817 Mar 21	8 53	0.67374	100.79	13.89
2819 Apr 24	12 35	−11 15	2819 Apr 30	5 34	0.61629	92.20	15.19
2821 Jun 7	6 31	−24 45	2821 Jun 15	17 27	0.50112	74.97	18.68
2823 Aug 12	11 10	−21 02	2823 Aug 16	11 19	0.38322	57.33	24.42
2825 Oct 30	3 10	+11 56	2825 Oct 22	15 46	0.41605	62.24	22.50
2827 Dec 24	0 58	+26 20	2827 Dec 16	22 27	0.54693	81.82	17.11
2830 Jan 31	18 22	+21 31	2830 Jan 28	7 07	0.64406	96.35	14.53
2832 Mar 6	5 47	+ 9 21	2832 Mar 6	15 54	0.67850	101.50	13.80
2834 Apr 9	21 39	− 5 23	2834 Apr 14	3 58	0.64669	96.74	14.47
2836 May 19	4 35	−20 12	2836 May 26	20 59	0.55182	82.55	16.96
2838 Jul 14	3 38	−26 36	2838 Jul 21	12 13	0.42100	62.98	22.23
2840 Oct 1	4 40	− 0 57	2840 Sep 26	18 46	0.38034	56.90	24.61
2842 Dec 5	6 52	+23 53	2842 Nov 27	5 30	0.49559	74.14	18.89
2845 Jan 17	7 59	+24 42	2845 Jan 12	8 52	0.61304	91.71	15.27

Opposition   *Nearest to Earth*

Date	TD	Decl.	Date	TD	Least distance		diam.
	h m	° ′		h m	a.u.	m.k.	″
2847 Feb 22	2 42	+ 14 27	2847 Feb 21	0 27	0.67294	100.67	13.91
2849 Mar 27	3 05	+ 0 25	2849 Mar 29	21 58	0.66691	99.77	14.03
2851 May 3	9 20	− 14 43	2851 May 9	20 50	0.59579	89.13	15.71
2853 Jun 19	11 29	− 26 29	2853 Jun 27	22 27	0.47192	70.60	19.83
2855 Aug 30	17 38	− 14 46	2855 Aug 31	18 11	0.37311	55.82	25.09
2857 Nov 13	9 39	+ 17 37	2857 Nov 5	6 29	0.44283	66.25	21.14
2860 Jan 2	16 57	+ 26 22	2859 Dec 27	7 20	0.57256	85.65	16.35
2862 Feb 8	16 32	+ 19 05	2862 Feb 6	1 42	0.65699	98.29	14.25
2864 Mar 13	19 27	+ 6 03	2864 Mar 15	2 04	0.67715	101.30	13.82
2866 Apr 18	3 41	− 8 55	2866 Apr 23	5 50	0.63102	94.40	14.83
2868 May 29	16 12	− 23 02	2868 Jun 6	20 33	0.52437	78.44	17.85
2870 Jul 30	0 51	− 24 11	2870 Aug 4	21 59	0.39764	59.49	23.54
2872 Oct 18	1 16	+ 6 52	2872 Oct 11	15 14	0.39734	59.44	23.56
2874 Dec 16	6 38	+ 25 42	2874 Dec 8	16 02	0.52457	78.47	17.84
2877 Jan 25	17 32	+ 22 55	2877 Jan 21	14 31	0.63138	94.45	14.82
2879 Mar 1	16 09	+ 11 24	2879 Mar 1	10 25	0.67725	101.32	13.82
2881 Apr 3	22 47	− 3 04	2881 Apr 7	13 45	0.65703	98.29	14.25
2883 May 12	17 43	− 18 03	2883 May 19	21 59	0.57253	85.65	16.35
2885 Jul 2	20 18	− 27 04	2885 Jul 10	22 06	0.44297	66.27	21.13
2887 Sep 18	4 02	− 7 02	2887 Sep 16	0 18	0.37292	55.79	25.10
2889 Nov 26	5 43	+ 21 46	2889 Nov 17	23 17	0.47173	70.57	19.84
2892 Jan 11	19 24	+ 25 34	2892 Jan 6	5 02	0.59608	89.17	15.70
2894 Feb 16	10 59	+ 16 20	2894 Feb 14	16 51	0.66714	99.80	14.03
2896 Mar 21	10 16	+ 2 39	2896 Mar 23	13 25	0.67281	100.65	13.91
2898 Apr 26	16 51	− 12 24	2898 May 2	14 03	0.61275	91.67	15.28
2900 Jun 11	1 07	− 25 19	2900 Jun 19	11 21	0.49575	74.16	18.88
2902 Aug 17	11 22	− 19 31	2902 Aug 20	22 26	0.38058	56.93	24.59
2904 Nov 3	12 51	+ 13 38	2904 Oct 26	21 02	0.42080	62.95	22.24
2906 Dec 27	11 40	+ 26 24	2906 Dec 20	12 23	0.55189	82.56	16.96
2909 Feb 3	20 02	+ 20 41	2909 Jan 31	13 20	0.64672	96.75	14.47
2911 Mar 10	5 31	+ 8 11	2911 Mar 10	20 14	0.67843	101.49	13.80
2913 Apr 13	0 03	− 6 35	2913 Apr 17	10 58	0.64377	96.31	14.54
2915 May 23	15 46	− 21 06	2915 May 31	10 18	0.54664	81.78	17.12
2917 Jul 18	12 45	− 26 05	2917 Jul 25	17 41	0.41623	62.27	22.49
2919 Oct 7	0 07	+ 1 09	2919 Oct 2	4 46	0.38287	57.28	24.45
2921 Dec 8	23 47	+ 24 32	2921 Nov 30	23 41	0.50127	74.99	18.67
2924 Jan 21	11 55	+ 24 08	2924 Jan 16	17 17	0.61674	92.26	15.18
2926 Feb 25	1 59	+ 13 23	2926 Feb 24	4 11	0.67394	100.82	13.89
2928 Mar 30	2 59	− 0 47	2928 Apr 2	2 08	0.66540	99.54	14.07
2930 May 6	15 43	− 15 48	2930 May 13	6 23	0.59166	88.51	15.82
2932 Jun 23	10 29	− 26 43	2932 Jul 1	20 05	0.46635	69.76	20.07
2934 Sep 4	17 39	− 12 52	2934 Sep 5	3 01	0.37217	55.68	25.15
2936 Nov 17	10 34	+ 18 54	2936 Nov 9	5 00	0.44807	67.03	20.89

Opposition				Nearest to Earth				
Date		TD	Decl.	Date		TD	Least distance		diam.
		h m	° ′			h m	a.u.	m.k.	″
2939 Jan	6	0 14	+26 09	2938 Dec	30	19 11	0.57730	86.36	16.21
2941 Feb	11	17 33	+18 08	2941 Feb	9	7 25	0.65934	98.64	14.20
2943 Mar	17	19 17	+ 4 51	2943 Mar	19	6 20	0.67664	101.22	13.83
2945 Apr	21	6 59	−10 06	2945 Apr	26	13 05	0.62784	93.92	14.91
2947 Jun	3	7 33	−23 45	2947 Jun	11	12 11	0.51919	77.67	18.03
2949 Aug	3	20 09	−23 05	2949 Aug	9	7 43	0.39381	58.91	23.77
2951 Oct	23	17 06	+ 8 49	2951 Oct	16	23 43	0.40131	60.03	23.32
2953 Dec	19	20 08	+26 01	2953 Dec	12	8 40	0.52989	79.27	17.66
2956 Jan	29	19 52	+22 11	2956 Jan	25	21 14	0.63444	94.91	14.75
2958 Mar	4	15 46	+10 15	2958 Mar	4	14 16	0.67764	101.37	13.81
2960 Apr	7	0 15	− 4 18	2960 Apr	10	19 18	0.65464	97.93	14.30
2962 May	16	2 21	−19 01	2962 May	23	8 57	0.56765	84.92	16.49
2964 Jul	6	23 58	−26 54	2964 Jul	15	0 31	0.43769	65.48	21.39
2966 Sep	23	2 33	− 4 56	2966 Sep	20	10 53	0.37404	55.96	25.02
2968 Nov	30	2 02	+22 41	2968 Nov	21	19 59	0.47739	71.42	19.61
2971 Jan	15	1 03	+25 08	2971 Jan	9	14 58	0.60019	89.79	15.60
2973 Feb	19	10 50	+15 18	2973 Feb	17	20 48	0.66862	100.02	14.00
2975 Mar	25	9 51	+ 1 27	2975 Mar	27	16 48	0.67169	100.48	13.94
2977 Apr	29	21 53	−13 32	2977 May	5	21 56	0.60902	91.11	15.37
2979 Jun	14	20 27	−25 46	2979 Jun	23	6 47	0.49023	73.34	19.09
2981 Aug	21	9 38	−17 55	2981 Aug	24	6 42	0.37811	56.56	24.75
2983 Nov	7	19 16	+15 12	2983 Oct	30	22 26	0.42556	63.66	21.99
2985 Dec	29	20 42	+26 23	2985 Dec	23	1 44	0.55695	83.32	16.81
2988 Feb	6	21 24	+19 49	2988 Feb	3	18 56	0.64952	97.17	14.41
2990 Mar	12	5 14	+ 7 00	2990 Mar	12	23 50	0.67841	101.49	13.80
2992 Apr	15	2 01	− 7 47	2992 Apr	19	16 17	0.64110	95.91	14.60
2994 May	26	3 35	−21 56	2994 Jun	2	23 18	0.54177	81.05	17.28
2996 Jul	22	2 02	−25 24	2996 Jul	29	1 04	0.41160	61.57	22.74
2998 Oct	10	21 00	+ 3 16	2998 Oct	5	15 43	0.38579	57.71	24.26
3000 Dec	12	16 23	+25 04	3000 Dec	4	18 51	0.50679	75.81	18.47

3

Equinoxes and Solstices

1 − 3000

Equinoxes and Solstices

The list on the next pages gives the times of the equinoxes and solstices on the planet Earth for the years 1 to 3000. These are the instants when the apparent longitude of the Sun's center (that is, calculated by including the effects of aberration and nutation) is an exact multiple of 90 degrees.

The instants are given to the nearest second of time, and they are expressed in *Dynamical Time*. For each equinox and solstice, the day of the month (d) is given first, followed by the hours, minutes, and seconds. The tabulated times can be converted into Universal Time by *subtracting* the quantity $\Delta T = \text{TD} - \text{UT}$, as explained on pages 6 − 8.

For the calculation of the Sun's longitude, use was made of the VSOP87 theory of P. Bretagnon.

The table below gives the durations of the four astronomical seasons for some epochs. In A.D. 1246 the Earth was in perihelion at the time of the winter solstice; then the spring had the same duration as the summer, and the autumn the same duration as the winter. Since the year +1246, the winter is the shortest season; it will reach its minimum value (88.71 days) by about A.D. 3500, and remain the shortest season till about A.D. 6430, when the Earth will be in perihelion at the March equinox.

Duration of the astronomical seasons, in days

Year	Spring	Summer	Autumn	Winter
−4000	93.55	89.18	89.07	93.44
−3000	94.04	89.92	88.61	92.67
−2000	94.28	90.76	88.39	91.81
−1000	94.25	91.63	88.42	90.94
0	93.96	92.45	88.69	90.13
+1000	93.44	93.15	89.18	89.47
2000	92.76	93.65	89.84	88.99
3000	91.97	93.92	90.61	88.74
4000	91.17	93.93	91.40	88.73
5000	90.44	93.70	92.15	88.96
6000	89.82	93.25	92.79	89.38

It appears that the first winter (since several thousands of years) having a duration less than 89 days was that of 1971−1972.

After four years, the equinoxes and the solstices occur approximately 45 minutes earlier. However, there is a jump of +1 day when the 4-year period contains a non-bissextile century year. For example:

1969 March 20	$19^h 09^m$ TD		1895 March 20	$20^h 49^m$ TD
1973 March 20	18 13		1899 March 20	19 46
1977 March 20	17 43		1903 March 21	19 15

After 400 years, in the Gregorian calendar, the equinoxes and the solstices occur approximately 3 hours earlier. This is a *mean* value, however. After a lapse of time of four centuries, the eccentricity and the longitude of perihelion of the Earth's orbit have sensibly changed, influencing the duration of the seasons. If we compare the times for the years 1600 and 2000, we see that in the year 2000 the June solstice occurs 8 hours earlier than in 1600, but the December solstice 3 hours *later*.

In 2044 the *March equinox* will take place on March 19, for the first time since 1796. It did not occur on March 21 from 1616 to 1701, from 1744 to 1800, and from 1876 to 1899. It will not occur on March 21 from 2008 to 2101.

A *June solstice* on June 20 occurred for the last time in 1896, and it will take place for the next time in 2012 (if we use TD, but in 2008 if the dates are based on UT!). The June solstice occurred for the last time on June 22 in 1975; it will take place again on that date in 2203, 2207, 2211, and 2215, and then not before 2302. The June solstice will take place on June 19 in 2488, for the first time since the Gregorian calendar reform in A.D. 1582.

In 1968, for the first time since 1897, the *September equinox* took place on September 22. Since the Gregorian calendar reform, the September equinox occurred on September 24 only in 1803, 1807, 1903, 1907, 1911, 1915, 1919, 1923, 1927, and 1931. The next time that will occur in 2303, but then not again before many centuries. (In 2307, if ΔT is larger than 8 ½ minutes, the equinox will take place on September 23 *Universal Time*). The September equinox will occur on September 21 in 2092, for the first time since the Gregorian calendar reform. It will take place on that date again in 2096, but then not again before A.D. 2464 — or maybe in 2460, Universal Time, if ΔT is then larger than 22 minutes.

The *December solstice* will occur on December 20 in A.D. 2080, for the first time since 1697. It will occur on that date again in 2084, 2088, 2092, and 2096, but then not again before 2492 (for dates based on Dynamical Time) or 2488 (for dates based on UT). Since the Gregorian calendar reform, the December solstice occurred only once on December 23, namely in 1903. It will take place on that date again in 2303, 2307, 2311, and 2315, and then not again before 2703.

Of course, these statements are not valid if the instants are expressed in standard times of zones other than that of Greenwich.

Year	March Equinox				June Solstice				September Equinox				December Solstice			
	d	h	m	s	d	h	m	s	d	h	m	s	d	h	m	s
1	23	0	40	48	24	23	50	30	25	10	36	41	23	3	16	18
2	23	6	37	03	25	5	40	31	25	16	27	52	23	9	06	58
3	23	12	15	00	25	11	18	38	25	22	21	10	23	15	01	52
4	22	18	05	20	24	17	17	05	25	4	14	48	22	20	57	09
5	23	0	03	15	24	23	10	16	25	10	02	12	23	2	50	07
6	23	5	54	49	25	4	50	39	25	15	50	08	23	8	44	29
7	23	11	51	43	25	10	46	04	25	21	46	08	23	14	31	32
8	22	17	39	28	24	16	36	14	25	3	35	10	22	20	18	57
9	22	23	28	51	24	22	29	12	25	9	22	19	23	2	11	57
10	23	5	24	27	25	4	22	48	25	15	11	25	23	8	01	38
11	23	11	01	56	25	10	00	36	25	21	00	54	23	13	47	35
12	22	16	47	18	24	15	54	50	25	2	53	34	22	19	33	45
13	22	22	40	11	24	21	44	07	25	8	41	58	23	1	21	36
14	23	4	22	15	25	3	15	51	25	14	24	59	23	7	11	22
15	23	10	06	44	25	9	04	01	25	20	16	49	23	13	01	06
16	22	15	50	23	24	14	49	35	25	2	00	25	22	18	48	59
17	22	21	37	21	24	20	34	55	25	7	42	24	23	0	40	14
18	23	3	35	39	25	2	26	23	25	13	32	58	23	6	30	12
19	23	9	19	04	25	8	04	00	25	19	21	52	23	12	15	59
20	22	15	07	39	24	14	00	57	25	1	14	45	22	18	06	25
21	22	21	04	50	24	19	58	03	25	7	03	39	22	23	59	15
22	23	2	52	04	25	1	37	20	25	12	47	41	23	5	50	24
23	23	8	41	14	25	7	33	26	25	18	45	42	23	11	39	37
24	22	14	30	21	24	13	26	22	25	0	37	29	22	17	26	44
25	22	20	19	53	24	19	14	02	25	6	26	27	22	23	18	30
26	23	2	14	56	25	1	07	58	25	12	23	42	23	5	14	22
27	23	7	56	15	25	6	46	54	25	18	13	22	23	11	05	35
28	22	13	42	09	24	12	40	31	25	0	04	44	22	16	58	57
29	22	19	39	06	24	18	34	03	25	5	52	24	22	22	51	41
30	23	1	28	38	25	0	06	48	25	11	33	14	23	4	38	54
31	23	7	17	23	25	5	56	46	25	17	27	13	23	10	24	21
32	22	13	04	35	24	11	47	25	24	23	12	08	22	16	10	21
33	22	18	51	28	24	17	31	55	25	4	51	09	22	21	57	21
34	23	0	41	02	24	23	24	10	25	10	42	19	23	3	47	59
35	23	6	21	53	25	5	04	06	25	16	29	22	23	9	31	12
36	22	12	05	33	24	10	54	43	24	22	20	04	22	15	16	03
37	22	17	59	09	24	16	49	18	25	4	13	40	22	21	09	26
38	22	23	46	15	24	22	23	16	25	9	56	49	23	2	58	56
39	23	5	29	46	25	4	13	46	25	15	53	29	23	8	51	53
40	22	11	19	05	24	10	08	35	24	21	43	18	22	14	46	25

Year	March Equinox				June Solstice				September Equinox				December Solstice			
	d	h	m	s	d	h	m	s	d	h	m	s	d	h	m	s
41	22	17	12	12	24	15	53	33	25	3	25	27	22	20	36	23
42	22	23	08	30	24	21	47	37	25	9	23	34	23	2	31	41
43	23	4	58	24	25	3	33	00	25	15	15	47	23	8	20	21
44	22	10	46	40	24	9	26	15	24	21	05	09	22	14	08	55
45	22	16	41	52	24	15	25	39	25	2	59	09	22	20	07	20
46	22	22	32	12	24	21	02	59	25	8	39	11	23	1	53	03
47	23	4	15	02	25	2	52	12	25	14	33	38	23	7	39	08
48	22	10	03	49	24	8	46	49	24	20	25	35	22	13	27	43
49	22	15	52	38	24	14	26	54	25	2	05	39	22	19	11	57
50	22	21	39	06	24	20	15	31	25	8	01	04	23	1	06	46
51	23	3	21	03	25	1	56	55	25	13	48	17	23	6	56	31
52	22	9	05	09	24	7	43	33	24	19	31	09	22	12	42	51
53	22	14	58	31	24	13	37	30	25	1	22	49	22	18	38	06
54	22	20	51	50	24	19	12	46	25	7	02	56	23	0	21	44
55	23	2	36	56	25	1	00	20	25	12	55	56	23	6	06	39
56	22	8	24	54	24	6	58	01	24	18	47	41	22	11	59	09
57	22	14	16	20	24	12	42	44	25	0	25	43	22	17	44	57
58	22	20	03	40	24	18	35	04	25	6	22	29	22	23	38	56
59	23	1	49	05	25	0	23	02	25	12	15	51	23	5	28	29
60	22	7	37	44	24	6	12	40	24	18	05	37	22	11	13	48
61	22	13	30	03	24	12	08	27	25	0	04	29	22	17	13	13
62	22	19	22	23	24	17	47	01	25	5	48	58	22	23	06	00
63	23	1	07	40	24	23	34	25	25	11	41	34	23	4	56	17
64	22	6	56	02	24	5	32	03	24	17	34	34	22	10	53	50
65	22	12	54	24	24	11	16	16	24	23	12	35	22	16	38	49
66	22	18	45	15	24	17	04	24	25	5	07	51	22	22	29	41
67	23	0	32	15	24	22	52	50	25	10	59	54	23	4	19	32
68	22	6	20	34	24	4	41	07	24	16	41	14	22	10	00	47
69	22	12	08	10	24	10	35	52	24	22	33	57	22	15	56	38
70	22	17	58	09	24	16	15	21	25	4	14	33	22	21	41	40
71	22	23	39	26	24	21	59	04	25	10	03	44	23	3	20	57
72	22	5	22	15	24	3	53	43	24	15	59	31	22	9	13	02
73	22	11	14	56	24	9	35	24	24	21	38	45	22	14	56	47
74	22	16	57	36	24	15	19	15	25	3	32	42	22	20	50	36
75	22	22	39	15	24	21	07	10	25	9	26	44	23	2	47	49
76	22	4	29	33	24	2	54	10	24	15	07	26	22	8	32	15
77	22	10	21	44	24	8	46	57	24	21	03	36	22	14	32	19
78	22	16	20	19	24	14	29	14	25	2	49	29	22	20	22	01
79	22	22	09	20	24	20	16	07	25	8	39	05	23	2	05	45
80	22	3	57	25	24	2	17	27	24	14	36	10	22	8	04	11

Year	March Equinox				June Solstice				September Equinox				December Solstice			
	d	h	m	s	d	h	m	s	d	h	m	s	d	h	m	s
81	22	9	55	59	24	8	07	08	24	20	15	17	22	13	50	53
82	22	15	43	29	24	13	55	48	25	2	09	08	22	19	40	58
83	22	21	26	55	24	19	47	37	25	8	07	25	23	1	33	50
84	22	3	19	15	24	1	36	48	24	13	52	02	22	7	14	08
85	22	9	06	29	24	7	26	52	24	19	48	33	22	13	11	38
86	22	14	56	02	24	13	07	30	25	1	34	26	22	19	04	54
87	22	20	40	05	24	18	49	52	25	7	19	07	23	0	47	36
88	22	2	22	13	24	0	43	35	24	13	11	53	22	6	43	23
89	22	8	20	42	24	6	27	56	24	18	48	55	22	12	27	03
90	22	14	08	28	24	12	09	36	25	0	38	40	22	18	11	47
91	22	19	48	21	24	17	58	31	25	6	33	34	23	0	04	51
92	22	1	40	04	23	23	48	34	24	12	13	31	22	5	46	30
93	22	7	25	39	24	5	37	44	24	18	04	15	22	11	40	54
94	22	13	14	15	24	11	21	03	24	23	51	25	22	17	33	14
95	22	19	03	31	24	17	06	52	25	5	39	26	22	23	11	39
96	22	0	46	16	23	23	01	43	24	11	38	24	22	5	07	54
97	22	6	45	31	24	4	51	22	24	17	23	03	22	10	59	12
98	22	12	33	37	24	10	35	33	24	23	14	54	22	16	50	37
99	22	18	14	12	24	16	28	02	25	5	14	06	22	22	52	18
100	22	0	12	47	23	22	21	47	24	10	56	54	22	4	37	45
101	22	6	05	50	24	4	11	04	24	16	50	28	22	10	32	42
102	22	11	59	25	24	9	56	18	24	22	42	04	22	16	26	36
103	22	17	51	20	24	15	44	03	25	4	27	40	22	22	05	22
104	21	23	32	21	23	21	38	18	24	10	20	51	22	4	00	19
105	22	5	27	57	24	3	28	00	24	16	01	00	22	9	47	56
106	22	11	13	24	24	9	09	17	24	21	47	01	22	15	28	53
107	22	16	48	14	24	14	56	34	25	3	45	12	22	21	21	15
108	21	22	40	24	23	20	45	37	24	9	27	14	22	3	00	58
109	22	4	24	42	24	2	27	32	24	15	16	51	22	8	54	44
110	22	10	08	45	24	8	07	41	24	21	06	10	22	14	51	45
111	22	15	57	28	24	13	52	33	25	2	48	24	22	20	33	28
112	21	21	40	48	23	19	43	23	24	8	40	10	22	2	28	51
113	22	3	41	22	24	1	32	51	24	14	24	09	22	8	19	20
114	22	9	35	59	24	7	17	47	24	20	12	56	22	14	02	58
115	22	15	15	49	24	13	09	52	25	2	12	02	22	20	00	16
116	21	21	13	47	23	19	09	11	24	7	57	18	22	1	47	26
117	22	3	05	04	24	0	59	14	24	13	47	57	22	7	40	27
118	22	8	51	31	24	6	46	08	24	19	43	22	22	13	36	46
119	22	14	45	35	24	12	37	32	25	1	33	16	22	19	16	44
120	21	20	28	25	23	18	28	07	24	7	29	12	22	1	10	15

Year	March Equinox				June Solstice				September Equinox				December Solstice			
	d	h	m	s	d	h	m	s	d	h	m	s	d	h	m	s
121	22	2	22	42	24	0	17	04	24	13	16	37	22	7	05	22
122	22	8	12	39	24	5	59	51	24	19	03	29	22	12	51	27
123	22	13	46	31	24	11	45	01	25	0	57	04	22	18	47	57
124	21	19	42	29	23	17	38	15	24	6	39	49	22	0	34	30
125	22	1	35	25	23	23	20	03	24	12	24	40	22	6	20	11
126	22	7	18	32	24	4	59	37	24	18	14	58	22	12	13	17
127	22	13	11	35	24	10	50	16	24	23	58	55	22	17	52	15
128	21	18	51	49	23	16	38	25	24	5	45	50	21	23	41	47
129	22	0	43	00	23	22	28	04	24	11	30	22	22	5	33	46
130	22	6	34	10	24	4	13	03	24	17	16	47	22	11	13	41
131	22	12	08	42	24	9	58	50	24	23	14	05	22	17	05	48
132	21	18	03	28	23	15	55	05	24	5	04	00	21	22	54	39
133	21	23	55	44	23	21	40	59	24	10	53	36	22	4	47	14
134	22	5	37	28	24	3	23	15	24	16	48	25	22	10	48	58
135	22	11	32	50	24	9	18	12	24	22	37	33	22	16	36	34
136	21	17	21	23	23	15	08	03	24	4	28	00	21	22	29	13
137	21	23	19	10	23	20	58	28	24	10	18	27	22	4	24	29
138	22	5	17	05	24	2	47	07	24	16	07	11	22	10	08	06
139	22	10	54	39	24	8	34	11	24	22	01	33	22	16	02	09
140	21	16	47	49	23	14	31	43	24	3	47	31	21	21	50	51
141	21	22	40	15	23	20	18	05	24	9	31	42	22	3	38	07
142	22	4	20	12	24	1	56	55	24	15	22	54	22	9	29	23
143	22	10	10	58	24	7	48	33	24	21	12	01	22	15	09	45
144	21	15	54	36	23	13	33	46	24	2	59	58	21	20	56	31
145	21	21	41	17	23	19	17	21	24	8	46	36	22	2	51	31
146	22	3	32	01	24	1	03	13	24	14	32	19	22	8	37	33
147	22	9	06	53	24	6	44	58	24	20	21	02	22	14	29	21
148	21	15	00	10	23	12	39	41	24	2	08	22	21	20	19	02
149	21	20	59	11	23	18	26	49	24	7	54	55	22	2	05	51
150	22	2	41	31	24	0	06	10	24	13	46	42	22	7	59	19
151	22	8	34	55	24	6	04	12	24	19	38	13	22	13	46	36
152	21	14	23	19	23	11	56	19	24	1	26	03	21	19	35	49
153	21	20	12	53	23	17	45	00	24	7	15	10	22	1	31	20
154	22	2	08	30	23	23	37	46	24	13	08	47	22	7	17	16
155	22	7	47	05	24	5	22	00	24	19	03	00	22	13	06	29
156	21	13	38	07	23	11	17	28	24	0	56	46	21	19	01	18
157	21	19	35	37	23	17	05	56	24	6	44	54	22	0	53	38
158	22	1	14	13	23	22	41	33	24	12	33	29	22	6	50	11
159	22	7	06	46	24	4	37	13	24	18	23	59	22	12	39	12
160	21	12	58	54	23	10	25	19	24	0	08	39	21	18	24	44

Year	March Equinox				June Solstice				September Equinox				December Solstice			
	d	h	m	s	d	h	m	s	d	h	m	s	d	h	m	s
161	21	18	50	00	23	16	08	00	24	5	54	23	22	0	15	41
162	22	0	44	16	23	21	58	32	24	11	43	37	22	5	59	48
163	22	6	21	23	24	3	42	00	24	17	29	57	22	11	45	46
164	21	12	07	40	23	9	35	52	23	23	16	47	21	17	35	13
165	21	18	02	04	23	15	25	49	24	5	02	28	21	23	22	07
166	21	23	39	43	23	21	00	01	24	10	50	42	22	5	09	52
167	22	5	27	21	24	2	54	01	24	16	45	44	22	10	55	54
168	21	11	16	53	23	8	43	07	23	22	34	07	21	16	44	19
169	21	17	03	13	23	14	24	05	24	4	21	08	21	22	40	51
170	21	22	54	40	23	20	15	55	24	10	13	42	22	4	33	40
171	22	4	37	26	24	2	01	13	24	16	02	48	22	10	25	15
172	21	10	30	34	23	7	54	13	23	21	53	00	21	16	17	17
173	21	16	33	08	23	13	47	30	24	3	44	00	21	22	10	05
174	21	22	19	24	23	19	26	01	24	9	32	52	22	4	01	00
175	22	4	09	22	24	1	24	31	24	15	27	03	22	9	51	23
176	21	10	03	32	23	7	20	37	23	21	14	12	21	15	40	21
177	21	15	51	39	23	13	02	26	24	2	58	00	21	21	30	05
178	21	21	42	04	23	18	55	17	24	8	53	17	22	3	16	37
179	22	3	23	25	24	0	39	45	24	14	42	33	22	9	00	33
180	21	9	08	15	23	6	28	00	23	20	31	18	21	14	50	33
181	21	15	01	25	23	12	18	05	24	2	20	36	21	20	44	58
182	21	20	39	57	23	17	50	28	24	8	03	50	22	2	34	21
183	22	2	25	13	23	23	41	51	24	13	54	36	22	8	23	03
184	21	8	20	23	23	5	33	28	23	19	41	15	21	14	09	52
185	21	14	10	19	23	11	10	15	24	1	22	22	21	19	57	12
186	21	20	00	00	23	17	02	45	24	7	17	17	22	1	46	57
187	22	1	42	59	23	22	50	52	24	13	03	23	22	7	33	03
188	21	7	28	58	23	4	41	10	23	18	49	41	21	13	23	55
189	21	13	25	43	23	10	36	45	24	0	42	58	21	19	17	51
190	21	19	09	31	23	16	13	45	24	6	31	35	22	1	04	59
191	22	0	57	19	23	22	08	23	24	12	29	51	22	6	55	27
192	21	6	52	32	23	4	04	45	23	18	23	01	21	12	50	04
193	21	12	42	53	23	9	43	51	24	0	05	57	21	18	44	01
194	21	18	31	48	23	15	36	52	24	6	01	54	22	0	38	57
195	22	0	20	05	23	21	27	12	24	11	49	55	22	6	27	46
196	21	6	11	58	23	3	14	37	23	17	35	22	21	12	14	38
197	21	12	09	02	23	9	08	40	23	23	29	07	21	18	07	59
198	21	17	53	42	23	14	46	07	24	5	13	22	21	23	53	29
199	21	23	36	40	23	20	37	37	24	11	03	29	22	5	39	47
200	21	5	26	58	23	2	33	13	23	16	51	27	21	11	29	57

Year	March Equinox				June Solstice				September Equinox				December Solstice			
	d	h	m	s	d	h	m	s	d	h	m	s	d	h	m	s
201	21	11	14	41	23	8	08	58	23	22	29	50	21	17	13	31
202	21	16	57	21	23	13	56	18	24	4	24	49	21	22	59	29
203	21	22	40	36	23	19	44	09	24	10	15	06	22	4	46	49
204	21	4	26	28	23	1	26	29	23	15	57	51	21	10	33	29
205	21	10	15	34	23	7	17	28	23	21	51	08	21	16	33	14
206	21	16	01	34	23	12	56	17	24	3	35	59	21	22	23	56
207	21	21	48	59	23	18	46	22	24	9	26	47	22	4	12	32
208	21	3	47	23	23	0	45	23	23	15	21	32	21	10	07	45
209	21	9	45	02	23	6	25	54	23	21	03	04	21	15	55	16
210	21	15	33	27	23	12	19	22	24	3	00	20	21	21	47	30
211	21	21	22	06	23	18	16	52	24	8	52	19	22	3	39	24
212	21	3	13	01	23	0	05	58	23	14	35	16	21	9	25	07
213	21	9	04	42	23	6	01	21	23	20	32	43	21	15	20	56
214	21	14	51	30	23	11	42	51	24	2	22	34	21	21	06	55
215	21	20	35	37	23	17	30	11	24	8	14	29	22	2	51	54
216	21	2	24	50	22	23	25	25	23	14	09	39	21	8	48	48
217	21	8	14	42	23	5	00	43	23	19	46	21	21	14	36	56
218	21	13	56	18	23	10	45	13	24	1	37	43	21	20	27	32
219	21	19	41	52	23	16	35	13	24	7	25	31	22	2	16	11
220	21	1	33	43	22	22	16	15	23	13	03	11	21	7	57	15
221	21	7	24	46	23	4	06	17	23	18	56	20	21	13	50	45
222	21	13	10	19	23	9	48	24	24	0	41	47	21	19	38	13
223	21	18	55	09	23	15	37	19	24	6	27	45	22	1	21	44
224	21	0	43	57	22	21	35	34	23	12	22	12	21	7	17	21
225	21	6	38	26	23	3	16	43	23	18	02	54	21	13	02	59
226	21	12	23	02	23	9	03	41	23	23	59	29	21	18	50	18
227	21	18	09	02	23	14	59	48	24	5	57	29	22	0	45	18
228	21	0	02	52	22	20	46	19	23	11	40	35	21	6	33	21
229	21	5	51	54	23	2	38	10	23	17	37	16	21	12	35	07
230	21	11	40	45	23	8	24	23	23	23	27	54	21	18	30	11
231	21	17	32	35	23	14	13	05	24	5	15	34	22	0	13	54
232	20	23	26	04	22	20	10	11	23	11	12	57	21	6	10	30
233	21	5	24	49	23	1	52	34	23	16	53	58	21	11	57	43
234	21	11	08	29	23	7	37	27	23	22	43	23	21	17	42	40
235	21	16	50	01	23	13	32	45	24	4	36	06	21	23	36	36
236	20	22	42	48	22	19	17	44	23	10	12	10	21	5	16	01
237	21	4	26	51	23	1	03	26	23	16	03	38	21	11	06	45
238	21	10	10	48	23	6	46	42	23	21	55	01	21	16	55	06
239	21	15	56	35	23	12	30	31	24	3	39	45	21	22	34	26
240	20	21	40	53	22	18	23	04	23	9	34	55	21	4	33	01

Year	March Equinox				June Solstice				September Equinox				December Solstice			
	d	h	m	s	d	h	m	s	d	h	m	s	d	h	m	s
241	21	3	34	03	23	0	03	44	23	15	14	11	21	10	24	11
242	21	9	18	15	23	5	46	30	23	21	02	06	21	16	10	20
243	21	15	03	30	23	11	41	37	24	2	58	22	21	22	05	24
244	20	21	03	54	22	17	29	58	23	8	39	20	21	3	48	23
245	21	2	54	31	22	23	19	11	23	14	33	30	21	9	43	16
246	21	8	41	10	23	5	09	47	23	20	28	32	21	15	39	00
247	21	14	32	58	23	11	01	49	24	2	14	07	21	21	21	38
248	20	20	21	09	22	16	59	07	23	8	11	57	21	3	19	34
249	21	2	17	30	22	22	44	33	23	13	57	25	21	9	09	15
250	21	8	03	58	23	4	28	07	23	19	49	53	21	14	52	41
251	21	13	44	42	23	10	21	20	24	1	48	25	21	20	48	33
252	20	19	39	25	22	16	07	42	23	7	27	38	21	2	36	04
253	21	1	25	51	22	21	51	15	23	13	16	02	21	8	30	39
254	21	7	08	04	23	3	35	52	23	19	07	27	21	14	25	34
255	21	13	02	28	23	9	23	10	24	0	49	05	21	20	03	22
256	20	18	50	48	22	15	13	03	23	6	41	25	21	1	55	45
257	21	0	45	29	22	20	57	03	23	12	23	36	21	7	45	15
258	21	6	30	56	23	2	40	27	23	18	08	03	21	13	24	41
259	21	12	08	09	23	8	34	39	24	0	02	32	21	19	18	12
260	20	18	03	39	22	14	25	02	23	5	42	33	21	1	01	43
261	20	23	50	16	22	20	08	51	23	11	33	05	21	6	49	38
262	21	5	30	32	23	1	55	16	23	17	32	36	21	12	45	11
263	21	11	24	07	23	7	46	00	23	23	20	15	21	18	28	25
264	20	17	09	53	22	13	36	18	23	5	14	59	21	0	28	33
265	20	23	04	07	22	19	22	47	23	11	02	45	21	6	28	29
266	21	4	55	54	23	1	07	49	23	16	49	06	21	12	11	39
267	21	10	40	00	23	7	01	18	23	22	46	45	21	18	08	36
268	20	16	44	03	22	12	54	27	23	4	30	29	20	23	55	46
269	20	22	35	34	22	18	39	50	23	10	18	14	21	5	44	39
270	21	4	16	09	23	0	29	14	23	16	14	55	21	11	41	22
271	21	10	10	21	23	6	23	12	23	21	57	54	21	17	21	40
272	20	15	55	57	22	12	12	18	23	3	48	01	20	23	12	34
273	20	21	45	56	22	17	55	56	23	9	35	47	21	5	03	29
274	21	3	33	55	22	23	38	54	23	15	22	32	21	10	40	18
275	21	9	09	28	23	5	26	50	23	21	17	25	21	16	33	37
276	20	15	02	20	22	11	15	54	23	2	59	16	20	22	23	52
277	20	20	49	21	22	16	56	45	23	8	42	59	21	4	12	38
278	21	2	26	39	22	22	40	18	23	14	38	15	21	10	08	36
279	21	8	24	01	23	4	32	42	23	20	23	11	21	15	50	05
280	20	14	14	51	22	10	20	17	23	2	13	40	20	21	41	42

Year	March Equinox				June Solstice				September Equinox				December Solstice			
	d	h	m	s	d	h	m	s	d	h	m	s	d	h	m	s
281	20	20	05	38	22	16	05	46	23	8	02	24	21	3	38	04
282	21	1	57	52	22	21	55	23	23	13	49	20	21	9	21	06
283	21	7	38	03	23	3	47	35	23	19	42	57	21	15	14	22
284	20	13	34	33	22	9	42	16	23	1	29	23	20	21	05	29
285	20	19	29	01	22	15	28	56	23	7	19	05	21	2	51	32
286	21	1	07	07	22	21	14	21	23	13	20	02	21	8	48	07
287	21	7	03	14	23	3	10	50	23	19	10	04	21	14	36	05
288	20	12	52	19	22	8	57	57	23	0	58	49	20	20	30	54
289	20	18	40	05	22	14	41	26	23	6	47	24	21	2	30	22
290	21	0	34	52	22	20	29	32	23	12	32	50	21	8	11	06
291	21	6	17	18	23	2	16	36	23	18	23	42	21	13	59	27
292	20	12	13	27	22	8	08	02	23	0	08	43	20	19	48	37
293	20	18	05	39	22	13	53	03	23	5	52	06	21	1	31	40
294	20	23	38	37	22	19	35	46	23	11	45	10	21	7	24	24
295	21	5	29	59	23	1	31	53	23	17	31	21	21	13	08	04
296	20	11	17	25	22	7	17	07	22	23	16	47	20	18	54	10
297	20	17	01	06	22	12	57	03	23	5	08	09	21	0	48	40
298	20	22	52	19	22	18	44	35	23	10	57	06	21	6	29	21
299	21	4	30	42	23	0	29	48	23	16	47	56	21	12	22	13
300	20	10	23	48	22	6	21	22	22	22	35	15	20	18	19	51
301	20	16	19	05	22	12	08	49	23	4	20	41	21	0	09	37
302	20	21	59	41	22	17	52	52	23	10	16	45	21	6	05	22
303	21	3	59	34	22	23	52	26	23	16	09	14	21	11	53	45
304	20	9	57	33	22	5	43	41	22	21	58	11	20	17	43	55
305	20	15	45	25	22	11	27	59	23	3	49	20	20	23	41	39
306	20	21	39	21	22	17	23	34	23	9	39	23	21	5	26	18
307	21	3	21	45	22	23	13	12	23	15	28	22	21	11	14	03
308	20	9	12	34	22	5	04	30	22	21	16	37	20	17	05	15
309	20	15	06	31	22	10	51	40	23	3	04	37	20	22	48	43
310	20	20	40	24	22	16	29	59	23	8	58	59	21	4	38	43
311	21	2	28	34	22	22	23	44	23	14	49	03	21	10	28	17
312	20	8	18	53	22	4	09	14	22	20	32	51	20	16	18	59
313	20	13	59	45	22	9	43	50	23	2	17	09	20	22	13	57
314	20	19	51	38	22	15	33	16	23	8	05	19	21	3	58	24
315	21	1	38	28	22	21	18	01	23	13	51	03	21	9	42	12
316	20	7	29	19	22	3	05	52	22	19	36	34	20	15	34	26
317	20	13	25	46	22	8	57	32	23	1	23	20	20	21	21	48
318	20	19	02	31	22	14	39	59	23	7	14	15	21	3	12	05
319	21	0	53	39	22	20	40	06	23	13	07	00	21	9	02	28
320	20	6	50	46	22	2	33	00	22	18	56	11	20	14	51	15

Year	March Equinox				June Solstice				September Equinox				December Solstice			
	d	h	m	s	d	h	m	s	d	h	m	s	d	h	m	s
321	20	12	35	53	22	8	12	04	23	0	48	39	20	20	45	37
322	20	18	28	26	22	14	07	34	23	6	46	20	21	2	34	51
323	21	0	15	01	22	19	57	16	23	12	36	37	21	8	25	59
324	20	6	04	33	22	1	45	39	22	18	24	22	20	14	24	56
325	20	12	01	22	22	7	37	39	23	0	13	24	20	20	16	24
326	20	17	42	50	22	13	17	32	23	6	03	34	21	2	04	40
327	20	23	35	46	22	19	13	22	23	11	56	11	21	7	53	16
328	20	5	33	25	22	1	04	34	22	17	41	24	20	13	41	00
329	20	11	15	16	22	6	39	39	22	23	25	35	20	19	32	34
330	20	17	01	46	22	12	32	15	23	5	15	37	21	1	16	54
331	20	22	45	22	22	18	19	43	23	10	58	50	21	6	59	46
332	20	4	32	02	22	0	03	01	22	16	43	02	20	12	48	14
333	20	10	23	23	22	5	51	53	22	22	33	52	20	18	34	42
334	20	16	00	55	22	11	30	16	23	4	24	10	21	0	21	45
335	20	21	46	27	22	17	23	05	23	10	15	50	21	6	14	39
336	20	3	42	17	21	23	16	01	22	16	03	08	20	12	09	54
337	20	9	27	46	22	4	51	06	22	21	47	45	20	18	03	15
338	20	15	19	56	22	10	46	07	23	3	44	15	20	23	52	29
339	20	21	14	53	22	16	39	50	23	9	34	12	21	5	40	03
340	20	3	07	32	21	22	26	53	22	15	20	43	20	11	33	24
341	20	9	02	21	22	4	23	30	22	21	14	54	20	17	27	08
342	20	14	45	01	22	10	09	12	23	3	05	18	20	23	15	14
343	20	20	32	28	22	16	04	30	23	8	57	07	21	5	05	10
344	20	2	29	58	21	21	59	39	22	14	49	01	20	10	57	09
345	20	8	14	10	22	3	31	56	22	20	34	10	20	16	43	55
346	20	13	58	10	22	9	22	08	23	2	30	01	20	22	33	28
347	20	19	46	53	22	15	12	29	23	8	16	09	21	4	22	23
348	20	1	32	33	21	20	51	00	22	13	54	37	20	10	13	49
349	20	7	23	09	22	2	41	13	22	19	45	06	20	16	05	17
350	20	13	07	24	22	8	21	42	23	1	31	27	20	21	48	12
351	20	18	55	48	22	14	11	37	23	7	19	29	21	3	34	31
352	20	0	52	24	21	20	06	38	22	13	08	33	20	9	27	13
353	20	6	37	12	22	1	42	02	22	18	49	57	20	15	14	25
354	20	12	20	58	22	7	35	39	23	0	44	15	20	21	02	23
355	20	18	11	25	22	13	32	08	23	6	34	00	21	2	50	18
356	20	0	01	19	21	19	14	01	22	12	17	47	20	8	38	23
357	20	5	51	17	22	1	07	05	22	18	17	29	20	14	31	47
358	20	11	35	45	22	6	52	57	23	0	10	41	20	20	22	10
359	20	17	22	59	22	12	43	50	23	6	01	37	21	2	16	28
360	19	23	18	49	21	18	40	05	22	11	54	15	20	8	17	21

Year	March Equinox				June Solstice				September Equinox				December Solstice			
	d	h	m	s	d	h	m	s	d	h	m	s	d	h	m	s
361	20	5	09	19	22	0	15	54	22	17	37	41	20	14	08	19
362	20	10	59	11	22	6	06	26	22	23	32	18	20	19	55	14
363	20	16	53	57	22	12	02	38	23	5	22	24	21	1	44	47
364	19	22	47	16	21	17	44	24	22	11	00	49	20	7	31	19
365	20	4	33	40	21	23	36	17	22	16	54	02	20	13	22	26
366	20	10	16	41	22	5	24	05	22	22	41	47	20	19	07	45
367	20	16	01	26	22	11	11	14	23	4	26	03	21	0	50	41
368	19	21	52	10	21	17	04	22	22	10	19	28	20	6	43	45
369	20	3	38	01	21	22	37	41	22	16	03	02	20	12	28	03
370	20	9	18	23	22	4	23	35	22	21	55	52	20	18	14	57
371	20	15	05	45	22	10	18	09	23	3	46	19	21	0	09	47
372	19	20	56	28	21	15	56	52	22	9	22	35	20	5	58	28
373	20	2	43	40	21	21	45	46	22	15	16	58	20	11	51	35
374	20	8	33	12	22	3	35	10	22	21	09	48	20	17	40	08
375	20	14	25	06	22	9	23	46	23	2	55	31	20	23	25	46
376	19	20	19	31	21	15	21	28	22	8	52	07	20	5	26	35
377	20	2	10	25	21	21	02	19	22	14	36	06	20	11	15	25
378	20	7	54	58	22	2	52	38	22	20	28	10	20	17	02	56
379	20	13	47	25	22	8	53	05	23	2	23	31	20	22	56	35
380	19	19	42	48	21	14	34	50	22	8	04	00	20	4	41	22
381	20	1	29	05	21	20	23	50	22	14	02	01	20	10	33	08
382	20	7	13	39	22	2	13	50	22	19	56	11	20	16	25	16
383	20	13	01	23	22	7	59	58	23	1	38	08	20	22	12	13
384	19	18	50	36	21	13	52	10	22	7	30	38	20	4	11	46
385	20	0	41	07	21	19	29	53	22	13	13	03	20	9	58	34
386	20	6	27	36	22	1	14	08	22	19	01	21	20	15	39	27
387	20	12	16	47	22	7	10	17	23	0	54	11	20	21	31	32
388	19	18	10	53	21	12	51	31	22	6	29	21	20	3	15	28
389	19	23	52	55	21	18	38	18	22	12	20	36	20	9	03	45
390	20	5	34	30	22	0	29	58	22	18	13	03	20	14	53	31
391	20	11	23	54	22	6	16	56	22	23	55	23	20	20	34	38
392	19	17	11	32	21	12	07	37	22	5	51	52	20	2	30	24
393	19	23	00	17	21	17	47	42	22	11	41	28	20	8	22	18
394	20	4	46	21	21	23	33	06	22	17	31	47	20	14	08	37
395	20	10	33	08	22	5	30	18	22	23	27	48	20	20	10	27
396	19	16	32	56	21	11	15	31	22	5	06	26	20	2	01	39
397	19	22	22	30	21	17	02	08	22	10	59	35	20	7	51	49
398	20	4	12	09	21	22	57	11	22	16	57	44	20	13	45	25
399	20	10	09	42	22	4	47	56	22	22	39	55	20	19	28	00
400	19	15	59	26	21	10	41	25	22	4	34	04	20	1	25	41

Year	March Equinox				June Solstice				September Equinox				December Solstice			
	d	h	m	s	d	h	m	s	d	h	m	s	d	h	m	s
401	19	21	48	11	21	16	26	17	22	10	21	38	20	7	16	33
402	20	3	33	30	21	22	12	50	22	16	07	25	20	12	55	33
403	20	9	17	20	22	4	07	49	22	22	02	53	20	18	48	44
404	19	15	12	26	21	9	49	45	22	3	42	17	20	0	32	28
405	19	20	54	01	21	15	29	44	22	9	32	39	20	6	17	26
406	20	2	32	06	21	21	19	01	22	15	28	49	20	12	12	41
407	20	8	22	27	22	3	04	31	22	21	05	55	20	17	56	27
408	19	14	08	19	21	8	50	27	22	2	55	54	19	23	54	13
409	19	19	57	51	21	14	31	19	22	8	43	48	20	5	44	51
410	20	1	48	43	21	20	16	05	22	14	28	51	20	11	23	41
411	20	7	37	05	22	2	12	21	22	20	24	48	20	17	20	18
412	19	13	35	43	21	8	00	39	22	2	04	59	19	23	10	32
413	19	19	22	55	21	13	47	47	22	7	54	59	20	4	57	22
414	20	1	04	59	21	19	44	01	22	13	54	35	20	10	53	21
415	20	7	02	43	22	1	37	52	22	19	38	45	20	16	36	32
416	19	12	52	32	21	7	26	54	22	1	34	30	19	22	31	22
417	19	18	39	40	21	13	12	13	22	7	30	44	20	4	27	23
418	20	0	29	42	21	19	00	04	22	13	17	57	20	10	10	50
419	20	6	13	12	22	0	53	05	22	19	12	26	20	16	10	27
420	19	12	10	22	21	6	38	44	22	0	52	30	19	22	02	30
421	19	17	59	42	21	12	19	51	22	6	39	25	20	3	43	50
422	19	23	40	28	21	18	09	27	22	12	36	00	20	9	36	03
423	20	5	37	03	22	0	00	22	22	18	15	30	20	15	18	20
424	19	11	22	36	21	5	44	14	22	0	00	49	19	21	08	08
425	19	17	02	59	21	11	26	53	22	5	50	45	20	3	01	31
426	19	22	52	54	21	17	14	40	22	11	33	17	20	8	37	27
427	20	4	34	19	21	23	03	59	22	17	26	05	20	14	29	00
428	19	10	29	04	21	4	49	59	21	23	11	21	19	20	19	47
429	19	16	16	17	21	10	31	44	22	5	00	06	20	2	02	53
430	19	21	53	09	21	16	22	16	22	10	59	12	20	8	02	40
431	20	3	50	28	21	22	16	39	22	16	41	49	20	13	53	26
432	19	9	42	44	21	4	03	08	22	22	30	30	19	19	47	56
433	19	15	31	16	21	9	49	08	22	4	27	48	20	1	45	14
434	19	21	30	37	21	15	42	57	22	10	16	11	20	7	26	03
435	20	3	18	56	21	21	36	33	22	16	10	07	20	13	21	32
436	19	9	14	54	21	3	27	41	21	21	56	27	19	19	16	33
437	19	15	04	50	21	9	14	16	22	3	42	55	20	0	57	48
438	19	20	41	39	21	15	04	39	22	9	40	33	20	6	51	14
439	20	2	36	36	21	20	57	26	22	15	24	09	20	12	35	11
440	19	8	24	50	21	2	38	48	21	21	11	46	19	18	23	17

ASTRONOMICAL TABLES

Year	March Equinox				June Solstice				September Equinox				December Solstice			
	d	h	m	s	d	h	m	s	d	h	m	s	d	h	m	s
441	19	14	03	12	21	8	17	48	22	3	06	08	20	0	18	09
442	19	19	52	51	21	14	06	15	22	8	49	07	20	6	00	30
443	20	1	35	52	21	19	52	22	22	14	35	33	20	11	53	57
444	19	7	27	42	21	1	36	30	21	20	18	26	19	17	47	33
445	19	13	21	17	21	7	20	23	22	2	03	33	19	23	25	54
446	19	19	00	44	21	13	07	43	22	7	58	34	20	5	17	23
447	20	0	56	59	21	19	04	06	22	13	43	00	20	11	06	12
448	19	6	48	55	21	0	51	10	21	19	28	31	19	16	55	04
449	19	12	28	00	21	6	35	47	22	1	24	36	19	22	51	27
450	19	18	23	08	21	12	32	46	22	7	14	07	20	4	34	19
451	20	0	11	10	21	18	22	51	22	13	07	01	20	10	25	01
452	19	6	03	18	21	0	10	20	21	18	59	24	19	16	22	20
453	19	11	56	52	21	5	58	46	22	0	50	54	19	22	07	33
454	19	17	33	51	21	11	45	44	22	6	46	16	20	4	04	52
455	19	23	28	28	21	17	41	15	22	12	32	52	20	10	00	31
456	19	5	24	21	20	23	25	28	21	18	16	28	19	15	48	00
457	19	11	05	24	21	5	03	59	22	0	10	14	19	21	41	59
458	19	17	03	08	21	10	59	08	22	5	57	35	20	3	23	59
459	19	22	50	53	21	16	46	23	22	11	42	05	20	9	11	25
460	19	4	38	20	20	22	31	43	21	17	27	00	19	15	05	30
461	19	10	28	58	21	4	20	14	21	23	12	01	19	20	44	19
462	19	16	04	16	21	10	04	49	22	5	02	37	20	2	31	13
463	19	21	54	29	21	15	57	33	22	10	50	13	20	8	19	08
464	19	3	46	50	20	21	41	52	21	16	36	42	19	14	04	34
465	19	9	21	54	21	3	18	41	21	22	31	30	19	20	00	27
466	19	15	12	58	21	9	14	04	22	4	21	23	20	1	50	13
467	19	21	02	50	21	15	02	07	22	10	07	10	20	7	42	40
468	19	2	54	01	20	20	46	07	21	15	55	49	19	13	40	13
469	19	8	52	45	21	2	38	00	21	21	47	19	19	19	24	03
470	19	14	37	13	21	8	26	22	22	3	41	01	20	1	15	05
471	19	20	30	52	21	14	23	35	22	9	29	55	20	7	09	09
472	19	2	27	43	20	20	15	04	21	15	16	36	19	12	59	36
473	19	8	07	00	21	1	55	26	21	21	09	41	19	18	52	29
474	19	13	58	53	21	7	53	15	22	3	01	54	20	0	38	43
475	19	19	51	37	21	13	42	46	22	8	50	00	20	6	25	07
476	19	1	37	37	20	19	22	31	21	14	38	22	19	12	19	08
477	19	7	28	35	21	1	12	26	21	20	29	08	19	18	05	01
478	19	13	06	45	21	6	55	38	22	2	16	02	19	23	54	26
479	19	18	54	24	21	12	46	13	22	8	00	42	20	5	47	32
480	19	0	51	26	20	18	33	30	21	13	45	04	19	11	33	36

Year	March Equinox				June Solstice				September Equinox				December Solstice			
	d	h	m	s	d	h	m	s	d	h	m	s	d	h	m	s
481	19	6	30	56	21	0	08	05	21	19	35	08	19	17	21	22
482	19	12	21	44	21	6	04	00	22	1	26	27	19	23	08	03
483	19	18	14	05	21	11	54	49	22	7	10	57	20	4	55	18
484	18	23	58	19	20	17	35	00	21	12	55	24	19	10	49	08
485	19	5	49	20	20	23	28	51	21	18	48	13	19	16	35	00
486	19	11	31	28	21	5	15	26	22	0	37	38	19	22	20	33
487	19	17	20	25	21	11	06	55	22	6	29	04	20	4	14	17
488	18	23	18	04	20	16	57	52	21	12	20	35	19	10	05	16
489	19	4	56	29	20	22	34	21	21	18	12	48	19	15	59	57
490	19	10	47	02	21	4	32	26	22	0	07	11	19	21	55	17
491	19	16	44	10	21	10	25	32	22	5	53	29	20	3	47	55
492	18	22	35	28	20	16	04	35	21	11	38	56	19	9	41	48
493	19	4	31	30	20	21	58	28	21	17	34	00	19	15	28	11
494	19	10	19	07	21	3	47	38	21	23	21	35	19	21	13	39
495	19	16	07	13	21	9	38	28	22	5	06	32	20	3	04	38
496	18	22	01	21	20	15	31	27	21	10	54	09	19	8	53	25
497	19	3	38	40	20	21	07	45	21	16	41	20	19	14	38	27
498	19	9	23	17	21	3	01	24	21	22	34	00	19	20	23	54
499	19	15	15	44	21	8	51	52	22	4	21	46	20	2	10	36
500	18	20	59	09	20	14	24	52	21	10	05	38	19	8	00	59
501	19	2	44	01	20	20	14	05	21	15	59	40	19	13	51	07
502	19	8	27	32	21	2	00	56	21	21	45	43	19	19	41	14
503	19	14	14	51	21	7	46	07	22	3	27	45	20	1	33	04
504	18	20	12	23	20	13	37	37	21	9	18	34	19	7	25	46
505	19	1	59	35	20	19	16	22	21	15	08	07	19	13	12	33
506	19	7	49	16	21	1	13	02	21	21	01	52	19	19	03	18
507	19	13	47	49	21	7	12	43	22	2	52	13	20	0	57	18
508	18	19	36	56	20	12	52	29	21	8	35	39	19	6	48	41
509	19	1	25	54	20	18	48	29	21	14	33	55	19	12	38	29
510	19	7	15	17	21	0	41	39	21	20	25	05	19	18	25	00
511	19	13	04	06	21	6	28	03	22	2	11	53	20	0	14	17
512	18	18	57	19	20	12	20	29	21	8	07	28	19	6	08	47
513	19	0	38	25	20	17	58	26	21	13	55	52	19	11	57	59
514	19	6	22	03	20	23	49	26	21	19	45	26	19	17	49	23
515	19	12	17	01	21	5	43	19	22	1	33	08	19	23	42	13
516	18	18	06	29	20	11	15	36	21	7	12	00	19	5	27	42
517	18	23	53	42	20	17	04	29	21	13	07	07	19	11	13	57
518	19	5	40	54	20	22	55	15	21	18	52	45	19	16	59	26
519	19	11	26	47	21	4	38	25	22	0	31	12	19	22	47	19
520	18	17	16	35	20	10	30	05	21	6	22	17	19	4	39	04

Year	March Equinox				June Solstice				September Equinox				December Solstice			
	d	h	m	s	d	h	m	s	d	h	m	s	d	h	m	s
521	18	22	58	19	20	16	09	59	21	12	08	54	19	10	23	35
522	19	4	43	59	20	22	01	17	21	18	00	18	19	16	09	36
523	19	10	38	36	21	3	57	32	21	23	55	02	19	22	03	56
524	18	16	28	47	20	9	34	00	21	5	38	54	19	3	54	27
525	18	22	13	59	20	15	25	00	21	11	36	59	19	9	48	37
526	19	4	03	59	20	21	21	29	21	17	28	19	19	15	44	03
527	19	9	57	54	21	3	06	58	21	23	09	34	19	21	34	08
528	18	15	53	48	20	9	00	36	21	5	06	58	19	3	29	17
529	18	21	44	24	20	14	45	46	21	10	58	10	19	9	17	03
530	19	3	32	59	20	20	38	13	21	16	47	08	19	15	04	03
531	19	9	26	15	21	2	37	03	21	22	39	48	19	21	00	18
532	18	15	16	16	20	8	15	03	21	4	19	21	19	2	45	56
533	18	20	57	55	20	14	02	23	21	10	12	45	19	8	30	32
534	19	2	43	38	20	19	55	35	21	16	04	33	19	14	19	03
535	19	8	31	51	21	1	34	12	21	21	42	19	19	20	00	42
536	18	14	15	40	20	7	19	43	21	3	35	54	19	1	54	57
537	18	19	58	26	20	13	01	13	21	9	23	26	19	7	44	54
538	19	1	42	25	20	18	46	49	21	15	06	01	19	13	31	18
539	19	7	35	17	21	0	41	43	21	20	58	22	19	19	28	21
540	18	13	31	04	20	6	18	56	21	2	38	56	19	1	13	34
541	18	19	18	18	20	12	07	09	21	8	33	16	19	6	59	48
542	19	1	07	49	20	18	06	40	21	14	27	14	19	12	53	56
543	19	7	01	31	20	23	53	58	21	20	06	06	19	18	40	17
544	18	12	49	44	20	5	46	33	21	2	02	59	19	0	35	45
545	18	18	36	40	20	11	36	14	21	7	57	40	19	6	26	16
546	19	0	26	14	20	17	26	09	21	13	46	41	19	12	10	27
547	19	6	18	06	20	23	22	33	21	19	46	27	19	18	09	41
548	18	12	11	01	20	5	01	51	21	1	30	24	19	0	00	30
549	18	17	54	16	20	10	47	00	21	7	22	46	19	5	50	23
550	18	23	39	41	20	16	42	06	21	13	14	30	19	11	46	22
551	19	5	35	03	20	22	23	58	21	18	49	23	19	17	30	35
552	18	11	25	00	20	4	08	41	21	0	41	37	18	23	19	56
553	18	17	10	27	20	9	54	39	21	6	32	08	19	5	08	32
554	18	22	59	18	20	15	42	40	21	12	12	40	19	10	49	00
555	19	4	45	53	20	21	36	14	21	18	04	36	19	16	44	35
556	18	10	36	43	20	3	17	07	20	23	45	08	18	22	30	43
557	18	16	19	37	20	9	00	58	21	5	34	22	19	4	11	00
558	18	22	02	52	20	14	56	37	21	11	31	13	19	10	03	50
559	19	3	57	45	20	20	41	08	21	17	11	11	19	15	48	14
560	18	9	42	25	20	2	26	07	20	23	06	12	18	21	42	50

Year	March Equinox				June Solstice				September Equinox				December Solstice			
	d	h	m	s	d	h	m	s	d	h	m	s	d	h	m	s
561	18	15	25	15	20	8	15	41	21	5	02	13	19	3	41	35
562	18	21	17	56	20	14	05	52	21	10	45	27	19	9	27	48
563	19	3	10	29	20	19	59	36	21	16	41	55	19	15	27	55
564	18	9	09	58	20	1	44	06	20	22	29	49	18	21	19	25
565	18	15	00	32	20	7	30	52	21	4	19	52	19	3	02	18
566	18	20	46	13	20	13	30	04	21	10	16	54	19	9	00	55
567	19	2	44	19	20	19	19	02	21	15	54	01	19	14	45	42
568	18	8	29	17	20	1	03	42	20	21	44	08	18	20	34	14
569	18	14	11	23	20	6	53	33	21	3	40	46	19	2	25	56
570	18	20	02	34	20	12	41	25	21	9	23	14	19	8	03	51
571	19	1	48	03	20	18	29	30	21	15	17	46	19	13	59	33
572	18	7	36	49	20	0	09	27	20	21	02	51	18	19	52	03
573	18	13	20	41	20	5	51	07	21	2	47	08	19	1	34	24
574	18	19	01	25	20	11	43	30	21	8	39	31	19	7	30	44
575	19	1	00	02	20	17	29	22	21	14	16	24	19	13	14	48
576	18	6	49	10	19	23	10	56	20	20	05	33	18	19	00	18
577	18	12	30	51	20	5	01	03	21	2	02	50	19	0	54	43
578	18	18	24	48	20	10	54	09	21	7	44	21	19	6	36	59
579	19	0	12	03	20	16	45	31	21	13	37	38	19	12	33	50
580	18	6	02	31	19	22	31	21	20	19	26	58	18	18	27	47
581	18	11	52	50	20	4	18	31	21	1	16	55	19	0	08	36
582	18	17	35	54	20	10	12	23	21	7	15	38	19	6	04	25
583	18	23	34	10	20	16	01	39	21	12	59	21	19	11	55	31
584	18	5	23	42	19	21	44	54	20	18	49	41	18	17	45	57
585	18	11	02	38	20	3	34	34	21	0	47	31	18	23	46	24
586	18	17	00	26	20	9	28	34	21	6	29	14	19	5	31	08
587	18	22	52	07	20	15	15	28	21	12	19	54	19	11	24	19
588	18	4	44	26	19	20	59	18	20	18	09	56	18	17	16	59
589	18	10	36	28	20	2	46	19	20	23	54	29	18	22	54	33
590	18	16	16	04	20	8	38	36	21	5	46	11	19	4	47	59
591	18	22	10	26	20	14	28	21	21	11	25	19	19	10	35	10
592	18	3	56	34	19	20	09	50	20	17	10	29	18	16	16	03
593	18	9	31	08	20	1	56	19	20	23	08	08	18	22	08	20
594	18	15	24	40	20	7	48	28	21	4	52	35	19	3	49	01
595	18	21	10	24	20	13	31	50	21	10	42	57	19	9	41	51
596	18	2	54	34	19	19	13	32	20	16	35	16	18	15	41	13
597	18	8	44	57	20	1	00	24	20	22	19	54	18	21	24	31
598	18	14	27	29	20	6	50	16	21	4	12	19	19	3	22	30
599	18	20	29	08	20	12	40	44	21	9	56	43	19	9	13	57
600	18	2	25	01	19	18	25	27	20	15	44	52	18	14	58	05

Year	March Equinox				June Solstice				September Equinox				December Solstice			
	d	h	m	s	d	h	m	s	d	h	m	s	d	h	m	s
601	18	8	06	25	20	0	17	01	20	21	44	14	18	20	55	50
602	18	14	04	39	20	6	17	01	21	3	29	42	19	2	42	37
603	18	19	55	58	20	12	06	48	21	9	18	54	19	8	34	50
604	18	1	41	45	19	17	52	04	20	15	12	42	18	14	30	24
605	18	7	35	09	19	23	43	06	20	21	01	26	18	20	09	06
606	18	13	16	34	20	5	31	11	21	2	54	51	19	2	00	25
607	18	19	08	52	20	11	19	00	21	8	40	45	19	7	53	07
608	18	0	57	53	19	17	00	45	20	14	25	42	18	13	37	54
609	18	6	30	53	19	22	44	02	20	20	18	53	18	19	34	01
610	18	12	25	21	20	4	37	36	21	2	01	09	19	1	19	43
611	18	18	18	20	20	10	20	24	21	7	46	30	19	7	06	22
612	18	0	02	22	19	15	59	42	20	13	37	20	18	12	59	27
613	18	5	55	28	19	21	51	14	20	19	23	31	18	18	40	44
614	18	11	36	54	20	3	39	06	21	1	10	25	19	0	30	30
615	18	17	27	22	20	9	28	28	21	6	54	54	19	6	23	52
616	17	23	21	29	19	15	15	50	20	12	42	08	18	12	05	54
617	18	4	57	45	19	21	01	10	20	18	39	34	18	17	58	52
618	18	10	53	50	20	2	59	51	21	0	31	11	18	23	48	40
619	18	16	48	14	20	8	47	23	21	6	21	06	19	5	41	09
620	17	22	30	49	19	14	29	28	20	12	15	49	18	11	42	55
621	18	4	26	16	19	20	25	01	20	18	05	36	18	17	31	01
622	18	10	14	16	20	2	14	17	20	23	54	28	18	23	22	32
623	18	16	10	29	20	8	03	16	21	5	42	52	19	5	16	14
624	17	22	08	27	19	13	51	57	20	11	30	53	18	10	58	35
625	18	3	45	12	19	19	36	39	20	17	23	08	18	16	50	19
626	18	9	36	40	20	1	34	23	20	23	09	55	18	22	38	38
627	18	15	28	27	20	7	20	55	21	4	53	14	19	4	24	05
628	17	21	06	24	19	12	57	38	20	10	43	39	18	10	15	48
629	18	2	56	06	19	18	48	27	20	16	32	40	18	15	55	35
630	18	8	38	18	20	0	32	02	20	22	19	17	18	21	42	19
631	18	14	25	24	20	6	14	54	21	4	05	17	19	3	36	42
632	17	20	16	05	19	12	00	56	20	9	51	21	18	9	23	41
633	18	1	52	31	19	17	43	04	20	15	40	52	18	15	17	13
634	18	7	46	10	19	23	38	44	20	21	29	24	18	21	08	27
635	18	13	47	13	20	5	28	25	21	3	17	02	19	2	56	52
636	17	19	32	27	19	11	08	04	20	9	09	20	18	8	51	32
637	18	1	27	16	19	17	07	43	20	15	02	51	18	14	40	01
638	18	7	17	11	19	23	01	12	20	20	50	50	18	20	29	48
639	18	13	07	16	20	4	50	01	21	2	39	18	19	2	24	51
640	17	19	02	46	19	10	43	12	20	8	32	05	18	8	10	46

Year	March Equinox				June Solstice				September Equinox				December Solstice			
	d	h	m	s	d	h	m	s	d	h	m	s	d	h	m	s
641	18	0	42	11	19	16	27	16	20	14	25	56	18	13	59	13
642	18	6	31	48	19	22	21	54	20	20	18	40	18	19	50	54
643	18	12	27	35	20	4	10	42	21	2	06	58	19	1	42	32
644	17	18	05	19	19	9	43	45	20	7	53	43	18	7	37	09
645	17	23	54	07	19	15	36	32	20	13	43	41	18	13	26	47
646	18	5	45	14	19	21	23	05	20	19	26	00	18	19	10	58
647	18	11	34	20	20	3	02	28	21	1	08	31	19	1	00	22
648	17	17	29	04	19	8	53	11	20	6	57	43	18	6	44	48
649	17	23	07	14	19	14	36	14	20	12	43	36	18	12	30	39
650	18	4	53	37	19	20	30	46	20	18	31	07	18	18	20	43
651	18	10	49	11	20	2	22	45	21	0	17	50	19	0	09	06
652	17	16	29	06	19	7	57	59	20	6	06	21	18	5	58	06
653	17	22	18	03	19	13	53	17	20	12	03	14	18	11	45	48
654	18	4	09	37	19	19	45	16	20	17	53	12	18	17	34	46
655	18	9	57	33	20	1	27	03	20	23	40	15	18	23	31	25
656	17	15	50	10	19	7	20	52	20	5	35	13	18	5	25	47
657	17	21	34	01	19	13	07	16	20	11	24	26	18	11	17	21
658	18	3	26	51	19	19	01	02	20	17	15	41	18	17	10	05
659	18	9	29	13	20	0	54	56	20	23	07	01	18	23	01	45
660	17	15	14	37	19	6	31	34	20	4	55	01	18	4	52	23
661	17	21	02	39	19	12	27	01	20	10	47	51	18	10	40	56
662	18	2	53	34	19	18	20	33	20	16	32	27	18	16	28	49
663	18	8	41	07	19	23	59	49	20	22	12	53	18	22	16	33
664	17	14	29	13	19	5	49	57	20	4	05	58	18	4	02	06
665	17	20	10	33	19	11	33	59	20	9	54	00	18	9	45	14
666	18	1	54	47	19	17	20	47	20	15	41	24	18	15	33	31
667	18	7	47	10	19	23	11	20	20	21	31	01	18	21	27	49
668	17	13	27	03	19	4	43	51	20	3	14	02	18	3	17	57
669	17	19	12	14	19	10	35	09	20	9	05	41	18	9	07	56
670	18	1	08	01	19	16	28	38	20	14	53	07	18	14	56	33
671	18	7	00	47	19	22	07	04	20	20	34	53	18	20	44	31
672	17	12	52	12	19	4	01	01	20	2	31	16	18	2	35	54
673	17	18	38	24	19	9	52	51	20	8	20	38	18	8	24	30
674	18	0	26	17	19	15	44	54	20	14	08	04	18	14	15	01
675	18	6	22	37	19	21	42	15	20	20	03	39	18	20	10	59
676	17	12	08	08	19	3	19	48	20	1	51	53	18	1	57	55
677	17	17	54	21	19	9	11	58	20	7	49	24	18	7	48	30
678	17	23	49	01	19	15	07	38	20	13	41	26	18	13	41	26
679	18	5	38	00	19	20	44	12	20	19	21	10	18	19	32	47
680	17	11	25	17	19	2	35	07	20	1	15	41	18	1	27	01

Year	March Equinox				June Solstice				September Equinox				December Solstice			
	d	h	m	s	d	h	m	s	d	h	m	s	d	h	m	s
681	17	17	12	05	19	8	24	01	20	7	01	54	18	7	14	49
682	17	23	02	36	19	14	09	45	20	12	44	54	18	12	59	46
683	18	4	58	22	19	20	02	25	20	18	37	34	18	18	51	57
684	17	10	43	35	19	1	39	49	20	0	20	50	18	0	36	33
685	17	16	26	10	19	7	29	59	20	6	10	32	18	6	22	36
686	17	22	15	37	19	13	26	29	20	11	59	15	18	12	12	59
687	18	4	04	21	19	19	03	12	20	17	36	44	18	17	56	52
688	17	9	48	08	19	0	51	28	19	23	33	20	17	23	44	39
689	17	15	32	52	19	6	41	36	20	5	25	12	18	5	32	28
690	17	21	20	26	19	12	26	34	20	11	10	31	18	11	20	30
691	18	3	10	34	19	18	19	00	20	17	06	10	18	17	21	14
692	17	8	57	16	18	23	58	57	19	22	52	52	17	23	14	52
693	17	14	45	35	19	5	48	03	20	4	43	35	18	5	03	59
694	17	20	42	14	19	11	46	01	20	10	37	44	18	10	59	41
695	18	2	42	07	19	17	27	10	20	16	18	12	18	16	46	34
696	17	8	30	17	18	23	18	23	19	22	14	06	17	22	38	04
697	17	14	18	20	19	5	15	35	20	4	05	37	18	4	29	43
698	17	20	08	52	19	11	03	32	20	9	46	21	18	10	13	27
699	18	1	58	34	19	16	56	53	20	15	41	42	18	16	07	55
700	17	7	44	59	18	22	37	32	19	21	29	32	17	21	52	53
701	17	13	28	21	19	4	22	58	20	3	19	34	18	3	35	34
702	17	19	15	25	19	10	17	06	20	9	13	40	18	9	31	02
703	18	1	05	45	19	15	53	17	20	14	49	56	18	15	18	12
704	17	6	46	49	18	21	36	42	19	20	40	30	17	21	08	45
705	17	12	32	15	19	3	28	24	20	2	30	41	18	2	59	45
706	17	18	25	45	19	9	11	25	20	8	09	08	18	8	40	25
707	18	0	16	38	19	15	01	38	20	14	04	10	18	14	35	58
708	17	6	04	15	18	20	45	14	19	19	51	01	17	20	24	45
709	17	11	49	42	19	2	33	52	20	1	37	34	18	2	10	03
710	17	17	39	47	19	8	33	05	20	7	33	03	18	8	06	54
711	17	23	35	47	19	14	15	42	20	13	13	32	18	13	53	12
712	17	5	22	16	18	20	03	10	19	19	10	18	17	19	41	34
713	17	11	08	37	19	1	59	37	20	1	09	01	18	1	36	31
714	17	17	03	13	19	7	47	21	20	6	51	47	18	7	23	10
715	17	22	51	35	19	13	37	32	20	12	47	22	18	13	23	56
716	17	4	39	21	18	19	23	00	19	18	37	03	17	19	18	26
717	17	10	30	05	19	1	10	01	20	0	22	33	18	1	00	56
718	17	16	21	25	19	7	05	12	20	6	18	20	18	6	55	35
719	17	22	19	27	19	12	47	12	20	11	57	19	18	12	40	56
720	17	4	03	09	18	18	31	12	19	17	46	23	17	18	25	30

Year	March Equinox				June Solstice				September Equinox				December Solstice			
	d	h	m	s	d	h	m	s	d	h	m	s	d	h	m	s
721	17	9	43	06	19	0	25	47	19	23	39	04	18	0	18	08
722	17	15	34	57	19	6	12	02	20	5	15	58	18	5	57	59
723	17	21	18	41	19	11	56	26	20	11	06	49	18	11	48	31
724	17	3	01	30	18	17	38	48	19	16	58	42	17	17	38	54
725	17	8	49	08	18	23	22	40	19	22	43	23	17	23	18	01
726	17	14	32	29	19	5	13	55	20	4	38	11	18	5	16	48
727	17	20	27	51	19	10	57	01	20	10	18	40	18	11	09	36
728	17	2	13	52	18	16	40	17	19	16	07	08	17	16	57	30
729	17	7	59	52	18	22	36	32	19	22	04	46	17	22	54	44
730	17	14	02	37	19	4	27	22	20	3	46	35	18	4	38	14
731	17	19	54	58	19	10	16	44	20	9	40	43	18	10	33	17
732	17	1	42	27	18	16	07	41	19	15	36	11	17	16	29	40
733	17	7	35	13	18	22	00	29	19	21	21	35	17	22	11	27
734	17	13	22	03	19	3	56	11	20	3	17	29	18	4	08	07
735	17	19	18	10	19	9	42	06	20	9	02	07	18	9	56	54
736	17	1	04	17	18	15	24	40	19	14	52	19	17	15	38	01
737	17	6	43	11	18	21	16	56	19	20	51	09	17	21	33	10
738	17	12	36	56	19	3	03	51	20	2	30	03	18	3	18	06
739	17	18	20	54	19	8	44	52	20	8	17	31	18	9	12	34
740	17	0	01	15	18	14	27	08	19	14	07	49	17	15	07	05
741	17	5	53	56	18	20	12	54	19	19	47	59	17	20	45	03
742	17	11	42	41	19	2	01	36	20	1	39	21	18	2	37	02
743	17	17	37	48	19	7	45	46	20	7	21	32	18	8	26	48
744	16	23	25	33	18	13	30	58	19	13	06	58	17	14	07	55
745	17	5	03	45	18	19	25	28	19	19	02	39	17	20	02	50
746	17	11	00	41	19	1	18	57	20	0	44	21	18	1	47	42
747	17	16	49	36	19	7	03	59	20	6	35	21	18	7	37	02
748	16	22	31	12	18	12	51	20	19	12	36	02	17	13	33	39
749	17	4	26	27	18	18	44	07	19	18	24	46	17	19	17	20
750	17	10	13	01	19	0	34	39	20	0	19	49	18	1	16	55
751	17	16	06	54	19	6	21	25	20	6	07	18	18	7	16	30
752	16	21	59	27	18	12	07	38	19	11	54	24	17	13	00	42
753	17	3	42	27	18	17	59	27	19	17	50	51	17	18	55	49
754	17	9	44	16	18	23	52	21	19	23	34	39	18	0	42	09
755	17	15	35	11	19	5	35	35	20	5	20	13	18	6	28	50
756	16	21	12	35	18	11	20	52	19	11	14	49	17	12	25	15
757	17	3	06	06	18	17	13	38	19	16	56	22	17	18	04	11
758	17	8	49	31	18	22	59	30	19	22	43	21	17	23	53	27
759	17	14	38	56	19	4	42	33	20	4	29	55	18	5	44	23
760	16	20	27	49	18	10	25	43	19	10	16	06	17	11	21	12

Year	March Equinox				June Solstice				September Equinox				December Solstice			
	d	h	m	s	d	h	m	s	d	h	m	s	d	h	m	s
761	17	2	03	39	18	16	13	08	19	16	10	46	17	17	14	25
762	17	7	57	10	18	22	03	49	19	21	53	47	17	23	04	56
763	17	13	45	55	19	3	46	18	20	3	38	11	18	4	54	52
764	16	19	24	20	18	9	30	10	19	9	34	14	17	10	52	59
765	17	1	23	29	18	15	25	27	19	15	21	22	17	16	36	02
766	17	7	16	19	18	21	14	28	19	21	12	20	17	22	28	11
767	17	13	09	21	19	3	02	14	20	3	03	17	18	4	26	09
768	16	19	03	44	18	8	54	32	19	8	51	37	17	10	09	40
769	17	0	44	17	18	14	47	23	19	14	46	43	17	16	04	13
770	17	6	40	28	18	20	42	49	19	20	33	40	17	21	54	22
771	17	12	33	51	19	2	28	31	20	2	22	18	18	3	40	38
772	16	18	11	07	18	8	10	32	19	8	20	40	17	9	34	56
773	17	0	04	12	18	14	04	29	19	14	08	45	17	15	20	41
774	17	5	52	19	18	19	49	58	19	19	55	14	17	21	13	09
775	17	11	37	42	19	1	30	21	20	1	41	27	18	3	11	08
776	16	17	31	11	18	7	18	09	19	7	26	00	17	8	52	12
777	16	23	12	58	18	13	03	08	19	13	15	01	17	14	39	42
778	17	5	08	12	18	18	54	16	19	18	59	43	17	20	27	56
779	17	11	01	33	19	0	40	04	20	0	42	53	18	2	11	28
780	16	16	35	17	18	6	22	14	19	6	35	50	17	8	04	33
781	16	22	26	55	18	12	19	43	19	12	23	24	17	13	49	22
782	17	4	16	16	18	18	07	15	19	18	10	04	17	19	36	32
783	17	10	01	26	18	23	48	10	20	0	01	46	18	1	31	53
784	16	15	54	57	18	5	39	28	19	5	54	28	17	7	15	39
785	16	21	35	39	18	11	26	20	19	11	46	59	17	13	08	34
786	17	3	28	05	18	17	18	43	19	17	36	29	17	19	07	36
787	17	9	24	19	18	23	06	49	19	23	22	04	18	0	57	59
788	16	15	03	49	18	4	47	58	19	5	16	06	17	6	54	32
789	16	21	03	45	18	10	46	45	19	11	07	58	17	12	42	02
790	17	3	01	12	18	16	36	39	19	16	54	39	17	18	30	04
791	17	8	48	49	18	22	19	16	19	22	43	41	18	0	26	31
792	16	14	41	40	18	4	13	55	19	4	32	49	17	6	10	25
793	16	20	23	01	18	10	02	12	19	10	19	50	17	11	56	40
794	17	2	12	12	18	15	51	47	19	16	06	11	17	17	46	14
795	17	8	05	31	18	21	38	52	19	21	52	51	17	23	28	55
796	16	13	39	24	18	3	15	32	19	3	45	21	17	5	17	57
797	16	19	26	58	18	9	09	29	19	9	36	08	17	11	06	36
798	17	1	17	26	18	14	56	16	19	15	19	54	17	16	56	56
799	17	6	59	30	18	20	31	52	19	21	05	22	17	22	53	31
800	16	12	52	02	18	2	23	26	19	2	55	40	17	4	39	19

Year	March Equinox				June Solstice				September Equinox				December Solstice			
	d	h	m	s	d	h	m	s	d	h	m	s	d	h	m	s
801	16	18	40	01	18	8	10	28	19	8	43	58	17	10	25	26
802	17	0	32	24	18	13	59	16	19	14	31	08	17	16	18	00
803	17	6	29	17	18	19	52	05	19	20	19	42	17	22	08	27
804	16	12	08	35	18	1	34	25	19	2	10	38	17	3	59	26
805	16	17	58	59	18	7	33	37	19	8	03	25	17	9	50	08
806	16	23	57	21	18	13	28	11	19	13	52	32	17	15	39	15
807	17	5	43	10	18	19	05	56	19	19	42	59	17	21	32	42
808	16	11	35	02	18	1	01	13	19	1	40	39	17	3	21	31
809	16	17	21	24	18	6	50	18	19	7	29	39	17	9	10	51
810	16	23	09	09	18	12	36	36	19	13	15	28	17	15	07	21
811	17	5	03	58	18	18	27	14	19	19	02	58	17	20	58	29
812	16	10	44	32	18	0	05	12	19	0	50	47	17	2	45	21
813	16	16	35	33	18	5	58	47	19	6	41	41	17	8	32	19
814	16	22	33	02	18	11	50	39	19	12	26	59	17	14	19	17
815	17	4	15	20	18	17	25	04	19	18	09	26	17	20	09	33
816	16	10	01	21	17	23	18	42	19	0	01	52	17	1	55	57
817	16	15	45	57	18	5	07	50	19	5	46	14	17	7	39	10
818	16	21	32	02	18	10	50	43	19	11	30	56	17	13	28	59
819	17	3	24	12	18	16	40	07	19	17	22	26	17	19	16	37
820	16	9	02	43	17	22	18	12	18	23	12	26	17	1	04	45
821	16	14	50	13	18	4	11	43	19	5	05	21	17	6	58	17
822	16	20	46	49	18	10	06	05	19	10	53	45	17	12	54	30
823	17	2	34	37	18	15	42	33	19	16	38	26	17	18	49	27
824	16	8	27	37	17	21	37	57	18	22	35	43	17	0	40	04
825	16	14	23	41	18	3	33	12	19	4	26	01	17	6	27	34
826	16	20	17	15	18	9	19	49	19	10	11	18	17	12	19	30
827	17	2	11	17	18	15	16	09	19	16	05	01	17	18	12	36
828	16	7	53	43	17	21	01	05	18	21	53	43	16	23	59	36
829	16	13	39	52	18	2	54	41	19	3	44	17	17	5	47	32
830	16	19	35	13	18	8	49	05	19	9	34	25	17	11	37	41
831	17	1	19	38	18	14	21	18	19	15	18	38	17	17	23	43
832	16	7	02	29	17	20	10	18	18	21	14	14	16	23	11	24
833	16	12	49	04	18	2	00	48	19	3	01	24	17	5	00	24
834	16	18	34	06	18	7	37	58	19	8	38	43	17	10	50	19
835	17	0	21	58	18	13	26	11	19	14	28	47	17	16	43	48
836	16	6	07	40	17	19	06	47	18	20	14	42	16	22	27	33
837	16	11	56	08	18	0	55	16	19	2	02	18	17	4	13	40
838	16	17	53	56	18	6	52	10	19	7	53	03	17	10	07	55
839	16	23	41	37	18	12	29	15	19	13	34	56	17	15	56	33
840	16	5	26	53	17	18	23	49	18	19	30	51	16	21	46	36

Year	March Equinox				June Solstice				September Equinox				December Solstice			
	d	h	m	s	d	h	m	s	d	h	m	s	d	h	m	s
841	16	11	18	55	18	0	22	26	19	1	22	12	17	3	36	07
842	16	17	10	45	18	6	05	40	19	7	05	45	17	9	24	01
843	16	23	01	15	18	11	58	52	19	13	05	27	17	15	18	10
844	16	4	46	56	17	17	45	46	18	18	58	50	16	21	07	57
845	16	10	33	49	17	23	35	32	19	0	48	28	17	3	00	19
846	16	16	28	15	18	5	31	44	19	6	41	33	17	9	00	44
847	16	22	18	24	18	11	07	17	19	12	22	53	17	14	50	02
848	16	4	06	20	17	16	56	04	18	18	17	12	16	20	37	07
849	16	9	59	43	17	22	51	16	19	0	06	48	17	2	24	56
850	16	15	51	26	18	4	31	06	19	5	43	25	17	8	10	25
851	16	21	36	39	18	10	20	17	19	11	34	38	17	14	00	33
852	16	3	17	57	17	16	06	17	18	17	20	47	16	19	45	33
853	16	9	03	21	17	21	52	58	18	23	04	03	17	1	27	50
854	16	14	53	12	18	3	45	34	19	4	57	11	17	7	20	58
855	16	20	41	00	18	9	20	57	19	10	40	57	17	13	06	23
856	16	2	23	03	17	15	07	07	18	16	34	43	16	18	54	12
857	16	8	11	06	17	21	03	40	18	22	27	20	17	0	50	15
858	16	14	04	09	18	2	44	47	19	4	04	32	17	6	40	01
859	16	19	52	30	18	8	34	34	19	9	59	45	17	12	35	03
860	16	1	43	29	17	14	25	43	18	15	53	45	16	18	25	07
861	16	7	37	51	17	20	16	02	18	21	40	47	17	0	10	28
862	16	13	32	18	18	2	14	19	19	3	37	41	17	6	10	37
863	16	19	24	29	18	7	57	32	19	9	23	07	17	12	00	44
864	16	1	09	09	17	13	47	02	18	15	15	04	16	17	47	11
865	16	6	58	18	17	19	45	53	18	21	10	32	16	23	40	55
866	16	12	52	50	18	1	26	13	19	2	48	17	17	5	23	03
867	16	18	35	54	18	7	10	38	19	8	42	46	17	11	13	35
868	16	0	19	33	17	12	58	43	18	14	35	19	16	17	04	01
869	16	6	05	43	17	18	42	30	18	20	14	56	16	22	48	16
870	16	11	52	58	18	0	33	07	19	2	06	16	17	4	47	22
871	16	17	43	16	18	6	10	44	19	7	47	41	17	10	34	39
872	15	23	30	12	17	11	54	22	18	13	35	33	16	16	15	52
873	16	5	19	27	17	17	50	32	18	19	29	13	16	22	08	29
874	16	11	15	27	17	23	34	04	19	1	04	56	17	3	52	49
875	16	16	58	55	18	5	21	09	19	6	56	25	17	9	42	58
876	15	22	42	06	17	11	14	44	18	12	51	05	16	15	34	38
877	16	4	33	48	17	17	04	13	18	18	34	13	16	21	16	20
878	16	10	23	15	17	22	56	56	19	0	32	42	17	3	13	54
879	16	16	14	18	18	4	39	51	19	6	23	49	17	9	06	03
880	15	22	01	21	17	10	26	31	18	12	16	15	16	14	53	35

Year	March Equinox				June Solstice				September Equinox				December Solstice			
	d	h	m	s	d	h	m	s	d	h	m	s	d	h	m	s
881	16	3	47	33	17	16	23	18	18	18	13	05	16	20	54	46
882	16	9	45	55	17	22	07	59	18	23	50	54	17	2	46	27
883	16	15	35	12	18	3	51	56	19	5	41	12	17	8	35	49
884	15	21	22	18	17	9	43	36	18	11	37	01	16	14	27	54
885	16	3	20	01	17	15	34	04	18	17	17	31	16	20	08	14
886	16	9	08	08	17	21	24	48	18	23	09	05	17	2	03	47
887	16	14	55	47	18	3	09	14	19	4	55	29	17	7	54	07
888	15	20	40	59	17	8	54	50	18	10	40	06	16	13	32	14
889	16	2	23	06	17	14	48	28	18	16	34	43	16	19	24	29
890	16	8	18	11	17	20	31	17	18	22	13	28	17	1	07	49
891	16	14	00	31	18	2	11	07	19	4	03	15	17	6	52	40
892	15	19	38	25	17	8	00	19	18	10	00	11	16	12	48	19
893	16	1	30	32	17	13	48	47	18	15	39	39	16	18	32	54
894	16	7	17	02	17	19	35	30	18	21	30	06	17	0	31	16
895	16	13	07	43	18	1	19	14	19	3	21	07	17	6	25	36
896	15	19	01	29	17	7	06	21	18	9	08	30	16	12	05	34
897	16	0	49	40	17	13	02	24	18	15	06	16	16	18	04	01
898	16	6	49	39	17	18	51	54	18	20	46	55	16	23	54	24
899	16	12	36	42	18	0	37	31	19	2	35	25	17	5	41	57
900	15	18	19	01	17	6	32	36	18	8	34	38	16	11	37	54
901	16	0	16	24	17	12	26	05	18	14	17	26	16	17	19	36
902	16	6	05	35	17	18	13	32	18	20	10	48	16	23	13	04
903	16	11	51	50	17	23	57	11	19	2	05	17	17	5	07	42
904	15	17	41	17	17	5	44	22	18	7	51	14	16	10	49	18
905	15	23	22	51	17	11	34	48	18	13	43	46	16	16	47	03
906	16	5	18	12	17	17	19	58	18	19	22	29	16	22	37	58
907	16	11	06	53	17	22	59	59	19	1	07	08	17	4	19	11
908	15	16	47	15	17	4	48	13	18	7	03	38	16	10	10	52
909	15	22	44	10	17	10	40	27	18	12	43	20	16	15	52	14
910	16	4	31	02	17	16	25	46	18	18	29	51	16	21	43	21
911	16	10	12	31	17	22	09	43	19	0	21	21	17	3	37	38
912	15	16	03	17	17	3	59	51	18	6	06	44	16	9	16	20
913	15	21	45	43	17	9	49	20	18	12	00	03	16	15	08	23
914	16	3	40	24	17	15	36	02	18	17	46	03	16	21	01	04
915	16	9	30	47	17	21	19	21	18	23	35	12	17	2	45	09
916	15	15	07	57	17	3	08	40	18	5	34	25	16	8	45	00
917	15	21	06	13	17	9	05	15	18	11	18	30	16	14	36	17
918	16	2	59	07	17	14	51	37	18	17	06	14	16	20	30	52
919	16	8	47	21	17	20	36	56	18	23	02	44	17	2	28	23
920	15	14	47	19	17	2	30	52	18	4	50	53	16	8	08	23

Year	March Equinox				June Solstice				September Equinox				December Solstice			
	d	h	m	s	d	h	m	s	d	h	m	s	d	h	m	s
921	15	20	34	52	17	8	22	48	18	10	43	00	16	14	01	34
922	16	2	29	12	17	14	12	50	18	16	27	41	16	19	55	08
923	16	8	18	31	17	19	58	59	18	22	12	36	17	1	35	16
924	15	13	53	28	17	1	46	31	18	4	07	59	16	7	27	02
925	15	19	47	12	17	7	39	51	18	9	51	57	16	13	10	14
926	16	1	35	06	17	13	20	59	18	15	38	02	16	18	56	20
927	16	7	12	24	17	18	59	22	18	21	33	12	17	0	52	03
928	15	13	02	19	17	0	48	36	18	3	17	46	16	6	34	21
929	15	18	44	02	17	6	33	24	18	9	04	21	16	12	29	04
930	16	0	36	02	17	12	17	09	18	14	47	00	16	18	23	31
931	16	6	30	04	17	18	01	15	18	20	31	40	17	0	03	44
932	15	12	11	59	16	23	48	44	18	2	27	23	16	5	56	15
933	15	18	09	34	17	5	46	45	18	8	13	28	16	11	45	49
934	16	0	03	51	17	11	36	19	18	13	59	52	16	17	36	17
935	16	5	44	46	17	17	21	04	18	19	56	17	16	23	34	02
936	15	11	40	55	16	23	19	38	18	1	47	07	16	5	17	45
937	15	17	29	50	17	5	09	51	18	7	38	58	16	11	07	52
938	15	23	21	45	17	10	56	55	18	13	30	10	16	17	03	44
939	16	5	15	14	17	16	45	28	18	19	20	22	16	22	48	00
940	15	10	51	47	16	22	30	47	18	1	14	41	16	4	43	20
941	15	16	44	00	17	4	25	24	18	7	00	11	16	10	36	32
942	15	22	38	43	17	10	10	13	18	12	43	24	16	16	24	15
943	16	4	19	01	17	15	46	41	18	18	35	42	16	22	16	48
944	15	10	14	29	16	21	40	46	18	0	23	58	16	3	59	11
945	15	16	01	55	17	3	26	48	18	6	07	01	16	9	44	56
946	15	21	47	21	17	9	09	40	18	11	50	07	16	15	39	08
947	16	3	39	09	17	14	59	00	18	17	35	17	16	21	19	07
948	15	9	14	56	16	20	42	30	17	23	25	08	16	3	06	00
949	15	15	05	46	17	2	36	48	18	5	13	58	16	8	54	47
950	15	21	00	14	17	8	23	28	18	11	01	20	16	14	41	24
951	16	2	37	26	17	14	01	06	18	16	56	59	16	20	38	19
952	15	8	29	47	16	19	58	18	17	22	49	36	16	2	29	38
953	15	14	20	59	17	1	48	23	18	4	36	25	16	8	22	49
954	15	20	12	43	17	7	32	29	18	10	24	33	16	14	21	29
955	16	2	12	27	17	13	25	45	18	16	16	59	16	20	06	21
956	15	7	57	54	16	19	13	36	17	22	09	43	16	1	56	04
957	15	13	51	22	17	1	11	18	18	3	59	29	16	7	49	30
958	15	19	47	55	17	7	03	29	18	9	45	29	16	13	38	20
959	16	1	25	36	17	12	41	54	18	15	37	39	16	19	31	01
960	15	7	15	20	16	18	37	38	17	21	29	13	16	1	15	44

Year	March Equinox				June Solstice				September Equinox				December Solstice			
	d	h	m	s	d	h	m	s	d	h	m	s	d	h	m	s
961	15	13	05	15	17	0	24	50	18	3	15	00	16	7	01	14
962	15	18	50	32	17	6	01	27	18	9	00	06	16	12	53	06
963	16	0	39	17	17	11	49	06	18	14	49	09	16	18	38	09
964	15	6	17	49	16	17	31	38	17	20	35	07	16	0	26	49
965	15	12	04	29	16	23	21	07	18	2	19	07	16	6	19	28
966	15	18	01	41	17	5	09	56	18	8	03	39	16	12	07	07
967	15	23	43	21	17	10	44	39	18	13	53	28	16	17	55	45
968	15	5	35	22	16	16	41	43	17	19	46	55	15	23	43	22
969	15	11	29	49	16	22	35	21	18	1	32	49	16	5	32	00
970	15	17	16	19	17	4	16	58	18	7	17	57	16	11	26	57
971	15	23	08	30	17	10	12	45	18	13	12	25	16	17	15	02
972	15	4	53	22	16	16	01	55	17	19	03	40	15	23	01	49
973	15	10	43	11	16	21	54	20	18	0	55	10	16	4	54	29
974	15	16	41	14	17	3	47	41	18	6	48	39	16	10	46	49
975	15	22	20	57	17	9	24	05	18	12	40	32	16	16	40	05
976	15	4	08	47	16	15	19	58	17	18	35	21	15	22	35	14
977	15	10	04	15	16	21	11	46	18	0	19	50	16	4	26	02
978	15	15	52	44	17	2	46	47	18	6	01	02	16	10	18	18
979	15	21	47	28	17	8	38	38	18	11	54	35	16	16	03	50
980	15	3	34	17	16	14	25	55	17	17	40	17	15	21	47	10
981	15	9	21	15	16	20	15	19	17	23	23	37	16	3	36	42
982	15	15	14	25	17	2	07	58	18	5	10	37	16	9	25	32
983	15	20	52	20	17	7	44	04	18	10	56	48	16	15	10	29
984	15	2	36	45	16	13	37	03	17	16	49	38	15	20	56	13
985	15	8	29	28	16	19	29	15	17	22	38	04	16	2	43	06
986	15	14	14	28	17	1	02	57	18	4	21	25	16	8	33	49
987	15	20	00	30	17	6	53	36	18	10	17	53	16	14	25	34
988	15	1	45	38	16	12	42	43	17	16	05	35	15	20	16	34
989	15	7	34	32	16	18	30	05	17	21	49	59	16	2	10	38
990	15	13	33	10	17	0	23	43	18	3	42	42	16	8	04	56
991	15	19	21	51	17	6	03	38	18	9	33	47	16	13	53	38
992	15	1	12	27	16	11	59	21	17	15	27	54	15	19	43	50
993	15	7	09	27	16	17	58	19	17	21	17	44	16	1	38	16
994	15	12	59	50	16	23	37	11	18	2	58	48	16	7	28	30
995	15	18	46	48	17	5	30	15	18	8	55	03	16	13	17	18
996	15	0	35	35	16	11	23	01	17	14	44	49	15	19	02	42
997	15	6	23	28	16	17	07	18	17	20	28	43	16	0	49	17
998	15	12	14	49	16	22	58	26	18	2	22	47	16	6	42	32
999	15	17	56	03	17	4	35	35	18	8	09	42	16	12	30	17
1000	14	23	38	23	16	10	24	42	17	13	58	18	15	18	20	12

Year	March Equinox				June Solstice				September Equinox				December Solstice			
	d	h	m	s	d	h	m	s	d	h	m	s	d	h	m	s
1001	15	5	31	34	16	16	18	29	17	19	45	27	16	0	13	06
1002	15	11	21	51	16	21	51	22	18	1	23	10	16	5	58	29
1003	15	17	09	11	17	3	39	50	18	7	18	20	16	11	45	12
1004	14	22	58	11	16	9	33	19	17	13	06	48	15	17	32	09
1005	15	4	45	55	16	15	18	34	17	18	46	11	15	23	19	40
1006	15	10	35	57	16	21	12	22	18	0	40	28	16	5	14	45
1007	15	16	20	19	17	2	54	39	18	6	29	07	16	11	00	55
1008	14	22	06	17	16	8	45	28	17	12	21	40	15	16	48	55
1009	15	4	01	45	16	14	42	23	17	18	17	00	15	22	43	28
1010	15	9	52	47	16	20	18	50	17	23	59	31	16	4	33	38
1011	15	15	38	52	17	2	09	22	18	5	57	34	16	10	27	53
1012	14	21	28	07	16	8	05	24	17	11	48	55	15	16	22	43
1013	15	3	21	24	16	13	50	39	17	17	28	28	15	22	11	48
1014	15	9	15	44	16	19	42	10	17	23	24	00	16	4	06	17
1015	15	15	05	58	17	1	26	23	18	5	13	48	16	9	52	33
1016	14	20	53	51	16	7	16	30	17	11	00	50	15	15	37	28
1017	15	2	44	52	16	13	14	01	17	16	52	17	15	21	32	07
1018	15	8	34	24	16	18	51	38	17	22	29	37	16	3	16	47
1019	15	14	15	38	17	0	37	42	18	4	22	25	16	9	01	12
1020	14	20	00	21	16	6	30	58	17	10	14	07	15	14	48	49
1021	15	1	49	21	16	12	11	35	17	15	52	50	15	20	30	59
1022	15	7	33	36	16	17	57	18	17	21	47	19	16	2	25	22
1023	15	13	16	18	16	23	39	39	18	3	37	04	16	8	17	35
1024	14	19	01	32	16	5	25	05	17	9	20	14	15	14	04	21
1025	15	0	52	57	16	11	19	11	17	15	12	40	15	20	03	30
1026	15	6	51	55	16	16	58	41	17	20	53	29	16	1	50	26
1027	15	12	41	02	16	22	46	24	18	2	47	58	16	7	37	22
1028	14	18	32	03	16	4	47	43	17	8	43	46	15	13	32	40
1029	15	0	27	50	16	10	37	02	17	14	22	42	15	19	18	55
1030	15	6	16	21	16	16	29	26	17	20	19	02	16	1	14	42
1031	15	12	03	39	16	22	19	18	18	2	13	32	16	7	05	14
1032	14	17	53	13	16	4	08	40	17	8	01	10	15	12	47	29
1033	14	23	43	01	16	10	03	16	17	13	59	02	15	18	44	51
1034	15	5	35	39	16	15	42	38	17	19	41	37	16	0	33	51
1035	15	11	17	37	16	21	25	38	18	1	31	53	16	6	21	34
1036	14	17	01	06	16	3	20	11	17	7	24	23	15	12	17	22
1037	14	22	55	42	16	9	02	48	17	12	58	15	15	17	59	48
1038	15	4	43	44	16	14	46	01	17	18	50	13	15	23	49	56
1039	15	10	29	06	16	20	31	20	18	0	41	22	16	5	38	13
1040	14	16	17	26	16	2	18	30	17	6	21	51	15	11	19	22

Year	March Equinox			June Solstice			September Equinox			December Solstice		
	d	h m s		d	h m s		d	h m s		d	h m s	
1041	14	22 04 09		16	8 11 13		17	12 13 32		15	17 15 15	
1042	15	3 55 16		16	13 52 31		17	17 53 50		15	23 02 41	
1043	15	9 40 27		16	19 37 31		17	23 43 44		16	4 44 44	
1044	14	15 24 40		16	1 33 43		17	5 41 48		15	10 38 46	
1045	14	21 21 46		16	7 21 31		17	11 22 57		15	16 24 01	
1046	15	3 08 29		16	13 07 03		17	17 18 16		15	22 19 23	
1047	15	8 52 11		16	18 57 30		17	23 15 54		16	4 18 45	
1048	14	14 45 50		16	0 48 46		17	4 59 27		15	10 05 08	
1049	14	20 37 34		16	6 41 40		17	10 54 56		15	16 04 50	
1050	15	2 36 41		16	12 26 01		17	16 41 28		15	21 55 54	
1051	15	8 28 11		16	18 12 25		17	22 31 10		16	3 37 54	
1052	14	14 12 23		16	0 09 56		17	4 27 01		15	9 34 08	
1053	14	20 09 10		16	5 59 57		17	10 04 15		15	15 18 39	
1054	15	1 53 22		16	11 42 46		17	15 52 28		15	21 05 44	
1055	15	7 32 35		16	17 29 53		17	21 48 27		16	2 57 44	
1056	14	13 23 30		15	23 16 43		17	3 29 12		15	8 33 41	
1057	14	19 06 35		16	5 01 33		17	9 21 00		15	14 27 56	
1058	15	0 55 52		16	10 41 52		17	15 05 53		15	20 20 32	
1059	15	6 40 31		16	16 23 25		17	20 50 04		16	2 03 03	
1060	14	12 21 08		15	22 15 57		17	2 43 19		15	8 00 49	
1061	14	18 20 59		16	4 04 14		17	8 21 05		15	13 46 28	
1062	15	0 12 24		16	9 47 09		17	14 10 40		15	19 33 20	
1063	15	5 56 17		16	15 38 18		17	20 09 48		16	1 29 16	
1064	14	11 52 44		15	21 34 36		17	1 53 23		15	7 12 17	
1065	14	17 41 02		16	3 26 46		17	7 46 48		15	13 09 46	
1066	14	23 32 45		16	9 14 28		17	13 37 38		15	19 04 54	
1067	15	5 24 20		16	15 02 56		17	19 27 33		16	0 45 25	
1068	14	11 07 16		15	20 56 50		17	1 27 09		15	6 41 26	
1069	14	17 05 08		16	2 47 00		17	7 10 53		15	12 30 31	
1070	14	22 52 59		16	8 28 46		17	13 00 16		15	18 20 21	
1071	15	4 29 50		16	14 14 55		17	18 56 33		16	0 18 47	
1072	14	10 24 11		15	20 06 26		17	0 36 11		15	6 02 43	
1073	14	16 14 29		16	1 50 29		17	6 23 27		15	11 54 05	
1074	14	22 04 41		16	7 31 22		17	12 11 08		15	17 45 27	
1075	15	3 57 16		16	13 18 29		17	17 55 16		15	23 22 38	
1076	14	9 36 32		15	19 09 06		16	23 46 01		15	5 15 30	
1077	14	15 30 29		16	1 00 01		17	5 25 34		15	11 03 13	
1078	14	21 18 12		16	6 42 24		17	11 10 31		15	16 45 16	
1079	15	2 53 41		16	12 28 41		17	17 08 46		15	22 38 31	
1080	14	8 48 42		15	18 23 10		16	22 54 39		15	4 20 21	

Year	March Equinox				June Solstice				September Equinox				December Solstice			
	d	h	m	s	d	h	m	s	d	h	m	s	d	h	m	s
1081	14	14	36	28	16	0	08	16	17	4	45	45	15	10	13	30
1082	14	20	22	01	16	5	51	18	17	10	39	04	15	16	13	57
1083	15	2	15	01	16	11	41	39	17	16	27	02	15	21	59	32
1084	14	7	58	33	15	17	32	17	16	22	20	13	15	3	57	37
1085	14	13	59	32	15	23	24	37	17	4	06	31	15	9	50	58
1086	14	19	57	02	16	5	09	55	17	9	54	33	15	15	34	48
1087	15	1	37	20	16	10	58	25	17	15	53	05	15	21	32	36
1088	14	7	35	11	15	16	57	28	16	21	37	26	15	3	17	43
1089	14	13	24	22	15	22	44	30	17	3	23	04	15	9	07	45
1090	14	19	09	08	16	4	27	50	17	9	14	31	15	15	02	15
1091	15	1	01	32	16	10	17	52	17	15	01	43	15	20	39	35
1092	14	6	41	51	15	16	04	20	16	20	53	07	15	2	29	07
1093	14	12	32	51	15	21	51	25	17	2	37	59	15	8	20	20
1094	14	18	21	48	16	3	33	31	17	8	22	19	15	14	04	24
1095	14	23	54	32	16	9	15	08	17	14	14	59	15	20	00	29
1096	14	5	48	18	15	15	09	40	16	19	58	17	15	1	46	42
1097	14	11	41	53	15	20	53	31	17	1	43	04	15	7	33	58
1098	14	17	27	51	16	2	33	44	17	7	35	19	15	13	28	25
1099	14	23	23	03	16	8	28	02	17	13	23	57	15	19	10	44
1100	14	5	06	27	15	14	18	41	16	19	13	48	15	1	02	56
1101	14	10	58	12	15	20	10	15	17	1	00	32	15	6	57	25
1102	14	16	52	53	16	1	59	17	17	6	49	35	15	12	42	23
1103	14	22	30	30	16	7	43	16	17	12	46	28	15	18	35	17
1104	14	4	25	02	15	13	40	43	16	18	37	36	15	0	24	52
1105	14	10	20	37	15	19	28	31	17	0	26	12	15	6	16	03
1106	14	16	02	22	16	1	07	43	17	6	18	34	15	12	16	08
1107	14	21	56	41	16	7	02	47	17	12	08	06	15	18	03	52
1108	14	3	43	25	15	12	50	18	16	17	54	29	14	23	53	55
1109	14	9	37	30	15	18	37	23	16	23	40	32	15	5	45	58
1110	14	15	35	04	16	0	25	19	17	5	27	06	15	11	27	22
1111	14	21	11	36	16	6	07	52	17	11	17	40	15	17	17	22
1112	14	3	01	27	15	12	04	47	16	17	03	50	14	23	04	59
1113	14	8	53	20	15	17	52	18	16	22	46	48	15	4	50	09
1114	14	14	31	34	15	23	28	12	17	4	35	46	15	10	41	45
1115	14	20	22	13	16	5	21	09	17	10	27	33	15	16	23	34
1116	14	2	06	16	15	11	06	56	16	16	15	14	14	22	09	55
1117	14	7	53	24	15	16	51	02	16	22	03	16	15	4	05	54
1118	14	13	45	43	15	22	39	05	17	3	51	25	15	9	54	24
1119	14	19	22	36	16	4	20	34	17	9	41	40	15	15	50	09
1120	14	1	17	13	15	10	16	49	16	15	31	00	14	21	42	34

Year	March Equinox				June Solstice				September Equinox				December Solstice			
	d	h	m	s	d	h	m	s	d	h	m	s	d	h	m	s
1121	14	7	18	45	15	16	07	12	16	21	18	10	15	3	31	29
1122	14	13	06	26	15	21	46	57	17	3	10	01	15	9	26	33
1123	14	19	01	56	16	3	46	46	17	9	04	24	15	15	15	06
1124	14	0	52	18	15	9	41	13	16	14	51	54	14	21	04	18
1125	14	6	41	52	15	15	28	49	16	20	38	31	15	2	58	00
1126	14	12	36	14	15	21	21	34	17	2	30	15	15	8	43	03
1127	14	18	15	02	16	3	03	50	17	8	21	47	15	14	29	37
1128	14	0	02	39	15	8	56	42	16	14	13	09	14	20	18	43
1129	14	5	56	40	15	14	45	01	16	19	59	47	15	2	08	24
1130	14	11	34	17	15	20	16	58	17	1	45	42	15	8	02	14
1131	14	17	21	22	16	2	08	58	17	7	35	59	15	13	51	11
1132	13	23	11	26	15	7	56	53	16	13	19	25	14	19	36	57
1133	14	5	00	50	15	13	35	47	16	19	01	31	15	1	26	01
1134	14	10	55	01	15	19	26	13	17	0	52	15	15	7	12	42
1135	14	16	35	29	16	1	09	21	17	6	38	18	15	12	58	53
1136	13	22	21	17	15	7	02	56	16	12	25	44	14	18	49	44
1137	14	4	18	40	15	12	57	24	16	18	13	45	15	0	40	03
1138	14	10	00	58	15	18	33	07	17	0	02	02	15	6	30	13
1139	14	15	51	16	16	0	29	38	17	6	00	41	15	12	19	38
1140	13	21	44	42	15	6	23	43	16	11	51	27	14	18	09	02
1141	14	3	33	48	15	12	05	49	16	17	37	46	15	0	04	54
1142	14	9	26	28	15	17	59	39	16	23	33	00	15	5	59	30
1143	14	15	10	34	15	23	45	58	17	5	21	26	15	11	50	13
1144	13	21	01	41	15	5	37	58	16	11	10	27	14	17	41	30
1145	14	3	02	56	15	11	31	38	16	17	01	03	14	23	31	57
1146	14	8	48	26	15	17	06	59	16	22	46	39	15	5	20	07
1147	14	14	35	21	15	23	01	40	17	4	40	20	15	11	08	38
1148	13	20	25	17	15	4	55	44	16	10	24	52	14	16	55	13
1149	14	2	11	00	15	10	33	36	16	16	04	07	14	22	42	56
1150	14	7	58	01	15	16	22	11	16	21	56	24	15	4	28	17
1151	14	13	38	30	15	22	04	55	17	3	43	19	15	10	11	42
1152	13	19	23	40	15	3	51	11	16	9	29	55	14	15	59	01
1153	14	1	15	32	15	9	41	50	16	15	19	54	14	21	53	20
1154	14	6	57	42	15	15	16	11	16	21	03	44	15	3	44	50
1155	14	12	43	48	15	21	07	47	17	2	56	56	15	9	36	38
1156	13	18	40	21	15	3	03	57	16	8	46	07	14	15	27	25
1157	14	0	35	49	15	8	43	54	16	14	28	11	14	21	16	11
1158	14	6	28	50	15	14	38	45	16	20	26	05	15	3	08	41
1159	14	12	16	51	15	20	32	13	17	2	16	25	15	8	58	11
1160	13	18	05	29	15	2	24	41	16	8	03	39	14	14	48	02

Year	March Equinox				June Solstice				September Equinox				December Solstice			
	d	h	m	s	d	h	m	s	d	h	m	s	d	h	m	s
1161	14	0	00	47	15	8	22	03	16	13	58	39	14	20	43	30
1162	14	5	47	25	15	14	00	34	16	19	46	38	15	2	30	22
1163	14	11	32	49	15	19	51	07	17	1	43	11	15	8	18	31
1164	13	17	24	44	15	1	46	30	16	7	35	42	14	14	10	56
1165	13	23	12	55	15	7	21	35	16	13	13	20	14	19	59	36
1166	14	4	56	39	15	13	08	29	16	19	06	19	15	1	53	56
1167	14	10	42	19	15	18	55	25	17	0	50	49	15	7	40	55
1168	13	16	30	35	15	0	38	02	16	6	30	23	14	13	24	15
1169	13	22	25	52	15	6	29	56	16	12	22	21	14	19	16	17
1170	14	4	12	44	15	12	07	37	16	18	05	05	15	1	00	51
1171	14	9	56	17	15	17	57	55	16	23	55	21	15	6	47	43
1172	13	15	46	15	14	23	56	04	16	5	45	34	14	12	39	36
1173	13	21	37	09	15	5	35	11	16	11	23	25	14	18	24	28
1174	14	3	22	35	15	11	24	09	16	17	20	56	15	0	14	14
1175	14	9	09	23	15	17	16	42	16	23	14	49	15	6	03	19
1176	13	14	58	57	14	23	03	26	16	5	00	26	14	11	51	09
1177	13	20	50	19	15	4	57	31	16	10	58	04	14	17	52	46
1178	14	2	38	38	15	10	39	29	16	16	45	29	14	23	46	25
1179	14	8	27	06	15	16	29	17	16	22	37	36	15	5	37	00
1180	13	14	22	41	14	22	27	27	16	4	32	12	14	11	32	06
1181	13	20	21	40	15	4	07	54	16	10	11	40	14	17	18	26
1182	14	2	09	08	15	9	55	49	16	16	05	23	14	23	08	00
1183	14	7	54	10	15	15	49	59	16	21	54	53	15	4	58	34
1184	13	13	44	15	14	21	36	47	16	3	33	07	14	10	40	09
1185	13	19	31	30	15	3	27	05	16	9	25	54	14	16	33	02
1186	14	1	17	41	15	9	07	44	16	15	12	38	14	22	17	42
1187	14	7	01	08	15	14	51	53	16	21	01	28	15	3	59	12
1188	13	12	46	54	14	20	45	55	16	2	55	46	14	9	54	05
1189	13	18	38	16	15	2	23	31	16	8	31	59	14	15	41	13
1190	14	0	19	45	15	8	06	40	16	14	23	02	14	21	32	40
1191	14	6	05	02	15	13	59	14	16	20	14	33	15	3	25	50
1192	13	12	00	30	14	19	44	47	16	1	54	07	14	9	07	25
1193	13	17	52	47	15	1	35	35	16	7	49	55	14	15	03	45
1194	13	23	43	30	15	7	22	36	16	13	40	14	14	20	55	22
1195	14	5	31	36	15	13	13	17	16	19	28	37	15	2	41	03
1196	13	11	20	46	14	19	13	21	16	1	26	16	14	8	40	00
1197	13	17	18	01	15	0	57	13	16	7	06	33	14	14	26	15
1198	13	23	03	37	15	6	42	08	16	13	01	33	14	20	15	04
1199	14	4	49	39	15	12	36	54	16	18	59	15	15	2	08	44
1200	13	10	42	58	14	18	23	10	16	0	39	17	14	7	52	27

Year	March Equinox				June Solstice				September Equinox				December Solstice			
	d	h	m	s	d	h	m	s	d	h	m	s	d	h	m	s
1201	13	16	30	01	15	0	11	02	16	6	32	39	14	13	51	29
1202	13	22	16	22	15	5	54	44	16	12	20	45	14	19	45	02
1203	14	4	05	57	15	11	40	33	16	18	04	21	15	1	26	28
1204	13	9	55	24	14	17	33	42	15	23	58	17	14	7	19	58
1205	13	15	53	29	14	23	16	01	16	5	36	19	14	13	04	01
1206	13	21	37	46	15	4	59	07	16	11	24	16	14	18	48	17
1207	14	3	17	17	15	10	53	41	16	17	17	46	15	0	41	12
1208	13	9	09	41	14	16	41	59	15	22	54	40	14	6	21	07
1209	13	14	54	46	14	22	27	31	16	4	46	39	14	12	13	12
1210	13	20	39	06	15	4	11	55	16	10	40	18	14	18	04	30
1211	14	2	28	44	15	9	59	00	16	16	28	11	14	23	45	48
1212	13	8	13	21	14	15	51	24	15	22	24	57	14	5	45	09
1213	13	14	09	04	14	21	35	56	16	4	07	23	14	11	40	12
1214	13	19	56	54	15	3	19	10	16	9	55	51	14	17	28	49
1215	14	1	41	19	15	9	13	33	16	15	52	46	14	23	26	49
1216	13	7	45	12	14	15	05	54	15	21	34	16	14	5	09	54
1217	13	13	37	52	14	20	53	23	16	3	26	39	14	11	03	32
1218	13	19	24	56	15	2	43	37	16	9	21	45	14	16	59	37
1219	14	1	17	37	15	8	36	03	16	15	06	00	14	22	40	08
1220	13	7	02	18	14	14	29	31	15	20	59	41	14	4	35	02
1221	13	12	57	06	14	20	14	31	16	2	42	22	14	10	22	41
1222	13	18	42	45	15	1	55	51	16	8	30	31	14	16	02	09
1223	14	0	19	50	15	7	45	48	16	14	27	44	14	21	55	47
1224	13	6	13	11	14	13	33	48	15	20	06	53	14	3	39	18
1225	13	11	56	47	14	19	14	05	16	1	53	14	14	9	32	35
1226	13	17	36	42	15	0	57	00	16	7	45	42	14	15	29	30
1227	13	23	30	24	15	6	44	45	16	13	27	21	14	21	08	22
1228	13	5	18	31	14	12	32	58	15	19	19	42	14	3	02	14
1229	13	11	15	21	14	18	17	56	16	1	02	59	14	8	52	52
1230	13	17	04	32	15	0	03	30	16	6	48	44	14	14	36	03
1231	13	22	44	50	15	5	58	18	16	12	45	05	14	20	32	23
1232	13	4	42	36	14	11	53	28	15	18	27	24	14	2	17	49
1233	13	10	33	41	14	17	39	59	16	0	18	20	14	8	08	03
1234	13	16	16	24	14	23	27	11	16	6	19	07	14	14	04	59
1235	13	22	12	35	15	5	21	31	16	12	08	21	14	19	48	05
1236	13	3	59	02	14	11	11	06	15	18	02	02	14	1	46	05
1237	13	9	51	42	14	16	57	07	15	23	48	42	14	7	44	21
1238	13	15	43	25	14	22	42	42	16	5	34	06	14	13	27	58
1239	13	21	24	34	15	4	32	04	16	11	28	30	14	19	21	39
1240	13	3	24	36	14	10	24	21	15	17	10	42	14	1	05	43

Year	March Equinox				June Solstice				September Equinox				December Solstice			
	d	h	m	s	d	h	m	s	d	h	m	s	d	h	m	s
1241	13	9	15	57	14	16	07	39	15	22	55	51	14	6	51	32
1242	13	14	52	49	14	21	51	45	16	4	50	03	14	12	46	31
1243	13	20	44	55	15	3	45	36	16	10	33	27	14	18	26	33
1244	13	2	27	59	14	9	30	50	15	16	19	51	14	0	15	20
1245	13	8	15	49	14	15	12	35	15	22	06	17	14	6	07	52
1246	13	14	06	39	14	20	56	31	16	3	52	32	14	11	45	21
1247	13	19	42	20	15	2	42	15	16	9	46	26	14	17	38	33
1248	13	1	37	05	14	8	35	09	15	15	31	09	13	23	29	53
1249	13	7	27	51	14	14	19	11	15	21	16	18	14	5	20	57
1250	13	13	07	24	14	20	03	36	16	3	13	18	14	11	21	14
1251	13	19	07	49	15	2	00	55	16	9	02	02	14	17	05	53
1252	13	1	01	59	14	7	50	41	15	14	53	00	13	22	57	46
1253	13	6	55	57	14	13	38	17	15	20	43	52	14	4	55	42
1254	13	12	51	19	14	19	31	36	16	2	32	27	14	10	39	10
1255	13	18	31	10	15	1	22	47	16	8	25	35	14	16	32	39
1256	13	0	26	10	14	7	18	10	15	14	12	14	13	22	21	43
1257	13	6	19	05	14	13	03	37	15	19	58	44	14	4	05	50
1258	13	11	55	10	14	18	43	59	16	1	56	24	14	9	59	40
1259	13	17	47	03	15	0	37	32	16	7	44	39	14	15	43	29
1260	12	23	32	45	14	6	21	11	15	13	29	59	13	21	35	03
1261	13	5	16	46	14	11	58	48	15	19	14	20	14	3	31	46
1262	13	11	08	05	14	17	44	58	16	0	57	33	14	9	13	38
1263	13	16	50	28	14	23	29	02	16	6	45	09	14	15	01	20
1264	12	22	45	31	14	5	19	41	15	12	29	45	13	20	49	09
1265	13	4	41	13	14	11	07	51	15	18	13	56	14	2	34	03
1266	13	10	17	15	14	16	50	26	16	0	07	22	14	8	28	28
1267	13	16	10	05	14	22	50	24	16	5	57	24	14	14	15	02
1268	12	22	01	38	14	4	40	35	15	11	45	00	13	20	03	30
1269	13	3	48	37	14	10	22	36	15	17	37	15	14	1	59	43
1270	13	9	43	36	14	16	16	02	15	23	31	25	14	7	44	47
1271	13	15	26	02	14	22	04	01	16	5	24	40	14	13	37	18
1272	12	21	17	59	14	3	56	39	15	11	14	11	13	19	34	49
1273	13	3	14	34	14	9	46	30	15	17	01	08	14	1	26	09
1274	13	8	53	41	14	15	26	17	15	22	53	53	14	7	21	27
1275	13	14	50	27	14	21	23	35	16	4	46	01	14	13	08	57
1276	12	20	46	57	14	3	12	02	15	10	30	47	13	18	54	26
1277	13	2	32	10	14	8	50	26	15	16	16	43	14	0	49	39
1278	13	8	24	10	14	14	43	37	15	22	04	41	14	6	32	51
1279	13	14	03	42	14	20	29	26	16	3	49	01	14	12	17	32
1280	12	19	51	39	14	2	18	01	15	9	34	04	13	18	06	22

Year	March Equinox				June Solstice				September Equinox				December Solstice			
	d	h	m	s	d	h	m	s	d	h	m	s	d	h	m	s
1281	13	1	45	04	14	8	05	16	15	15	20	09	13	23	49	14
1282	13	7	19	59	14	13	41	37	15	21	11	57	14	5	38	17
1283	13	13	08	00	14	19	35	59	16	3	04	00	14	11	27	14
1284	12	18	59	36	14	1	24	57	15	8	48	57	13	17	17	59
1285	13	0	43	01	14	7	01	00	15	14	34	34	13	23	16	00
1286	13	6	36	33	14	12	54	33	15	20	27	08	14	5	04	22
1287	13	12	26	18	14	18	43	27	16	2	15	56	14	10	51	37
1288	12	18	20	54	14	0	33	57	15	8	05	06	13	16	45	04
1289	13	0	19	42	14	6	29	10	15	13	55	20	13	22	35	59
1290	13	6	00	17	14	12	12	43	15	19	47	40	14	4	28	39
1291	13	11	50	35	14	18	11	43	16	1	41	08	14	10	18	56
1292	12	17	47	16	14	0	05	39	15	7	29	28	13	16	08	18
1293	12	23	33	12	14	5	40	46	15	13	16	28	13	21	59	31
1294	13	5	22	20	14	11	32	51	15	19	11	52	14	3	46	36
1295	13	11	08	14	14	17	21	17	16	0	58	56	14	9	33	45
1296	12	16	54	06	13	23	04	49	15	6	42	13	13	15	27	26
1297	12	22	47	00	14	4	54	55	15	12	29	00	13	21	18	36
1298	13	4	27	18	14	10	31	46	15	18	14	57	14	3	05	11
1299	13	10	17	13	14	16	24	07	16	0	05	19	14	8	51	23
1300	12	16	15	02	13	22	16	46	15	5	50	53	13	14	38	08
1301	12	21	59	02	14	3	51	39	15	11	32	51	13	20	28	16
1302	13	3	45	21	14	9	46	01	15	17	26	34	14	2	16	14
1303	13	9	31	47	14	15	38	08	15	23	13	03	14	8	01	07
1304	12	15	19	14	13	21	22	30	15	4	58	12	13	13	50	55
1305	12	21	12	51	14	3	15	08	15	10	53	18	13	19	41	50
1306	13	2	54	20	14	8	55	42	15	16	45	15	14	1	30	37
1307	13	8	41	32	14	14	49	07	15	22	40	13	14	7	25	29
1308	12	14	38	15	13	20	43	57	15	4	29	25	13	13	21	37
1309	12	20	25	24	14	2	18	35	15	10	11	58	13	19	16	56
1310	13	2	18	09	14	8	12	34	15	16	08	14	14	1	07	48
1311	13	8	13	11	14	14	06	25	15	21	56	41	14	6	53	47
1312	12	14	06	35	13	19	51	36	15	3	39	44	13	12	43	19
1313	12	19	59	25	14	1	46	12	15	9	32	06	13	18	35	24
1314	13	1	41	13	14	7	30	25	15	15	19	05	14	0	21	15
1315	13	7	25	58	14	13	21	45	15	21	07	46	14	6	07	41
1316	12	13	19	27	13	19	15	32	15	2	56	54	13	11	56	45
1317	12	19	04	10	14	0	47	11	15	8	38	52	13	17	41	54
1318	13	0	46	50	14	6	35	22	15	14	34	32	13	23	29	16
1319	13	6	33	18	14	12	26	54	15	20	22	03	14	5	17	49
1320	12	12	19	40	13	18	05	51	15	2	00	38	13	11	08	18

Year	March Equinox				June Solstice				September Equinox				December Solstice			
	d	h	m	s	d	h	m	s	d	h	m	s	d	h	m	s
1321	12	18	07	57	13	23	55	21	15	7	52	28	13	17	03	07
1322	12	23	54	25	14	5	38	21	15	13	40	50	13	22	50	12
1323	13	5	44	44	14	11	27	16	15	19	29	44	14	4	36	51
1324	12	11	42	28	13	17	24	33	15	1	22	14	13	10	33	14
1325	12	17	33	42	13	23	03	11	15	7	04	06	13	16	22	41
1326	12	23	19	09	14	4	56	35	15	12	59	41	13	22	13	36
1327	13	5	12	02	14	10	56	54	15	18	51	47	14	4	03	59
1328	12	11	04	56	13	16	40	21	15	0	33	46	13	9	50	33
1329	12	16	55	00	13	22	33	00	15	6	32	41	13	15	44	17
1330	12	22	40	59	14	4	19	48	15	12	25	12	13	21	32	42
1331	13	4	26	49	14	10	07	56	15	18	13	16	14	3	22	10
1332	12	10	18	33	13	16	02	24	15	0	05	14	13	9	21	03
1333	12	16	07	41	13	21	37	21	15	5	44	35	13	15	09	10
1334	12	21	53	36	14	3	23	29	15	11	36	36	13	20	55	27
1335	13	3	45	56	14	9	18	30	15	17	26	42	14	2	42	50
1336	12	9	38	11	13	14	58	41	14	23	02	13	13	8	26	02
1337	12	15	22	51	13	20	48	07	15	4	54	49	13	14	18	03
1338	12	21	04	58	14	2	35	26	15	10	42	32	13	20	03	59
1339	13	2	49	54	14	8	21	48	15	16	26	27	14	1	47	39
1340	12	8	39	54	13	14	14	05	14	22	19	58	13	7	41	19
1341	12	14	28	36	13	19	49	55	15	4	03	12	13	13	27	58
1342	12	20	13	14	14	1	36	38	15	9	57	46	13	19	16	52
1343	13	2	01	46	14	7	33	53	15	15	51	54	14	1	13	40
1344	12	7	56	49	13	13	17	32	14	21	29	54	13	7	04	11
1345	12	13	46	01	13	19	07	09	15	3	24	55	13	13	00	49
1346	12	19	37	31	14	0	59	06	15	9	19	16	13	18	51	42
1347	13	1	33	06	14	6	49	20	15	15	05	25	14	0	35	40
1348	12	7	26	52	13	12	46	32	14	21	01	35	13	6	34	11
1349	12	13	18	58	13	18	29	46	15	2	45	20	13	12	23	29
1350	12	19	03	09	14	0	17	52	15	8	36	01	13	18	08	45
1351	13	0	49	48	14	6	15	16	15	14	30	01	14	0	00	43
1352	12	6	44	04	13	11	56	43	14	20	07	20	13	5	42	01
1353	12	12	26	34	13	17	39	38	15	2	00	28	13	11	30	55
1354	12	18	08	09	13	23	26	53	15	7	53	51	13	17	21	44
1355	12	23	54	19	14	5	10	07	15	13	32	47	13	23	04	10
1356	12	5	38	53	13	10	58	11	14	19	23	10	13	5	03	51
1357	12	11	30	15	13	16	36	38	15	1	04	17	13	10	52	26
1358	12	17	17	39	13	22	19	25	15	6	51	27	13	16	34	17
1359	12	23	07	52	14	4	16	37	15	12	46	51	13	22	28	07
1360	12	5	06	40	13	10	02	56	14	18	23	48	13	4	13	13

Year	March Equinox				June Solstice				September Equinox				December Solstice			
	d	h	m	s	d	h	m	s	d	h	m	s	d	h	m	s
1361	12	10	52	14	13	15	51	19	15	0	16	13	13	10	05	22
1362	12	16	37	02	13	21	46	30	15	6	12	54	13	15	59	03
1363	12	22	30	41	14	3	38	23	15	11	56	58	13	21	40	54
1364	12	4	20	30	13	9	31	05	14	17	55	02	13	3	38	37
1365	12	10	12	33	13	15	15	16	14	23	46	30	13	9	30	36
1366	12	15	59	57	13	21	01	40	15	5	37	57	13	15	16	40
1367	12	21	44	51	14	2	57	46	15	11	35	26	13	21	17	01
1368	12	3	42	08	13	8	42	49	14	17	12	19	13	3	06	28
1369	12	9	29	07	13	14	25	05	14	23	01	29	13	8	56	29
1370	12	15	14	17	13	20	14	17	15	4	56	20	13	14	47	29
1371	12	21	10	09	14	2	02	45	15	10	35	33	13	20	26	39
1372	12	2	57	29	13	7	50	15	14	16	24	25	13	2	20	08
1373	12	8	43	18	13	13	32	35	14	22	08	51	13	8	10	07
1374	12	14	29	23	13	19	18	24	15	3	52	44	13	13	48	27
1375	12	20	10	45	14	1	10	48	15	9	46	44	13	19	40	39
1376	12	2	06	47	13	6	55	53	14	15	26	05	13	1	24	45
1377	12	7	51	19	13	12	36	35	14	21	15	55	13	7	10	42
1378	12	13	30	18	13	18	26	35	15	3	14	40	13	13	07	26
1379	12	19	24	32	14	0	17	54	15	8	55	59	13	18	52	53
1380	12	1	12	18	13	6	05	33	14	14	47	09	13	0	52	16
1381	12	7	03	57	13	11	50	49	14	20	38	55	13	6	48	43
1382	12	13	00	18	13	17	40	20	15	2	27	55	13	12	29	47
1383	12	18	49	13	13	23	36	27	15	8	25	48	13	18	27	07
1384	12	0	49	59	13	5	28	31	14	14	08	31	13	0	18	21
1385	12	6	37	54	13	11	14	31	14	19	56	36	13	6	05	21
1386	12	12	17	34	13	17	07	06	15	1	55	36	13	12	01	37
1387	12	18	13	52	13	22	59	32	15	7	36	40	13	17	41	02
1388	12	0	00	11	13	4	42	58	14	13	26	03	12	23	32	26
1389	12	5	45	43	13	10	24	40	14	19	18	26	13	5	25	30
1390	12	11	34	10	13	16	10	24	15	1	02	44	13	11	04	58
1391	12	17	14	12	13	21	59	09	15	6	54	03	13	17	01	24
1392	11	23	08	26	13	3	44	20	14	12	32	24	12	22	52	07
1393	12	4	57	09	13	9	24	45	14	18	16	12	13	4	34	16
1394	12	10	37	43	13	15	11	56	15	0	12	52	13	10	26	36
1395	12	16	35	49	13	21	06	04	15	5	54	13	13	16	08	03
1396	11	22	24	15	13	2	52	23	14	11	40	54	12	22	00	06
1397	12	4	07	23	13	8	37	49	14	17	34	11	13	3	56	29
1398	12	10	00	08	13	14	30	45	14	23	21	16	13	9	36	44
1399	12	15	44	24	13	20	22	13	15	5	16	38	13	15	30	37
1400	11	21	40	37	13	2	11	22	14	11	04	26	12	21	23	24

Year	March Equinox				June Solstice				September Equinox				December Solstice			
	d	h	m	s	d	h	m	s	d	h	m	s	d	h	m	s
1401	12	3	32	00	13	7	56	34	14	16	55	15	13	3	09	18
1402	12	9	09	37	13	13	44	29	14	22	54	37	13	9	08	25
1403	12	15	05	50	13	19	39	58	15	4	38	41	13	14	59	39
1404	11	20	58	29	13	1	24	38	14	10	23	46	12	20	52	59
1405	12	2	44	09	13	7	06	05	14	16	16	53	13	2	49	25
1406	12	8	43	47	13	12	59	23	14	22	03	46	13	8	28	18
1407	12	14	30	30	13	18	48	39	15	3	53	12	13	14	18	50
1408	11	20	23	30	13	0	37	51	14	9	36	33	12	20	11	08
1409	12	2	12	50	13	6	24	01	14	15	20	05	13	1	51	00
1410	12	7	46	59	13	12	09	43	14	21	14	07	13	7	41	51
1411	12	13	39	54	13	18	03	29	15	2	58	07	13	13	24	38
1412	11	19	28	35	12	23	45	33	14	8	43	36	12	19	10	16
1413	12	1	06	23	13	5	23	39	14	14	38	30	13	1	06	16
1414	12	6	58	01	13	11	15	39	14	20	26	13	13	6	50	06
1415	12	12	40	59	13	17	02	07	15	2	13	54	13	12	44	45
1416	11	18	33	27	12	22	48	26	14	7	59	37	12	18	42	30
1417	12	0	29	49	13	4	35	32	14	13	46	21	13	0	25	08
1418	12	6	12	26	13	10	22	15	14	19	43	14	13	6	19	29
1419	12	12	11	23	13	16	20	55	15	1	30	33	13	12	08	40
1420	11	18	05	47	12	22	10	16	14	7	15	46	12	17	59	08
1421	11	23	47	47	13	3	54	14	14	13	11	09	12	23	57	06
1422	12	5	43	14	13	9	52	14	14	19	01	41	13	5	40	19
1423	12	11	31	37	13	15	41	59	15	0	51	44	13	11	29	00
1424	11	17	22	28	12	21	27	17	14	6	40	50	12	17	22	59
1425	11	23	15	03	13	3	15	27	14	12	29	41	12	23	05	56
1426	12	4	50	33	13	8	58	12	14	18	21	52	13	4	59	03
1427	12	10	40	21	13	14	51	29	15	0	06	43	13	10	50	20
1428	11	16	33	31	12	20	35	55	14	5	48	03	12	16	37	34
1429	11	22	13	29	13	2	10	51	14	11	39	04	12	22	30	06
1430	12	4	08	30	13	8	04	56	14	17	28	03	13	4	12	01
1431	12	9	57	13	13	13	52	54	14	23	12	47	13	9	58	44
1432	11	15	43	46	12	19	36	38	14	4	56	54	12	15	53	21
1433	11	21	35	40	13	1	27	46	14	10	44	37	12	21	36	54
1434	12	3	13	20	13	7	11	25	14	16	34	48	13	3	24	36
1435	12	9	03	45	13	13	05	26	14	22	24	11	13	9	14	51
1436	11	15	01	06	12	18	54	22	14	4	12	03	12	15	02	26
1437	11	20	39	51	13	0	31	13	14	10	06	55	12	20	59	21
1438	12	2	33	10	13	6	29	53	14	16	01	33	13	2	51	31
1439	12	8	25	15	13	12	21	01	14	21	48	21	13	8	44	28
1440	11	14	16	41	12	18	04	36	14	3	35	12	12	14	43	07

Year	March Equinox				June Solstice				September Equinox				December Solstice			
	d	h	m	s	d	h	m	s	d	h	m	s	d	h	m	s
1441	11	20	16	11	12	23	57	46	14	9	26	58	12	20	28	20
1442	12	2	01	49	13	5	44	18	14	15	18	02	13	2	15	52
1443	12	7	53	56	13	11	40	20	14	21	06	14	13	8	07	04
1444	11	13	49	36	12	17	32	38	14	2	51	01	12	13	54	36
1445	11	19	26	21	12	23	09	07	14	8	40	22	12	19	45	36
1446	12	1	14	47	13	5	04	35	14	14	32	51	13	1	30	15
1447	12	7	04	07	13	10	52	34	14	20	18	01	13	7	13	48
1448	11	12	48	27	12	16	28	45	14	2	03	11	12	13	06	01
1449	11	18	37	08	12	22	16	43	14	7	53	30	12	18	51	31
1450	12	0	15	12	13	3	58	25	14	13	39	51	13	0	41	10
1451	12	6	02	07	13	9	47	10	14	19	23	54	13	6	33	50
1452	11	11	58	41	12	15	36	15	14	1	08	14	12	12	23	13
1453	11	17	43	00	12	21	12	04	14	6	58	04	12	18	13	24
1454	11	23	36	33	13	3	09	34	14	12	53	23	13	0	02	00
1455	12	5	33	01	13	9	06	16	14	18	41	05	13	5	51	52
1456	11	11	21	42	12	14	48	48	14	0	26	12	12	11	47	42
1457	11	17	14	21	12	20	45	41	14	6	21	56	12	17	37	15
1458	11	23	00	24	13	2	35	30	14	12	12	51	12	23	23	57
1459	12	4	49	56	13	8	27	04	14	18	03	12	13	5	14	56
1460	11	10	47	01	12	14	20	06	13	23	55	22	12	11	06	09
1461	11	16	27	12	12	19	55	49	14	5	45	53	12	16	57	42
1462	11	22	13	04	13	1	49	41	14	11	39	47	12	22	50	15
1463	12	4	06	23	13	7	42	04	14	17	24	46	13	4	41	07
1464	11	9	53	56	12	13	15	39	13	23	03	51	12	10	32	03
1465	11	15	45	42	12	19	05	05	14	4	57	09	12	16	18	40
1466	11	21	32	45	13	0	50	57	14	10	41	41	12	22	00	25
1467	12	3	18	11	13	6	37	34	14	16	22	49	13	3	48	59
1468	11	9	11	53	12	12	30	44	13	22	09	47	12	9	38	44
1469	11	14	51	09	12	18	06	57	14	3	55	04	12	15	24	07
1470	11	20	36	31	13	0	00	41	14	9	49	01	12	21	11	07
1471	12	2	30	47	13	5	55	22	14	15	38	54	13	2	59	10
1472	11	8	18	20	12	11	31	13	13	21	22	43	12	8	50	34
1473	11	14	06	03	12	17	23	17	14	3	21	11	12	14	44	05
1474	11	19	53	00	12	23	14	48	14	9	10	53	12	20	35	50
1475	12	1	42	38	13	5	02	51	14	14	55	18	13	2	30	18
1476	11	7	41	20	12	10	57	39	13	20	48	54	12	8	25	52
1477	11	13	31	22	12	16	38	11	14	2	39	05	12	14	13	49
1478	11	19	22	32	12	22	33	40	14	8	34	15	12	20	03	37
1479	12	1	18	50	13	4	33	10	14	14	24	21	13	1	56	13
1480	11	7	07	44	12	10	11	26	13	20	04	22	12	7	46	06

Year	March Equinox				June Solstice				September Equinox				December Solstice			
	d	h	m	s	d	h	m	s	d	h	m	s	d	h	m	s
1481	11	12	52	47	12	16	01	49	14	1	59	15	12	13	33	51
1482	11	18	38	37	12	21	52	03	14	7	47	12	12	19	18	38
1483	12	0	25	55	13	3	33	58	14	13	27	59	13	1	02	35
1484	11	6	14	36	12	9	22	13	13	19	19	50	12	6	54	27
1485	11	11	56	24	12	14	59	31	14	1	05	56	12	12	41	37
1486	11	17	38	21	12	20	46	58	14	6	54	00	12	18	30	51
1487	11	23	30	29	13	2	41	36	14	12	42	01	13	0	24	48
1488	11	5	22	22	12	8	15	51	13	18	19	10	12	6	11	23
1489	11	11	10	46	12	14	04	28	14	0	15	17	12	11	59	25
1490	11	17	01	41	12	19	59	51	14	6	05	46	12	17	47	41
1491	11	22	51	50	13	1	47	16	14	11	46	13	12	23	35	39
1492	11	4	42	40	12	7	42	16	13	17	41	32	12	5	32	34
1493	11	10	29	47	12	13	27	35	13	23	32	20	12	11	20	51
1494	11	16	17	14	12	19	19	04	14	5	25	18	12	17	08	20
1495	11	22	12	11	13	1	17	34	14	11	22	45	12	23	04	10
1496	11	4	04	50	12	6	55	10	13	17	04	49	12	4	53	15
1497	11	9	49	08	12	12	42	59	13	23	02	03	12	10	47	31
1498	11	15	37	07	12	18	37	03	14	4	52	14	12	16	40	31
1499	11	21	27	41	13	0	19	28	14	10	28	03	12	22	27	18
1500	11	3	20	16	12	6	08	36	13	16	20	59	12	4	20	57
1501	11	9	09	46	12	11	51	06	13	22	08	49	12	10	05	37
1502	11	14	57	24	12	17	39	51	14	3	54	31	12	15	48	57
1503	11	20	47	10	12	23	36	36	14	9	45	23	12	21	42	58
1504	11	2	37	11	12	5	15	23	13	15	21	57	12	3	27	33
1505	11	8	18	45	12	11	00	49	13	21	14	13	12	9	12	36
1506	11	14	03	36	12	16	55	00	14	3	07	09	12	15	00	40
1507	11	19	54	01	12	22	37	39	14	8	45	55	12	20	42	44
1508	11	1	39	41	12	4	24	35	13	14	41	59	12	2	38	18
1509	11	7	24	07	12	10	09	20	13	20	33	49	12	8	31	24
1510	11	13	11	21	12	15	57	32	14	2	20	20	12	14	20	25
1511	11	19	03	11	12	21	53	13	14	8	15	03	12	20	20	58
1512	11	1	02	35	12	3	34	31	13	13	57	15	12	2	10	35
1513	11	6	53	10	12	9	21	06	13	19	51	22	12	7	57	28
1514	11	12	42	53	12	15	20	26	14	1	46	54	12	13	52	48
1515	11	18	39	52	12	21	10	17	14	7	24	16	12	19	37	35
1516	11	0	26	42	12	2	59	40	13	13	17	44	12	1	32	00
1517	11	6	13	06	12	8	48	45	13	19	11	09	12	7	21	53
1518	11	12	02	01	12	14	36	33	14	0	56	34	12	13	01	54
1519	11	17	49	29	12	20	29	01	14	6	52	34	12	18	57	21
1520	10	23	41	39	12	2	07	59	13	12	33	30	12	0	45	02

Year	March Equinox				June Solstice				September Equinox				December Solstice			
	d	h	m	s	d	h	m	s	d	h	m	s	d	h	m	s
1521	11	5	23	00	12	7	49	17	13	18	22	16	12	6	31	19
1522	11	11	04	39	12	13	42	24	14	0	14	24	12	12	26	48
1523	11	16	59	09	12	19	26	20	14	5	47	53	12	18	09	04
1524	10	22	47	02	12	1	08	50	13	11	38	27	11	23	59	31
1525	11	4	34	00	12	6	55	49	13	17	32	17	12	5	49	32
1526	11	10	24	52	12	12	45	27	13	23	14	33	12	11	30	31
1527	11	16	11	37	12	18	39	36	14	5	08	54	12	17	29	02
1528	10	22	04	51	12	0	23	31	13	10	51	09	11	23	18	33
1529	11	3	51	03	12	6	08	39	13	16	41	47	12	5	03	12
1530	11	9	36	44	12	12	04	43	13	22	40	22	12	10	57	48
1531	11	15	34	24	12	17	53	25	14	4	20	49	12	16	42	30
1532	10	21	22	38	11	23	39	09	13	10	15	13	11	22	37	40
1533	11	3	06	19	12	5	28	52	13	16	12	55	12	4	36	19
1534	11	8	59	50	12	11	20	48	13	21	56	07	12	10	21	45
1535	11	14	49	52	12	17	11	46	14	3	49	32	12	16	20	19
1536	10	20	47	41	11	22	55	21	13	9	34	30	11	22	10	19
1537	11	2	38	57	12	4	40	14	13	15	22	20	12	3	50	35
1538	11	8	21	32	12	10	35	29	13	21	17	02	12	9	44	50
1539	11	14	16	58	12	16	25	40	14	2	52	50	12	15	27	54
1540	10	20	00	55	11	22	07	53	13	8	39	59	11	21	14	43
1541	11	1	39	13	12	3	54	15	13	14	35	45	12	3	05	58
1542	11	7	30	25	12	9	43	17	13	20	18	27	12	8	43	00
1543	11	13	13	51	12	15	28	11	14	2	10	40	12	14	36	50
1544	10	19	02	29	11	21	08	55	13	7	57	17	11	20	31	36
1545	11	0	48	57	12	2	51	03	13	13	42	12	12	2	14	48
1546	11	6	28	32	12	8	41	45	13	19	35	29	12	8	14	02
1547	11	12	30	02	12	14	32	26	14	1	14	10	12	14	01	28
1548	10	18	23	15	11	20	15	33	13	7	03	07	11	19	49	12
1549	11	0	09	02	12	2	07	16	13	13	03	35	12	1	46	17
1550	11	6	07	39	12	8	05	37	13	18	48	17	12	7	29	30
1551	11	11	56	19	12	13	57	53	14	0	40	57	12	13	26	37
1552	10	17	47	59	11	19	45	32	13	6	31	18	11	19	21	54
1553	10	23	39	49	12	1	34	08	13	12	20	13	12	1	01	29
1554	11	5	21	19	12	7	25	37	13	18	17	34	12	6	55	40
1555	11	11	18	01	12	13	15	41	14	0	00	21	12	12	42	47
1556	10	17	05	04	11	18	56	20	13	5	47	10	11	18	30	13
1557	10	22	40	46	12	0	41	05	13	11	43	42	12	0	28	17
1558	11	4	34	08	12	6	33	07	13	17	23	40	12	6	10	49
1559	11	10	22	02	12	12	15	59	13	23	10	07	12	12	02	41
1560	10	16	11	45	11	17	55	39	13	4	57	14	11	17	53	56

Year	March Equinox				June Solstice				September Equinox				December Solstice			
	d	h	m	s	d	h	m	s	d	h	m	s	d	h	m	s
1561	10	22	04	10	11	23	42	08	13	10	41	18	11	23	32	04
1562	11	3	44	46	12	5	31	54	13	16	31	35	12	5	24	52
1563	11	9	38	19	12	11	23	20	13	22	11	23	12	11	13	11
1564	10	15	28	23	11	17	08	11	13	3	56	57	11	16	57	08
1565	10	21	05	40	11	22	54	33	13	9	56	01	11	22	51	44
1566	11	3	02	25	12	4	52	00	13	15	44	12	12	4	34	52
1567	11	8	52	20	12	10	38	48	13	21	35	41	12	10	28	13
1568	10	14	39	03	11	16	22	44	13	3	29	56	11	16	29	05
1569	10	20	33	10	11	22	14	43	13	9	19	04	11	22	15	17
1570	11	2	16	31	12	4	04	47	13	15	11	47	12	4	12	52
1571	11	8	15	48	12	9	56	50	13	20	57	04	12	10	05	32
1572	10	14	13	44	11	15	42	37	13	2	44	36	11	15	48	59
1573	10	19	53	32	11	21	28	42	13	8	41	41	11	21	44	12
1574	11	1	49	20	12	3	27	47	13	14	27	00	12	3	29	14
1575	11	7	37	29	12	9	13	46	13	20	10	55	12	9	17	25
1576	10	13	19	23	11	14	53	43	13	2	00	29	11	15	12	07
1577	10	19	11	19	11	20	42	25	13	7	46	23	11	20	48	29
1578	11	0	49	45	12	2	25	42	13	13	34	49	12	2	36	21
1579	11	6	40	25	12	8	12	40	13	19	19	09	12	8	26	58
1580	10	12	30	05	11	13	55	16	13	1	03	03	11	14	11	06
1581	10	18	03	38	11	19	36	32	13	6	56	08	11	20	08	05
1582	10	23	57	54	12	1	32	29	13	12	41	11	22	1	55	59
1583	21	5	52	56	22	7	18	32	23	18	26	24	22	7	44	33
1584	20	11	41	10	21	12	59	19	23	0	19	09	21	13	40	33
1585	20	17	38	55	21	18	56	09	23	6	10	19	21	19	24	15
1586	20	23	24	11	22	0	48	17	23	12	00	34	22	1	16	56
1587	21	5	16	56	22	6	41	37	23	17	48	40	22	7	12	08
1588	20	11	12	42	21	12	32	36	22	23	37	52	21	12	57	13
1589	20	16	51	11	21	18	16	47	23	5	35	25	21	18	50	47
1590	20	22	45	09	22	0	14	24	23	11	27	15	22	0	38	50
1591	21	4	38	54	22	6	01	37	23	17	15	03	22	6	29	12
1592	20	10	19	35	21	11	37	32	22	23	04	40	21	12	26	58
1593	20	16	10	38	21	17	29	32	23	4	52	35	21	18	13	56
1594	20	21	56	28	21	23	15	24	23	10	36	16	22	0	02	17
1595	21	3	47	57	22	4	59	34	23	16	19	22	22	5	52	32
1596	20	9	45	21	21	10	47	46	22	22	05	22	21	11	33	41
1597	20	15	22	46	21	16	29	00	23	3	54	56	21	17	22	49
1598	20	21	11	55	21	22	26	15	23	9	42	04	21	23	10	35
1599	21	3	04	34	22	4	15	31	23	15	25	23	22	4	56	32
1600	20	8	44	01	21	9	51	34	22	21	14	12	21	10	49	04

Year	March Equinox				June Solstice				September Equinox				December Solstice			
	d	h	m	s	d	h	m	s	d	h	m	s	d	h	m	s
1601	20	14	35	36	21	15	45	57	23	3	07	50	21	16	32	48
1602	20	20	21	52	21	21	34	14	23	8	56	54	21	22	19	52
1603	21	2	10	11	22	3	19	35	23	14	45	24	22	4	16	03
1604	20	8	04	35	21	9	10	59	22	20	36	55	21	10	07	02
1605	20	13	43	27	21	14	53	43	23	2	28	32	21	16	02	48
1606	20	19	36	50	21	20	50	36	23	8	20	07	21	21	57	11
1607	21	1	38	39	22	2	41	44	23	14	07	14	22	3	46	07
1608	20	7	25	46	21	8	18	32	22	19	57	00	21	9	41	10
1609	20	13	21	10	21	14	16	20	23	1	50	29	21	15	28	30
1610	20	19	09	47	21	20	08	43	23	7	35	03	21	21	15	33
1611	21	0	58	09	22	1	54	27	23	13	18	45	22	3	07	29
1612	20	6	51	03	21	7	45	41	22	19	08	39	21	8	51	34
1613	20	12	29	25	21	13	26	46	23	0	58	12	21	14	36	38
1614	20	18	15	51	21	19	18	01	23	6	48	16	21	20	24	03
1615	21	0	09	01	22	1	06	47	23	12	34	25	22	2	12	34
1616	20	5	47	11	21	6	38	02	22	18	18	59	21	8	06	00
1617	20	11	33	49	21	12	29	58	23	0	10	35	21	13	55	38
1618	20	17	24	05	21	18	19	30	23	5	54	14	21	19	42	22
1619	20	23	15	01	21	23	59	44	23	11	36	59	22	1	32	33
1620	20	5	10	56	21	5	52	14	22	17	29	45	21	7	20	18
1621	20	10	54	12	21	11	38	41	22	23	19	07	21	13	09	05
1622	20	16	41	46	21	17	34	00	23	5	08	50	21	19	00	53
1623	20	22	38	54	21	23	30	06	23	10	59	03	22	0	53	51
1624	20	4	23	27	21	5	05	52	22	16	46	33	21	6	44	12
1625	20	10	12	41	21	11	00	24	22	22	44	41	21	12	33	49
1626	20	16	07	03	21	16	55	25	23	4	35	00	21	18	22	27
1627	20	21	55	45	21	22	35	47	23	10	18	40	22	0	15	55
1628	20	3	47	08	21	4	28	44	22	16	13	22	21	6	09	54
1629	20	9	30	42	21	10	14	01	22	22	00	16	21	11	59	15
1630	20	15	19	29	21	16	03	49	23	3	47	02	21	17	48	44
1631	20	21	18	47	21	21	56	19	23	9	36	10	21	23	38	04
1632	20	3	04	24	21	3	30	29	22	15	19	46	21	5	24	25
1633	20	8	50	06	21	9	23	16	22	21	12	39	21	11	12	13
1634	20	14	39	26	21	15	18	11	23	2	57	41	21	16	58	41
1635	20	20	25	12	21	20	56	06	23	8	35	27	21	22	45	34
1636	20	2	12	32	21	2	45	49	22	14	29	51	21	4	33	20
1637	20	7	55	00	21	8	30	35	22	20	18	26	21	10	17	12
1638	20	13	40	09	21	14	17	23	23	2	06	37	21	16	05	56
1639	20	19	32	55	21	20	09	10	23	7	58	17	21	22	01	10
1640	20	1	16	13	21	1	43	52	22	13	42	18	21	3	54	32

Year	March Equinox				June Solstice				September Equinox				December Solstice			
	d	h	m	s	d	h	m	s	d	h	m	s	d	h	m	s
1641	20	7	04	04	21	7	35	19	22	19	36	18	21	9	47	42
1642	20	13	00	19	21	13	32	01	23	1	25	39	21	15	39	30
1643	20	18	57	49	21	19	13	19	23	7	07	01	21	21	28	26
1644	20	0	51	53	21	1	07	34	22	13	04	55	21	3	21	04
1645	20	6	40	54	21	7	01	55	22	18	55	24	21	9	10	19
1646	20	12	29	29	21	12	53	38	23	0	40	56	21	14	58	34
1647	20	18	23	03	21	18	50	03	23	6	34	51	21	20	52	44
1648	20	0	09	30	21	0	28	02	22	12	20	28	21	2	38	14
1649	20	5	53	46	21	6	16	26	22	18	15	24	21	8	24	23
1650	20	11	43	27	21	12	11	03	23	0	06	42	21	14	14	55
1651	20	17	31	36	21	17	46	41	23	5	43	35	21	20	02	15
1652	19	23	14	13	20	23	32	23	22	11	36	18	21	1	55	29
1653	20	4	58	38	21	5	20	09	22	17	22	47	21	7	44	17
1654	20	10	47	04	21	11	02	52	22	23	02	05	21	13	27	25
1655	20	16	40	55	21	16	53	35	23	4	54	41	21	19	21	11
1656	19	22	30	36	20	22	32	11	22	10	37	36	21	1	06	15
1657	20	4	14	32	21	4	21	22	22	16	27	42	21	6	53	46
1658	20	10	05	36	21	10	21	42	22	22	19	51	21	12	47	28
1659	20	15	58	38	21	16	02	54	23	3	57	56	21	18	33	10
1660	19	21	45	33	20	21	52	41	22	9	56	24	21	0	24	52
1661	20	3	33	58	21	3	46	42	22	15	51	44	21	6	15	01
1662	20	9	24	48	21	9	34	35	22	21	37	10	21	12	01	53
1663	20	15	15	53	21	15	28	08	23	3	34	31	21	18	03	01
1664	19	21	04	38	20	21	10	28	22	9	21	31	20	23	55	57
1665	20	2	51	56	21	2	58	25	22	15	11	42	21	5	45	19
1666	20	8	45	23	21	8	55	33	22	21	05	44	21	11	39	34
1667	20	14	43	45	21	14	35	41	23	2	42	50	21	17	23	11
1668	19	20	30	26	20	20	22	11	22	8	36	18	20	23	12	38
1669	20	2	14	30	21	2	15	41	22	14	26	09	21	5	02	17
1670	20	8	02	35	21	8	01	33	22	20	03	36	21	10	43	38
1671	20	13	48	32	21	13	49	22	23	1	54	40	21	16	35	46
1672	19	19	33	42	20	19	28	32	22	7	39	54	20	22	20	58
1673	20	1	18	50	21	1	12	24	22	13	27	50	21	4	02	08
1674	20	7	04	00	21	7	05	52	22	19	22	08	21	9	56	47
1675	20	12	57	23	21	12	46	27	23	0	59	16	21	15	44	47
1676	19	18	40	43	20	18	30	10	22	6	51	08	20	21	37	45
1677	20	0	26	49	21	0	24	31	22	12	45	02	21	3	33	15
1678	20	6	24	28	21	6	12	46	22	18	25	26	21	9	15	56
1679	20	12	18	13	21	12	04	06	23	0	21	55	21	15	12	53
1680	19	18	10	54	20	17	52	46	22	6	13	15	20	21	05	31

Year	March Equinox				June Solstice				September Equinox				December Solstice			
	d	h	m	s	d	h	m	s	d	h	m	s	d	h	m	s
1681	20	0	00	50	20	23	44	40	22	12	02	13	21	2	51	01
1682	20	5	48	59	21	5	44	31	22	17	59	26	21	8	49	08
1683	20	11	46	47	21	11	30	22	22	23	40	12	21	14	35	28
1684	19	17	32	15	20	17	14	14	22	5	33	53	20	20	22	23
1685	19	23	15	29	20	23	07	27	22	11	32	06	21	2	15	52
1686	20	5	07	49	21	4	53	09	22	17	10	39	21	7	56	50
1687	20	10	51	36	21	10	36	51	22	23	01	25	21	13	54	48
1688	19	16	36	58	20	16	18	28	22	4	47	59	20	19	47	24
1689	19	22	24	40	20	22	01	51	22	10	29	05	21	1	27	39
1690	20	4	12	45	21	3	53	43	22	16	21	46	21	7	21	00
1691	20	10	11	30	21	9	36	30	22	21	59	24	21	13	04	46
1692	19	15	57	15	20	15	20	08	22	3	47	24	20	18	49	41
1693	19	21	37	27	20	21	15	06	22	9	42	19	21	0	43	58
1694	20	3	31	15	21	3	06	17	22	15	20	34	21	6	24	42
1695	20	9	17	49	21	8	52	30	22	21	12	53	21	12	18	13
1696	19	15	03	57	20	14	38	35	22	3	08	25	20	18	11	18
1697	19	20	55	47	20	20	27	50	22	8	57	17	20	23	53	14
1698	20	2	41	39	21	2	21	23	22	14	55	31	21	5	53	20
1699	20	8	38	28	21	8	08	00	22	20	38	56	21	11	47	57
1700	20	14	27	00	21	13	52	32	23	2	28	43	21	17	38	09
1701	20	20	10	45	21	19	45	56	23	8	25	53	21	23	36	03
1702	21	2	12	43	22	1	37	37	23	14	06	39	22	5	18	58
1703	21	8	05	07	22	7	22	04	23	19	55	58	22	11	10	06
1704	20	13	49	46	21	13	08	34	23	1	48	28	21	17	04	43
1705	20	19	42	21	21	19	00	34	23	7	31	04	21	22	43	40
1706	21	1	24	50	22	0	51	02	23	13	21	51	22	4	36	34
1707	21	7	18	34	22	6	35	59	23	19	03	29	22	10	23	45
1708	20	13	04	50	21	12	17	03	23	0	50	13	21	16	02	41
1709	20	18	41	17	21	18	05	57	23	6	47	03	21	21	55	47
1710	21	0	35	00	21	23	55	39	23	12	26	49	22	3	38	56
1711	21	6	19	31	22	5	36	34	23	18	13	15	22	9	32	07
1712	20	11	59	44	21	11	19	55	23	0	06	52	21	15	30	39
1713	20	17	55	10	21	17	10	36	23	5	50	49	21	21	11	26
1714	20	23	44	19	21	22	59	46	23	11	43	38	22	3	05	49
1715	21	5	43	22	22	4	47	56	23	17	30	26	22	8	59	04
1716	20	11	35	47	21	10	36	31	22	23	18	28	21	14	43	16
1717	20	17	15	48	21	16	31	13	23	5	16	38	21	20	41	46
1718	20	23	13	58	21	22	27	24	23	10	59	40	22	2	27	04
1719	21	5	04	13	22	4	12	13	23	16	48	32	22	8	17	17
1720	20	10	47	12	21	9	57	25	22	22	47	37	21	14	13	13

Year	March Equinox				June Solstice				September Equinox				December Solstice			
	d	h	m	s	d	h	m	s	d	h	m	s	d	h	m	s
1721	20	16	42	05	21	15	50	16	23	4	35	03	21	19	54	09
1722	20	22	27	14	21	21	38	09	23	10	26	20	22	1	49	54
1723	21	4	18	08	22	3	22	09	23	16	11	17	22	7	46	27
1724	20	10	08	44	21	9	07	06	22	21	55	18	21	13	29	19
1725	20	15	48	13	21	14	53	39	23	3	47	23	21	19	22	06
1726	20	21	46	47	21	20	45	28	23	9	28	48	22	1	04	49
1727	21	3	38	35	22	2	28	26	23	15	12	27	22	6	49	46
1728	20	9	15	49	21	8	11	45	22	21	06	31	21	12	44	47
1729	20	15	08	04	21	14	07	05	23	2	50	52	21	18	25	18
1730	20	20	52	20	21	19	54	05	23	8	38	30	22	0	15	22
1731	21	2	41	17	22	1	37	19	23	14	26	21	22	6	08	38
1732	20	8	33	44	21	7	24	29	22	20	15	46	21	11	49	24
1733	20	14	11	33	21	13	10	52	23	2	10	57	21	17	43	28
1734	20	20	06	10	21	19	04	35	23	7	57	30	21	23	36	31
1735	21	1	59	26	22	0	49	48	23	13	42	42	22	5	27	52
1736	20	7	38	41	21	6	32	13	22	19	38	23	21	11	28	36
1737	20	13	39	42	21	12	30	49	23	1	27	45	21	17	13	51
1738	20	19	34	07	21	18	20	00	23	7	16	59	21	23	04	13
1739	21	1	27	49	22	0	06	47	23	13	06	54	22	5	01	07
1740	20	7	23	21	21	6	00	12	22	18	54	45	21	10	43	56
1741	20	13	01	58	21	11	49	29	23	0	45	48	21	16	35	37
1742	20	18	54	53	21	17	43	36	23	6	31	02	21	22	23	03
1743	21	0	47	09	21	23	28	44	23	12	15	50	22	4	05	35
1744	20	6	22	10	21	5	06	40	22	18	11	01	21	9	57	55
1745	20	12	13	17	21	11	00	31	23	0	00	12	21	15	40	57
1746	20	17	58	32	21	16	44	37	23	5	44	44	21	21	30	37
1747	20	23	42	07	21	22	22	44	23	11	30	38	22	3	28	48
1748	20	5	34	02	21	4	10	40	22	17	16	04	21	9	12	22
1749	20	11	15	49	21	9	54	21	22	23	04	27	21	15	02	26
1750	20	17	11	38	21	15	44	56	23	4	49	52	21	20	50	40
1751	20	23	08	25	21	21	33	38	23	10	34	17	22	2	37	08
1752	20	4	47	29	21	3	16	41	22	16	27	49	21	8	32	48
1753	20	10	40	58	21	9	17	22	22	22	19	12	21	14	20	18
1754	20	16	34	08	21	15	09	59	23	4	07	22	21	20	09	31
1755	20	22	22	28	21	20	51	50	23	9	58	54	22	2	05	57
1756	20	4	17	57	21	2	46	15	22	15	53	49	21	7	51	26
1757	20	10	00	52	21	8	33	47	22	21	45	45	21	13	42	41
1758	20	15	51	24	21	14	25	09	23	3	34	09	21	19	37	56
1759	20	21	46	50	21	20	14	25	23	9	19	34	22	1	28	12
1760	20	3	25	03	21	1	52	03	22	15	09	58	21	7	22	09

Year	March Equinox				June Solstice				September Equinox				December Solstice			
	d	h	m	s	d	h	m	s	d	h	m	s	d	h	m	s
1761	20	9	19	24	21	7	47	36	22	21	00	36	21	13	08	00
1762	20	15	15	30	21	13	36	43	23	2	45	13	21	18	52	38
1763	20	21	00	50	21	19	13	59	23	8	29	52	22	0	45	53
1764	20	2	51	08	21	1	07	21	22	14	19	34	21	6	30	43
1765	20	8	30	55	21	6	53	02	22	20	03	30	21	12	15	07
1766	20	14	17	08	21	12	39	52	23	1	47	53	21	18	04	39
1767	20	20	12	03	21	18	27	59	23	7	34	02	21	23	48	25
1768	20	1	48	03	21	0	03	30	22	13	24	39	21	5	37	36
1769	20	7	37	23	21	5	59	20	22	19	18	30	21	11	27	37
1770	20	13	30	45	21	11	50	47	23	1	04	48	21	17	19	02
1771	20	19	16	00	21	17	28	11	23	6	51	03	21	23	18	16
1772	20	1	10	39	20	23	23	22	22	12	45	37	21	5	08	54
1773	20	7	01	45	21	5	13	56	22	18	34	57	21	10	56	36
1774	20	12	57	02	21	11	04	18	23	0	23	44	21	16	49	18
1775	20	18	56	24	21	17	00	22	23	6	14	34	21	22	40	04
1776	20	0	37	13	20	22	43	21	22	12	05	11	21	4	31	39
1777	20	6	26	21	21	4	41	30	22	17	58	52	21	10	21	10
1778	20	12	21	34	21	10	35	22	22	23	46	02	21	16	08	19
1779	20	18	06	15	21	16	09	23	23	5	31	41	21	21	59	03
1780	19	23	54	00	20	21	59	54	22	11	26	49	21	3	45	05
1781	20	5	37	36	21	3	46	38	22	17	12	51	21	9	31	38
1782	20	11	22	23	21	9	27	24	22	22	53	42	21	15	22	50
1783	20	17	12	38	21	15	15	09	23	4	39	08	21	21	14	00
1784	19	22	53	48	20	20	51	55	22	10	23	50	21	3	01	04
1785	20	4	43	15	21	2	42	46	22	16	13	42	21	8	47	15
1786	20	10	41	59	21	8	37	16	22	22	00	31	21	14	34	58
1787	20	16	28	52	21	14	13	26	23	3	42	28	21	20	26	02
1788	19	22	16	35	20	20	09	17	22	9	38	14	21	2	16	10
1789	20	4	05	07	21	2	04	09	22	15	26	22	21	8	02	49
1790	20	9	54	21	21	7	49	56	22	21	11	56	21	13	52	53
1791	20	15	48	58	21	13	44	01	23	3	08	07	21	19	45	12
1792	19	21	32	56	20	19	26	33	22	9	00	53	21	1	34	17
1793	20	3	20	23	21	1	19	45	22	14	55	41	21	7	27	39
1794	20	9	16	38	21	7	16	13	22	20	46	45	21	13	24	17
1795	20	15	04	17	21	12	50	55	23	2	28	18	21	19	18	14
1796	19	20	54	04	20	18	42	33	22	8	24	16	21	1	09	54
1797	20	2	47	49	21	0	34	52	22	14	11	01	21	6	54	10
1798	20	8	38	58	21	6	16	29	22	19	50	29	21	12	41	16
1799	20	14	31	04	21	12	09	13	23	1	41	07	21	18	32	09
1800	20	20	11	52	21	17	51	54	23	7	25	52	22	0	16	30

Year	March Equinox				June Solstice				September Equinox				December Solstice			
	d	h	m	s	d	h	m	s	d	h	m	s	d	h	m	s
1801	21	1	55	47	21	23	42	06	23	13	13	28	22	6	02	04
1802	21	7	48	17	22	5	35	46	23	19	02	33	22	11	50	58
1803	21	13	34	07	22	11	08	27	24	0	43	51	22	17	36	15
1804	20	19	17	31	21	16	56	28	23	6	40	06	21	23	24	27
1805	21	1	04	46	21	22	49	48	23	12	29	19	22	5	13	37
1806	21	6	52	33	22	4	30	20	23	18	08	15	22	11	04	16
1807	21	12	41	46	22	10	21	11	24	0	02	02	22	17	01	12
1808	20	18	29	49	21	16	06	28	23	5	51	30	21	22	50	03
1809	21	0	22	09	21	21	57	01	23	11	42	27	22	4	38	03
1810	21	6	20	50	22	3	55	55	23	17	37	04	22	10	34	21
1811	21	12	13	22	22	9	36	43	23	23	20	24	22	16	25	31
1812	20	17	59	17	21	15	29	12	23	5	16	10	21	22	16	40
1813	20	23	50	00	21	21	27	55	23	11	07	56	22	4	07	40
1814	21	5	43	05	22	3	09	52	23	16	46	47	22	9	51	40
1815	21	11	30	32	22	8	58	32	23	22	42	20	22	15	43	36
1816	20	17	16	07	21	14	44	35	23	4	33	02	21	21	30	27
1817	20	23	00	42	21	20	30	12	23	10	18	26	22	3	16	52
1818	21	4	49	54	22	2	23	16	23	16	09	31	22	9	14	30
1819	21	10	38	49	22	7	58	05	23	21	47	14	22	15	02	09
1820	20	16	23	49	21	13	42	36	23	3	37	59	21	20	48	24
1821	20	22	15	33	21	19	37	25	23	9	28	18	22	2	36	00
1822	21	4	09	26	22	1	19	01	23	15	03	41	22	8	18	20
1823	21	9	54	46	22	7	08	39	23	20	56	23	22	14	11	26
1824	20	15	38	52	21	12	58	42	23	2	46	40	21	19	59	28
1825	20	21	25	37	21	18	47	10	23	8	31	40	22	1	43	08
1826	21	3	16	20	22	0	41	59	23	14	28	32	22	7	39	40
1827	21	9	08	05	22	6	21	12	23	20	13	49	22	13	27	43
1828	20	14	53	18	21	12	07	53	23	2	09	51	21	19	18	40
1829	20	20	42	13	21	18	05	05	23	8	04	57	22	1	15	23
1830	21	2	36	53	21	23	48	23	23	13	41	35	22	7	05	05
1831	21	8	26	20	22	5	36	44	23	19	35	01	22	13	01	43
1832	20	14	16	38	21	11	27	03	23	1	27	54	21	18	51	50
1833	20	20	11	59	21	17	16	43	23	7	12	23	22	0	33	39
1834	21	2	04	22	21	23	11	44	23	13	06	52	22	6	30	08
1835	21	7	55	45	22	4	54	53	23	18	49	15	22	12	18	19
1836	20	13	39	03	21	10	41	12	23	0	38	05	21	18	02	29
1837	20	19	23	23	21	16	37	01	23	6	31	24	21	23	53	24
1838	21	1	17	02	21	22	18	44	23	12	07	07	22	5	33	25
1839	21	6	59	28	22	4	00	34	23	17	59	12	22	11	22	06
1840	20	12	40	44	21	9	47	49	22	23	52	56	21	17	12	38

Year	March Equinox				June Solstice				September Equinox				December Solstice			
	d	h	m	s	d	h	m	s	d	h	m	s	d	h	m	s
1841	20	18	28	16	21	15	33	23	23	5	33	48	21	22	55	39
1842	21	0	13	05	21	21	22	02	23	11	25	36	22	4	55	32
1843	21	6	04	33	22	3	02	32	23	17	09	18	22	10	47	39
1844	20	11	53	51	21	8	46	02	22	22	57	21	21	16	31	04
1845	20	17	43	41	21	14	42	14	23	4	53	45	21	22	26	52
1846	20	23	45	38	21	20	30	45	23	10	31	12	22	4	12	15
1847	21	5	32	12	22	2	18	15	23	16	22	26	22	10	05	02
1848	20	11	17	54	21	8	14	17	22	22	20	04	21	15	59	53
1849	20	17	12	46	21	14	07	07	23	4	03	39	21	21	40	51
1850	20	23	02	02	21	19	58	53	23	10	00	17	22	3	37	35
1851	21	4	54	12	22	1	42	54	23	15	50	37	22	9	28	44
1852	20	10	41	19	21	7	28	28	22	21	40	25	21	15	12	35
1853	20	16	23	59	21	13	22	16	23	3	36	17	21	21	10	52
1854	20	22	19	53	21	19	07	19	23	9	12	02	22	2	58	21
1855	21	4	05	16	22	0	47	43	23	14	58	25	22	8	47	25
1856	20	9	49	10	21	6	35	51	22	20	53	28	21	14	38	39
1857	20	15	45	13	21	12	25	06	23	2	32	33	21	20	15	41
1858	20	21	31	51	21	18	12	34	23	8	22	09	22	2	09	56
1859	21	3	18	29	21	23	55	57	23	14	07	54	22	8	00	56
1860	20	9	04	38	21	5	42	07	22	19	52	46	21	13	41	20
1861	20	14	46	35	21	11	33	45	23	1	46	51	21	19	33	55
1862	20	20	42	44	21	17	19	35	23	7	26	19	22	1	18	46
1863	21	2	30	03	21	23	01	59	23	13	16	26	22	7	06	04
1864	20	8	10	07	21	4	51	56	22	19	16	20	21	13	03	36
1865	20	14	05	59	21	10	46	10	23	0	59	34	21	18	49	38
1866	20	19	54	44	21	16	34	04	23	6	50	35	22	0	49	27
1867	21	1	46	29	21	22	19	52	23	12	42	43	22	6	47	00
1868	20	7	43	46	21	4	09	59	22	18	31	16	21	12	28	02
1869	20	13	32	09	21	10	04	24	23	0	28	00	21	18	23	34
1870	20	19	32	03	21	15	56	19	23	6	09	35	22	0	13	11
1871	21	1	19	47	21	21	41	53	23	11	56	10	22	5	59	09
1872	20	6	57	23	21	3	31	56	22	17	53	30	21	11	53	32
1873	20	12	52	26	21	9	25	04	22	23	35	10	21	17	32	36
1874	20	18	37	59	21	15	07	12	23	5	22	55	21	23	22	08
1875	21	0	21	16	21	20	47	01	23	11	15	01	22	5	15	54
1876	20	6	09	58	21	2	32	14	22	16	58	44	21	10	54	22
1877	20	11	47	55	21	8	18	01	22	22	48	11	21	16	50	20
1878	20	17	42	39	21	14	03	46	23	4	26	35	21	22	41	18
1879	20	23	31	57	21	19	44	24	23	10	09	19	22	4	24	29
1880	20	5	13	37	21	1	31	44	22	16	06	46	21	10	18	18

Year	March Equinox			June Solstice			September Equinox			December Solstice		
	d	h m s		d	h m s		d	h m s		d	h m s	
1881	20	11 13 54		21	7 28 12		22	21 50 10		21	16 00 33	
1882	20	17 04 39		21	13 16 41		23	3 37 45		21	21 53 40	
1883	20	22 49 48		21	19 03 31		23	9 32 20		22	3 52 06	
1884	20	4 44 34		21	0 59 06		22	15 21 17		21	9 33 35	
1885	20	10 29 39		21	6 50 54		22	21 16 12		21	15 27 32	
1886	20	16 26 26		21	12 41 15		23	3 04 34		21	21 19 56	
1887	20	22 18 32		21	18 27 25		23	8 54 19		22	3 04 56	
1888	20	3 56 00		21	0 14 27		22	14 53 48		21	9 03 15	
1889	20	9 50 54		21	6 10 04		22	20 38 14		21	14 51 58	
1890	20	15 41 05		21	11 53 59		23	2 22 38		21	20 44 59	
1891	20	21 24 56		21	17 32 39		23	8 13 53		22	2 40 36	
1892	20	3 22 04		20	23 23 33		22	13 59 39		21	8 19 23	
1893	20	9 08 28		21	5 10 06		22	19 46 20		21	14 07 26	
1894	20	14 59 23		21	10 56 54		23	1 27 34		21	19 58 19	
1895	20	20 49 19		21	16 44 09		23	7 10 31		22	1 38 45	
1896	20	2 23 27		20	22 28 17		22	13 03 36		21	7 29 26	
1897	20	8 16 14		21	4 23 27		22	18 49 02		21	13 12 51	
1898	20	14 06 33		21	10 07 12		23	0 34 38		21	18 59 13	
1899	20	19 45 48		21	15 45 40		23	6 30 11		22	0 56 16	
1900	21	1 39 05		21	21 39 52		23	12 20 16		22	6 41 39	
1901	21	7 23 40		22	3 27 53		23	18 09 02		22	12 36 42	
1902	21	13 16 40		22	9 15 15		23	23 55 27		22	18 35 39	
1903	21	19 14 54		22	15 05 04		24	5 43 49		23	0 20 34	
1904	21	0 58 42		21	20 51 30		23	11 40 22		22	6 14 06	
1905	21	6 57 41		22	2 51 33		23	17 30 07		22	12 03 53	
1906	21	12 52 59		22	8 41 59		23	23 15 11		22	17 53 29	
1907	21	18 33 12		22	14 23 12		24	5 09 06		22	23 51 45	
1908	21	0 27 27		21	20 19 17		23	10 58 30		22	5 33 40	
1909	21	6 13 12		22	2 05 48		23	16 44 46		22	11 20 03	
1910	21	12 03 09		22	7 48 58		23	22 31 00		22	17 12 00	
1911	21	17 54 38		22	13 35 49		24	4 17 49		22	22 53 29	
1912	20	23 29 38		21	19 17 11		23	10 08 19		22	4 45 00	
1913	21	5 18 16		22	1 09 48		23	15 53 04		22	10 35 12	
1914	21	11 11 03		22	6 55 22		23	21 34 08		22	16 22 45	
1915	21	16 51 37		22	12 29 43		24	3 24 09		22	22 16 08	
1916	20	22 47 14		21	18 24 47		23	9 15 07		22	3 58 56	
1917	21	4 37 37		22	0 14 42		23	15 00 32		22	9 46 04	
1918	21	10 26 01		22	6 00 01		23	20 46 02		22	15 41 54	
1919	21	16 19 32		22	11 53 59		24	2 35 46		22	21 27 29	
1920	20	21 59 42		21	17 40 13		23	8 28 32		22	3 17 24	

Year	March Equinox				June Solstice				September Equinox				December Solstice			
	d	h	m	s	d	h	m	s	d	h	m	s	d	h	m	s
1921	21	3	51	27	21	23	36	04	23	14	20	10	22	9	07	51
1922	21	9	49	00	22	5	27	07	23	20	09	59	22	14	57	21
1923	21	15	29	10	22	11	03	10	24	2	03	59	22	20	53	41
1924	20	21	20	33	21	16	59	48	23	7	58	42	22	2	45	54
1925	21	3	12	36	21	22	50	23	23	13	43	48	22	8	37	04
1926	21	9	01	36	22	4	30	27	23	19	27	02	22	14	33	46
1927	21	14	59	31	22	10	22	37	24	1	17	21	22	20	18	55
1928	20	20	44	40	21	16	06	52	23	7	05	55	22	2	04	05
1929	21	2	35	16	21	22	01	02	23	12	52	44	22	7	53	10
1930	21	8	30	11	22	3	53	15	23	18	36	22	22	13	39	59
1931	21	14	06	43	22	9	28	29	24	0	23	43	22	19	30	00
1932	20	19	54	00	21	15	23	03	23	6	16	18	22	1	14	40
1933	21	1	43	31	21	21	12	15	23	12	01	34	22	6	57	55
1934	21	7	28	20	22	2	48	20	23	17	45	36	22	12	49	51
1935	21	13	18	10	22	8	38	20	23	23	38	35	22	18	37	32
1936	20	18	58	16	21	14	22	03	23	5	26	23	22	0	27	06
1937	21	0	45	29	21	20	12	25	23	11	13	21	22	6	22	05
1938	21	6	43	30	22	2	04	01	23	16	59	56	22	12	13	50
1939	21	12	28	53	22	7	39	52	23	22	49	52	22	18	06	24
1940	20	18	24	10	21	13	36	51	23	4	46	00	21	23	55	10
1941	21	0	20	49	21	19	33	44	23	10	33	12	22	5	44	36
1942	21	6	11	04	22	1	16	42	23	16	16	54	22	11	40	00
1943	21	12	03	04	22	7	12	46	23	22	12	11	22	17	29	33
1944	20	17	49	03	21	13	02	45	23	4	02	05	21	23	15	16
1945	20	23	37	40	21	18	52	32	23	9	50	15	22	5	04	02
1946	21	5	33	09	22	0	44	48	23	15	41	06	22	10	53	49
1947	21	11	13	09	22	6	19	18	23	21	29	06	22	16	43	15
1948	20	16	57	16	21	12	11	05	23	3	22	11	21	22	33	46
1949	20	22	48	32	21	18	03	15	23	9	06	20	22	4	23	23
1950	21	4	35	39	21	23	36	33	23	14	44	04	22	10	13	51
1951	21	10	26	14	22	5	25	21	23	20	37	22	22	16	00	35
1952	20	16	14	16	21	11	13	04	23	2	24	11	21	21	43	40
1953	20	22	01	01	21	17	00	28	23	8	06	25	22	3	31	59
1954	21	3	53	57	21	22	54	34	23	13	55	48	22	9	24	53
1955	21	9	35	38	22	4	31	55	23	19	41	25	22	15	11	26
1956	20	15	20	50	21	10	24	17	23	1	35	35	21	21	00	01
1957	20	21	17	02	21	16	21	04	23	7	26	37	22	2	49	10
1958	21	3	06	22	21	21	57	28	23	13	09	24	22	8	40	16
1959	21	8	55	05	22	3	50	20	23	19	09	00	22	14	34	54
1960	20	14	43	14	21	9	42	52	23	0	59	26	21	20	26	30

Year	March Equinox				June Solstice				September Equinox				December Solstice			
	d	h	m	s	d	h	m	s	d	h	m	s	d	h	m	s
1961	20	20	32	41	21	15	30	41	23	6	43	01	22	2	20	03
1962	21	2	30	08	21	21	24	42	23	12	35	48	22	8	15	51
1963	21	8	20	17	22	3	04	37	23	18	24	06	22	14	02	29
1964	20	14	10	28	21	8	57	25	23	0	17	16	21	19	50	09
1965	20	20	05	22	21	14	56	18	23	6	06	37	22	1	41	03
1966	21	1	53	33	21	20	34	01	23	11	43	47	22	7	28	48
1967	21	7	37	27	22	2	23	30	23	17	38	42	22	13	16	58
1968	20	13	22	40	21	8	13	57	22	23	26	51	21	19	00	26
1969	20	19	08	46	21	13	55	44	23	5	07	35	22	0	44	23
1970	21	0	57	01	21	19	43	20	23	10	59	40	22	6	36	22
1971	21	6	38	51	22	1	20	18	23	16	45	36	22	12	24	38
1972	20	12	22	08	21	7	06	54	22	22	33	29	21	18	13	38
1973	20	18	13	12	21	13	01	20	23	4	21	53	22	0	08	28
1974	21	0	07	24	21	18	38	23	23	9	59	12	22	5	56	44
1975	21	5	57	27	22	0	27	13	23	15	55	59	22	11	46	21
1976	20	11	50	24	21	6	25	00	22	21	48	59	21	17	35	57
1977	20	17	43	05	21	12	14	35	23	3	30	04	21	23	23	58
1978	20	23	34	23	21	18	10	24	23	9	26	14	22	5	21	48
1979	21	5	22	47	21	23	57	02	23	15	17	15	22	11	10	37
1980	20	11	10	33	21	5	47	53	22	21	09	33	21	16	56	58
1981	20	17	03	43	21	11	45	33	23	3	06	05	21	22	51	25
1982	20	22	56	44	21	17	23	53	23	8	47	05	22	4	39	03
1983	21	4	39	39	21	23	09	35	23	14	42	31	22	10	30	50
1984	20	10	25	15	21	5	03	09	22	20	33	49	21	16	23	44
1985	20	16	14	39	21	10	45	03	23	2	08	22	21	22	08	35
1986	20	22	03	38	21	16	30	54	23	7	59	48	22	4	03	04
1987	21	3	52	54	21	22	11	41	23	13	46	13	22	9	46	50
1988	20	9	39	33	21	3	57	29	22	19	29	47	21	15	28	50
1989	20	15	29	12	21	9	53	58	23	1	20	33	21	21	22	58
1990	20	21	20	14	21	15	33	46	23	6	56	28	22	3	07	58
1991	21	3	02	55	21	21	19	46	23	12	49	04	22	8	54	38
1992	20	8	49	02	21	3	15	08	22	18	43	46	21	14	44	14
1993	20	14	41	38	21	9	00	44	23	0	23	29	21	20	26	49
1994	20	20	29	01	21	14	48	33	23	6	20	14	22	2	23	44
1995	21	2	15	27	21	20	35	24	23	12	14	01	22	8	17	50
1996	20	8	04	07	21	2	24	46	22	18	01	08	21	14	06	56
1997	20	13	55	42	21	8	20	59	22	23	56	49	21	20	08	05
1998	20	19	55	35	21	14	03	38	23	5	38	15	22	1	57	31
1999	21	1	46	53	21	19	50	11	23	11	32	34	22	7	44	52
2000	20	7	36	19	21	1	48	46	22	17	28	40	21	13	38	30

Year	March Equinox				June Solstice				September Equinox				December Solstice			
	d	h	m	s	d	h	m	s	d	h	m	s	d	h	m	s
2001	20	13	31	47	21	7	38	48	22	23	05	32	21	19	22	34
2002	20	19	17	13	21	13	25	29	23	4	56	28	22	1	15	26
2003	21	1	00	50	21	19	11	32	23	10	47	53	22	7	04	53
2004	20	6	49	42	21	0	57	57	22	16	30	54	21	12	42	40
2005	20	12	34	29	21	6	47	12	22	22	24	14	21	18	36	01
2006	20	18	26	39	21	12	26	56	23	4	04	27	22	0	23	11
2007	21	0	08	30	21	18	07	30	23	9	52	18	22	6	08	54
2008	20	5	49	23	21	0	00	27	22	15	45	34	21	12	04	51
2009	20	11	44	44	21	5	46	37	22	21	19	41	21	17	47	53
2010	20	17	33	18	21	11	29	31	23	3	10	07	21	23	39	33
2011	20	23	21	50	21	17	17	36	23	9	05	44	22	5	31	09
2012	20	5	15	32	20	23	09	55	22	14	50	05	21	11	12	43
2013	20	11	03	02	21	5	05	04	22	20	45	15	21	17	12	07
2014	20	16	58	12	21	10	52	21	23	2	30	11	21	23	04	08
2015	20	22	46	16	21	16	39	02	23	8	21	40	22	4	49	04
2016	20	4	31	19	20	22	35	18	22	14	22	15	21	10	45	18
2017	20	10	29	46	21	4	25	17	22	20	02	56	21	16	29	05
2018	20	16	16	36	21	10	08	26	23	1	55	14	21	22	23	52
2019	20	21	59	34	21	15	55	23	23	7	51	18	22	4	20	34
2020	20	3	50	45	20	21	44	49	22	13	31	47	21	10	03	28
2021	20	9	38	37	21	3	33	18	22	19	22	13	21	16	00	26
2022	20	15	34	33	21	9	14	59	23	1	04	50	21	21	49	21
2023	20	21	25	35	21	14	58	58	23	6	51	07	22	3	28	30
2024	20	3	07	32	20	20	52	08	22	12	44	47	21	9	21	42
2025	20	9	02	37	21	2	43	24	22	18	20	28	21	15	04	13
2026	20	14	47	05	21	8	25	38	23	0	06	21	21	20	51	22
2027	20	20	25	49	21	14	11	58	23	6	02	51	22	2	43	17
2028	20	2	18	16	20	20	03	08	22	11	46	26	21	8	20	48
2029	20	8	03	06	21	1	49	25	22	17	39	38	21	14	15	14
2030	20	13	53	14	21	7	32	26	22	23	28	01	21	20	10	46
2031	20	19	42	06	21	13	18	16	23	5	16	25	22	1	56	41
2032	20	1	23	01	20	19	09	54	22	11	12	00	21	7	57	04
2033	20	7	23	51	21	1	02	16	22	16	52	48	21	13	47	08
2034	20	13	18	38	21	6	45	19	22	22	40	42	21	19	35	08
2035	20	19	03	52	21	12	34	16	23	4	40	04	22	1	32	01
2036	20	1	03	58	20	18	33	22	22	10	24	27	21	7	14	01
2037	20	6	51	24	21	0	23	35	22	16	14	13	21	13	08	53
2038	20	12	41	47	21	6	10	32	22	22	03	25	21	19	03	28
2039	20	18	33	11	21	11	58	34	23	3	50	45	22	0	41	45
2040	20	0	12	51	20	17	47	32	22	9	46	04	21	6	34	00

Year	March Equinox				June Solstice				September Equinox				December Solstice			
	d	h	m	s	d	h	m	s	d	h	m	s	d	h	m	s
2041	20	6	07	58	20	23	37	01	22	15	27	43	21	12	19	30
2042	20	11	54	29	21	5	17	00	22	21	12	43	21	18	05	14
2043	20	17	28	58	21	10	59	33	23	3	08	07	22	0	02	25
2044	19	23	21	44	20	16	52	19	22	8	49	03	21	5	44	47
2045	20	5	08	49	20	22	35	06	22	14	34	07	21	11	36	19
2046	20	10	59	04	21	4	15	52	22	20	22	56	21	17	29	42
2047	20	16	53	53	21	10	04	43	23	2	09	19	21	23	08	28
2048	19	22	35	04	20	15	55	10	22	8	01	53	21	5	03	31
2049	20	4	29	52	20	21	48	34	22	13	43	52	21	10	53	26
2050	20	10	20	51	21	3	34	18	22	19	29	48	21	16	40	00
2051	20	16	00	28	21	9	19	57	23	1	28	40	21	22	35	27
2052	19	21	57	25	20	15	17	33	22	7	17	02	21	4	18	33
2053	20	3	48	44	20	21	05	30	22	13	07	40	21	10	11	18
2054	20	9	35	52	21	2	48	35	22	19	00	54	21	16	11	22
2055	20	15	30	03	21	8	41	18	23	0	50	11	21	21	57	03
2056	19	21	12	32	20	14	29	41	22	6	40	59	21	3	53	06
2057	20	3	09	25	20	20	20	34	22	12	24	47	21	9	44	22
2058	20	9	06	32	21	2	05	35	22	18	09	56	21	15	26	31
2059	20	14	45	48	21	7	48	49	23	0	05	05	21	21	19	31
2060	19	20	40	05	20	13	47	13	22	5	49	47	21	3	03	02
2061	20	2	27	52	20	19	33	50	22	11	33	00	21	8	50	28
2062	20	8	09	08	21	1	13	02	22	17	21	33	21	14	44	20
2063	20	14	00	51	21	7	03	35	22	23	09	54	21	20	22	48
2064	19	19	40	20	20	12	47	23	22	4	58	45	21	2	10	29
2065	20	1	29	55	20	18	34	13	22	10	44	21	21	8	02	31
2066	20	7	21	34	21	0	18	10	22	16	28	47	21	13	47	19
2067	20	12	55	25	21	5	57	48	22	22	21	27	21	19	44	55
2068	19	18	50	48	20	11	55	36	22	4	08	36	21	1	34	31
2069	20	0	46	54	20	17	43	09	22	9	53	44	21	7	23	56
2070	20	6	36	47	20	23	24	29	22	15	46	42	21	13	21	14
2071	20	12	36	35	21	5	22	44	22	21	39	39	21	19	05	48
2072	19	18	22	59	20	11	15	44	22	3	29	46	21	0	57	58
2073	20	0	15	18	20	17	08	58	22	9	17	16	21	6	52	41
2074	20	6	10	51	20	23	00	27	22	15	05	48	21	12	37	16
2075	20	11	48	37	21	4	42	28	22	21	00	52	21	18	29	08
2076	19	17	40	54	20	10	38	51	22	2	52	19	21	0	15	29
2077	19	23	33	12	20	16	25	34	22	8	37	59	21	6	02	57
2078	20	5	12	59	20	22	00	06	22	14	26	41	21	12	00	08
2079	20	11	02	52	21	3	51	31	22	20	15	22	21	17	46	14
2080	19	16	46	21	20	9	36	20	22	1	58	40	20	23	35	04

Year	March Equinox				June Solstice				September Equinox				December Solstice			
	d	h	m	s	d	h	m	s	d	h	m	s	d	h	m	s
2081	19	22	36	32	20	15	18	40	22	7	40	06	21	5	24	49
2082	20	4	32	55	20	21	05	36	22	13	25	17	21	11	07	00
2083	20	10	12	52	21	2	46	15	22	19	14	00	21	16	55	45
2084	19	16	01	51	20	8	42	58	22	1	01	31	20	22	43	42
2085	19	21	55	59	20	14	35	17	22	6	45	57	21	4	31	12
2086	20	3	37	46	20	20	12	06	22	12	34	43	21	10	25	11
2087	20	9	30	51	21	2	08	31	22	18	30	52	21	16	11	04
2088	19	15	19	37	20	7	59	18	22	0	20	53	20	21	58	48
2089	19	21	09	13	20	13	45	37	22	6	09	34	21	3	54	42
2090	20	3	04	35	20	19	38	36	22	12	02	03	21	9	46	18
2091	20	8	44	36	21	1	21	41	22	17	53	36	21	15	41	19
2092	19	14	36	21	20	7	17	45	21	23	44	36	20	21	34	43
2093	19	20	37	21	20	13	09	53	22	5	31	44	21	3	23	35
2094	20	2	24	25	20	18	45	10	22	11	19	25	21	9	16	11
2095	20	8	17	50	21	0	41	44	22	17	14	00	21	15	03	43
2096	19	14	05	49	20	6	33	42	21	22	57	38	20	20	49	07
2097	19	19	51	19	20	12	16	18	22	4	38	46	21	2	40	22
2098	20	1	43	23	20	18	05	55	22	10	27	10	21	8	23	47
2099	20	7	20	40	20	23	44	36	22	16	14	03	21	14	07	19
2100	20	13	06	41	21	5	35	12	22	22	03	29	21	19	53	54
2101	20	18	59	38	21	11	24	31	23	3	49	39	22	1	42	02
2102	21	0	39	05	21	16	56	35	23	9	34	05	22	7	35	54
2103	21	6	26	15	21	22	48	55	23	15	27	37	22	13	27	12
2104	20	12	17	20	21	4	40	46	22	21	12	44	21	19	15	36
2105	20	18	09	37	21	10	21	59	23	2	55	18	22	1	06	59
2106	21	0	07	22	21	16	15	55	23	8	50	07	22	6	56	20
2107	21	5	53	20	21	22	04	09	23	14	40	22	22	12	45	35
2108	20	11	42	16	21	4	00	41	22	20	31	24	21	18	37	58
2109	20	17	39	27	21	9	58	26	23	2	22	27	22	0	30	48
2110	20	23	25	02	21	15	35	33	23	8	10	22	22	6	22	15
2111	21	5	14	03	21	21	29	09	23	14	08	58	22	12	11	15
2112	20	11	06	18	21	3	23	07	22	19	58	52	21	17	59	36
2113	20	16	54	16	21	9	00	57	23	1	39	07	21	23	50	03
2114	20	22	42	28	21	14	50	01	23	7	31	11	22	5	42	44
2115	21	4	25	49	21	20	34	09	23	13	16	03	22	11	30	21
2116	20	10	12	10	21	2	20	49	22	18	59	39	21	17	17	44
2117	20	16	09	41	21	8	12	57	23	0	47	57	21	23	06	43
2118	20	21	56	34	21	13	46	52	23	6	30	04	22	4	52	11
2119	21	3	42	25	21	19	39	01	23	12	23	28	22	10	40	03
2120	20	9	32	11	21	1	35	39	22	18	09	35	21	16	27	27

Year	March Equinox				June Solstice				September Equinox				December Solstice			
	d	h	m	s	d	h	m	s	d	h	m	s	d	h	m	s
2121	20	15	19	11	21	7	15	08	22	23	47	11	21	22	14	39
2122	20	21	07	16	21	13	05	47	23	5	42	44	22	4	04	25
2123	21	2	52	20	21	18	53	17	23	11	33	28	22	9	49	43
2124	20	8	39	02	21	0	41	33	22	17	22	16	21	15	38	17
2125	20	14	33	10	21	6	36	19	22	23	17	24	21	21	35	32
2126	20	20	19	13	21	12	13	46	23	5	03	05	22	3	29	03
2127	21	2	06	26	21	18	05	12	23	10	59	15	22	9	24	20
2128	20	8	02	09	21	0	02	18	22	16	49	22	21	15	16	42
2129	20	13	58	31	21	5	41	58	22	22	28	21	21	21	05	17
2130	20	19	52	46	21	11	33	37	23	4	24	29	22	2	56	35
2131	21	1	40	37	21	17	25	41	23	10	12	50	22	8	43	50
2132	20	7	28	09	20	23	16	08	22	15	55	50	21	14	30	00
2133	20	13	19	24	21	5	10	27	22	21	47	50	21	20	22	42
2134	20	19	05	20	21	10	48	09	23	3	31	34	22	2	07	06
2135	21	0	48	50	21	16	34	24	23	9	24	47	22	7	51	57
2136	20	6	36	56	20	22	28	32	22	15	15	55	21	13	41	21
2137	20	12	25	18	21	4	04	37	22	20	51	25	21	19	27	27
2138	20	18	07	42	21	9	49	31	23	2	44	20	22	1	20	54
2139	20	23	52	02	21	15	38	03	23	8	31	48	22	7	10	43
2140	20	5	41	31	20	21	22	34	22	14	12	02	21	12	55	09
2141	20	11	35	59	21	3	14	12	22	20	05	57	21	18	49	42
2142	20	17	28	20	21	8	56	06	23	1	52	01	22	0	37	41
2143	20	23	15	08	21	14	46	37	23	7	44	04	22	6	26	27
2144	20	5	05	43	20	20	47	41	22	13	38	25	21	12	22	48
2145	20	11	01	02	21	2	30	27	22	19	15	58	21	18	08	31
2146	20	16	47	34	21	8	18	08	23	1	12	46	22	0	00	27
2147	20	22	36	57	21	14	12	39	23	7	08	13	22	5	50	30
2148	20	4	27	53	20	19	59	57	22	12	51	34	21	11	34	52
2149	20	10	17	35	21	1	52	15	22	18	47	43	21	17	34	29
2150	20	16	06	09	21	7	34	08	23	0	33	23	21	23	26	08
2151	20	21	51	53	21	13	20	14	23	6	21	39	22	5	13	49
2152	20	3	42	34	20	19	15	37	22	12	14	08	21	11	07	07
2153	20	9	40	18	21	0	55	40	22	17	49	26	21	16	48	45
2154	20	15	26	27	21	6	40	07	22	23	41	06	21	22	36	58
2155	20	21	10	13	21	12	33	38	23	5	32	02	22	4	26	37
2156	20	2	58	15	20	18	20	43	22	11	09	00	21	10	06	48
2157	20	8	43	52	21	0	09	17	22	17	01	40	21	16	00	50
2158	20	14	30	43	21	5	50	28	22	22	48	51	21	21	47	17
2159	20	20	16	17	21	11	35	00	23	4	38	27	22	3	30	27
2160	20	2	02	09	20	17	28	42	22	10	34	02	21	9	25	49

Year	March Equinox				June Solstice				September Equinox				December Solstice			
	d	h	m	s	d	h	m	s	d	h	m	s	d	h	m	s
2161	20	7	56	15	20	23	10	26	22	16	11	29	21	15	14	58
2162	20	13	42	01	21	4	54	33	22	22	03	31	21	21	08	56
2163	20	19	27	51	21	10	48	28	23	3	58	14	22	3	05	31
2164	20	1	26	32	20	16	38	57	22	9	38	47	21	8	48	45
2165	20	7	20	52	20	22	29	10	22	15	34	19	21	14	45	23
2166	20	13	14	25	21	4	18	04	22	21	25	56	21	20	37	52
2167	20	19	05	08	21	10	09	43	23	3	13	44	22	2	22	13
2168	20	0	51	24	20	16	07	56	22	9	09	33	21	8	18	52
2169	20	6	48	09	20	21	53	41	22	14	48	19	21	14	03	47
2170	20	12	32	56	21	3	35	49	22	20	39	39	21	19	49	02
2171	20	18	14	06	21	9	26	55	23	2	36	31	22	1	40	44
2172	20	0	06	00	20	15	13	40	22	8	15	01	21	7	20	22
2173	20	5	48	55	20	20	56	03	22	14	04	38	21	13	16	17
2174	20	11	32	45	21	2	37	42	22	19	52	54	21	19	10	35
2175	20	17	20	57	21	8	21	30	23	1	34	18	22	0	51	13
2176	19	23	07	01	20	14	11	21	22	7	26	24	21	6	46	14
2177	20	5	07	44	20	19	55	02	22	13	04	07	21	12	30	18
2178	20	10	54	57	21	1	38	13	22	18	51	19	21	18	15	30
2179	20	16	36	53	21	7	34	31	23	0	47	56	22	0	11	31
2180	19	22	32	31	20	13	28	38	22	6	27	26	21	5	53	08
2181	20	4	20	36	20	19	16	19	22	12	20	18	21	11	47	55
2182	20	10	08	28	21	1	03	43	22	18	17	11	21	17	42	28
2183	20	16	02	02	21	6	54	51	23	0	06	54	21	23	24	12
2184	19	21	47	52	20	12	47	52	22	6	04	27	21	5	23	24
2185	20	3	44	35	20	18	35	06	22	11	47	55	21	11	16	50
2186	20	9	32	35	21	0	18	59	22	17	36	00	21	17	05	43
2187	20	15	14	28	21	6	10	34	22	23	32	35	21	23	03	01
2188	19	21	14	34	20	12	02	07	22	5	11	55	21	4	43	34
2189	20	3	05	42	20	17	45	22	22	11	00	26	21	10	34	00
2190	20	8	49	46	20	23	30	16	22	16	52	50	21	16	27	36
2191	20	14	40	29	21	5	21	21	22	22	35	20	21	22	06	52
2192	19	20	21	42	20	11	09	16	22	4	23	54	21	3	58	35
2193	20	2	13	24	20	16	52	23	22	10	03	57	21	9	45	49
2194	20	8	01	26	20	22	34	07	22	15	49	45	21	15	25	00
2195	20	13	37	51	21	4	21	23	22	21	45	55	21	21	17	53
2196	19	19	32	36	20	10	13	39	22	3	27	18	21	3	01	35
2197	20	1	19	06	20	15	55	53	22	9	14	08	21	8	55	19
2198	20	7	00	23	20	21	40	08	22	15	09	33	21	14	55	48
2199	20	12	57	27	21	3	33	26	22	20	55	16	21	20	38	38
2200	20	18	47	34	21	9	23	12	23	2	48	10	22	2	33	41

Year	March Equinox				June Solstice				September Equinox				December Solstice			
	d	h	m	s	d	h	m	s	d	h	m	s	d	h	m	s
2201	21	0	47	50	21	15	12	22	23	8	35	45	22	8	27	30
2202	21	6	42	49	21	21	02	51	23	14	24	36	22	14	11	59
2203	21	12	22	50	22	2	56	42	23	20	21	53	22	20	09	40
2204	20	18	20	37	21	8	54	36	23	2	06	15	22	1	55	14
2205	21	0	10	56	21	14	39	28	23	7	53	47	22	7	43	32
2206	21	5	51	40	21	20	22	23	23	13	52	14	22	13	39	28
2207	21	11	45	42	22	2	14	17	23	19	38	48	22	19	18	28
2208	20	17	28	00	21	7	58	40	23	1	26	53	22	1	12	16
2209	20	23	17	52	21	13	40	37	23	7	09	54	22	7	07	02
2210	21	5	06	54	21	19	24	00	23	12	51	54	22	12	48	51
2211	21	10	45	18	22	1	09	02	23	18	42	22	22	18	41	36
2212	20	16	43	12	21	7	00	50	23	0	23	41	22	0	24	00
2213	20	22	36	25	21	12	45	02	23	6	07	23	22	6	08	57
2214	21	4	15	22	21	18	28	25	23	12	01	55	22	12	04	52
2215	21	10	08	52	22	0	26	15	23	17	48	44	22	17	46	45
2216	20	15	54	22	21	6	15	06	22	23	36	57	21	23	37	49
2217	20	21	44	41	21	11	59	55	23	5	26	23	22	5	32	35
2218	21	3	39	04	21	17	49	43	23	11	17	16	22	11	14	49
2219	21	9	18	48	21	23	37	16	23	17	13	54	22	17	09	55
2220	20	15	13	52	21	5	32	33	22	23	02	07	21	23	02	08
2221	20	21	07	33	21	11	19	45	23	4	48	57	22	4	54	41
2222	21	2	46	56	21	17	00	50	23	10	44	09	22	10	55	16
2223	21	8	45	16	21	22	57	39	23	16	33	20	22	16	41	16
2224	20	14	38	54	21	4	44	34	22	22	19	22	21	22	29	04
2225	20	20	30	13	21	10	26	57	23	4	05	46	22	4	23	26
2226	21	2	25	35	21	16	19	50	23	9	52	11	22	10	05	12
2227	21	8	02	49	21	22	06	22	23	15	40	14	22	15	54	55
2228	20	13	53	39	21	3	59	39	22	21	24	31	21	21	41	26
2229	20	19	45	59	21	9	44	58	23	3	08	00	22	3	23	26
2230	21	1	21	05	21	15	21	33	23	9	01	42	22	9	15	10
2231	21	7	12	14	21	21	15	51	23	14	51	48	22	14	58	15
2232	20	12	58	20	21	3	01	27	22	20	36	31	21	20	47	17
2233	20	18	42	25	21	8	39	54	23	2	22	33	22	2	46	00
2234	21	0	35	43	21	14	30	30	23	8	10	49	22	8	32	06
2235	21	6	18	45	21	20	15	39	23	13	59	45	22	14	23	06
2236	20	12	15	41	21	2	08	34	22	19	48	20	21	20	13	45
2237	20	18	15	33	21	8	00	17	23	1	35	16	22	2	01	14
2238	20	23	55	42	21	13	43	22	23	7	29	45	22	7	58	54
2239	21	5	49	47	21	19	44	17	23	13	22	13	22	13	46	47
2240	20	11	42	02	21	1	36	24	22	19	08	48	21	19	35	39

Year	March Equinox				June Solstice				September Equinox				December Solstice			
	d	h	m	s	d	h	m	s	d	h	m	s	d	h	m	s
2241	20	17	31	05	21	7	17	01	23	0	57	53	22	1	30	49
2242	20	23	25	21	21	13	09	53	23	6	51	26	22	7	14	55
2243	21	5	07	39	21	18	56	39	23	12	41	28	22	13	04	01
2244	20	10	56	23	21	0	45	52	22	18	28	04	21	18	56	36
2245	20	16	50	11	21	6	34	40	23	0	12	35	22	0	45	40
2246	20	22	27	23	21	12	10	16	23	6	00	31	22	6	38	35
2247	21	4	19	32	21	18	04	13	23	11	50	31	22	12	23	47
2248	20	10	14	55	20	23	53	24	22	17	34	04	21	18	07	22
2249	20	16	00	59	21	5	29	56	22	23	17	48	21	23	59	49
2250	20	21	51	18	21	11	24	03	23	5	08	37	22	5	45	26
2251	21	3	32	25	21	17	12	08	23	10	54	22	22	11	31	34
2252	20	9	19	18	20	23	00	06	22	16	40	04	21	17	21	17
2253	20	15	14	41	21	4	50	45	22	22	29	26	21	23	08	33
2254	20	20	53	33	21	10	27	25	23	4	20	53	22	4	58	53
2255	21	2	42	53	21	16	22	37	23	10	16	13	22	10	50	42
2256	20	8	38	17	20	22	15	35	22	16	03	02	21	16	42	04
2257	20	14	23	48	21	3	51	39	22	21	47	24	21	22	40	50
2258	20	20	18	36	21	9	47	24	23	3	42	38	22	4	32	44
2259	21	2	09	51	21	15	37	51	23	9	30	34	22	10	19	50
2260	20	8	04	35	20	21	27	00	22	15	17	46	21	16	10	45
2261	20	14	03	51	21	3	22	27	22	21	07	52	21	22	00	32
2262	20	19	44	33	21	9	04	27	23	2	56	19	22	3	50	20
2263	21	1	31	46	21	15	00	28	23	8	48	23	22	9	38	20
2264	20	7	25	22	20	20	54	03	22	14	34	22	21	15	23	52
2265	20	13	09	18	21	2	26	42	22	20	16	56	21	21	12	45
2266	20	18	56	22	21	8	16	44	23	2	12	32	22	2	58	44
2267	21	0	39	54	21	14	04	15	23	7	58	25	22	8	43	38
2268	20	6	24	13	20	19	45	46	22	13	40	21	21	14	35	12
2269	20	12	14	43	21	1	34	41	22	19	28	01	21	20	27	41
2270	20	17	55	47	21	7	11	52	23	1	13	30	22	2	17	21
2271	20	23	45	58	21	13	01	54	23	7	03	39	22	8	04	20
2272	20	5	44	42	20	18	56	29	22	12	51	02	21	13	52	59
2273	20	11	35	19	21	0	34	35	22	18	33	01	21	19	44	41
2274	20	17	24	18	21	6	30	45	23	0	29	50	22	1	36	15
2275	20	23	14	17	21	12	28	24	23	6	19	36	22	7	24	17
2276	20	5	04	46	20	18	14	56	22	12	04	36	21	13	13	59
2277	20	10	59	21	21	0	09	30	22	18	01	23	21	19	06	42
2278	20	16	44	22	21	5	52	29	22	23	53	13	22	0	54	56
2279	20	22	30	57	21	11	43	57	23	5	46	50	22	6	46	01
2280	20	4	25	01	20	17	39	27	22	11	36	56	21	12	40	56

ASTRONOMICAL TABLES

Year	March Equinox				June Solstice				September Equinox				December Solstice			
	d	h	m	s	d	h	m	s	d	h	m	s	d	h	m	s
2281	20	10	11	57	20	23	13	22	22	17	16	29	21	18	33	10
2282	20	15	59	09	21	5	02	25	22	23	10	32	22	0	23	42
2283	20	21	51	21	21	10	54	40	23	4	57	31	22	6	08	10
2284	20	3	42	20	20	16	35	20	22	10	35	35	21	11	52	33
2285	20	9	32	33	20	22	26	42	22	16	26	56	21	17	44	30
2286	20	15	14	06	21	4	09	25	22	22	11	15	21	23	28	52
2287	20	20	56	39	21	9	57	23	23	3	57	37	22	5	14	53
2288	20	2	49	33	20	15	51	23	22	9	46	58	21	11	04	21
2289	20	8	36	40	20	21	24	31	22	15	26	58	21	16	49	47
2290	20	14	21	42	21	3	13	18	22	21	24	13	21	22	39	30
2291	20	20	10	39	21	9	08	31	23	3	15	16	22	4	29	44
2292	20	2	00	25	20	14	51	30	22	8	55	02	21	10	20	34
2293	20	7	50	52	20	20	43	24	22	14	50	06	21	16	19	19
2294	20	13	40	21	21	2	30	40	22	20	40	33	21	22	09	26
2295	20	19	33	25	21	8	21	13	23	2	30	54	22	3	57	04
2296	20	1	31	59	20	14	20	11	22	8	26	13	21	9	52	32
2297	20	7	25	23	20	20	01	44	22	14	08	08	21	15	42	22
2298	20	13	10	59	21	1	53	26	22	20	03	37	21	21	33	22
2299	20	18	59	59	21	7	51	39	23	1	55	04	22	3	22	44
2300	21	0	51	29	21	13	33	45	23	7	33	07	22	9	05	52
2301	21	6	37	48	21	19	20	19	23	13	27	21	22	14	56	42
2302	21	12	21	27	22	1	04	24	23	19	17	06	22	20	43	18
2303	21	18	05	46	22	6	47	52	24	1	00	14	23	2	26	57
2304	20	23	51	49	21	12	38	03	23	6	49	47	22	8	23	11
2305	21	5	41	15	21	18	14	01	23	12	26	49	22	14	11	09
2306	21	11	25	59	21	23	57	13	23	18	16	34	22	19	58	07
2307	21	17	17	29	22	5	52	48	24	0	08	25	23	1	47	09
2308	20	23	13	55	21	11	36	33	23	5	44	25	22	7	29	36
2309	21	5	00	58	21	17	27	00	23	11	38	11	22	13	24	20
2310	21	10	47	04	21	23	19	20	23	17	30	29	22	19	14	36
2311	21	16	35	46	22	5	09	52	23	23	16	33	23	0	58	41
2312	20	22	26	43	21	11	05	27	23	5	14	09	22	6	55	59
2313	21	4	20	40	21	16	47	14	23	11	00	37	22	12	44	56
2314	21	10	06	51	21	22	33	54	23	16	56	16	22	18	34	55
2315	21	15	54	56	22	4	31	39	23	22	53	28	23	0	32	19
2316	20	21	49	53	21	10	15	52	23	4	29	54	22	6	20	09
2317	21	3	36	32	21	16	01	16	23	10	21	52	22	12	17	17
2318	21	9	25	14	21	21	49	02	23	16	12	49	22	18	06	32
2319	21	15	18	19	22	3	35	22	23	21	53	44	22	23	45	55
2320	20	21	09	51	21	9	27	44	23	3	45	42	22	5	40	03

Year	March Equinox				June Solstice				September Equinox				December Solstice			
	d	h	m	s	d	h	m	s	d	h	m	s	d	h	m	s
2321	21	3	00	30	21	15	09	41	23	9	25	53	22	11	26	43
2322	21	8	43	34	21	20	55	33	23	15	13	24	22	17	10	28
2323	21	14	26	44	22	2	50	25	23	21	06	33	22	23	01	18
2324	20	20	20	49	21	8	34	00	23	2	42	14	22	4	41	04
2325	21	2	04	23	21	14	15	52	23	8	33	44	22	10	30	28
2326	21	7	46	40	21	20	04	06	23	14	29	10	22	16	21	54
2327	21	13	36	01	22	1	52	06	23	20	11	09	22	22	04	59
2328	20	19	21	44	21	7	41	52	23	2	04	29	22	4	05	47
2329	21	1	14	28	21	13	24	57	23	7	49	48	22	9	59	46
2330	21	7	05	51	21	19	11	06	23	13	40	03	22	15	45	41
2331	21	12	56	33	22	1	08	02	23	19	38	25	22	21	41	40
2332	20	18	59	16	21	6	58	55	23	1	18	10	22	3	28	12
2333	21	0	47	12	21	12	46	07	23	7	08	49	22	9	21	05
2334	21	6	31	03	21	18	39	51	23	13	06	02	22	15	16	44
2335	21	12	26	14	22	0	32	09	23	18	47	43	22	20	55	34
2336	20	18	12	44	21	6	19	59	23	0	40	29	22	2	49	40
2337	21	0	04	00	21	12	03	11	23	6	29	04	22	8	39	36
2338	21	5	50	28	21	17	47	10	23	12	16	31	22	14	21	05
2339	21	11	30	47	21	23	38	46	23	18	11	07	22	20	17	36
2340	20	17	25	37	21	5	23	56	22	23	46	07	22	2	03	50
2341	20	23	10	07	21	11	03	18	23	5	30	49	22	7	52	43
2342	21	4	52	51	21	16	50	02	23	11	25	30	22	13	44	38
2343	21	10	49	45	21	22	40	23	23	17	05	14	22	19	21	06
2344	20	16	36	58	21	4	27	37	22	22	54	17	22	1	15	05
2345	20	22	25	24	21	10	13	09	23	4	42	19	22	7	08	22
2346	21	4	13	43	21	16	02	04	23	10	28	45	22	12	49	43
2347	21	9	56	14	21	21	55	25	23	16	25	38	22	18	45	14
2348	20	15	54	35	21	3	44	24	22	22	07	41	22	0	31	30
2349	20	21	43	01	21	9	27	26	23	3	58	29	22	6	20	57
2350	21	3	24	26	21	15	16	24	23	9	58	37	22	12	18	29
2351	21	9	19	56	21	21	10	21	23	15	41	19	22	18	03	34
2352	20	15	09	04	21	2	57	45	22	21	30	35	22	0	02	21
2353	20	20	59	38	21	8	41	52	23	3	21	02	22	5	59	23
2354	21	2	56	30	21	14	32	06	23	9	08	19	22	11	39	39
2355	21	8	43	50	21	20	23	55	23	15	02	43	22	17	32	54
2356	20	14	42	37	21	2	15	21	22	20	43	22	21	23	20	30
2357	20	20	30	03	21	8	00	26	23	2	27	45	22	5	05	19
2358	21	2	06	04	21	13	48	20	23	8	23	52	22	10	58	35
2359	21	7	59	49	21	19	41	43	23	14	05	04	22	16	36	32
2360	20	13	45	11	21	1	23	42	22	19	51	47	21	22	25	13

Year	March Equinox				June Solstice				September Equinox				December Solstice			
	d	h	m	s	d	h	m	s	d	h	m	s	d	h	m	s
2361	20	19	28	13	21	7	03	13	23	1	43	43	22	4	18	41
2362	21	1	18	14	21	12	51	05	23	7	30	11	22	9	58	44
2363	21	6	56	58	21	18	37	30	23	13	20	54	22	15	54	31
2364	20	12	50	56	21	0	24	45	22	19	02	15	21	21	48	09
2365	20	18	42	05	21	6	07	06	23	0	45	44	22	3	33	23
2366	21	0	23	37	21	11	52	41	23	6	43	06	22	9	29	20
2367	21	6	26	21	21	17	50	49	23	12	28	03	22	15	12	00
2368	20	12	18	21	20	23	39	21	22	18	14	45	21	21	04	40
2369	20	18	04	39	21	5	26	50	23	0	09	39	22	3	04	19
2370	21	0	00	26	21	11	23	29	23	5	59	04	22	8	46	09
2371	21	5	45	07	21	17	14	38	23	11	52	35	22	14	39	06
2372	20	11	41	14	20	23	04	23	22	17	39	54	21	20	30	15
2373	20	17	33	11	21	4	50	22	22	23	27	55	22	2	13	38
2374	20	23	09	20	21	10	34	34	23	5	24	58	22	8	09	40
2375	21	5	02	21	21	16	29	17	23	11	09	03	22	13	56	09
2376	20	10	50	33	20	22	12	09	22	16	51	01	21	19	47	08
2377	20	16	33	00	21	3	49	22	22	22	41	21	22	1	43	14
2378	20	22	29	28	21	9	40	15	23	4	27	23	22	7	21	17
2379	21	4	15	16	21	15	26	34	23	10	14	29	22	13	09	44
2380	20	10	06	53	20	21	13	36	22	15	56	29	21	19	00	47
2381	20	15	57	03	21	3	01	34	22	21	40	01	22	0	43	41
2382	20	21	32	50	21	8	44	59	23	3	32	28	22	6	34	45
2383	21	3	25	18	21	14	40	12	23	9	18	31	22	12	18	40
2384	20	9	18	04	20	20	26	52	22	15	04	41	21	18	06	02
2385	20	14	59	19	21	2	05	31	22	21	00	17	22	0	03	54
2386	20	20	54	11	21	8	02	17	23	2	53	03	22	5	50	42
2387	21	2	40	23	21	13	51	48	23	8	42	17	22	11	45	45
2388	20	8	33	24	20	19	39	40	22	14	29	07	21	17	44	54
2389	20	14	32	03	21	1	30	40	22	20	17	19	21	23	30	46
2390	20	20	15	54	21	7	15	41	23	2	12	19	22	5	22	52
2391	21	2	13	41	21	13	14	37	23	8	01	32	22	11	10	28
2392	20	8	08	49	20	19	05	35	22	13	45	45	21	16	58	32
2393	20	13	47	48	21	0	44	47	22	19	37	15	21	22	54	47
2394	20	19	40	05	21	6	40	50	23	1	28	03	22	4	37	25
2395	21	1	24	58	21	12	26	58	23	7	13	21	22	10	22	07
2396	20	7	12	17	20	18	07	41	22	12	58	30	21	16	14	05
2397	20	13	03	41	20	23	53	44	22	18	44	33	21	21	55	10
2398	20	18	37	35	21	5	32	19	23	0	32	47	22	3	45	44
2399	21	0	26	19	21	11	24	46	23	6	17	49	22	9	35	32
2400	20	6	19	03	20	17	11	01	22	11	58	33	21	15	23	33

Year	March Equinox				June Solstice				September Equinox				December Solstice			
	d	h	m	s	d	h	m	s	d	h	m	s	d	h	m	s
2401	20	12	00	44	20	22	45	54	22	17	48	18	21	21	18	38
2402	20	17	57	42	21	4	41	59	22	23	41	27	22	3	03	11
2403	20	23	50	17	21	10	34	20	23	5	28	19	22	8	50	59
2404	20	5	40	51	20	16	20	49	22	11	14	25	21	14	47	57
2405	20	11	35	56	20	22	17	05	22	17	06	00	21	20	35	32
2406	20	17	17	38	21	4	04	16	22	22	58	34	22	2	25	47
2407	20	23	09	46	21	10	00	31	23	4	50	48	22	8	15	51
2408	20	5	07	27	20	15	52	37	22	10	40	03	21	14	04	22
2409	20	10	48	21	20	21	28	34	22	16	33	18	21	20	00	09
2410	20	16	38	41	21	3	24	18	22	22	28	35	22	1	50	26
2411	20	22	28	14	21	9	14	36	23	4	13	34	22	7	40	59
2412	20	4	15	41	20	14	52	10	22	9	54	15	21	13	36	03
2413	20	10	10	23	20	20	41	23	22	15	42	47	21	19	21	54
2414	20	15	55	48	21	2	23	34	22	21	28	39	22	1	04	56
2415	20	21	44	35	21	8	14	31	23	3	13	03	22	6	51	33
2416	20	3	39	30	20	14	07	48	22	8	56	28	21	12	38	22
2417	20	9	16	57	20	19	42	48	22	14	42	17	21	18	28	12
2418	20	15	04	02	21	1	37	52	22	20	36	21	22	0	13	56
2419	20	20	54	48	21	7	29	24	23	2	22	37	22	5	57	58
2420	20	2	41	12	20	13	06	40	22	8	06	53	21	11	50	14
2421	20	8	32	23	20	18	58	26	22	14	02	01	21	17	39	55
2422	20	14	14	48	21	0	44	28	22	19	51	38	21	23	30	02
2423	20	20	02	35	21	6	35	36	23	1	39	30	22	5	25	17
2424	20	2	01	23	20	12	29	43	22	7	28	30	21	11	19	18
2425	20	7	48	14	20	18	06	34	22	13	17	58	21	17	11	45
2426	20	13	43	13	21	0	03	37	22	19	16	15	21	23	01	51
2427	20	19	40	34	21	6	01	22	23	1	04	12	22	4	50	24
2428	20	1	29	28	20	11	42	20	22	6	45	54	21	10	45	21
2429	20	7	20	08	20	17	35	56	22	12	39	32	21	16	34	03
2430	20	13	03	49	20	23	22	58	22	18	26	14	21	22	17	41
2431	20	18	51	18	21	5	10	30	23	0	11	24	22	4	03	59
2432	20	0	44	50	20	11	00	45	22	6	00	13	21	9	52	08
2433	20	6	24	53	20	16	34	47	22	11	46	16	21	15	40	08
2434	20	12	08	14	20	22	24	48	22	17	38	40	21	21	29	38
2435	20	17	58	26	21	4	17	40	22	23	23	19	22	3	19	01
2436	19	23	45	57	20	9	51	24	22	4	59	45	21	9	09	42
2437	20	5	36	28	20	15	40	16	22	10	53	51	21	14	58	01
2438	20	11	26	07	20	21	29	32	22	16	41	44	21	20	41	48
2439	20	17	15	08	21	3	18	29	22	22	25	17	22	2	30	31
2440	19	23	09	10	20	9	14	46	22	4	16	39	21	8	24	59

Year	March Equinox				June Solstice				September Equinox				December Solstice			
	d	h	m	s	d	h	m	s	d	h	m	s	d	h	m	s
2441	20	4	53	46	20	14	55	54	22	10	04	55	21	14	14	24
2442	20	10	40	50	20	20	49	27	22	16	01	07	21	20	03	57
2443	20	16	36	34	21	2	47	44	22	21	54	32	22	1	55	10
2444	19	22	27	49	20	8	24	47	22	3	36	14	21	7	45	24
2445	20	4	15	16	20	14	15	17	22	9	34	42	21	13	40	07
2446	20	10	04	02	20	20	07	36	22	15	24	19	21	19	30	11
2447	20	15	51	39	21	1	52	55	22	21	04	51	22	1	20	49
2448	19	21	46	39	20	7	45	36	22	2	56	16	21	7	16	10
2449	20	3	36	19	20	13	24	21	22	8	42	05	21	13	01	16
2450	20	9	25	16	20	19	14	36	22	14	33	35	21	18	46	51
2451	20	15	18	55	21	1	12	43	22	20	22	19	22	0	36	29
2452	19	21	07	01	20	6	50	42	22	1	57	45	21	6	22	48
2453	20	2	49	51	20	12	39	00	22	7	51	48	21	12	11	14
2454	20	8	34	58	20	18	30	09	22	13	40	57	21	17	54	55
2455	20	14	21	18	21	0	12	37	22	19	20	59	21	23	37	58
2456	19	20	09	59	20	6	01	27	22	1	15	14	21	5	31	50
2457	20	1	54	02	20	11	40	51	22	7	02	55	21	11	20	32
2458	20	7	37	55	20	17	28	34	22	12	53	32	21	17	11	53
2459	20	13	29	43	20	23	24	31	22	18	44	37	21	23	08	38
2460	19	19	24	39	20	5	02	40	22	0	22	09	21	4	59	18
2461	20	1	16	30	20	10	50	37	22	6	18	20	21	10	49	54
2462	20	7	09	35	20	16	47	23	22	12	11	19	21	16	39	08
2463	20	13	04	00	20	22	38	15	22	17	51	32	21	22	25	59
2464	19	18	54	43	20	4	32	47	21	23	46	26	21	4	23	18
2465	20	0	43	02	20	10	19	48	22	5	36	41	21	10	11	45
2466	20	6	30	23	20	16	09	02	22	11	26	53	21	15	56	21
2467	20	12	21	18	20	22	05	03	22	17	22	06	21	21	49	03
2468	19	18	13	56	20	3	43	11	21	23	00	43	21	3	34	34
2469	19	23	55	47	20	9	26	39	22	4	54	06	21	⁻9	24	29
2470	20	5	39	27	20	15	18	59	22	10	44	51	21	15	15	59
2471	20	11	28	25	20	21	01	41	22	16	18	54	21	20	59	42
2472	19	17	15	57	20	2	46	32	21	22	09	22	21	2	53	53
2473	19	23	05	31	20	8	28	57	22	3	57	55	21	8	39	51
2474	20	4	54	08	20	14	15	38	22	9	42	48	21	14	21	48
2475	20	10	42	52	20	20	12	01	22	15	35	33	21	20	18	09
2476	19	16	36	13	20	1	53	53	21	21	12	03	21	2	04	18
2477	19	22	19	21	20	7	39	03	22	3	04	03	21	7	52	37
2478	20	4	07	25	20	13	35	52	22	9	00	23	21	13	43	40
2479	20	10	01	48	20	19	23	22	22	14	40	08	21	19	25	49
2480	19	15	50	21	20	1	12	01	21	20	37	11	21	1	23	27

Year	March Equinox				June Solstice				September Equinox				December Solstice			
	d	h	m	s	d	h	m	s	d	h	m	s	d	h	m	s
2481	19	21	38	14	20	6	59	56	22	2	32	03	21	7	17	48
2482	20	3	27	21	20	12	49	55	22	8	19	17	21	13	05	56
2483	20	9	17	20	20	18	44	58	22	14	14	16	21	19	06	31
2484	19	15	16	24	20	0	27	41	21	19	54	20	21	0	55	12
2485	19	21	06	42	20	6	12	04	22	1	45	59	21	6	41	05
2486	20	2	54	50	20	12	08	34	22	7	41	34	21	12	33	00
2487	20	8	49	18	20	17	58	39	22	13	16	39	21	18	14	23
2488	19	14	33	29	19	23	44	21	21	19	06	40	21	0	07	15
2489	19	20	16	19	20	5	29	53	22	0	58	22	21	5	55	58
2490	20	2	03	52	20	11	16	08	22	6	41	47	21	11	34	12
2491	20	7	47	53	20	17	03	50	22	12	34	29	21	17	27	03
2492	19	13	39	25	19	22	43	03	21	18	14	18	20	23	15	13
2493	19	19	23	20	20	4	23	37	22	0	01	22	21	5	01	00
2494	20	1	03	35	20	10	15	05	22	5	54	49	21	10	57	19
2495	20	7	00	22	20	16	04	24	22	11	30	11	21	16	41	40
2496	19	12	50	23	19	21	47	49	21	17	20	11	20	22	34	51
2497	19	18	40	40	20	3	37	17	21	23	17	43	21	4	28	09
2498	20	0	37	26	20	9	32	02	22	5	03	28	21	10	10	01
2499	20	6	25	45	20	15	27	45	22	10	58	55	21	16	09	27
2500	20	12	21	41	20	21	16	41	22	16	44	20	21	22	02	10
2501	20	18	10	37	21	3	03	58	22	22	35	16	22	3	46	43
2502	20	23	54	21	21	8	58	34	23	4	34	45	22	9	41	38
2503	21	5	52	30	21	14	49	35	23	10	15	20	22	15	24	09
2504	20	11	38	30	20	20	31	25	22	16	05	14	21	21	16	09
2505	20	17	19	27	21	2	17	08	22	22	01	51	22	3	12	48
2506	20	23	09	58	21	8	06	26	23	3	42	11	22	8	53	59
2507	21	4	54	17	21	13	51	39	23	9	30	36	22	14	50	42
2508	20	10	49	10	20	19	31	43	22	15	11	34	21	20	39	02
2509	20	16	39	17	21	1	13	36	22	20	55	36	22	2	17	12
2510	20	22	21	39	21	7	05	40	23	2	48	51	22	8	10	00
2511	21	4	17	02	21	12	58	01	23	8	24	52	22	13	52	28
2512	20	10	02	39	20	18	41	51	22	14	10	49	21	19	40	54
2513	20	15	42	48	21	0	28	43	22	20	08	36	22	1	34	35
2514	20	21	37	05	21	6	22	42	23	1	54	19	22	7	13	16
2515	21	3	23	22	21	12	10	11	23	7	47	51	22	13	08	20
2516	20	9	15	01	20	17	55	00	22	13	37	58	21	19	04	52
2517	20	15	05	39	20	23	42	53	22	19	27	19	22	0	51	24
2518	20	20	47	05	21	5	34	25	23	1	23	50	22	6	52	08
2519	21	2	47	15	21	11	27	53	23	7	05	05	22	12	41	48
2520	20	8	42	12	20	17	11	47	22	12	52	53	21	18	30	47

Year	March Equinox				June Solstice				September Equinox				December Solstice			
	d	h	m	s	d	h	m	s	d	h	m	s	d	h	m	s
2521	20	14	27	29	20	22	58	51	22	18	51	53	22	0	26	23
2522	20	20	25	59	21	4	57	01	23	0	36	31	22	6	07	58
2523	21	2	12	15	21	10	44	52	23	6	23	11	22	12	00	23
2524	20	7	59	36	20	16	28	20	22	12	09	30	21	17	54	09
2525	20	13	51	03	20	22	15	42	22	17	54	44	21	23	31	06
2526	20	19	28	50	21	4	01	06	22	23	46	53	22	5	21	07
2527	21	1	22	50	21	9	50	51	23	5	28	12	22	11	05	46
2528	20	7	10	04	20	15	31	08	22	11	11	56	21	16	50	49
2529	20	12	44	38	20	21	12	52	22	17	07	23	21	22	48	05
2530	20	18	38	02	21	3	07	14	22	22	49	59	22	4	31	17
2531	21	0	25	41	21	8	51	22	23	4	35	03	22	10	23	22
2532	20	6	16	54	20	14	33	05	22	10	24	35	21	16	18	33
2533	20	12	14	30	20	20	24	50	22	16	13	27	21	21	58	52
2534	20	17	57	28	21	2	16	19	22	22	06	48	22	3	54	11
2535	20	23	53	36	21	8	12	56	23	3	52	05	22	9	46	11
2536	20	5	46	36	20	14	01	28	22	9	39	16	21	15	33	17
2537	20	11	26	15	20	19	46	59	22	15	39	28	21	21	30	36
2538	20	17	23	40	21	1	45	07	22	21	29	06	22	3	13	33
2539	20	23	13	40	21	7	31	31	23	3	17	45	22	9	05	23
2540	20	5	00	33	20	13	12	06	22	9	08	25	21	15	03	22
2541	20	10	52	40	20	19	02	32	22	14	55	35	21	20	47	01
2542	20	16	33	34	21	0	48	59	22	20	43	48	22	2	40	57
2543	20	22	27	37	21	6	37	41	23	2	25	32	22	8	30	39
2544	20	4	23	36	20	12	22	25	22	8	09	04	21	14	12	07
2545	20	10	02	53	20	18	03	11	22	14	02	22	21	20	03	50
2546	20	15	56	24	21	0	01	21	22	19	47	43	22	1	46	43
2547	20	21	44	14	21	5	48	49	23	1	30	05	22	7	33	47
2548	20	3	25	47	20	11	27	44	22	7	18	20	21	13	28	12
2549	20	9	18	11	20	17	19	45	22	13	08	01	21	19	08	02
2550	20	14	59	24	20	23	05	11	22	18	58	00	22	0	56	33
2551	20	20	49	52	21	4	53	36	23	0	45	05	22	6	48	46
2552	20	2	43	29	20	10	41	22	22	6	32	53	21	12	36	39
2553	20	8	19	52	20	16	21	57	22	12	27	12	21	18	35	15
2554	20	14	14	26	20	22	20	04	22	18	16	59	22	0	27	20
2555	20	20	11	34	21	4	08	24	23	0	01	27	22	6	16	52
2556	20	2	00	55	20	9	46	59	22	5	51	43	21	12	13	59
2557	20	8	02	01	20	15	45	20	22	11	44	39	21	17	58	08
2558	20	13	48	23	20	21	37	04	22	17	32	21	21	23	47	55
2559	20	19	39	14	21	3	29	32	22	23	18	19	22	5	41	09
2560	20	1	34	06	20	9	20	48	22	5	05	25	21	11	24	49

Year	March Equinox				June Solstice				September Equinox				December Solstice			
	d	h	m	s	d	h	m	s	d	h	m	s	d	h	m	s
2561	20	7	10	59	20	15	00	54	22	10	58	08	21	17	14	46
2562	20	13	01	29	20	20	55	44	22	16	48	41	21	22	59	35
2563	20	18	52	45	21	2	42	32	22	22	33	01	22	4	45	01
2564	20	0	31	53	20	8	15	31	22	4	19	28	21	10	40	54
2565	20	6	21	13	20	14	07	04	22	10	09	33	21	16	27	15
2566	20	12	04	24	20	19	52	37	22	15	52	35	21	22	15	29
2567	20	17	54	22	21	1	36	05	22	21	35	18	22	4	06	59
2568	19	23	52	16	20	7	25	09	22	3	22	47	21	9	50	33
2569	20	5	33	33	20	13	06	43	22	9	13	30	21	15	41	18
2570	20	11	24	03	20	19	04	19	22	15	03	16	21	21	30	31
2571	20	17	18	24	21	0	57	55	22	20	48	37	22	3	20	07
2572	19	23	02	57	20	6	35	14	22	2	36	39	21	9	14	59
2573	20	4	56	06	20	12	31	07	22	8	33	23	21	15	01	42
2574	20	10	46	16	20	18	23	41	22	14	23	32	21	20	49	20
2575	20	16	36	12	21	0	09	18	22	20	10	42	22	2	43	54
2576	19	22	31	14	20	6	02	33	22	2	03	18	21	8	35	11
2577	20	4	11	26	20	11	44	46	22	7	53	11	21	14	28	29
2578	20	10	00	54	20	17	38	45	22	13	42	57	21	20	20	07
2579	20	15	59	41	20	23	30	06	22	19	28	13	22	2	07	43
2580	19	21	46	11	20	5	03	20	22	1	13	04	21	7	58	20
2581	20	3	37	59	20	10	57	46	22	7	06	48	21	13	44	33
2582	20	9	25	41	20	16	50	25	22	12	50	30	21	19	29	24
2583	20	15	10	25	20	22	32	25	22	18	30	09	22	1	19	01
2584	19	21	01	25	20	4	22	52	22	0	20	29	21	7	04	54
2585	20	2	40	29	20	10	02	26	22	6	07	50	21	12	48	47
2586	20	8	25	42	20	15	51	50	22	11	57	44	21	18	36	34
2587	20	14	19	59	20	21	42	16	22	17	44	34	22	0	25	00
2588	19	20	00	34	20	3	14	06	21	23	27	49	21	6	19	02
2589	20	1	49	15	20	9	07	37	22	5	23	15	21	12	12	06
2590	20	7	41	28	20	15	01	19	22	11	09	26	21	18	01	37
2591	20	13	34	59	20	20	44	04	22	16	51	55	21	23	53	58
2592	19	19	34	17	20	2	38	56	21	22	47	55	21	5	44	38
2593	20	1	22	25	20	8	28	35	22	4	38	40	21	11	33	28
2594	20	7	11	35	20	14	24	50	22	10	29	02	21	17	24	53
2595	20	13	07	46	20	20	22	57	22	16	20	04	21	23	16	50
2596	19	18	53	02	20	1	59	36	21	22	05	32	21	5	06	40
2597	20	0	40	46	20	7	51	24	22	4	03	36	21	10	54	37
2598	20	6	31	15	20	13	44	48	22	9	52	25	21	16	40	33
2599	20	12	18	04	20	19	22	21	22	15	31	40	21	22	29	53
2600	20	18	05	03	21	1	10	10	22	21	23	50	22	4	21	48

Year	March Equinox				June Solstice				September Equinox				December Solstice			
	d	h	m	s	d	h	m	s	d	h	m	s	d	h	m	s
2601	20	23	46	33	21	6	53	16	23	3	08	56	22	10	09	53
2602	21	5	31	50	21	12	37	51	23	8	50	49	22	15	56	18
2603	21	11	26	49	21	18	27	59	23	14	38	04	22	21	45	55
2604	20	17	16	11	21	0	02	35	22	20	19	04	22	3	31	16
2605	20	23	02	34	21	5	53	24	23	2	12	22	22	9	19	17
2606	21	4	53	16	21	11	52	31	23	8	00	32	22	15	08	21
2607	21	10	42	20	21	17	34	05	23	13	38	20	22	20	56	23
2608	20	16	31	38	20	23	25	54	22	19	35	37	22	2	48	28
2609	20	22	19	15	21	5	15	38	23	1	27	56	22	8	35	13
2610	21	4	07	30	21	11	04	51	23	7	16	57	22	14	23	29
2611	21	10	01	50	21	17	00	26	23	13	12	47	22	20	20	54
2612	20	15	49	37	20	22	39	25	22	18	58	24	22	2	13	55
2613	20	21	36	08	21	4	29	36	23	0	53	42	22	8	08	07
2614	21	3	30	24	21	10	27	06	23	6	44	44	22	14	00	54
2615	21	9	26	12	21	16	06	36	23	12	21	44	22	19	47	05
2616	20	15	18	15	20	21	55	56	22	18	17	29	22	1	38	35
2617	20	21	06	02	21	3	46	50	23	0	05	14	22	7	24	29
2618	21	2	51	20	21	9	34	34	23	5	45	33	22	13	09	21
2619	21	8	41	17	21	15	26	49	23	11	35	37	22	19	00	47
2620	20	14	26	24	20	21	03	12	22	17	16	47	22	0	43	57
2621	20	20	10	01	21	2	48	43	22	23	08	53	22	6	28	22
2622	21	1	57	37	21	8	42	49	23	5	00	25	22	12	17	32
2623	21	7	47	08	21	14	21	13	23	10	36	03	22	18	03	29
2624	20	13	30	41	20	20	06	29	22	16	29	47	21	23	58	11
2625	20	19	16	06	21	1	56	53	22	22	19	46	22	5	49	40
2626	21	1	06	58	21	7	43	27	23	4	00	42	22	11	35	07
2627	21	7	02	15	21	13	36	03	23	9	56	06	22	17	31	06
2628	20	12	57	07	20	19	20	16	22	15	43	20	21	23	20	01
2629	20	18	46	12	21	1	12	10	22	21	36	52	22	5	09	38
2630	21	0	36	40	21	7	14	19	23	3	32	41	22	11	05	51
2631	21	6	32	22	21	12	59	43	23	9	11	28	22	16	52	37
2632	20	12	19	08	20	18	46	24	22	15	07	47	21	22	44	22
2633	20	18	06	21	21	0	38	47	22	21	03	15	22	4	34	48
2634	20	23	56	46	21	6	24	21	23	2	43	46	22	10	16	05
2635	21	5	43	03	21	12	12	06	23	8	36	19	22	16	13	19
2636	20	11	31	17	20	17	53	06	22	14	20	02	21	22	03	26
2637	20	17	15	25	20	23	36	24	22	20	05	28	22	3	49	01
2638	20	23	03	33	21	5	30	25	23	1	57	04	22	9	41	59
2639	21	5	01	25	21	11	10	59	23	7	31	04	22	15	22	49
2640	20	10	48	00	20	16	54	30	22	13	21	54	21	21	10	51

Year	March Equinox				June Solstice				September Equinox				December Solstice			
	d	h	m	s	d	h	m	s	d	h	m	s	d	h	m	s
2641	20	16	32	20	20	22	48	23	22	19	14	05	22	3	01	21
2642	20	22	21	33	21	4	37	39	23	0	51	39	22	8	41	36
2643	21	4	07	29	21	10	26	38	23	6	44	37	22	14	36	48
2644	20	9	56	35	20	16	10	19	22	12	34	00	21	20	25	25
2645	20	15	44	20	20	21	56	47	22	18	24	33	22	2	09	01
2646	20	21	31	24	21	3	52	38	23	0	23	09	22	8	06	36
2647	21	3	27	59	21	9	37	49	23	6	02	44	22	13	56	00
2648	20	9	13	59	20	15	22	18	22	11	55	54	21	19	51	51
2649	20	14	59	58	20	21	15	42	22	17	51	38	22	1	48	57
2650	20	20	57	26	21	3	05	30	22	23	30	37	22	7	31	51
2651	21	2	51	44	21	8	53	30	23	5	23	25	22	13	26	50
2652	20	8	44	16	20	14	40	00	22	11	12	57	21	19	17	29
2653	20	14	34	44	20	20	31	16	22	16	59	02	22	0	59	58
2654	20	20	18	54	21	2	27	05	22	22	52	43	22	6	54	47
2655	21	2	14	19	21	8	13	02	23	4	30	20	22	12	38	28
2656	20	7	58	46	20	13	53	58	22	10	19	28	21	18	22	36
2657	20	13	38	37	20	19	43	31	22	16	16	05	22	0	13	25
2658	20	19	30	14	21	1	31	17	22	21	54	11	22	5	51	47
2659	21	1	13	00	21	7	13	11	23	3	43	16	22	11	47	03
2660	20	6	56	40	20	12	55	16	22	9	32	33	21	17	42	06
2661	20	12	46	09	20	18	41	27	22	15	16	05	21	23	24	33
2662	20	18	32	20	21	0	31	46	22	21	09	12	22	5	20	25
2663	21	0	34	01	21	6	18	20	23	2	50	12	22	11	07	40
2664	20	6	24	39	20	12	03	23	22	8	38	57	21	16	54	18
2665	20	12	06	54	20	17	59	03	22	14	37	10	21	22	52	49
2666	20	18	04	30	20	23	55	00	22	20	17	16	22	4	34	33
2667	20	23	52	15	21	5	41	07	23	2	08	01	22	10	28	49
2668	20	5	40	52	20	11	28	35	22	8	04	46	21	16	23	43
2669	20	11	34	54	20	17	19	30	22	13	53	16	21	22	03	59
2670	20	17	19	26	20	23	10	58	22	19	48	58	22	4	01	26
2671	20	23	15	19	21	4	57	35	23	1	31	09	22	9	53	20
2672	20	5	02	32	20	10	40	33	22	7	17	13	21	15	40	25
2673	20	10	42	16	20	16	29	22	22	13	11	51	21	21	36	30
2674	20	16	40	40	20	22	20	40	22	18	50	06	22	3	15	28
2675	20	22	30	51	21	4	02	24	23	0	36	02	22	9	04	00
2676	20	4	14	57	20	9	46	33	22	6	28	42	21	14	57	28
2677	20	10	06	00	20	15	38	55	22	12	11	41	21	20	36	00
2678	20	15	46	52	20	21	27	47	22	18	01	25	22	2	29	15
2679	20	21	39	28	21	3	12	42	22	23	43	35	22	8	17	47
2680	20	3	28	18	20	8	55	48	22	5	30	52	21	13	59	36

Year	March Equinox			June Solstice			September Equinox			December Solstice		
	d	h m s		d	h m s		d	h m s		d	h m s	
2681	20	9 06 20		20	14 42 32		22	11 27 31		21	19 53 19	
2682	20	15 01 22		20	20 35 38		22	17 09 48		22	1 37 53	
2683	20	20 50 39		21	2 19 47		22	22 56 40		22	7 31 51	
2684	20	2 32 40		20	8 03 36		22	4 52 23		21	13 32 59	
2685	20	8 30 45		20	13 59 22		22	10 39 50		21	19 17 03	
2686	20	14 21 14		20	19 48 59		22	16 31 44		22	1 11 53	
2687	20	20 21 25		21	1 38 13		22	22 19 35		22	7 05 21	
2688	20	2 17 34		20	7 29 17		22	4 07 43		21	12 49 06	
2689	20	7 56 41		20	13 21 07		22	10 03 22		21	18 45 12	
2690	20	13 52 29		20	19 18 28		22	15 46 44		22	0 29 14	
2691	20	19 41 54		21	1 02 40		22	21 32 16		22	6 15 40	
2692	20	1 20 54		20	6 43 00		22	3 28 25		21	12 09 51	
2693	20	7 14 03		20	12 35 19		22	9 15 53		21	17 48 34	
2694	20	12 55 31		20	18 18 56		22	15 02 46		21	23 40 01	
2695	20	18 43 12		21	0 00 09		22	20 46 43		22	5 35 46	
2696	20	0 32 51		20	5 44 04		22	2 29 16		21	11 18 15	
2697	20	6 09 54		20	11 26 35		22	8 18 36		21	17 12 25	
2698	20	12 08 30		20	17 18 34		22	14 00 09		21	22 55 20	
2699	20	18 02 51		20	23 02 56		22	19 42 58		22	4 40 13	
2700	20	23 44 12		21	4 47 08		23	1 37 59		22	10 37 29	
2701	21	5 39 35		21	10 47 07		23	7 26 55		22	16 20 52	
2702	21	11 26 32		21	16 38 22		23	13 15 57		22	22 12 47	
2703	21	17 18 19		21	22 24 18		23	19 06 08		23	4 08 43	
2704	20	23 14 23		21	4 16 09		23	0 58 10		22	9 51 49	
2705	21	4 55 00		21	10 03 22		23	6 53 51		22	15 46 17	
2706	21	10 49 34		21	15 58 44		23	12 42 19		22	21 37 06	
2707	21	16 42 53		21	21 46 15		23	18 27 44		23	3 27 53	
2708	20	22 21 33		21	3 25 43		23	0 21 34		22	9 27 32	
2709	21	4 17 51		21	9 21 38		23	6 10 24		22	15 11 55	
2710	21	10 09 30		21	15 08 25		23	11 55 42		22	20 59 10	
2711	21	16 00 12		21	20 49 00		23	17 40 59		23	2 52 03	
2712	20	21 53 55		21	2 40 41		22	23 27 25		22	8 34 27	
2713	21	3 31 02		21	8 25 17		23	5 13 16		22	14 22 46	
2714	21	9 19 20		21	14 16 14		23	10 56 07		22	20 08 31	
2715	21	15 12 46		21	20 02 56		23	16 39 15		23	1 50 47	
2716	20	20 48 41		21	1 38 22		22	22 31 31		22	7 42 24	
2717	21	2 40 29		21	7 34 10		23	4 23 50		22	13 26 38	
2718	21	8 28 23		21	13 22 13		23	10 09 25		22	19 16 02	
2719	21	14 13 44		21	19 01 48		23	15 56 27		23	1 15 46	
2720	20	20 08 31		21	0 54 48		22	21 47 14		22	7 04 18	

Year	March Equinox				June Solstice				September Equinox				December Solstice			
	d	h	m	s	d	h	m	s	d	h	m	s	d	h	m	s
2721	21	1	52	55	21	6	41	19	23	3	36	52	22	12	56	19
2722	21	7	50	01	21	12	34	40	23	9	26	03	22	18	47	23
2723	21	13	51	42	21	18	28	13	23	15	14	18	23	0	35	19
2724	20	19	32	59	21	0	10	53	22	21	07	31	22	6	31	58
2725	21	1	26	28	21	6	12	21	23	3	01	52	22	12	20	43
2726	21	7	18	25	21	12	05	10	23	8	48	00	22	18	07	57
2727	21	13	05	11	21	17	43	23	23	14	35	22	23	0	02	47
2728	20	18	58	24	20	23	34	21	22	20	27	51	22	5	45	54
2729	21	0	38	30	21	5	17	57	23	2	14	35	22	11	33	05
2730	21	6	26	08	21	11	04	46	23	7	58	34	22	17	23	04
2731	21	12	18	05	21	16	51	50	23	13	41	10	22	23	10	32
2732	20	17	55	03	20	22	26	40	22	19	27	15	22	5	02	58
2733	20	23	46	12	21	4	19	29	23	1	17	04	22	10	48	22
2734	21	5	41	47	21	10	10	06	23	7	00	56	22	16	32	05
2735	21	11	29	45	21	15	47	06	23	12	44	07	22	22	24	40
2736	20	17	21	30	20	21	42	53	22	18	37	13	22	4	11	59
2737	20	23	04	29	21	3	33	34	23	0	24	07	22	9	59	19
2738	21	4	52	38	21	9	23	06	23	6	11	13	22	15	49	54
2739	21	10	49	17	21	15	16	14	23	12	02	28	22	21	38	47
2740	20	16	30	57	20	20	55	18	22	17	55	26	22	3	30	39
2741	20	22	21	11	21	2	51	14	22	23	52	30	22	9	22	09
2742	21	4	16	20	21	8	46	21	23	5	41	37	22	15	14	43
2743	21	10	02	48	21	14	22	14	23	11	25	04	22	21	12	41
2744	20	15	55	17	20	20	15	40	22	17	20	05	22	3	05	31
2745	20	21	46	06	21	2	04	44	22	23	05	37	22	8	50	50
2746	21	3	38	06	21	7	49	47	23	4	48	47	22	14	38	33
2747	21	9	36	27	21	13	44	13	23	10	37	40	22	20	27	00
2748	20	15	16	46	20	19	24	39	22	16	23	21	22	2	14	45
2749	20	21	02	00	21	1	18	57	22	22	14	30	22	8	01	47
2750	21	2	54	26	21	7	12	33	23	3	59	59	22	13	46	44
2751	21	8	38	27	21	12	44	52	23	9	40	49	22	19	34	48
2752	20	14	25	19	20	18	34	24	22	15	36	40	22	1	21	27
2753	20	20	09	42	21	0	23	14	22	21	23	23	22	7	06	18
2754	21	1	54	25	21	6	05	19	23	3	05	00	22	12	57	16
2755	21	7	45	50	21	11	56	04	23	8	55	25	22	18	51	50
2756	20	13	28	44	20	17	35	27	22	14	41	59	22	0	42	31
2757	20	19	19	40	20	23	26	57	22	20	34	57	22	6	32	31
2758	21	1	20	05	21	5	23	50	23	2	25	10	22	12	22	34
2759	21	7	12	12	21	11	02	46	23	8	07	42	22	18	15	53
2760	20	13	02	33	20	16	58	20	22	14	04	45	22	0	08	11

Year	March Equinox				June Solstice				September Equinox				December Solstice			
	d	h	m	s	d	h	m	s	d	h	m	s	d	h	m	s
2761	20	18	51	52	20	22	55	12	22	19	53	35	22	5	56	13
2762	21	0	43	04	21	4	41	29	23	1	36	11	22	11	44	06
2763	21	6	36	17	21	10	33	57	23	7	31	10	22	17	35	27
2764	20	12	21	15	20	16	16	54	22	13	21	18	21	23	22	07
2765	20	18	06	47	20	22	06	02	22	19	12	39	22	5	10	36
2766	20	23	58	29	21	4	00	36	23	1	02	03	22	11	03	45
2767	21	5	44	54	21	9	34	05	23	6	39	18	22	16	54	18
2768	20	11	30	22	20	15	21	06	22	12	32	12	21	22	44	23
2769	20	17	21	13	20	21	13	03	22	18	18	40	22	4	28	37
2770	20	23	12	36	21	2	53	56	22	23	55	51	22	10	11	37
2771	21	5	02	49	21	8	45	17	23	5	47	38	22	16	03	38
2772	20	10	46	13	20	14	30	55	22	11	34	21	21	21	50	05
2773	20	16	30	02	20	20	20	10	22	17	22	18	22	3	36	22
2774	20	22	22	12	21	2	15	53	22	23	15	08	22	9	29	00
2775	21	4	12	15	21	7	51	28	23	4	56	13	22	15	15	50
2776	20	9	57	48	20	13	39	14	22	10	53	59	21	21	07	48
2777	20	15	48	42	20	19	35	34	22	16	46	13	22	2	58	38
2778	20	21	39	01	21	1	18	40	22	22	24	38	22	8	48	02
2779	21	3	29	42	21	7	10	46	23	4	19	51	22	14	47	27
2780	20	9	19	34	20	12	58	12	22	10	09	47	21	20	37	35
2781	20	15	11	57	20	18	47	51	22	15	58	39	22	2	23	54
2782	20	21	09	12	21	0	45	18	22	21	53	08	22	8	17	53
2783	21	3	02	38	21	6	26	41	23	3	33	15	22	14	05	46
2784	20	8	46	57	20	12	16	08	22	9	26	28	21	19	55	29
2785	20	14	33	52	20	18	13	01	22	15	17	21	22	1	43	43
2786	20	20	24	07	20	23	54	41	22	20	53	01	22	7	24	28
2787	21	2	09	35	21	5	39	59	23	2	46	30	22	13	15	18
2788	20	7	53	13	20	11	24	11	22	8	36	21	21	19	00	59
2789	20	13	37	16	20	17	08	25	22	14	20	25	22	0	45	02
2790	20	19	23	09	20	22	58	49	22	20	11	25	22	6	41	40
2791	21	1	12	40	21	4	35	37	23	1	49	19	22	12	31	50
2792	20	6	58	49	20	10	18	20	22	7	38	41	21	18	19	57
2793	20	12	49	39	20	16	12	59	22	13	30	55	22	0	10	21
2794	20	18	49	13	20	21	59	43	22	19	07	32	22	5	53	00
2795	21	0	38	14	21	3	50	15	23	1	01	15	22	11	48	37
2796	20	6	26	06	20	9	44	59	22	6	55	35	21	17	40	42
2797	20	12	16	39	20	15	37	16	22	12	41	54	21	23	24	37
2798	20	18	07	13	20	21	33	01	22	18	39	44	22	5	21	46
2799	21	0	02	07	21	3	16	01	23	0	25	44	22	11	10	23
2800	20	5	48	21	20	9	01	47	22	6	19	58	21	16	58	46

Year	March Equinox				June Solstice				September Equinox				December Solstice			
	d	h	m	s	d	h	m	s	d	h	m	s	d	h	m	s
2801	20	11	34	20	20	14	57	52	22	12	16	21	21	22	54	27
2802	20	17	28	28	20	20	42	31	22	17	51	48	22	4	39	54
2803	20	23	13	07	21	2	25	35	22	23	41	21	22	10	35	09
2804	20	4	59	55	20	8	12	44	22	5	32	57	21	16	25	19
2805	20	10	52	35	20	13	58	49	22	11	13	12	21	22	02	47
2806	20	16	41	36	20	19	48	50	22	17	04	54	22	3	57	01
2807	20	22	32	58	21	1	30	49	22	22	44	50	22	9	43	38
2808	20	4	15	21	20	7	15	09	22	4	30	54	21	15	27	55
2809	20	9	58	45	20	13	09	38	22	10	24	19	21	21	19	20
2810	20	15	53	20	20	18	54	18	22	15	59	44	22	2	59	01
2811	20	21	38	31	21	0	37	14	22	21	51	24	22	8	49	52
2812	20	3	22	33	20	6	26	23	22	3	48	34	21	14	42	52
2813	20	9	14	01	20	12	17	14	22	9	32	16	21	20	26	20
2814	20	15	00	49	20	18	07	41	22	15	26	12	22	2	27	49
2815	20	20	54	44	20	23	52	35	22	21	12	56	22	8	23	06
2816	20	2	47	05	20	5	39	39	22	3	02	50	21	14	09	33
2817	20	8	37	24	20	11	35	42	22	9	01	18	21	20	05	11
2818	20	14	40	14	20	17	27	19	22	14	40	27	22	1	49	55
2819	20	20	28	31	20	23	14	20	22	20	30	10	22	7	42	30
2820	20	2	11	08	20	5	06	39	22	2	27	02	21	13	36	50
2821	20	8	04	28	20	10	59	08	22	8	08	52	21	19	15	16
2822	20	13	49	38	20	16	44	38	22	13	59	25	22	1	07	39
2823	20	19	38	47	20	22	25	20	22	19	46	24	22	6	57	35
2824	20	1	25	46	20	4	08	07	22	1	31	33	21	12	37	03
2825	20	7	03	38	20	9	56	17	22	7	23	39	21	18	31	37
2826	20	12	58	18	20	15	42	50	22	12	58	47	22	0	17	16
2827	20	18	43	10	20	21	22	06	22	18	42	15	22	6	06	24
2828	20	0	25	41	20	3	08	53	22	0	38	00	21	12	00	01
2829	20	6	24	25	20	9	01	29	22	6	19	15	21	17	37	11
2830	20	12	13	15	20	14	49	46	22	12	08	57	21	23	31	37
2831	20	18	03	45	20	20	37	26	22	17	58	50	22	5	27	00
2832	19	23	54	34	20	2	29	24	21	23	47	14	21	11	09	39
2833	20	5	37	37	20	8	23	18	22	5	44	39	21	17	05	52
2834	20	11	37	14	20	14	14	58	22	11	28	57	21	22	53	04
2835	20	17	27	00	20	19	59	13	22	17	19	27	22	4	41	52
2836	19	23	08	04	20	1	48	01	21	23	21	06	21	10	40	26
2837	20	5	03	43	20	7	42	46	22	5	04	45	21	16	24	09
2838	20	10	50	12	20	13	27	43	22	10	52	18	21	22	22	13
2839	20	16	39	11	20	19	08	57	22	16	40	33	22	4	18	13
2840	19	22	33	33	20	0	56	18	21	22	24	44	21	9	57	08

Year	March Equinox				June Solstice				September Equinox				December Solstice			
	d	h	m	s	d	h	m	s	d	h	m	s	d	h	m	s
2841	20	4	19	47	20	6	45	10	22	4	15	56	21	15	48	01
2842	20	10	17	08	20	12	34	46	22	9	54	49	21	21	33	22
2843	20	16	04	34	20	18	20	14	22	15	37	46	22	3	17	38
2844	19	21	40	06	20	0	06	31	21	21	32	56	21	9	10	45
2845	20	3	33	28	20	6	01	02	22	3	15	14	21	14	48	45
2846	20	9	19	32	20	11	43	41	22	9	01	13	21	20	37	31
2847	20	15	03	30	20	17	23	32	22	14	53	53	22	2	32	04
2848	19	20	55	18	19	23	13	41	21	20	41	54	21	8	13	13
2849	20	2	35	29	20	5	01	16	22	2	33	38	21	14	09	34
2850	20	8	30	15	20	10	50	34	22	8	16	54	21	20	03	59
2851	20	14	23	37	20	16	36	33	22	14	02	49	22	1	52	09
2852	19	20	06	34	19	22	22	31	21	20	01	15	21	7	48	57
2853	20	2	09	08	20	4	21	56	22	1	49	06	21	13	33	12
2854	20	8	02	47	20	10	10	58	22	7	35	17	21	19	25	04
2855	20	13	48	05	20	15	56	03	22	13	28	33	22	1	25	06
2856	19	19	44	24	19	21	52	28	21	19	16	57	21	7	05	59
2857	20	1	26	57	20	3	40	32	22	1	06	43	21	12	56	09
2858	20	7	21	29	20	9	29	17	22	6	52	23	21	18	45	38
2859	20	13	13	03	20	15	14	41	22	12	38	12	22	0	27	16
2860	19	18	47	55	19	20	56	39	21	18	33	15	21	6	21	21
2861	20	0	39	34	20	2	50	53	22	0	17	21	21	12	06	17
2862	20	6	26	48	20	8	33	51	22	5	58	17	21	17	55	53
2863	20	12	08	18	20	14	09	48	22	11	47	26	21	23	52	22
2864	19	18	04	42	19	20	01	25	21	17	34	47	21	5	31	19
2865	19	23	50	57	20	1	48	01	21	23	21	32	21	11	19	17
2866	20	5	43	53	20	7	36	40	22	5	05	42	21	17	11	48
2867	20	11	36	13	20	13	27	46	22	10	51	19	21	22	56	14
2868	19	17	13	06	19	19	12	51	21	16	46	15	21	4	50	19
2869	19	23	07	10	20	1	10	10	21	22	35	31	21	10	35	57
2870	20	5	00	37	20	6	58	03	22	4	22	36	21	16	25	13
2871	20	10	43	59	20	12	35	53	22	10	17	10	21	22	23	11
2872	19	16	38	23	19	18	31	38	21	16	09	54	21	4	09	48
2873	19	22	25	17	20	0	21	38	21	21	57	56	21	10	03	29
2874	20	4	17	05	20	6	07	38	22	3	42	54	21	16	01	05
2875	20	10	14	45	20	11	58	41	22	9	30	08	21	21	46	59
2876	19	15	57	55	19	17	41	23	21	15	22	15	21	3	37	18
2877	19	21	54	01	19	23	38	28	21	21	10	32	21	9	22	37
2878	20	3	48	34	20	5	29	16	22	2	52	55	21	15	08	49
2879	20	9	26	52	20	11	06	40	22	8	41	59	21	21	03	39
2880	19	15	17	30	19	17	02	02	21	14	32	44	21	2	45	55

Year	March Equinox				June Solstice				September Equinox				December Solstice			
	d	h	m	s	d	h	m	s	d	h	m	s	d	h	m	s
2881	19	21	02	07	19	22	48	45	21	20	17	36	21	8	29	53
2882	20	2	48	56	20	4	28	57	22	2	01	53	21	14	20	51
2883	20	8	41	04	20	10	17	16	22	7	50	30	21	20	04	22
2884	19	14	16	57	19	15	57	03	21	13	39	47	21	1	54	59
2885	19	20	04	58	19	21	49	53	21	19	27	23	21	7	46	45
2886	20	1	59	20	20	3	38	13	22	1	09	29	21	13	36	09
2887	20	7	41	33	20	9	12	18	22	6	58	09	21	19	33	03
2888	19	13	40	29	19	15	09	29	21	12	53	00	21	1	19	08
2889	19	19	34	29	19	21	02	47	21	18	39	50	21	7	06	22
2890	20	1	26	31	20	2	50	19	22	0	25	56	21	13	03	30
2891	20	7	22	31	20	8	47	47	22	6	18	21	21	18	52	08
2892	19	13	04	42	19	14	35	20	21	12	10	20	21	0	41	42
2893	19	18	56	01	19	20	30	34	21	18	01	42	21	6	30	24
2894	20	0	52	49	20	2	22	36	21	23	50	00	21	12	17	24
2895	20	6	33	03	20	7	56	49	22	5	40	09	21	18	10	58
2896	19	12	21	34	19	13	50	43	21	11	35	07	20	23	59	35
2897	19	18	09	03	19	19	40	30	21	17	18	36	21	5	47	40
2898	19	23	55	03	20	1	17	05	21	22	58	12	21	11	42	11
2899	20	5	48	15	20	7	05	46	22	4	47	01	21	17	28	16
2900	20	11	32	45	20	12	47	45	22	10	33	26	21	23	12	19
2901	20	17	22	04	20	18	37	39	22	16	18	07	22	4	58	29
2902	20	23	16	35	21	0	30	48	22	22	02	09	22	10	47	07
2903	21	4	56	35	21	6	06	19	23	3	46	45	22	16	37	20
2904	20	10	43	25	20	12	00	22	22	9	41	08	21	22	23	58
2905	20	16	35	55	20	17	54	54	22	15	28	51	22	4	09	19
2906	20	22	24	26	20	23	33	09	22	21	12	43	22	10	01	53
2907	21	4	16	58	21	5	26	44	23	3	10	01	22	15	53	36
2908	20	10	01	45	20	11	14	45	22	9	00	33	21	21	44	02
2909	20	15	49	59	20	17	06	04	22	14	48	38	22	3	38	38
2910	20	21	48	25	20	23	00	57	22	20	38	00	22	9	33	06
2911	21	3	35	47	21	4	37	47	23	2	25	42	22	15	24	44
2912	20	9	29	29	20	10	32	46	22	8	22	44	21	21	13	21
2913	20	15	26	21	20	16	30	42	22	14	10	39	22	3	00	23
2914	20	21	14	38	20	22	10	51	22	19	49	49	22	8	52	24
2915	21	3	03	09	21	4	03	55	23	1	44	12	22	14	42	15
2916	20	8	46	28	20	9	50	56	22	7	30	44	21	20	24	52
2917	20	14	31	32	20	15	35	56	22	13	14	27	22	2	10	40
2918	20	20	24	19	20	21	24	51	22	19	02	27	22	7	58	12
2919	21	2	04	06	21	2	57	24	23	0	46	01	22	13	45	13
2920	20	7	48	04	20	8	46	50	22	6	38	12	21	19	34	22

Year	March Equinox				June Solstice				September Equinox				December Solstice			
	d	h	m	s	d	h	m	s	d	h	m	s	d	h	m	s
2921	20	13	37	56	20	14	40	13	22	12	23	31	22	1	23	46
2922	20	19	26	28	20	20	16	01	22	18	00	04	22	7	15	17
2923	21	1	17	54	21	2	05	38	22	23	55	28	22	13	06	03
2924	20	7	09	28	20	7	57	08	22	5	45	34	21	18	51	14
2925	20	13	00	50	20	13	47	32	22	11	29	58	22	0	40	13
2926	20	18	55	58	20	19	45	32	22	17	23	21	22	6	36	15
2927	21	0	42	29	21	1	28	49	22	23	11	54	22	12	26	30
2928	20	6	30	14	20	7	22	17	22	5	08	37	21	18	16	19
2929	20	12	25	13	20	13	20	58	22	11	02	21	22	0	06	46
2930	20	18	17	04	20	18	59	13	22	16	43	35	22	5	56	26
2931	21	0	03	54	21	0	48	18	22	22	41	30	22	11	49	51
2932	20	5	50	23	20	6	39	28	22	4	31	28	21	17	39	38
2933	20	11	36	47	20	12	22	43	22	10	09	38	21	23	27	37
2934	20	17	27	53	20	18	11	29	22	15	58	40	22	5	22	49
2935	20	23	17	35	20	23	48	58	22	21	41	38	22	11	06	38
2936	20	5	05	00	20	5	35	24	22	3	30	03	21	16	49	59
2937	20	10	57	55	20	11	33	24	22	9	18	49	21	22	39	06
2938	20	16	47	14	20	17	12	29	22	14	52	56	22	4	24	49
2939	20	22	30	11	20	23	01	00	22	20	47	13	22	10	14	30
2940	20	4	16	29	20	4	53	47	22	2	38	06	21	15	59	31
2941	20	10	04	27	20	10	38	16	22	8	18	44	21	21	42	31
2942	20	15	54	06	20	16	28	15	22	14	14	17	22	3	37	54
2943	20	21	40	52	20	22	10	29	22	20	03	52	22	9	27	47
2944	20	3	26	12	20	3	59	17	22	1	55	20	21	15	19	20
2945	20	9	18	26	20	9	57	16	22	7	49	24	21	21	17	53
2946	20	15	14	49	20	15	37	57	22	13	27	28	22	3	08	29
2947	20	21	06	14	20	21	25	58	22	19	24	33	22	9	01	19
2948	20	3	00	02	20	3	22	34	22	1	18	49	21	14	50	18
2949	20	8	53	41	20	9	12	20	22	6	57	36	21	20	35	51
2950	20	14	43	30	20	15	04	21	22	12	50	04	22	2	31	33
2951	20	20	29	48	20	20	48	56	22	18	37	32	22	8	18	18
2952	20	2	16	17	20	2	36	41	22	0	25	17	21	14	00	50
2953	20	8	04	39	20	8	30	06	22	6	18	38	21	19	51	42
2954	20	13	56	48	20	14	08	42	22	11	55	55	22	1	35	51
2955	20	19	38	21	20	19	50	41	22	17	47	34	22	7	24	58
2956	20	1	21	01	20	1	42	33	21	23	39	10	21	13	16	11
2957	20	7	10	11	20	7	26	34	22	5	13	01	21	18	59	19
2958	20	12	57	15	20	13	11	10	22	11	03	26	22	0	54	33
2959	20	18	47	36	20	18	54	45	22	16	53	02	22	6	42	04
2960	20	0	38	35	20	0	43	02	21	22	39	39	21	12	24	44

Year	March Equinox				June Solstice				September Equinox				December Solstice			
	d	h	m	s	d	h	m	s	d	h	m	s	d	h	m	s
2961	20	6	28	13	20	6	40	28	22	4	34	13	21	18	21	48
2962	20	12	23	52	20	12	26	37	22	10	13	53	22	0	11	06
2963	20	18	09	23	20	18	13	27	22	16	07	19	22	6	01	09
2964	19	23	57	02	20	0	10	16	21	22	06	03	21	11	54	47
2965	20	5	53	16	20	5	59	03	22	3	45	32	21	17	36	08
2966	20	11	40	55	20	11	45	08	22	9	40	04	21	23	33	14
2967	20	17	29	43	20	17	32	41	22	15	34	01	22	5	26	32
2968	19	23	17	55	19	23	20	57	21	21	18	46	21	11	11	42
2969	20	5	05	19	20	5	14	09	22	3	12	07	21	17	10	43
2970	20	11	03	09	20	10	56	14	22	8	50	07	21	22	58	13
2971	20	16	52	12	20	16	38	51	22	14	39	16	22	4	42	47
2972	19	22	39	12	19	22	33	08	21	20	33	58	21	10	33	19
2973	20	4	33	34	20	4	23	52	22	2	08	20	21	16	12	44
2974	20	10	16	58	20	10	08	47	22	7	56	27	21	22	05	07
2975	20	15	59	44	20	15	54	40	22	13	49	04	22	3	54	26
2976	19	21	47	48	19	21	42	16	21	19	32	44	21	9	32	11
2977	20	3	32	02	20	3	30	59	22	1	27	06	21	15	26	24
2978	20	9	25	21	20	9	12	55	22	7	09	14	21	21	15	23
2979	20	15	10	23	20	14	55	33	22	12	58	47	22	3	03	52
2980	19	20	52	03	19	20	47	31	21	18	54	44	21	9	01	51
2981	20	2	49	08	20	2	38	14	22	0	31	31	21	14	48	07
2982	20	8	40	55	20	8	21	49	22	6	20	23	21	20	42	07
2983	20	14	31	11	20	14	09	49	22	12	17	27	22	2	35	39
2984	19	20	29	49	19	20	06	09	21	18	03	36	21	8	16	46
2985	20	2	18	00	20	2	00	42	21	23	57	28	21	14	14	52
2986	20	8	13	16	20	7	50	11	22	5	42	42	21	20	07	24
2987	20	14	02	03	20	13	37	08	22	11	31	54	22	1	50	56
2988	19	19	43	48	19	19	29	05	21	17	29	51	21	7	44	12
2989	20	1	40	35	20	1	19	40	21	23	09	00	21	13	24	41
2990	20	7	25	41	20	6	59	54	22	4	56	15	21	19	14	30
2991	20	13	05	02	20	12	43	28	22	10	51	30	22	1	09	40
2992	19	18	55	08	19	18	33	25	21	16	32	25	21	6	50	01
2993	20	0	37	53	20	0	17	35	21	22	19	34	21	12	45	23
2994	20	6	31	40	20	5	58	44	22	4	02	21	21	18	36	19
2995	20	12	23	35	20	11	41	58	22	9	47	24	22	0	15	31
2996	19	18	05	47	19	17	32	25	21	15	41	39	21	6	10	01
2997	20	0	03	07	19	23	26	24	21	21	18	52	21	11	53	14
2998	20	5	49	36	20	5	10	31	22	3	03	46	21	17	42	45
2999	20	11	32	04	20	10	58	26	22	9	02	26	21	23	38	11
3000	20	17	28	11	20	16	54	15	22	14	49	21	22	5	17	20

4

Phases of the Moon

1970 − 2050

Phases of the Moon

The list in this Part gives the instants of the phases of the Moon for the years 1970 to 2050, to the nearest second of time. These instants are expressed in *Dynamical Time*. The tabulated times can be converted to Universal Time by *subtracting* the quantity ΔT, as explained on pages 6 − 8. See Example 1 on page 8.

By definition, the times of New Moon, First Quarter, Full Moon and Last Quarter are the times at which the excess of the apparent longitude of the Moon (i.e. affected by the nutation and the effect of light-time) over the apparent longitude of the Sun is 0, 90, 180, and 270 degrees, respectively.

Cycles

The Methonic Cycle is a cycle of 19 years, or 235 lunations (synodic months), after which time the phases of the Moon are repeated on the same days of the year, or approximately so, forming a nearly accurate prediction technique. For example:

New Moon	1942 Aug	12
New Moon	1961 Aug	11
New Moon	1980 Aug	10
New Moon	1999 Aug	11

There are other interesting cycles. After 2 years, the *preceding* lunar phase occurs at nearly the same calendar date. For example:

New Moon	1983 Aug	8
Last Quarter	1985 Aug	8
Full Moon	1987 Aug	9
First Quarter	1989 Aug	9
New Moon	1991 Aug	10
Last Quarter	1993 Aug	10
Full Moon	1995 Aug	10

Hence, after 8 years the same lunar phases repeat, occurring one or two days later in the year. The Greek called this 8-year cycle the *octaeteris*:

New Moon	1983 Aug	8
New Moon	1991 Aug	10
New Moon	1999 Aug	11
New Moon	2007 Aug	12

Combining the cycle of 19 years with that of 2 years, we obtain cycles of 17 and 21 years, after which respectively the following and the preceding lunar phase occurs near the same calendar date.

In the Gregorian calendar, 372 years is an excellent cycle of long period:

New Moon	1627 Aug	11
New Moon	1999 Aug	11
New Moon	2371 Aug	11
New Moon	2743 Aug	12

During the period 1800 − 2100, the following lunar phases take place on bissextile day, February 29:

1820	Full Moon		1972	Full Moon
1824	New Moon		1976	New Moon
1856	Last Quarter		2008	Last Quarter
1860	First Quarter		2048	Full Moon
1936	First Quarter		2088	First Quarter

In some years there are only three lunar phases in February. Here is the complete list of these cases during the period 1800 − 2100. For each case the missing phase is given. For instance, there was no First Quarter in February 1993. The sign *B* refers to a bissextile year. The list is based on the *Universal Time*; different results are obtained for dates based on another standard time.

1805	NM		1870	NM		1925	FQ		1993	FQ	*B*	2052	NM
1807	LQ		1879	FQ		1934	FM		1995	NM		2054	LQ
1809	FM		1883	LQ		1938	NM		1997	LQ		2058	FQ
1811	FQ		1885	FM	*B*	1940	LQ		1999	FM		2067	FM
1826	LQ		1889	NM		1955	FQ	*B*	2012	FQ		2071	NM
1830	FQ		1900	NM		1957	NM		2014	NM		2077	FQ
1843	NM		1902	LQ		1959	LQ		2018	FM		2081	LQ
1845	LQ		1911	NM		1961	FM		2031	FQ		2090	NM
1847	FM		1915	FM		1970	LQ		2033	NM	*B*	2092	LQ
1849	FQ		1919	NM		1974	FQ		2035	LQ		2094	FM
1866	FM		1921	LQ		1978	LQ		2037	FM			

	New Moon	First Quarter	Full Moon	Last Quarter

1970

	New Moon (h m s)	First Quarter (h m s)	Full Moon (h m s)	Last Quarter (h m s)
	Jan 7 — 20 36 09	Jan 14 — 13 18 51	Jan 22 — 12 56 00	Jan 30 — 14 39 08
	Feb 6 — 7 13 31	Feb 13 — 4 11 02	Feb 21 — 8 19 21	Mar 1 — 2 33 48
	Mar 7 — 17 43 07	Mar 14 — 21 16 17	Mar 23 — 1 53 06	Mar 30 — 11 05 12
	Apr 6 — 4 10 01	Apr 13 — 15 44 16	Apr 21 — 16 21 58	Apr 28 — 17 18 53
	May 5 — 14 51 35	May 13 — 10 26 43	May 21 — 3 38 15	May 27 — 22 32 19
	Jun 4 — 2 21 53	Jun 12 — 4 07 05	Jun 19 — 12 28 05	Jun 26 — 4 01 56
	Jul 3 — 15 18 29	Jul 11 — 19 43 40	Jul 18 — 19 59 09	Jul 25 — 11 00 13
	Aug 2 — 5 58 48	Aug 10 — 8 50 33	Aug 17 — 3 16 01	Aug 23 — 20 35 01
	Aug 31 — 22 01 53	Sep 8 — 19 38 57	Sep 15 — 11 10 05	Sep 22 — 9 42 44
	Sep 30 — 14 32 11	Oct 8 — 4 43 31	Oct 14 — 20 21 55	Oct 22 — 2 47 58
	Oct 30 — 6 28 35	Nov 6 — 12 47 35	Nov 13 — 7 28 35	Nov 20 — 23 13 45
	Nov 28 — 21 14 50	Dec 5 — 20 36 16	Dec 12 — 21 04 01	Dec 20 — 21 09 33
	Dec 28 — 10 43 21			

1971

	New Moon (h m s)	First Quarter (h m s)	Full Moon (h m s)	Last Quarter (h m s)
		Jan 4 — 4 55 36	Jan 11 — 13 20 58	Jan 19 — 18 08 57
	Jan 26 — 22 55 48	Feb 2 — 14 31 16	Feb 10 — 7 42 04	Feb 18 — 12 14 17
	Feb 25 — 9 49 14	Mar 4 — 2 01 43	Mar 12 — 2 34 16	Mar 20 — 2 30 39
	Mar 26 — 19 24 10	Apr 2 — 15 46 29	Apr 10 — 20 10 38	Apr 18 — 12 58 31
	Apr 25 — 4 02 21	May 2 — 7 34 35	May 10 — 11 24 13	May 17 — 20 15 34
	May 24 — 12 32 28	Jun 1 — 0 42 55	Jun 9 — 0 04 18	Jun 16 — 1 24 48
	Jun 22 — 21 58 00	Jun 30 — 18 11 40	Jul 8 — 10 37 17	Jul 15 — 5 47 24
	Jul 22 — 9 15 40	Jul 30 — 11 07 44	Aug 6 — 19 43 02	Aug 13 — 10 55 48
	Aug 20 — 22 54 02	Aug 29 — 2 56 47	Sep 5 — 4 03 06	Sep 11 — 18 23 50
	Sep 19 — 14 42 54	Sep 27 — 17 18 01	Oct 4 — 12 20 08	Oct 11 — 5 29 57
	Oct 19 — 7 59 37	Oct 27 — 5 54 57	Nov 2 — 21 20 04	Nov 9 — 20 52 01
	Nov 18 — 1 46 29	Nov 25 — 16 37 27	Dec 2 — 7 48 49	Dec 9 — 16 03 05
	Dec 17 — 19 03 28	Dec 25 — 1 35 35	Dec 31 — 20 20 14	

1972

	New Moon (h m s)	First Quarter (h m s)	Full Moon (h m s)	Last Quarter (h m s)
				Jan 8 — 13 31 26
	Jan 16 — 10 53 05	Jan 23 — 9 29 24	Jan 30 — 10 58 54	Feb 7 — 11 12 12
	Feb 15 — 0 29 28	Feb 21 — 17 20 58	Feb 29 — 3 12 37	Mar 8 — 7 06 09
	Mar 15 — 11 35 23	Mar 22 — 2 12 23	Mar 29 — 20 06 11	Apr 6 — 23 45 07
	Apr 13 — 20 31 34	Apr 20 — 12 45 53	Apr 28 — 12 45 10	May 6 — 12 27 03
	May 13 — 4 08 42	May 20 — 1 17 01	May 28 — 4 28 19	Jun 4 — 21 22 14
	Jun 11 — 11 30 39	Jun 18 — 15 41 41	Jun 26 — 18 46 52	Jul 4 — 3 26 09
	Jul 10 — 19 39 28	Jul 18 — 7 46 25	Jul 26 — 7 24 15	Aug 2 — 8 03 01
	Aug 9 — 5 26 31	Aug 17 — 1 09 51	Aug 24 — 18 22 22	Aug 31 — 12 48 52
	Sep 7 — 17 29 08	Sep 15 — 19 13 38	Sep 23 — 4 07 27	Sep 29 — 19 16 44
	Oct 7 — 8 08 39	Oct 15 — 12 55 28	Oct 22 — 13 25 45	Oct 29 — 4 41 30
	Nov 6 — 1 22 00	Nov 14 — 5 01 27	Nov 20 — 23 07 14	Nov 27 — 17 45 19
	Dec 5 — 20 24 46	Dec 13 — 18 36 20	Dec 20 — 9 45 48	Dec 27 — 10 28 08

	New Moon		First Quarter		Full Moon		Last Quarter

1973

New Moon	h m s	First Quarter	h m s	Full Moon	h m s	Last Quarter	h m s
Jan 4	15 43 12	Jan 12	5 27 45	Jan 18	21 29 08	Jan 26	6 05 36
Feb 3	9 23 33	Feb 10	14 06 06	Feb 17	10 07 41	Feb 25	3 11 01
Mar 5	0 08 08	Mar 11	21 26 27	Mar 18	23 34 08	Mar 26	23 46 54
Apr 3	11 45 47	Apr 10	4 28 43	Apr 17	13 51 14	Apr 25	17 59 18
May 2	20 55 37	May 9	12 07 23	May 17	4 58 46	May 25	8 40 34
Jun 1	4 34 59	Jun 7	21 11 31	Jun 15	20 35 18	Jun 23	19 45 59
Jun 30	11 39 29	Jul 7	8 26 25	Jul 15	11 56 37	Jul 23	3 58 16
Jul 29	18 59 31	Aug 5	22 27 27	Aug 14	2 17 13	Aug 21	10 22 53
Aug 28	3 26 02	Sep 4	15 22 39	Sep 12	15 17 00	Sep 19	16 11 16
Sep 26	13 54 34	Oct 4	10 32 42	Oct 12	3 09 41	Oct 18	22 33 25
Oct 26	3 17 14	Nov 3	6 30 03	Nov 10	14 27 27	Nov 17	6 35 09
Nov 24	19 55 50	Dec 3	1 29 21	Dec 10	1 35 20	Dec 16	17 13 22
Dec 24	15 07 46						

1974

New Moon	h m s	First Quarter	h m s	Full Moon	h m s	Last Quarter	h m s
		Jan 1	18 07 04	Jan 8	12 37 04	Jan 15	7 04 17
Jan 23	11 02 38	Jan 31	7 39 58	Feb 6	23 24 57	Feb 14	0 04 25
Feb 22	5 34 29	Mar 1	18 03 16	Mar 8	10 03 39	Mar 15	19 15 57
Mar 23	21 24 42	Mar 31	1 44 50	Apr 6	21 01 03	Apr 14	14 58 11
Apr 22	10 17 14	Apr 29	7 40 01	May 6	8 55 17	May 14	9 29 25
May 21	20 35 01	May 28	13 03 56	Jun 4	22 10 23	Jun 13	1 45 59
Jun 20	4 56 24	Jun 26	19 20 47	Jul 4	12 41 03	Jul 12	15 28 43
Jul 19	12 07 15	Jul 26	3 51 40	Aug 3	3 57 45	Aug 11	2 46 35
Aug 17	19 02 19	Aug 24	15 39 11	Sep 1	19 25 35	Sep 9	12 01 41
Sep 16	2 45 59	Sep 23	7 08 36	Oct 1	10 38 42	Oct 8	19 46 19
Oct 15	12 25 33	Oct 23	1 53 49	Oct 31	1 19 44	Nov 7	2 47 56
Nov 14	0 53 58	Nov 21	22 40 14	Nov 29	15 10 49	Dec 6	10 10 42
Dec 13	16 25 30	Dec 21	19 44 06	Dec 29	3 51 36		

1975

New Moon	h m s	First Quarter	h m s	Full Moon	h m s	Last Quarter	h m s
						Jan 4	19 04 48
Jan 12	10 20 22	Jan 20	15 15 05	Jan 27	15 10 12	Feb 3	6 23 31
Feb 11	5 17 37	Feb 19	7 39 14	Feb 26	1 15 16	Mar 4	20 20 31
Mar 12	23 48 11	Mar 20	20 05 22	Mar 27	10 36 48	Apr 3	12 25 36
Apr 11	16 39 45	Apr 19	4 41 55	Apr 25	19 55 41	May 3	5 44 25
May 11	7 05 39	May 18	10 29 44	May 25	5 51 21	Jun 1	23 23 13
Jun 9	18 49 52	Jun 16	14 59 03	Jun 23	16 54 58	Jul 1	16 37 55
Jul 9	4 10 53	Jul 15	19 47 28	Jul 23	5 29 13	Jul 31	8 49 10
Aug 7	11 58 08	Aug 14	2 24 16	Aug 21	19 48 24	Aug 29	23 20 21
Sep 5	19 19 28	Sep 12	11 59 37	Sep 20	11 51 08	Sep 28	11 46 36
Oct 5	3 24 03	Oct 12	1 16 03	Oct 20	5 06 32	Oct 27	22 07 38
Nov 3	13 05 32	Nov 10	18 21 30	Nov 18	22 29 01	Nov 26	6 52 36
Dec 3	0 50 49	Dec 10	14 39 50	Dec 18	14 40 15	Dec 25	14 52 39

New Moon		First Quarter		Full Moon		Last Quarter	

1976

	New Moon h m s		First Quarter h m s		Full Moon h m s		Last Quarter h m s
Jan 1	14 40 49	Jan 9	12 40 22	Jan 17	4 47 43	Jan 23	23 05 07
Jan 31	6 20 57	Feb 8	10 05 37	Feb 15	16 44 02	Feb 22	8 16 37
Feb 29	23 25 32	Mar 9	4 38 43	Mar 16	2 53 31	Mar 22	18 55 09
Mar 30	17 08 42	Apr 7	19 02 30	Apr 14	11 49 32	Apr 21	7 14 49
Apr 29	10 20 16	May 7	5 18 02	May 13	20 04 49	May 20	21 22 48
May 29	1 47 33	Jun 5	12 20 41	Jun 12	4 15 45	Jun 19	13 16 06
Jun 27	14 50 44	Jul 4	17 29 14	Jul 11	13 09 39	Jul 19	6 30 06
Jul 27	1 39 31	Aug 2	22 07 17	Aug 9	23 44 16	Aug 18	0 13 17
Aug 25	11 01 21	Sep 1	3 36 11	Sep 8	12 52 51	Sep 16	17 20 59
Sep 23	19 55 39	Sep 30	11 13 00	Oct 8	4 56 11	Oct 16	8 59 24
Oct 23	5 10 25	Oct 29	22 05 50	Nov 6	23 15 22	Nov 14	22 39 33
Nov 21	15 11 25	Nov 28	12 59 32	Dec 6	18 15 19	Dec 14	10 15 00
Dec 21	2 08 33	Dec 28	7 48 29				

1977

	New Moon h m s		First Quarter h m s		Full Moon h m s		Last Quarter h m s
				Jan 5	12 11 09	Jan 12	19 55 47
Jan 19	14 11 53	Jan 27	5 12 14	Feb 4	3 57 02	Feb 11	4 07 58
Feb 18	3 37 40	Feb 26	2 50 49	Mar 5	17 14 00	Mar 12	11 35 22
Mar 19	18 33 26	Mar 27	22 27 26	Apr 4	4 09 40	Apr 10	19 15 25
Apr 18	10 36 13	Apr 26	14 42 59	May 3	13 04 12	May 10	4 09 03
May 18	2 52 10	May 26	3 20 54	Jun 1	20 31 40	Jun 8	15 07 49
Jun 16	18 23 30	Jun 24	12 44 33	Jul 1	3 24 50	Jul 8	4 40 02
Jul 16	8 37 17	Jul 23	19 39 02	Jul 30	10 53 02	Aug 6	20 40 55
Aug 14	21 31 48	Aug 22	1 05 12	Aug 28	20 10 51	Sep 5	14 33 40
Sep 13	9 23 35	Sep 20	6 18 48	Sep 27	8 18 12	Oct 5	9 21 30
Oct 12	20 31 30	Oct 19	12 46 32	Oct 26	23 36 09	Nov 4	3 59 08
Nov 11	7 10 11	Nov 17	21 53 14	Nov 25	17 32 13	Dec 3	21 16 49
Dec 10	17 33 32	Dec 17	10 37 33	Dec 25	12 49 50		

1978

	New Moon h m s		First Quarter h m s		Full Moon h m s		Last Quarter h m s
						Jan 2	12 08 05
Jan 9	4 00 31	Jan 16	3 03 40	Jan 24	7 56 20	Jan 31	23 51 49
Feb 7	14 54 49	Feb 14	22 11 26	Feb 23	1 27 09	Mar 2	8 34 49
Mar 9	2 37 02	Mar 16	18 21 44	Mar 24	16 20 48	Mar 31	15 11 35
Apr 7	15 15 58	Apr 15	13 56 23	Apr 23	4 11 46	Apr 29	21 02 48
May 7	4 47 31	May 15	7 40 16	May 22	13 17 30	May 29	3 31 02
Jun 5	19 02 16	Jun 13	22 44 48	Jun 20	20 31 20	Jun 27	11 44 48
Jul 5	9 51 08	Jul 13	10 49 46	Jul 20	3 05 39	Jul 26	22 31 51
Aug 4	1 01 32	Aug 11	20 07 15	Aug 18	10 14 48	Aug 25	12 18 23
Sep 2	16 09 43	Sep 10	3 21 02	Sep 16	19 01 55	Sep 24	5 08 19
Oct 2	6 41 28	Oct 9	9 38 45	Oct 16	6 10 20	Oct 24	0 34 41
Oct 31	20 07 21	Nov 7	16 18 56	Nov 14	20 00 57	Nov 22	21 25 02
Nov 30	8 20 16	Dec 7	0 34 41	Dec 14	12 31 28	Dec 22	17 42 20
Dec 29	19 36 43						

New Moon	First Quarter	Full Moon	Last Quarter

1979

	New Moon				First Quarter				Full Moon				Last Quarter		
	h	m	s		h	m	s		h	m	s		h	m	s
				Jan	5	11	15	39	Jan	13	7	09	27	Jan	21	11	24	09
Jan	28	6	20	22	Feb	4	0	37	12	Feb	12	2	40	13	Feb	20	1	17	58
Feb	26	16	46	03	Mar	5	16	23	38	Mar	13	21	15	05	Mar	21	11	22	57
Mar	28	3	00	18	Apr	4	9	58	04	Apr	12	13	15	51	Apr	19	18	31	06
Apr	26	13	15	34	May	4	4	26	09	May	12	2	01	41	May	18	23	57	40
May	26	0	01	13	Jun	2	22	38	17	Jun	10	11	56	06	Jun	17	5	01	49
Jun	24	11	58	40	Jul	2	15	24	25	Jul	9	20	00	06	Jul	16	10	59	32
Jul	24	1	41	28	Aug	1	5	58	09	Aug	8	3	21	56	Aug	14	19	02	42
Aug	22	17	11	16	Aug	30	18	09	53	Sep	6	10	59	25	Sep	13	6	16	12
Sep	21	9	47	25	Sep	29	4	20	34	Oct	5	19	36	10	Oct	12	21	24	38
Oct	21	2	23	52	Oct	28	13	06	42	Nov	4	5	47	53	Nov	11	16	24	49
Nov	19	18	04	28	Nov	26	21	09	25	Dec	3	18	08	29	Dec	11	13	59	42
Dec	19	8	24	13	Dec	26	5	11	46									

1980

	New Moon				First Quarter				Full Moon				Last Quarter		
	h	m	s		h	m	s		h	m	s		h	m	s
									Jan	2	9	02	58	Jan	10	11	50	22
Jan	17	21	20	14	Jan	24	13	58	58	Feb	1	2	22	04	Feb	9	7	35	59
Feb	16	8	51	44	Feb	23	0	14	50	Mar	1	21	00	30	Mar	9	23	49	31
Mar	16	18	56	43	Mar	23	12	32	08	Mar	31	15	14	42	Apr	8	12	07	16
Apr	15	3	47	05	Apr	22	3	00	16	Apr	30	7	36	08	May	7	20	51	27
May	14	12	01	02	May	21	19	16	48	May	29	21	28	26	Jun	6	2	54	02
Jun	12	20	39	09	Jun	20	12	32	36	Jun	28	9	02	57	Jul	5	7	27	59
Jul	12	6	46	30	Jul	20	5	51	24	Jul	27	18	54	29	Aug	3	12	01	09
Aug	10	19	10	09	Aug	18	22	28	43	Aug	26	3	43	03	Sep	1	18	08	24
Sep	9	10	01	02	Sep	17	13	55	09	Sep	24	12	08	34	Oct	1	3	18	45
Oct	9	2	50	29	Oct	17	3	48	02	Oct	23	20	52	44	Oct	30	16	33	39
Nov	7	20	43	23	Nov	15	15	47	38	Nov	22	6	39	42	Nov	29	9	59	24
Dec	7	14	35	54	Dec	15	1	47	42	Dec	21	18	08	51	Dec	29	6	32	48

1981

	New Moon				First Quarter				Full Moon				Last Quarter		
	h	m	s		h	m	s		h	m	s		h	m	s
Jan	6	7	24	52	Jan	13	10	10	51	Jan	20	7	39	49	Jan	28	4	19	38
Feb	4	22	14	37	Feb	11	17	50	00	Feb	18	22	59	05	Feb	27	1	14	52
Mar	6	10	31	50	Mar	13	1	51	22	Mar	20	15	23	19	Mar	28	19	34	54
Apr	4	20	20	17	Apr	11	11	11	25	Apr	19	7	59	46	Apr	27	10	15	19
May	4	4	20	01	May	10	22	22	40	May	19	0	04	26	May	26	21	01	01
Jun	2	11	32	33	Jun	9	11	34	07	Jun	17	15	05	19	Jun	25	4	25	55
Jul	1	19	03	58	Jul	9	2	40	12	Jul	17	4	39	41	Jul	24	9	40	46
Jul	31	3	52	49	Aug	7	19	26	55	Aug	15	16	37	29	Aug	22	14	16	25
Aug	29	14	44	17	Sep	6	13	26	25	Sep	14	3	09	34	Sep	20	19	48	16
Sep	28	4	08	04	Oct	6	7	46	04	Oct	13	12	50	17	Oct	20	3	41	21
Oct	27	20	14	08	Nov	5	1	09	47	Nov	11	22	27	22	Nov	18	14	54	35
Nov	26	14	39	12	Dec	4	16	23	05	Dec	11	8	42	14	Dec	18	5	48	08
Dec	26	10	10	52														

New Moon   First Quarter   Full Moon   Last Quarter

1982

New Moon	h m s	First Quarter	h m s	Full Moon	h m s	Last Quarter	h m s
		Jan 3	4 46 22	Jan 9	19 53 42	Jan 16	23 58 51
Jan 25	4 56 49	Feb 1	14 28 41	Feb 8	7 57 59	Feb 15	20 21 46
Feb 23	21 13 58	Mar 2	22 15 55	Mar 9	20 46 00	Mar 17	17 15 30
Mar 25	10 18 21	Apr 1	5 09 04	Apr 8	10 19 11	Apr 16	12 43 05
Apr 23	20 29 33	Apr 30	12 08 09	May 8	0 45 32	May 16	5 12 06
May 23	4 41 11	May 29	20 07 27	Jun 6	16 00 14	Jun 14	18 06 47
Jun 21	11 52 38	Jun 28	5 57 17	Jul 6	7 32 29	Jul 14	3 47 38
Jul 20	18 57 31	Jul 27	18 22 32	Aug 4	22 34 37	Aug 12	11 09 20
Aug 19	2 45 45	Aug 26	9 50 13	Sep 3	12 29 05	Sep 10	17 19 32
Sep 17	12 09 58	Sep 25	4 07 52	Oct 3	1 09 24	Oct 9	23 27 15
Oct 17	0 04 59	Oct 25	0 08 40	Nov 1	12 57 33	Nov 8	6 38 37
Nov 15	15 10 47	Nov 23	20 06 26	Dec 1	0 21 35	Dec 7	15 54 04
Dec 15	9 18 56	Dec 23	14 17 27	Dec 30	11 33 28		

1983

New Moon	h m s	First Quarter	h m s	Full Moon	h m s	Last Quarter	h m s
						Jan 6	4 00 51
Jan 14	5 08 40	Jan 22	5 34 17	Jan 28	22 27 10	Feb 4	19 17 38
Feb 13	0 32 42	Feb 20	17 32 53	Feb 27	8 59 00	Mar 6	13 16 31
Mar 14	17 44 19	Mar 22	2 26 20	Mar 28	19 27 33	Apr 5	8 39 10
Apr 13	7 59 12	Apr 20	8 58 38	Apr 27	6 31 33	May 5	3 43 49
May 12	19 25 58	May 19	14 17 52	May 26	18 48 34	Jun 3	21 08 17
Jun 11	4 38 24	Jun 17	19 46 43	Jun 25	8 32 39	Jul 3	12 12 52
Jul 10	12 19 23	Jul 17	2 51 20	Jul 24	23 27 46	Aug 2	0 53 12
Aug 8	19 18 49	Aug 15	12 47 52	Aug 23	15 00 11	Aug 31	11 23 11
Sep 7	2 35 52	Sep 14	2 24 38	Sep 22	6 37 03	Sep 29	20 06 08
Oct 6	11 16 33	Oct 13	19 43 07	Oct 21	21 54 09	Oct 29	3 37 40
Nov 4	22 22 01	Nov 12	15 50 04	Nov 20	12 30 15	Nov 27	10 50 58
Dec 4	12 26 45	Dec 12	13 09 55	Dec 20	2 01 10	Dec 26	18 53 20

1984

New Moon	h m s	First Quarter	h m s	Full Moon	h m s	Last Quarter	h m s
Jan 3	5 16 33	Jan 11	9 49 11	Jan 18	14 06 04	Jan 25	4 48 39
Feb 1	23 47 19	Feb 10	4 00 30	Feb 17	0 41 58	Feb 23	17 12 49
Mar 2	18 31 46	Mar 10	18 28 23	Mar 17	10 10 37	Mar 24	7 59 11
Apr 1	12 10 27	Apr 9	4 52 18	Apr 15	19 11 40	Apr 23	0 27 05
May 1	3 46 12	May 8	11 50 35	May 15	4 29 30	May 22	17 45 45
May 30	16 48 45	Jun 6	16 42 34	Jun 13	14 42 42	Jun 21	11 10 32
Jun 29	3 19 21	Jul 5	21 05 01	Jul 13	2 20 38	Jul 21	4 02 17
Jul 28	11 52 08	Aug 4	2 33 42	Aug 11	15 44 06	Aug 19	19 41 26
Aug 26	19 26 25	Sep 2	10 30 30	Sep 10	7 01 58	Sep 18	9 31 53
Sep 25	3 11 32	Oct 1	21 53 22	Oct 9	23 59 04	Oct 17	21 14 44
Oct 24	12 09 03	Oct 31	13 08 20	Nov 8	17 43 31	Nov 16	7 00 02
Nov 22	22 57 35	Nov 30	8 01 25	Dec 8	10 54 16	Dec 15	15 26 09
Dec 22	11 47 30	Dec 30	5 28 13				

New Moon	First Quarter	Full Moon	Last Quarter

1985

New Moon	h m s	First Quarter	h m s	Full Moon	h m s	Last Quarter	h m s
				Jan 7	2 17 03	Jan 13	23 27 38
Jan 21	2 29 10	Jan 29	3 29 55	Feb 5	15 19 41	Feb 12	7 57 41
Feb 19	18 43 36	Feb 27	23 41 32	Mar 7	2 14 01	Mar 13	17 34 55
Mar 21	11 59 36	Mar 29	16 12 02	Apr 5	11 33 20	Apr 12	4 42 23
Apr 20	5 22 52	Apr 28	4 26 04	May 4	19 53 43	May 11	17 34 41
May 19	21 42 02	May 27	12 56 39	Jun 3	3 51 21	Jun 10	8 19 59
Jun 18	11 58 51	Jun 25	18 53 55	Jul 2	12 09 15	Jul 10	0 50 17
Jul 17	23 57 16	Jul 24	23 39 55	Jul 31	21 41 51	Aug 8	18 29 42
Aug 16	10 06 26	Aug 23	4 36 56	Aug 30	9 28 05	Sep 7	12 16 38
Sep 14	19 20 49	Sep 21	11 03 51	Sep 29	0 09 23	Oct 7	5 05 01
Oct 14	4 34 13	Oct 20	20 13 36	Oct 28	17 38 27	Nov 5	20 07 26
Nov 12	14 21 15	Nov 19	9 04 26	Nov 27	12 42 39	Dec 5	9 02 10
Dec 12	0 55 21	Dec 19	1 58 44	Dec 27	7 31 13		

1986

New Moon	h m s	First Quarter	h m s	Full Moon	h m s	Last Quarter	h m s
						Jan 3	19 48 08
Jan 10	12 22 36	Jan 17	22 14 12	Jan 26	0 31 58	Feb 2	4 41 42
Feb 9	0 56 14	Feb 16	19 56 02	Feb 24	15 03 09	Mar 3	12 18 11
Mar 10	14 52 29	Mar 18	16 39 30	Mar 26	3 02 42	Apr 1	19 30 49
Apr 9	6 09 03	Apr 17	10 35 44	Apr 24	12 47 17	May 1	3 22 50
May 8	22 10 31	May 17	1 00 36	May 23	20 45 39	May 30	12 55 26
Jun 7	14 01 19	Jun 15	12 00 44	Jun 22	3 42 41	Jun 29	0 54 03
Jul 7	4 55 45	Jul 14	20 10 51	Jul 21	10 41 08	Jul 28	15 35 06
Aug 5	18 36 40	Aug 13	2 22 23	Aug 19	18 55 06	Aug 27	8 39 26
Sep 4	7 11 18	Sep 11	7 41 46	Sep 18	5 34 39	Sep 26	3 18 16
Oct 3	18 55 41	Oct 10	13 29 16	Oct 17	19 22 34	Oct 25	22 26 27
Nov 2	6 03 03	Nov 8	21 11 28	Nov 16	12 12 28	Nov 24	16 51 17
Dec 1	16 43 31	Dec 8	8 02 23	Dec 16	7 05 20	Dec 24	9 17 59
Dec 31	3 10 42						

1987

New Moon	h m s	First Quarter	h m s	Full Moon	h m s	Last Quarter	h m s
		Jan 6	22 35 22	Jan 15	2 31 20	Jan 22	22 46 18
Jan 29	13 45 26	Feb 5	16 21 40	Feb 13	20 58 50	Feb 21	8 56 45
Feb 28	0 51 34	Mar 7	11 59 02	Mar 15	13 13 36	Mar 22	16 22 35
Mar 29	12 46 28	Apr 6	7 48 35	Apr 14	2 31 54	Apr 20	22 16 06
Apr 28	1 35 18	May 6	2 26 33	May 13	12 51 09	May 20	4 03 19
May 27	15 14 16	Jun 4	18 53 35	Jun 11	20 49 44	Jun 18	11 03 26
Jun 26	5 37 46	Jul 4	8 35 05	Jul 11	3 33 32	Jul 17	20 17 44
Jul 25	20 38 27	Aug 2	19 24 54	Aug 9	10 18 14	Aug 16	8 25 56
Aug 24	11 59 33	Sep 1	3 48 33	Sep 7	18 13 47	Sep 14	23 45 24
Sep 23	3 09 09	Sep 30	10 39 58	Oct 7	4 13 11	Oct 14	18 06 31
Oct 22	17 28 48	Oct 29	17 11 12	Nov 5	16 46 46	Nov 13	14 39 07
Nov 21	6 33 49	Nov 28	0 37 56	Dec 5	8 01 55	Dec 13	11 42 10
Dec 20	18 26 11	Dec 27	10 01 39				

New Moon   First Quarter   Full Moon   Last Quarter

1988

New Moon	h m s	First Quarter	h m s	Full Moon	h m s	Last Quarter	h m s
				Jan 4	1 41 18	Jan 12	7 04 45
Jan 19	5 26 32	Jan 25	21 54 28	Feb 2	20 52 26	Feb 10	23 01 44
Feb 17	15 55 11	Feb 24	12 16 04	Mar 3	16 02 00	Mar 11	10 57 14
Mar 18	2 03 15	Mar 25	4 42 27	Apr 2	9 22 13	Apr 9	19 21 54
Apr 16	12 00 56	Apr 23	22 32 54	May 1	23 41 43	May 9	1 23 43
May 15	22 11 29	May 23	16 49 48	May 31	10 54 26	Jun 7	6 22 28
Jun 14	9 14 45	Jun 22	10 24 06	Jun 29	19 46 42	Jul 6	11 37 12
Jul 13	21 53 58	Jul 22	2 15 14	Jul 29	3 26 09	Aug 4	18 23 17
Aug 12	12 31 55	Aug 20	15 52 24	Aug 27	10 56 47	Sep 3	3 51 17
Sep 11	4 50 04	Sep 19	3 19 08	Sep 25	19 08 01	Oct 2	16 59 25
Oct 10	21 49 36	Oct 18	13 01 49	Oct 25	4 36 28	Nov 1	10 12 24
Nov 9	14 20 40	Nov 16	21 36 13	Nov 23	15 54 01	Dec 1	6 50 23
Dec 9	5 36 55	Dec 16	5 41 15	Dec 23	5 29 50	Dec 31	4 57 31

1989

New Moon	h m s	First Quarter	h m s	Full Moon	h m s	Last Quarter	h m s
Jan 7	19 23 10	Jan 14	13 59 18	Jan 21	21 34 28	Jan 30	2 03 04
Feb 6	7 37 54	Feb 12	23 15 39	Feb 20	15 32 48	Feb 28	20 08 52
Mar 7	18 19 36	Mar 14	10 11 45	Mar 22	9 58 52	Mar 30	10 22 23
Apr 6	3 33 38	Apr 12	23 13 44	Apr 21	3 14 13	Apr 28	20 46 55
May 5	11 47 18	May 12	14 20 28	May 20	18 17 11	May 28	4 01 40
Jun 3	19 53 43	Jun 11	6 59 49	Jun 19	6 58 15	Jun 26	9 09 50
Jul 3	4 59 59	Jul 11	0 19 54	Jul 18	17 42 33	Jul 25	13 32 11
Aug 1	16 06 32	Aug 9	17 29 29	Aug 17	3 07 33	Aug 23	18 41 07
Aug 31	5 45 28	Sep 8	9 50 09	Sep 15	11 51 30	Sep 22	2 10 46
Sep 29	21 47 48	Oct 8	0 53 06	Oct 14	20 32 48	Oct 21	13 19 44
Oct 29	15 28 00	Nov 6	14 11 51	Nov 13	5 52 20	Nov 20	4 44 34
Nov 28	9 41 37	Dec 6	1 26 42	Dec 12	16 30 49	Dec 19	23 55 27
Dec 28	3 20 32						

1990

New Moon	h m s	First Quarter	h m s	Full Moon	h m s	Last Quarter	h m s
		Jan 4	10 41 16	Jan 11	4 57 46	Jan 18	21 18 13
Jan 26	19 20 59	Feb 2	18 33 22	Feb 9	19 16 41	Feb 17	18 48 34
Feb 25	8 55 21	Mar 4	2 06 00	Mar 11	10 59 27	Mar 19	14 31 23
Mar 26	19 49 16	Apr 2	10 24 57	Apr 10	3 19 26	Apr 18	7 03 30
Apr 25	4 28 24	May 1	20 18 46	May 9	19 31 43	May 17	19 45 54
May 24	11 48 04	May 31	8 11 40	Jun 8	11 02 04	Jun 16	4 48 39
Jun 22	18 55 33	Jun 29	22 08 25	Jul 8	1 24 26	Jul 15	11 04 43
Jul 22	2 55 15	Jul 29	14 02 19	Aug 6	14 20 23	Aug 13	15 55 21
Aug 20	12 39 57	Aug 28	7 35 05	Sep 5	1 46 33	Sep 11	20 54 11
Sep 19	0 47 18	Sep 27	2 06 36	Oct 4	12 02 47	Oct 11	3 32 20
Oct 18	15 37 38	Oct 26	20 27 22	Nov 2	21 49 14	Nov 9	13 02 45
Nov 17	9 05 32	Nov 25	13 12 28	Dec 2	7 50 37	Dec 9	2 04 54
Dec 17	4 22 38	Dec 25	3 16 45	Dec 31	18 36 08		

	New Moon		First Quarter		Full Moon		Last Quarter	

1991

New Moon	h m s	First Quarter	h m s	Full Moon	h m s	Last Quarter	h m s
						Jan 7	18 36 27
Jan 15	23 50 37	Jan 23	14 22 30	Jan 30	6 10 40	Feb 6	13 53 11
Feb 14	17 32 40	Feb 21	22 59 11	Feb 28	18 25 32	Mar 8	10 32 53
Mar 16	8 11 27	Mar 23	6 03 42	Mar 30	7 18 21	Apr 7	6 46 17
Apr 14	19 38 34	Apr 21	12 39 38	Apr 28	20 59 28	May 7	0 47 09
May 14	4 36 47	May 20	19 46 47	May 28	11 37 32	Jun 5	15 30 57
Jun 12	12 06 53	Jun 19	4 20 19	Jun 27	2 59 23	Jul 5	2 51 13
Jul 11	19 07 03	Jul 18	15 11 40	Jul 26	18 25 19	Aug 3	11 26 28
Aug 10	2 28 40	Aug 17	5 01 43	Aug 25	9 08 07	Sep 1	18 17 21
Sep 8	11 01 46	Sep 15	22 02 18	Sep 23	22 40 56	Oct 1	0 30 32
Oct 7	21 39 47	Oct 15	17 33 50	Oct 23	11 09 08	Oct 30	7 11 30
Nov 6	11 11 58	Nov 14	14 02 31	Nov 21	22 57 16	Nov 28	15 22 02
Dec 6	3 57 16	Dec 14	9 32 54	Dec 21	10 24 09	Dec 28	1 55 59

1992

New Moon	h m s	First Quarter	h m s	Full Moon	h m s	Last Quarter	h m s
Jan 4	23 10 33	Jan 13	2 32 56	Jan 19	21 29 27	Jan 26	15 28 05
Feb 3	19 00 37	Feb 11	16 15 42	Feb 18	8 05 05	Feb 25	7 56 43
Mar 4	13 23 07	Mar 12	2 36 52	Mar 18	18 18 46	Mar 26	2 30 57
Apr 3	5 02 25	Apr 10	10 06 57	Apr 17	4 43 26	Apr 24	21 40 53
May 2	17 45 26	May 9	15 44 29	May 16	16 03 32	May 24	15 54 26
Jun 1	3 57 34	Jun 7	20 47 46	Jun 15	4 50 43	Jun 23	8 12 04
Jun 30	12 19 00	Jul 7	2 44 27	Jul 14	19 07 16	Jul 22	22 13 17
Jul 29	19 36 09	Aug 5	10 59 32	Aug 13	10 28 10	Aug 21	10 02 12
Aug 28	2 42 51	Sep 3	22 39 40	Sep 12	2 17 38	Sep 19	19 53 56
Sep 26	10 41 08	Oct 3	14 12 51	Oct 11	18 04 00	Oct 19	4 13 03
Oct 25	20 34 44	Nov 2	9 12 12	Nov 10	9 21 06	Nov 17	11 40 00
Nov 24	9 12 17	Dec 2	6 17 59	Dec 9	23 41 40	Dec 16	19 13 57
Dec 24	0 43 50						

1993

New Moon	h m s	First Quarter	h m s	Full Moon	h m s	Last Quarter	h m s
		Jan 1	3 39 20	Jan 8	12 38 14	Jan 15	4 02 15
Jan 22	18 27 51	Jan 30	23 20 49	Feb 6	23 56 23	Feb 13	14 57 45
Feb 21	13 06 04	Mar 1	15 47 32	Mar 8	9 46 52	Mar 15	4 17 32
Mar 23	7 15 25	Mar 31	4 10 41	Apr 6	18 44 15	Apr 13	19 39 38
Apr 21	23 49 57	Apr 29	12 41 30	May 6	3 34 44	May 13	12 20 40
May 21	14 07 31	May 28	18 22 23	Jun 4	13 03 12	Jun 12	5 36 45
Jun 20	1 53 27	Jun 26	22 44 21	Jul 3	23 46 06	Jul 11	22 49 35
Jul 19	11 25 02	Jul 26	3 26 01	Aug 2	12 10 37	Aug 10	15 20 23
Aug 17	19 29 17	Aug 24	9 58 20	Sep 1	2 33 55	Sep 9	6 27 23
Sep 16	3 11 17	Sep 22	19 33 04	Sep 30	18 54 49	Oct 8	19 36 21
Oct 15	11 36 56	Oct 22	8 53 02	Oct 30	12 38 38	Nov 7	6 36 50
Nov 13	21 35 20	Nov 21	2 04 13	Nov 29	6 31 45	Dec 6	15 49 51
Dec 13	9 27 54	Dec 20	22 27 01	Dec 28	23 06 23		

New Moon	First Quarter	Full Moon	Last Quarter

1994

New Moon (h m s)	First Quarter (h m s)	Full Moon (h m s)	Last Quarter (h m s)
			Jan 5 — 0 01 32
Jan 11 — 23 11 21	Jan 19 — 20 27 37	Jan 27 — 13 23 51	Feb 3 — 8 07 07
Feb 10 — 14 30 54	Feb 18 — 17 48 19	Feb 26 — 1 16 02	Mar 4 — 16 54 19
Mar 12 — 7 05 36	Mar 20 — 12 15 19	Mar 27 — 11 10 32	Apr 3 — 2 55 48
Apr 11 — 0 18 06	Apr 19 — 2 35 12	Apr 25 — 19 45 58	May 2 — 14 33 25
May 10 — 17 07 34	May 18 — 12 50 54	May 25 — 3 40 22	Jun 1 — 4 03 24
Jun 9 — 8 27 28	Jun 16 — 19 57 31	Jun 23 — 11 34 05	Jun 30 — 19 31 49
Jul 8 — 21 38 18	Jul 16 — 1 12 36	Jul 22 — 20 16 53	Jul 30 — 12 41 01
Aug 7 — 8 46 09	Aug 14 — 5 58 21	Aug 21 — 6 47 53	Aug 29 — 6 41 41
Sep 5 — 18 33 46	Sep 12 — 11 34 59	Sep 19 — 20 01 31	Sep 28 — 0 24 23
Oct 5 — 3 56 09	Oct 11 — 19 18 19	Oct 19 — 12 18 46	Oct 27 — 16 45 18
Nov 3 — 13 36 30	Nov 10 — 6 14 54	Nov 18 — 6 58 01	Nov 26 — 7 04 37
Dec 2 — 23 55 03	Dec 9 — 21 07 15	Dec 18 — 2 18 04	Dec 25 — 19 07 26

1995

New Moon (h m s)	First Quarter (h m s)	Full Moon (h m s)	Last Quarter (h m s)
Jan 1 — 10 56 39	Jan 8 — 15 47 27	Jan 16 — 20 27 22	Jan 24 — 4 59 03
Jan 30 — 22 48 44	Feb 7 — 12 55 07	Feb 15 — 12 16 24	Feb 22 — 13 04 47
Mar 1 — 11 48 48	Mar 9 — 10 14 38	Mar 17 — 1 26 38	Mar 23 — 20 10 52
Mar 31 — 2 09 38	Apr 8 — 5 35 53	Apr 15 — 12 09 13	Apr 22 — 3 19 13
Apr 29 — 17 37 20	May 7 — 21 44 50	May 14 — 20 49 08	May 21 — 11 36 41
May 29 — 9 28 08	Jun 6 — 10 26 39	Jun 13 — 4 04 30	Jun 19 — 22 01 48
Jun 28 — 0 50 58	Jul 5 — 20 03 25	Jul 12 — 10 50 19	Jul 19 — 11 10 37
Jul 27 — 15 13 58	Aug 4 — 3 17 01	Aug 10 — 18 16 34	Aug 18 — 3 04 54
Aug 26 — 4 31 59	Sep 2 — 9 04 11	Sep 9 — 3 37 40	Sep 16 — 21 10 23
Sep 24 — 16 55 46	Oct 1 — 14 36 36	Oct 8 — 15 52 52	Oct 16 — 16 26 54
Oct 24 — 4 37 14	Oct 30 — 21 17 41	Nov 7 — 7 21 31	Nov 15 — 11 40 49
Nov 22 — 15 43 49	Nov 29 — 6 29 20	Dec 7 — 1 27 55	Dec 15 — 5 32 08
Dec 22 — 2 23 26	Dec 28 — 19 07 28		

1996

New Moon (h m s)	First Quarter (h m s)	Full Moon (h m s)	Last Quarter (h m s)
		Jan 5 — 20 51 56	Jan 13 — 20 46 06
Jan 20 — 12 51 30	Jan 27 — 11 14 43	Feb 4 — 15 58 40	Feb 12 — 8 38 08
Feb 18 — 23 31 10	Feb 26 — 5 53 27	Mar 5 — 9 23 44	Mar 12 — 17 15 55
Mar 19 — 10 45 45	Mar 27 — 1 31 56	Apr 4 — 0 07 58	Apr 10 — 23 36 45
Apr 17 — 22 49 48	Apr 25 — 20 41 22	May 3 — 11 49 15	May 10 — 5 04 43
May 17 — 11 47 13	May 25 — 14 14 01	Jun 1 — 20 47 56	Jun 8 — 11 06 31
Jun 16 — 1 36 49	Jun 24 — 5 24 18	Jul 1 — 3 59 10	Jul 7 — 18 55 48
Jul 15 — 16 15 53	Jul 23 — 17 49 58	Jul 30 — 10 36 12	Aug 6 — 5 25 47
Aug 14 — 7 34 51	Aug 22 — 3 37 28	Aug 28 — 17 53 13	Sep 4 — 19 07 01
Sep 12 — 23 08 19	Sep 20 — 11 23 45	Sep 27 — 2 51 48	Oct 4 — 12 05 08
Oct 12 — 14 15 28	Oct 19 — 18 10 06	Oct 26 — 14 12 12	Nov 3 — 7 51 11
Nov 11 — 4 17 10	Nov 18 — 1 09 53	Nov 25 — 4 10 56	Dec 3 — 5 06 36
Dec 10 — 16 57 20	Dec 17 — 9 31 56	Dec 24 — 20 42 01	

	New Moon		First Quarter		Full Moon		Last Quarter	
1997	*h m s*		*h m s*		*h m s*		*h m s*	
							Jan 2	1 46 07
Jan 9	4 26 45	Jan 15	20 03 02	Jan 23	15 11 45	Jan 31	19 41 18
Feb 7	15 07 18	Feb 14	8 58 35	Feb 22	10 27 42	Mar 2	9 38 35
Mar 9	1 15 37	Mar 16	0 07 18	Mar 24	4 46 17	Mar 31	19 39 19
Apr 7	11 02 55	Apr 14	17 00 57	Apr 22	20 34 32	Apr 30	2 38 05
May 6	20 47 33	May 14	10 56 02	May 22	9 14 26	May 29	7 52 16
Jun 5	7 04 35	Jun 13	4 52 32	Jun 20	19 09 51	Jun 27	12 43 10
Jul 4	18 40 48	Jul 12	21 44 40	Jul 20	3 21 20	Jul 26	18 29 09
Aug 3	8 14 57	Aug 11	12 43 28	Aug 18	10 56 29	Aug 25	2 24 33
Sep 1	23 52 37	Sep 10	1 32 09	Sep 16	18 51 32	Sep 23	13 36 13
Oct 1	16 52 34	Oct 9	12 23 05	Oct 16	3 46 43	Oct 23	4 49 19
Oct 31	10 02 06	Nov 7	21 44 25	Nov 14	14 12 40	Nov 21	23 59 02
Nov 30	2 15 03	Dec 7	6 10 21	Dec 14	2 38 08	Dec 21	21 44 08
Dec 29	16 57 35						

	New Moon		First Quarter		Full Moon		Last Quarter	
1998	*h m s*		*h m s*		*h m s*		*h m s*	
			Jan 5	14 19 09	Jan 12	17 24 45	Jan 20	19 41 12
Jan 28	6 01 54	Feb 3	22 54 29	Feb 11	10 23 42	Feb 19	15 27 53
Feb 26	17 26 58	Mar 5	8 41 55	Mar 13	4 35 12	Mar 21	7 38 58
Mar 28	3 14 40	Apr 3	20 19 25	Apr 11	22 24 31	Apr 19	19 53 44
Apr 26	11 42 25	May 3	10 04 45	May 11	14 30 24	May 19	4 36 21
May 25	19 33 17	Jun 2	1 46 04	Jun 10	4 19 20	Jun 17	10 39 10
Jun 24	3 51 20	Jul 1	18 43 45	Jul 9	16 01 56	Jul 16	15 14 30
Jul 23	13 44 50	Jul 31	12 06 14	Aug 8	2 10 41	Aug 14	19 49 31
Aug 22	2 04 09	Aug 30	5 07 33	Sep 6	11 22 25	Sep 13	1 58 58
Sep 20	17 02 32	Sep 28	21 11 54	Oct 5	20 12 58	Oct 12	11 11 56
Oct 20	10 10 26	Oct 28	11 47 12	Nov 4	5 19 15	Nov 11	0 29 05
Nov 19	4 27 49	Nov 27	0 23 38	Dec 3	15 20 13	Dec 10	17 54 35
Dec 18	22 43 24	Dec 26	10 47 03				

	New Moon		First Quarter		Full Moon		Last Quarter	
1999	*h m s*		*h m s*		*h m s*		*h m s*	
					Jan 2	2 50 36	Jan 9	14 22 40
Jan 17	15 47 08	Jan 24	19 16 12	Jan 31	16 07 34	Feb 8	11 58 50
Feb 16	6 39 45	Feb 23	2 43 53	Mar 2	6 59 31	Mar 10	8 41 18
Mar 17	18 49 00	Mar 24	10 18 53	Mar 31	22 49 58	Apr 9	2 51 39
Apr 16	4 22 51	Apr 22	19 02 36	Apr 30	14 55 40	May 8	17 29 35
May 15	12 06 04	May 22	5 35 04	May 30	6 40 57	Jun 7	4 21 00
Jun 13	19 03 55	Jun 20	18 13 54	Jun 28	21 38 31	Jul 6	11 57 50
Jul 13	2 25 01	Jul 20	9 01 15	Jul 28	11 25 51	Aug 4	17 27 39
Aug 11	11 09 34	Aug 19	1 47 56	Aug 26	23 48 53	Sep 2	22 18 21
Sep 9	22 03 19	Sep 17	20 06 50	Sep 25	10 52 08	Oct 2	4 03 02
Oct 9	11 35 29	Oct 17	15 00 49	Oct 24	21 03 27	Oct 31	12 04 39
Nov 8	3 54 05	Nov 16	9 04 10	Nov 23	7 04 43	Nov 29	23 19 37
Dec 7	22 32 41	Dec 16	0 51 20	Dec 22	17 32 22	Dec 29	14 05 23

New Moon   First Quarter   Full Moon   Last Quarter

2000

New Moon			First Quarter			Full Moon			Last Quarter		
	h	m s		h	m s		h	m s		h	m s
Jan 6	18	14 42	Jan 14	13	35 15	Jan 21	4	41 30	Jan 28	7	57 45
Feb 5	13	04 20	Feb 12	23	22 28	Feb 19	16	27 44	Feb 27	3	54 34
Mar 6	5	17 46	Mar 13	6	59 53	Mar 20	4	45 25	Mar 28	0	21 43
Apr 4	18	13 04	Apr 11	13	31 25	Apr 18	17	42 35	Apr 26	19	31 07
May 4	4	13 08	May 10	20	01 34	May 18	7	35 30	May 26	11	56 05
Jun 2	12	15 01	Jun 9	3	30 17	Jun 16	22	28 07	Jun 25	1	01 08
Jul 1	19	20 59	Jul 8	12	53 51	Jul 16	13	56 17	Jul 24	11	03 07
Jul 31	2	26 13	Aug 7	1	02 43	Aug 15	5	13 42	Aug 22	18	51 44
Aug 29	10	20 22	Sep 5	16	28 26	Sep 13	19	37 54	Sep 21	1	29 25
Sep 27	19	54 01	Oct 5	11	00 16	Oct 13	8	54 02	Oct 20	8	00 09
Oct 27	7	59 05	Nov 4	7	27 54	Nov 11	21	15 41	Nov 18	15	25 32
Nov 25	23	12 22	Dec 4	3	56 26	Dec 11	9	03 52	Dec 18	0	42 22
Dec 25	17	22 41									

2001

New Moon			First Quarter			Full Moon			Last Quarter		
	h	m s		h	m s		h	m s		h	m s
			Jan 2	22	32 36	Jan 9	20	25 28	Jan 16	12	35 40
Jan 24	13	07 54	Feb 1	14	03 28	Feb 8	7	12 43	Feb 15	3	24 36
Feb 23	8	22 10	Mar 3	2	04 16	Mar 9	17	24 08	Mar 16	20	46 13
Mar 25	1	22 04	Apr 1	10	50 19	Apr 8	3	22 56	Apr 15	15	32 10
Apr 23	15	26 41	Apr 30	17	08 39	May 7	13	53 36	May 15	10	11 40
May 23	2	47 06	May 29	22	10 12	Jun 6	1	40 25	Jun 14	3	29 21
Jun 21	11	58 49	Jun 28	3	20 36	Jul 5	15	04 51	Jul 13	18	46 15
Jul 20	19	45 29	Jul 27	10	09 14	Aug 4	5	56 48	Aug 12	7	54 14
Aug 19	2	56 20	Aug 25	19	55 53	Sep 2	21	44 05	Sep 10	19	00 36
Sep 17	10	28 25	Sep 24	9	31 55	Oct 2	13	49 53	Oct 10	4	20 42
Oct 16	19	24 23	Oct 24	2	59 25	Nov 1	5	42 05	Nov 8	12	22 19
Nov 15	6	41 03	Nov 22	23	21 48	Nov 30	20	50 07	Dec 7	19	52 45
Dec 14	20	48 27	Dec 22	20	57 30	Dec 30	10	41 37			

2002

New Moon			First Quarter			Full Moon			Last Quarter		
	h	m s		h	m s		h	m s		h	m s
									Jan 6	3	55 46
Jan 13	13	29 40	Jan 21	17	47 35	Jan 28	22	51 34	Feb 4	13	34 06
Feb 12	7	41 57	Feb 20	12	02 48	Feb 27	9	17 45	Mar 6	1	25 37
Mar 14	2	03 36	Mar 22	2	29 17	Mar 28	18	25 57	Apr 4	15	30 05
Apr 12	19	22 12	Apr 20	12	49 23	Apr 27	3	00 59	May 4	7	17 03
May 12	10	46 09	May 19	19	43 07	May 26	11	52 19	Jun 3	0	06 12
Jun 10	23	47 35	Jun 18	0	30 21	Jun 24	21	43 25	Jul 2	17	20 21
Jul 10	10	27 06	Jul 17	4	48 11	Jul 24	9	08 04	Aug 1	10	23 08
Aug 8	19	16 15	Aug 15	10	13 21	Aug 22	22	30 19	Aug 31	2	32 19
Sep 7	3	11 23	Sep 13	18	09 06	Sep 21	14	00 20	Sep 29	17	03 56
Oct 6	11	18 38	Oct 13	5	34 06	Oct 21	7	21 06	Oct 29	5	28 44
Nov 4	20	35 31	Nov 11	20	53 13	Nov 20	1	34 44	Nov 27	15	47 24
Dec 4	7	35 26	Dec 11	15	49 38	Dec 19	19	11 14	Dec 27	0	32 04

	New Moon			First Quarter			Full Moon			Last Quarter		

2003

	New Moon h m s			First Quarter h m s			Full Moon h m s			Last Quarter h m s		
Jan 2	20 23 54		Jan 10	13 15 52		Jan 18	10 48 43		Jan 25	8 34 13	
Feb 1	10 49 27		Feb 9	11 12 14		Feb 16	23 52 15		Feb 23	16 46 58	
Mar 3	2 36 03		Mar 11	7 16 21		Mar 18	10 35 38		Mar 25	1 52 13	
Apr 1	19 19 47		Apr 9	23 41 15		Apr 16	19 36 46		Apr 23	12 19 27	
May 1	12 15 52		May 9	11 54 17		May 16	3 37 02		May 23	0 31 45	
May 31	4 20 57		Jun 7	20 28 44		Jun 14	11 16 59		Jun 21	14 46 03	
Jun 29	18 39 45		Jul 7	2 33 25		Jul 13	19 22 28		Jul 21	7 02 18	
Jul 29	6 53 48		Aug 5	7 28 40		Aug 12	4 49 17		Aug 20	0 49 16	
Aug 27	17 27 25		Sep 3	12 35 10		Sep 10	16 37 19		Sep 18	19 03 57	
Sep 26	3 10 16		Oct 2	19 10 22		Oct 10	7 28 33		Oct 18	12 32 21	
Oct 25	12 51 23		Nov 1	4 25 39		Nov 9	1 14 26		Nov 17	4 15 59	
Nov 23	23 00 01		Nov 30	17 17 03		Dec 8	20 37 46		Dec 16	17 43 13	
Dec 23	9 44 04		Dec 30	10 04 22							

2004

	New Moon h m s			First Quarter h m s			Full Moon h m s			Last Quarter h m s		
							Jan 7	15 41 14		Jan 15	4 46 42	
Jan 21	21 05 58		Jan 29	6 04 18		Feb 6	8 47 58		Feb 13	13 40 38	
Feb 20	9 18 45		Feb 28	3 25 06		Mar 6	23 15 21		Mar 13	21 01 46	
Mar 20	22 42 25		Mar 28	23 48 56		Apr 5	11 03 49		Apr 12	3 47 07	
Apr 19	13 22 16		Apr 27	17 33 32		May 4	20 34 31		May 11	11 05 09	
May 19	4 53 00		May 27	7 58 05		Jun 3	4 20 39		Jun 9	20 03 26	
Jun 17	20 27 51		Jun 25	19 09 02		Jul 2	11 09 59		Jul 9	7 34 39	
Jul 17	11 24 51		Jul 25	3 38 15		Jul 31	18 06 15		Aug 7	22 02 18	
Aug 16	1 24 58		Aug 23	10 12 54		Aug 30	2 23 20		Sep 6	15 11 37	
Sep 14	14 30 07		Sep 21	15 54 39		Sep 28	13 10 21		Oct 6	10 12 54	
Oct 14	2 49 20		Oct 20	21 59 45		Oct 28	3 08 27		Nov 5	5 54 30	
Nov 12	14 28 15		Nov 19	5 51 26		Nov 26	20 08 20		Dec 5	0 53 48	
Dec 12	1 30 06		Dec 18	16 40 41		Dec 26	15 07 24				

2005

	New Moon h m s			First Quarter h m s			Full Moon h m s			Last Quarter h m s		
										Jan 3	17 46 45	
Jan 10	12 03 53		Jan 17	6 58 33		Jan 25	10 33 25		Feb 2	7 27 54	
Feb 8	22 29 07		Feb 16	0 17 04		Feb 24	4 54 48		Mar 3	17 37 31	
Mar 10	9 11 26		Mar 17	19 20 07		Mar 25	20 59 36		Apr 2	0 51 29	
Apr 8	20 33 05		Apr 16	14 38 33		Apr 24	10 07 33		May 1	6 25 15	
May 8	8 46 31		May 16	8 57 39		May 23	20 19 14		May 30	11 48 24	
Jun 6	21 56 10		Jun 15	1 23 20		Jun 22	4 14 54		Jun 28	18 24 31	
Jul 6	12 03 34		Jul 14	15 20 57		Jul 21	11 01 19		Jul 28	3 20 00	
Aug 5	3 05 50		Aug 13	2 39 37		Aug 19	17 54 04		Aug 26	15 19 08	
Sep 3	18 46 30		Sep 11	11 37 43		Sep 18	2 01 51		Sep 25	6 41 50	
Oct 3	10 28 57		Oct 10	19 01 53		Oct 17	12 14 43		Oct 25	1 17 47	
Nov 2	1 25 40		Nov 9	1 58 08		Nov 16	0 58 37		Nov 23	22 12 20	
Dec 1	15 02 00		Dec 8	9 37 20		Dec 15	16 16 37		Dec 23	19 37 12	
Dec 31	3 12 49										

New Moon	First Quarter	Full Moon	Last Quarter

2006

New Moon	h m s	First Quarter	h m s	Full Moon	h m s	Last Quarter	h m s
		Jan 6	18 57 34	Jan 14	9 49 12	Jan 22	15 14 52
Jan 29	14 15 43	Feb 5	6 29 54	Feb 13	4 45 18	Feb 21	7 17 47
Feb 28	0 31 51	Mar 6	20 16 52	Mar 14	23 36 31	Mar 22	19 11 35
Mar 29	10 16 20	Apr 5	12 01 45	Apr 13	16 41 12	Apr 21	3 29 31
Apr 27	19 44 58	May 5	5 14 06	May 13	6 52 08	May 20	9 21 38
May 27	5 26 41	Jun 3	23 06 50	Jun 11	18 04 12	Jun 18	14 09 22
Jun 25	16 06 21	Jul 3	16 37 49	Jul 11	3 02 58	Jul 17	19 13 38
Jul 25	4 32 00	Aug 2	8 46 52	Aug 9	10 55 03	Aug 16	1 51 50
Aug 23	19 10 51	Aug 31	22 57 36	Sep 7	18 43 08	Sep 14	11 16 26
Sep 22	11 46 08	Sep 30	11 04 56	Oct 7	3 13 54	Oct 14	0 26 40
Oct 22	5 15 08	Oct 29	21 26 27	Nov 5	12 59 21	Nov 12	17 46 26
Nov 20	22 19 05	Nov 28	6 30 04	Dec 5	0 25 54	Dec 12	14 32 46
Dec 20	14 01 52	Dec 27	14 48 52				

2007

New Moon	h m s	First Quarter	h m s	Full Moon	h m s	Last Quarter	h m s
				Jan 3	13 58 27	Jan 11	12 45 40
Jan 19	4 01 47	Jan 25	23 02 31	Feb 2	5 46 23	Feb 10	9 52 08
Feb 17	16 15 23	Feb 24	7 56 51	Mar 3	23 18 12	Mar 12	3 55 12
Mar 19	2 43 39	Mar 25	18 17 12	Apr 2	17 16 07	Apr 10	18 05 17
Apr 17	11 37 06	Apr 24	6 36 36	May 2	10 10 30	May 10	4 28 07
May 16	19 28 19	May 23	21 03 37	Jun 1	1 04 45	Jun 8	11 43 44
Jun 15	3 14 14	Jun 22	13 16 22	Jun 30	13 49 48	Jul 7	16 54 42
Jul 14	12 04 52	Jul 22	6 30 11	Jul 30	0 48 50	Aug 5	21 20 42
Aug 12	23 03 36	Aug 20	23 55 18	Aug 28	10 36 11	Sep 4	2 33 28
Sep 11	12 45 19	Sep 19	16 49 00	Sep 26	19 46 14	Oct 3	10 06 57
Oct 11	5 01 45	Oct 19	8 34 04	Oct 26	4 52 37	Nov 1	21 19 16
Nov 9	23 04 10	Nov 17	22 33 39	Nov 24	14 30 55	Dec 1	12 45 08
Dec 9	17 41 30	Dec 17	10 18 36	Dec 24	1 16 41	Dec 31	7 51 54

2008

New Moon	h m s	First Quarter	h m s	Full Moon	h m s	Last Quarter	h m s
Jan 8	11 38 15	Jan 15	19 46 46	Jan 22	13 35 46	Jan 30	5 03 59
Feb 7	3 45 36	Feb 14	3 34 33	Feb 21	3 31 36	Feb 29	2 19 25
Mar 7	17 15 19	Mar 14	10 46 40	Mar 21	18 41 02	Mar 29	21 48 24
Apr 6	3 56 23	Apr 12	18 32 50	Apr 20	10 26 28	Apr 28	14 13 15
May 5	12 19 19	May 12	3 48 01	May 20	2 12 28	May 28	2 57 40
Jun 3	19 23 39	Jun 10	15 04 41	Jun 18	17 31 34	Jun 26	12 10 52
Jul 3	2 19 40	Jul 10	4 35 57	Jul 18	8 00 10	Jul 25	18 42 39
Aug 1	10 13 39	Aug 8	20 21 19	Aug 16	21 17 33	Aug 23	23 50 38
Aug 30	19 59 08	Sep 7	14 05 11	Sep 15	9 14 28	Sep 22	5 05 26
Sep 29	8 13 20	Oct 7	9 05 18	Oct 14	20 03 32	Oct 21	11 55 43
Oct 28	23 14 56	Nov 6	4 04 28	Nov 13	6 18 25	Nov 19	21 31 49
Nov 27	16 55 40	Dec 5	21 26 42	Dec 12	16 38 14	Dec 19	10 30 20
Dec 27	12 23 29						

	New Moon	First Quarter	Full Moon	Last Quarter

2009

New Moon	h m s	First Quarter	h m s	Full Moon	h m s	Last Quarter	h m s
		Jan 4	11 57 19	Jan 11	3 27 52	Jan 18	2 46 49
Jan 26	7 56 23	Feb 2	23 14 12	Feb 9	14 50 15	Feb 16	21 38 16
Feb 25	1 36 10	Mar 4	7 46 55	Mar 11	2 38 50	Mar 18	17 48 27
Mar 26	16 07 01	Apr 2	14 34 48	Apr 9	14 56 54	Apr 17	13 37 28
Apr 25	3 23 39	May 1	20 45 20	May 9	4 02 31	May 17	7 27 08
May 24	12 12 06	May 31	3 23 18	Jun 7	18 12 50	Jun 15	22 15 41
Jun 22	19 36 04	Jun 29	11 29 31	Jul 7	9 22 31	Jul 15	9 54 15
Jul 22	2 35 42	Jul 28	22 00 53	Aug 6	0 55 58	Aug 13	18 56 17
Aug 20	10 02 41	Aug 27	11 43 04	Sep 4	16 03 42	Sep 12	2 16 50
Sep 18	18 45 25	Sep 26	4 50 47	Oct 4	6 11 16	Oct 11	8 56 55
Oct 18	5 34 11	Oct 26	0 43 20	Nov 2	19 15 01	Nov 9	15 56 54
Nov 16	19 14 50	Nov 24	21 40 20	Dec 2	7 31 33	Dec 9	0 14 26
Dec 16	12 03 12	Dec 24	17 37 03	Dec 31	19 13 51		

2010

New Moon	h m s	First Quarter	h m s	Full Moon	h m s	Last Quarter	h m s
						Jan 7	10 40 34
Jan 15	7 12 28	Jan 23	10 54 29	Jan 30	6 18 41	Feb 5	23 49 32
Feb 14	2 52 25	Feb 22	0 43 28	Feb 28	16 38 58	Mar 7	15 42 53
Mar 15	21 02 11	Mar 23	11 01 10	Mar 30	2 26 31	Apr 6	9 37 55
Apr 14	12 30 01	Apr 21	18 20 52	Apr 28	12 19 34	May 6	4 16 01
May 14	1 05 29	May 20	23 43 46	May 27	23 08 25	Jun 4	22 14 13
Jun 12	11 15 41	Jun 19	4 30 35	Jun 26	11 31 28	Jul 4	14 36 19
Jul 11	19 41 33	Jul 18	10 11 39	Jul 26	1 37 39	Aug 3	4 59 43
Aug 10	3 09 15	Aug 16	18 15 08	Aug 24	17 05 40	Sep 1	17 22 52
Sep 8	10 30 55	Sep 15	5 50 51	Sep 23	9 18 19	Oct 1	3 53 13
Oct 7	18 45 35	Oct 14	21 28 30	Oct 23	1 37 37	Oct 30	12 46 58
Nov 6	4 52 53	Nov 13	16 39 40	Nov 21	17 28 26	Nov 28	20 37 32
Dec 5	17 36 48	Dec 13	13 59 48	Dec 21	8 14 33	Dec 28	4 19 32

2011

New Moon	h m s	First Quarter	h m s	Full Moon	h m s	Last Quarter	h m s
Jan 4	9 03 43	Jan 12	11 32 32	Jan 19	21 22 31	Jan 26	12 58 20
Feb 3	2 31 46	Feb 11	7 19 22	Feb 18	8 36 46	Feb 24	23 27 27
Mar 4	20 46 58	Mar 12	23 46 04	Mar 19	18 11 09	Mar 26	12 08 26
Apr 3	14 33 25	Apr 11	12 06 29	Apr 18	2 45 05	Apr 25	2 47 59
May 3	6 51 48	May 10	20 34 01	May 17	11 09 45	May 24	18 53 16
Jun 1	21 03 43	Jun 9	2 11 40	Jun 15	20 14 41	Jun 23	11 49 22
Jul 1	8 55 01	Jul 8	6 30 29	Jul 15	6 40 43	Jul 23	5 03 03
Jul 30	18 40 54	Aug 6	11 09 24	Aug 13	18 58 36	Aug 21	21 55 34
Aug 29	3 05 12	Sep 4	17 40 25	Sep 12	9 27 46	Sep 20	13 39 41
Sep 27	11 09 47	Oct 4	3 16 12	Oct 12	2 06 50	Oct 20	3 31 30
Oct 26	19 56 54	Nov 2	16 39 16	Nov 10	20 17 13	Nov 18	15 10 10
Nov 25	6 10 47	Dec 2	9 53 21	Dec 10	14 37 29	Dec 18	0 48 46
Dec 24	18 07 30						

New Moon	First Quarter	Full Moon	Last Quarter

2012

New Moon		First Quarter		Full Moon		Last Quarter	
	h m s		h m s		h m s		h m s
		Jan 1	6 15 41	Jan 9	7 31 12	Jan 16	9 09 01
Jan 23	7 40 23	Jan 31	4 10 45	Feb 7	21 54 51	Feb 14	17 04 58
Feb 21	22 35 42	Mar 1	1 22 36	Mar 8	9 40 35	Mar 15	1 26 14
Mar 22	14 38 13	Mar 30	19 41 46	Apr 6	19 19 47	Apr 13	10 50 43
Apr 21	7 19 30	Apr 29	9 58 36	May 6	3 36 11	May 12	21 47 50
May 20	23 48 08	May 28	20 17 07	Jun 4	11 12 40	Jun 11	10 42 28
Jun 19	15 03 13	Jun 27	3 31 29	Jul 3	18 52 59	Jul 11	1 48 58
Jul 19	4 25 09	Jul 26	8 57 17	Aug 2	3 28 34	Aug 9	18 56 06
Aug 17	15 55 33	Aug 24	13 54 41	Aug 31	13 59 13	Sep 8	13 16 08
Sep 16	2 11 45	Sep 22	19 41 56	Sep 30	3 19 44	Oct 8	7 34 23
Oct 15	12 03 37	Oct 22	3 33 05	Oct 29	19 50 34	Nov 7	0 36 46
Nov 13	22 09 07	Nov 20	14 32 33	Nov 28	14 47 03	Dec 6	15 32 33
Dec 13	8 42 43	Dec 20	5 20 11	Dec 28	10 22 18		

2013

New Moon		First Quarter		Full Moon		Last Quarter	
	h m s		h m s		h m s		h m s
						Jan 5	3 58 52
Jan 11	19 44 44	Jan 18	23 46 09	Jan 27	4 39 29	Feb 3	13 57 25
Feb 10	7 21 13	Feb 17	20 31 42	Feb 25	20 27 11	Mar 4	21 53 55
Mar 11	19 52 07	Mar 19	17 27 41	Mar 27	9 28 25	Apr 3	4 37 41
Apr 10	9 36 25	Apr 18	12 31 59	Apr 25	19 58 14	May 2	11 15 17
May 10	0 29 30	May 18	4 35 39	May 25	4 26 03	May 31	18 59 15
Jun 8	15 57 27	Jun 16	17 24 49	Jun 23	11 33 22	Jun 30	4 54 39
Jul 8	7 15 24	Jul 16	3 19 25	Jul 22	18 16 39	Jul 29	17 44 28
Aug 6	21 51 49	Aug 14	10 57 08	Aug 21	1 45 44	Aug 28	9 36 04
Sep 5	11 37 15	Sep 12	17 09 31	Sep 19	11 13 58	Sep 27	3 56 35
Oct 5	0 35 38	Oct 11	23 03 26	Oct 18	23 38 47	Oct 26	23 41 36
Nov 3	12 51 04	Nov 10	5 58 18	Nov 17	15 16 51	Nov 25	19 28 50
Dec 3	0 23 29	Dec 9	15 12 51	Dec 17	9 29 11	Dec 25	13 48 47

2014

New Moon		First Quarter		Full Moon		Last Quarter	
	h m s		h m s		h m s		h m s
Jan 1	11 15 17	Jan 8	3 40 24	Jan 16	4 53 17	Jan 24	5 20 05
Jan 30	21 39 39	Feb 6	19 23 10	Feb 14	23 54 08	Feb 22	17 16 22
Mar 1	8 00 47	Mar 8	13 27 52	Mar 16	17 09 27	Mar 24	1 47 12
Mar 30	18 45 48	Apr 7	8 31 45	Apr 15	7 43 25	Apr 22	7 52 49
Apr 29	6 15 28	May 7	3 16 00	May 14	19 17 02	May 21	13 00 18
May 28	18 41 20	Jun 5	20 39 54	Jun 13	4 12 36	Jun 19	18 39 51
Jun 27	8 09 35	Jul 5	11 59 56	Jul 12	11 26 02	Jul 19	2 09 28
Jul 26	22 42 53	Aug 4	0 50 48	Aug 10	18 10 29	Aug 17	12 26 54
Aug 25	14 13 53	Sep 2	11 12 17	Sep 9	1 39 18	Sep 16	2 06 01
Sep 24	6 14 53	Oct 1	19 33 41	Oct 8	10 51 43	Oct 15	19 13 10
Oct 23	21 57 47	Oct 31	2 49 26	Nov 6	22 23 57	Nov 14	15 16 38
Nov 22	12 33 23	Nov 29	10 07 26	Dec 6	12 27 52	Dec 14	12 52 04
Dec 22	1 36 58	Dec 28	18 32 31				

2015

New Moon	h m s	First Quarter	h m s	Full Moon	h m s	Last Quarter	h m s
				Jan 5	4 54 23	Jan 13	9 47 35
Jan 20	13 14 49	Jan 27	4 49 31	Feb 3	23 10 03	Feb 12	3 50 57
Feb 18	23 48 23	Feb 25	17 15 07	Mar 5	18 06 31	Mar 13	17 49 02
Mar 20	9 37 18	Mar 27	7 43 44	Apr 4	12 06 41	Apr 12	3 45 35
Apr 18	18 57 58	Apr 25	23 56 17	May 4	3 43 12	May 11	10 37 16
May 18	4 14 21	May 25	17 19 59	Jun 2	16 20 09	Jun 9	15 42 53
Jun 16	14 06 28	Jun 24	11 03 40	Jul 2	2 20 43	Jul 8	20 25 05
Jul 16	1 25 29	Jul 24	4 05 07	Jul 31	10 44 02	Aug 7	2 03 50
Aug 14	14 54 32	Aug 22	19 32 08	Aug 29	18 36 20	Sep 5	9 55 10
Sep 13	6 42 24	Sep 21	9 00 12	Sep 28	2 51 38	Oct 4	21 07 12
Oct 13	0 06 51	Oct 20	20 32 29	Oct 27	12 06 15	Nov 3	12 25 00
Nov 11	17 48 16	Nov 19	6 28 30	Nov 25	22 45 23	Dec 3	7 41 29
Dec 11	10 30 32	Dec 18	15 15 26	Dec 25	11 12 36		

2016

New Moon	h m s	First Quarter	h m s	Full Moon	h m s	Last Quarter	h m s
						Jan 2	5 31 34
Jan 10	1 31 40	Jan 16	23 27 29	Jan 24	1 46 52	Feb 1	3 28 57
Feb 8	14 40 03	Feb 15	7 47 36	Feb 22	18 21 00	Mar 1	23 11 49
Mar 9	1 55 37	Mar 15	17 04 02	Mar 23	12 01 58	Mar 31	15 17 59
Apr 7	11 24 47	Apr 14	4 00 28	Apr 22	5 24 44	Apr 30	3 29 51
May 6	19 30 38	May 13	17 03 16	May 21	21 15 33	May 29	12 13 07
Jun 5	3 00 43	Jun 12	8 10 57	Jun 20	11 03 27	Jun 27	18 19 47
Jul 4	11 02 09	Jul 12	0 52 57	Jul 19	22 57 43	Jul 26	23 00 43
Aug 2	20 45 41	Aug 10	18 21 58	Aug 18	9 27 42	Aug 25	3 41 58
Sep 1	9 04 14	Sep 9	11 50 00	Sep 16	19 06 14	Sep 23	9 57 11
Oct 1	0 12 30	Oct 9	4 34 08	Oct 16	4 24 14	Oct 22	19 14 55
Oct 30	17 39 18	Nov 7	19 52 15	Nov 14	13 53 11	Nov 21	8 34 20
Nov 29	12 19 21	Dec 7	9 04 09	Dec 14	0 06 41	Dec 21	1 56 52
Dec 29	6 54 19						

2017

New Moon	h m s	First Quarter	h m s	Full Moon	h m s	Last Quarter	h m s
		Jan 5	19 48 08	Jan 12	11 35 06	Jan 19	22 14 34
Jan 28	0 08 10	Feb 4	4 20 01	Feb 11	0 34 01	Feb 18	19 34 16
Feb 26	14 59 31	Mar 5	11 33 30	Mar 12	14 54 56	Mar 20	15 59 21
Mar 28	2 58 21	Apr 3	18 40 34	Apr 11	6 09 14	Apr 19	9 57 53
Apr 26	12 17 17	May 3	2 48 00	May 10	21 43 38	May 19	0 33 57
May 25	19 45 36	Jun 1	12 43 15	Jun 9	13 10 44	Jun 17	11 33 51
Jun 24	2 31 51	Jul 1	0 52 14	Jul 9	4 07 43	Jul 16	19 26 47
Jul 23	9 46 43	Jul 30	15 24 14	Aug 7	18 11 46	Aug 15	1 16 12
Aug 21	18 31 19	Aug 29	8 14 08	Sep 6	7 03 57	Sep 13	6 26 09
Sep 20	5 31 00	Sep 28	2 54 40	Oct 5	18 41 15	Oct 12	12 26 34
Oct 19	19 13 12	Oct 27	22 23 13	Nov 4	5 24 03	Nov 10	20 37 32
Nov 18	11 43 16	Nov 26	17 04 04	Dec 3	15 48 07	Dec 10	7 52 29
Dec 18	6 31 34	Dec 26	9 21 15				

New Moon				First Quarter				Full Moon				Last Quarter			

2018

		h	m	s			h	m	s			h	m	s			h	m	s
										Jan	2	2	25	14	Jan	8	22	26	23
Jan	17	2	18	23	Jan	24	22	21	31	Jan	31	13	27	53	Feb	7	15	55	04
Feb	15	21	06	21	Feb	23	8	10	18	Mar	2	0	52	29	Mar	9	11	20	56
Mar	17	13	12	43	Mar	24	15	36	19	Mar	31	12	37	59	Apr	8	7	18	41
Apr	16	1	58	16	Apr	22	21	46	44	Apr	30	0	59	20	May	8	2	09	45
May	15	11	48	55	May	22	3	50	18	May	29	14	20	42	Jun	6	18	32	50
Jun	13	19	44	23	Jun	20	10	52	01	Jun	28	4	54	07	Jul	6	7	51	51
Jul	13	2	49	01	Jul	19	19	53	23	Jul	27	20	21	30	Aug	4	18	19	08
Aug	11	9	58	53	Aug	18	7	49	40	Aug	26	11	57	19	Sep	3	2	38	33
Sep	9	18	02	37	Sep	16	23	16	07	Sep	25	2	53	33	Oct	2	9	46	34
Oct	9	3	47	59	Oct	16	18	02	48	Oct	24	16	46	20	Oct	31	16	41	28
Nov	7	16	03	11	Nov	15	14	55	20	Nov	23	5	40	21	Nov	30	0	19	58
Dec	7	7	21	30	Dec	15	11	50	24	Dec	22	17	49	44	Dec	29	9	35	27

2019

		h	m	s			h	m	s			h	m	s			h	m	s
Jan	6	1	29	20	Jan	14	6	46	40	Jan	21	5	17	14	Jan	27	21	11	29
Feb	4	21	04	44	Feb	12	22	27	23	Feb	19	15	54	43	Feb	26	11	28	52
Mar	6	16	05	07	Mar	14	10	28	14	Mar	21	1	44	00	Mar	28	4	10	50
Apr	5	8	51	37	Apr	12	19	07	02	Apr	19	11	13	19	Apr	26	22	19	27
May	4	22	46	39	May	12	1	13	23	May	18	21	12	30	May	26	16	34	44
Jun	3	10	03	06	Jun	10	6	00	27	Jun	17	8	31	49	Jun	25	9	47	33
Jul	2	19	17	22	Jul	9	10	56	00	Jul	16	21	39	22	Jul	25	1	19	10
Aug	1	3	13	04	Aug	7	17	32	06	Aug	15	12	30	24	Aug	23	14	57	15
Aug	30	10	38	17	Sep	6	3	11	35	Sep	14	4	33	55	Sep	22	2	42	03
Sep	28	18	27	30	Oct	5	16	48	15	Oct	13	21	09	01	Oct	21	12	40	27
Oct	28	3	39	37	Nov	4	10	24	13	Nov	12	13	35	33	Nov	19	21	12	05
Nov	26	15	06	44	Dec	4	6	59	22	Dec	12	5	13	25	Dec	19	4	58	14
Dec	26	5	14	17															

2020

		h	m	s			h	m	s			h	m	s			h	m	s
					Jan	3	4	46	33	Jan	10	19	22	27	Jan	17	12	59	34
Jan	24	21	43	09	Feb	2	1	42	49	Feb	9	7	34	26	Feb	15	22	18	21
Feb	23	15	33	10	Mar	2	19	58	32	Mar	9	17	48	53	Mar	16	9	35	20
Mar	24	9	29	22	Apr	1	10	22	24	Apr	8	2	36	14	Apr	14	22	57	17
Apr	23	2	27	00	Apr	30	20	39	30	May	7	10	46	22	May	14	14	03	51
May	22	17	40	00	May	30	3	31	04	Jun	5	19	13	32	Jun	13	6	24	50
Jun	21	6	42	36	Jun	28	8	16	50	Jul	5	4	45	34	Jul	12	23	30	09
Jul	20	17	34	06	Jul	27	12	33	43	Aug	3	15	59	55	Aug	11	16	45	56
Aug	19	2	42	49	Aug	25	17	58	47	Sep	2	5	23	13	Sep	10	9	26	53
Sep	17	11	01	22	Sep	24	1	56	01	Oct	1	21	06	25	Oct	10	0	40	41
Oct	16	19	32	12	Oct	23	13	24	05	Oct	31	14	50	18	Nov	8	13	47	15
Nov	15	5	08	20	Nov	22	4	46	10	Nov	30	9	30	50	Dec	8	0	37	45
Dec	14	16	17	44	Dec	21	23	42	22	Dec	30	3	29	22					

	New Moon		*First Quarter*		*Full Moon*		*Last Quarter*

2021

New Moon	h m s	First Quarter	h m s	Full Moon	h m s	Last Quarter	h m s
						Jan 6	9 38 22
Jan 13	5 01 20	Jan 20	21 02 46	Jan 28	19 17 23	Feb 4	17 38 15
Feb 11	19 06 49	Feb 19	18 48 29	Feb 27	8 18 29	Mar 6	1 31 22
Mar 13	10 22 19	Mar 21	14 41 34	Mar 28	18 49 20	Apr 4	10 03 36
Apr 12	2 32 00	Apr 20	7 00 05	Apr 27	3 32 42	May 3	19 51 15
May 11	19 00 57	May 19	19 13 47	May 26	11 15 02	Jun 2	7 25 33
Jun 10	10 53 48	Jun 18	3 55 26	Jun 24	18 40 51	Jul 1	21 11 47
Jul 10	1 17 46	Jul 17	10 11 48	Jul 24	2 38 03	Jul 31	13 17 09
Aug 8	13 51 17	Aug 15	15 20 45	Aug 22	12 03 07	Aug 30	7 14 23
Sep 7	0 52 55	Sep 13	20 40 31	Sep 20	23 55 51	Sep 29	1 58 18
Oct 6	11 06 33	Oct 13	3 26 18	Oct 20	14 57 50	Oct 28	20 06 20
Nov 4	21 15 46	Nov 11	12 47 11	Nov 19	8 58 37	Nov 27	12 28 49
Dec 4	7 44 11	Dec 11	1 36 43	Dec 19	4 36 40	Dec 27	2 24 54

2022

New Moon	h m s	First Quarter	h m s	Full Moon	h m s	Last Quarter	h m s
Jan 2	18 34 39	Jan 9	18 12 26	Jan 17	23 49 35	Jan 25	13 42 04
Feb 1	5 47 10	Feb 8	13 51 16	Feb 16	16 57 40	Feb 23	22 33 36
Mar 2	17 35 56	Mar 10	10 46 33	Mar 18	7 18 44	Mar 25	5 38 26
Apr 1	6 25 34	Apr 9	6 48 47	Apr 16	18 56 12	Apr 23	11 57 30
Apr 30	20 29 15	May 9	0 22 33	May 16	4 15 18	May 22	18 44 15
May 30	11 31 26	Jun 7	14 49 39	Jun 14	11 52 55	Jun 21	3 11 59
Jun 29	2 53 26	Jul 7	2 15 18	Jul 13	18 38 47	Jul 20	14 19 47
Jul 28	17 56 11	Aug 5	11 07 42	Aug 12	1 36 54	Aug 19	4 37 15
Aug 27	8 18 17	Sep 3	18 08 53	Sep 10	10 00 12	Sep 17	21 53 09
Sep 25	21 55 43	Oct 3	0 15 12	Oct 9	20 56 08	Oct 17	17 16 16
Oct 25	10 49 51	Nov 1	6 38 15	Nov 8	11 03 18	Nov 16	13 28 13
Nov 23	22 58 23	Nov 30	14 37 42	Dec 8	4 09 19	Dec 16	8 57 15
Dec 23	10 18 02	Dec 30	1 21 42				

2023

New Moon	h m s	First Quarter	h m s	Full Moon	h m s	Last Quarter	h m s
				Jan 6	23 09 03	Jan 15	2 11 28
Jan 21	20 54 24	Jan 28	15 19 55	Feb 5	18 29 43	Feb 13	16 01 53
Feb 20	7 06 59	Feb 27	8 06 46	Mar 7	12 41 31	Mar 15	2 09 26
Mar 21	17 24 17	Mar 29	2 33 31	Apr 6	4 35 40	Apr 13	9 12 33
Apr 20	4 13 41	Apr 27	21 21 05	May 5	17 35 12	May 12	14 29 27
May 19	15 54 25	May 27	15 23 26	Jun 4	3 42 53	Jun 10	19 32 32
Jun 18	4 38 18	Jun 26	7 50 52	Jul 3	11 39 51	Jul 10	1 49 02
Jul 17	18 32 59	Jul 25	22 07 57	Aug 1	18 32 49	Aug 8	10 29 35
Aug 16	9 39 20	Aug 24	9 58 24	Aug 31	1 36 47	Sep 6	22 22 14
Sep 15	1 40 57	Sep 22	19 32 57	Sep 29	9 58 41	Oct 6	13 48 52
Oct 14	17 56 18	Oct 22	3 30 36	Oct 28	20 25 12	Nov 5	8 37 58
Nov 13	9 28 33	Nov 20	10 51 02	Nov 27	9 17 27	Dec 5	5 50 25
Dec 12	23 33 11	Dec 19	18 40 23	Dec 27	0 34 21		

2024

New Moon (h m s)	First Quarter (h m s)	Full Moon (h m s)	Last Quarter (h m s)
			Jan 4 3 31 36
Jan 11 11 58 33	Jan 18 3 53 45	Jan 25 17 55 09	Feb 2 23 19 08
Feb 9 23 00 20	Feb 16 15 02 05	Feb 24 12 31 35	Mar 3 15 24 38
Mar 10 9 01 35	Mar 17 4 11 52	Mar 25 7 01 28	Apr 2 3 15 53
Apr 8 18 22 00	Apr 15 19 14 16	Apr 23 23 50 08	May 1 11 28 25
May 8 3 23 05	May 15 11 49 09	May 23 13 54 17	May 30 17 13 49
Jun 6 12 38 53	Jun 14 5 19 36	Jun 22 1 09 02	Jun 28 21 54 34
Jul 5 22 58 33	Jul 13 22 49 58	Jul 21 10 18 18	Jul 28 2 52 43
Aug 4 11 14 13	Aug 12 15 19 57	Aug 19 18 26 58	Aug 26 9 27 00
Sep 3 1 56 44	Sep 11 6 06 48	Sep 18 2 35 37	Sep 24 18 51 02
Oct 2 18 50 26	Oct 10 18 56 17	Oct 17 11 27 33	Oct 24 8 04 14
Nov 1 12 48 17	Nov 9 5 56 37	Nov 15 21 29 40	Nov 23 1 29 05
Dec 1 6 22 34	Dec 8 15 27 46	Dec 15 9 02 50	Dec 22 22 19 20
Dec 30 22 27 57			

2025

New Moon (h m s)	First Quarter (h m s)	Full Moon (h m s)	Last Quarter (h m s)
	Jan 6 23 57 26	Jan 13 22 28 03	Jan 21 20 31 56
Jan 29 12 37 08	Feb 5 8 03 18	Feb 12 13 54 33	Feb 20 17 33 41
Feb 28 0 45 58	Mar 6 16 32 47	Mar 14 6 55 48	Mar 22 11 30 35
Mar 29 10 58 59	Apr 5 2 15 50	Apr 13 0 23 24	Apr 21 1 36 43
Apr 27 19 32 18	May 4 13 52 54	May 12 16 57 05	May 20 11 59 55
May 27 3 03 30	Jun 3 3 42 07	Jun 11 7 44 59	Jun 18 19 20 15
Jun 25 10 32 46	Jul 2 19 31 20	Jul 10 20 37 56	Jul 18 0 38 49
Jul 24 19 12 21	Aug 1 12 42 28	Aug 9 7 56 13	Aug 16 5 13 23
Aug 23 6 07 42	Aug 31 6 26 21	Sep 7 18 10 03	Sep 14 10 34 06
Sep 21 19 55 17	Sep 29 23 54 59	Oct 7 3 48 46	Oct 13 18 13 50
Oct 21 12 26 19	Oct 29 16 21 58	Nov 5 13 20 27	Nov 12 5 29 17
Nov 20 6 48 25	Nov 28 6 59 57	Dec 4 23 15 13	Dec 11 20 52 50
Dec 20 1 44 30	Dec 27 19 11 00		

2026

New Moon (h m s)	First Quarter (h m s)	Full Moon (h m s)	Last Quarter (h m s)
		Jan 3 10 04 03	Jan 10 15 49 32
Jan 18 19 53 08	Jan 26 4 48 32	Feb 1 22 10 23	Feb 9 12 44 15
Feb 17 12 02 18	Feb 24 12 28 45	Mar 3 11 39 02	Mar 11 9 39 39
Mar 19 1 24 37	Mar 25 19 18 51	Apr 2 2 13 07	Apr 10 4 52 48
Apr 17 11 52 57	Apr 24 2 32 54	May 1 17 24 19	May 9 21 11 37
May 16 20 02 12	May 23 11 12 06	May 31 8 46 21	Jun 8 10 01 40
Jun 15 2 55 19	Jun 21 21 56 33	Jun 29 23 57 50	Jul 7 19 30 08
Jul 14 9 44 46	Jul 21 11 06 45	Jul 29 14 36 52	Aug 6 2 22 38
Aug 12 17 37 54	Aug 20 2 47 29	Aug 28 4 19 41	Sep 4 7 52 23
Sep 11 3 28 09	Sep 18 20 44 56	Sep 26 16 50 11	Oct 3 13 26 12
Oct 10 15 51 14	Oct 18 16 13 50	Oct 26 4 12 58	Nov 1 20 29 36
Nov 9 7 03 16	Nov 17 11 48 58	Nov 24 14 54 42	Dec 1 6 09 49
Dec 9 0 53 00	Dec 17 5 43 49	Dec 24 1 29 23	Dec 30 19 00 38

| *New Moon* | | | | *First Quarter* | | | | *Full Moon* | | | | *Last Quarter* | | | |

2027

		h	m	s			h	m	s			h	m	s			h	m	s
Jan	7	20	25	32	Jan	15	20	35	40	Jan	22	12	18	32	Jan	29	10	56	37
Feb	6	15	57	16	Feb	14	7	59	38	Feb	20	23	24	48	Feb	28	5	17	39
Mar	8	9	30	38	Mar	15	16	26	21	Mar	22	10	44	57	Mar	30	0	55	05
Apr	6	23	52	18	Apr	13	22	57	48	Apr	20	22	28	18	Apr	28	20	19	01
May	6	10	59	47	May	13	4	45	03	May	20	11	00	10	May	28	13	59	05
Jun	4	19	41	30	Jun	11	10	57	17	Jun	19	0	45	30	Jun	27	4	55	30
Jul	4	3	03	14	Jul	10	18	40	10	Jul	18	15	46	04	Jul	26	16	56	04
Aug	2	10	06	23	Aug	9	4	55	19	Aug	17	7	29	50	Aug	25	2	28	30
Aug	31	17	42	20	Sep	7	18	32	28	Sep	15	23	04	40	Sep	23	10	21	37
Sep	30	2	37	14	Oct	7	11	48	33	Oct	15	13	48	09	Oct	22	17	30	19
Oct	29	13	37	43	Nov	6	8	01	10	Nov	14	3	27	05	Nov	21	0	49	21
Nov	28	3	25	35	Dec	6	5	23	17	Dec	13	16	09	57	Dec	20	9	12	03
Dec	27	20	13	29															

2028

		h	m	s			h	m	s			h	m	s			h	m	s
					Jan	5	1	41	37	Jan	12	4	04	14	Jan	18	19	27	08
Jan	26	15	13	40	Feb	3	19	11	40	Feb	10	15	04	55	Feb	17	8	09	09
Feb	25	10	38	34	Mar	4	9	03	34	Mar	11	1	07	14	Mar	17	23	24	01
Mar	26	4	32	28	Apr	2	19	16	38	Apr	9	10	27	47	Apr	16	16	38	05
Apr	24	19	48	05	May	2	2	26	54	May	8	19	50	06	May	16	10	44	16
May	24	8	17	28	May	31	7	37	46	Jun	7	6	09	58	Jun	15	4	28	34
Jun	22	18	28	42	Jun	29	12	11	47	Jul	6	18	11	58	Jul	14	20	57	44
Jul	22	3	02	52	Jul	28	17	41	18	Aug	5	8	10	58	Aug	13	11	46	29
Aug	20	10	45	00	Aug	27	1	37	02	Sep	3	23	48	45	Sep	12	0	46	56
Sep	18	18	24	54	Sep	25	13	11	11	Oct	3	16	26	08	Oct	11	11	58	01
Oct	18	2	57	58	Oct	25	4	54	20	Nov	2	9	18	31	Nov	9	21	27	02
Nov	16	13	19	11	Nov	24	0	15	48	Dec	2	1	41	23	Dec	9	5	40	05
Dec	16	2	07	29	Dec	23	21	46	07	Dec	31	16	49	40					

2029

		h	m	s			h	m	s			h	m	s			h	m	s
															Jan	7	13	27	32
Jan	14	17	25	40	Jan	22	19	24	23	Jan	30	6	04	46	Feb	5	21	53	22
Feb	13	10	32	40	Feb	21	15	11	04	Feb	28	17	11	24	Mar	7	7	52	46
Mar	15	4	20	24	Mar	23	7	34	18	Mar	30	2	27	34	Apr	5	19	52	45
Apr	13	21	41	20	Apr	21	19	51	18	Apr	28	10	37	58	May	5	9	49	16
May	13	13	43	18	May	21	4	17	19	May	27	18	38	40	Jun	4	1	20	01
Jun	12	3	51	42	Jun	19	9	55	17	Jun	26	3	23	30	Jul	3	17	58	40
Jul	11	15	52	13	Jul	18	14	15	29	Jul	25	13	36	55	Aug	2	11	16	33
Aug	10	1	56	57	Aug	16	18	56	29	Aug	24	1	52	23	Sep	1	4	34	14
Sep	8	10	45	32	Sep	15	1	30	27	Sep	22	16	30	28	Sep	30	20	58	01
Oct	7	19	15	43	Oct	14	11	10	02	Oct	22	9	28	44	Oct	30	11	33	27
Nov	6	4	25	17	Nov	13	0	36	27	Nov	21	4	04	07	Nov	28	23	48	58
Dec	5	14	53	17	Dec	12	17	50	39	Dec	20	22	47	40	Dec	28	9	50	16

New Moon	First Quarter	Full Moon	Last Quarter

2030

New Moon	h m s	First Quarter	h m s	Full Moon	h m s	Last Quarter	h m s
Jan 4	2 50 43	Jan 11	14 07 13	Jan 19	15 55 31	Jan 26	18 15 36
Feb 2	16 08 40	Feb 10	11 50 36	Feb 18	6 20 58	Feb 25	1 58 52
Mar 4	6 35 50	Mar 12	8 48 46	Mar 19	17 57 39	Mar 26	9 52 33
Apr 2	22 03 42	Apr 11	2 58 03	Apr 18	3 21 11	Apr 24	18 40 09
May 2	14 13 19	May 10	17 12 39	May 17	11 20 18	May 24	4 58 38
Jun 1	6 22 30	Jun 9	3 36 55	Jun 15	18 42 10	Jun 22	17 20 52
Jun 30	21 35 38	Jul 8	11 03 03	Jul 15	2 13 09	Jul 22	8 08 45
Jul 30	11 12 10	Aug 6	16 43 53	Aug 13	10 45 34	Aug 21	1 16 36
Aug 28	23 08 32	Sep 4	21 56 45	Sep 11	21 19 05	Sep 19	19 57 41
Sep 27	9 55 50	Oct 4	3 57 20	Oct 11	10 48 00	Oct 19	14 51 32
Oct 26	20 18 09	Nov 2	11 57 10	Nov 10	3 31 28	Nov 18	8 33 30
Nov 25	6 47 39	Dec 1	22 58 02	Dec 9	22 41 37	Dec 18	0 02 22
Dec 24	17 33 23	Dec 31	13 37 19				

2031

New Moon	h m s	First Quarter	h m s	Full Moon	h m s	Last Quarter	h m s
				Jan 8	18 27 00	Jan 16	12 48 25
Jan 23	4 32 10	Jan 30	7 44 16	Feb 7	12 47 22	Feb 14	22 50 59
Feb 21	15 50 01	Mar 1	4 03 24	Mar 9	4 30 49	Mar 16	6 36 58
Mar 23	3 50 15	Mar 31	0 33 25	Apr 7	17 22 29	Apr 14	12 59 02
Apr 21	16 58 15	Apr 29	19 20 37	May 7	3 41 02	May 13	19 08 07
May 21	7 18 22	May 29	11 20 45	Jun 5	11 59 42	Jun 12	2 21 49
Jun 19	22 25 52	Jun 28	0 20 11	Jul 4	19 02 32	Jul 11	11 51 00
Jul 19	13 41 25	Jul 27	10 35 57	Aug 3	1 46 47	Aug 10	0 24 50
Aug 18	4 33 28	Aug 25	18 40 53	Sep 1	9 21 41	Sep 8	16 15 35
Sep 16	18 48 03	Sep 24	1 20 56	Sep 30	18 59 02	Oct 8	10 51 24
Oct 16	8 21 54	Oct 23	7 37 37	Oct 30	7 33 52	Nov 7	7 03 26
Nov 14	21 10 47	Nov 21	14 46 04	Nov 28	23 19 38	Dec 7	3 20 52
Dec 14	9 06 58	Dec 21	0 01 42	Dec 28	17 34 11		

2032

New Moon	h m s	First Quarter	h m s	Full Moon	h m s	Last Quarter	h m s
						Jan 5	22 05 22
Jan 12	20 07 53	Jan 19	12 15 29	Jan 27	12 53 40	Feb 4	13 50 07
Feb 11	6 25 26	Feb 18	3 30 09	Feb 26	7 44 25	Mar 5	1 48 10
Mar 11	16 25 50	Mar 18	20 57 51	Mar 27	0 47 29	Apr 3	10 11 25
Apr 10	2 40 40	Apr 17	15 25 33	Apr 25	15 10 51	May 2	16 02 56
May 9	13 36 54	May 17	9 44 34	May 25	2 38 24	May 31	20 52 23
Jun 8	1 33 16	Jun 16	3 01 04	Jun 23	11 33 43	Jun 30	2 13 00
Jul 7	14 42 42	Jul 15	18 33 14	Jul 22	18 52 45	Jul 29	9 26 32
Aug 6	5 12 43	Aug 14	7 52 10	Aug 21	1 48 02	Aug 27	19 34 35
Sep 4	20 57 48	Sep 12	18 50 23	Sep 19	9 31 26	Sep 26	9 13 38
Oct 4	13 27 32	Oct 12	3 48 46	Oct 18	18 59 18	Oct 26	2 30 03
Nov 3	5 46 07	Nov 10	11 34 27	Nov 17	6 43 14	Nov 24	22 48 58
Dec 2	20 54 03	Dec 9	19 09 42	Dec 16	20 50 10	Dec 24	20 40 24

New Moon	First Quarter	Full Moon	Last Quarter

2033

New Moon	h m s	First Quarter	h m s	Full Moon	h m s	Last Quarter	h m s
Jan 1	10 18 12	Jan 8	3 35 33	Jan 15	13 08 17	Jan 23	17 46 55
Jan 30	22 01 02	Feb 6	13 35 23	Feb 14	7 05 23	Feb 22	11 54 17
Mar 1	8 24 42	Mar 8	1 28 15	Mar 16	1 38 36	Mar 24	1 50 58
Mar 30	17 52 49	Apr 6	15 15 06	Apr 14	19 18 33	Apr 22	11 43 28
Apr 29	2 47 21	May 6	6 46 28	May 14	10 43 54	May 21	18 30 10
May 28	11 37 43	Jun 4	23 40 05	Jun 12	23 20 23	Jun 19	23 30 43
Jun 26	21 08 13	Jul 4	17 13 18	Jul 12	9 29 45	Jul 19	4 08 25
Jul 26	8 13 45	Aug 3	10 26 54	Aug 10	18 08 53	Aug 17	9 44 07
Aug 24	21 40 59	Sep 2	2 24 51	Sep 9	2 21 45	Sep 15	17 34 45
Sep 23	13 40 57	Oct 1	16 33 57	Oct 8	10 59 19	Oct 15	4 48 39
Oct 23	7 29 36	Oct 31	4 47 32	Nov 6	20 33 16	Nov 13	20 09 59
Nov 22	1 40 17	Nov 29	15 16 24	Dec 6	7 23 15	Dec 13	15 29 10
Dec 21	18 47 39	Dec 29	0 21 16				

2034

New Moon	h m s	First Quarter	h m s	Full Moon	h m s	Last Quarter	h m s
				Jan 4	19 48 17	Jan 12	13 18 14
Jan 20	10 02 43	Jan 27	8 33 09	Feb 3	10 05 43	Feb 11	11 10 10
Feb 18	23 11 26	Feb 25	16 35 21	Mar 5	2 11 20	Mar 13	6 45 38
Mar 20	10 15 45	Mar 27	1 19 40	Apr 3	19 20 03	Apr 11	22 46 20
Apr 18	19 27 04	Apr 25	11 35 52	May 3	12 16 53	May 11	10 57 24
May 18	3 13 45	May 24	23 58 45	Jun 2	3 55 11	Jun 9	19 45 08
Jun 16	10 27 07	Jun 23	14 36 20	Jul 1	17 45 43	Jul 9	2 00 14
Jul 15	18 16 26	Jul 23	7 06 13	Jul 31	5 55 40	Aug 7	6 51 29
Aug 14	3 54 14	Aug 22	0 44 32	Aug 29	16 50 28	Sep 5	11 42 34
Sep 12	16 14 59	Sep 20	18 40 41	Sep 28	2 58 00	Oct 4	18 05 53
Oct 12	7 33 50	Oct 20	12 04 10	Oct 27	12 43 41	Nov 3	3 28 30
Nov 11	1 17 25	Nov 19	4 02 39	Nov 25	22 33 21	Dec 2	16 47 36
Dec 10	20 15 36	Dec 18	17 46 08	Dec 25	8 55 40		

2035

New Moon	h m s	First Quarter	h m s	Full Moon	h m s	Last Quarter	h m s
						Jan 1	10 02 11
Jan 9	15 04 17	Jan 17	4 46 35	Jan 23	20 17 47	Jan 31	6 03 41
Feb 8	8 23 19	Feb 15	13 18 10	Feb 22	8 55 06	Mar 2	3 02 14
Mar 9	23 10 39	Mar 16	20 16 11	Mar 23	22 43 18	Mar 31	23 07 49
Apr 8	10 58 59	Apr 15	2 56 02	Apr 22	13 21 56	Apr 30	16 55 08
May 7	20 05 05	May 14	10 29 39	May 22	4 26 58	May 30	7 31 54
Jun 6	3 21 54	Jun 12	19 51 13	Jun 20	19 38 40	Jun 28	18 43 57
Jul 5	10 00 30	Jul 12	7 34 04	Jul 20	10 38 03	Jul 28	2 56 47
Aug 3	17 13 00	Aug 10	21 53 40	Aug 19	1 01 29	Aug 26	9 09 10
Sep 2	2 00 43	Sep 9	14 48 25	Sep 17	14 24 43	Sep 24	14 40 48
Oct 1	13 08 00	Oct 9	9 50 29	Oct 17	2 36 45	Oct 23	20 58 00
Oct 31	2 59 54	Nov 8	5 51 39	Nov 15	13 50 04	Nov 22	5 17 33
Nov 29	19 38 47	Dec 8	1 06 24	Dec 15	0 34 22	Dec 21	16 29 43
Dec 29	14 32 07						

	New Moon	First Quarter	Full Moon	Last Quarter

2036

	New Moon h m s	First Quarter h m s	Full Moon h m s	Last Quarter h m s
		Jan 6 17 49 14	Jan 13 11 17 18	Jan 20 6 47 33
Jan 28	10 18 23	Feb 5 7 02 01	Feb 11 22 09 53	Feb 18 23 48 09
Feb 27	5 00 27	Mar 5 16 50 09	Mar 12 9 10 40	Mar 19 18 39 52
Mar 27	20 57 55	Apr 4 0 04 35	Apr 10 20 23 45	Apr 18 14 07 03
Apr 26	9 34 24	May 3 5 55 32	May 10 8 10 44	May 18 8 40 41
May 25	19 18 09	Jun 1 11 35 38	Jun 8 21 03 10	Jun 17 1 04 25
Jun 24	3 10 49	Jun 30 18 14 16	Jul 8 11 20 30	Jul 16 14 40 51
Jul 23	10 18 12	Jul 30 2 57 30	Aug 7 2 50 08	Aug 15 1 37 05
Aug 21	17 36 32	Aug 28 14 44 38	Sep 5 18 46 46	Sep 13 10 30 23
Sep 20	1 52 46	Sep 27 6 13 44	Oct 5 10 16 21	Oct 12 18 10 36
Oct 19	11 51 12	Oct 27 1 14 57	Nov 4 0 45 28	Nov 11 1 29 50
Nov 18	0 15 41	Nov 25 22 29 20	Dec 3 14 09 42	Dec 10 9 19 42
Dec 17	15 35 39	Dec 25 19 45 46		

2037

	New Moon h m s	First Quarter h m s	Full Moon h m s	Last Quarter h m s
			Jan 2 2 36 28	Jan 8 18 30 19
Jan 16	9 35 35	Jan 24 14 56 18	Jan 31 14 05 21	Feb 7 5 44 40
Feb 15	4 55 16	Feb 23 6 42 11	Mar 2 0 29 21	Mar 8 19 26 25
Mar 16	23 37 25	Mar 24 18 40 48	Mar 31 9 54 44	Apr 7 11 26 14
Apr 15	16 08 54	Apr 23 3 12 43	Apr 29 18 55 05	May 7 4 57 17
May 15	5 55 33	May 22 9 09 45	May 29 4 25 20	Jun 5 22 50 16
Jun 13	17 11 27	Jun 20 13 46 31	Jun 27 15 21 08	Jul 5 16 01 43
Jul 13	2 33 00	Jul 19 18 32 15	Jul 27 4 16 21	Aug 4 7 52 19
Aug 11	10 42 47	Aug 18 1 00 53	Aug 25 19 10 32	Sep 2 22 04 29
Sep 9	18 26 40	Sep 16 10 37 20	Sep 24 11 32 54	Oct 2 10 30 26
Oct 9	2 35 43	Oct 16 0 16 42	Oct 24 4 37 43	Oct 31 21 07 45
Nov 7	12 04 19	Nov 14 17 59 58	Nov 22 21 36 26	Nov 30 6 07 44
Dec 6	23 39 33	Dec 14 14 43 14	Dec 22 13 39 50	Dec 29 14 06 13

2038

	New Moon h m s	First Quarter h m s	Full Moon h m s	Last Quarter h m s
Jan 5	13 42 33	Jan 13 12 35 06	Jan 21 4 01 09	Jan 27 22 01 42
Feb 4	5 53 22	Feb 12 9 30 55	Feb 19 16 10 36	Feb 26 6 57 09
Mar 5	23 16 11	Mar 14 3 42 51	Mar 21 2 10 47	Mar 27 17 37 15
Apr 4	16 44 11	Apr 12 18 03 08	Apr 19 10 37 12	Apr 26 6 16 31
May 4	9 20 51	May 12 4 19 15	May 18 18 24 44	May 25 20 44 40
Jun 3	0 25 25	Jun 10 11 12 36	Jun 17 2 31 44	Jun 24 12 40 50
Jul 2	13 33 21	Jul 9 16 01 48	Jul 16 11 49 27	Jul 24 5 41 07
Aug 1	0 41 32	Aug 7 20 22 42	Aug 14 22 58 01	Aug 22 23 13 18
Aug 30	10 13 58	Sep 6 1 51 58	Sep 13 12 25 37	Sep 21 16 28 14
Sep 28	18 58 47	Oct 5 9 53 29	Oct 13 4 23 03	Oct 21 8 24 38
Oct 28	3 54 03	Nov 3 21 24 59	Nov 11 22 28 23	Nov 19 22 11 14
Nov 26	13 47 55	Dec 3 12 47 23	Dec 11 17 31 37	Dec 19 9 30 13
Dec 26	1 03 10			

New Moon	First Quarter	Full Moon	Last Quarter

2039

	h m s		h m s		h m s		h m s
		Jan 2	7 38 06	Jan 10	11 46 43	Jan 17	18 42 53
Jan 24	13 37 18	Feb 1	4 46 20	Feb 9	3 40 26	Feb 16	2 37 12
Feb 23	3 18 43	Mar 3	2 16 15	Mar 10	16 36 11	Mar 17	10 08 52
Mar 24	18 00 42	Apr 1	21 55 53	Apr 9	2 53 58	Apr 15	18 08 12
Apr 23	9 35 57	May 1	14 08 34	May 8	11 21 12	May 15	3 18 11
May 23	1 39 13	May 31	2 25 40	Jun 6	18 48 53	Jun 13	14 17 41
Jun 21	17 22 38	Jun 29	11 18 30	Jul 6	2 04 41	Jul 13	3 39 28
Jul 21	7 55 17	Jul 28	17 50 55	Aug 4	9 57 51	Aug 11	19 37 10
Aug 19	20 51 44	Aug 26	23 17 36	Sep 2	19 24 45	Sep 10	13 46 50
Sep 18	8 24 13	Sep 25	4 53 45	Oct 2	7 24 29	Oct 10	9 00 41
Oct 17	19 10 13	Oct 24	11 51 45	Oct 31	22 37 28	Nov 9	3 47 16
Nov 16	5 47 17	Nov 22	21 17 52	Nov 30	16 50 46	Dec 8	20 45 46
Dec 15	16 33 14	Dec 22	10 02 51	Dec 30	12 38 56		

2040

	h m s		h m s		h m s		h m s
						Jan 7	11 06 44
Jan 14	3 26 27	Jan 21	2 22 22	Jan 29	7 55 53	Feb 5	22 33 39
Feb 12	14 25 38	Feb 19	21 34 51	Feb 28	1 00 44	Mar 6	7 19 59
Mar 13	1 47 18	Mar 20	18 00 17	Mar 28	15 12 54	Apr 4	14 07 26
Apr 11	14 01 24	Apr 19	13 38 40	Apr 27	2 39 04	May 3	20 00 56
May 11	3 29 05	May 19	7 01 45	May 26	11 48 17	Jun 2	2 18 35
Jun 9	18 04 16	Jun 17	21 33 34	Jun 24	19 20 30	Jul 1	10 18 56
Jul 9	9 15 56	Jul 17	9 17 28	Jul 24	2 06 49	Jul 30	21 06 59
Aug 8	0 27 36	Aug 15	18 37 11	Aug 22	9 10 53	Aug 29	11 17 45
Sep 6	15 14 53	Sep 14	2 08 46	Sep 20	17 43 59	Sep 28	4 42 37
Oct 6	5 27 06	Oct 13	8 42 26	Oct 20	4 51 03	Oct 28	0 28 06
Nov 4	18 57 11	Nov 11	15 24 35	Nov 18	19 07 21	Nov 26	21 08 40
Dec 4	7 34 21	Dec 10	23 30 59	Dec 18	12 16 56	Dec 26	17 03 37

2041

	h m s		h m s		h m s		h m s
Jan 2	19 08 58	Jan 9	10 06 59	Jan 17	7 12 34	Jan 25	10 34 23
Feb 1	5 44 06	Feb 7	23 41 27	Feb 16	2 22 27	Feb 24	0 29 56
Mar 2	15 40 31	Mar 9	15 52 22	Mar 17	20 20 20	Mar 25	10 33 08
Apr 1	1 30 39	Apr 8	9 39 45	Apr 16	12 01 51	Apr 23	17 25 19
Apr 30	11 47 32	May 8	3 55 20	May 16	0 53 34	May 22	22 27 07
May 29	22 57 10	Jun 6	21 41 51	Jun 14	10 59 55	Jun 21	3 13 23
Jun 28	11 18 06	Jul 6	14 13 41	Jul 13	19 02 02	Jul 20	9 14 23
Jul 28	1 03 31	Aug 5	4 53 56	Aug 12	2 05 53	Aug 18	17 44 19
Aug 26	16 17 22	Sep 3	17 20 01	Sep 10	9 25 03	Sep 17	5 34 07
Sep 25	8 42 29	Oct 3	3 33 53	Oct 9	18 03 53	Oct 16	21 06 25
Oct 25	1 31 31	Nov 1	12 06 09	Nov 8	4 44 41	Nov 15	16 07 35
Nov 23	17 37 55	Nov 30	19 49 57	Dec 7	17 43 25	Dec 15	13 33 56
Dec 23	8 07 34	Dec 30	3 46 52				

	New Moon	First Quarter	Full Moon	Last Quarter
	h m s	h m s	h m s	h m s

2042

New Moon	First Quarter	Full Moon	Last Quarter
		Jan 6 — 8 55 08	Jan 14 — 11 25 39
Jan 21 — 20 43 18	Jan 28 — 12 49 47	Feb 5 — 1 58 59	Feb 13 — 7 17 34
Feb 20 — 7 40 07	Feb 26 — 23 30 38	Mar 6 — 20 11 09	Mar 14 — 23 22 19
Mar 21 — 17 24 13	Mar 28 — 12 00 59	Apr 5 — 14 17 08	Apr 13 — 11 10 22
Apr 20 — 2 20 31	Apr 27 — 2 20 31	May 5 — 6 49 47	May 12 — 19 19 17
May 19 — 10 56 03	May 26 — 18 19 19	Jun 3 — 20 49 29	Jun 11 — 1 01 24
Jun 17 — 19 49 23	Jun 25 — 11 30 03	Jul 3 — 8 10 36	Jul 10 — 5 39 28
Jul 17 — 5 53 05	Jul 25 — 5 02 51	Aug 1 — 17 34 31	Aug 8 — 10 35 56
Aug 15 — 18 02 31	Aug 23 — 21 56 41	Aug 31 — 2 03 33	Sep 6 — 17 10 09
Sep 14 — 8 51 23	Sep 22 — 13 21 54	Sep 29 — 10 35 31	Oct 6 — 2 36 04
Oct 14 — 2 04 20	Oct 22 — 2 54 20	Oct 28 — 19 49 36	Nov 4 — 15 52 34
Nov 12 — 20 29 38	Nov 20 — 14 32 33	Nov 27 — 6 07 13	Dec 4 — 9 20 02
Dec 12 — 14 30 52	Dec 20 — 0 28 46	Dec 26 — 17 43 57	

2043

New Moon	First Quarter	Full Moon	Last Quarter
			Jan 3 — 6 09 21
Jan 11 — 6 54 29	Jan 18 — 9 06 08	Jan 25 — 6 57 50	Feb 2 — 4 15 58
Feb 9 — 21 08 56	Feb 16 — 17 01 23	Feb 23 — 21 59 10	Mar 4 — 1 08 38
Mar 11 — 9 10 31	Mar 18 — 1 04 22	Mar 25 — 14 27 30	Apr 2 — 18 57 28
Apr 9 — 19 07 50	Apr 16 — 10 10 07	Apr 24 — 7 24 14	May 2 — 9 00 07
May 9 — 3 22 26	May 15 — 21 06 18	May 23 — 23 38 11	May 31 — 19 25 55
Jun 7 — 10 36 16	Jun 14 — 10 19 58	Jun 22 — 14 21 51	Jun 30 — 2 54 15
Jul 6 — 17 52 09	Jul 14 — 1 48 03	Jul 22 — 3 25 25	Jul 29 — 8 23 59
Aug 5 — 2 23 58	Aug 12 — 18 58 24	Aug 20 — 15 05 44	Aug 27 — 13 10 24
Sep 3 — 13 18 34	Sep 11 — 13 02 15	Sep 19 — 1 48 17	Sep 25 — 18 41 33
Oct 3 — 3 13 23	Oct 11 — 7 06 10	Oct 18 — 11 56 54	Oct 25 — 2 28 46
Nov 1 — 19 58 40	Nov 10 — 0 14 19	Nov 16 — 21 53 41	Nov 23 — 13 46 53
Dec 1 — 14 38 11	Dec 9 — 15 28 43	Dec 16 — 8 03 13	Dec 23 — 5 05 39
Dec 31 — 9 49 17			

2044

New Moon	First Quarter	Full Moon	Last Quarter
	Jan 8 — 4 03 08	Jan 14 — 18 52 21	Jan 21 — 23 48 16
Jan 30 — 4 05 44	Feb 6 — 13 47 24	Feb 13 — 6 43 00	Feb 20 — 20 21 22
Feb 28 — 20 13 35	Mar 6 — 21 18 25	Mar 13 — 19 42 24	Mar 21 — 16 53 39
Mar 29 — 9 27 13	Apr 5 — 3 46 14	Apr 12 — 9 40 22	Apr 20 — 11 49 45
Apr 27 — 19 43 20	May 4 — 10 29 17	May 12 — 0 17 53	May 20 — 4 03 13
May 27 — 3 40 49	Jun 2 — 18 34 43	Jun 10 — 15 17 21	Jun 18 — 17 01 06
Jun 25 — 10 25 33	Jul 2 — 4 49 44	Jul 10 — 6 23 15	Jul 18 — 2 47 57
Jul 24 — 17 11 43	Jul 31 — 17 41 37	Aug 8 — 21 15 07	Aug 16 — 10 04 32
Aug 23 — 1 07 13	Aug 30 — 9 19 52	Sep 7 — 11 25 38	Sep 14 — 15 58 58
Sep 21 — 11 04 42	Sep 29 — 3 31 50	Oct 7 — 0 31 17	Oct 13 — 21 53 45
Oct 20 — 23 37 39	Oct 28 — 23 28 54	Nov 5 — 12 27 58	Nov 12 — 5 10 40
Nov 19 — 14 59 01	Nov 27 — 19 37 37	Dec 4 — 23 35 08	Dec 11 — 14 53 31
Dec 19 — 8 54 24	Dec 27 — 14 01 16		

New Moon	First Quarter	Full Moon	Last Quarter

2045

New Moon	h m s	First Quarter	h m s	Full Moon	h m s	Last Quarter	h m s
				Jan 3	10 21 42	Jan 10	3 33 21
Jan 18	4 26 43	Jan 26	5 10 16	Feb 1	21 06 45	Feb 8	19 04 30
Feb 16	23 52 22	Feb 24	16 38 13	Mar 3	7 53 53	Mar 10	12 51 10
Mar 18	17 16 06	Mar 26	0 57 35	Apr 1	18 44 13	Apr 9	7 53 29
Apr 17	7 27 55	Apr 24	7 13 16	May 1	5 53 24	May 9	2 52 08
May 16	18 27 51	May 23	12 39 38	May 30	17 53 39	Jun 7	20 24 24
Jun 15	3 06 03	Jun 21	18 29 47	Jun 29	7 17 02	Jul 7	11 31 54
Jul 14	10 29 37	Jul 21	1 53 32	Jul 28	22 11 50	Aug 5	23 58 19
Aug 12	17 40 29	Aug 19	11 56 28	Aug 27	14 08 54	Sep 4	10 04 45
Sep 11	1 28 58	Sep 18	1 31 27	Sep 26	6 12 46	Oct 3	18 32 54
Oct 10	10 38 06	Oct 17	18 56 37	Oct 25	21 32 25	Nov 2	2 10 54
Nov 8	21 50 10	Nov 16	15 27 08	Nov 24	11 44 47	Dec 1	9 47 41
Dec 8	11 42 37	Dec 16	13 09 37	Dec 24	0 50 30	Dec 30	18 12 33

2046

New Moon	h m s	First Quarter	h m s	Full Moon	h m s	Last Quarter	h m s
Jan 7	4 25 17	Jan 15	9 43 48	Jan 22	12 52 32	Jan 29	4 12 41
Feb 5	23 10 56	Feb 14	3 21 42	Feb 20	23 45 31	Feb 27	16 24 11
Mar 7	18 16 35	Mar 15	17 14 04	Mar 22	9 28 11	Mar 29	6 58 30
Apr 6	11 52 54	Apr 14	3 22 46	Apr 20	18 22 20	Apr 27	23 31 39
May 6	2 57 19	May 13	10 26 07	May 20	3 16 31	May 27	17 07 50
Jun 4	15 23 41	Jun 11	15 28 55	Jun 18	13 11 16	Jun 26	10 41 14
Jul 4	1 40 08	Jul 10	19 54 39	Jul 18	0 56 16	Jul 26	3 20 40
Aug 2	10 26 43	Aug 9	1 16 56	Aug 16	14 51 21	Aug 24	18 37 31
Aug 31	18 26 40	Sep 7	9 08 13	Sep 15	6 40 36	Sep 23	8 16 58
Sep 30	2 26 45	Oct 6	20 42 20	Oct 14	23 42 32	Oct 22	20 08 57
Oct 29	11 18 17	Nov 5	12 29 52	Nov 13	17 05 41	Nov 21	6 11 39
Nov 27	21 51 21	Dec 5	7 57 38	Dec 13	9 56 41	Dec 20	14 44 18
Dec 27	10 40 08						

2047

New Moon	h m s	First Quarter	h m s	Full Moon	h m s	Last Quarter	h m s
		Jan 4	5 32 13	Jan 12	1 22 35	Jan 18	22 33 48
Jan 26	1 45 00	Feb 3	3 10 17	Feb 10	14 41 04	Feb 17	6 43 48
Feb 24	18 27 07	Mar 4	22 53 07	Mar 12	1 38 13	Mar 18	16 12 01
Mar 26	11 45 25	Apr 3	15 12 09	Apr 10	10 36 31	Apr 17	3 31 26
Apr 25	4 41 04	May 3	3 27 37	May 9	18 25 42	May 16	16 47 04
May 24	20 28 47	Jun 1	11 55 39	Jun 8	2 06 07	Jun 15	7 46 25
Jun 23	10 37 04	Jun 30	17 38 14	Jul 7	10 35 04	Jul 15	0 10 47
Jul 22	22 50 34	Jul 29	22 04 06	Aug 5	20 39 43	Aug 13	17 35 20
Aug 21	9 17 29	Aug 28	2 50 49	Sep 4	8 55 25	Sep 12	11 19 34
Sep 19	18 32 40	Sep 26	9 30 08	Oct 3	23 43 17	Oct 12	4 23 03
Oct 19	3 29 11	Oct 25	19 14 05	Nov 2	16 59 23	Nov 10	19 40 38
Nov 17	13 00 11	Nov 24	8 42 02	Dec 2	11 56 20	Dec 10	8 30 14
Dec 16	23 39 29	Dec 24	1 52 13				

2048

New Moon	h m s	First Quarter	h m s	Full Moon	h m s	Last Quarter	h m s
				Jan 1	6 58 19	Jan 8	18 50 45
Jan 15	11 33 40	Jan 22	21 57 24	Jan 31	0 15 32	Feb 7	3 17 56
Feb 14	0 32 45	Feb 21	19 23 41	Feb 29	14 39 21	Mar 7	10 46 21
Mar 14	14 28 54	Mar 22	16 04 44	Mar 30	2 05 37	Apr 5	18 11 50
Apr 13	5 20 57	Apr 21	10 03 25	Apr 28	11 14 12	May 5	2 23 39
May 12	20 59 25	May 21	0 17 28	May 27	18 58 28	Jun 3	12 05 58
Jun 11	12 51 10	Jun 19	10 50 49	Jun 26	2 09 20	Jul 2	23 59 05
Jul 11	4 05 26	Jul 18	18 32 56	Jul 25	9 35 09	Aug 1	14 31 45
Aug 9	18 00 09	Aug 17	0 33 07	Aug 23	18 08 24	Aug 31	7 43 09
Sep 8	6 25 49	Sep 15	6 05 09	Sep 22	4 47 47	Sep 30	2 46 44
Oct 7	17 46 27	Oct 14	12 21 20	Oct 21	18 26 11	Oct 29	22 15 43
Nov 6	4 39 35	Nov 12	20 30 05	Nov 20	11 20 44	Nov 28	16 34 48
Dec 5	15 31 21	Dec 12	7 30 18	Dec 20	6 40 26	Dec 28	8 32 48

2049

New Moon	h m s	First Quarter	h m s	Full Moon	h m s	Last Quarter	h m s
Jan 4	2 25 43	Jan 10	21 57 02	Jan 19	2 30 19	Jan 26	21 34 18
Feb 2	13 17 05	Feb 9	15 39 33	Feb 17	20 48 39	Feb 25	7 37 32
Mar 4	0 12 46	Mar 11	11 27 23	Mar 19	12 24 31	Mar 26	15 11 30
Apr 2	11 40 26	Apr 10	7 28 41	Apr 18	1 05 56	Apr 24	21 12 25
May 2	0 12 16	May 10	1 58 46	May 17	11 15 00	May 24	2 55 20
May 31	14 01 21	Jun 8	17 57 43	Jun 15	19 28 02	Jun 22	9 42 27
Jun 30	4 51 33	Jul 8	7 11 23	Jul 15	2 30 52	Jul 21	18 49 50
Jul 29	20 08 29	Aug 6	17 52 58	Aug 13	9 20 49	Aug 20	7 12 00
Aug 28	11 19 53	Sep 5	2 29 27	Sep 11	17 05 37	Sep 18	23 05 00
Sep 27	2 06 19	Oct 4	9 40 06	Oct 11	2 54 30	Oct 18	17 56 48
Oct 26	16 16 21	Nov 2	16 20 24	Nov 9	15 39 09	Nov 17	14 33 23
Nov 25	5 36 56	Dec 1	23 40 58	Dec 9	7 29 16	Dec 17	11 15 55
Dec 24	17 53 00	Dec 31	8 54 01				

2050

New Moon	h m s	First Quarter	h m s	Full Moon	h m s	Last Quarter	h m s
				Jan 8	1 40 15	Jan 16	6 18 50
Jan 23	4 58 11	Jan 29	20 49 07	Feb 6	20 48 43	Feb 14	22 11 58
Feb 21	15 04 44	Feb 28	11 30 55	Mar 8	15 24 28	Mar 16	10 09 12
Mar 23	0 42 09	Mar 30	4 18 38	Apr 7	8 13 20	Apr 14	18 25 07
Apr 21	10 26 59	Apr 28	22 09 45	May 6	22 27 19	May 14	0 05 04
May 20	20 52 15	May 28	16 05 29	Jun 5	9 52 22	Jun 12	4 40 36
Jun 19	8 23 04	Jun 27	9 18 04	Jul 4	18 52 10	Jul 11	9 47 12
Jul 18	21 18 02	Jul 27	1 06 34	Aug 3	2 21 42	Aug 9	16 49 34
Aug 17	11 48 41	Aug 25	14 57 34	Sep 1	9 32 04	Sep 8	2 52 15
Sep 16	3 50 32	Sep 24	2 35 22	Sep 30	17 33 00	Oct 7	16 33 21
Oct 15	20 49 50	Oct 23	12 11 52	Oct 30	3 17 08	Nov 6	9 58 17
Nov 14	13 42 37	Nov 21	20 26 35	Nov 28	15 10 59	Dec 6	6 28 44
Dec 14	5 19 19	Dec 21	4 16 24	Dec 28	5 16 47		

In Chapter 47 of our *Astronomical Algorithms* (Wilmann-Bell, ed.), formulae are given for the calculation of the times of the lunar phases with an accuracy better than 18 seconds. By means of the tables on the next pages, it is possible to find the instant of any lunar phase between the years -1500 and $+2999$ with an accuracy better than 10 minutes.

Write down, from Table 1, the values of 'Time', A, B, and C for the century of the given year. (Thus, for instance, for the year 1967 take the value for 1900). The time is expressed in days and decimals.

Write down, from Tables 2 and 3, the values of the same quantities for the year and for the date of the year.

Add up the numbers obtained, in order to find the date (of the *mean* lunar phase), and the values of the arguments A, B, C. If the value of any of the quantities A, B, C exceeds 1000, subtract 1000 or a convenient multiple.

Calculate the quantities $A + B$ and $A - B$. If $A + B$ is larger than 1000, subtract 1000. If $A - B < 0$, add 1000.

With the arguments A, B, C, $A + B$, $A - B$, find the five corrections from Tables 4 to 8, respectively. Note that Table 4 is in a certain sense a double-entry table, the value of the correction being given for the years -1500, 0, $+1500$, and $+3000$. Hence, a 'double' interpolation should be performed here.

From Tables $5 - 8$, take the correction from the appropriate column, according to the lunar phase (NM = New Moon, FQ = First Quarter, FM = Full Moon, LQ = Last Quarter).

If the final instant should be expressed in Universal Time instead of Dynamical Time, take the correction given by Table 9.

Add the five (or six) corrections to the instant of the mean phase, in order to obtain the time of the true phase, expressed in TD (or UT). The fraction of the day can then be converted into hours and minutes, if this is desired.

Remarks

● It will be clear to the reader that *no* interpolation is involved in Tables 1, 2, and 3. On the contrary, in Tables 4 to 9 the corrections *should* be interpolated; in these tables the tabular intervals have been taken so small that linear interpolation is always sufficient for our purpose.

● In the case of a negative year, the year number should be split in such a way as to have the last two figures positive. For example, the year -328 (this is the year 329 B.C. of the historians) should be split up as $-400 + 72$. Then Table 1 should be consulted for -400, and Table 2 for $+72$.

● After the addition of the several times and corrections, the final instant may exceed the number of days in the month. It must be remembered that, for instance, May 32 is the same as June 1. In Example 2, we obtain April 78; this is the same as May 48, or as June 17.

● Care must be given to choose the correct line in Table 3. In Example 2, for instance, the sum of the first two times (25.633 and 25.011 days) is already as large as 50 days. Consequently, in order to obtain the New Moon taking place in mid-June, the line corresponding to April 28 should be taken from Table 3. (The line June 26 would give the New Moon of August).

● In the case of a bissextile year, for the months January and February the lines indicated by *(B)* should be used in Table 3.

● If an accuracy of one hour is sufficient, it's not necessary to use Tables 6, 7, and 8. In that case, it is not needed to calculate the arguments C, $A + B$, $A - B$.

Example 1. — Find the time of the Full Moon of July 1953.

Table	Argument	Time	A	B	C
1	1900	1.259	998	850	118
2	53	14.066	36	44	777
3	July	10.949	526	966	108
4	560	−0.062			
5	860	+0.298	1560	1860	1003
6	3	+0.000	= 560	= 860	= 3
7	420	−0.002			
8	700	+0.007	$A + B$ =	1420 or 420	
9	1953	0.000	$A - B$ =	−300 or 700	
	Sum	26.515			

Hence, the Full Moon took place on 1953 July 26.515, that is, on 1953 July 26 at approximately 12^h22^m Universal Time. The correct value is 12^h21^m UT.

Example 2. — Find the time of the New Moon of June 1433.

Table	Argument	Time	A	B	C
1	1400	25.633	113	445	586
2	33	25.011	67	331	690
3	April	28.122	323	287	682
4	503	−0.003			
5	63	−0.146	503	1063	1958
6	958	−0.003		= 63	= 958
7	566	+0.002	$A + B$ = 566		
8	440	−0.003	$A - B$ = 440		
	Sum	78.613			

Hence, New Moon took place on 1433 April 78.613 TD that is, 1433 June 17, at approximately 14^h43^m Dynamical Time.

TABLE 1   —   *Values for beginning of centuries*

Julian Calendar					Gregorian Calendar				
Year	Time	A	B	C	Year	Time	A	B	C
−1500	d 18.108	169	180	797	1500	d 10.440	42	83	190
−1400	22.437	178	888	572	1600	14.777	51	793	965
−1300	26.765	187	596	347	1700	20.114	60	502	740
−1200	1.564	116	232	952	1800	25.452	69	212	514
−1100	5.893	125	940	727	1900	1.259	998	850	118
−1000	10.223	134	649	502	2000	5.598	7	560	893
−900	14.553	143	357	277	2100	10.936	16	270	667
−800	18.884	153	65	52	2200	16.275	26	980	442
−700	23.214	162	773	827	2300	21.614	35	690	216
−600	27.545	171	482	602	2400	25.953	44	400	991
−500	2.346	99	118	206	2500	1.762	972	38	595
−400	6.677	109	827	981	2600	7.102	982	748	369
−300	11.009	118	535	756	2700	12.442	991	459	144
−200	15.341	127	244	531	2800	16.783	0	169	918
−100	19.674	136	953	306	2900	22.123	9	879	692
0	24.007	146	661	8					
+100	28.340	155	370	856					
200	3.142	83	7	460					
300	7.476	93	716	235					
400	11.810	102	425	10					
500	16.144	111	134	784					
600	20.478	120	843	559					
700	24.813	130	552	334					
800	29.148	139	261	109					
900	3.953	67	898	713					
1000	8.288	76	608	488					
1100	12.624	86	317	262					
1200	16.960	95	26	37					
1300	21.297	104	736	812					
1400	25.633	113	445	586					
1500	0.440	42	83	190					

TABLE 2 — Values for additional years

Year	Time (d)	A	B	C
0 B?	0.000	0	0	0
1	18.898	51	932	215
2	8.265	21	793	260
3	27.162	72	725	475
4 B	15.529	42	586	520
5	4.896	13	446	564
6	23.794	64	379	779
7	13.161	34	239	824
8 B	1.528	4	100	869
9	20.426	55	32	84
10	9.793	25	892	129
11	28.691	76	825	344
12 B	17.058	46	685	388
13	6.425	17	546	433
14	25.322	68	478	648
15	14.689	38	339	693
16 B	3.057	8	199	738
17	21.954	59	132	953
18	11.321	29	992	997
19	0.688	999	853	42
20 B	18.586	50	785	257
21	7.953	21	646	302
22	26.851	72	578	517
23	16.218	42	438	562
24 B	4.585	12	299	606
25	23.482	63	231	821
26	12.850	33	92	866
27	2.217	3	952	911
28 B	20.114	54	885	126
29	9.481	25	745	171
30	28.379	76	677	386
31	17.746	46	538	430
32 B	6.113	16	399	475
33	25.011	67	331	690
34	14.378	37	191	735

Year	Time (d)	A	B	C
35	3.745	7	52	780
36 B	21.643	58	984	995
37	11.010	28	845	39
38	0.377	999	705	84
39	19.274	50	638	299
40 B	7.641	20	498	344
41	26.539	71	431	559
42	15.906	41	291	604
43	5.273	11	152	648
44 B	23.171	62	84	863
45	12.538	32	945	908
46	1.905	3	805	953
47	20.803	54	737	168
48 B	9.170	24	598	213
49	28.067	75	530	428
50	17.434	45	391	472
51	6.801	15	251	517
52 B	24.699	66	184	732
53	14.066	36	44	777
54	3.433	7	905	822
55	22.331	58	837	37
56 B	10.698	28	698	81
57	0.065	998	558	126
58	18.963	49	490	341
59	8.330	19	351	386
60 B	26.227	70	283	601
61	15.594	40	144	646
62	4.962	11	4	690
63	23.859	62	937	906
64 B	12.226	32	797	950
65	1.593	2	658	995
66	20.491	53	590	210
67	9.858	23	451	255
68 B	27.756	74	383	470
69	17.123	44	243	515

Year	Time (d)	A	B	C
70	6.490	15	104	559
71	25.387	66	36	774
72 B	13.755	36	897	819
73	3.122	6	757	864
74	22.019	57	690	79
75	11.386	27	550	123
76 B	29.284	78	483	339
77	18.651	48	343	383
78	8.018	19	204	428
79	26.916	70	136	643
80 B	15.283	40	997	688
81	4.650	10	857	732
82	23.548	61	789	948
83	12.915	31	650	992
84 B	1.282	1	510	37
85	20.179	52	443	252
86	9.546	22	303	297
87	28.444	74	236	512
88 B	16.811	44	96	557
89	6.178	14	957	601
90	25.076	65	889	816
91	14.443	35	750	861
92 B	2.810	5	610	906
93	21.708	56	542	121
94	11.075	26	403	166
95	0.442	997	264	210
96 B	18.339	48	196	425
97	7.706	18	56	470
98	26.604	69	989	685
99	15.971	39	849	730

B = Leap (Bissextile) Year

All the century years are bissextile, except 1700, 1800, 1900,
2100, 2200, 2300, 2500, etc., which are common years.

TABLE 3 — Values for dates of the year

Lunar Phase	Month	Time	A	B	C	Lunar Phase	Month	Time	A	B	C
		d						*d*			
NM	Jan.	0.000	0	0	0	NM	May	27.653	404	359	852
NM	Jan. *(B)*	1.000	0	0	0	FQ	June	4.036	424	626	395
FQ	Jan.	7.383	20	268	543	FM	June	11.418	445	894	937
FQ	Jan. *(B)*	8.383	20	268	543	LQ	June	18.801	465	162	480
FM	Jan.	14.765	40	536	85	NM	June	26.184	485	430	22
FM	Jan. *(B)*	15.765	40	536	85	FQ	July	3.566	505	698	565
LQ	Jan.	22.148	61	804	628	FM	July	10.949	526	966	108
LQ	Jan. *(B)*	23.148	61	804	628	LQ	July	18.331	546	234	650
NM	Jan.	29.531	81	72	170	NM	July	25.714	566	502	193
NM	Jan. *(B)*	30.531	81	72	170	FQ	Aug.	2.097	586	770	735
FQ	Feb.	5.913	101	340	713	FM	Aug.	9.479	606	38	278
FQ	Feb. *(B)*	6.913	101	340	713	LQ	Aug.	16.862	627	306	821
FM	Feb.	13.296	121	608	256	NM	Aug.	24.245	647	574	363
FM	Feb. *(B)*	14.296	121	608	256	FQ	Sept.	0.627	667	842	906
LQ	Feb.	20.679	141	875	798	FM	Sept.	8.010	687	110	448
LQ	Feb. *(B)*	21.679	141	875	798	LQ	Sept.	15.393	707	377	991
NM	March	0.061	162	143	341	NM	Sept.	22.775	728	645	534
FQ	March	7.444	182	411	883	FQ	Oct.	0.158	748	913	76
FM	March	14.826	202	679	426	FM	Oct.	7.541	768	181	619
LQ	March	22.209	222	947	969	LQ	Oct.	14.923	788	449	161
NM	March	29.592	243	215	511	NM	Oct.	22.306	808	717	704
FQ	April	5.974	263	483	54	FQ	Oct.	29.689	829	985	247
FM	April	13.357	283	751	596	FM	Nov.	6.071	849	253	789
LQ	April	20.740	303	19	139	LQ	Nov.	13.454	869	521	332
NM	April	28.122	323	287	682	NM	Nov.	20.836	889	789	874
FQ	May	5.505	344	555	224	FQ	Nov.	28.219	910	57	417
FM	May	12.888	364	823	767	FM	Dec.	5.602	930	325	960
LQ	May	20.270	384	91	309	LQ	Dec.	12.984	950	593	502
						NM	Dec.	20.367	970	861	45
						FQ	Dec.	27.750	990	128	587
						FM	Dec.	35.132	11	396	130

(B) = Leap (Bissextile) Year

NM = New Moon
FQ = First Quarter
FM = Full Moon
LQ = Last Quarter

TABLE 4 — First Correction

A	Year -1500	Year 0	Year 1500	Year 3000	A	Year -1500	Year 0	Year 1500	Year 3000
0	d 0.000	d 0.000	d 0.000	d 0.000	500	d 0.000	d 0.000	d 0.000	d 0.000
10	+0.012	+0.012	+0.011	+0.011	510	−0.011	−0.011	−0.011	−0.010
20	+0.024	+0.023	+0.022	+0.022	520	−0.023	−0.022	−0.021	−0.021
30	+0.036	+0.035	+0.034	+0.032	530	−0.034	−0.033	−0.032	−0.031
40	+0.047	+0.046	+0.044	+0.043	540	−0.045	−0.044	−0.042	−0.041
50	+0.059	+0.057	+0.055	+0.053	550	−0.056	−0.054	−0.053	−0.051
60	+0.070	+0.068	+0.066	+0.063	560	−0.067	−0.065	−0.063	−0.061
70	+0.081	+0.079	+0.076	+0.073	570	−0.077	−0.075	−0.073	−0.070
80	+0.092	+0.089	+0.086	+0.083	580	−0.088	−0.085	−0.082	−0.079
90	+0.102	+0.099	+0.095	+0.092	590	−0.098	−0.095	−0.092	−0.088
100	+0.112	+0.108	+0.105	+0.101	600	−0.107	−0.104	−0.101	−0.097
110	+0.121	+0.117	+0.113	+0.109	610	−0.116	−0.113	−0.109	−0.105
120	+0.130	+0.126	+0.122	+0.117	620	−0.125	−0.121	−0.117	−0.113
130	+0.138	+0.134	+0.129	+0.124	630	−0.133	−0.129	−0.125	−0.120
140	+0.146	+0.141	+0.137	+0.131	640	−0.141	−0.137	−0.132	−0.127
150	+0.153	+0.148	+0.143	+0.138	650	−0.148	−0.144	−0.139	−0.134
160	+0.159	+0.155	+0.149	+0.144	660	−0.155	−0.150	−0.145	−0.140
170	+0.165	+0.160	+0.155	+0.149	670	−0.161	−0.156	−0.151	−0.146
180	+0.170	+0.165	+0.160	+0.153	680	−0.166	−0.162	−0.156	−0.150
190	+0.175	+0.169	+0.164	+0.157	690	−0.171	−0.166	−0.161	−0.155
200	+0.178	+0.173	+0.167	+0.161	700	−0.175	−0.170	−0.165	−0.159
210	+0.181	+0.176	+0.170	+0.164	710	−0.179	−0.174	−0.168	−0.162
220	+0.184	+0.178	+0.172	+0.166	720	−0.182	−0.177	−0.171	−0.164
230	+0.185	+0.180	+0.174	+0.167	730	−0.184	−0.179	−0.173	−0.166
240	+0.186	+0.180	+0.174	+0.168	740	−0.185	−0.180	−0.174	−0.167
250	+0.186	+0.181	+0.175	+0.168	750	−0.186	−0.181	−0.175	−0.168
260	+0.185	+0.180	+0.174	+0.167	760	−0.186	−0.180	−0.174	−0.168
270	+0.184	+0.179	+0.173	+0.166	770	−0.185	−0.180	−0.174	−0.167
280	+0.182	+0.177	+0.171	+0.164	780	−0.184	−0.178	−0.172	−0.166
290	+0.179	+0.174	+0.168	+0.162	790	−0.181	−0.176	−0.170	−0.164
300	+0.175	+0.170	+0.165	+0.159	800	−0.178	−0.173	−0.167	−0.161
310	+0.171	+0.166	+0.161	+0.155	810	−0.175	−0.169	−0.164	−0.157
320	+0.166	+0.162	+0.156	+0.150	820	−0.170	−0.165	−0.160	−0.153
330	+0.161	+0.156	+0.151	+0.146	830	−0.165	−0.160	−0.155	−0.149
340	+0.155	+0.150	+0.145	+0.140	840	−0.159	−0.155	−0.149	−0.144
350	+0.148	+0.144	+0.139	+0.134	850	−0.153	−0.148	−0.143	−0.138
360	+0.141	+0.137	+0.132	+0.127	860	−0.146	−0.141	−0.137	−0.131
370	+0.133	+0.129	+0.125	+0.120	870	−0.138	−0.134	−0.129	−0.124
380	+0.125	+0.121	+0.117	+0.113	880	−0.130	−0.126	−0.122	−0.117
390	+0.116	+0.113	+0.109	+0.105	890	−0.121	−0.117	−0.113	−0.109
400	+0.107	+0.104	+0.101	+0.097	900	−0.112	−0.108	−0.105	−0.101
410	+0.098	+0.095	+0.092	+0.088	910	−0.102	−0.099	−0.095	−0.092
420	+0.088	+0.085	+0.082	+0.079	920	−0.092	−0.089	−0.086	−0.083
430	+0.077	+0.075	+0.073	+0.070	930	−0.081	−0.079	−0.076	−0.073
440	+0.067	+0.065	+0.063	+0.061	940	−0.070	−0.068	−0.066	−0.063
450	+0.056	+0.054	+0.053	+0.051	950	−0.059	−0.057	−0.055	−0.053
460	+0.045	+0.044	+0.042	+0.041	960	−0.047	−0.046	−0.044	−0.043
470	+0.034	+0.033	+0.032	+0.031	970	−0.036	−0.035	−0.034	−0.032
480	+0.023	+0.022	+0.021	+0.021	980	−0.024	−0.023	−0.022	−0.022
490	+0.011	+0.011	+0.011	+0.010	990	−0.012	−0.012	−0.011	−0.011
500	0.000	0.000	0.000	0.000	1000	0.000	0.000	0.000	0.000

TABLE 5 — Second Correction

B	NM/FM	FQ/LQ	B	NM/FM	FQ/LQ
	d	d		d	d
0	0.000	0.000	500	0.000	0.000
10	−0.024	−0.038	510	+0.028	+0.041
20	−0.047	−0.077	520	+0.055	+0.081
30	−0.070	−0.115	530	+0.082	+0.121
40	−0.094	−0.152	540	+0.109	+0.161
50	−0.117	−0.189	550	+0.135	+0.199
60	−0.139	−0.226	560	+0.161	+0.237
70	−0.161	−0.261	570	+0.186	+0.274
80	−0.183	−0.296	580	+0.210	+0.310
90	−0.204	−0.329	590	+0.233	+0.345
100	−0.224	−0.361	600	+0.255	+0.378
110	−0.244	−0.392	610	+0.275	+0.409
120	−0.263	−0.422	620	+0.295	+0.439
130	−0.281	−0.449	630	+0.313	+0.467
140	−0.298	−0.476	640	+0.329	+0.493
150	−0.314	−0.500	650	+0.344	+0.516
160	−0.329	−0.522	660	+0.358	+0.538
170	−0.343	−0.543	670	+0.370	+0.558
180	−0.355	−0.561	680	+0.380	+0.575
190	−0.367	−0.578	690	+0.389	+0.590
200	−0.377	−0.592	700	+0.396	+0.602
210	−0.386	−0.604	710	+0.401	+0.612
220	−0.393	−0.613	720	+0.405	+0.620
230	−0.399	−0.621	730	+0.407	+0.625
240	−0.403	−0.625	740	+0.407	+0.627
250	−0.406	−0.628	750	+0.406	+0.628
260	−0.407	−0.627	760	+0.403	+0.625
270	−0.407	−0.625	770	+0.399	+0.621
280	−0.405	−0.620	780	+0.393	+0.613
290	−0.401	−0.612	790	+0.386	+0.604
300	−0.396	−0.602	800	+0.377	+0.592
310	−0.389	−0.590	810	+0.367	+0.578
320	−0.380	−0.575	820	+0.355	+0.561
330	−0.370	−0.558	830	+0.343	+0.543
340	−0.358	−0.538	840	+0.329	+0.522
350	−0.344	−0.516	850	+0.314	+0.500
360	−0.329	−0.493	860	+0.298	+0.476
370	−0.313	−0.467	870	+0.281	+0.449
380	−0.295	−0.439	880	+0.263	+0.422
390	−0.275	−0.409	890	+0.244	+0.392
400	−0.255	−0.378	900	+0.224	+0.361
410	−0.233	−0.345	910	+0.204	+0.329
420	−0.210	−0.310	920	+0.183	+0.296
430	−0.186	−0.274	930	+0.161	+0.261
440	−0.161	−0.237	940	+0.139	+0.226
450	−0.135	−0.199	950	+0.117	+0.189
460	−0.109	−0.161	960	+0.094	+0.152
470	−0.082	−0.121	970	+0.070	+0.115
480	−0.055	−0.081	980	+0.047	+0.077
490	−0.028	−0.041	990	+0.024	+0.038
500	0.000	0.000	1000	0.000	0.000

TABLE 6 — Third Correction

C	NM/FM d	FQ/LQ d
0	0.000	0.000
20	+0.001	+0.001
40	+0.003	+0.002
60	+0.004	+0.003
80	+0.005	+0.004
100	+0.006	+0.005
120	+0.007	+0.006
140	+0.008	+0.006
160	+0.009	+0.007
180	+0.009	+0.007
200	+0.010	+0.008
220	+0.010	+0.008
240	+0.010	+0.008
260	+0.010	+0.008
280	+0.010	+0.008
300	+0.010	+0.008
320	+0.009	+0.007
340	+0.009	+0.007
360	+0.008	+0.006
380	+0.007	+0.006
400	+0.006	+0.005
420	+0.005	+0.004
440	+0.004	+0.003
460	+0.003	+0.002
480	+0.001	+0.001
500	0.000	0.000
520	−0.001	−0.001
540	−0.003	−0.002
560	−0.004	−0.003
580	−0.005	−0.004
600	−0.006	−0.005
620	−0.007	−0.006
640	−0.008	−0.006
660	−0.009	−0.007
680	−0.009	−0.007
700	−0.010	−0.008
720	−0.010	−0.008
740	−0.010	−0.008
760	−0.010	−0.008
780	−0.010	−0.008
800	−0.010	−0.008
820	−0.009	−0.007
840	−0.009	−0.007
860	−0.008	−0.006
880	−0.007	−0.006
900	−0.006	−0.005
920	−0.005	−0.004
940	−0.004	−0.003
960	−0.003	−0.002
980	−0.001	−0.001
1000	0.000	0.000

TABLE 7 — Fourth Correction

A + B	NM/FM d	FQ d	LQ d
0	0.000	+0.003	−0.003
20	−0.001	+0.001	−0.005
40	−0.001	+0.000	−0.006
60	−0.002	−0.001	−0.007
80	−0.002	−0.003	−0.009
100	−0.003	−0.004	−0.010
120	−0.004	−0.005	−0.011
140	−0.004	−0.006	−0.012
160	−0.004	−0.007	−0.013
180	−0.005	−0.008	−0.014
200	−0.005	−0.008	−0.014
220	−0.005	−0.009	−0.014
240	−0.005	−0.009	−0.015
260	−0.005	−0.009	−0.015
280	−0.005	−0.009	−0.015
300	−0.005	−0.008	−0.014
320	−0.005	−0.008	−0.014
340	−0.004	−0.007	−0.013
360	−0.004	−0.006	−0.012
380	−0.004	−0.005	−0.011
400	−0.003	−0.004	−0.010
420	−0.002	−0.003	−0.009
440	−0.002	−0.001	−0.007
460	−0.001	+0.000	−0.006
480	−0.001	+0.002	−0.005
500	0.000	+0.003	−0.003
520	+0.001	+0.005	−0.002
540	+0.001	+0.006	−0.000
560	+0.002	+0.007	+0.001
580	+0.002	+0.009	+0.003
600	+0.003	+0.010	+0.004
620	+0.004	+0.011	+0.005
640	+0.004	+0.012	+0.006
660	+0.004	+0.013	+0.007
680	+0.005	+0.014	+0.008
700	+0.005	+0.014	+0.009
720	+0.005	+0.015	+0.009
740	+0.005	+0.015	+0.009
760	+0.005	+0.015	+0.009
780	+0.005	+0.015	+0.009
800	+0.005	+0.014	+0.008
820	+0.005	+0.014	+0.008
840	+0.004	+0.013	+0.007
860	+0.004	+0.012	+0.006
880	+0.004	+0.011	+0.005
900	+0.003	+0.010	+0.004
920	+0.002	+0.009	+0.003
940	+0.002	+0.007	+0.001
960	+0.001	+0.006	−0.000
980	+0.001	+0.005	−0.002
1000	0.000	+0.003	−0.003

TABLE 8 — Fifth Correction

A − B	NM/FM	FQ/LQ
	d	d
0	0.000	0.000
20	−0.001	−0.001
40	−0.002	−0.001
60	−0.003	−0.002
80	−0.004	−0.002
100	−0.004	−0.003
120	−0.005	−0.003
140	−0.006	−0.003
160	−0.006	−0.004
180	−0.007	−0.004
200	−0.007	−0.004
220	−0.007	−0.004
240	−0.007	−0.005
260	−0.007	−0.005
280	−0.007	−0.004
300	−0.007	−0.004
320	−0.007	−0.004
340	−0.006	−0.004
360	−0.006	−0.003
380	−0.005	−0.003
400	−0.004	−0.003
420	−0.004	−0.002
440	−0.003	−0.002
460	−0.002	−0.001
480	−0.001	−0.001
500	0.000	0.000
520	+0.001	+0.001
540	+0.002	+0.001
560	+0.003	+0.002
580	+0.004	+0.002
600	+0.004	+0.003
620	+0.005	+0.003
640	+0.006	+0.003
660	+0.006	+0.004
680	+0.007	+0.004
700	+0.007	+0.004
720	+0.007	+0.004
740	+0.007	+0.005
760	+0.007	+0.005
780	+0.007	+0.004
800	+0.007	+0.004
820	+0.007	+0.004
840	+0.006	+0.004
860	+0.006	+0.003
880	+0.005	+0.003
900	+0.004	+0.003
920	+0.004	+0.002
940	+0.003	+0.002
960	+0.002	+0.001
980	+0.001	+0.001
1000	0.000	0.000

TABLE 9

Correction for reduction from Dynamical Time to Universal Time

Year	Corr.
	d
−1500	−0.412
−1400	−0.387
−1300	−0.364
−1200	−0.341
−1100	−0.318
−1000	−0.297
− 900	−0.276
− 800	−0.256
− 700	−0.237
− 600	−0.218
− 500	−0.201
− 400	−0.184
− 300	−0.167
− 200	−0.152
− 100	−0.137
0	−0.123
+ 100	−0.110
200	−0.097
300	−0.086
400	−0.075
500	−0.064
600	−0.055
700	−0.046
800	−0.038
900	−0.031
1000	−0.025
1100	−0.019
1200	−0.014
1300	−0.010
1400	−0.006
1500	−0.003
1620	−0.001
1700	0.000
1970	0.000
1975	−0.001
2000	−0.001
2050	−0.002
2100	−0.003
2200	−0.006
2300	−0.009
2400	−0.013
2500	−0.018
2600	−0.023
2700	−0.030
2800	−0.037
2900	−0.045
3000	−0.053

5

Occultations
of Planets and Bright Stars
by the Moon

1990 – 2020

Occultations 1990 − 2020

General Data

Table I, which begins on page 237, gives the list of all occultations of first-magnitude stars and of planets by the Moon taking place during the period 1990 − 2020. These events are listed chronologically. For reason of completeness, *all* these occultations are given, even those which occur at a small angular distance from the Sun and thus are not observable. For example, the occultation of Uranus on 2000 February 5 will not be visible.

The second column of the Table gives the nearest integer hour T_0 (Dynamical Time) of the least geocentric angular distance between the center of the Moon and that of the body.

The visual stellar magnitude of the occulted star or planet is given in the fourth column. For the stars, the magnitudes are those of the RHP (Revised Harvard Photometry). For the planets, they have been calculated from the classical formulae by G. Müller (1893) — see Chapter 40 of our *Astronomical Algorithms* (Willmann-Bell, ed.).

The fifth column contains the elongation of the occulted body from the Sun, at the time T_0, in degrees:

E = East from the Sun = visible in the *evening*;
W = West from the Sun = visible in the *morning*.

In the next column we give the least distance γ from the axis of the lunar 'shadow' to the center of the Earth, expressed in units of the Earth's equatorial radius. In fact, γ is the minimum value of the expression

$$\sqrt{x^2 + y^2}$$

where x and y are the Besselian elements which will be defined further. The quantity γ is positive or negative, according to whether the axis of the shadow passes north or south of the center of the Earth. There is an occultation for some places on the Earth's surface only when $|\gamma| < 1.270$.

In the last column, the region of visibility on the Earth's surface is briefly described. However, it is important to note that no distinction is made between events taking place during the night and events in daylight. For example, the occultation of Aldebaran on 1999 July 10 is said to be visible in Europe; in fact, it will take place in daylight there. The descriptions are, of course, necessarily short and not many details can be given. For instance, an occultation stated to be visible in Africa is not necessarily visible from each point of this continent. The letters N, E, S and W signify 'northern part of', etc.

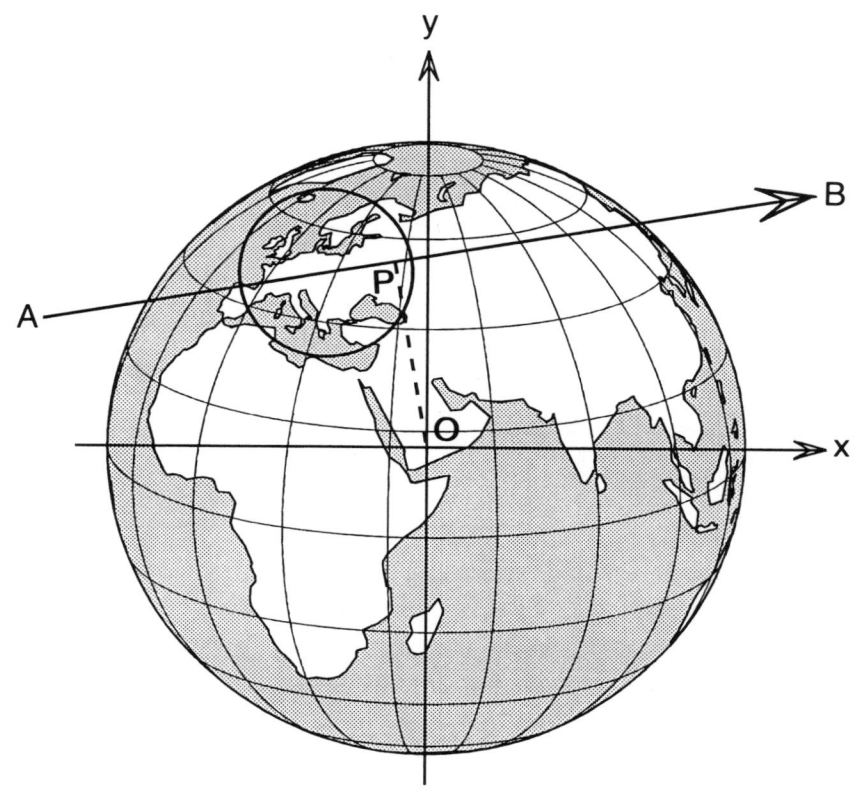

The Earth and the Moon as seen from the occulted body

The Moon's center is moving from *A* towards *B*. The position of the Moon
(the smaller circle) is indicated for the reference time T_0.

O is the center of the Earth's disk. In the fundamental plane, the coordinates
of the center of the Moon (x, y) and those of the observer (ξ, η) are referred
to the axes *Ox, Oy*.

OP is the least distance γ of the Moon's center to the center of the Earth.

While the Moon is moving from *A* to *B*, the Earth's globe is rotating from
West to East. The calculation of local circumstances of the occultation takes
both motions into account.

Besselian Elements

The Besselian elements, needed for the calculation of local circumstances, are given in Table II for the occultations of the stars, and in Table III for the occultations of the planets. Their geometric significance is as follows.

Let us imagine a plane passing through the center of the Earth, perpendicular to the line joining the star (or the planet) and the center of the Moon. The latter line is the axis of the Moon's 'shadow', and the plane is called the *fundamental plane*, or plane of *xy*. We take the intersection of this plane with that of the Earth's equator as the *x*-axis, and the center of the Earth as the origin of rectangular coordinates. The *y*-axis is perpendicular to the *x*-axis, and is directed towards the North. Hence, as seen from the occulted body, the *x*-axis passes through the two points of intersection of the Earth's equator with the 'horizon' (the Earth's limb), while the poles project on the *y*-axis — see the drawing on the preceding page.

Then, at any instant, *x* and *y* are the coordinates of the point in which the axis of the 'shadow' intersects the fundamental plane. In other words, they are the coordinates of the center of the Moon's disk with respect to the disk of the Earth, as seen from the occulted body (star or planet). These coordinates are here expressed in terms of the Earth's equatorial radius taken as unity.

The angle *d* is the declination of the occulted body, while *HO* denotes the Greenwich hour angle of this body. [The latter statement is not rigorously exact. *HO* would be the body's hour angle for the meridian of Greenwich if the quantity ΔT was exactly zero. Actually, *HO* is the hour angle of the body for the so-called *ephemeris meridian*, whose position in space is conceived as being where the Greenwich meridian would have been if the Earth had rotated uniformly at the rate implicit in the definition of the Dynamical Time].

In Tables II and III, T_0 is an *integer* hour (Dynamical Time), the same as in Table I. The values of *d* and of *HO* are given for that instant, and they are expressed in degrees and decimals. In the case of the occultation of a star, the declination may be considered as constant during the whole event, while the hourly variation of *HO* is +15.04107. This is not the case for the planets, since their right ascension and declination are varying: therefore, the hourly variations *D1* and *H1* are given in Table III (in degrees), and the value of *d* at time T_0 is designated *D0*.

In the case of an occultation of a star, the 'shadow' of the Moon is a cylinder, whose intersection with the fundamental plane is a circle. The radius of this circle is equal to the radius of the Moon, for which we adopted the value *k* = 0.272495, again the equatorial radius of the Earth being taken as unity.

In the case of an occultation of a planet, the 'shadow' is not a cylinder but a cone. Its intersection with the fundamental plane is a circle, whose radius *k* is a little larger than 0.272495. In Table III, the value of *k* is given, together with a quantity *F* which is needed in the accurate calculation of the occultation. In fact, the quantity *F* takes into account the fact that the planet is not at infinite distance. The quantities *k* and *F* can be considered as constants during the whole event.

The values of *x* and *y* at time T_0 are denoted by *XO* and *YO*. They are given in Tables II and III, together with their hourly variations *X1* and *Y1*, and the coefficients *X2* and *Y2* of their second-order terms. At any instant *t* during the occultation, measured in hours from T_0, the values of *x* and *y* can be calculated from

$$x = XO + X1\,t + X2\,t^2 \qquad \text{and} \qquad y = YO + Y1\,t + Y2\,t^2$$

Important Remarks

● The Besselian elements are given in the uniform time scale known as *Dynamical Time* (TD). The Universal Time (UT), necessary for civil life and for astronomical calculations where hour angles are involved, is based on the rotation of the Earth. Because this rotation is slowing down — and, moreover, with unpredictable irregularities — TD is not a uniform time. The exact value of the difference $\Delta T = \text{TD} - \text{UT}$ can be deduced only from observations. For more information, see pages $5 - 7$.

● In the case of the occultation of a planet, all results calculated by using the formulae given in this work refer to the *center* of that planet's disk.

● The *geographical longitudes* are supposed to be measured *positively westward* from the meridian of Greenwich, and negatively to the east. This convention, which is in conformity with the longitude systems on the other planets, has been followed by most astronomers during more than one century. For instance, the longitude of Washington, D.C., is $+77°04'$. That of Vienna, Austria, is $-16°23'$.

● Most programming languages can calculate trigonometric functions (SIN, COS, TAN) only if their argument is expressed in *radians*, not degrees. In such a case, one has to convert the degrees into radians before calculating a trigonometric function of an angle given in degrees. To convert degrees into radians, multiply by 0.01745329252, which is equal to $\text{ATN}(1)/45$.

Least Geocentric Distance

The least geocentric distance between the occulted body and the center of the Moon occurs when $x^2 + y^2$ is a minimum. The instant of this least separation can be obtained, with sufficient accuracy, from

$$t = - \frac{X0\,X1 + Y0\,Y1 + 3t^2\,(X1\,X2 + Y1\,Y2)}{X1^2 + Y1^2 + 2\,(X0\,X2 + Y0\,Y2)} \tag{1}$$

where t is measured in hours from the instant T_0. This equation can be solved by iteration (that is, repetition), first putting $t = 0$ in the second member. Once t is known, the value of the least distance γ between the axis of the shadow and the center of the Earth, expressed in units of the Earth's equatorial radius, is found from

$$\gamma = \pm\sqrt{x^2 + y^2}$$

where x and y are the values for time t, to be calculated from

$$x = X0 + X1\,t + X2\,t^2 \qquad\qquad y = Y0 + Y1\,t + Y2\,t^2$$

and where γ has the same sign as y.

If no high accuracy is required, one may use the approximate formulae

$$t = - \frac{X0\,X1 + Y0\,Y1}{X1^2 + Y1^2} \qquad \text{and} \qquad \gamma = \frac{Y0\,X1 - X0\,Y1}{\sqrt{X1^2 + Y1^2}} \tag{2}$$

Example 1. — Occultation of Mercury, 1999 August 10.

From Table III we take the following values :

$T_0 = 3h$ $X0 = +0.216049$ $Y0 = +1.139263$
                  $X1 = +0.579536$ $Y1 = -0.078132$
                  $X2 = -0.000112$ $Y2 = -0.000105$

Putting these values in formula (1), we find

$$t = -\frac{0.0361953 - 0.0001701\,t^2}{0.3416789}$$

Putting $t = 0$ in the second member, we find $t = -0.1059337$. Putting now this latter value in the second member, we obtain $t = -0.1059281$ which is the correct value, as another iteration gives the value -0.1059281 again. Consequently, the time of least geocentric separation is

$$\begin{aligned} T_0 + t &= 3^h - 0^h.1059281 \\ &= 2^h.8940719 \\ &= 2^h53^m39^s \text{ TD} \\ &= 2^h52^m34^s \text{ UT} \quad \text{if we adopt the value } \Delta T = 65 \text{ seconds} \end{aligned}$$

For $t = -0.1059281$, we find $x = +0.154659$, $y = +1.147538$, whence $\gamma = +1.157913$. The approximate formulae (2) give $t = -0.1058445$ and $\gamma = +1.157915$.

Geocentric Conjunction in Right Ascension

At the time of the geocentric conjunction in right ascension between the occulted body and the center of the Moon, we have $x = 0$. The instant of the geocentric conjunction in right ascension can be found from

$$t = -\frac{X0 + X2\,t^2}{X1} \tag{3}$$

or, when no high accuracy is required, from $t = -X0/X1$. Again, equation (3) can be solved by iteration, by first putting $t = 0$ in the second member.

Example 2. — Occultation of Mercury, 1999 August 10. As in Example 1, we have

$T_0 = 3h$ $X0 = +0.216049$ $Y0 = +1.139263$
                  $X1 = +0.579536$ $Y1 = -0.078132$
                  $X2 = -0.000112$ $Y2 = -0.000105$

Putting these values into formula (3), we find

$$t = -\frac{0.216049 - 0.000112\,t^2}{0.579536}$$

Putting $t = 0$ in the second member, this formula gives $t = -0.372797$. Putting this latter value in the second member, we find $t = -0.372770$. A new iteration gives the value -0.372770 again. Hence, this is the correct value for t, and the time of geocentric conjunction in right ascension is

$$
\begin{aligned}
T_0 + t \; = \; & 3^{\text{h}} - 0^{\text{h}}.372770 \\
= \; & 2^{\text{h}}.627230 \\
= \; & 2^{\text{h}} 37^{\text{m}} 38^{\text{s}} \text{ TD} \\
= \; & 2^{\text{h}} 36^{\text{m}} 33^{\text{s}} \text{ UT}
\end{aligned}
$$

The value of y at the instant of conjunction in right ascension is found by putting the value $t = -0.372770$ in the formula $y = Y0 + Y1\,t + Y2\,t^2$. This gives $y_a = +1.168374$.

An approximate value of y_a is given by $\quad y_a \approx (Y0\,X1 - X0\,Y1)/X1$, which gives $y_a = +1.168390$ in this case.

Local Circumstances

Let λ be the observer's geographical longitude, in degrees, measured *westward* from Greenwich, and φ his geographical latitude, positive in the northern hemisphere, negative in the southern. Let z be the observer's height above sea-level, expressed in *meters*. (The factor for converting feet into meters is 0.3048).

If not already available, calculate the geocentric rectangular coordinates of the observer as follows :

$$
\kappa = 0.99664719 \qquad\qquad \tan w = \kappa \tan \varphi
$$

$$
\rho \sin \varphi' = \kappa \sin w + \frac{z}{6378\,140} \sin \varphi
$$

$$
\rho \cos \varphi' = \cos w + \frac{z}{6378\,140} \cos \varphi
$$

For an assumed value of t (take $t = 0$ as a first approximation), calculate

$$
d = D0 + D1\,t \qquad \text{(in the case of an occultation of a \emph{planet} only)}
$$

$$
H = H0 + H1\,t - \lambda - \Delta T/239.345
$$

where ΔT is the difference $\text{TD} - \text{UT}$ expressed in *seconds of time*. Remember that, in the case of the occultation of a star, we have $H1 = +15.04107$. H, the local hour angle of the occulted body, is then expressed in degrees. Further, calculate

$$
x = X0 + X1\,t + X2\,t^2 \qquad\qquad y = Y0 + Y1\,t + Y2\,t^2
$$

and the hourly variations

$$
x' = X1 + 2\,X2\,t \qquad\qquad y' = Y1 + 2\,Y2\,t
$$

Then calculate the observer's coordinates ξ, η and ζ, referred to the fundamental plane, by the following expressions. (In the case of an occultation of a star, ζ is not needed).

$$\xi = \rho \cos \varphi' \sin H$$
$$\eta = \rho \sin \varphi' \cos d - \rho \cos \varphi' \cos H \sin d$$
$$\zeta = \rho \sin \varphi' \sin d + \rho \cos \varphi' \cos H \cos d$$

The hourly variations of ξ and η are

$$\xi' = 0.01745329 \ H1 \ \rho \cos \varphi' \cos H$$
$$\eta' = 0.01745329 \ (H1 \ \xi \sin d - \zeta D1)$$

Next, calculate

$$u = x - \xi \qquad\qquad u' = x' - \xi'$$
$$v = y - \eta \qquad\qquad v' = y' - \eta'$$
$$n^2 = u'^2 + v'^2 \qquad (n > 0)$$

$$\tau = - \ \frac{uu' + vv'}{n^2}$$

Then, τ is the correction to t. That is, add τ algebraically to t in order to obtain a better value of t.

Repeat the whole calculation again and again, until t no longer varies, that is, until τ is negligible, say less than 0.000001. In practice, two or three iterations will suffice.

The time of least distance between the body and the Moon's center, as seen from the given place, is then $T_0 + t$, in hours (Dynamical Time), where t is the final value for t. Subtract ΔT in order to obtain the instant in Universal Time.

The position angle P of the star or planet at the time of nearest approach may be found from

$$\tan P = u/v, \tag{4}$$

$\cos P$ having the opposite sign of v, and where u and v are taken from the last iteration. The position angle P is measured in the usual way, namely from the North Point (*not* the north pole) of the Moon's disk, towards the East, South and West.

In the case of an occultation of a star take $L = k = 0.272495$. In the case of an occultation of a planet, find L from

$$L = k - \zeta F / 1000000$$

The value of the least distance Δ between the body and the center of the Moon's disk as seen from the given place, the Moon's radius being taken as unity, is then

$$\Delta = \frac{uv' - u'v}{nL} \tag{5}$$

where, again, u, v, u', etc. are taken from the last iteration.

If $|\Delta| > 1$, there is no occultation for the given place.

If there is an occultation for the given place, *approximate* values of t corresponding to the immersion and to the emersion of the star or planet are obtained by subtracting or adding, to the final value of t found above, the quantity

$$\frac{\sqrt{1 - \Delta^2}}{n} \, L \tag{6}$$

To calculate the exact times of immersion or emersion, start from the approximate value of t just found, for the immersion and the emersion *separately*. Then calculate, as above, the values of

$$d, \; H, \; x, \; y, \; x', \; y', \; \xi, \; \eta, \; \zeta, \; \xi', \; \eta', \; u, \; v, \; u', \; v', \; n, \; L, \; \text{and} \; \Delta.$$

(In the case of the occultation of a star, d is directly given in Table II, the quantity ζ is not needed, and $L = 0.272\,495$).

The correction τ to t is then

$$\tau = - \frac{uu' + vv'}{n^2} \mp \frac{\sqrt{1 - \Delta^2}}{n} \, L \tag{7}$$

where the upper sign is to be taken for immersion (disappearance), the lower for emersion (reappearance).

Repeat the calculation until t no longer varies. The time (TD) of immersion or emersion is then $T_0 + t$, in hours, where t is the final value for t. Again, subtract ΔT in order to obtain the instant in Universal Time.

The position angle P at ingress or egress can be found by means of formula (4).

The altitude h of the star or planet above the observer's horizon, at any instant t, is given by

$$\sin h = \sin d \sin \varphi + \cos d \cos \varphi \cos H, \tag{8}$$

where d and H are obtained as a function of t, as indicated above. Of course, for the altitude at the instant of immersion, least distance, or emersion, take t from the last iteration. Evidently, the event is not visible when h is negative, which indicates that the body is below the horizon. Moreover, the event is not visible if it occurs in daylight, except when the body is bright enough and is not too close to the Sun.

Example 3. — Occultation of Saturn, 1997 November 12.

Calculate the circumstances at the Uccle Observatory, Belgium, for which we have

$$\lambda = -\ 4°21'29'' = -\ 4°.3581$$
$$\varphi = +50°47'55'' = +50°.7986 \qquad \text{height} = 104 \text{ meters}$$

$$\rho \sin \varphi' = +0.771\,306$$
$$\rho \cos \varphi' = +0.633\,333$$

We take the elements of the occultation from Table III :

$T_0 = 1\,h$	$X0 = +0.065\,485$	$Y0 = +0.467\,324$
$D0 = +3°.23676$	$X1 = +0.582\,727$	$Y1 = +0.190\,910$
$D1 = -0°.00085$	$X2 = +0.000\,014$	$Y2 = -0.000\,015$
$H0 = 51°.74585$		
$H1 = 15°.04346$	$k = 0.272\,572$	$F = 1.36$

We will use the value $\Delta T = +63$ seconds.

Starting with the value $t = 0$, we find successively:

d	$+3°.23676$	ξ	$+0.524070$	u	-0.458585
H	$+55°.84073$	η	$+0.749997$	v	-0.282673
x	$+0.065485$	ζ	$+0.398596$	u'	$+0.489358$
y	$+0.467324$	ξ'	$+0.093369$	v'	$+0.183135$
x'	$+0.582727$	η'	$+0.007775$	n^2	0.2730096
y'	$+0.190910$			τ	$+1.011612$

Hence, the improved value of t is

$$t = 0 + 1.011612 = +1.011612$$

Starting with this new value of t, we obtain:

d	$+3°.23590$	ξ	$+0.599040$	u	$+0.055953$
H	$+71°.05888$	η	$+0.758472$	v	-0.098036
x	$+0.654993$	ζ	$+0.248788$	u'	$+0.528779$
y	$+0.660435$	ξ'	$+0.053976$	v'	$+0.181998$
x'	$+0.582755$	η'	$+0.008882$	n^2	0.3127308
y'	$+0.190880$			τ	-0.037555

Consequently, the new improved value of t is

$$t = +1.011612 - 0.037555 = +0.974057$$

A new calculation, performed with this new value of t, gives the correction $\tau = +0.000142$, so that the final value of t is

$$t = +0.974057 + 0.000142 = +0.974199$$

(A new iteration gives the negligibly small correction $\tau = -0.000001$). Using the value of ζ resulting from the last iteration, we find $L = 0.272572$. Then formula (5) gives $\Delta = 0.406936$, and expression (6) is equal to 0.446391.

Hence, approximate values of t for the immersion and the emersion of Saturn are:

immersion : $\quad t = +0.974199 - 0.446391 = +0.527808$
emersion : $\quad t = +0.974199 + 0.446391 = +1.420590$

Let us calculate the circumstances of the immersion. We thus start with the value $t = +0.527808$, and we find successively:

d	$+3°.23631$	ξ	$+0.568170$	v	-0.186196
H	$+63°.78079$	η	$+0.754279$	u'	$+0.509275$
x	$+0.373057$	ζ	$+0.322908$	v'	$+0.182468$
y	$+0.568084$	ξ'	$+0.073467$	n^2	0.2926558
x'	$+0.582742$	η'	$+0.008426$	n	0.540977
y'	$+0.190894$	u	-0.195113	Δ	0.401635

Formula (7) then gives $\quad \tau = +0.455623 - 0.461426 = -0.005803$, whence

$$t = +0.527808 - 0.005803 = +0.522005$$

Starting with this improved value of t, and proceeding in the same manner as above, we obtain the correction $\tau = -0.000001$. Consequently, the correct value of t is

$$t = +0.522005 - 0.000001 = +0.522004$$

and the time of immersion is

$$T_0 + t = 1^h + 0^h.522004 = 1^h.522004$$
$$= 1^h 31^m 19^s \text{ TD}$$
$$= 1^h 30^m 16^s \text{ UT}$$

The values from the last iteration are

$$d = +3°.23632 \qquad H = +63°.6935 \qquad u = -0.198067 \qquad v = -0.187255$$

Using formulae (4) and (8), we find that Saturn's position angle and altitude, at the instant of immersion, are

$$P = 46°.61 \qquad \text{and} \qquad h = 18°.87$$

Starting from $t = +1.420590$ we find, in a similar way, the circumstances at the emersion of the center of Saturn's disk:

$$2^h 23^m 43^s \text{ UT} \qquad P = 275°.28 \qquad h = 10°.64$$

Hence, at Uccle the immersion and the emersion of Saturn take place above the horizon, since the altitude h is positive in both cases.

Example 4. — Same occultation as in Example 3, but now for Sofia, Bulgaria, for which we have

$$\lambda = 23°20' \text{ East}, \qquad \varphi = 42°42' \text{ North}.$$

We find :

immersion :	$1^h 40^m 37^s$ UT	$P = 75°$	$h = +6°$
emersion :	$2^h 33^m 41^s$ UT	$P = 251°$	$h = -4°$

Hence, at Sofia the immersion occurs above the horizon, though at the low altitude of 6 degrees, while the emersion of Saturn takes place below the horizon.

Example 5. — Occultation of Regulus, 1999 March 1.

Calculate the circumstances at Palomar Mountain Observatory, for which we have

$$\lambda = +116°.8640 \qquad \rho \sin \varphi' = +0.546862$$
$$\varphi = +33°.3562 \qquad \rho \cos \varphi' = +0.836338$$
$$z = 1706 \text{ meters}$$

The elements of the occultation are (Table II)

T_0	$10h$	X0	$+0.22151$	Y0	$+0.19947$
d	$+11°.9694$	X1	$+0.55549$	Y1	-0.15258
H0	$156°.6836$	X2	-0.00000	Y2	-0.00001
H1	$15°.04107$			L	0.272495

We will adopt the value $\Delta T = +65$ seconds.

Proceeding as in Example 3, we find that the value of t corresponding to the least topocentric distance is $t = +0.455608$, and that the least separation is $\Delta = +1.1526$. Because $|\Delta|$ is larger than 1, there is no occultation of Regulus for the given place. The *positive* sign of Δ indicates that, as seen from the given place, Regulus passes to the *north* of the center of the Moon's disk.

The least distance of the star to the Moon's *limb* is $|\Delta| - 1$, that is 0.1526 in this case, the radius of the Moon's disk being taken as unity. As the apparent radius of the Moon is approximately a quarter of a degree, the least distance of the star from the Moon's limb, as seen from the given place, is approximately 2 minutes of arc.

The time of least distance is

$$
\begin{aligned}
T_0 + t \ &= \ 10^h.455\,608 \\
&= \ 10^h 27^m 20^s \ \text{TD} \\
&= \ 10^h 26^m 15^s \ \text{UT}
\end{aligned}
$$

The values taken from the last iteration are

$$
\begin{array}{lll}
u = -0.131\,07 & u' = +0.404\,08 & H = +46°.4009 \\
v = -0.285\,41 & v' = -0.185\,56 & n = 0.444\,655
\end{array}
$$

whence, by formula (4), $P = 24°.67$. Finally, formula (8) gives $h = +43°$, which indicates that the conjunction is visible from Palomar.

Grazing Occultations

If γ is between $+0.725$ and -0.725, the region of visibility of the occultation on the Earth's surface is bordered by a *northern limit* and a *southern limit*. An observer situated exactly on one of these lines will see the star grazing the northern or the southern limb of the Moon. These limits run over the Earth's surface approximately from West to East, over thousands of kilometers. At one side of such a line an occultation of the star or planet takes place, while at the other side the body is not occulted by the Moon.

If γ is between $+1.270$ and $+0.725$, only the southern limit does exist, the northern one being replaced by the curve 'star on the horizon'. Similarly, only the northern limit does exist if γ is between -0.725 and -1.270.

Points of the northern or southern limit of an occultation can be calculated as follows, using the Besselian elements given in the present publication.

For a given geographical longitude λ (positive westward from Greenwich!), start from approximate values for the latitude φ and for the time t, the latter being, as before, measured in hours from the reference time T_0. Good choices are $\varphi = 0°$ and $t = 0$. Then calculate

$$
\begin{aligned}
H &= HO + H1\,t - \lambda - \Delta T / 239.345 \\
d &= DO + D1\,t
\end{aligned}
$$

where, as before, ΔT is the difference $\text{TD} - \text{UT}$ expressed in seconds of time. In the case of an occultation of a star, d is already given in Table II, and $H1 = 15.041\,07$. Then calculate

$$
\begin{aligned}
x &= XO + X1\,t + X2\,t^2 \\
y &= YO + Y1\,t + Y2\,t^2 \\
x' &= X1 + 2X2\,t \\
y' &= Y1 + 2Y2\,t \\
\tan w &= 0.996\,647\,19 \tan \varphi \\
\rho \sin \varphi' &= 0.996\,647\,19 \sin w \\
\rho \cos \varphi' &= \cos w \\
\xi &= \rho \cos \varphi' \sin H \\
\eta &= \rho \sin \varphi' \cos d - \rho \cos \varphi' \cos H \sin d \\
\zeta &= \rho \sin \varphi' \sin d + \rho \cos \varphi' \cos H \cos d \qquad \text{(not needed for a star)}
\end{aligned}
$$

$$\xi' = 0.017\,453\,29\ HI\ \rho\cos\varphi'\cos H$$
$$\eta' = 0.017\,453\,29\ (HI\ \xi\sin d - \zeta DI) \qquad \text{(in the case of a star, } DI = 0)$$

$$u = x - \xi \qquad\qquad u' = x' - \xi'$$
$$v = y - \eta \qquad\qquad v' = y' - \eta'$$

$$n^2 = u'^2 + v'^2 \qquad (n > 0)$$

$$L = k - \zeta F/1\,000\,000$$
(in the case of a star, $L = 0.272\,495$)

$$\Delta = \frac{uv' - u'v}{nL}$$

$$A = v'\,\rho\sin\varphi'\,\sin H + u'\,\rho\sin\varphi'\,\cos H\sin d + 0.99328\ u'\,\rho\cos\varphi'\,\cos d$$

$$q = \frac{0.064\,311}{n}\,A$$

Then the correction to the assumed value of t is

$$\tau = -\,\frac{uu' + vv'}{n^2} \qquad \text{hours}$$

and the correction to the latitude φ is

$$\Delta\varphi = \frac{\Delta_0 - \Delta}{q} \qquad \text{degrees}$$

where $\Delta_0 = +1$ for a northern limit, and $\Delta_0 = -1$ for a southern limit.

Thus, the new values for the instant and for the latitude are $t + \tau$ and $\varphi + \Delta\varphi$, respectively.

The whole calculation is then repeated with these new values, until both corrections τ and $\Delta\varphi$ can be neglected, for instance until they are, in absolute value, less than 0.0001 hour and 0.0001 degree, respectively.

After the last iteration, the position angle P of the point of graze is given by

$$\tan P = -v'/u' \tag{9}$$

$\cos P$ having the same sign as Δ_0, while the altitude above the horizon can be found by means of formula (8). For the calculation of these quantities, use the values of u', v', φ, d and H resulting from the last iteration.

Similar calculations can then be performed for other longitudes λ, that is, for other points of the northern or southern limit of the area of visibility.

In the case of an occultation of a planet, the limit calculated in this manner refers to the *center* of this body.

It should further be pointed out that the calculated limits refer to sea level.

Example 6. — Southern limit of the occultation of Venus, 2009 April 22.

Let us calculate the point of this limit at longitude 100° West = +100°. The elements of the occultation are (Table III):

$$T_0 = 13\,h \qquad X0 = -0.658606 \qquad Y0 = +0.793889$$
$$D0 = +2°.72523 \qquad X1 = +0.511259 \qquad Y1 = +0.263955$$
$$D1 = -0°.00695 \qquad X2 = +0.000025 \qquad Y2 = -0.000054$$
$$H0 = 47°.30376$$
$$H1 = 15°.03406 \qquad k = 0.274398 \qquad F = 31.63$$

We will adopt the value $\Delta T = +75$ seconds. Starting with $t = 0$ and $\varphi = 0°$, we find:

H	$-53°.00960$	ξ	-0.798736	n^2	0.1998707
d	$+2°.72523$	η	-0.028608	n	0.447069
x	-0.658606	ζ	$+0.601001$	L	0.274379
y	$+0.793889$	ξ'	$+0.157877$	Δ	-2.056643
x'	$+0.511259$	η'	-0.009892	A	$+0.3506099$
y'	$+0.263955$	u	$+0.140130$	q	$+0.0504353$
w	$+0.000000$	v	$+0.822497$	Δ_0	-1
$\rho \sin \varphi'$	$+0.000000$	u'	$+0.353382$	τ	-1.374677
$\rho \cos \varphi'$	$+1.000000$	v'	$+0.273847$	$\Delta\varphi$	$+20°.9505$

Hence, improved values for the time and for the latitude are

$$t = 0 - 1.374677 = -1^h.374677$$
$$\varphi = 0 + 20.9505 = +20°.9505$$

Starting with these new values, still better ones are obtained, and so on. The values obtained after the successive iterations are as follows:

t	φ
0^h	$0°$
-1.374677	$+20.9505$
-0.707136	$+25.2528$
-0.584914	$+26.3885$
-0.554007	$+26.4426$
-0.553368	$+26.4447$
-0.553325	$+26.4448$
-0.553324	$+26.4448$

Hence, for $\lambda = +100°$ we have $t = -0^h.553324$, which corresponds to

$$T_0 + t = 12^h.446676 = 12^h 26^m 48^s \text{ TD} = 12^h 25^m 33^s \text{ UT},$$

and $\varphi = +26°.4448$.

The values taken from the last iteration are

$$u' = +0.39844 \qquad H = -61°.3283$$
$$v' = +0.27378 \qquad d = +2°.7291$$

and formulae (9) and (8) then give $P = 145°.51$ and $h = 26°.76$.

In a similar manner, other points of the southern limit can be calculated. For instance, the following values are found:

Longitude West	Latitude North	Universal Time	Altitude of Venus	Pos. Angle of contact
°	°	h m s	°	°
100	26.4448	12 25 33	26.76	145.51
96	28.7860	12 32 20	31.14	144.40
92	31.6349	12 40 46	35.25	143.39
88	35.0062	12 50 51	38.74	142.57
84	38.8322	13 02 17	41.17	142.07
80	42.9295	13 14 21	42.18	141.98

Calculation of the Besselian Elements

If α' and δ' are the apparent right ascension and declination of the star, α and δ the apparent geocentric right ascension and declination of the Moon, and π the Moon's equatorial horizontal parallax, then we have

$$x = \frac{\cos \delta \sin(\alpha - \alpha')}{\sin \pi}$$

$$y = \frac{\sin \delta \cos \delta' - \cos \delta \sin \delta' \cos(\alpha - \alpha')}{\sin \pi}$$

$$H = \theta - \alpha'$$

where θ is the apparent *ephemeris sidereal time*, that is the hour angle of the vernal equinox referred to the *ephemeris* meridian; its value at time T Dynamical Time is numerically equal to the value of the sidereal time at time T Universal Time for the Greenwich meridian.

If the occulted body is a planet, the Besselian elements are calculated as follows. Let

α, δ = geocentric apparent right ascension and declination of the Moon;
α', δ' = geocentric apparent right ascension and declination of the planet;
Δ = geocentric distance of the planet, in astronomical units;
r = geocentric distance of the Moon, in astronomical units, given by

$$r = \frac{0.000\,042\,635\,2327}{\sin \pi}$$

where π is the Moon's equatorial horizontal parallax. Then we have

$$G \cos d \cos a = \Delta \cos \delta' \cos \alpha' - r \cos \delta \cos \alpha$$
$$G \cos d \sin a = \Delta \cos \delta' \sin \alpha' - r \cos \delta \sin \alpha$$
$$G \sin d = \Delta \sin \delta' - r \sin \delta$$

where a and d designate the right ascension and declination of the point on the celestial sphere towards which the axis of the shadow is directed, and G is the distance between the centers of the planet and the Moon in astronomical units. The rectangular coordinates x, y, z of the Moon with respect to the fundamental plane, in units of the equatorial radius of the Earth, are obtained from

$$x = \cos \delta \sin (\alpha - a) / \sin \pi$$
$$y = (\sin \delta \cos d - \cos \delta \sin d \cos (\alpha - a)) / \sin \pi$$
$$z = (\sin \delta \sin d + \cos \delta \cos d \cos (\alpha - a)) / \sin \pi$$

Further, we have

$$H = \theta - a$$
$$\sin f = k_0 \sin \pi_0 / G$$

where, as above, θ is the apparent ephemeris sidereal time; k_0 is the Moon's radius expressed in units of the equatorial radius of the Earth, and π_0 is the mean parallax of the Sun. With $k_0 = 0.272\,495$ and $\pi_0 = 8''.794\,148$, the last formula becomes

$$\sin f = 0.000\,011\,617\,888 / G$$

where G is, as above, expressed in astronomical units. The angle f is the angle which the elements of the 'umbral' cone make with the axis of the shadow. The vertex of this cone is at the planet's center and, because f is a very small angle, never exceeding $10''$ in the case of an occultation of a planet by the Moon, we can take $\tan f$ equal to $\sin f$. The radius of the umbral cone in the fundamental plane is then, with sufficient accuracy,

$$k = k_0 + z \sin f$$

where, as above, z is in units of the Earth's equatorial radius.

In Table III, the quantity F is equal to $\sin f$ (at the instant T_0) multiplied by $1\,000\,000$, in order to avoid the tabulation of a very small quantity.

For the present work, the values of x and y have been calculated for three instants at intervals of two hours, namely for $T_0 - 2$ hours, T_0, and $T_0 + 2$ hours. If the values of x for these three instants are designated by A, B, C, respectively, the coefficients $X0$, $X1$ and $X2$ are calculated from

$$X0 = B \qquad\qquad X1 = \frac{C - A}{4} \qquad\qquad X2 = \frac{A + C - 2B}{8}$$

with similar expressions for $Y0$, $Y1$, $Y2$.

BASIC Program

The BASIC program given below computes local circumstances or points of the limits of the region of visibility of an occultation. The elements should be entered manually, and taken from either Table II (occultation of a star) or Table III (occultation of a planet). To have the results for the graze lines printed on the printer, change the instruction PRINT into LPRINT on the lines 528, 530, 532 (twice), 534, 544, and 562.

```
100 R = ATN(1) / 45: A$ = "    ###.##"
102 J1$ = "     Least distance   = &#.####"
104 J2$ = "     Immersion   ## & &        ###       ###"
106 J3$ = "     Emersion    ## & &        ###       ###"
108 J4$ = "     ####.###    ###.####   ## & &" + A$ + A$
110 J5$ = "     ####.###        No limit"
112 DEF FNFP (X) = X - FIX(X)
114 CLS : PRINT
116 PRINT TAB(26); "Occultation by the Moon"
118 PRINT TAB(26); "using Besselian elements"
120 PRINT : PRINT : PRINT : PRINT
122 PRINT TAB(25); "1 - Occultation of a Star"
124 PRINT TAB(25); "2 - Occultation of a Planet": PRINT
126 PRINT : PRINT " Your Choice = "; : INPUT K5
128 K5 = INT(K5): IF K5 < 1 OR K5 > 2 THEN GOSUB 1000: GOTO 126
130 PRINT
132 PRINT " TO (integer hours Dynamical Time) = "; : INPUT TO
134 IF FNFP(TO) <> 0 THEN GOSUB 1000: GOTO 130
136 ON K5 GOSUB 1100, 1200
138 PRINT
140 PRINT " Delta T  (TD-UT)  (seconds) = "; : INPUT Z9
142 PRINT : PRINT
144 PRINT , "1 - Local Circumstances"
146 PRINT , "2 - Graze"
148 PRINT , "3 - Other Occultation"
150 PRINT , "4 - Stop": PRINT
152 PRINT : PRINT " Your Choice = "; : INPUT J
154 J = INT(J)
156 IF J < 1 OR J > 4 THEN GOSUB 1000: GOTO 152
158 ON J GOTO 160, 500, 120, 9000
160 REM ----- Local Circumstances
162 PRINT : PRINT
164 PRINT " Longitude of place (degrees)"
166 PRINT , "(negative if East of Greenwich !)  = "; : INPUT LO
168 PRINT : PRINT " Latitude of place (degrees)"
170 PRINT , "(negative if South)  = "; : INPUT F
172 PRINT : PRINT " Height (meters) = "; : INPUT Y
174 F = R * F: U = ATN(0.99664719 * TAN(F))
176 Y = Y / 6378140: S7 = SIN(F): C7 = COS(F)
178 S8 = 0.99664719 * SIN(U) + Y * S7: C8 = COS(U) + Y * C7
180 T = 0
182 FOR J2 = 1 TO 4
184 GOSUB 1400
186 NEXT J2
188 CLS : PRINT
190 S$ = "+": IF D5 < 0 THEN S$ = "-"
192 IF ABS(D5) < 1 THEN 212
```

```
194 PRINT : PRINT "        NO OCCULTATION"
196 PRINT USING J1$; S$; ABS(D5)
198 GOSUB 1500
200 PRINT "       at "; H$; " UT"
202 IF J5 = 1 THEN PRINT "        on the previous day"
204 PRINT
206 PRINT USING "      Position Angle   =  ###°"; N9
208 PRINT USING "      Star's Altitude  =  ###°"; Q0
210 GOTO 142
212 PRINT USING J1$; S$; ABS(D5)
214 PRINT : C = L * SQR(1 - D5 * D5) / N
216 T1 = T - C: T2 = T + C
218 T = T1
220 FOR J2 = 1 TO 4
222 GOSUB 1400
224 T = T - L * SQR(1 - D5 * D5) / N
226 NEXT J2
228 GOSUB 1500
230 PRINT TAB(24); "UT"; TAB(36); "P.A."; TAB(46); "Alt."
232 PRINT TAB(21); "h  m   s"; TAB(38); "°"; TAB(48); "°"
234 PRINT USING J2$; H9; V$; W$; N9; Q0
236 IF J5 = 1 THEN PRINT TAB(21); "on the previous day": PRINT
238 T = T2
240 FOR J2 = 1 TO 4
242 GOSUB 1400
244 T = T + L * SQR(1 - D5 * D5) / N
246 NEXT J2
248 GOSUB 1500
250 PRINT USING J3$; H9; V$; W$; N9; Q0
252 IF J5 = 1 THEN PRINT TAB(21); "on the previous day"
254 GOTO 142
500 REM ----- Grazing Occultation
502 PRINT : PRINT : PRINT
504 PRINT : PRINT " Limit   (N or S)  = "; : INPUT L$
506 IF L$ = "N" OR L$ = "n" THEN Q9 = 1: GOTO 512
508 IF L$ = "S" OR L$ = "s" THEN Q9 = -1: GOTO 512
510 GOSUB 1000: GOTO 504
512 PRINT : PRINT
514 PRINT "      Longitudes EAST of Greenwich are NEGATIVE !"
516 PRINT : PRINT " First longitude (degrees) = "; : INPUT A1
518 PRINT : PRINT " Last longitude (degrees) = "; : INPUT A2
520 PRINT
522 PRINT " Step (degrees) (negative towards the East) = ";
524 INPUT S2
526 CLS : PRINT
528 PRINT TAB(8); "Long."; TAB(21); "Latit.";
530 PRINT TAB(35); "UT"; TAB(46); "Alt."; TAB(56); "P.A."
532 PRINT : PRINT TAB(10); "°"; TAB(23); "°";
534 PRINT TAB(33); "h  m   s"; TAB(47); "°"; TAB(57); "°"
536 FOR L0 = A1 TO A2 STEP S2
538 F = 0: T = 0: J3 = 0
540 Z = ATN(0.99664719 * TAN(R * F))
542 J3 = J3 + 1: IF J3 < 20 THEN 546
544 PRINT USING J5$; L0: GOTO 564
546 S8 = 0.99664719 * SIN(Z): C8 = COS(Z)
548 GOSUB 1400
550 Q = S8 * (V1 * SIN(H) + U1 * COS(H) * S9)
552 Q = Q + 0.99328 * U1 * C8 * C9
554 Q = 0.064311 * Q / N
556 F6 = (Q9 - D5) / Q: F = F + F6
558 IF ABS(F6) > 0.00001 THEN 540
```

```
560 S7 = SIN(R * F): C7 = COS(R * F): GOSUB 1500
562 PRINT USING J4$; L0; F; H9; V$; W$; Q0; N9
564 NEXT L0
566 GOTO 142
1000 PRINT : PRINT TAB(9); "ERROR !"
1002 RETURN
1100 PRINT : PRINT " d (degrees) = "; : INPUT D0
1102 PRINT : PRINT " H0 (degrees) = "; : INPUT H0
1104 D1 = 0: H1 = 15.04107: K = 0.272495: F5 = 0
1106 GOSUB 1300
1108 RETURN
1200 PRINT : PRINT " D0 (degrees) = "; : INPUT D0
1202 PRINT : PRINT " D1 (degrees) = "; : INPUT D1
1204 PRINT : PRINT " H0 (degrees) = "; : INPUT H0
1206 PRINT : PRINT " H1 (degrees) = "; : INPUT H1
1208 GOSUB 1300
1210 PRINT : PRINT " k = "; : INPUT K
1212 PRINT : PRINT " F = "; : INPUT F5
1214 RETURN
1300 PRINT : PRINT " X0 = "; : INPUT X0
1302 PRINT " X1 = "; : INPUT X1
1304 PRINT " X2 = "; : INPUT X2
1306 PRINT : PRINT " Y0 = "; : INPUT Y0
1308 PRINT " Y1 = "; : INPUT Y1
1310 PRINT " Y2 = "; : INPUT Y2
1312 RETURN
1400 H = R * (H0 + H1 * T - L0 - Z9 / 239.345)
1402 X = X0 + T * (X1 + T * X2): Y = Y0 + T * (Y1 + T * Y2)
1404 X4 = X1 + 2 * T * X2: Y4 = Y1 + 2 * T * Y2
1406 D = R * (D0 + T * D1): S9 = SIN(D): C9 = COS(D)
1408 K0 = C8 * SIN(H): E0 = S8 * C9 - C8 * S9 * COS(H)
1410 Z = S8 * S9 + C8 * C9 * COS(H)
1412 K1 = C8 * H1 * R * COS(H)
1414 E1 = R * (K0 * S9 * H1 - Z * D1)
1416 U = X - K0: V = Y - E0: U1 = X4 - K1: V1 = Y4 - E1
1418 N2 = U1 * U1 + V1 * V1: N = SQR(N2)
1420 L = K - Z * F5 / 1000000: D5 = (U * V1 - V * U1) / (N * L)
1422 T = T - (U * U1 + V * V1) / N2: RETURN
1500 B = S9 * S7 + C9 * C7 * COS(H)
1502 Q0 = ATN(B / SQR(1 - B * B)) / R
1504 T7 = T0 + T - Z9 / 3600: J5 = 0
1506 IF T7 < 0 THEN T7 = T7 + 24: J5 = 1
1508 H9 = INT(T7): T5 = 60 * FNFP(T7): M9 = INT(T5)
1510 G9 = INT(60 * FNFP(T5) + .5)
1512 IF G9 = 60 THEN G9 = 0: M9 = M9 + 1
1514 IF M9 = 60 THEN M9 = 0: H9 = H9 + 1
1516 U$ = STR$(H9): V$ = STR$(M9): W$ = STR$(G9)
1518 N$ = "0": P$ = ":"
1520 IF H9 < 10 THEN U$ = N$ + RIGHT$(U$, 1): GOTO 1524
1522 U$ = RIGHT$(U$, 2)
1524 IF M9 < 10 THEN V$ = N$ + RIGHT$(V$, 1): GOTO 1528
1526 V$ = RIGHT$(V$, 2)
1528 IF G9 < 10 THEN W$ = N$ + RIGHT$(W$, 1): GOTO 1532
1530 W$ = RIGHT$(W$, 2)
1532 H$ = U$ + P$ + V$ + P$ + W$
1534 N9 = ATN(U / V) / R + 180
1536 IF V < 0 THEN N9 = N9 + 180
1538 IF N9 > 360 THEN N9 = N9 - 360
1540 RETURN
9000 END
```

Table I

Date			T_0	Body	Magn.	Elong.	γ	Area of Visibility
			h					
1990 Jan	22		8	Antares	+1.2	53°W	−0.2994	South America, S Atlantic Ocean, S Africa, Madagascar, S Indian Ocean
1990 Feb	18		16	Antares	+1.2	80°W	−0.3056	Australia, New Zealand, Pacific, SW South America
1990 Mar	18		0	Antares	+1.2	107°W	−0.1611	Africa, Madagascar, S Indian Ocean, Australia
1990 Mar	22		18	Mars	+1.3	54°W	+0.4098	Pacific, North America
1990 Apr	14		7	Antares	+1.2	134°W	+0.0548	E Pacific Ocean, South America, Atlantic Ocean
1990 May	11		13	Antares	+1.2	160°W	+0.2091	New Guinea, Pacific Ocean
1990 Jun	7		19	Antares	+1.2	172°E	+0.2312	Africa, Madagascar, Indian Ocean, Indonesia
1990 Jul	5		2	Antares	+1.2	147°E	+0.1620	E Pacific, South America, Atlantic Ocean, W Africa
1990 Jul	21		18	Jupiter	−1.4	5°W	+0.9811	Greenland, Iceland, W and N Europe, N Asia, Arctic
1990 Aug	1		9	Antares	+1.2	121°E	+0.1135	Indonesia, New Guinea, Australia, Pacific Ocean
1990 Aug	18		13	Jupiter	−1.4	25°W	+0.3833	North America, Greenland, Atlantic, SW Europe, Africa
1990 Aug	18		24	Venus	−3.3	20°W	−0.4723	Indonesia, New Guinea, Australia, New Zealand, Pacific
1990 Aug	22		11	Mercury	+1.0	24°E	−0.1584	Atlantic Ocean, Africa, Madagascar, Indian Ocean
1990 Aug	28		17	Antares	+1.2	94°E	+0.1837	NE South America, Atlantic, Africa, Madagascar, Indian Ocean
1990 Sep	15		6	Jupiter	−1.5	47°W	−0.2651	Africa, Madagascar, Arabia, Indian Ocean
1990 Sep	25		1	Antares	+1.2	68°E	+0.3818	Pacific Ocean, Hawaii, Central America, NW South America
1990 Oct	12		19	Jupiter	−1.6	69°W	−0.9476	Australia, New Zealand

Table I   (cont.)

Date	T_0	Body	Magn.	Elong.	γ	Area of Visibility
	h					
1990 Oct 22	9	Antares	+1.2	41°E	+0.6133	Asia, Indonesia, Japan.
1990 Oct 25	18	Saturn	+0.8	78°E	−1.1694	S Indian Ocean, Antarctica
1990 Nov 18	15	Antares	+1.2	14°E	+0.7488	E North America, Atlantic, W Europe, NW Africa
1990 Nov 22	4	Saturn	+0.8	52°E	−0.6909	S Indian Ocean, S Pacific Ocean, Antarctica
1990 Dec 15	21	Antares	+1.2	15°W	+0.7402	Pacific Ocean, Hawaii, W North America
1990 Dec 18	6	Venus	−3.4	11°E	−1.0721	S Pacific Ocean, Antarctica
1990 Dec 19	16	Saturn	+0.8	27°E	−0.2508	SE Pacific, South America, Atlantic Ocean, Africa
1991 Jan 12	4	Antares	+1.2	42°W	+0.6754	Asia
1991 Jan 16	4	Saturn	+0.7	2°E	+0.1178	Indian Ocean, Indonesia, Australia, New Guinea, Pacific
1991 Feb 8	11	Antares	+1.2	70°W	+0.7002	E North America, N South America, Atlantic Ocean, Iberia, NW Africa
1991 Feb 11	5	Uranus	+6.1	40°W	−1.2265	Part of Antarctica
1991 Feb 12	18	Saturn	+0.8	23°W	+0.4807	Pacific Ocean, Hawaii, North and Central America
1991 Mar 7	19	Antares	+1.2	97°W	+0.8767	E Asia, Japan, NW Pacific Ocean
1991 Mar 10	15	Uranus	+6.1	66°W	−0.9552	Antarctica, S Atlantic Ocean
1991 Mar 12	7	Saturn	+0.9	48°W	+0.9126	Europe, N Africa, Turkey, W Siberia
1991 Mar 22	17	Mars	+1.1	83°E	+0.6602	NE North America, Greenland, Iceland, N Atlantic, Europe except S, Asia, Arctic
1991 Apr 4	3	Antares	+1.2	124°W	+1.1290	Europe, extreme N Africa, SW Asia
1991 Apr 7	0	Uranus	+6.0	93°W	−0.6172	S Indian Ocean, Australia, New Zealand, Antarctica

T a b l e   I   (cont.)

Date	T_0	Body	Magn.	Elong.	γ	Area of Visibility
	h					
1991 Apr 19	24	Mars	+1.5	70°E	−0.6017	Pacific Ocean, South America
1991 May 4	8	Uranus	+6.0	120°W	−0.3342	SE Pacific, S South America, Atlantic Ocean, SW Africa
1991 May 31	13	Uranus	+5.9	147°W	−0.2178	Australia, New Zealand, Pacific Ocean
1991 May 31	20	Neptune	+7.7	144°W	−1.2075	S Pacific, part of Antarctica
1991 Jun 27	17	Uranus	+5.9	173°W	−0.2739	SE Africa, Madagascar, Indian Ocean, Australia, New Guinea
1991 Jun 28	1	Neptune	+7.7	170°W	−1.2322	Part of Antarctica
1991 Jul 24	21	Uranus	+5.9	160°E	−0.3896	South America, S Atlantic Ocean, S Africa, Madagascar, Indian Ocean
1991 Aug 11	9	Mercury	+1.7	18°E	+0.5632	Europe, Svalbard, Asia, Borneo, W Pacific Ocean
1991 Aug 21	2	Uranus	+6.0	133°E	−0.4100	S Pacific Ocean, South America, Atlantic Ocean
1991 Sep 17	9	Uranus	+6.0	106°E	−0.2467	Indian Ocean, Australia, New Zealand, Pacific Ocean
1991 Sep 17	18	Neptune	+7.7	110°E	−1.1165	SE Indian Ocean, Antarctica
1991 Oct 4	15	Venus	−4.3	42°W	−0.2126	Pacific, Hawaii, South America
1991 Oct 14	18	Uranus	+6.1	79°E	+0.0647	South America, Atlantic Ocean, Africa, Arabia
1991 Oct 15	2	Neptune	+7.8	83°E	−0.8009	S Pacific, South America, Antarctica
1991 Nov 8	4	Mercury	−0.2	20°E	−0.7091	S Indian Ocean, New Zealand, S Pacific Ocean, Antarctica
1991 Nov 11	3	Uranus	+6.1	53°E	+0.4001	New Guinea, Pacific, Hawaii
1991 Nov 11	11	Neptune	+7.8	56°E	−0.4768	S Africa, S Indian Ocean, Australia, New Guinea, part of Antarctica
1991 Dec 8	13	Uranus	+6.1	26°E	+0.6523	Atlantic Ocean, Europe, Africa, SW Asia

ASTRONOMICAL TABLES

T a b l e   I   (cont.)

Date			T_0	Body	Magn.	Elong.	γ	Area of Visibility
			h					
1991 Dec	8		20	Neptune	+7.8	29°E	−0.2571	Pacific Ocean, South America, Atlantic Ocean
1992 Jan	3		10	Mars	+1.6	17°W	−0.9070	S South America, S Indian Ocean, Antarctica
1992 Jan	4		23	Uranus	+6.2	0°E	+0.8204	Japan, Pacific, Hawaii, Kamchatka, NE North America
1992 Jan	5		4	Neptune	+7.8	2°E	−0.1433	Indian Ocean, Australia, New Guinea, Pacific Ocean
1992 Jan	31		17	Venus	−3.4	33°W	−1.1267	Antarctica, S Atlantic Ocean
1992 Feb	1		8	Uranus	+6.1	26°W	+1.0011	Europe, N Africa, W Asia
1992 Feb	1		12	Neptune	+7.8	25°W	−0.0293	South America, Atlantic Ocean, Africa
1992 Feb	28		21	Neptune	+7.8	51°W	+0.1871	Australia, New Guinea, Pacific Ocean, Hawaii
1992 Mar	27		5	Neptune	+7.8	78°W	+0.4961	Atlantic, Africa, SW Asia
1992 Apr	23		13	Neptune	+7.7	105°W	+0.7827	Pacific Ocean, Hawaii, North America
1992 May	20		21	Neptune	+7.7	131°W	+0.9318	Asia
1992 Jun	1		5	Mercury	−1.9	1°E	+1.0237	Greenland, Iceland, Europe except S, N Asia, Alaska, Arctic regions
1992 Jun	17		3	Neptune	+7.7	158°W	+0.9222	E North America, Atlantic Ocean, W Europe
1992 Jul	14		9	Neptune	+7.7	175°E	+0.8432	Pacific, Hawaii, North America
1992 Aug	10		13	Neptune	+7.7	149°E	+0.8317	Asia, Japan, NW Pacific Ocean
1992 Sep	6		18	Neptune	+7.7	122°E	+0.9758	Europe, N Africa, W Asia
1992 Sep	20		9	Mars	+0.6	83°W	−0.8784	South America, S Atlantic
1992 Oct	4		1	Neptune	+7.7	95°E	+1.2494	Canada
1992 Oct	27		15	Mercury	−0.0	23°E	+0.5474	E North America, Cuba, Haiti, Atlantic Ocean, Africa, S Italy, Greece

T a b l e I (cont.)

Date	T_0	Body	Magn.	Elong.	γ	Area of Visibility
	h					
1992 Oct 28	15	Venus	−3.4	35°E	+0.4048	N South America, Atlantic Ocean, Africa, Arabia
1993 Feb 25	5	Venus	−4.3	40°E	−0.5246	SE Indian Ocean, Australia, Pacific Ocean
1993 Apr 19	16	Venus	−4.0	26°W	+0.4969	E Pacific Ocean, Hawaii, North America, N Atlantic Ocean, Greenland, Iceland
1993 May 22	5	Mercury	−1.5	7°E	−0.9186	S Indian Ocean, Australia
1993 Oct 15	14	Spica	+1.2	3°E	−1.2321	Part of Antarctica
1993 Nov 14	18	Mars	+1.6	12°E	+0.5104	E Pacific Ocean, the Americas, Atlantic Ocean
1993 Dec 12	11	Mercury	−0.5	12°W	+0.1507	N South America, Atlantic, Africa, Madagascar, Arabia
1993 Dec 12	18	Venus	−3.4	8°W	+0.4492	Pacific Ocean, Hawaii, the Americas, W Atlantic Ocean
1994 Jan 5	16	Spica	+1.2	81°W	−1.1805	Antarctica
1994 Feb 1	21	Spica	+1.2	109°W	−0.9421	S Indian Ocean, Antarctica
1994 Mar 1	5	Spica	+1.2	137°W	−0.7026	SE Pacific, South America, S Atlantic, Antarctica
1994 Mar 28	14	Spica	+1.2	164°W	−0.5791	E Indian Ocean, Java, Australia, New Zealand, S Pacific Ocean, part of Antarctica
1994 Apr 12	23	Venus	−3.3	21°E	+1.1257	NE Asia, Greenland, Arctic
1994 Apr 25	2	Spica	+1.2	169°E	−0.5740	SE Pacific, South America, S Atlantic, part of Antarctica
1994 May 22	12	Spica	+1.2	143°E	−0.5931	Indian Ocean, Australia, New Zealand, SW Pacific Ocean, part of Antarctica
1994 Jun 18	20	Spica	+1.2	116°E	−0.5309	South America, S Atlantic, S Indian Ocean, part Antarctica
1994 Jul 5	5	Mars	+1.4	42°W	−0.3730	Africa, Madagascar, Indian Ocean, extreme S Asia, Indonesia, W Australia

Table I (cont.)

Date			T_0	Body	Magn.	Elong.	γ	Area of Visibility
			h					
1994 Jul	16		2	Spica	+1.2	90°E	−0.3507	Pacific Ocean, South America
1994 Aug	12		7	Spica	+1.2	64°E	−0.1118	Indian Ocean, S Asia, Indonesia, Australia, New Zealand, SW Pacific
1994 Sep	8		14	Spica	+1.2	38°E	+0.0780	N South America, Atlantic Ocean, Africa, Madagascar, Indian Ocean
1994 Oct	5		23	Spica	+1.2	11°E	+0.1521	Japan, Pacific Ocean
1994 Oct	7		12	Jupiter	−1.3	32°E	−0.7149	Extreme E South America, S Atlantic Ocean, S Indian Ocean, Antarctica
1994 Nov	2		10	Spica	+1.2	16°W	+0.1442	Atlantic Ocean, Africa, Madagascar, Indian Ocean
1994 Nov	4		8	Jupiter	−1.3	11°E	−0.1057	Africa, Madagascar, Indian Ocean, Australia, New Guinea
1994 Nov	29		21	Spica	+1.2	44°W	+0.1670	Japan, E China, Pacific Ocean
1994 Dec	2		5	Jupiter	−1.3	11°W	+0.4644	Asia, Indonesia, New Guinea, W Pacific Ocean
1994 Dec	27		6	Spica	+1.2	71°W	+0.3191	Atlantic Ocean, Iberia, Africa, Arabia, Madagascar, Indian Ocean, India, Ceylon
1994 Dec	30		1	Jupiter	−1.3	34°W	+1.0502	NE Asia, Japan, NW Pacific
1995 Jan	23		12	Spica	+1.2	99°W	+0.5839	NE Pacific, North America, Central America, N South America, Atlantic Ocean
1995 Jan	27		12	Venus	−3.9	46°W	+0.1759	E Pacific Ocean, Central America, South America, Atlantic Ocean, W Africa
1995 Feb	19		17	Spica	+1.2	127°W	+0.8330	NE Asia, Japan, Pacific, Hawaii, Alaska
1995 Mar	19		0	Spica	+1.2	154°W	+0.9562	Iceland, Europe exc. SW, western half of Asia
1995 Apr	15		10	Spica	+1.2	178°E	+0.9652	NE Siberia, NE Pacific Ocean, North America

T a b l e I (cont.)

Date	T_0	Body	Magn.	Elong.	γ	Area of Visibility
	h					
1995 May 12	21	Spica	+1.2	152°E	+0.9672	Iceland, Europe except SW, western half of Asia
1995 May 27	7	Venus	−3.3	23°W	+0.9172	Greenland, Iceland, Europe except SE, extreme NW of Africa, N Asia, Arctic
1995 Jun 9	6	Spica	+1.2	126°E	+1.0673	Extreme NE Siberia, North America except E
1995 Jun 26	2	Mercury	+1.0	22°W	+0.6668	NE Europe, Asia, Japan, N Pacific, Alaska, Arctic regions
1995 Aug 30	3	Mars	+1.6	48°E	−0.1545	Indonesia, New Guinea, NE Australia, N New Zealand, Pacific Ocean
1996 Feb 22	5	Venus	−3.7	42°E	−0.0575	Indonesia, Australia, New Guinea, Pacific, Hawaii
1996 Apr 17	5	Mars	+1.4	9°W	+0.5960	SE Europe, NE Africa, Asia, Japan
1996 Jun 14	0	Mercury	+0.5	23°W	−0.4655	Australia, New Guinea, Pacific Ocean
1996 Jul 12	9	Venus	−4.2	37°W	+0.4106	Atlantic Ocean, Europe, N Africa, Asia
1996 Aug 8	7	Aldebaran	+1.1	66°W	+1.1261	NE North America, Greenland, Arctic, N Asia
1996 Aug 16	18	Mercury	+0.4	27°E	−0.3421	Pacific Ocean, South America, S Atlantic Ocean
1996 Sep 4	14	Aldebaran	+1.1	93°W	+0.9314	NE Asia, Arctic, Alaska, Canada, Greenland, Iceland
1996 Oct 1	22	Aldebaran	+1.1	119°W	+0.8590	Europe, N Africa, northern half of Asia, Greenland, Iceland, Arctic regions
1996 Oct 29	8	Aldebaran	+1.1	146°W	+0.9139	NE Siberia, N Pacific Ocean, N North America, Greenland, Iceland, W Europe except Iberia, N Europe, Arctic
1996 Nov 25	16	Aldebaran	+1.1	172°W	+0.9928	N and E Europe, N Asia, Alaska, Arctic regions

Table I (cont.)

Date	T_0	Body	Magn.	Elong.	γ	Area of Visibility
	h					
1996 Dec 22	24	Aldebaran	+1.1	158°E	+0.9666	North America, Greenland, Iceland, Scandinavia, NW Asia, Arctic regions
1997 Jan 19	5	Aldebaran	+1.1	130°E	+0.7991	NE Asia, N Pacific Ocean, N North America, Greenland, Iceland, Arctic regions
1997 Feb 15	11	Aldebaran	+1.1	103°E	+0.5909	E Europe, Asia, N Pacific Ocean, Alaska
1997 Mar 14	19	Aldebaran	+1.1	76°E	+0.4828	SE North America, Central America, Atlantic Ocean, S Greenland, Iceland, Europe, N Africa, W Asia
1997 Apr 7	1	Saturn	+1.0	6°W	+1.0429	NE Asia, N Alaska, Arctic regions
1997 Apr 7	13	Venus	−3.5	2°E	−0.6483	S South America, S Atlantic Ocean, S Africa, Madagascar, part of Antarctica
1997 Apr 11	4	Aldebaran	+1.1	49°E	+0.5179	SE and E Asia, N Pacific Ocean, North America
1997 May 4	15	Saturn	+1.0	30°W	+0.7580	NE Pacific, North America, N Atlantic, Greenland, Iceland, British Isles, Arctic
1997 May 5	17	Mercury	+2.0	15°W	−1.1534	Extreme S South America, S Atlantic Ocean, part of Antarctica
1997 May 8	14	Aldebaran	+1.1	22°E	+0.6171	E North America, Atlantic, Greenland, Iceland, Europe, Svalbard, western half of Asia
1997 Jun 1	3	Saturn	+1.0	53°W	+0.4510	Africa, Asia, Japan
1997 Jun 4	22	Aldebaran	+1.1	7°W	+0.6624	E Asia, N Pacific Ocean, North America, Greenland
1997 Jun 13	16	Mars	+0.5	95°E	−0.2907	Atlantic Ocean, Africa, Madagascar, Indian Ocean
1997 Jun 28	12	Saturn	+0.9	77°W	+0.1605	Pacific, Central America, Florida, NW South America, Atlantic, Iberia, NW Africa

Table I (cont.) ·

Date		T_0	Body	Magn.	Elong.	γ	Area of Visibility
		h					
1997 Jul	2	5	Aldebaran	+1.1	31°W	+0.5894	Europe, N Africa, Asia except S, Japan
1997 Jul	25	19	Saturn	+0.8	103°W	−0.0185	Indian Ocean, Indonesia, Australia, New Guinea, Pacific Ocean, Hawaii
1997 Jul	29	11	Aldebaran	+1.1	57°W	+0.4302	North and Central America, Atlantic, Europe except N, N Africa, Turkey
1997 Aug	5	20	Mercury	+0.7	27°E	+1.0596	N Canada, N Atlantic Ocean, Greenland, Iceland, British Isles, Arctic regions
1997 Aug	22	2	Saturn	+0.6	129°W	−0.0072	E South America, Atlantic Ocean, Africa, S Asia
1997 Aug	25	17	Aldebaran	+1.1	83°W	+0.2887	Pacific Ocean, Hawaii, North and Central America
1997 Sep	18	10	Saturn	+0.5	157°W	+0.1740	Pacific Ocean, Hawaii, North and Central America
1997 Sep	21	24	Aldebaran	+1.1	109°W	+0.2609	Africa, Asia, Japan
1997 Oct	15	18	Saturn	+0.4	173°E	+0.3811	Africa, Indian Ocean, Asia, Japan
1997 Oct	19	9	Aldebaran	+1.1	136°W	+0.3521	Pacific Ocean, Hawaii, North America, Atlantic Ocean
1997 Nov	12	1	Saturn	+0.6	145°E	+0.4237	The Americas, Atlantic Ocean, Europe, N Africa
1997 Nov	15	20	Aldebaran	+1.1	163°W	+0.4672	Africa, S and SE Europe, Asia, Japan, W Pacific Ocean
1997 Dec	9	6	Saturn	+0.8	116°E	+0.2046	Australia, New Guinea, Pacific Ocean, Hawaii, North America
1997 Dec	13	5	Aldebaran	+1.1	167°E	+0.4886	Pacific Ocean, Hawaii, North America, S Greenland, Atlantic Ocean, extreme W Europe
1998 Jan	5	12	Saturn	+0.9	89°E	−0.1998	S Africa, Madagascar, Indian Ocean, SE Asia, Indonesia, New Guinea

Table I (cont.)

Date			T_0	Body	Magn.	Elong.	γ	Area of Visibility
			h					
1998 Jan	9		13	Aldebaran	+1.1	140°E	+0.3809	Indian Ocean, Asia, Japan, Pacific Ocean
1998 Feb	1		21	Saturn	+0.9	63°E	−0.6150	SE Pacific, South America, Atlantic Ocean
1998 Feb	5		18	Aldebaran	+1.1	113°E	+0.2325	Atlantic Ocean, N Africa, S Europe, SW Asia
1998 Feb	27		23	Mars	+1.4	17°E	−0.6954	New Zealand, Pacific Ocean, South America, Antarctica
1998 Mar	1		10	Saturn	+0.9	38°E	−0.9301	S Indian Ocean, Australia, Antarctica
1998 Mar	5		0	Aldebaran	+1.1	86°E	+0.1744	Pacific Ocean, Hawaii, S North America, Central America, Atlantic Ocean
1998 Mar	24		19	Venus	−4.1	46°W	+0.0915	NE Australia, Pacific, S North America, Central America
1998 Mar	26		11	Jupiter	−1.6	24°W	+0.7756	NE North America, Atlantic, Greenland, Iceland, Europe, NW Africa
1998 Mar	29		1	Saturn	+0.8	13°E	−1.1606	S Pacific, part of Antarctica
1998 Apr	1		8	Aldebaran	+1.1	59°E	+0.2509	Indian Ocean, S and SE Asia, Japan, Pacific Ocean
1998 Apr	23		7	Jupiter	−1.7	45°W	+0.2051	Atlantic Ocean, Africa, SW Asia
1998 Apr	23		8	Venus	−3.8	45°W	−0.0754	E South America, Atlantic, Africa, N Madagascar, Indian Ocean, Arabia, India
1998 Apr	24		19	Mercury	+1.2	24°W	−0.8115	S Pacific Ocean, South America, part of Antarctica
1998 Apr	28		18	Aldebaran	+1.1	32°E	+0.3863	North and Central America, Atlantic Ocean, Europe except N, N Africa
1998 May	20		23	Jupiter	−1.8	67°W	−0.3416	Indian Ocean, Australia, New Guinea, Pacific Ocean
1998 May	26		4	Aldebaran	+1.1	7°E	+0.4683	Asia, Japan, N Pacific Ocean, extreme SW Alaska

Table I (cont.)

Date			T_0	Body	Magn.	Elong.	γ	Area of Visibility
			h					
1998 Jun	1		3	Regulus	+1.3	79°E	−1.0417	S Pacific Ocean, New Zealand
1998 Jun	17		11	Jupiter	−2.0	89°W	−0.7695	S Pacific, South America, Atlantic Ocean
1998 Jun	22		14	Aldebaran	+1.1	22°W	+0.4401	North and Central America, Atlantic Ocean, S Greenland, S Iceland, Europe, N Africa, Turkey
1998 Jun	28		12	Regulus	+1.3	53°E	−0.7956	S Africa, Madagascar, S Indian Ocean, part Antarctica
1998 Jul	14		19	Jupiter	−2.2	114°W	−0.9527	New Zealand, S Pacific Oc., Antarctica
1998 Jul	19		21	Aldebaran	+1.1	47°W	+0.3345	SE Asia, Japan, Pacific Ocean, SW North America
1998 Jul	25		20	Regulus	+1.3	27°E	−0.6740	Pacific Ocean, S South America
1998 Aug	11		0	Jupiter	−2.4	141°W	−0.8262	S Atlantic Ocean, Indian Oc., Antarctica
1998 Aug	16		3	Aldebaran	+1.1	73°W	+0.2470	Africa, Asia, Japan
1998 Aug	22		4	Regulus	+1.3	1°E	−0.6636	Indian Ocean, Australia, New Zealand
1998 Sep	7		4	Jupiter	−2.5	170°W	−0.4880	SE Pacific, South America, Atlantic Ocean, W Africa
1998 Sep	12		8	Aldebaran	+1.1	100°W	+0.2645	S and SE North America, Central America, NW South America, Atlantic, W and S Europe, Africa, Turkey
1998 Sep	18		11	Regulus	+1.3	26°W	−0.6642	South America, Atlantic Oc., S Africa
1998 Sep	19		18	Venus	−3.4	11°W	−0.0798	Pacific, Hawaii, South America
1998 Sep	20		7	Mercury	−1.3	5°W	+0.1380	Italy, Greece, NE Africa, S Asia, Indian Ocean, Indonesia, Australia
1998 Oct	4		10	Jupiter	−2.4	160°E	−0.1994	Australia, New Zealand, Pacific Ocean, Mexico

Table I   (cont.)

Date		T_0	Body	Magn.	Elong.	γ	Area of Visibility
		h					
1998 Oct	9	16	Aldebaran	+1.1	126°W	+0.3914	SE Asia, Borneo, Japan, Pacific, W North America
1998 Oct	15	16	Regulus	+1.3	52°W	−0.5580	Pacific Ocean, S South America
1998 Oct	16	3	Mars	+1.8	47°W	−1.0387	S Indian Ocean, Antarctica
1998 Oct	31	16	Jupiter	−2.3	130°E	−0.2141	SE Atlantic Ocean, S Africa, Madagascar, Indian Ocean, SE Asia, Indonesia
1998 Nov	6	2	Aldebaran	+1.1	153°W	+0.5389	E North America, NW South America, Atlantic Ocean, S Greenland, Iceland, Europe, extreme NW Africa, Asia
1998 Nov	11	22	Regulus	+1.3	79°W	−0.3095	Indian Ocean, S Asia, Indonesia, New Guinea, Australia, New Zealand
1998 Nov	13	18	Mars	+1.7	59°W	+0.5510	NE Asia, Japan, Pacific Oc., North and Central America
1998 Nov	28	1	Jupiter	−2.1	103°E	−0.5874	S Pacific Ocean, South America, Atlantic Ocean
1998 Dec	3	13	Aldebaran	+1.1	174°E	+0.5998	Asia, Japan, N Pacific Ocean, North America
1998 Dec	9	6	Regulus	+1.3	107°W	−0.0104	Florida, N South America, Cuba, Atlantic, Africa
1998 Dec	25	11	Jupiter	−1.9	78°E	−1.1690	S New Zealand, Antarctica
1998 Dec	30	23	Aldebaran	+1.1	150°E	+0.5448	North and Central America, Atlantic Ocean, Greenland, Iceland, Europe, extreme N Africa, W Asia
1999 Jan	5	15	Regulus	+1.3	135°W	+0.1974	E Asia, Japan, Pacific, Hawaii
1999 Jan	27	7	Aldebaran	+1.1	123°E	+0.4543	Japan, Pacific, North America
1999 Feb	2	1	Regulus	+1.3	163°W	+0.2581	E North America, Atlantic, SW Europe, Africa, Arabia, Indian Ocean
1999 Feb	17	2	Mercury	−1.1	10°E	−0.2398	Australia, New Zealand, Pacific Ocean

T a b l e   I   (cont.)

Date	T_0	Body	Magn.	Elong.	γ	Area of Visibility
	h					
1999 Feb 23	12	Aldebaran	+1.1	95°E	+0.4473	SE Europe, Africa, Asia, Japan, NW Pacific Ocean
1999 Mar 1	10	Regulus	+1.3	169°E	+0.2510	Japan, Pacific Ocean, Hawaii, Mexico, Central America
1999 Mar 22	18	Aldebaran	+1.1	68°E	+0.5636	North and Central America, Atlantic Ocean, Greenland, Iceland, Europe, extreme N Africa, W Asia
1999 Mar 28	16	Regulus	+1.3	142°E	+0.3092	Europe, N Africa, Asia, Borneo, New Guinea
1999 Apr 10	9	Neptune	+7.8	76°W	+1.1949	Labrador, N Atlantic Ocean, Greenland, Iceland
1999 Apr 11	7	Uranus	+6.2	65°W	+1.0663	NE Atlantic Ocean, Europe, extreme NW Africa
1999 Apr 14	5	Mercury	+0.7	27°W	−1.0279	SE Australia, Tasmania, Antarctica
1999 Apr 19	2	Aldebaran	+1.1	41°E	+0.7296	Asia, Japan, N Pacific Ocean, North America, Arctic regions
1999 Apr 24	22	Regulus	+1.3	116°E	+0.4972	North America, Greenland, Iceland, Atlantic Ocean, Europe except N and NE, Africa, Turkey, Arabia
1999 May 7	16	Neptune	+7.7	102°W	+0.9154	Japan, extreme E Siberia, N Pacific, NW North America
1999 May 8	16	Uranus	+6.1	91°W	+0.7738	Pacific Ocean, Hawaii, North America
1999 May 16	12	Aldebaran	+1.1	16°E	+0.8391	NE North America, Atlantic, Greenland, Iceland, British Isles, N Europe, N Asia, Arctic regions
1999 May 22	4	Regulus	+1.3	89°E	+0.7596	NE Asia, North America, Greenland, Arctic regions
1999 Jun 3	22	Neptune	+7.7	129°W	+0.6979	NE Africa, Asia, Japan
1999 Jun 4	22	Uranus	+6.1	117°W	+0.5414	E Africa, Indian Ocean, Asia, Japan

T a b l e   I   (cont.)

Date			T_0	Body	Magn.	Elong.	γ	Area of Visibility
			h					
1999 Jun	12		23	Aldebaran	+1.1	13°W	+0.8448	Asia, Japan, N North America, Arctic regions
1999 Jun	18		12	Regulus	+1.3	63°E	+0.9814	Greenland, Iceland, Scandinavia, N and NE Asia, Japan, Arctic regions
1999 Jul	1		2	Neptune	+7.7	155°W	+0.6280	Newfoundland, Atlantic Oc., Europe, Africa, Turkey, Arabia
1999 Jul	2		3	Uranus	+6.0	144°W	+0.4624	N South America, Atlantic, Africa, SE Europe, Turkey, Arabia
1999 Jul	10		9	Aldebaran	+1.1	38°W	+0.7854	NE North America, Atlantic, Greenland, Iceland, Europe except S, Asia, Arctic regions
1999 Jul	15		21	Regulus	+1.3	37°E	+1.0882	Arctic regions, Greenland, Iceland, N Atlantic Ocean, British Isles, W Scandinavia
1999 Jul	28		8	Neptune	+7.7	178°E	+0.6866	Pacific, Hawaii, North America
1999 Jul	29		7	Uranus	+6.0	171°W	+0.5329	Pacific, North and Central America, W Atlantic Ocean
1999 Aug	6		16	Aldebaran	+1.1	64°W	+0.7515	NE Asia, N Pacific, N North America, Greenland, Iceland, Arctic regions
1999 Aug	10		3	Mercury	+0.9	18°W	+1.1579	NW North America, Arctic
1999 Aug	12		7	Regulus	+1.3	11°E	+1.0979	Arctic regions, NE Siberia
1999 Aug	24		14	Neptune	+7.7	151°E	+0.7593	Asia, Japan, W Pacific Ocean
1999 Aug	25		13	Uranus	+6.0	162°E	+0.6406	E Asia, Japan, N Pacific
1999 Sep	2		22	Aldebaran	+1.1	90°W	+0.8183	N and E Europe, Asia, Alaska, Arctic regions
1999 Sep	8		15	Regulus	+1.3	16°W	+1.1028	Arctic regions, Greenland, Iceland, Europe except SW
1999 Sep	20		22	Neptune	+7.7	124°E	+0.7169	E North America, Atlantic, Iceland, Europe except NE, NW Africa

T a b l e I (cont.)

Date			T_0	Body	Magn.	Elong.	γ	Area of Visibility
			h					
1999 Sep	21		20	Uranus	+6.1	135°E	+0.6393	Atlantic Ocean, Europe except NW and N, Africa, W Asia
1999 Sep	30		4	Aldebaran	+1.1	117°W	+0.9816	NE North America, Greenland, Iceland, Scandinavia, N Asia, Arctic regions
1999 Oct	5		22	Regulus	+1.3	42°W	+1.2040	Canada, Arctic regions
1999 Oct	18		7	Neptune	+7.7	97°E	+0.5067	S Japan, Pacific Ocean, Hawaii
1999 Oct	19		4	Uranus	+6.1	108°E	+0.4502	Pacific Ocean, Hawaii, North America
1999 Oct	27		11	Aldebaran	+1.1	143°W	+1.1537	E Siberia, Arctic regions
1999 Nov	14		15	Neptune	+7.8	70°E	+0.1964	Atlantic Ocean, Africa, Madagascar, Indian Ocean, S Asia
1999 Nov	15		13	Uranus	+6.2	80°E	+0.1208	Africa, Madagascar, Indian Ocean, SE Asia, Indonesia
1999 Nov	23		22	Aldebaran	+1.1	170°W	+1.2376	NE Canada
1999 Dec	11		22	Neptune	+7.8	43°E	−0.0777	Pacific Ocean, South America
1999 Dec	12		18	Mars	+1.1	52°E	+0.6068	The Americas, Atlantic Ocean, W Europe, NW Africa
1999 Dec	12		21	Uranus	+6.2	54°E	−0.2111	Pacific Ocean, South America, Atlantic Ocean
1999 Dec	21		9	Aldebaran	+1.1	160°E	+1.2170	Siberia, Arctic regions
2000 Jan	8		6	Neptune	+7.8	16°E	−0.2339	Madagascar, Indian Ocean, Australia, New Guinea, SW Pacific
2000 Jan	9		5	Uranus	+6.2	27°E	−0.4392	Indian Ocean, Australia, New Zealand, Pacific Ocean
2000 Jan	17		19	Aldebaran	+1.1	133°E	+1.1773	NE Canada, Greenland, Arctic
2000 Feb	4		14	Neptune	+7.8	11°W	−0.3216	Pacific Ocean, South America, Atlantic Ocean, Africa
2000 Feb	5		14	Uranus	+6.2	1°E	−0.5821	SE Pacific, S South America, S Atlantic Ocean, S Africa, part of Antarctica

Table I   (cont.)

Date	T_0	Body	Magn.	Elong.	γ	Area of Visibility
	h					
2000 Feb 14	3	Aldebaran	+1.1	105°E	+1.2295	Siberia
2000 Mar 2	24	Neptune	+7.8	37°W	−0.4607	Indian Ocean, S Australia, New Zealand, Pacific Ocean
2000 Mar 4	1	Venus	−3.3	25°W	−0.6784	S Indian Ocean, Tasmania, New Zealand, S Pacific, Antarctica
2000 Mar 4	1	Uranus	+6.2	25°W	−0.7462	S Indian Ocean, New Zealand, S Pacific Ocean, Antarctica
2000 Mar 30	10	Neptune	+7.8	64°W	−0.7164	SE Pacific, S South America, S Atlantic Ocean, S Africa, Madagascar, Antarctica
2000 Mar 31	12	Uranus	+6.2	51°W	−1.0069	S Atlantic Ocean, Antarctica
2000 Apr 26	19	Neptune	+7.7	90°W	−1.0360	S Pacific Ocean, Antarctica
2000 Jul 29	17	Mercury	+0.1	20°W	+0.8077	N Pacific Ocean, Alaska, Canada, Atlantic Ocean, Greenland, Iceland, W and N Europe, E Siberia, Arctic regions
2000 Jul 30	12	Mars	+1.9	9°W	−0.6139	South America, Atlantic, S Africa, Madagascar
2000 Aug 1	2	Venus	−3.3	14°E	+1.0384	N Siberia, NW North America, Arctic regions
2000 Aug 13	17	Neptune	+7.7	163°E	−1.2511	Part of Antarctica
2000 Aug 28	3	Mars	+2.0	18°W	+0.8928	N and NE Europe, N Asia, Pacific, Alaska, Arctic regions
2001 May 23	6	Saturn	+0.3	3°E	−1.1318	SE Indian Ocean
2001 May 24	8	Jupiter	−1.5	15°E	−1.2733	Extreme SE Indian Ocean
2001 Jun 19	22	Saturn	+0.3	21°W	−0.8781	New Zealand, S Pacific Ocean
2001 Jun 21	4	Jupiter	−1.5	5°W	−0.7329	SE Indian Ocean, New Guinea, Australia, New Zealand
2001 Jul 17	13	Saturn	+0.4	44°W	−0.5859	SE Pacific, South America, Atlantic Ocean

Table I   (cont.)

Date			T_0	Body	Magn.	Elong.	γ	Area of Visibility
			h					
2001 Jul	17		18	Venus	−3.6	42°W	+0.2709	Pacific Ocean, Hawaii, North America, N Mexico, Cuba, Haiti
2001 Jul	19		0	Jupiter	−1.5	25°W	−0.1903	Indonesia, extreme NW Australia, New Guinea, Pacific
2001 Jul	19		13	Mercury	−0.5	18°W	+0.9835	NE North America, Greenland, NE Europe
2001 Aug	14		3	Saturn	+0.4	68°W	−0.2157	Africa, Madagascar, Indian Ocean, S Asia, Indonesia
2001 Aug	15		20	Jupiter	−1.6	46°W	+0.3824	Extreme E Asia, Japan, N Pacific, North America
2001 Sep	10		13	Saturn	+0.3	93°W	+0.2003	Pacific Ocean, Hawaii, North America, Mexico, Cuba, Haiti, W Atlantic Ocean
2001 Sep	12		12	Jupiter	−1.7	68°W	+0.9707	NW North America, Greenland, Iceland, Arctic regions, Central and N Europe
2001 Oct	7		19	Saturn	+0.1	120°W	+0.5381	Asia, Japan, Pacific Ocean
2001 Oct	23		20	Mars	+0.1	87°E	−0.1323	South America, Atlantic, Africa
2001 Nov	3		22	Saturn	−0.1	148°W	+0.6460	NW Africa, Europe, Siberia, Japan
2001 Dec	1		2	Saturn	−0.2	177°W	+0.4888	S and E North America, Central America, N Atlantic, Greenland, Iceland, Europe, extreme N Africa
2001 Dec	14		6	Venus	−3.4	7°W	+0.8625	Asia
2001 Dec	28		8	Saturn	−0.1	153°E	+0.2222	Pacific, Hawaii, North America, Central America
2001 Dec	30		14	Jupiter	−2.3	178°W	+1.2609	Greenland
2002 Jan	24		16	Saturn	+0.1	124°E	+0.0791	Africa, Arabia, Asia, Japan
2002 Jan	26		19	Jupiter	−2.2	151°E	+0.9220	Greenland, Iceland, N Europe, Arctic regions, Siberia

Table I   (cont.)

Date	T_0	Body	Magn.	Elong.	γ	Area of Visibility
	h					
2002 Feb 21	0	Saturn	+0.3	96°E	+0.1820	E Pacific, North America, Mexico, Cuba, N Atlantic
2002 Feb 23	2	Jupiter	−2.0	121°E	+0.8880	N North America, Arctic, Greenland, Iceland, W Europe
2002 Mar 20	10	Saturn	+0.3	70°E	+0.4768	NE Africa, Arabia, Asia, Japan, NW Pacific Ocean
2002 Mar 22	11	Jupiter	−1.8	95°E	+1.1839	Iceland, Greenland, Arctic
2002 Apr 16	20	Saturn	+0.4	45°E	+0.8275	NE Pacific, NW North America, Greenland, Iceland, Europe except SW
2002 May 14	8	Saturn	+0.3	22°E	+1.1451	NW Europe, Greenland, Iceland, Arctic regions
2002 May 14	19	Mars	+1.9	27°E	−0.6659	SE Pacific, South America
2002 May 14	23	Venus	−3.4	29°E	−0.8941	S Pacific Ocean
2002 Jun 12	12	Mars	+1.9	19°E	+0.9836	NE North America, Greenland, Iceland, Arctic, Siberia
2002 Dec 5	4	Mercury	−0.5	12°E	+0.6018	Central and SE Asia, S Japan, N Indonesia, New Guinea, Pacific Ocean
2002 Dec 30	2	Mars	+1.7	50°W	+1.1090	E Siberia, Japan
2003 Jan 27	15	Mars	+1.5	61°W	−0.4193	Pacific, S South America
2003 May 29	4	Venus	−3.3	22°W	+0.1287	E Africa, Madagascar, Indian Ocean, SE Asia, Japan
2003 Jul 17	8	Mars	−1.7	135°W	+0.3304	Pacific, Central America, Cuba, Haiti, NW South America, Atlantic Ocean
2003 Sep 9	12	Mars	−2.5	164°E	+1.2268	NE Siberia
2003 Oct 6	16	Mars	−1.8	138°E	−1.0518	Antarctica, extreme SE Australia, Tasmania, New Zealand
2003 Oct 25	13	Mercury	−1.0	1°E	+1.0696	Europe except SW, NE Africa, N Arabia
2003 Oct 26	20	Venus	−3.3	18°E	−0.0703	Pacific, Hawaii, South America

Table I   (cont.)

Date	T_0	Body	Magn.	Elong.	γ	Area of Visibility
	h					
2003 Nov 25	3	Mercury	−0.4	17°E	−0.2603	Indian Ocean, SW Indonesia, Australia, New Zealand, S Pacific
2004 Feb 26	2	Mars	+1.2	68°E	−0.9086	S Pacific
2004 Mar 25	23	Mars	+1.5	57°E	+0.8672	NE Siberia, N Pacific Ocean, N North America, Arctic, Greenland, Iceland
2004 May 21	12	Venus	−3.9	25°E	+0.3695	Atlantic Ocean, NW Africa, Europe, Asia
2004 Oct 13	9	Mars	+2.0	9°W	+1.1266	Central Asia
2004 Oct 14	14	Mercury	−0.7	6°E	+0.1503	W North America, NE South America, Atlantic, S Africa
2004 Nov 9	17	Jupiter	−1.3	38°W	+0.9287	N and E North America, W Atlantic Ocean
2004 Nov 10	2	Venus	−3.5	33°W	−0.1699	S Asia, Indonesia, Australia, New Zealand
2004 Nov 11	4	Mars	+1.9	19°W	−0.4193	E Africa, Madagascar, Indian Oc., S Australia, New Zealand
2004 Nov 14	3	Mercury	−0.2	21°E	−0.8908	S Indian Ocean, Antarctica
2004 Dec 7	11	Jupiter	−1.4	61°W	+0.3170	North America, Atlantic, NE South America, W Africa
2005 Jan 4	1	Jupiter	−1.6	86°W	−0.3352	Africa, Madagascar, Indian Ocean, SW Australia
2005 Jan 7	20	Antares	+1.2	38°W	+1.2517	N Pacific Ocean
2005 Jan 31	10	Jupiter	−1.8	113°W	−0.8727	SE Pacific, Antarctica
2005 Feb 4	5	Antares	+1.2	66°W	+1.0526	Central and E Europe, N Arabia, W Asia
2005 Feb 27	14	Jupiter	−1.9	141°W	−1.0908	S Australia, Antarctica
2005 Mar 3	12	Antares	+1.2	93°W	+0.8404	North America, Central America, N South America, W Atlantic Ocean

ASTRONOMICAL TABLES

T a b l e   I   (cont.)

Date	T_0	Body	Magn.	Elong.	γ	Area of Visibility
	h					
2005 Mar 26	15	Jupiter	−2.0	171°W	−0.9282	Indian Ocean, extreme SW Australia, Antarctica
2005 Mar 30	17	Antares	+1.2	120°W	+0.7245	E Asia, Japan, Pacific, Hawaii
2005 Apr 9	1	Venus	−3.5	3°E	+1.0443	E Asia, Arctic regions
2005 Apr 22	17	Jupiter	−2.0	159°E	−0.5761	Africa, Madagascar, Indian Ocean, Antarctica
2005 Apr 26	24	Antares	+1.2	147°W	+0.7247	Europe, N and NE Africa, Arabia, W and S Asia
2005 May 19	22	Jupiter	−1.9	130°E	−0.3361	Cuba, Haiti, South America, Atlantic, extreme S Africa
2005 May 24	8	Antares	+1.2	172°W	+0.7610	NE Pacific, North America, Central America
2005 May 31	10	Mars	+0.5	78°W	−0.4687	S South America, Atlantic Ocean, Africa
2005 Jun 16	6	Jupiter	−1.7	104°E	−0.3890	E Indonesia, New Guinea, N Australia, New Zealand, Pacific Ocean
2005 Jun 20	18	Antares	+1.2	160°E	+0.7332	SE Europe, Arabia, Asia except NE
2005 Jul 13	18	Jupiter	−1.5	80°E	−0.7239	South America, S Atlantic, Antarctica
2005 Jul 18	4	Antares	+1.2	134°E	+0.5979	NE Pacific, S North America, Central America, NW South America
2005 Aug 8	5	Venus	−3.4	34°E	+1.2059	Alaska
2005 Aug 10	7	Jupiter	−1.4	57°E	−1.2222	S Indian Ocean, Antarctica
2005 Aug 14	13	Antares	+1.2	108°E	+0.3972	NE Africa, Arabia, S Asia, Indonesia, New Guinea, NW Australia
2005 Sep 7	7	Spica	+1.2	39°E	+1.2166	NE Asia
2005 Sep 7	8	Venus	−3.5	40°E	−0.6062	Africa, Madagascar, S Indian Ocean, Antarctica

T a b l e   I   (cont.)

Date	T_0	Body	Magn.	Elong.	γ	Area of Visibility
	h					
2005 Sep 10	20	Antares	+1.2	82°E	+0.2312	SE North America, Central America, South America, Atlantic Ocean, W Africa
2005 Oct 4	11	Mercury	−0.4	12°E	−0.8620	South America, S Atlantic, Antarctica
2005 Oct 4	13	Spica	+1.2	13°E	+1.1851	NE Europe, W Asia
2005 Oct 8	1	Antares	+1.2	55°E	+0.1819	Pacific Ocean
2005 Oct 31	20	Spica	+1.2	15°W	+1.1848	North America
2005 Nov 4	7	Antares	+1.2	28°E	+0.2355	NE Africa, Arabia, Indian Ocean, S Asia, Indonesia, New Guinea, Australia
2005 Nov 28	5	Spica	+1.2	42°W	+1.0962	E Asia
2005 Dec 1	16	Antares	+1.2	5°E	+0.2826	The Americas, Atlantic, W Africa
2005 Dec 12	4	Mars	−1.1	138°E	+1.2168	E Siberia
2005 Dec 25	15	Spica	+1.2	70°W	+0.8688	North America, Central America, N South America, W Atlantic
2005 Dec 29	2	Antares	+1.2	28°W	+0.2143	S Asia, Indonesia, New Guinea, Australia, W Pacific
2006 Jan 21	23	Spica	+1.2	98°W	+0.5782	Asia, Japan, Indonesia, New Guinea, W Pacific Ocean
2006 Jan 25	12	Antares	+1.2	56°W	+0.0251	E Pacific, SW Central America, South America, Atlantic
2006 Feb 18	5	Spica	+1.2	125°W	+0.3652	E North America, Atlantic, W and Central Africa
2006 Feb 21	21	Antares	+1.2	83°W	−0.1760	Indian Ocean, S Indonesia, Australia, New Zealand, S Pacific Ocean
2006 Mar 17	11	Spica	+1.2	153°W	+0.2993	Pacific, Hawaii, W South America
2006 Mar 21	3	Antares	+1.2	110°W	−0.2642	E South America, Atlantic, S Africa, S Indian Ocean

ASTRONOMICAL TABLES

Table I   (cont.)

Date	T_0	Body	Magn.	Elong.	γ	Area of Visibility
	h					
2006 Mar 27	16	Uranus	+6.3	25°W	−1.2397	Part of Antarctica, S Atlantic
2006 Apr 13	17	Spica	+1.2	178°W	+0.3152	E Europe, Asia exc. N and E, Indonesia, New Guinea, N Australia
2006 Apr 17	9	Antares	+1.2	137°W	−0.2171	Pacific Ocean, S South America, SW Atlantic
2006 Apr 24	3	Uranus	+6.2	50°W	−1.0920	Antarctica, SE Australia
2006 Apr 24	14	Venus	−3.7	44°W	−0.4005	S Pacific, South America except NW, Atlantic Ocean, extreme W Africa
2006 May 11	1	Spica	+1.2	154°E	+0.2846	North America, NE South America, Atlantic, W Africa
2006 May 14	15	Antares	+1.2	163°W	−0.1283	Indonesia, Australia, New Zealand, S Pacific Ocean
2006 May 21	11	Uranus	+6.2	76°W	−0.8534	Antarctica, S Atlantic Ocean
2006 Jun 7	9	Spica	+1.2	127°E	+0.1298	E Asia, Japan, Pacific Ocean
2006 Jun 10	23	Antares	+1.2	169°E	−0.1143	N South America, Atlantic, S Africa, Madagascar
2006 Jun 17	17	Uranus	+6.2	102°W	−0.5682	Antarctica, New Zealand, Pacific
2006 Jul 4	17	Spica	+1.2	101°E	−0.1234	Atlantic, W and S Africa, Madagascar, S Indian Ocean
2006 Jul 8	8	Antares	+1.2	144°E	−0.2218	New Guinea, N and E Australia, New Zealand, S Pacific
2006 Jul 14	23	Uranus	+6.1	128°W	−0.3463	Extreme S Africa, Indian Ocean, Indonesia
2006 Jul 27	18	Mars	+2.0	29°E	+1.0624	Greenland, Iceland, Europe, N Africa
2006 Aug 1	1	Spica	+1.2	75°E	−0.3726	Pacific, S South America
2006 Aug 4	17	Antares	+1.2	118°E	−0.4018	E South America, Atlantic, S Africa, S Indian Ocean
2006 Aug 11	6	Uranus	+6.1	155°W	−0.2746	SE Pacific, South America, Atlantic, extreme W Africa

Table I   (cont.)

Date	T_0	Body	Magn.	Elong.	γ	Area of Visibility
	h					
2006 Aug 25	13	Mars	+2.0	19°E	−0.5395	South America, Atlantic, extreme S Africa, Antarctica
2006 Aug 28	8	Spica	+1.2	49°E	−0.5150	E Africa, Madagascar, Indian Ocean, SW Australia, New Zealand, Antarctica
2006 Sep 1	2	Antares	+1.2	91°E	−0.5457	E Australia, New Zealand, S Pacific Ocean, Antarctica, S South America
2006 Sep 7	15	Uranus	+6.1	178°E	−0.3422	Indian Ocean, Australia, New Guinea, Pacific Ocean
2006 Sep 21	15	Venus	−3.4	10°W	−0.9207	SE Pacific, S South America, part of Antarctica
2006 Sep 24	14	Spica	+1.2	23°E	−0.5292	S South America, S Atlantic, Africa, Madagascar
2006 Sep 28	8	Antares	+1.2	65°E	−0.5634	Madagascar, S Indian Ocean, Antarctica, New Zealand
2006 Oct 5	0	Uranus	+6.1	150°E	−0.4381	S South America, Atlant., Africa
2006 Oct 21	20	Spica	+1.2	5°W	−0.5019	Pacific, S South America, part of Antarctica
2006 Oct 25	14	Antares	+1.2	38°E	−0.4631	South America, S Atlantic, S Indian Ocean, Antarctica
2006 Nov 1	9	Uranus	+6.1	122°E	−0.4220	Part of Antarctica, S Australia, New Zealand, Pacific Ocean
2006 Nov 18	3	Spica	+1.2	32°W	−0.5643	E Africa, Madagascar, S Indian Ocean, Antarctica
2006 Nov 21	20	Antares	+1.2	11°E	−0.3598	Pacific Ocean, S South America
2006 Nov 28	15	Uranus	+6.2	95°E	−0.2296	S Africa, Madagascar, Indian Ocean, SE Asia, Indonesia
2006 Dec 10	12	Saturn	+0.6	113°W	+1.1569	Greenland, Iceland, N Atlantic
2006 Dec 15	10	Spica	+1.2	59°W	−0.7762	SE Pacific Ocean, S South America, Antarctica
2006 Dec 19	4	Antares	+1.2	18°W	−0.3759	E Africa, Madagasc., S Indian Ocean, Australia, New Zealand

T a b l e   I   (cont.)

Date	T_0	Body	Magn.	Elong.	γ	Area of Visibility
	h					
2006 Dec 25	21	Uranus	+6.2	67°E	+0.0694	E Pacific, South America, Atlantic, extreme NE Africa
2007 Jan 6	19	Saturn	+0.4	142°W	+0.8949	N Europe, N and E Siberia, NW Pacific Ocean
2007 Jan 7	6	Regulus	+1.3	137°W	+1.2004	Arctic regions, NE Europe
2007 Jan 11	19	Spica	+1.2	87°W	−1.0610	S Indian Ocean, Antarctica
2007 Jan 15	13	Antares	+1.2	45°W	−0.5250	S Pacific, S South America, Antarctica, S Atlantic Ocean
2007 Jan 19	19	Mercury	−0.9	8°E	−1.2505	Part of Antarctica
2007 Jan 20	17	Venus	−3.3	21°E	−0.7013	SE Pacific, extreme S South America, Antarctica, S Atlantic Ocean
2007 Jan 22	5	Uranus	+6.3	41°E	+0.3396	S India, Indonesia, extreme NW Australia, New Guinea, Japan, NW Pacific Ocean
2007 Feb 2	24	Saturn	+0.2	171°W	+0.9032	N North America, Greenland, Arctic, NE Europe, Asia
2007 Feb 3	15	Regulus	+1.3	164°W	+1.0944	Arctic, NW North America
2007 Feb 11	22	Antares	+1.2	73°W	−0.6948	Madagascar, S Indian Ocean, Antarctica
2007 Feb 18	17	Uranus	+6.3	14°E	+0.5166	E Pacific, the Americas, Atlantic Ocean, Greenland, Iceland, W Europe
2007 Mar 2	2	Saturn	+0.2	159°E	+1.0825	Arctic regions, Greenland, Europe except SW
2007 Mar 2	22	Regulus	+1.3	168°E	+1.1048	Arctic regions, part of Asia
2007 Mar 11	6	Antares	+1.2	100°W	−0.7512	S South America, S Atlantic Ocean, Antarctica
2007 Mar 17	4	Mercury	+0.6	27°W	−1.2325	Ocean S of New Zealand
2007 Mar 18	6	Uranus	+6.3	12°W	+0.6529	SE Europe, Africa, Arabia, Asia
2007 Mar 29	5	Saturn	+0.4	130°E	+1.2036	Greenland, Iceland

Table I (cont.)

Date	T_0	Body	Magn.	Elong.	γ	Area of Visibility
	h					
2007 Mar 30	4	Regulus	+1.3	141°E	+1.0950	Arctic regions, Greenland, Iceland, W Europe
2007 Apr 7	13	Antares	+1.2	127°W	−0.6630	Australia, New Zealand, Antarctica, S South America
2007 Apr 14	1	Mars	+1.3	48°W	+0.4608	Extreme E Africa, N Madagascar, Indian Ocean, Indonesia, S and E Asia, Japan
2007 Apr 14	19	Uranus	+6.3	38°W	+0.8449	NW Pacific, extreme NE of Siberia, NW North America
2007 Apr 25	10	Saturn	+0.5	103°E	+1.1030	Arctic regions, Alaska
2007 Apr 26	9	Regulus	+1.3	114°E	+0.9553	N Siberia, NE Pacific, NW North America
2007 May 4	18	Antares	+1.2	154°W	−0.5243	S Africa, Madagascar, S Indian Ocean, Antarctica, Tasmania, New Zealand
2007 May 12	6	Uranus	+6.2	63°W	+1.1360	N Atlantic Ocean, Iceland, NW Europe
2007 May 22	20	Saturn	+0.7	78°E	+0.7889	NW North America, Greenland, Iceland, Europe, N Africa
2007 May 23	16	Regulus	+1.3	88°E	+0.6961	N Atlantic Ocean, Greenland, Iceland, Arctic, N and E Europe, E Arabia, Asia except E
2007 Jun 1	0	Antares	+1.2	175°E	−0.4605	South America, S Atlantic, part of Antarctica
2007 Jun 18	15	Venus	−4.0	45°E	+0.5666	NE North Amer., Greenland, Iceland, Europe except SW, NE Africa, Arabia, SW Asia
2007 Jun 19	8	Saturn	+0.7	53°E	+0.3738	E Europe, Asia, Japan, Pacific
2007 Jun 20	1	Regulus	+1.3	62°E	+0.4248	Extreme E Asia, N Pacific, W and S North America, Central America, NW South America
2007 Jun 28	8	Antares	+1.2	153°E	−0.5239	New Guinea, Australia, New Zealand, S Pacific, S South America, Antarctica

Table I   (cont.)

Date			T_0	Body	Magn.	Elong.	γ	Area of Visibility
			h					
2007 Jul	3	20		Neptune	+7.7	140°W	−1.2486	SW Pacific Ocean
2007 Jul	16	23		Saturn	+0.8	30°E	−0.0402	Pacific, Hawaii, W South America
2007 Jul	17	10		Regulus	+1.3	35°E	+0.2496	Europe except N, N Africa, Central and S Asia, Indonesia, Australia
2007 Jul	25	16		Antares	+1.2	127°E	−0.6630	S Africa, Antarctica, S Indian Ocean, SE Australia
2007 Jul	31	2		Neptune	+7.7	167°W	−1.2107	Antarctica, S Indian Ocean
2007 Aug	12	16		Mercury	−1.6	4°W	+0.2203	North America, NE South America, Atlantic, W Africa
2007 Aug	13	13		Saturn	+0.8	7°E	−0.4185	South America, Atlantic, S Africa
2007 Aug	13	18		Regulus	+1.3	9°E	+0.2004	Pacific, Hawaii, the Americas
2007 Aug	22	0		Antares	+1.2	101°E	−0.7662	New Zealand, S Pacific, Antarctica, S Atlantic
2007 Aug	27	10		Neptune	+7.7	166°E	−1.2699	Extreme S Atlantic Ocean
2007 Sep	10	1		Regulus	+1.3	17°W	+0.2105	Asia, Japan, Pacific Ocean
2007 Sep	10	3		Saturn	+0.8	16°W	−0.7937	Indian Ocean, W Australia
2007 Sep	18	8		Antares	+1.2	75°E	−0.7433	Madagascar, S Indian Ocean, Antarctica
2007 Oct	7	7		Regulus	+1.3	44°W	+0.1603	Europe except N and E, N and E Africa, Arabia, Indian Oc.
2007 Oct	7	15		Saturn	+0.9	40°W	−1.2274	SE Pacific Ocean
2007 Oct	15	15		Antares	+1.2	48°E	−0.5998	South America, S Atlantic, Antarctica
2007 Oct	21	4		Neptune	+7.7	112°E	−1.2106	Antarctica, SW Atlantic Ocean
2007 Nov	3	13		Regulus	+1.3	71°W	−0.0294	Pacific, Hawaii, the Americas
2007 Nov	11	21		Antares	+1.2	21°E	−0.4458	New Zealand, S Pacific, S South America

Table I (cont.)

Date	T_0	Body	Magn.	Elong.	γ	Area of Visibility
	h					
2007 Nov 17	12	Neptune	+7.7	85°E	−0.9579	Antarctica, SE Australia, New Zealand
2007 Nov 30	20	Regulus	+1.3	98°W	−0.3184	S Asia, Indonesia, New Guinea, N and E Australia, New Zealand
2007 Dec 9	3	Antares	+1.2	8°W	−0.4078	Madagascar, Indian Ocean, S Australia, New Zealand, part of Antarctica
2007 Dec 14	18	Neptune	+7.8	57°E	−0.6478	S South America, Antarctica, S Atlantic, SW Africa
2007 Dec 24	3	Mars	−1.4	177°W	+0.8913	NW North America, Greenland, Iceland, Arctic regions, Central and E Europe
2007 Dec 28	5	Regulus	+1.3	126°W	−0.5723	South America, S Atlantic Oc.
2008 Jan 5	10	Antares	+1.2	35°W	−0.5068	South America, S Atlantic, Antarctica
2008 Jan 9	16	Mercury	−0.8	14°E	−0.2964	SE Pacific, S South America, S Atlantic, Africa
2008 Jan 11	2	Neptune	+7.8	30°E	−0.4109	S Australia, part of Antarctica, New Zealand, Pacific
2008 Jan 19	24	Mars	−0.8	145°E	+1.1268	NW North America, Arctic regions, Siberia
2008 Jan 24	15	Regulus	+1.3	154°W	−0.6900	Indonesia, New Guinea, Australia, New Zealand
2008 Feb 1	18	Antares	+1.2	62°W	−0.6311	Australia, New Zealand, S Pacific, S South America, Antarctica
2008 Feb 7	11	Neptune	+7.8	4°E	−0.2822	S Atlantic, S and SE Africa, Madagascar, Indian Oc., India
2008 Feb 20	24	Regulus	+1.3	178°E	−0.6990	South America, S Atlantic
2008 Feb 29	2	Antares	+1.2	90°W	−0.6428	S Atlantic, S Indian Ocean, Antarctica, S Australia
2008 Mar 5	14	Mercury	+0.3	27°W	−0.1784	S and E South America, Atlantic Ocean, W Africa

T a b l e   I   (cont.)

Date	T_0	Body	Magn.	Elong.	γ	Area of Visibility
	h					
2008 Mar 5	19	Venus	−3.3	24°W	+0.2279	Pacific, North America, N Central America, Cuba
2008 Mar 5	22	Neptune	+7.8	23°W	−0.1793	Australia except N, New Zealand, Pacific Ocean
2008 Mar 19	7	Regulus	+1.3	151°E	−0.7210	New Zealand, Pacific Ocean
2008 Mar 27	10	Antares	+1.2	117°W	−0.5083	N New Zealand, S Pacific, Antarctica, S South America
2008 Apr 2	9	Neptune	+7.8	50°W	+0.0022	South America, Atlantic, Africa, Arabia
2008 Apr 12	6	Mars	+1.2	83°E	+1.2004	Arctic regions, Greenland
2008 Apr 15	13	Regulus	+1.3	124°E	−0.8564	Madagascar, S Indian Ocean
2008 Apr 23	17	Antares	+1.2	144°W	−0.3214	Indian Ocean, S Australia, New Zealand
2008 Apr 29	19	Neptune	+7.7	76°W	+0.2873	Indonesia, New Guinea, Australia, Pacific, Hawaii
2008 May 10	14	Mars	+1.5	70°E	+0.2391	Europe except N, N Africa, Arabia, Central and S Asia
2008 May 12	19	Regulus	+1.3	98°E	−1.0966	S South America, part of Antarctica
2008 May 20	23	Antares	+1.2	169°W	−0.2104	E South America, Atlantic Ocean, S Africa, S Madagascar, S Indian Ocean
2008 May 27	3	Neptune	+7.7	102°W	+0.5964	Africa, SE Europe, Arabia, Central Asia
2008 Jun 8	1	Mars	+1.7	59°E	−1.0269	S Pacific, New Zealand
2008 Jun 17	5	Antares	+1.2	163°E	−0.2305	Pacific, South America
2008 Jun 23	8	Neptune	+7.7	128°W	+0.8070	North America, Central America, Cuba
2008 Jul 14	12	Antares	+1.2	137°E	−0.3280	Indian Ocean, S Australia, New Zealand
2008 Jul 20	13	Neptune	+7.7	155°W	+0.8533	Japan, extreme NE Siberia, NW Pacific, Alaska

Table I   (cont.)

Date			T_0	Body	Magn.	Elong.	γ	Area of Visibility
			h					
2008 Aug	1		16	Mercury	−1.5	4°E	−1.2517	SE Pacific Ocean
2008 Aug	10		19	Antares	+1.2	111°E	−0.3882	South America, S Atlantic, extr. S Africa, S Madagascar
2008 Aug	16		18	Neptune	+7.7	179°E	+0.7852	NE Africa, Arabia, Asia
2008 Sep	7		3	Antares	+1.2	85°E	−0.3218	New Guinea, Australia, New Zealand, S Pacific Ocean, W South America
2008 Sep	13		1	Neptune	+7.7	151°E	+0.7375	North America except NW, Central America, NW South America, N Atlantic, Greenland, Iceland, British Isles
2008 Sep	30		10	Mercury	+1.7	13°E	−1.1219	S Atlantic, Antarctica
2008 Oct	4		11	Antares	+1.2	58°E	−0.1357	Africa, Madagascar, Indian Ocean, S Indonesia, Australia
2008 Oct	10		10	Neptune	+7.7	124°E	+0.8351	E and SE Asia, Japan, NW Pacific Ocean
2008 Oct	31		19	Antares	+1.2	31°E	+0.0637	Pacific Ocean, South America, Atlantic Ocean
2008 Nov	6		18	Neptune	+7.7	97°E	+1.0991	N Atlantic Ocean, Iceland, NW Europe
2008 Nov	28		1	Antares	+1.2	6°E	+0.1522	SE Asia, Indonesia, New Guinea, NE Australia, Pacific
2008 Dec	1		16	Venus	−3.7	43°E	+0.8651	Atlantic, NW Africa, Europe
2008 Dec	25		7	Antares	+1.2	24°W	+0.1017	Africa, Madagascar, Indian Ocean, W Indonesia
2008 Dec	29		4	Mercury	−0.6	18°E	+0.7037	E and SE Asia, Japan, N Indonesia, N Pacific
X 2008 Dec	29		9	Jupiter	−1.5	20°E	−0.6629	Antarctica, S Indian Ocean, Australia
2009 Jan	21		13	Antares	+1.2	52°W	+0.0182	Pacific Ocean, South America, Atlantic Ocean
2009 Jan	25		3	Mars	+1.5	13°W	−0.7664	Antarctica, New Zealand

Table I (cont.)

Date	T_0	Body	Magn.	Elong.	γ	Area of Visibility
	h					
2009 Jan 26	5	Jupiter	−1.5	2°W	+0.0361	Madagascar, Indian Ocean, NW Australia, Indonesia, New Guinea, W Pacific
2009 Feb 17	21	Antares	+1.2	79°W	+0.0435	SE Asia, Indonesia, New Guinea, Australia, N New Zealand, Pacific Ocean
2009 Feb 22	21	Mercury	+0.1	25°W	+1.0841	Extreme E Asia, NW Pacific
2009 Feb 23	1	Jupiter	−1.5	23°W	+0.7454	SE and E Asia, Japan, N Indonesia, NW Pacific
2009 Feb 28	0	Venus	−4.3	35°E	−1.2336	S Pacific Ocean
2009 Mar 17	5	Antares	+1.2	107°W	+0.2167	NE South America, Atlantic, Africa, Madagascar, S Arabia
2009 Apr 13	13	Antares	+1.2	134°W	+0.4452	Pacific, Hawaii, NW Central America
2009 Apr 22	13	Venus	−4.2	33°W	+1.0075	North America, Greenland
2009 May 10	21	Antares	+1.2	160°W	+0.6004	SE Europe, N and E Africa, Arabia, S Asia, NW Indonesia
2009 Jun 7	4	Antares	+1.2	172°E	+0.6247	The Americas, Atlantic Ocean, W Africa
2009 Jul 4	10	Antares	+1.2	147°E	+0.5672	Japan, Pacific, Hawaii
2009 Jul 31	16	Antares	+1.2	121°E	+0.5406	SE Europe, N and E Africa, Arabia, S and SE Asia
2009 Aug 27	23	Antares	+1.2	95°E	+0.6343	The Americas, Atlantic Ocean, W Africa
2009 Sep 13	16	Mars	+1.1	69°W	+1.1103	N Siberia, Arctic, Greenland
2009 Sep 18	24	Mercury	+2.8	4°E	−1.0786	SW Australia, Antarctica
2009 Sep 24	7	Antares	+1.2	69°E	+0.8446	E Asia, Japan, N Pacific
2009 Oct 12	1	Mars	+0.9	81°W	−1.1232	S Indian Ocean
2009 Oct 21	15	Antares	+1.2	42°E	+1.0732	N Atlant., Europe, NW Africa
2009 Nov 17	24	Antares	+1.2	15°E	+1.2044	N Pacific Ocean

Table I (cont.)

Date	T_0	Body	Magn.	Elong.	γ	Area of Visibility
	h					
2009 Dec 15	7	Antares	+1.2	14°W	+1.2052	Central Asia
2010 Jan 11	13	Antares	+1.2	41°W	+1.1642	NE North America, N Atlantic
2010 Feb 7	19	Antares	+1.2	69°W	+1.2141	N Pacific Ocean
2010 May 16	10	Venus	−3.4	30°E	+0.0860	Africa, Arabia, S Asia, Indonesia
2010 Sep 11	13	Venus	−4.2	44°E	−0.3221	E South America, Atlantic, S Africa, S Indian Ocean
2010 Nov 5	8	Venus	−3.5	12°W	−0.1641	Atlantic, Africa, Madagascar, Indian Ocean
2010 Dec 6	22	Mars	+1.5	15°E	+0.5510	Pacific Ocean, Hawaii, North America, N Central America
2011 Jun 30	8	Venus	−3.3	13°W	+0.0900	Africa, SE Europe, Arabia, S Asia, Indonesia
2011 Jul 27	17	Mars	+1.6	39°W	−0.5012	Pacific, South America
2011 Oct 28	2	Mercury	−0.2	18°E	−0.2197	Indonesia, Australia, New Zealand, Pacific Ocean
2012 Jun 17	8	Jupiter	−1.6	25°W	+1.2388	Arctic regions
2012 Jul 15	3	Jupiter	−1.7	46°W	+0.5502	Europe except NW, N Africa, N Arabia, Asia exc. S, Japan
2012 Jul 20	8	Mercury	+2.1	14°E	−0.5622	S Africa, Madagascar, Indian Ocean, SW Australia
2012 Jul 25	16	Spica	+1.2	81°E	−1.1862	S South America, Antarctica
2012 Aug 11	21	Jupiter	−1.8	68°W	−0.1217	Indonesia, extreme NW Australia, New Guinea, Pacific Ocean, Hawaii
2012 Aug 13	20	Venus	−4.0	46°W	+0.6033	E Asia, Japan, Arctic regions, N Pacific, North America
2012 Aug 21	22	Spica	+1.2	55°E	−0.9550	New Zealand, S Pacific, Antarctica
2012 Sep 8	11	Jupiter	−2.0	91°W	−0.6882	SE Pacific Ocean, South America

Table I   (cont.)

Date	T_0	Body	Magn.	Elong.	γ	Area of Visibility
	h					
2012 Sep 18	5	Spica	+1.2	28°E	−0.7814	Madagascar, S Indian Ocean, Antarctica
2012 Sep 19	21	Mars	+1.4	51°E	−0.1484	Pacific Ocean, South America, SW Atlantic Ocean
2012 Oct 5	21	Jupiter	−2.1	117°W	−1.0121	S Indian Ocean, S Australia
2012 Oct 15	15	Spica	+1.2	2°E	−0.7284	SE Pacific Ocean, S South America, Antarctica
2012 Oct 17	2	Mercury	−0.0	22°E	+1.2683	Alaska
2012 Nov 2	1	Jupiter	−2.3	145°W	−0.9907	S Atlantic Ocean, S Africa
2012 Nov 12	2	Spica	+1.2	26°W	−0.7533	Madagascar, S Indian Ocean, Antarctica
2012 Nov 14	11	Mercury	+2.1	7°E	+1.0149	Europe except SW, extreme N Africa, N Arabia, SW Asia
2012 Nov 29	1	Jupiter	−2.4	175°W	−0.7058	Central South America, Atlantic Ocean, S Africa
2012 Dec 9	12	Spica	+1.2	54°W	−0.7375	S Pacific, S South America, S Atlantic, Antarctica
2012 Dec 12	1	Mercury	−0.4	19°W	−1.0915	Antarctica
✗ 2012 Dec 26	0	Jupiter	−2.3	154°E	−0.4619	South America, S Atlantic, SW Africa
⅄ 2013 Jan 5	20	Spica	+1.2	82°W	−0.5868	Indian Ocean, S Indonesia, Australia, New Zealand, part of Antarctica
2013 Jan 22	3	Jupiter	−2.2	124°E	−0.5478	Pacific, South America
2013 Feb 2	2	Spica	+1.2	109°W	−0.3298	Atlantic, Central and S Africa, Madagascar, S Indian Ocean, W Australia
2013 Feb 18	12	Jupiter	−2.0	97°E	−0.9926	S Indian Ocean, S Australia
2013 Mar 1	7	Spica	+1.2	137°W	−0.0988	Pacific, SW Central America, S America, S Atlantic Ocean
2013 Mar 28	15	Spica	+1.2	164°W	+0.0053	SE Asia, Indonesia, NE Australia, Pacific Ocean

Table I (cont.)

Date	T_0	Body	Magn.	Elong.	γ	Area of Visibility
	h					
2013 Apr 25	1	Spica	+1.2	169°E	+0.0040	Central America, N and NE South America, Atlantic, S Africa, Madagascar
2013 May 9	14	Mars	+1.5	5°W	+0.4580	S and E North America, Central America, Atlantic, Iceland, Europe, N Africa
2013 May 9	19	Mercury	−1.7	3°W	+0.3143	Pacific, Hawaii, North America, Mexico, Cuba, Haiti
2013 May 22	11	Spica	+1.2	142°E	+0.0046	SE Asia, Indonesia, New Guinea, NE Australia, Pacific
2013 Jun 18	20	Spica	+1.2	116°E	+0.1059	N South America, Atlantic, W and S Africa, Madagascar
2013 Jul 8	12	Mercury	+3.0	5°E	+0.1060	NW South America, Atlantic, N and Centr. Africa, SW Europe, SW Arabia, Indian Ocean
2013 Jul 16	4	Spica	+1.2	90°E	+0.3253	Pacific Ocean, Hawaii, Central America
2013 Aug 12	9	Spica	+1.2	64°E	+0.5854	Asia, Japan, N Indonesia, W Pacific Ocean
2013 Sep 8	15	Spica	+1.2	38°E	+0.7721	NE North America, N Atlantic, Greenland, Iceland, Europe, N Africa, Arabia
2013 Sep 8	21	Venus	−3.6	41°E	−0.4344	Pacific, S South America
2013 Oct 5	22	Spica	+1.2	11°E	+0.8299	N Pacific, North America, N Central America
2013 Nov 2	7	Spica	+1.2	16°W	+0.8156	Europe except SW, Asia
2013 Nov 3	7	Mercury	+2.7	3°W	+0.0194	Central Africa, SW Arabia, N Madagascar, Indian Ocean, S Indonesia, Australia
2013 Nov 29	18	Spica	+1.2	44°W	+0.8586	N Pacific, North America, N Central America
2013 Dec 1	10	Saturn	+0.8	22°W	−1.2482	Antarctica
2013 Dec 1	23	Mercury	−0.6	15°W	+0.4262	Extreme E Asia, Japan, Pacific Ocean, Hawaii

Table I (cont.)

Date	T_0	Body	Magn.	Elong.	γ	Area of Visibility
	h					
2013 Dec 27	3	Spica	+1.2	71°W	+1.0516	N and NE Europe, Siberia, N China
2013 Dec 29	1	Saturn	+0.8	47°W	−0.9170	Extreme SE Africa, S Indian Ocean, Antarctica
2014 Jan 25	14	Saturn	+0.8	73°W	−0.5727	New Zealand, S Pacific, S South America, part of Antarctica
2014 Feb 21	22	Saturn	+0.7	100°W	−0.3133	E Africa, Madagascar, Indian Oc., Australia, New Zealand
2014 Feb 26	5	Venus	−4.3	44°W	+0.3439	Africa, S Arabia, Indian Ocean, S Asia
2014 Mar 21	3	Saturn	+0.5	127°W	−0.2523	South America, Atlantic, S Africa, Madagascar
2014 Apr 17	7	Saturn	+0.4	155°W	−0.3896	Pacific, S South America, SW Atlantic
2014 May 14	12	Saturn	+0.3	175°E	−0.5790	Australia, New Zealand, Antarctica, S Pacific
2014 Jun 10	19	Saturn	+0.4	148°E	−0.6328	S Atlantic Ocean, S Africa, S Indian Ocean, Antarctica, extreme SW Australia
2014 Jun 26	12	Mercury	+2.4	11°W	+0.2933	SE North America, Central America, NW South America, Atlantic Ocean, N Africa, Central and S Europe, Arabia
2014 Jul 6	1	Mars	+0.3	97°E	+0.2171	Pacific, Hawaii, SW Central America, South America
2014 Jul 8	2	Saturn	+0.6	121°E	−0.4520	S Pacific, S South America, S Atlantic Ocean
2014 Aug 4	11	Saturn	+0.8	95°E	−0.0739	S India, Indian Ocean, S Indonesia, Australia
2014 Aug 14	16	Uranus	+6.1	126°W	+1.1208	E Siberia, Arctic regions
2014 Aug 31	19	Saturn	+0.8	70°E	+0.3731	SE North America, Central America, N South America, Atlantic Ocean, W Africa

Table I   (cont.)

Date	T_0	Body	Magn.	Elong.	γ	Area of Visibility
	h					
2014 Sep 11	1	Uranus	+6.1	153°W	+1.0568	NE North America, N Atlantic, Greenland, Iceland, NW Europe
2014 Sep 28	5	Saturn	+0.8	45°E	+0.7634	E Asia, Japan, NW Pacific, Hawaii
2014 Oct 8	10	Uranus	+6.1	179°E	+1.1354	E Asia, Arctic regions
2014 Oct 22	22	Mercury	+1.3	12°W	−0.7399	Australia, New Zealand, S Pacific, Antarctica
2014 Oct 23	21	Venus	−3.5	1°W	−0.0702	Pacific Ocean
2014 Oct 25	16	Saturn	+0.8	21°E	+1.0537	NE North America, N Atlantic Ocean, SW Europe
2014 Nov 4	18	Uranus	+6.1	151°E	+1.2303	N Atlantic Ocean, Iceland
2014 Dec 1	23	Uranus	+6.1	123°E	+1.1847	NW North America
2014 Dec 29	4	Uranus	+6.1	95°E	+0.9404	NE Asia, Japan, NW Pacific, N North America
2015 Jan 25	12	Uranus	+6.2	68°E	+0.5947	SE Europe, Africa, Arabia, Asia
2015 Jan 29	17	Aldebaran	+1.1	120°E	+1.2537	Arctic regions
2015 Feb 21	22	Uranus	+6.2	41°E	+0.2971	Pacific Ocean, Hawaii, North America, Mexico, Cuba
2015 Feb 25	23	Aldebaran	+1.1	93°E	+1.0266	Alaska, NW Canada, Greenland, Iceland, N Europe
2015 Mar 21	11	Uranus	+6.3	15°E	+0.1058	Atlantic Ocean, Africa, Arabia, S Asia
2015 Mar 21	23	Mars	+1.5	22°E	−0.9197	S Pacific Ocean
2015 Mar 25	7	Aldebaran	+1.1	66°E	+0.9104	Siberia, Alaska
2015 Apr 18	1	Uranus	+6.3	11°W	−0.0344	Indonesia, New Guinea, Australia, Pacific, Hawaii
2015 Apr 21	17	Aldebaran	+1.1	39°E	+0.9319	NW and N North America, Greenland, Iceland, N and E Europe

Table I   (cont.)

Date	T_0	Body	Magn.	Elong.	γ	Area of Visibility
	h					
2015 May 15	12	Uranus	+6.2	36°W	−0.2138	South America, Atlantic, Africa
2015 May 19	3	Aldebaran	+1.1	13°E	+1.0042	Siberia, Alaska, Arctic regions
2015 Jun 11	21	Uranus	+6.2	61°W	−0.4717	Australia, New Zealand, Pacific Ocean
2015 Jun 15	2	Mercury	+1.5	19°W	−0.0456	Indian Ocean, SE Asia, Indonesia, Pacific Ocean
2015 Jun 15	11	Aldebaran	+1.1	15°W	+1.0137	NE North America, Greenland, Iceland, N Europe, NW Siberia
2015 Jul 9	3	Uranus	+6.2	86°W	−0.7596	Indian Ocean, Antarctica
2015 Jul 12	18	Aldebaran	+1.1	40°W	+0.9058	Japan, NE Siberia, N Pacific Ocean, N North America
2015 Jul 19	1	Venus	−4.1	34°E	−0.4409	NE Australia, New Guinea, Pacific Ocean
2015 Aug 5	9	Uranus	+6.1	112°W	−0.9716	S South America, Antarctica, S Atlantic Ocean
2015 Aug 8	24	Aldebaran	+1.1	66°W	+0.7213	N Arabia, E Europe, Asia except SE, NW Pacific
2015 Sep 1	16	Uranus	+6.1	139°W	−1.0279	Antarctica, New Zealand, S Pacific Ocean
2015 Sep 5	5	Aldebaran	+1.1	92°W	+0.5665	E North America, N Atlantic, Greenland, Iceland, Europe, extreme N Africa, W and Central Asia
2015 Sep 29	1	Uranus	+6.0	167°W	−0.9490	S Atlantic Ocean, S Africa, S Madagascar
2015 Oct 2	13	Aldebaran	+1.1	119°W	+0.5295	Japan, N Pacific, North America
2015 Oct 8	20	Venus	−4.2	45°W	−0.7351	Australia, New Zealand, S Pacific
2015 Oct 11	11	Mercury	+0.4	17°W	−1.0008	S South America, S Atlantic, Antarctica
2015 Oct 26	11	Uranus	+6.0	165°E	−0.8574	Antarctica, New Zealand, S Pacific Ocean

Table I (cont.)

Date	T_0	Body	Magn.	Elong.	γ	Area of Visibility
	h					
2015 Oct 29	23	Aldebaran	+1.1	146°W	+0.6020	Europe, NW Africa, Asia except S
2015 Nov 22	19	Uranus	+6.1	137°E	−0.8956	Antarctica, S Indian Ocean
2015 Nov 26	10	Aldebaran	+1.1	172°W	+0.6817	NE Asia, N Pacific Ocean, North America except S
2015 Dec 6	3	Mars	+1.7	60°W	−0.1023	Central and E Africa, Arabia, Indian Ocean, S Indonesia, Australia
2015 Dec 7	17	Venus	−3.7	42°W	+0.7037	NE Pacific, North and Central America, Cuba, Haiti
2015 Dec 20	1	Uranus	+6.1	109°E	−1.1111	Antarctica, S South America
X 2015 Dec 23	19	Aldebaran	+1.1	158°E	+0.6575	Europe, NW Africa, Asia except S
2016 Jan 20	3	Aldebaran	+1.1	130°E	+0.5126	North America, Greenland, Iceland, extreme W Europe, extreme NW Africa
2016 Feb 16	8	Aldebaran	+1.1	103°E	+0.3472	SE Asia, Japan, N Indonesia, Pacific, Hawaii, California
2016 Mar 14	14	Aldebaran	+1.1	76°E	+0.2841	SE Europe, N Africa, Asia
2016 Apr 6	8	Venus	−3.3	16°W	+0.6503	Europe, N Africa, N Arabia, W and Central Asia
2016 Apr 10	22	Aldebaran	+1.1	49°E	+0.3478	Pacific Ocean, Hawaii, North America except NW and N, Atlantic Ocean
2016 May 8	9	Aldebaran	+1.1	22°E	+0.4534	S and E Europe, N Africa, Arabia, Asia, Japan
2016 Jun 3	10	Mercury	+0.9	24°W	−0.7181	S Atlantic Ocean, S Africa, Madagascar
2016 Jun 4	19	Aldebaran	+1.1	7°W	+0.4948	Pacific Ocean, Hawaii, North America, NE Atlantic Ocean
2016 Jun 25	24	Neptune	+7.7	113°W	+1.1658	Central and NE Europe, NW Asia
2016 Jul 2	4	Aldebaran	+1.1	31°W	+0.4278	SE Europe, Africa, Asia

T a b l e   I   (cont.)

Date	T_0	Body	Magn.	Elong.	γ	Area of Visibility
	h					
2016 Jul 9	10	Jupiter	−1.4	61°E	−0.8791	SE Africa, Madagascar, S Indian Ocean, Antarctica
2016 Jul 23	5	Neptune	+7.7	139°W	+1.0366	E North America, NW Atlantic, Greenland, Iceland
2016 Jul 29	11	Aldebaran	+1.1	57°W	+0.2953	S and SE North America, Central America, Atlantic, SW Europe, NW Africa
2016 Aug 4	22	Mercury	+0.2	25°E	−0.5794	New Zealand, S Pacific, S South America
2016 Aug 6	3	Jupiter	−1.3	39°E	−0.2171	Indonesia, New Guinea, extr. N Australia, Pacific Ocean
2016 Aug 19	12	Neptune	+7.6	166°W	+1.0414	Extreme E Asia, N Pacific, Alaska
2016 Aug 25	17	Aldebaran	+1.1	83°W	+0.1959	Indonesia, Pacific, Hawaii, S North America, Mexico
2016 Sep 2	22	Jupiter	−1.2	18°E	+0.3864	Pacific, Hawaii, SW North America, Central America, NW South America
2016 Sep 3	11	Venus	−3.3	24°E	+1.1677	Siberia
2016 Sep 15	20	Neptune	+7.6	167°E	+1.1126	Europe except S
2016 Sep 21	23	Aldebaran	+1.1	109°W	+0.2082	Africa, Arabia, S and SE Asia, Japan
2016 Sep 29	10	Mercury	−0.3	18°W	−0.7322	South America, S Atlantic, extreme S Africa, Antarctica
2016 Sep 30	17	Jupiter	−1.2	4°W	+0.9419	N North America, N Atlantic, W Europe, NW Africa
2016 Oct 13	5	Neptune	+7.7	139°E	+1.1236	N Pacific Ocean, Alaska
2016 Oct 19	7	Aldebaran	+1.1	136°W	+0.3209	S and E North America, Central America, Atlantic Ocean, S and W Europe, NW Africa
2016 Nov 9	14	Neptune	+7.7	112°E	+0.9786	NE Africa, E Europe, Asia
2016 Nov 15	17	Aldebaran	+1.1	163°W	+0.4352	NE Africa, Arabia, Asia, Japan, NW Pacific Ocean

Table I (cont.)

Date			T_0	Body	Magn.	Elong.	γ	Area of Visibility
			h					
2016 Dec	6		22	Neptune	+7.7	84°E	+0.6947	North and Central America, N Atlantic, W Europe
2016 Dec	13		5	Aldebaran	+1.1	167°E	+0.4497	Pacific, Hawaii, North America, S Greenland, Atlantic, W Europe, NW Africa
2016 Dec	18		18	Regulus	+1.3	117°W	−1.0281	S Australia, Tasmania, part of Antarctica
2017 Jan	3		4	Neptune	+7.8	57°E	+0.3962	Indonesia, New Guinea, Pacific Ocean, Hawaii, extreme W North America
2017 Jan	3		7	Mars	+1.1	58°E	+0.2448	Indian Ocean, S Asia, Indonesia, New Guinea, NW Australia, Pacific
2017 Jan	9		14	Aldebaran	+1.1	140°E	+0.3530	E and NE Africa, Arabia, Asia, Japan, NW Pacific
2017 Jan	15		4	Regulus	+1.3	145°W	−0.8354	Central and S South America, S Atlantic Ocean
2017 Jan	30		11	Neptune	+7.8	30°E	+0.1982	Africa, S Arabia, S Asia
2017 Feb	5		22	Aldebaran	+1.1	113°E	+0.2413	Central America, NW South America, Florida, Atlantic, S Europe, N Africa
2017 Feb	11		14	Regulus	+1.3	173°W	−0.7934	Australia, New Zealand, part of Antarctica
2017 Feb	26		21	Neptune	+7.8	3°E	+0.0975	Pacific, Central America, S North America, NW South America
2017 Mar	5		3	Aldebaran	+1.1	85°E	+0.2295	E Indonesia, Pacific, Hawaii, North America except Canada, Central America
2017 Mar	10		22	Regulus	+1.3	160°E	−0.8051	Central and S South America, S Atlantic, extreme S Africa
2017 Mar	26		8	Neptune	+7.8	23°W	−0.0049	Atlantic, Africa, S Arabia, Madagascar, Indian Oc., India
2017 Apr	1		9	Aldebaran	+1.1	58°E	+0.3378	E and NE Africa, Arabia, Asia exc. NW, Japan, NW Pacific

Table I (cont.)

Date	T_0	Body	Magn.	Elong.	γ	Area of Visibility
	h					
2017 Apr 7	5	Regulus	+1.3	132°E	−0.7345	S Pacific, S South America
2017 Apr 22	20	Neptune	+7.8	49°W	−0.1955	Australia, New Zealand, Pacific
2017 Apr 28	18	Aldebaran	+1.1	32°E	+0.4798	North America except NW, Mexico, N Atlantic, S Greenland, Icel., N Africa, Europe
2017 May 4	10	Regulus	+1.3	106°E	−0.5285	Indian Ocean, Indonesia, Australia, New Zealand
2017 May 20	6	Neptune	+7.7	75°W	−0.4703	S Atlantic Ocean, S Africa, Madagascar, Indian Ocean
2017 May 26	4	Aldebaran	+1.1	7°E	+0.5523	NE Africa, Arabia, Asia, Japan, N Pacific, Alaska
2017 May 31	16	Regulus	+1.3	80°E	−0.2553	NE South America, Africa, Madagascar
2017 Jun 16	13	Neptune	+7.7	101°W	−0.7364	S Pacific, South America, Antarctica
2017 Jun 22	15	Aldebaran	+1.1	22°W	+0.5197	North America, S Greenland, Iceland, Europe, NW Africa
2017 Jun 28	1	Regulus	+1.3	54°E	−0.0336	Pacific Ocean, Hawaii, W South America
2017 Jul 13	18	Neptune	+7.7	127°W	−0.8819	Antarctica, New Zealand, S Pacific Ocean
2017 Jul 19	24	Aldebaran	+1.1	48°W	+0.4310	Arabia, Asia, N Pacific
2017 Jul 25	9	Mercury	+0.5	27°E	+0.8543	N Europe, Siberia, E Asia, Japan
2017 Jul 25	11	Regulus	+1.3	27°E	+0.0676	S Europe, N half of Africa, Arabia, S India, Indonesia
2017 Aug 9	23	Neptune	+7.6	154°W	−0.8691	Antarctica, S Indian Ocean
2017 Aug 16	7	Aldebaran	+1.1	74°W	+0.3818	Florida, Central America, N South America, Europe except N, Central Asia, N Africa, Arabia except SW
2017 Aug 21	21	Regulus	+1.3	1°E	+0.0746	Pacific, Hawaii, SW Central America, W South America

Table I (cont.)

Date	T_0	Body	Magn.	Elong.	γ	Area of Visibility
	h					
2017 Sep 6	5	Neptune	+7.6	179°E	−0.7757	SE Pacific, Antarctica, S South America, S Atlantic Ocean
2017 Sep 12	12	Aldebaran	+1.1	100°W	+0.4427	Pacific, Hawaii, North America, NW Atlantic Ocean
2017 Sep 18	1	Venus	−3.4	28°W	−0.5459	Indonesia, Australia, New Zealand
2017 Sep 18	5	Regulus	+1.3	25°W	+0.0868	N and Central Africa, Arabia, S and SE Asia, Indonesia, NW Australia
2017 Sep 18	20	Mars	+2.0	18°W	+0.1374	Pacific, Hawaii, SW Central America, NW South America
2017 Sep 18	23	Mercury	−0.8	16°W	−0.0299	SE Asia, S Japan, N Indonesia, Pacific Ocean
2017 Oct 3	13	Neptune	+7.6	152°E	−0.7455	Antarctica, extreme SE Australia, New Zealand, S Pacific
2017 Oct 9	18	Aldebaran	+1.1	126°W	+0.5966	Asia, Japan, N Pacific, Alaska
2017 Oct 15	11	Regulus	+1.3	52°W	+0.2055	North America except N, Mexico, Atlantic, W Africa
2017 Oct 30	21	Neptune	+7.7	124°E	−0.8837	Antarctica, S Atlantic Ocean, S Africa, S Madagascar
2017 Nov 6	3	Aldebaran	+1.1	153°W	+0.7445	Mexico, North America except NW, N Atlantic, Greenland, Iceland, Europe except S
2017 Nov 11	17	Regulus	+1.3	79°W	+0.4479	E Asia, N Pacific, SW Alaska, W and S North America, Central America
2017 Nov 27	6	Neptune	+7.7	96°E	−1.1720	Antarctica, S Pacific
2017 Dec 3	13	Aldebaran	+1.1	174°E	+0.7910	Asia except SE, N Pacific, NW North America, Arctic
2017 Dec 8	23	Regulus	+1.3	107°W	+0.7156	Europe except W, Siberia, China, Japan, NW Pacific
2017 Dec 31	1	Aldebaran	+1.1	150°E	+0.7335	North America, N Atlantic, Greenland, Iceland, Arctic regions, Europe except S

ASTRONOMICAL TABLES

T a b l e   I   (cont.)

Date	T_0	Body	Magn.	Elong.	γ	Area of Visibility
	h					
2018 Jan 5	8	Regulus	+1.3	135°W	+0.8812	N North America, N Atlantic, NW Africa, W Europe
2018 Jan 27	10	Aldebaran	+1.1	122°E	+0.6710	Asia except SE, N Japan, Arctic, N Pacific, Alaska
2018 Feb 1	19	Regulus	+1.3	163°W	+0.9156	N Europe, Siberia, Japan except S
2018 Feb 15	18	Mercury	−1.1	2°W	+1.1565	NW North America
2018 Feb 16	17	Venus	−3.4	9°E	−0.5774	SE Pacific, S South America, part of Antarctica, S Atlantic
2018 Feb 23	17	Aldebaran	+1.1	95°E	+0.7139	NE North America, Atlantic, Greenland, Iceland, Arctic, Europe except S, W Asia
2018 Mar 1	6	Regulus	+1.3	169°E	+0.9120	N North America, Greenland, Iceland, N Atlantic Ocean
2018 Mar 22	23	Aldebaran	+1.1	68°E	+0.8746	Extreme NE Asia, N Pacific, N North America, Arctic, Greenland, Iceland, W Europe
2018 Mar 28	15	Regulus	+1.3	142°E	+0.9909	N Europe, N and NE Siberia, Sakhalin
2018 Apr 19	5	Aldebaran	+1.1	41°E	+1.0583	W and N Siberia, Arctic, Alaska
2018 Apr 24	21	Regulus	+1.3	115°E	+1.1905	Arctic regions, NE Europe
2018 May 16	13	Aldebaran	+1.1	15°E	+1.1595	N Canada, NW Greenland, Arctic regions
2018 Jun 12	23	Aldebaran	+1.1	13°W	+1.1517	N Siberia, Arctic regions
2018 Jul 10	10	Aldebaran	+1.1	38°W	+1.0969	Canada, Greenland, Arctic
2018 Aug 6	19	Aldebaran	+1.1	64°W	+1.0955	E Siberia, Arctic regions
2018 Sep 3	2	Aldebaran	+1.1	90°W	+1.2128	Greenland, Arctic regions
2018 Sep 8	23	Mercury	−1.2	11°W	+0.8821	E Siberia, N Pacific Ocean, North America
2018 Nov 16	5	Mars	−0.1	96°E	−1.0511	Antarctica, SE Pacific

T a b l e   I   (cont.)

Date	T_0	Body	Magn.	Elong.	γ	Area of Visibility
	h					
2018 Dec 9	5	Saturn	+0.7	22°E	+1.2357	Extreme NE Asia
2019 Jan 5	19	Saturn	+0.7	3°W	+0.9580	North America
2019 Jan 31	18	Venus	−3.8	45°W	+0.0985	Pacific, NW South America
2019 Feb 2	7	Saturn	+0.8	28°W	+0.6861	Europe, Africa, Arabia, SW Asia
2019 Feb 5	7	Mercury	−1.1	5°E	−0.2094	S Africa, Madagascar, Indian Oc., Indonesia, NW Australia
2019 Mar 1	18	Saturn	+0.8	53°W	+0.3439	Pacific, SW North America, NE Central America
2019 Mar 29	5	Saturn	+0.8	79°W	−0.0583	Atlantic, S Africa, Madagascar, Indian Ocean, S India
2019 Apr 25	15	Saturn	+0.7	105°W	−0.4061	E Australia, New Zealand, Pacific, SW South America
2019 May 22	22	Saturn	+0.5	131°W	−0.5629	S Africa, S Indian Ocean, Antarctica, Australia
2019 Jun 19	4	Saturn	+0.4	159°W	−0.4778	SE Pacific, S South America, S Atlantic, S Africa
2019 Jul 4	6	Mars	+2.0	19°E	+0.0871	Arabia, Central and S Asia, Japan, Pacific Ocean
2019 Jul 16	7	Saturn	+0.3	173°E	−0.2414	S Pacific, South America
2019 Jul 31	21	Venus	−3.5	4°W	+0.5813	NE Asia, N Pacific Ocean, North America, Cuba, Haiti
2019 Aug 12	10	Saturn	+0.4	146°E	−0.0423	E Indonesia, New Guinea, Australia, N New Zealand, Pacific Ocean
2019 Sep 8	14	Saturn	+0.6	118°E	−0.0438	Africa, Madagascar, Indian Ocean, S Indonesia, Australia, New Guinea
2019 Oct 5	21	Saturn	+0.7	92°E	−0.2750	South America, S Atlantic Ocean, SW Africa
2019 Nov 2	7	Saturn	+0.8	66°E	−0.6326	Indian Ocean, Antarctica, New Zealand, S Pacific Ocean

ASTRONOMICAL TABLES

Table I (cont.)

Date	T_0	Body	Magn.	Elong.	γ	Area of Visibility
	h					
2019 Nov 28	11	Jupiter	−1.4	23°E	+0.7551	N Africa, Europe, Arabia, Asia
2019 Nov 29	21	Saturn	+0.8	41°E	−0.9851	S New Zealand, Antarctica
2019 Dec 26	8	Jupiter	−1.4	1°E	+0.1907	Africa, Madagascar, Indian Ocean, Indonesia
2019 Dec 29	2	Venus	−3.4	34°E	−1.0646	Antarctica
2020 Jan 23	3	Jupiter	−1.4	21°W	−0.3820	Madagascar, Indian Ocean, Australia, New Zealand
2020 Feb 18	13	Mars	+1.4	58°W	+0.7989	North America, Central America, extreme N South America, Cuba, Haiti, Atlantic Ocean
2020 Feb 19	20	Jupiter	−1.5	43°W	−0.9984	Extreme S South America, Antarctica
2020 Mar 18	8	Mars	+1.1	67°W	−0.7947	S South America, S Atlantic Ocean, Antarctica
2020 Jun 19	9	Venus	−3.8	23°W	+0.7759	NE North America, Greenland, Iceland, Atlantic Ocean, Europe except SE, Siberia
2020 Aug 9	9	Mars	−1.1	115°W	−0.7647	S and SE South America, Atlantic, part of Antarctica
2020 Sep 6	5	Mars	−1.7	136°W	+0.0273	South America, Atlantic, S Europe, Africa
2020 Oct 3	4	Mars	−2.3	166°W	−0.7385	S South America, S Atlantic, SW Africa
2020 Dec 12	21	Venus	−3.4	25°W	+0.7446	Pacific, Hawaii, extreme W and NW North America
2020 Dec 14	11	Mercury	−0.7	3°W	+0.9515	Europe, N Africa, N Arabia, SW Asia

Notes

1990 Jul	21	First of a series of occultations of Jupiter
1990 Oct	12	Last of a series of occultations of Jupiter.
1990 Oct	25	First of a series of occultations of Saturn
1991 Feb	11	First of a series of occultations of Uranus.
1991 Mar	12	Last of a series of occultations of Saturn.
1991 Apr	4	Last of a series of occultations of Antares, which began on 1986 March 30. The next series for this star will begin on 2005 January 7.
1991 May	31	First of a series of occultations of Neptune.
1992 Feb	1	Last of a series of occultations of Uranus.
1992 Oct	4	Last of a series of occultations of Neptune.
1993 Oct	15	First of a series of occultations of Spica, though this series was interrupted in November - December 1993.
1994 Oct	7	First of a series of occultations of Jupiter.
1994 Dec	30	Last of a series of occultations of Jupiter.
1995 Apr	15	In a part of North America, the occultation of Spica occurs during the initial *penumbral* phase preceding the partial lunar eclipse of that date.
1995 Jun	9	Last of a series of occultations of Spica.
1996 Aug	8	First of a series of occultations of Aldebaran, which will end on 2000 Febr. 14.
1997 Apr	7	First of a series of occultations of Saturn.
1998 Mar	26	First of a series of occultations of Jupiter.
1998 Mar	29	Last of a series of occultations of Saturn.
1998 Apr	23	In a part of the Atlantic Ocean, of Africa, of Arabia, of the Indian Ocean and of India, Jupiter and Venus will simultaneously be hidden by the Moon.
1998 Jun	1	First of a series of occultations of Regulus.
1998 Dec	25	Last of a series of occultations of Jupiter.
1999 Apr	10	First of a series of occultations of Neptune.
1999 Apr	11	First of a series of occultations of Uranus.
1999 Jul	28	In a part of the U.S.A., the emersion of Neptune occurs during the *penumbral* phase preceding the partial lunar eclipse of that date.
1999 Oct	5	Last of a series of occultations of Regulus.
2000 Feb	14	Last of a series of occultations of Aldebaran, which began on 1996 August 8. There will be no other occultation of a first-magnitude star by the Moon until 2005 January 7, when a new series of occultations of Antares will begin. The next occultation of Aldebaran will take place on 2015 January 29.
2000 Mar	4	In a part of the southern Indian Ocean, in New Zealand, in a part of the southern Pacific Ocean and in Antarctica, Venus and Uranus will simultaneously be hidden by the Moon.
2000 Mar	31	Last of a series of occultations of Uranus.
2000 Aug	13	Last of a series of occultations of Neptune, though this series was interrupted from May to July 2000.
2001 May	23	First of a series of occultations of Saturn.
2001 May	24	First of a series of occultations of Jupiter. — This is a rare, grazing event. Nowhere on the Earth's surface will the *center* of Jupiter's disk be occulted by the Moon. Only a partial occultation will take place, at low altitude.

2002 Mar 22	Last of a series of occultations of Jupiter, though this series was interrupted in October and November 2001.
2002 May 14	Last of a series of occultations of Saturn. — On 2002 May 14 (UT date), *three* planets will be occulted by the Moon (Saturn, Mars, Venus), but nowhere simultaneously.
2004 Nov 9	First of a series of occultations of Jupiter.
2005 Jan 7	First of a series of occultations of Antares, which will end on 2010 February 7. First occultation of Antares since 1991 April 4.
2005 Aug 10	Last of a series of occultations of Jupiter.
2005 Sep 7	First of a series of occultations of Spica.
2006 Mar 27	First of a series of occultations of Uranus.
2006 Dec 10	First of a series of occultations of Saturn.
2007 Jan 7	First of a series of occultations of Regulus. — In a time interval of eight days, three first-magnitude stars will be occulted by the Moon, though not all in the same countries : Regulus on January 7, Spica on January 11, and Antares on January 15. This happens for the same time since June 1988. Next case : five times, from July to November 2025.
2007 Jan 11	Last of a series of occultations of Spica.
2007 May 12	Last of a series of occultations of Uranus.
2007 Jul 3	First of a series of occultations of Neptune.
2007 Aug 27	This is a grazing case.
2007 Oct 7	Last of a series of occultations of Saturn.
2008 Mar 5	On 2008 March 5 (UT date), *three* planets will be occulted by the Moon (Mercury, Venus, Neptune), but nowhere simultaneously.
2008 May 12	Last of a series of occultations of Regulus.
2008 Aug 16	In a part of Asia, occultation of Neptune during the initial *penumbral* phase of the partial lunar eclipse of that date.
2008 Nov 6	Last of a series of occultations of Neptune.
2008 Dec 29	First of a short series of occultations of Jupiter (only three events).
2010 Feb 7	Last of a series of occultations of Antares, which began on 2005 January 7. There will be no other occultation of Antares by the Moon until the beginning of a new series on 2023 August 25.
2012 Jun 17	First of a series of occultations of Jupiter.
2012 Jul 25	First of a series of occultations of Spica.
2013 Feb 18	Last of a series of occultations of Jupiter.
2013 Dec 1	First of a series of occultations of Saturn.
2013 Dec 27	Last of a series of occultations of Spica.
2014 Aug 14	First of a series of occultations of Uranus.
2014 Oct 8	In northeastern Asia and in the northern polar regions, Uranus will be occulted by the Moon during either the penumbral or the partial phase preceding the total phase, or during the total phase itself, of the lunar eclipse of that date.
2014 Oct 25	Last of a series of occultations of Saturn.
2015 Jan 29	First of a series of occultations of Aldebaran, which will end on 2018 Sept. 3.
2015 Dec 20	Last of a series of occultations of Uranus.
2016 Jun 25	First of a series of occultations of Neptune.

2016 Jul 9   First of a series of occultations of Jupiter.

2016 Sep 30   Last of a series of occultations of Jupiter.

2016 Dec 18   First of a series of occultations of Regulus.

2017 Sep 18   On 2017 September 18 (UT date), three planets (Venus, Mars, Mercury) and a first-magnitude star (Regulus) will be occulted by the Moon, but nowhere simultaneously.

2017 Nov 27   Last of a series of occultations of Neptune.

2018 Apr 24   Last of a series of occultations of Regulus.

2018 Sep 3   Last of a series of occultations of Aldebaran, which began on 2015 January 29. There will be no other occultation of a first-magnitude star by the Moon until 2023 August 25, when a new series of occultations of Antares will begin. The next occultation of Aldebaran will take place on 2033 Aug. 18.

2018 Dec 9   First of a series of occultations of Saturn.

2019 Nov 28   First of a series of occultations of Jupiter.

2019 Nov 29   Last of a series of occultations of Saturn.

2020 Feb 19   Last of a series of occultations of Jupiter.

During the period 1990 − 2020, the calendar year with the least number of occultations of planets by the Moon is 2011, with only three events (one occultation each of Mercury, Venus, and Mars). Moreover, in 2011 there will be no occultation of a first-magnitude star. — The next years with only three occultations of planets will be 2087, 2118, and 2200. Until A.D. 2200, we found no year with less than three occultations of planets.

At the other extreme, there will be 29 occultations of planets in 2007. This number includes occultations of Uranus and Neptune, and the events occurring close to the Sun.

Table II

Date	T_0	d	$H0$	$X0$ $X1$ $X2$	$Y0$ $Y1$ $Y2$
Aldebaran					
1996 Aug 8	7	+16.5006	353.1375	−0.04426 +0.55186 +0.00002	+1.12731 +0.05566 −0.00004
1996 Sep 4	14	+16.5013	125.0338	−0.00509 +0.55569 −0.00000	+0.93568 +0.05647 −0.00004
1996 Oct 1	22	+16.5016	271.9712	−0.26161 +0.56240 −0.00001	+0.83685 +0.05719 −0.00004
1996 Oct 29	8	+16.5017	88.9913	+0.12091 +0.56790 −0.00003	+0.93111 +0.05822 −0.00004
1996 Nov 25	16	+16.5016	235.9302	−0.35389 +0.56851 +0.00001	+0.96104 +0.05996 −0.00004
1996 Dec 22	24	+16.5014	22.8705	+0.08034 +0.56409 +0.00001	+0.98119 +0.06174 −0.00004
1997 Jan 19	5	+16.5012	124.6888	−0.33716 +0.55921 +0.00004	+0.76610 +0.06325 −0.00003
1997 Feb 15	11	+16.5010	241.5492	−0.19917 +0.55985 +0.00002	+0.57196 +0.06427 −0.00002
1997 Mar 14	19	+16.5008	28.4920	+0.15136 +0.56703 −0.00002	+0.50336 +0.06523 −0.00002
1997 Apr 11	4	+16.5007	190.4755	+0.04411 +0.57582 −0.00002	+0.52645 +0.06660 −0.00003
1997 May 8	14	+16.5006	7.4992	+0.19445 +0.58030 −0.00002	+0.64431 +0.06840 −0.00003
1997 Jun 4	22	+16.5009	154.4395	−0.24157 +0.57841 +0.00001	+0.63789 +0.07033 −0.00003

Table II   (cont.)

Date	T_0	d	H0	X0 X1 X2	Y0 Y1 Y2

Aldebaran

Date	T_0	d	H0		
1997 Jul 2	5	+16.5014	286.3376	−0.22 703 +0.57 266 +0.00 003	+0.56 557 +0.07 181 −0.00 002
1997 Jul 29	11	+16.5021	43.1937	−0.00 130 +0.56 824 +0.00 002	+0.43 351 +0.07 271 −0.00 002
1997 Aug 25	17	+16.5028	160.0490	+0.20 988 +0.56 965 −0.00 000	+0.31 819 +0.07 341 −0.00 001
1997 Sep 21	24	+16.5033	291.9453	+0.10 016 +0.57 750 −0.00 002	+0.27 600 +0.07 456 −0.00 001
1997 Oct 19	9	+16.5035	93.9240	−0.08 570 +0.58 737 −0.00 002	+0.34 388 +0.07 656 −0.00 002
1997 Nov 15	20	+16.5035	285.9858	+0.21 761 +0.59 251 −0.00 002	+0.50 043 +0.07 909 −0.00 002
1997 Dec 13	5	+16.5033	87.9666	−0.22 296 +0.58 938 +0.00 002	+0.46 244 +0.08 127 −0.00 002
1998 Jan 9	13	+16.5031	234.9077	+0.05 765 +0.58 108 +0.00 002	+0.39 290 +0.08 230 −0.00 001
1998 Feb 5	18	+16.5030	336.7267	−0.29 205 +0.57 573 +0.00 003	+0.19 294 +0.08 263 −0.00 001
1998 Mar 5	0	+16.5028	93.5873	−0.15 248 +0.57 940 +0.00 000	+0.15 426 +0.08 339 −0.00 001
1998 Apr 1	8	+16.5026	240.5300	−0.04 056 +0.58 985 −0.00 002	+0.24 769 +0.08 534 −0.00 001
1998 Apr 28	18	+16.5025	57.5540	+0.02 362 +0.59 933 −0.00 002	+0.39 397 +0.08 819 −0.00 002

T a b l e   II   (cont.)

Date	T_0	d	$H0$	$X0$ $X1$ $X2$	$Y0$ $Y1$ $Y2$
A l d e b a r a n					
1998 May 26	4	+16.5027	234.5770	−0.26 088 +0.60 168 +0.00 001	+0.43 424 +0.09 090 −0.00 002
1998 Jun 22	14	+16.5031	51.5987	+0.14 540 +0.59 635 +0.00 001	+0.46 794 +0.09 243 −0.00 002
1998 Jul 19	21	+16.5038	183.4961	−0.00 718 +0.58 795 +0.00 002	+0.33 748 +0.09 270 −0.00 001
1998 Aug 16	3	+16.5045	300.3517	+0.21 317 +0.58 318 +0.00 000	+0.28 387 +0.09 250 −0.00 001
1998 Sep 12	8	+16.5051	42.1658	−0.16 677 +0.58 657 +0.00 000	+0.24 136 +0.09 321 −0.00 001
1998 Oct 9	16	+16.5055	189.1033	+0.10 953 +0.59 685 −0.00 003	+0.41 391 +0.09 557 −0.00 002
1998 Nov 6	2	+16.5055	6.1236	+0.09 441 +0.60 700 −0.00 002	+0.56 146 +0.09 912 −0.00 003
1998 Dec 3	13	+16.5054	198.1861	−0.08 288 +0.60 924 +0.00 001	+0.59 432 +0.10 216 −0.00 003
1998 Dec 30	23	+16.5053	15.2089	−0.24 289 +0.60 170 +0.00 003	+0.51 105 +0.10 318 −0.00 002
1999 Jan 27	7	+16.5051	162.1507	−0.02 074 +0.59 087 +0.00 002	+0.45 748 +0.10 240 −0.00 002
1999 Feb 23	12	+16.5050	263.9701	−0.35 084 +0.58 630 +0.00 002	+0.39 306 +0.10 187 −0.00 002
1999 Mar 22	18	+16.5048	20.8308	−0.26 994 +0.59 203 −0.00 000	+0.52 502 +0.10 328 −0.00 003

Table II (cont.)

Date	T_0	d	$H0$	$X0$ $X1$ $X2$	$Y0$ $Y1$ $Y2$

Aldebaran

Date	T_0	d	$H0$	$X0$/$X1$/$X2$	$Y0$/$Y1$/$Y2$
1999 Apr 19	2	+16.5047	167.7730	−0.30233 +0.60298 −0.00000	+0.68751 +0.10661 −0.00004
1999 May 16	12	+16.5048	344.7964	−0.40864 +0.61060 +0.00001	+0.77889 +0.11027 −0.00004
1999 Jun 12	23	+16.5051	176.8596	−0.26701 +0.60983 +0.00002	+0.80978 +0.11240 −0.00004
1999 Jul 10	9	+16.5058	353.8806	−0.02593 +0.60165 +0.00002	+0.79409 +0.11231 −0.00004
1999 Aug 6	16	+16.5065	125.7775	−0.28200 +0.59178 +0.00003	+0.71177 +0.11095 −0.00003
1999 Sep 2	22	+16.5072	242.6328	−0.08468 +0.58737 +0.00000	+0.81673 +0.11017 −0.00004
1999 Sep 30	4	+16.5077	359.4880	+0.09673 +0.59198 −0.00002	+1.01715 +0.11163 −0.00006
1999 Oct 27	11	+16.5079	131.3848	−0.33708 +0.60243 −0.00000	+1.11014 +0.11540 −0.00006
1999 Nov 23	22	+16.5078	323.4468	+0.05953 +0.61026 −0.00001	+1.27266 +0.11921 −0.00007
1999 Dec 21	9	+16.5077	155.5102	−0.27117 +0.60852 +0.00003	+1.18688 +0.12091 −0.00006
2000 Jan 17	19	+16.5076	332.5338	−0.44907 +0.59787 +0.00005	+1.11074 +0.11964 −0.00006
2000 Feb 14	3	+16.5074	119.4762	−0.16957 +0.58673 +0.00002	+1.21997 +0.11735 −0.00006

ASTRONOMICAL TABLES

T a b l e II (cont.)

Date	T_0	d	$H0$	$X0$ $X1$ $X2$	$Y0$ $Y1$ $Y2$
A l d e b a r a n					
2015 Jan 29	17	+16.5362	314.4256	−0.30 230 +0.56 536 +0.00 003	+1.23 036 +0.05 446 −0.00 005
2015 Feb 25	23	+16.5360	71.2861	−0.25 435 +0.56 829 +0.00 001	+1.00 667 +0.05 547 −0.00 005
2015 Mar 25	7	+16.5358	218.2288	−0.17 808 +0.57 741 −0.00 001	+0.89 736 +0.05 635 −0.00 004
2015 Apr 21	17	+16.5357	35.2531	+0.02 279 +0.58 624 −0.00 002	+0.93 862 +0.05 748 −0.00 005
2015 May 19	3	+16.5357	212.2764	+0.05 916 +0.58 917 −0.00 001	+1.01 515 +0.05 904 −0.00 005
2015 Jun 15	11	+16.5360	359.2163	−0.33 108 +0.58 536 +0.00 003	+0.98 473 +0.06 078 −0.00 004
2015 Jul 12	18	+16.5366	131.1140	−0.18 128 +0.57 869 +0.00 003	+0.89 155 +0.06 211 −0.00 004
2015 Aug 8	24	+16.5373	247.9698	+0.12 939 +0.57 499 +0.00 001	+0.73 980 +0.06 296 −0.00 003
2015 Sep 5	5	+16.5379	349.7839	−0.32 646 +0.57 834 +0.00 001	+0.53 400 +0.06 375 −0.00 003
2015 Oct 2	13	+16.5383	136.7213	−0.14 598 +0.58 795 −0.00 001	+0.51 657 +0.06 487 −0.00 003
2015 Oct 29	23	+16.5384	313.7414	−0.08 562 +0.59 780 −0.00 002	+0.59 621 +0.06 673 −0.00 003
2015 Nov 26	10	+16.5383	145.8035	+0.03 395 +0.60 093 −0.00 001	+0.69 004 +0.06 902 −0.00 003

Table II (cont.)

Date	T_0	d	$H0$	$X0$ $X1$ $X2$	$Y0$ $Y1$ $Y2$
Aldebaran					
2015 Dec 23	19	+16.5381	307.7849	−0.33018 +0.59519 +0.00003	+0.62286 +0.07093 −0.00003
2016 Jan 20	3	+16.5379	94.7264	+0.18835 +0.58601 +0.00001	+0.53950 +0.07180 −0.00002
2016 Feb 16	8	+16.5377	196.5457	−0.05462 +0.58224 +0.00001	+0.34304 +0.07223 −0.00002
2016 Mar 14	14	+16.5375	313.4063	−0.07980 +0.58825 −0.00001	+0.27641 +0.07314 −0.00002
2016 Apr 10	22	+16.5374	100.3488	−0.27860 +0.59955 −0.00001	+0.31567 +0.07509 −0.00002
2016 May 8	9	+16.5374	292.4135	+0.16522 +0.60777 −0.00002	+0.47817 +0.07769 −0.00003
2016 Jun 4	19	+16.5376	109.4361	−0.15231 +0.60785 +0.00001	+0.47902 +0.08011 −0.00002
2016 Jul 2	4	+16.5381	271.4163	−0.20920 +0.60064 +0.00003	+0.40334 +0.08146 −0.00002
2016 Jul 29	11	+16.5388	43.3134	−0.16922 +0.59185 +0.00003	+0.27475 +0.08169 −0.00001
2016 Aug 25	17	+16.5395	160.1687	+0.14275 +0.58838 −0.00000	+0.21756 +0.08165 −0.00001
2016 Sep 21	23	+16.5400	277.0239	+0.22100 +0.59374 −0.00003	+0.24092 +0.08256 −0.00001
2016 Oct 19	7	+16.5402	63.9616	+0.18668 +0.60494 −0.00003	+0.35025 +0.08502 −0.00002

T a b l e   II   (cont.)

Date	T_0	d	$H0$	$X0$ $X1$ $X2$	$Y0$ $Y1$ $Y2$
Aldebaran					
2016 Nov 15	17	+16.5402	240.9823	−0.13747 +0.61376 −0.00000	+0.41990 +0.08841 −0.00002
2016 Dec 13	5	+16.5401	88.0863	+0.22663 +0.61307 −0.00000	+0.48825 +0.09105 −0.00002
2017 Jan 9	14	+16.5399	250.0685	−0.32295 +0.60319 +0.00004	+0.30795 +0.09182 −0.00001
2017 Feb 5	22	+16.5397	37.0107	+0.20299 +0.59253 +0.00001	+0.27539 +0.09114 −0.00001
2017 Mar 5	3	+16.5396	138.8302	−0.03762 +0.59013 −0.00000	+0.22636 +0.09098 −0.00001
2017 Apr 1	9	+16.5394	255.6907	−0.13995 +0.59760 −0.00001	+0.32015 +0.09268 −0.00002
2017 Apr 28	18	+16.5393	57.6737	+0.17175 +0.60822 −0.00003	+0.51282 +0.09593 −0.00003
2017 May 26	4	+16.5395	234.6967	−0.06454 +0.61384 −0.00000	+0.54908 +0.09928 −0.00003
2017 Jun 22	15	+16.5399	66.7595	+0.13895 +0.61069 +0.00001	+0.54974 +0.10110 −0.00003
2017 Jul 19	24	+16.5406	228.7390	−0.02041 +0.60107 +0.00002	+0.43358 +0.10096 −0.00002
2017 Aug 16	7	+16.5413	0.6356	−0.04703 +0.59151 +0.00001	+0.37928 +0.09979 −0.00002
2017 Sep 12	12	+16.5420	102.4498	−0.34638 +0.58880 +0.00001	+0.39047 +0.09949 −0.00002

Table II   (cont.)

Date	T_0	d	$H0$	$X0$ $X1$ $X2$	$Y0$ $Y1$ $Y2$
Aldebaran					
2017 Oct 9	18	+16.5423	219.3051	−0.30991 +0.59509 −0.00000	+0.55239 +0.10137 −0.00003
2017 Nov 6	3	+16.5424	21.2844	+0.16084 +0.60535 −0.00003	+0.78354 +0.10503 −0.00004
2017 Dec 3	13	+16.5423	198.3058	−0.25720 +0.61085 +0.00002	+0.75765 +0.10863 −0.00004
2017 Dec 31	1	+16.5421	45.4107	+0.09775 +0.60613 +0.00001	+0.76319 +0.10981 −0.00004
2018 Jan 27	10	+16.5420	207.3936	−0.35034 +0.59414 +0.00004	+0.61814 +0.10854 −0.00003
2018 Feb 23	17	+16.5419	339.2952	−0.33595 +0.58423 +0.00002	+0.66428 +0.10682 −0.00003
2018 Mar 22	23	+16.5417	96.1558	−0.00213 +0.58341 −0.00002	+0.88886 +0.10705 −0.00005
2018 Apr 19	5	+16.5416	213.0159	−0.11450 +0.59052 −0.00002	+1.05513 +0.10982 −0.00006
2018 May 16	13	+16.5417	359.9572	−0.29515 +0.59851 +0.00001	+1.12420 +0.11353 −0.00006
2018 Jun 12	23	+16.5421	176.9793	−0.33115 +0.60079 +0.00003	+1.10907 +0.11602 −0.00006
2018 Jul 10	10	+16.5427	9.0414	+0.04193 +0.59537 +0.00002	+1.12576 +0.11603 −0.00006
2018 Aug 6	19	+16.5434	171.0203	−0.02168 +0.58519 +0.00002	+1.11191 +0.11423 −0.00005

Table II   (cont.)

Date	T_0	d	$H0$	$X0$ $X1$ $X2$	$Y0$ $Y1$ $Y2$
Aldebaran					
2018 Sep 3	2	+16.5441	302.9167	−0.02421 +0.57660 +0.00000	+1.23093 +0.11249 −0.00006
Regulus					
1998 Jun 1	3	+11.9750	142.3308	−0.35644 +0.53634 +0.00001	−0.98203 −0.14974 +0.00004
1998 Jun 28	12	+11.9754	304.3141	−0.03275 +0.54196 −0.00002	−0.81657 −0.15033 +0.00004
1998 Jul 25	20	+11.9756	91.2557	−0.34623 +0.54550 −0.00000	−0.60379 −0.15007 +0.00003
1998 Aug 22	4	+11.9755	238.1963	−0.23096 +0.54503 +0.00001	−0.62482 −0.14925 +0.00003
1998 Sep 18	11	+11.9752	10.0948	+0.03635 +0.54143 +0.00002	−0.69871 −0.14865 +0.00002
1998 Oct 15	16	+11.9743	111.9100	−0.28871 +0.53810 +0.00003	−0.49913 −0.14889 +0.00001
1998 Nov 11	22	+11.9731	228.7656	−0.25125 +0.53938 +0.00002	−0.25137 −0.15009 +0.00001
1998 Dec 9	6	+11.9717	15.7031	−0.01508 +0.54630 −0.00001	−0.00664 −0.15175 +0.00001
1999 Jan 5	15	+11.9705	177.6819	−0.20567 +0.55467 −0.00001	+0.26143 −0.15295 −0.00000
1999 Feb 2	1	+11.9697	354.7027	−0.08055 +0.55842 −0.00001	+0.28975 −0.15311 −0.00001

Table II (cont.)

Date	T_0	d	$H0$	$X0$ $X1$ $X2$	$Y0$ $Y1$ $Y2$
Regulus					
1999 Mar 1	10	+11.9694	156.6836	+0.22 151 +0.55 549 −0.00 000	+0.19 947 −0.15 258 −0.00 001
1999 Mar 28	16	+11.9696	273.5425	−0.02 968 +0.54 960 +0.00 002	+0.32 907 −0.15 215 −0.00 002
1999 Apr 24	22	+11.9701	30.4024	+0.31 667 +0.54 694 +0.00 000	+0.42 784 −0.15 262 −0.00 003
1999 May 22	4	+11.9706	147.2627	+0.30 604 +0.55 081 −0.00 001	+0.70 318 −0.15 399 −0.00 003
1999 Jun 18	12	+11.9710	294.2052	+0.41 139 +0.55 923 −0.00 004	+0.90 419 −0.15 571 −0.00 004
1999 Jul 15	21	+11.9713	96.1881	+0.19 375 +0.56 693 −0.00 003	+1.07 554 −0.15 698 −0.00 005
1999 Aug 12	7	+11.9714	273.2113	+0.35 818 +0.56 929 −0.00 003	+1.04 006 −0.15 759 −0.00 005
1999 Sep 8	15	+11.9711	60.1512	+0.03 112 +0.56 535 +0.00 001	+1.13 615 −0.15 744 −0.00 006
1999 Oct 5	22	+11.9705	192.0489	+0.22 925 +0.55 850 +0.00 001	+1.18 622 −0.15 713 −0.00 006
2007 Jan 7	6	+11.9319	44.2377	+0.32 852 +0.50 394 −0.00 004	+1.17 287 −0.23 921 −0.00 004
2007 Feb 3	15	+11.9310	206.2174	+0.66 216 +0.50 486 −0.00 004	+0.89 640 −0.24 156 −0.00 003
2007 Mar 2	22	+11.9307	338.1163	+0.68 031 +0.50 254 −0.00 002	+0.89 891 −0.24 139 −0.00 004

Table II   (cont.)

Date	T_0	d	$H0$	$X0$ $X1$ $X2$	$Y0$ $Y1$ $Y2$
Regulus					
2007 Mar 30	4	+11.9308	94.9753	+0.71 809 +0.49 960 −0.00 001	+0.87 007 −0.23 963 −0.00 004
2007 Apr 26	9	+11.9312	196.7941	+0.19 452 +0.49 945 +0.00 001	+0.96 579 −0.23 854 −0.00 005
2007 May 23	16	+11.9316	328.6956	+0.08 310 +0.50 311 −0.00 001	+0.73 161 −0.23 993 −0.00 003
2007 Jun 20	1	+11.9319	130.6791	+0.32 908 +0.50 841 −0.00 003	+0.31 351 −0.24 325 −0.00 001
2007 Jul 17	10	+11.9321	292.6620	+0.32 198 +0.51 216 −0.00 003	+0.12 202 −0.24 652 +0.00 000
2007 Aug 13	18	+11.9320	79.6029	+0.10 385 +0.51 240 −0.00 001	+0.17 241 −0.24 784 −0.00 001
2007 Sep 10	1	+11.9317	211.5017	+0.04 687 +0.50 943 +0.00 001	+0.21 113 −0.24 658 −0.00 001
2007 Oct 7	7	+11.9309	328.3584	+0.08 519 +0.50 599 +0.00 002	+0.13 687 −0.24 410 −0.00 001
2007 Nov 3	13	+11.9297	85.2142	+0.17 080 +0.50 594 +0.00 001	−0.11 469 −0.24 310 −0.00 000
2007 Nov 30	20	+11.9282	217.1106	+0.06 831 +0.51 116 −0.00 001	−0.38 605 −0.24 555 +0.00 002
2007 Dec 28	5	+11.9267	19.0892	−0.02 111 +0.51 902 −0.00 002	−0.62 520 −0.25 043 +0.00 004
2008 Jan 24	15	+11.9257	196.1097	−0.10 184 +0.52 404 −0.00 001	−0.71 753 −0.25 418 +0.00 004

T a b l e II (cont.)

Date	T_0	d	$H0$	$X0$ $X1$ $X2$	$Y0$ $Y1$ $Y2$

R e g u l u s

Date	T_0	d	$H0$	$X0$ $X1$ $X2$	$Y0$ $Y1$ $Y2$
2008 Feb 20	24	+11.9251	358.0902	−0.26 523 +0.52 306 +0.00 001	−0.64 821 −0.25 410 +0.00 003
2008 Mar 19	7	+11.9251	129.9899	−0.43 385 +0.51 807 +0.00 004	−0.59 098 −0.25 086 +0.00 002
2008 Apr 15	13	+11.9254	246.8495	−0.32 031 +0.51 445 +0.00 003	−0.79 628 −0.24 784 +0.00 003
2008 May 12	19	+11.9258	3.7098	−0.24 606 +0.51 625 +0.00 002	−1.09 847 −0.24 808 +0.00 005
2016 Dec 18	18	+11.8822	205.4577	−0.36 482 +0.55 507 −0.00 000	−0.96 551 −0.15 553 +0.00 005
2017 Jan 15	4	+11.8811	22.4778	−0.30 265 +0.56 353 −0.00 001	−0.78 303 −0.15 632 +0.00 005
2017 Feb 11	14	+11.8805	199.4990	−0.28 408 +0.56 545 +0.00 001	−0.74 472 −0.15 566 +0.00 004
2017 Mar 10	22	+11.8804	346.4393	−0.43 451 +0.56 054 +0.00 003	−0.71 538 −0.15 440 +0.00 003
2017 Apr 7	5	+11.8807	118.3397	+0.03 066 +0.55 433 +0.00 002	−0.77 070 −0.15 368 +0.00 003
2017 May 4	10	+11.8812	220.1588	−0.14 338 +0.55 309 +0.00 002	−0.50 869 −0.15 434 +0.00 002
2017 May 31	16	+11.8817	337.0192	−0.31 619 +0.55 897 +0.00 001	−0.17 681 −0.15 601 +0.00 001
2017 Jun 28	1	+11.8821	139.0025	+0.08 928 +0.56 845 −0.00 002	−0.05 965 −0.15 765 +0.00 001

Table II (cont.)

Date	T_0	d	$H0$	$X0$ $X1$ $X2$	$Y0$ $Y1$ $Y2$
Regulus					
2017 Jul 25	11	+11.8823	316.0263	+0.21 325 +0.57 559 −0.00 003	+0.01 136 −0.15 849 +0.00 001
2017 Aug 21	21	+11.8823	133.0490	+0.28 690 +0.57 621 −0.00 002	−0.00 143 −0.15 833 +0.00 000
2017 Sep 18	5	+11.8819	279.9885	+0.02 599 +0.57 046 +0.00 001	+0.08 288 −0.15 754 −0.00 001
2017 Oct 15	11	+11.8811	36.8449	−0.18 483 +0.56 312 +0.00 003	+0.26 481 −0.15 686 −0.00 002
2017 Nov 11	17	+11.8799	153.7005	+0.24 930 +0.56 134 −0.00 000	+0.39 527 −0.15 722 −0.00 002
2017 Dec 8	23	+11.8785	270.5558	+0.08 516 +0.56 863 −0.00 002	+0.71 923 −0.15 889 −0.00 003
2018 Jan 5	8	+11.8772	72.5346	+0.10 590 +0.58 055 −0.00 003	+0.88 511 −0.16 115 −0.00 004
2018 Feb 1	19	+11.8764	264.5964	+0.10 605 +0.58 803 −0.00 002	+0.92 070 −0.16 273 −0.00 005
2018 Mar 1	6	+11.8761	96.6595	+0.25 528 +0.58 586 −0.00 001	+0.87 566 −0.16 298 −0.00 005
2018 Mar 28	15	+11.8764	258.6416	+0.54 449 +0.57 704 −0.00 001	+0.87 620 −0.16 225 −0.00 005
2018 Apr 24	21	+11.8768	15.5015	+0.51 740 +0.56 957 −0.00 001	+1.09 068 −0.16 151 −0.00 006

Table II (cont.)

Date	T_0	d	$H0$	$X0$ $X1$ $X2$	$Y0$ $Y1$ $Y2$
Spica					
1993 Oct 15	14	−11.1281	32.8977	−0.55 873 +0.59 156 +0.00 003	−1.11 457 −0.19 038 +0.00 007
1994 Jan 5	16	−11.1308	143.7952	−0.38 160 +0.56 863 +0.00 003	−1.11 733 −0.18 177 +0.00 006
1994 Feb 1	21	−11.1323	245.6095	−0.52 927 +0.57 125 +0.00 001	−0.81 988 −0.18 253 +0.00 005
1994 Mar 1	5	−11.1335	32.5474	+0.00 155 +0.58 204 −0.00 002	−0.73 750 −0.18 434 +0.00 005
1994 Mar 28	14	−11.1343	194.5273	−0.44 514 +0.59 195 +0.00 002	−0.46 749 −0.18 521 +0.00 003
1994 Apr 25	2	−11.1347	41.6315	+0.07 982 +0.59 363 +0.00 000	−0.62 565 −0.18 391 +0.00 003
1994 May 22	12	−11.1347	218.6547	−0.12 853 +0.58 651 +0.00 003	−0.58 108 −0.18 130 +0.00 003
1994 Jun 18	20	−11.1344	5.5966	−0.19 365 +0.57 610 +0.00 003	−0.49 583 −0.17 864 +0.00 002
1994 Jul 16	2	−11.1340	122.4569	−0.13 379 +0.56 980 +0.00 001	−0.32 568 −0.17 717 +0.00 002
1994 Aug 12	7	−11.1334	224.2761	−0.26 651 +0.57 199 −0.00 000	−0.03 439 −0.17 735 +0.00 001
1994 Sep 8	14	−11.1329	356.1770	−0.06 265 +0.58 093 −0.00 002	+0.10 081 −0.17 853 +0.00 000
1994 Oct 5	23	−11.1327	158.1591	−0.16 517 +0.58 974 +0.00 000	+0.20 919 −0.17 940 −0.00 001

T a b l e   II   (cont.)

Date	T_0	d	$H0$	X0 X1 X2	Y0 Y1 Y2
Spica					
1994 Nov 2	10	−11.1328	350.2222	−0.18 634 +0.59 106 +0.00 002	+0.20 703 −0.17 886 −0.00 001
1994 Nov 29	21	−11.1336	182.2841	−0.06 432 +0.58 284 +0.00 003	+0.19 406 −0.17 674 −0.00 002
1994 Dec 27	6	−11.1349	344.2630	+0.25 743 +0.57 086 −0.00 000	+0.25 514 −0.17 404 −0.00 002
1995 Jan 23	12	−11.1363	101.1183	+0.35 166 +0.56 472 −0.00 002	+0.50 316 −0.17 237 −0.00 002
1995 Feb 19	17	−11.1377	202.9328	+0.17 655 +0.56 882 −0.00 003	+0.81 697 −0.17 250 −0.00 003
1995 Mar 19	0	−11.1387	334.8302	+0.07 564 +0.57 820 −0.00 002	+0.97 567 −0.17 368 −0.00 004
1995 Apr 15	10	−11.1392	151.8519	+0.32 809 +0.58 430 −0.00 002	+0.90 938 −0.17 457 −0.00 005
1995 May 12	21	−11.1393	343.9157	+0.53 135 +0.58 228 −0.00 001	+0.85 055 −0.17 429 −0.00 005
1995 Jun 9	6	−11.1391	145.8985	+0.04 325 +0.57 341 +0.00 002	+1.10 166 −0.17 266 −0.00 006
2005 Sep 7	7	−11.1898	250.0907	+0.68 838 +0.50 190 −0.00 005	+1.01 479 −0.25 523 −0.00 003
2005 Oct 4	13	−11.1896	6.9496	+0.66 563 +0.50 575 −0.00 004	+0.99 092 −0.25 773 −0.00 004
2005 Oct 31	20	−11.1899	138.8485	+0.37 497 +0.50 869 −0.00 001	+1.13 848 −0.25 869 −0.00 005

T a b l e II (cont.)

Date	T_0	d	$H0$	X0 X1 X2	Y0 Y1 Y2

S p i c a

Date	T_0	d	$H0$		
2005 Nov 28	5	-11.1907	300.8283	$+0.31594$ $+0.50731$ $+0.00001$	$+1.06860$ -0.25654 -0.00006
2005 Dec 25	15	-11.1920	117.8483	$+0.57883$ $+0.50188$ -0.00001	$+0.68138$ -0.25236 -0.00004
2006 Jan 21	23	-11.1936	264.7858	$+0.39394$ $+0.49625$ -0.00002	$+0.44921$ -0.24900 -0.00002
2006 Feb 18	5	-11.1951	21.6413	$+0.03065$ $+0.49435$ -0.00002	$+0.39330$ -0.24854 -0.00001
2006 Mar 17	11	-11.1962	138.4976	$+0.08916$ $+0.49630$ -0.00002	$+0.29028$ -0.25012 -0.00000
2006 Apr 13	17	-11.1968	255.3550	-0.04176 $+0.49921$ $+0.00000$	$+0.37395$ -0.25126 -0.00001
2006 May 11	1	-11.1970	42.2956	$+0.27761$ $+0.50031$ $+0.00001$	$+0.17934$ -0.25043 -0.00001
2006 Jun 7	9	-11.1969	189.2372	$+0.10664$ $+0.49864$ $+0.00002$	$+0.09199$ -0.24786 -0.00001
2006 Jul 4	17	-11.1966	336.1794	-0.18004 $+0.49517$ $+0.00002$	-0.04859 -0.24502 -0.00000
2006 Aug 1	1	-11.1962	123.1219	-0.14414 $+0.49201$ -0.00000	-0.34437 -0.24340 $+0.00002$
2006 Aug 28	8	-11.1958	255.0230	-0.05201 $+0.49104$ -0.00002	-0.54910 -0.24347 $+0.00003$
2006 Sep 24	14	-11.1955	11.8824	-0.04720 $+0.49255$ -0.00001	-0.56739 -0.24440 $+0.00003$

T a b l e   II   (cont.)

Date	T_0	d	$H0$	$X0$ $X1$ $X2$	$Y0$ $Y1$ $Y2$
S p i c a					
2006 Oct 21	20	−11.1955	128.7406	−0.07337 +0.49472 +0.00001	−0.52361 −0.24453 +0.00002
2006 Nov 18	3	−11.1961	260.6387	−0.05387 +0.49526 +0.00002	−0.60211 −0.24288 +0.00002
2006 Dec 15	10	−11.1973	32.5358	−0.57440 +0.49355 +0.00004	−0.58364 −0.24023 +0.00002
2007 Jan 11	19	−11.1988	194.5144	−0.35114 +0.49143 +0.00002	−1.00894 −0.23835 +0.00004
2012 Jul 25	16	−11.2276	342.2971	−0.58900 +0.56556 +0.00002	−1.05692 −0.18112 +0.00006
2012 Aug 21	22	−11.2271	99.1572	−0.21479 +0.56938 −0.00001	−0.93387 −0.18152 +0.00006
2012 Sep 18	5	−11.2267	231.0578	−0.22611 +0.57854 −0.00000	−0.74808 −0.18255 +0.00005
2012 Oct 15	15	−11.2265	48.0806	+0.04759 +0.58570 −0.00000	−0.77779 −0.18263 +0.00004
2012 Nov 12	2	−11.2269	240.1432	+0.00572 +0.58443 +0.00002	−0.79043 −0.18104 +0.00004
2012 Dec 9	12	−11.2279	57.1637	−0.22005 +0.57455 +0.00003	−0.70394 −0.17814 +0.00003
2013 Jan 5	20	−11.2293	204.1013	−0.12730 +0.56339 +0.00001	−0.57497 −0.17533 +0.00003
2013 Feb 2	2	−11.2307	320.9567	+0.13208 +0.55972 −0.00002	−0.38647 −0.17408 +0.00002

Table II (cont.)

Date	T_0	d	$H0$	$X0$ $X1$ $X2$	$Y0$ $Y1$ $Y2$
Spica					
2013 Mar 1	7	−11.2320	62.7714	−0.15 928 +0.56 535 −0.00 001	−0.05 428 −0.17 450 +0.00 001
2013 Mar 28	15	−11.2328	209.7102	+0.10 489 +0.57 392 −0.00 002	−0.02 652 −0.17 526 +0.00 001
2013 Apr 25	1	−11.2331	26.7323	+0.28 976 +0.57 783 −0.00 001	−0.08 369 −0.17 524 +0.00 000
2013 May 22	11	−11.2331	203.7555	+0.05 205 +0.57 389 +0.00 002	−0.01 093 −0.17 405 −0.00 001
2013 Jun 18	20	−11.2328	5.7385	−0.15 965 +0.56 457 +0.00 002	+0.15 938 −0.17 200 −0.00 001
2013 Jul 16	4	−11.2324	152.6809	+0.23 685 +0.55 582 −0.00 001	+0.26 778 −0.16 987 −0.00 001
2013 Aug 12	9	−11.2318	254.5001	−0.06 892 +0.55 302 −0.00 001	+0.63 299 −0.16 850 −0.00 002
2013 Sep 8	15	−11.2313	11.3599	+0.26 331 +0.55 719 −0.00 003	+0.72 697 −0.16 839 −0.00 003
2013 Oct 5	22	−11.2311	143.2599	+0.26 647 +0.56 415 −0.00 002	+0.78 649 −0.16 907 −0.00 003
2013 Nov 2	7	−11.2312	305.2408	+0.15 746 +0.56 729 +0.00 000	+0.80 423 −0.16 955 −0.00 004
2013 Nov 29	18	−11.2320	137.3027	+0.52 809 +0.56 269 −0.00 000	+0.73 786 −0.16 899 −0.00 004
2013 Dec 27	3	−11.2333	299.2816	+0.26 051 +0.55 276 +0.00 000	+1.01 984 −0.16 704 −0.00 005

T a b l e II (cont.)

Date			T_0	d	$H0$	X0 X1 X2	Y0 Y1 Y2
Antares							
1990 Jan	22	8	−26.4118	354.2152	+0.03 506 +0.54 923 +0.00 002	−0.30 576 −0.06 578 +0.00 001
1990 Feb	18	16	−26.4124	141.1523	−0.27 989 +0.54 520 +0.00 002	−0.27 434 −0.06 512 +0.00 001
1990 Mar	18	0	−26.4130	288.0893	−0.23 818 +0.54 291 −0.00 000	−0.13 392 −0.06 463 +0.00 001
1990 Apr	14	7	−26.4135	59.9856	−0.15 373 +0.54 339 −0.00 002	+0.07 315 −0.06 360 +0.00 000
1990 May	11	13	−26.4139	176.8418	−0.19 322 +0.54 530 −0.00 001	+0.23 235 −0.06 188 −0.00 000
1990 Jun	7	19	−26.4142	293.6992	−0.20 262 +0.54 658 +0.00 001	+0.25 483 −0.05 995 −0.00 001
1990 Jul	5	2	−26.4144	65.5990	+0.00 857 +0.54 622 +0.00 002	+0.16 198 −0.05 853 −0.00 001
1990 Aug	1	9	−26.4145	197.4998	−0.23 341 +0.54 450 +0.00 003	+0.13 897 −0.05 790 −0.00 001
1990 Aug	28	17	−26.4144	344.4424	−0.22 047 +0.54 280 +0.00 001	+0.20 822 −0.05 775 −0.00 001
1990 Sep	25	1	−26.4141	131.3852	−0.15 071 +0.54 247 −0.00 001	+0.39 989 −0.05 733 −0.00 001
1990 Oct	22	9	−26.4137	278.3275	+0.27 871 +0.54 380 −0.00 003	+0.58 775 −0.05 613 −0.00 002
1990 Nov	18	15	−26.4133	35.1865	+0.03 065 +0.54 551 −0.00 001	+0.74 942 −0.05 419 −0.00 002

Table II (cont.)

Date	T_0	d	$H0$	$X0$ $X1$ $X2$	$Y0$ $Y1$ $Y2$
Antares					
1990 Dec 15	21	−26.4131	152.0441	−0.03513 +0.54553 +0.00001	+0.74695 −0.05231 −0.00003
1991 Jan 12	4	−26.4134	283.9414	+0.27439 +0.54391 +0.00001	+0.65251 −0.05127 −0.00003
1991 Feb 8	11	−26.4139	55.8377	+0.15268 +0.54290 +0.00001	+0.68894 −0.05112 −0.00003
1991 Mar 7	19	−26.4144	202.7746	+0.15798 +0.54435 −0.00001	+0.86571 −0.05119 −0.00003
1991 Apr 4	3	−26.4150	349.7118	+0.06078 +0.54742 −0.00002	+1.12819 −0.05054 −0.00004
2005 Jan 7	20	−26.4434	160.0576	+0.24321 +0.59564 +0.00001	+1.22826 −0.13375 −0.00007
2005 Feb 4	5	−26.4439	322.0361	+0.07339 +0.58410 +0.00001	+1.06187 −0.12931 −0.00006
2005 Mar 3	12	−26.4446	93.9320	+0.28872 +0.57594 −0.00002	+0.79694 −0.12680 −0.00004
2005 Mar 30	17	−26.4452	195.7460	+0.04334 +0.57684 −0.00002	+0.73236 −0.12699 −0.00003
2005 Apr 26	24	−26.4457	327.6427	+0.37453 +0.58442 −0.00004	+0.65963 −0.12853 −0.00003
2005 May 24	8	−26.4461	114.5816	+0.07256 +0.59187 −0.00001	+0.76308 −0.12900 −0.00004
2005 Jun 20	18	−26.4465	291.6039	+0.01841 +0.59330 +0.00001	+0.74594 −0.12729 −0.00004

Table II (cont.)

Date			T_0	d	$H0$	$X0$ $X1$ $X2$	$Y0$ $Y1$ $Y2$

Antares

Date			T_0	d	$H0$		
2005 Jul	18	4	−26.4467	108.6274	−0.10326	+0.63286
					+0.58731	−0.12389
					+0.00003	−0.00003
2005 Aug	14	13	−26.4468	270.6108	+0.01343	+0.40299
					+0.57735	−0.12045
					+0.00001	−0.00002
2005 Sep	10	20	−26.4467	42.5125	+0.17644	+0.19943
					+0.56980	−0.11852
					−0.00001	−0.00001
2005 Oct	8	1	−26.4464	144.3319	−0.07307	+0.20099
					+0.56973	−0.11851
					−0.00002	−0.00001
2005 Nov	4	7	−26.4460	261.1916	−0.13009	+0.26743
					+0.57642	−0.11929
					−0.00001	−0.00001
2005 Dec	1	16	−26.4458	63.1731	+0.29133	+0.22904
					+0.58336	−0.11894
					−0.00001	−0.00001
2005 Dec	29	2	−26.4459	240.1943	+0.15716	+0.18719
					+0.58363	−0.11647
					+0.00001	−0.00001
2006 Jan	25	12	−26.4463	57.2142	−0.06678	+0.03866
					+0.57590	−0.11282
					+0.00003	−0.00000
2006 Feb	21	21	−26.4469	219.1924	+0.14490	−0.20741
					+0.56584	−0.10991
					−0.00000	+0.00001
2006 Mar	21	3	−26.4476	336.0472	−0.09691	−0.25027
					+0.56068	−0.10882
					−0.00001	+0.00001
2006 Apr	17	9	−26.4481	92.9026	+0.17972	−0.25590
					+0.56278	−0.10888
					−0.00003	+0.00002
2006 May	14	15	−26.4485	209.7589	+0.00896	−0.13235
					+0.56864	−0.10868
					−0.00001	+0.00001

Table II   (cont.)

Date		T_0	d	$H0$	$X0$ $X1$ $X2$	$Y0$ $Y1$ $Y2$

A n t a r e s

Date		T_0	d	$H0$	$X0$ / $X1$ / $X2$	$Y0$ / $Y1$ / $Y2$
2006 Jun	10	23	−26.4489	356.6986	+0.03959 +0.57274 +0.00001	−0.12365 −0.10724 +0.00001
2006 Jul	8	8	−26.4491	158.6806	−0.10277 +0.57154 +0.00003	−0.20669 −0.10473 +0.00001
2006 Aug	4	17	−26.4493	320.6637	−0.32977 +0.56511 +0.00003	−0.34880 −0.10206 +0.00001
2006 Sep	1	2	−26.4492	122.6475	+0.07799 +0.55715 −0.00000	−0.56846 −0.10011 +0.00003
2006 Sep	28	8	−26.4489	239.5081	−0.23372 +0.55267 −0.00001	−0.53039 −0.09927 +0.00003
2006 Oct	25	14	−26.4486	356.3681	+0.01884 +0.55427 −0.00002	−0.47382 −0.09888 +0.00003
2006 Nov	21	20	−26.4482	113.2269	−0.03242 +0.55942 −0.00001	−0.35955 −0.09797 +0.00002
2006 Dec	19	4	−26.4482	260.1665	+0.09524 +0.56247 +0.00001	−0.39757 −0.09593 +0.00002
2007 Jan	15	13	−26.4485	62.1458	−0.01594 +0.55986 +0.00003	−0.52958 −0.09331 +0.00002
2007 Feb	11	22	−26.4490	224.1241	−0.15398 +0.55321 +0.00002	−0.67876 −0.09117 +0.00003
2007 Mar	11	6	−26.4497	11.0610	−0.13785 +0.54746 −0.00000	−0.73858 −0.09003 +0.00003
2007 Apr	7	13	−26.4502	142.9572	+0.16974 +0.54611 −0.00003	−0.69967 −0.08940 +0.00003

Table II   (cont.)

Date	T_0	d	$H0$	X0 X1 X2	Y0 Y1 Y2
Antares					
2007 May 4	18	−26.4507	244.7721	−0.23144 +0.54868 −0.00001	−0.49375 −0.08852 +0.00003
2007 Jun 1	0	−26.4510	1.6292	−0.31834 +0.55189 +0.00001	−0.41602 −0.08686 +0.00002
2007 Jun 28	8	−26.4513	148.5696	+0.09967 +0.55286 +0.00001	−0.54532 −0.08474 +0.00002
2007 Jul 25	16	−26.4515	295.5113	−0.03239 +0.55053 +0.00002	−0.66562 −0.08291 +0.00003
2007 Aug 22	0	−26.4515	82.4538	−0.30597 +0.54610 +0.00002	−0.72892 −0.08176 +0.00003
2007 Sep 18	8	−26.4513	229.3966	−0.21221 +0.54240 −0.00000	−0.71988 −0.08109 +0.00003
2007 Oct 15	15	−26.4509	1.2980	−0.03669 +0.54181 −0.00002	−0.60091 −0.08032 +0.00003
2007 Nov 11	21	−26.4505	118.1573	+0.01313 +0.54412 −0.00002	−0.45236 −0.07892 +0.00002
2007 Dec 9	3	−26.4503	235.0153	−0.04622 +0.54638 +0.00001	−0.40534 −0.07682 +0.00002
2008 Jan 5	10	−26.4504	6.9129	−0.08450 +0.54598 +0.00003	−0.49995 −0.07468 +0.00002
2008 Feb 1	18	−26.4509	153.8505	−0.10314 +0.54335 +0.00002	−0.62291 −0.07333 +0.00002
2008 Feb 29	2	−26.4515	300.7874	−0.28009 +0.54106 +0.00001	−0.61087 −0.07285 +0.00002

Table II (cont.)

Date	T_0	d	$H0$	$X0$ $X1$ $X2$	$Y0$ $Y1$ $Y2$
Antares					
2008 Mar 27	10	−26.4521	87.7245	−0.17669 +0.54088 −0.00001	−0.48914 −0.07243 +0.00002
2008 Apr 23	17	−26.4525	219.6211	−0.11485 +0.54236 −0.00002	−0.30909 −0.07131 +0.00002
2008 May 20	23	−26.4529	336.4777	−0.20074 +0.54383 +0.00000	−0.18654 −0.06940 +0.00001
2008 Jun 17	5	−26.4532	93.3356	−0.21907 +0.54403 +0.00002	−0.20511 −0.06732 +0.00001
2008 Jul 14	12	−26.4534	225.2358	+0.03240 +0.54306 +0.00002	−0.33436 −0.06584 +0.00001
2008 Aug 10	19	−26.4534	357.1369	−0.14929 +0.54192 +0.00002	−0.37304 −0.06529 +0.00001
2008 Sep 7	3	−26.4533	144.0797	−0.10837 +0.54185 +0.00000	−0.31111 −0.06522 +0.00001
2008 Oct 4	11	−26.4529	291.0223	−0.07175 +0.54332 −0.00002	−0.12813 −0.06479 +0.00001
2008 Oct 31	19	−26.4525	77.9643	+0.28028 +0.54543 −0.00003	+0.03160 −0.06338 +0.00001
2008 Nov 28	1	−26.4522	194.8228	−0.02803 +0.54636 −0.00000	+0.15627 −0.06121 −0.00000
2008 Dec 25	7	−26.4522	311.6799	−0.07870 +0.54503 +0.00002	+0.11085 −0.05914 −0.00000
2009 Jan 21	13	−26.4525	68.5357	−0.25554 +0.54322 +0.00003	+0.04563 −0.05807 −0.00000

Table II   (cont.)

Date	T_0	d	$H0$	$X0$ $X1$ $X2$	$Y0$ $Y1$ $Y2$
Antares					
2009 Feb 17	21	−26.4531	215.4728	+0.17 121 +0.54 390 +0.00 000	+0.02 551 −0.05 806 −0.00 000
2009 Mar 17	5	−26.4537	2.4097	+0.08 923 +0.54 753 −0.00 002	+0.20 843 −0.05 816 −0.00 001
2009 Apr 13	13	−26.4541	149.3471	−0.13 221 +0.55 169 −0.00 002	+0.46 133 −0.05 738 −0.00 001
2009 May 10	21	−26.4545	296.2854	−0.00 482 +0.55 357 −0.00 001	+0.60 393 −0.05 560 −0.00 002
2009 Jun 7	4	−26.4548	68.1839	+0.15 614 +0.55 246 −0.00 000	+0.61 252 −0.05 356 −0.00 002
2009 Jul 4	10	−26.4550	185.0425	+0.19 266 +0.54 996 +0.00 001	+0.55 153 −0.05 214 −0.00 002
2009 Jul 31	16	−26.4551	301.9023	+0.21 557 +0.54 878 +0.00 001	+0.52 264 −0.05 175 −0.00 002
2009 Aug 27	23	−26.4550	73.8039	+0.31 406 +0.55 095 −0.00 001	+0.60 748 −0.05 202 −0.00 003
2009 Sep 24	7	−26.4547	220.7466	+0.30 104 +0.55 612 −0.00 003	+0.82 012 −0.05 203 −0.00 003
2009 Oct 21	15	−26.4543	7.6889	−0.13 830 +0.56 155 −0.00 002	+1.09 024 −0.05 103 −0.00 004
2009 Nov 17	24	−26.4539	169.6712	+0.13 602 +0.56 360 −0.00 002	+1.19 706 −0.04 924 −0.00 005
2009 Dec 15	7	−26.4537	301.5699	−0.02 145 +0.56 102 +0.00 001	+1.21 126 −0.04 733 −0.00 005

Table II (cont.)

Date	T_0	d	$H0$	$X0$ $X1$ $X2$	$Y0$ $Y1$ $Y2$
Antares					
2010 Jan 11	13	−26.4539	58.4261	−0.00475 +0.55673 +0.00003	+1.16864 −0.04630 −0.00005
2010 Feb 7	19	−26.4544	175.2813	+0.12044 +0.55616 +0.00001	+1.20824 −0.04630 −0.00005

Table   III

Date	T_0	D0 D1	H0 H1	X0 X1 X2	Y0 Y1 Y2	k F

Mercury

Date	T_0	D0 / D1	H0 / H1	X0 / X1 / X2	Y0 / Y1 / Y2	k / F
1990 Aug 22	11	− 0.58 857 − 0.00 968	323.23 800 15.03 460	−0.316 234 +0.504 589 +0.000 065	−0.020 409 −0.249 341 −0.000 021	0.273 445 15.63
1991 Aug 11	9	+ 5.46 381 + 0.00 038	298.72 751 15.05 702	+0.506 538 +0.566 932 +0.000 045	+0.389 416 −0.255 702 −0.000 041	0.273 533 18.09
1991 Nov 8	4	−23.46 752 − 0.01 275	224.08 862 14.98 212	−0.295 091 +0.499 547 +0.000 010	−0.683 623 −0.048 927 +0.000 237	0.273 091 9.59
1992 Jun 1	5	+22.87 411 + 0.01 917	254.95 929 14.94 445	−0.223 618 +0.510 453 +0.000 051	+1.019 419 +0.010 370 −0.000 376	0.273 007 8.81
1992 Oct 27	15	−22.24 890 − 0.01 320	26.48 982 14.99 214	−0.087 458 +0.535 775 +0.000 013	+0.564 855 −0.074 563 +0.000 147	0.273 126 10.68
1993 May 22	5	+22.91 089 + 0.02 006	248.26 798 14.94 749	+0.080 054 +0.479 004 +0.000 056	−0.916 291 +0.017 321 −0.000 248	0.273 048 9.07
1993 Dec 12	11	−21.34 953 − 0.01 501	359.66 078 14.97 552	−0.159 630 +0.542 442 +0.000 006	+0.158 477 −0.025 901 +0.000 220	0.272 991 8.60
1995 Jun 26	2	+18.57 027 + 0.00 878	231.88 891 15.01 427	−0.102 771 +0.518 576 −0.000 050	+0.663 276 +0.020 635 −0.000 119	0.273 465 15.23
1996 Jun 14	0	+17.08 243 + 0.01 509	203.94 074 14.99 455	−0.176 148 +0.506 506 −0.000 031	−0.492 138 +0.065 619 −0.000 119	0.273 309 13.09
1996 Aug 16	18	+ 2.41 976 − 0.02 492	64.17 172 15.00 080	−0.202 842 +0.482 554 +0.000 032	−0.296 051 −0.146 167 −0.000 013	0.273 236 11.79
1997 May 5	17	+ 9.92 953 − 0.01 345	90.28 186 15.04 934	+0.512 044 +0.586 884 −0.000 079	−1.049 132 +0.180 769 +0.000 043	0.273 632 19.63
1997 Aug 5	20	+ 5.90 591 − 0.02 050	93.00 158 15.01 229	+0.528 121 +0.492 351 +0.000 002	+0.952 581 −0.137 716 −0.000 072	0.273 362 13.64

Table III (cont.)

Date	T_0	D0 D1	H0 H1	X0 X1 X2	Y0 Y1 Y2	k F

Mercury

1998 Apr 24	19	+ 2.51 054 − 0.00 162	127.11 928 15.02 631	+0.025 322 +0.571 380 −0.000 039	−0.850 850 +0.199 806 +0.000 013	0.273 442 16.84
1998 Sep 20	7	+ 4.63 420 − 0.03 202	290.29 369 14.97 079	+0.178 477 +0.449 981 −0.000 026	+0.088 875 −0.140 257 −0.000 040	0.273 038 8.63
1999 Feb 17	2	− 9.65 014 + 0.03 332	196.36 595 14.97 249	+0.205 552 +0.500 074 −0.000 011	−0.191 742 +0.139 101 +0.000 115	0.273 033 9.17
1999 Apr 14	5	− 3.22 874 + 0.00 876	279.40 406 15.00 750	+0.465 107 +0.536 939 −0.000 036	−0.925 231 +0.192 209 +0.000 053	0.273 308 14.14
1999 Aug 10	3	+17.92 673 + 0.00 773	242.05 593 15.02 136	+0.216 049 +0.579 536 −0.000 112	+1.139 263 −0.078 132 −0.000 105	0.273 349 14.76
2000 Jul 29	17	+20.82 542 + 0.00 440	94.04 586 14.99 084	−0.085 610 +0.572 746 −0.000 061	+0.812 539 −0.026 514 −0.000 213	0.273 199 12.52
2001 Jul 19	13	+22.39 481 + 0.00 445	32.34 042 14.96 834	−0.095 262 +0.543 925 −0.000 004	+0.982 142 +0.008 681 −0.000 307	0.273 112 10.81
2002 Dec 5	4	−25.34 986 − 0.00 496	230.02 598 14.97 146	−0.073 923 +0.534 832 +0.000 004	+0.617 816 −0.074 308 +0.000 247	0.272 987 8.52
2003 Oct 25	13	−11.52 930 − 0.02 774	18.67 885 14.97 730	+0.371 261 +0.497 971 +0.000 050	+1.007 689 −0.235 936 +0.000 046	0.272 954 8.15
2003 Nov 25	3	−25.26 704 − 0.00 669	210.64 433 14.97 590	−0.189 728 +0.554 023 +0.000 036	−0.231 713 −0.094 650 +0.000 282	0.273 008 9.14
2004 Oct 14	14	−10.44 166 − 0.02 842	27.58 249 14.97 879	−0.128 057 +0.472 029 +0.000 079	+0.231 390 −0.234 501 +0.000 070	0.272 974 8.18
2004 Nov 14	3	−24.88 452 − 0.00 789	207.30 640 14.98 415	−0.183 114 +0.550 366 +0.000 067	−0.871 788 −0.112 464 +0.000 273	0.273 069 10.11

T a b l e   III   (cont.)

Date	T_0	D0 D1	H0 H1	X0 X1 X2	Y0 Y1 Y2	k F

M e r c u r y

2005 Oct	4	11	− 9.14 586 − 0.02 872	336.68 057 14.98 120	−0.543 961 +0.442 622 +0.000 085	−0.690 210 −0.225 896 +0.000 093	0.273 014 8.42
2007 Jan	19	19	−20.37 739 + 0.01 935	93.37 311 14.96 918	+0.346 121 +0.499 557 −0.000 080	−1.203 834 +0.175 329 +0.000 280	0.272 998 8.66
2007 Mar	17	4	−12.10 560 + 0.00 273	263.07 195 15.01 024	+0.461 177 +0.520 013 −0.000 071	−1.147 384 +0.259 994 +0.000 104	0.273 288 13.89
2007 Aug	12	16	+17.58 749 − 0.02 349	61.59 333 14.95 558	−0.036 281 +0.445 607 −0.000 112	+0.256 208 −0.194 432 −0.000 196	0.273 026 8.75
2008 Jan	9	16	−21.55 272 + 0.01 746	43.44 992 14.97 242	+0.231 437 +0.474 859 −0.000 054	−0.235 724 +0.156 800 +0.000 219	0.273 062 9.27
2008 Mar	5	14	−15.98 905 + 0.00 755	52.70 015 14.99 531	−0.006 455 +0.488 981 −0.000 038	−0.198 788 +0.221 407 +0.000 111	0.273 206 11.88
2008 Aug	1	16	+18.63 052 − 0.02 339	54.72 543 14.95 562	−0.304 289 +0.477 330 −0.000 094	−1.230 060 −0.201 259 −0.000 152	0.272 994 8.64
2008 Sep	30	10	−11.29 936 + 0.01 764	321.83 854 15.07 006	−0.707 984 +0.545 531 +0.000 104	−0.907 295 −0.255 425 −0.000 085	0.273 554 17.19
2008 Dec	29	4	−22.86 991 + 0.01 435	220.00 238 14.98 019	−0.085 114 +0.467 053 −0.000 022	+0.707 660 +0.134 028 +0.000 161	0.273 142 10.20
2009 Feb	22	21	−18.71 818 + 0.00 919	155.43 883 14.98 533	−0.558 979 +0.461 674 −0.000 024	+0.943 143 +0.187 148 +0.000 103	0.273 151 10.46
2009 Sep	18	24	− 2.72 123 + 0.02 749	180.19 500 15.07 999	−0.405 204 +0.581 172 +0.000 032	−1.003 560 −0.292 191 −0.000 003	0.273 543 18.12
2011 Oct	28	2	−20.01 519 − 0.01 919	196.84 486 14.98 187	−0.141 708 +0.553 075 +0.000 018	−0.199 193 −0.091 593 +0.000 214	0.273 003 9.01

T a b l e   III   (cont.)

Date	T_0	D0 D1	H0 H1	X0 X1 X2	Y0 Y1 Y2	k F

M e r c u r y

2012 Jul 20	8	+13.31 125 − 0.00 211	285.78 698 15.06 240	+0.083 639 +0.568 713 +0.000 053	−0.612 259 −0.171 160 +0.000 049	0.273 678 19.43
2012 Oct 17	2	−18.81 762 − 0.01 929	192.80 839 14.98 869	+0.146 930 +0.553 737 +0.000 024	+1.263 207 −0.106 113 +0.000 091	0.273 055 9.90
2012 Nov 14	11	−20.78 091 + 0.02 399	341.45 215 15.09 120	+0.441 866 +0.663 100 +0.000 031	+0.959 795 −0.099 941 −0.000 293	0.273 447 16.99
2012 Dec 12	1	−18.84 598 − 0.01 611	216.63 326 14.98 574	+0.137 986 +0.565 511 +0.000 013	−1.109 350 −0.053 222 +0.000 238	0.273 058 10.03
2013 May 9	19	+16.58 003 + 0.03 227	108.33 651 14.95 442	−0.140 648 +0.454 945 +0.000 001	+0.295 798 +0.072 993 −0.000 223	0.273 047 8.78
2013 Jul 8	12	+17.57 092 − 0.00 114	357.21 367 15.07 160	+0.159 516 +0.573 361 +0.000 007	+0.081 568 −0.092 821 +0.000 069	0.273 804 20.58
2013 Nov 3	7	−14.10 587 + 0.03 081	292.33 680 15.08 885	+0.090 578 +0.629 424 −0.000 117	−0.004 770 −0.172 610 −0.000 088	0.273 492 17.12
2013 Dec 1	23	−17.94 175 − 0.01 961	182.80 588 14.97 927	+0.302 709 +0.542 162 +0.000 041	+0.388 832 −0.074 037 +0.000 148	0.273 015 9.08
2014 Jun 26	12	+18.75 556 − 0.00 235	9.46 030 15.05 918	+0.015 515 +0.576 769 −0.000 044	+0.293 227 −0.003 238 +0.000 026	0.273 737 19.84
2014 Oct 22	22	− 6.39 492 + 0.02 064	164.87 133 15.05 922	+0.027 979 +0.558 377 −0.000 158	−0.790 331 −0.188 321 +0.000 104	0.273 470 15.74
2015 Jun 15	2	+16.95 872 + 0.00 204	229.25 537 15.03 323	−0.260 690 +0.577 426 −0.000 041	−0.079 748 +0.074 746 −0.000 043	0.273 533 17.57
2015 Oct 11	11	+ 0.36 238 + 0.00 405	3.25 620 15.02 634	−0.512 354 +0.504 366 −0.000 112	−0.881 443 −0.175 256 +0.000 116	0.273 377 13.85

Table III (cont.)

Date	T_0	D0 D1	H0 H1	X0 X1 X2	Y0 Y1 Y2	k F

Mercury

Date	T_0	D0 / D1	H0 / H1	X0 / X1 / X2	Y0 / Y1 / Y2	k / F
2016 Jun 3	10	+ 14.03 452 + 0.01 068	354.25 414 15.00 908	+ 0.109 887 + 0.568 249 − 0.000 024	− 0.711 433 + 0.128 556 − 0.000 055	0.273 335 14.84
2016 Aug 4	22	+ 8.77 959 − 0.02 707	124.60 758 14.99 004	− 0.098 287 + 0.491 017 − 0.000 025	− 0.573 807 − 0.134 211 − 0.000 030	0.273 157 10.81
2016 Sep 29	10	+ 5.54 259 − 0.01 055	348.33 683 14.99 850	− 0.338 220 + 0.487 211 − 0.000 094	− 0.659 418 − 0.160 732 + 0.000 053	0.273 241 11.98
2017 Jul 25	9	+ 11.36 099 − 0.02 345	286.98 305 14.99 877	+ 0.094 132 + 0.534 830 − 0.000 029	+ 0.856 706 − 0.132 196 − 0.000 110	0.273 214 12.35
2017 Sep 18	23	+ 9.02 618 − 0.02 049	180.46 708 14.97 864	− 0.185 758 + 0.498 222 − 0.000 077	+ 0.026 157 − 0.154 037 − 0.000 048	0.273 114 10.43
2018 Feb 15	18	− 14.90 647 + 0.02 623	87.11 771 14.96 982	− 0.452 926 + 0.465 366 − 0.000 007	+ 1.080 326 + 0.112 391 + 0.000 104	0.273 018 8.36
2018 Sep 8	23	+ 11.16 611 − 0.02 612	175.27 844 14.96 738	+ 0.365 137 + 0.515 549 − 0.000 075	+ 0.811 327 − 0.154 793 − 0.000 156	0.273 026 9.36
2019 Feb 5	7	− 16.43 589 + 0.02 548	276.38 776 14.96 931	− 0.022 619 + 0.455 829 − 0.000 047	− 0.219 910 + 0.103 508 + 0.000 161	0.273 038 8.53
2020 Dec 14	11	− 23.97 957 − 0.00 899	349.80 558 14.97 109	+ 0.228 043 + 0.538 540 + 0.000 017	+ 0.926 117 − 0.094 067 + 0.000 218	0.272 954 8.05

Venus

Date	T_0	D0 / D1	H0 / H1	X0 / X1 / X2	Y0 / Y1 / Y2	k / F
1990 Aug 18	24	+ 19.27 002 − 0.01 083	198.35 019 14.98 916	− 0.055 751 + 0.509 938 − 0.000 056	− 0.483 117 − 0.189 295 − 0.000 128	0.272 925 7.33
1990 Dec 18	6	− 24.10 910 + 0.00 144	258.41 327 14.98 454	+ 0.220 058 + 0.487 750 − 0.000 041	− 1.050 712 + 0.076 300 + 0.000 238	0.272 939 6.99

Table III (cont.)

Date	T_0	D0 D1	H0 H1	X0 X1 X2	Y0 Y1 Y2	k F

Venus

1991 Oct 4	15	+ 8.93930 − 0.00165	88.28259 15.01626	−0.200062 +0.523264 −0.000007	−0.142322 −0.239271 −0.000025	0.273914 24.42
1992 Jan 31	17	−22.35317 + 0.00077	106.67640 14.98668	+0.022648 +0.486577 −0.000016	−1.138403 +0.080929 +0.000221	0.273061 8.91
1992 Oct 28	15	−23.47270 − 0.00899	13.22507 14.98798	+0.194137 +0.528482 −0.000051	+0.396936 −0.022360 +0.000186	0.273050 9.26
1993 Feb 25	5	+10.91689 + 0.01308	216.93441 15.02770	+0.302598 +0.499452 +0.000003	−0.449227 +0.177545 −0.000003	0.274241 27.55
1993 Apr 19	16	+ 5.54493 − 0.01242	85.90676 15.04553	−0.266445 +0.515318 −0.000011	+0.427054 +0.216488 −0.000028	0.274753 35.57
1993 Dec 12	18	−21.86731 − 0.00845	100.60184 14.98615	−0.009051 +0.552207 −0.000004	+0.449596 −0.015629 +0.000175	0.272894 6.91
1994 Apr 12	23	+15.87428 + 0.01687	144.62090 14.99168	−0.199486 +0.483050 +0.000020	+1.108183 +0.097500 −0.000161	0.272969 7.45
1995 Jan 27	12	−20.05941 − 0.00508	46.11353 14.99487	−0.116708 +0.562222 +0.000025	+0.174911 +0.004533 +0.000144	0.273353 14.94
1995 May 27	7	+14.18490 + 0.01599	308.83678 14.99243	−0.182762 +0.486671 +0.000033	+0.898797 +0.101145 −0.000135	0.272973 7.56
1996 Feb 22	5	+ 6.31212 + 0.02093	212.37602 14.99875	+0.192114 +0.515143 −0.000043	−0.002275 +0.154990 −0.000064	0.273187 11.67
1996 Jul 12	9	+17.68051 + 0.00060	352.69891 15.02921	+0.180838 +0.540168 −0.000016	+0.425855 +0.041748 −0.000055	0.274260 28.09
1997 Apr 7	13	+ 6.35599 + 0.02018	12.83743 14.99407	+0.094177 +0.536390 −0.000008	−0.648609 +0.161157 −0.000020	0.272880 6.74

Table III   (cont.)

Date	T_0	D0 D1	H0 H1	X0 X1 X2	Y0 Y1 Y2	k F

Venus

Date	T_0	D0 / D1	H0 / H1	X0 / X1 / X2	Y0 / Y1 / Y2	k / F
1998 Mar 24	19	− 13.72 116 + 0.00 723	147.72 645 15.00 222	− 0.020 964 + 0.548 695 + 0.000 020	+ 0.089 044 + 0.125 135 + 0.000 091	0.273 520 17.76
1998 Apr 23	8	− 5.28 943 + 0.01 487	341.59 628 14.99 905	+ 0.252 029 + 0.538 411 + 0.000 013	+ 0.001 950 + 0.173 408 + 0.000 050	0.273 242 13.12
1998 Sep 19	18	+ 6.79 593 − 0.01 954	100.74 185 14.99 359	− 0.166 970 + 0.478 670 − 0.000 016	− 0.031 637 − 0.148 991 − 0.000 043	0.272 929 6.92
2000 Mar 4	1	− 16.02 196 + 0.01 418	216.10 180 14.99 045	+ 0.102 312 + 0.495 249 + 0.000 006	− 0.672 389 + 0.111 491 + 0.000 161	0.272 981 7.84
2000 Aug 1	2	+ 15.21 561 − 0.01 646	194.09 901 14.99 170	+ 0.208 015 + 0.548 262 − 0.000 070	+ 1.018 496 − 0.138 345 − 0.000 172	0.272 892 7.01
2001 Jul 17	18	+ 20.07 860 + 0.00 694	133.44 928 14.99 350	+ 0.163 967 + 0.536 046 + 0.000 074	+ 0.312 098 + 0.114 614 − 0.000 161	0.273 168 11.52
2001 Dec 14	6	− 22.27 395 − 0.00 739	279.41 135 14.98 577	+ 0.070 074 + 0.515 933 + 0.000 021	+ 0.866 749 − 0.109 069 + 0.000 150	0.272 908 6.89
2002 May 14	23	+ 24.60 441 + 0.00 397	134.91 408 14.98 705	− 0.000 374 + 0.514 264 + 0.000 030	− 0.904 229 + 0.077 331 − 0.000 161	0.273 001 8.30
2003 May 29	4	+ 15.11 576 + 0.01 531	262.88 608 14.99 192	− 0.026 419 + 0.455 690 + 0.000 061	+ 0.128 290 + 0.187 761 − 0.000 101	0.272 971 7.47
2003 Oct 26	20	− 18.05 686 − 0.01 500	106.00 295 14.98 980	+ 0.035 899 + 0.532 330 + 0.000 038	− 0.088 799 − 0.201 848 + 0.000 157	0.272 904 7.27
2004 May 21	12	+ 26.73 904 − 0.00 619	334.06 277 15.05 203	− 0.146 230 + 0.555 766 + 0.000 024	+ 0.355 836 + 0.059 889 + 0.000 041	0.274 723 34.98
2004 Nov 10	2	− 4.15 968 − 0.01 873	245.39 769 14.99 465	+ 0.149 886 + 0.473 653 + 0.000 058	− 0.273 394 − 0.254 508 + 0.000 020	0.273 026 8.99

Table III (cont.)

Date	T_0	D0 D1	H0 H1	X0 X1 X2	Y0 Y1 Y2	k F
Venus						
2005 Apr 9	1	+ 7.35 910 + 0.01 986	192.03 171 14.99 382	−0.323 621 +0.475 846 +0.000 020	+1.009 330 +0.249 173 −0.000 108	0.272 895 6.75
2005 Aug 8	5	+ 5.23 290 − 0.02 050	221.12 674 14.99 657	+0.493 039 +0.439 296 −0.000 044	+1.103 705 −0.233 929 −0.000 073	0.273 061 8.96
2005 Sep 7	8	−10.08 041 − 0.01 976	263.55 242 14.99 720	−0.484 037 +0.457 253 +0.000 048	−0.434 692 −0.231 000 +0.000 081	0.273 151 10.63
2006 Apr 24	14	− 4.72 991 + 0.01 505	71.23 577 14.99 863	+0.190 529 +0.494 181 −0.000 004	−0.352 318 +0.272 932 +0.000 058	0.273 222 12.73
2006 Sep 21	15	+ 5.63 931 − 0.01 974	54.85 637 14.99 371	−0.274 928 +0.429 994 −0.000 003	−0.900 445 −0.238 595 −0.000 003	0.272 934 6.89
2007 Jan 20	17	−16.08 270 + 0.01 626	51.05 857 14.99 058	+0.061 974 +0.505 302 −0.000 063	−0.738 939 +0.222 374 +0.000 161	0.272 926 7.45
2007 Jun 18	15	+18.93 512 − 0.01 243	356.46 523 15.00 699	+0.099 395 +0.504 682 −0.000 073	+0.568 244 −0.191 266 −0.000 111	0.273 608 18.46
2008 Mar 5	19	−15.23 127 + 0.01 473	125.08 945 14.99 080	−0.164 972 +0.482 990 +0.000 001	+0.175 097 +0.221 484 +0.000 117	0.272 957 7.74
2008 Dec 1	16	−23.90 553 + 0.00 786	16.04 605 14.98 952	−0.024 972 +0.477 777 −0.000 059	+0.888 782 +0.127 798 +0.000 137	0.273 231 11.60
2009 Feb 28	0	+11.19 595 + 0.00 915	146.86 491 15.03 738	+0.501 203 +0.520 157 −0.000 007	−1.127 222 +0.238 051 +0.000 052	0.274 367 31.14
2009 Apr 22	13	+ 2.72 523 − 0.00 695	47.30 376 15.03 406	−0.658 606 +0.511 259 +0.000 025	+0.793 889 +0.263 955 −0.000 054	0.274 398 31.63
2010 May 16	10	+24.79 576 + 0.00 319	298.80 980 14.98 709	−0.152 465 +0.542 748 −0.000 011	+0.095 892 −0.034 360 −0.000 220	0.272 990 8.43

T a b l e   III   (cont.)

Date	T_0	D0 D1	H0 H1	X0 X1 X2	Y0 Y1 Y2	k F
Venus						
2010 Sep 11	13	− 16.27 729 − 0.01 450	337.03 354 15.01 645	−0.069 915 +0.547 592 +0.000 006	−0.315 920 −0.177 619 +0.000 096	0.273 814 22.90
2010 Nov 5	8	− 16.02 669 + 0.01 890	316.86 405 15.06 335	−0.304 869 +0.598 250 −0.000 014	−0.066 745 −0.210 395 −0.000 061	0.274 919 42.24
2011 Jun 30	8	+ 23.10 311 + 0.00 356	313.16 477 14.98 661	+0.225 695 +0.522 502 +0.000 012	+0.074 530 −0.036 342 −0.000 196	0.272 916 6.95
2012 Aug 13	20	+ 19.93 835 + 0.00 142	166.54 211 15.00 105	+0.147 538 +0.513 065 +0.000 009	+0.589 267 −0.066 033 −0.000 146	0.273 540 16.77
2013 Sep 8	21	− 11.06 254 − 0.01 964	98.03 069 14.99 735	−0.045 205 +0.515 535 +0.000 020	−0.438 024 −0.141 106 +0.000 097	0.273 146 10.90
2014 Feb 26	5	− 16.63 884 − 0.00 021	295.93 150 15.01 236	−0.234 906 +0.574 231 +0.000 010	+0.305 573 +0.108 742 +0.000 064	0.273 849 23.83
2014 Oct 23	21	− 10.48 853 − 0.01 914	138.90 550 14.99 247	−0.129 750 +0.497 698 +0.000 044	−0.038 732 −0.129 821 +0.000 071	0.272 911 6.78
2015 Jul 19	1	+ 9.23 953 − 0.01 031	160.25 032 15.03 829	−0.070 090 +0.527 841 +0.000 015	−0.437 840 −0.142 835 +0.000 017	0.274 392 30.11
2015 Oct 8	20	+ 9.51 288 − 0.00 604	165.69 564 15.00 877	−0.273 898 +0.492 921 −0.000 000	−0.685 759 −0.144 144 −0.000 017	0.273 809 20.76
2015 Dec 7	17	− 10.37 436 − 0.01 593	119.60 108 14.99 526	+0.026 970 +0.480 928 +0.000 049	+0.722 274 −0.131 789 +0.000 028	0.273 227 11.60
2016 Apr 6	8	− 1.06 306 + 0.02 008	313.73 073 14.99 461	−0.275 052 +0.541 700 +0.000 037	+0.594 736 +0.173 818 −0.000 014	0.272 896 7.13
2016 Sep 3	11	− 1.23 590 − 0.02 123	322.71 065 14.99 551	+0.219 358 +0.478 300 −0.000 039	+1.155 674 −0.152 609 −0.000 025	0.272 976 7.67

Table III (cont.)

Date	T_0	D0 D1	H0 H1	X0 X1 X2	Y0 Y1 Y2	k F

Venus

2017 Sep 18	1	+13.12 422 − 0.01 575	222.10 568 14.99 248	+0.031 564 +0.523 632 −0.000 024	−0.572 443 −0.136 857 −0.000 074	0.272 972 8.11
2018 Feb 16	17	−10.22 990 + 0.01 924	62.19 101 14.99 264	+0.402 807 +0.486 437 −0.000 028	−0.483 038 +0.143 751 +0.000 099	0.272 924 6.91
2019 Jan 31	18	−20.79 511 − 0.00 333	134.53 682 14.99 226	+0.202 914 +0.501 843 +0.000 016	+0.076 728 −0.055 237 +0.000 148	0.273 337 13.44
2019 Jul 31	21	+20.01 901 − 0.01 107	136.93 130 14.98 856	+0.214 957 +0.559 205 −0.000 021	+0.555 349 −0.084 683 −0.000 211	0.272 876 6.73
2019 Dec 29	2	−19.29 800 + 0.01 407	173.42 359 14.99 004	+0.228 715 +0.498 154 −0.000 081	−1.040 045 +0.096 636 +0.000 179	0.273 053 8.99
2020 Jun 19	9	+18.55 211 − 0.00 931	338.39 892 15.05 440	+0.045 549 +0.558 175 −0.000 041	+0.816 734 +0.152 314 −0.000 008	0.274 730 36.07
2020 Dec 12	21	−18.04 983 − 0.01 293	162.49 423 14.98 909	+0.170 056 +0.534 174 +0.000 057	+0.728 286 −0.178 993 +0.000 105	0.272 940 7.84

Mars

1990 Mar 22	18	−19.23 095 + 0.00 727	139.04 377 15.00 997	+0.091 573 +0.511 830 +0.000 004	+0.471 956 +0.191 966 +0.000 081	0.272 884 6.49
1991 Mar 22	17	+25.31 409 + 0.00 066	351.20 044 15.01 875	+0.277 577 +0.576 799 −0.000 025	+0.651 499 −0.018 909 −0.000 130	0.272 966 8.12
1991 Apr 19	24	+24.90 447 − 0.00 181	107.45 305 15.01 668	+0.139 928 +0.571 034 −0.000 044	−0.631 724 −0.091 414 −0.000 068	0.272 889 6.82
1992 Jan 3	10	−23.85 210 − 0.00 149	347.13 789 15.00 808	−0.172 434 +0.518 079 +0.000 006	−0.919 564 +0.032 308 +0.000 158	0.272 801 4.84

Table III   (cont.)

Date	T_0	D0 D1	H0 H1	X0 X1 X2	Y0 Y1 Y2	k F
Mars						
1992 Sep 20	9	+ 23.46 642 + 0.00 030	39.56 800 15.01 760	+ 0.006 165 + 0.562 774 + 0.000 009	− 0.886 182 − 0.071 052 − 0.000 048	0.273 060 9.62
1993 Nov 14	18	− 21.35 997 − 0.00 597	81.95 641 15.01 003	+ 0.181 582 + 0.578 169 − 0.000 021	+ 0.495 981 − 0.052 458 + 0.000 092	0.272 766 4.74
1994 Jul 5	5	+ 19.96 361 + 0.00 650	299.12 534 15.01 107	+ 0.059 495 + 0.516 131 + 0.000 038	− 0.368 619 + 0.058 001 − 0.000 071	0.272 858 5.76
1995 Aug 30	3	− 9.65 143 − 0.01 021	180.05 024 15.01 638	− 0.297 815 + 0.537 775 + 0.000 011	− 0.074 993 − 0.154 973 + 0.000 048	0.272 837 5.78
1996 Apr 17	5	+ 6.51 813 + 0.01 229	263.56 125 15.01 210	− 0.150 947 + 0.529 594 + 0.000 012	+ 0.576 954 + 0.159 346 − 0.000 059	0.272 786 4.88
1997 Jun 13	16	+ 1.45 229 − 0.00 734	323.97 811 15.02 706	+ 0.000 245 + 0.504 149 + 0.000 001	− 0.305 887 − 0.164 439 + 0.000 008	0.273 147 10.33
1998 Feb 27	23	− 2.13 273 + 0.01 305	145.64 416 15.01 205	+ 0.354 668 + 0.558 840 − 0.000 035	− 0.616 016 + 0.181 472 + 0.000 041	0.272 778 5.02
1998 Oct 16	3	+ 10.88 516 − 0.00 884	271.84 996 15.01 766	− 0.281 341 + 0.508 554 + 0.000 015	− 0.999 953 − 0.148 511 − 0.000 004	0.272 833 5.41
1998 Nov 13	18	+ 4.54 723 − 0.00 906	149.17 920 15.01 938	+ 0.040 082 + 0.498 155 + 0.000 006	+ 0.567 623 − 0.166 535 − 0.000 044	0.272 873 6.00
1999 Dec 12	18	− 18.25 820 + 0.00 945	35.47 803 15.00 945	− 0.340 036 + 0.509 635 − 0.000 017	+ 0.552 257 + 0.098 499 + 0.000 062	0.272 914 6.67
2000 Jul 30	12	+ 21.36 879 − 0.00 533	7.07 190 15.01 323	− 0.007 834 + 0.590 136 − 0.000 006	− 0.616 956 − 0.066 891 − 0.000 071	0.272 745 4.44
2000 Aug 28	3	+ 16.71 717 − 0.00 785	241.44 399 15.01 518	− 0.006 932 + 0.573 385 − 0.000 008	+ 0.917 250 − 0.130 629 − 0.000 129	0.272 751 4.51

Table III (cont.)

Date	T_0	$D0$ $D1$	$H0$ $H1$	$X0$ $X1$ $X2$	$Y0$ $Y1$ $Y2$	k F
Mars						
2001 Oct 23	20	−22.94 843 + 0.00 648	32.39 014 15.01 269	−0.082 001 +0.519 266 −0.000 024	−0.142 556 +0.059 680 +0.000 103	0.273 213 11.47
2002 May 14	19	+24.03 361 + 0.00 204	77.28 011 15.01 139	+0.202 699 +0.535 530 +0.000 003	−0.641 343 +0.089 137 −0.000 085	0.272 788 4.81
2002 Jun 12	12	+24.10 500 − 0.00 189	339.84 807 15.01 193	+0.062 739 +0.556 389 −0.000 004	+0.984 580 +0.007 909 −0.000 163	0.272 768 4.56
2002 Dec 30	2	−16.61 436 − 0.00 748	262.27 071 15.01 488	+0.604 034 +0.537 354 +0.000 018	+0.955 455 −0.207 645 +0.000 013	0.272 820 5.63
2003 Jan 27	15	−21.00 884 − 0.00 478	106.51 175 15.01 359	−0.099 829 +0.547 633 +0.000 046	−0.407 321 −0.146 658 +0.000 112	0.272 868 6.38
2003 Jul 17	8	−13.03 904 − 0.00 106	71.91 511 15.03 673	−0.139 816 +0.520 354 −0.000 010	+0.299 415 +0.228 933 −0.000 006	0.273 981 24.47
2003 Sep 9	12	−16.44 157 − 0.00 237	191.72 840 15.05 403	−0.455 889 +0.545 484 +0.000 041	+1.139 006 +0.221 560 −0.000 083	0.274 338 30.39
2003 Oct 6	16	−15.06 758 + 0.00 382	280.63 624 15.04 012	+0.560 140 +0.528 526 −0.000 017	−0.909 712 +0.208 576 +0.000 048	0.273 983 24.40
2004 Feb 26	2	+17.05 332 + 0.00 798	143.55 268 15.01 528	+0.221 749 +0.482 576 −0.000 008	−0.891 754 +0.200 542 −0.000 035	0.272 945 7.17
2004 Mar 25	23	+21.84 214 + 0.00 521	108.04 125 15.01 362	−0.480 373 +0.498 719 +0.000 018	+0.762 789 +0.146 349 −0.000 125	0.272 887 6.19
2004 Oct 13	9	− 3.61 765 − 0.01 062	326.81 203 15.01 657	+0.348 809 +0.494 336 +0.000 008	+1.091 435 −0.265 276 −0.000 048	0.272 758 4.44
2004 Nov 11	4	−10.88 483 − 0.00 977	262.54 003 15.01 529	−0.173 657 +0.513 731 +0.000 057	−0.381 862 −0.250 619 +0.000 053	0.272 763 4.61

ASTRONOMICAL TABLES

Table III (cont.)

Date	T_0	$D0$ $D1$	$H0$ $H1$	$X0$ $X1$ $X2$	$Y0$ $Y1$ $Y2$	k F
Mars						
2005 May 31	10	− 5.20 546 + 0.01 055	45.71 565 15.01 458	+0.354 257 +0.493 862 −0.000 030	−0.342 339 +0.262 970 +0.000 038	0.273 072 9.79
2005 Dec 12	4	+15.32 557 + 0.00 060	105.44 263 15.04 258	−0.695 331 +0.535 744 +0.000 042	+1.025 334 +0.231 014 −0.000 042	0.273 616 18.76
2006 Jul 27	18	+11.36 603 − 0.00 944	59.85 084 15.01 692	+0.466 532 +0.469 061 −0.000 061	+0.954 472 −0.230 040 −0.000 069	0.272 792 4.69
2006 Aug 25	13	+ 4.26 420 − 0.01 052	356.34 873 15.01 713	−0.314 355 +0.455 291 −0.000 011	−0.442 902 −0.249 498 +0.000 009	0.272 783 4.53
2007 Apr 14	1	−10.59 828 + 0.01 119	238.65 380 15.01 142	−0.451 680 +0.510 326 +0.000 035	+0.286 547 +0.260 985 +0.000 048	0.272 854 6.26
2007 Dec 24	3	+26.73 996 + 0.00 169	44.03 704 15.06 232	+0.021 594 +0.629 389 +0.000 011	+0.892 016 −0.045 051 +0.000 054	0.273 617 19.70
2008 Jan 19	24	+26.86 057 − 0.00 063	34.51 315 15.05 021	+0.200 108 +0.613 778 −0.000 001	+1.126 847 +0.000 044 −0.000 017	0.273 460 16.79
2008 Apr 12	6	+24.59 122 − 0.00 282	183.25 535 15.02 038	+0.222 140 +0.562 054 −0.000 045	+1.179 716 −0.107 392 −0.000 142	0.272 965 8.04
2008 May 10	14	+21.76 023 − 0.00 512	315.74 948 15.01 854	+0.128 285 +0.544 430 −0.000 059	+0.210 525 −0.169 196 −0.000 084	0.272 892 6.80
2008 Jun 8	1	+17.31 299 − 0.00 735	132.69 784 15.01 809	−0.652 378 +0.522 973 −0.000 029	−0.839 622 −0.219 929 −0.000 009	0.272 845 5.97
2009 Jan 25	3	−22.54 343 + 0.00 470	235.77 200 15.00 736	+0.320 612 +0.499 742 −0.000 025	−0.708 297 +0.130 950 +0.000 144	0.272 806 4.91
2009 Sep 13	16	+23.33 824 − 0.00 126	130.00 052 15.01 492	−0.009 108 +0.567 594 +0.000 005	+1.130 019 −0.102 838 −0.000 164	0.272 935 7.62

Table III (cont.)

Date	T_0	D0 D1	H0 H1	X0 X1 X2	Y0 Y1 Y2	k F
Mars						
2009 Oct 12	1	+21.62081 − 0.00304	275.80628 15.01915	−0.261018 +0.552410 +0.000002	−1.094429 −0.166505 −0.000024	0.273001 8.72
2010 Dec 6	22	−24.28843 − 0.00052	136.47226 15.00717	+0.119591 +0.545177 −0.000045	+0.564362 +0.050293 +0.000108	0.272790 4.91
2011 Jul 27	17	+23.59950 + 0.00152	114.94675 15.01097	+0.051328 +0.544570 +0.000030	−0.505341 −0.033510 −0.000090	0.272823 5.42
2012 Sep 19	21	−17.88699 − 0.00820	88.83659 15.01326	+0.200047 +0.567275 −0.000017	−0.189711 −0.109478 +0.000102	0.272848 6.14
2013 May 9	14	+15.92537 + 0.00940	35.78781 15.01136	−0.079575 +0.513847 +0.000015	+0.451239 +0.105433 −0.000093	0.272792 4.74
2014 Jul 6	1	− 8.52854 − 0.00710	100.29505 15.02600	−0.198498 +0.522465 +0.000023	+0.286938 −0.158192 +0.000004	0.273195 11.40
2015 Mar 21	23	+ 8.62386 + 0.01185	143.00587 15.01253	+0.418125 +0.559926 −0.000036	−0.836090 +0.163234 −0.000006	0.272779 5.01
2015 Dec 6	3	− 4.08106 − 0.00916	286.33371 15.01917	+0.142538 +0.495625 +0.000022	−0.152805 −0.157871 +0.000009	0.272879 6.05
2017 Jan 3	7	− 8.11894 + 0.01239	225.06906 15.01273	+0.108045 +0.522179 −0.000024	+0.288599 +0.158332 +0.000021	0.272917 7.03
2017 Sep 18	20	+ 9.46009 − 0.00975	137.32610 15.01676	+0.148586 +0.538852 −0.000015	+0.098717 −0.162186 −0.000045	0.272760 4.48
2018 Nov 16	5	−12.97389 + 0.01014	157.36768 15.01851	+0.346242 +0.500938 −0.000019	−0.994182 +0.143333 +0.000088	0.273315 12.99
2019 Jul 4	6	+20.98374 − 0.00582	247.99781 15.01399	+0.189876 +0.579277 −0.000032	+0.062884 −0.076074 −0.000100	0.272753 4.51

Table III   (cont.)

Date	T_0	D0 D1	H0 H1	X0 X1 X2	Y0 Y1 Y2	k F

Mars

2020 Feb 18	13	− 23.66 291	71.41 562	− 0.164 437	+ 0.815 014	0.272 887
		− 0.000 25	15.01 055	+ 0.538 866	− 0.044 011	6.44
				+ 0.000 024	+ 0.000 079	
2020 Mar 18	8	− 22.44 672	3.17 497	− 0.172 749	− 0.808 517	0.272 951
		+ 0.00 372	15.01 092	+ 0.531 021	+ 0.036 553	7.41
				+ 0.000 002	+ 0.000 143	
2020 Aug 9	9	+ 4.73 367	71.79 122	+ 0.474 930	− 0.625 587	0.273 732
		+ 0.00 428	15.02 813	+ 0.485 150	+ 0.215 978	19.49
				− 0.000 023	+ 0.000 013	
2020 Sep 6	5	+ 6.79 815	33.17 581	+ 0.106 544	+ 0.075 821	0.274 057
		− 0.00 021	15.04 159	+ 0.502 190	+ 0.217 396	24.55
				− 0.000 013	− 0.000 005	
2020 Oct 3	4	+ 6.11 507	48.35 368	+ 0.285 541	− 0.681 160	0.274 286
		− 0.00 383	15.05 607	+ 0.517 564	+ 0.227 266	28.12
				− 0.000 022	+ 0.000 041	

Jupiter

1990 Jul 21	18	+ 21.55 451	93.36 240	− 0.028 855	+ 1.024 385	0.272 602
		− 0.00 149	15.03 142	+ 0.574 717	− 0.155 934	1.86
				− 0.000 008	− 0.000 090	
1990 Aug 18	13	+ 20.42 258	39.33 719	− 0.079 748	+ 0.427 243	0.272 606
		− 0.00 173	15.03 206	+ 0.560 374	− 0.177 355	1.90
				+ 0.000 009	− 0.000 056	
1990 Sep 15	6	+ 19.20 972	315.90 428	+ 0.093 322	− 0.313 994	0.272 612
		− 0.00 171	15.03 340	+ 0.546 969	− 0.192 920	1.98
				+ 0.000 008	− 0.000 014	
1990 Oct 12	19	+ 18.12 671	173.47 458	− 0.344 552	− 0.882 771	0.272 620
		− 0.00 133	15.03 552	+ 0.539 288	− 0.204 372	2.10
				+ 0.000 025	+ 0.000 022	
1994 Oct 7	12	− 15.91 783	331.75 842	− 0.009 369	− 0.727 102	0.272 601
		− 0.00 244	15.03 296	+ 0.595 011	− 0.118 752	1.87
				− 0.000 016	+ 0.000 070	
1994 Nov 4	8	− 17.56 094	293.27 810	+ 0.258 450	− 0.150 174	0.272 597
		− 0.00 235	15.03 206	+ 0.604 632	− 0.100 701	1.83
				− 0.000 018	+ 0.000 039	

Table III (cont.)

Date	T_0	$D0$ $D1$	$H0$ $H1$	$X0$ $X1$ $X2$	$Y0$ $Y1$ $Y2$	k F
Jupiter						
1994 Dec 2	5	-19.05519 $-\,0.00201$	269.46047 15.03187	$+0.269910$ $+0.607580$ -0.000002	$+0.432648$ -0.080638 $+0.000004$	0.272598 1.83
1994 Dec 30	1	-20.26239 $-\,0.00153$	230.73952 15.03248	$+0.324211$ $+0.603071$ $+0.000008$	$+1.022874$ -0.060786 -0.000031	0.272602 1.88
1998 Mar 26	11	$-\,7.94817$ $+\,0.00352$	5.01009 15.03231	-0.331344 $+0.579959$ $+0.000033$	$+0.709554$ $+0.178345$ -0.000019	0.272606 1.97
1998 Apr 23	7	$-\,5.65438$ $+\,0.00313$	326.76502 15.03341	-0.120350 $+0.572345$ $+0.000028$	$+0.176683$ $+0.184665$ $+0.000009$	0.272612 2.06
1998 May 20	23	$-\,3.74670$ $+\,0.00239$	229.33154 15.03518	-0.164264 $+0.564196$ $+0.000025$	-0.414875 $+0.188156$ $+0.000033$	0.272621 2.19
1998 Jun 17	11	$-\,2.45826$ $+\,0.00128$	73.15675 15.03768	-0.015690 $+0.560690$ $+0.000003$	-0.817914 $+0.190192$ $+0.000045$	0.272633 2.38
1998 Jul 14	19	$-\,2.00209$ $-\,0.00012$	218.70521 15.04088	$+0.135340$ $+0.565282$ -0.000019	-0.960296 $+0.192453$ $+0.000045$	0.272646 2.60
1998 Aug 11	0	$-\,2.48152$ $-\,0.00152$	321.30148 15.04422	$+0.080253$ $+0.576694$ -0.000021	-0.845139 $+0.195403$ $+0.000037$	0.272656 2.80
1998 Sep 7	4	$-\,3.71153$ $-\,0.00232$	50.70890 15.04635	-0.138075 $+0.587102$ $+0.000002$	-0.561136 $+0.197108$ $+0.000026$	0.272661 2.93
1998 Oct 4	10	$-\,5.10460$ $-\,0.00197$	170.80404 15.04593	$+0.297848$ $+0.586931$ -0.000006	-0.111107 $+0.195096$ $+0.000007$	0.272659 2.90
1998 Oct 31	16	$-\,5.93119$ $-\,0.00068$	289.79482 15.04314	-0.162300 $+0.574821$ $+0.000026$	-0.279210 $+0.190219$ $+0.000020$	0.272652 2.73
1998 Nov 28	1	$-\,5.79878$ $+\,0.00087$	91.74778 15.03953	$+0.256042$ $+0.559502$ -0.000006	-0.533824 $+0.186287$ $+0.000032$	0.272641 2.50

T a b l e　　III　　(cont.)

Date	T_0	D0 D1	H0 H1	X0 X1 X2	Y0 Y1 Y2	k F
Jupiter						
1998 Dec 25	11	− 4.71 289 + 0.00 219	266.49 507 15.03 638	+0.247 498 +0.550 590 −0.000 023	−1.150 276 +0.185 870 +0.000 056	0.272 630 2.29
2001 May 24	8	+22.60 144 + 0.00 084	284.29 876 15.03 111	+0.454 466 +0.579 610 −0.000 024	−1.213 057 +0.101 470 +0.000 023	0.272 607 1.92
2001 Jun 21	4	+23.03 855 + 0.00 039	244.80 624 15.03 091	+0.365 927 +0.590 102 −0.000 006	−0.691 373 +0.076 940 −0.000 004	0.272 605 1.90
2001 Jul 19	0	+23.16 354 − 0.00 006	205.41 636 15.03 143	−0.091 988 +0.595 351 +0.000 030	−0.198 604 +0.049 455 −0.000 029	0.272 606 1.93
2001 Aug 15	20	+23.01 776 − 0.00 038	166.61 925 15.03 272	+0.112 303 +0.594 296 +0.000 023	+0.386 869 +0.022 195 −0.000 056	0.272 611 2.01
2001 Sep 12	12	+22.72 285 − 0.00 047	68.82 510 15.03 488	−0.240 740 +0.588 924 +0.000 032	+0.970 909 −0.000 460 −0.000 076	0.272 620 2.14
2001 Dec 30	14	+22.99 332 + 0.00 056	207.26 662 15.04 755	−0.004 135 +0.607 443 −0.000 004	+1.260 899 +0.003 717 −0.000 038	0.272 656 2.78
2002 Jan 26	19	+23.29 032 + 0.00 032	312.77 106 15.04 610	+0.005 892 +0.603 269 +0.000 014	+0.922 629 +0.018 517 −0.000 028	0.272 652 2.70
2002 Feb 23	2	+23.43 233 + 0.00 010	86.58 545 15.04 239	−0.157 944 +0.589 133 +0.000 024	+0.882 100 +0.024 920 −0.000 039	0.272 644 2.52
2002 Mar 22	11	+23.44 772 − 0.00 008	247.88 034 15.03 838	−0.313 599 +0.575 698 +0.000 018	+1.173 277 +0.020 962 −0.000 067	0.272 633 2.31
2004 Nov 9	17	− 2.70 313 − 0.00 294	115.06 361 15.03 390	+0.676 894 +0.509 002 −0.000 004	+0.691 184 −0.272 135 −0.000 043	0.272 606 1.88
2004 Dec 7	11	− 4.50 268 − 0.00 219	47.97 378 15.03 544	+0.165 149 +0.507 763 +0.000 018	+0.271 317 −0.267 453 −0.000 019	0.272 614 1.99

T a b l e III (cont.)

Date	T_0	D0 D1	H0 H1	X0 X1 X2	Y0 Y1 Y2	k F
Jupiter						
2005 Jan 4	1	− 5.68675 − 0.00111	282.04562 15.03787	−0.328935 +0.505320 +0.000024	−0.207149 −0.261772 +0.000010	0.272624 2.14
2005 Jan 31	10	− 6.07089 + 0.00021	82.78134 15.04104	−0.411481 +0.505488 +0.000005	−0.769756 −0.260192 +0.000040	0.272637 2.33
2005 Feb 27	14	− 5.58999 + 0.00149	170.45189 15.04426	−0.306813 +0.509818 −0.000009	−1.069783 −0.264594 +0.000053	0.272648 2.50
2005 Mar 26	15	− 4.44051 + 0.00220	214.73144 15.04618	−0.373045 +0.514138 +0.000006	−0.852630 −0.271016 +0.000037	0.272653 2.60
2005 Apr 22	17	− 3.14594 + 0.00191	274.55274 15.04572	−0.282825 +0.512735 +0.000022	−0.502022 −0.273324 +0.000015	0.272651 2.58
2005 May 19	22	− 2.33094 + 0.00079	18.50589 15.04326	−0.196122 +0.505313 +0.000025	−0.276309 −0.270206 +0.000004	0.272644 2.45
2005 Jun 16	6	− 2.34990 − 0.00057	165.69803 15.04013	−0.422163 +0.496106 +0.000026	−0.215522 −0.264899 +0.000006	0.272634 2.27
2005 Jul 13	18	− 3.20902 − 0.00174	11.07439 15.03732	−0.179509 +0.489369 +0.000001	−0.724432 −0.260324 +0.000036	0.272625 2.10
2005 Aug 10	7	− 4.72366 − 0.00256	229.86079 15.03518	−0.544622 +0.487483 +0.000007	−1.094558 −0.257221 +0.000058	0.272617 1.96
2008 Dec 29	9	−20.89568 + 0.00181	292.67481 15.03130	−0.045428 +0.517129 −0.000023	−0.706527 +0.156969 +0.000057	0.272617 1.93
2009 Jan 26	5	−19.48692 + 0.00224	253.33081 15.03112	+0.185808 +0.512924 −0.000014	+0.103413 +0.179995 +0.000027	0.272615 1.91
2009 Feb 23	1	−17.85228 + 0.00245	214.12627 15.03169	−0.023327 +0.509045 +0.000012	+0.791219 +0.198946 −0.000001	0.272617 1.94

Table III   (cont.)

Date	T_0	D0 D1	H0 H1	X0 X1 X2	Y0 Y1 Y2	k F

Jupiter

Date	T_0	D0 D1	H0 H1	X0 X1 X2	Y0 Y1 Y2	k F
2012 Jun 17	8	+19.65 557 + 0.00 189	326.67 131 15.03 158	−0.248 057 +0.535 792 +0.000 032	+1.218 817 +0.059 598 −0.000 074	0.272 620 1.97
2012 Jul 15	3	+20.74 388 + 0.00 133	272.89 718 15.03 267	−0.066 525 +0.538 887 +0.000 030	+0.546 809 +0.040 171 −0.000 045	0.272 625 2.05
2012 Aug 11	21	+21.45 559 + 0.00 079	205.09 559 15.03 466	+0.258 278 +0.541 088 +0.000 010	−0.110 337 +0.023 991 −0.000 015	0.272 633 2.18
2012 Sep 8	11	+21.82 741 + 0.00 033	78.79 830 15.03 768	−0.051 760 +0.543 010 +0.000 005	−0.689 624 +0.012 873 +0.000 016	0.272 645 2.37
2012 Oct 5	21	+21.92 033 − 0.00 005	254.64 052 15.04 158	+0.017 317 +0.546 310 −0.000 017	−1.011 974 +0.008 815 +0.000 040	0.272 659 2.58
2012 Nov 2	1	+21.77 216 − 0.00 041	342.84 443 15.04 546	−0.066 047 +0.551 468 −0.000 019	−0.992 600 +0.013 351 +0.000 053	0.272 671 2.77
2012 Nov 29	1	+21.41 143 − 0.00 069	12.87 476 15.04 745	+0.045 387 +0.554 748 −0.000 007	−0.704 419 +0.025 285 +0.000 050	0.272 677 2.86
2012 Dec 26	0	+20.98 204 − 0.00 060	28.13 166 15.04 624	−0.104 229 +0.551 886 +0.000 021	−0.470 351 +0.038 541 +0.000 040	0.272 673 2.79
2013 Jan 22	3	+20.76 544 − 0.00 007	101.93 202 15.04 267	−0.033 365 +0.545 547 +0.000 021	−0.552 572 +0.046 050 +0.000 031	0.272 661 2.61
2013 Feb 18	12	+20.95 030 + 0.00 056	263.41 913 15.03 862	+0.206 318 +0.541 173 −0.000 005	−0.978 998 +0.044 492 +0.000 031	0.272 646 2.39
2016 Jul 9	10	+ 5.71 871 − 0.00 248	268.08 977 15.03 552	−0.105 855 +0.525 824 −0.000 019	−0.889 587 −0.169 837 +0.000 042	0.272 618 1.98
2016 Aug 6	3	+ 3.80 115 − 0.00 310	186.04 348 15.03 405	−0.273 690 +0.525 022 −0.000 015	−0.139 053 −0.171 404 +0.000 012	0.272 611 1.88

Table III (cont.)

Date	T_0	D0 D1	H0 H1	X0 X1 X2	Y0 Y1 Y2	k F

Jupiter

Date	T_0	D0 / D1	H0 / H1	X0 / X1 / X2	Y0 / Y1 / Y2	k / F
2016 Sep 2	22	+ 1.55 801 − 0.00 344	133.31 504 15.03 323	+0.034 588 +0.522 380 −0.000 018	+0.395 454 −0.172 234 −0.000 012	0.272 608 1.82
2016 Sep 30	17	− 0.81 344 − 0.00 349	80.23 808 15.03 299	+0.416 729 +0.518 541 −0.000 016	+0.853 980 −0.172 459 −0.000 035	0.272 608 1.80
2019 Nov 28	11	−23.28 992 − 0.00 012	323.09 492 15.03 134	+0.092 336 +0.571 548 −0.000 024	+0.749 714 −0.055 842 +0.000 003	0.272 608 1.90
2019 Dec 26	8	−23.23 011 + 0.00 033	298.69 367 15.03 087	+0.273 307 +0.568 795 −0.000 015	+0.176 450 −0.030 150 +0.000 032	0.272 608 1.87
2020 Jan 23	3	−22.86 588 + 0.00 075	244.19 915 15.03 124	+0.165 739 +0.562 303 +0.000 007	−0.382 840 −0.002 948 +0.000 056	0.272 611 1.90
2020 Feb 19	20	−22.26 838 + 0.00 100	160.16 773 15.03 245	+0.214 422 +0.555 796 +0.000 012	−0.990 496 +0.022 575 +0.000 078	0.272 617 1.97

Saturn

Date	T_0	D0 / D1	H0 / H1	X0 / X1 / X2	Y0 / Y1 / Y2	k / F
1990 Oct 25	18	−22.08 039 + 0.00 023	12.67 483 15.03 888	+0.406 437 +0.533 595 −0.000 029	−1.105 721 +0.125 673 +0.000 051	0.272 567 1.14
1990 Nov 22	4	−21.81 948 + 0.00 050	187.62 429 15.03 731	+0.048 539 +0.529 220 −0.000 030	−0.700 266 +0.133 392 +0.000 037	0.272 564 1.10
1990 Dec 19	16	−21.39 155 + 0.00 073	31.80 369 15.03 632	+0.293 332 +0.527 812 −0.000 036	−0.179 419 +0.145 220 +0.000 020	0.272 563 1.07
1991 Jan 16	4	−20.82 996 + 0.00 089	235.54 562 15.03 598	−0.045 732 +0.527 672 −0.000 004	+0.109 233 +0.157 932 +0.000 012	0.272 562 1.06
1991 Feb 12	18	−20.19 733 + 0.00 093	109.35 038 15.03 629	+0.093 129 +0.526 618 +0.000 011	+0.534 600 +0.168 733 −0.000 003	0.272 562 1.07

Table III   (cont.)

Date	T_0	$D0$ $D1$	$H0$ $H1$	$X0$ $X1$ $X2$	$Y0$ $Y1$ $Y2$	k F

Saturn

Date	T_0	$D0$ / $D1$	$H0$ / $H1$	$X0$ / $X1$ / $X2$	$Y0$ / $Y1$ / $Y2$	k / F
1991 Mar 12	7	−19.59 197 + 0.00 080	328.52 259 15.03 720	−0.479 270 +0.523 840 +0.000 042	+0.801 659 +0.176 219 −0.000 015	0.272 563 1.09
1997 Apr 7	1	+ 2.39 130 + 0.00 197	199.24 791 15.03 640	−0.058 179 +0.578 546 +0.000 007	+1.076 793 +0.186 014 −0.000 058	0.272 558 1.11
1997 May 4	15	+ 3.65 271 + 0.00 172	73.34 483 15.03 676	−0.357 518 +0.571 617 +0.000 036	+0.681 510 +0.182 862 −0.000 031	0.272 560 1.13
1997 Jun 1	3	+ 4.68 174 + 0.00 128	277.77 550 15.03 763	−0.125 701 +0.564 224 +0.000 026	+0.433 340 +0.179 852 −0.000 016	0.272 563 1.16
1997 Jun 28	12	+ 5.35 884 + 0.00 067	77.82 339 15.03 897	−0.017 968 +0.561 876 +0.000 010	+0.162 711 +0.178 363 −0.000 006	0.272 566 1.21
1997 Jul 25	19	+ 5.59 840 − 0.00 004	208.80 116 15.04 069	−0.138 466 +0.567 612 −0.000 000	−0.063 163 +0.179 376 +0.000 000	0.272 569 1.27
1997 Aug 22	2	+ 5.36 893 − 0.00 074	340.94 524 15.04 249	−0.183 128 +0.579 477 −0.000 008	−0.065 339 +0.182 933 −0.000 001	0.272 571 1.33
1997 Sep 18	10	+ 4.73 948 − 0.00 123	129.20 228 15.04 387	−0.004 205 +0.590 306 −0.000 014	+0.181 248 +0.187 572 −0.000 011	0.272 572 1.37
1997 Oct 15	18	+ 3.92 384 − 0.00 128	278.06 500 15.04 427	+0.058 610 +0.592 007 −0.000 001	+0.419 243 +0.190 772 −0.000 018	0.272 573 1.38
1997 Nov 12	1	+ 3.23 676 − 0.00 085	51.74 585 15.04 346	+0.065 485 +0.582 727 +0.000 014	+0.467 324 +0.190 910 −0.000 015	0.272 572 1.36
1997 Dec 9	6	+ 2.95 237 − 0.00 010	154.50 782 15.04 180	−0.296 335 +0.570 011 +0.000 031	+0.117 297 +0.188 955 +0.000 001	0.272 571 1.31
1998 Jan 5	12	+ 3.18 829 + 0.00 071	271.11 857 15.03 989	−0.134 585 +0.564 484 +0.000 005	−0.255 159 +0.187 391 +0.000 011	0.272 567 1.25

Table III (cont.)

Date	T_0	D0 D1	H0 H1	X0 X1 X2	Y0 Y1 Y2	k F

S a t u r n

Date	T_0	D0 D1	H0 H1	X0 X1 X2	Y0 Y1 Y2	k F
1998 Feb 1	21	+ 3.90 999 + 0.00 136	71.65 885 15.03 821	+0.195 094 +0.569 502 −0.000 029	−0.583 231 +0.187 563 +0.000 021	0.272 563 1.19
1998 Mar 1	10	+ 4.98 904 + 0.00 177	291.40 153 15.03 702	+0.486 617 +0.579 427 −0.000 048	−0.819 671 +0.188 684 +0.000 034	0.272 560 1.15
1998 Mar 29	1	+ 6.25 622 + 0.00 192	180.61 125 15.03 638	+0.122 209 +0.586 386 −0.000 016	−1.180 036 +0.189 057 +0.000 060	0.272 558 1.13
2001 May 23	6	+19.29 175 + 0.00 100	268.50 923 15.03 560	+0.037 919 +0.567 050 +0.000 003	−1.161 870 +0.152 706 +0.000 041	0.272 563 1.15
2001 Jun 19	22	+19.91 701 + 0.00 080	172.08 512 15.03 580	+0.208 433 +0.572 749 +0.000 011	−0.853 054 +0.142 725 +0.000 029	0.272 563 1.16
2001 Jul 17	13	+20.38 410 + 0.00 054	60.94 797 15.03 659	−0.139 251 +0.573 252 +0.000 035	−0.632 668 +0.130 738 +0.000 020	0.272 565 1.19
2001 Aug 14	3	+20.67 265 + 0.00 027	295.49 519 15.03 798	+0.077 528 +0.569 452 +0.000 017	−0.204 176 +0.119 442 +0.000 002	0.272 568 1.24
2001 Sep 10	13	+20.78 701 + 0.00 002	110.98 897 15.03 990	+0.064 192 +0.565 038 −0.000 002	+0.216 934 +0.112 275 −0.000 014	0.272 573 1.30
2001 Oct 7	19	+20.74 684 − 0.00 020	227.69 139 15.04 210	+0.042 270 +0.564 826 −0.000 021	+0.556 960 +0.112 042 −0.000 024	0.272 577 1.36
2001 Nov 3	22	+20.57 840 − 0.00 037	300.63 813 15.04 400	−0.290 222 +0.569 820 −0.000 010	+0.599 321 +0.118 881 −0.000 019	0.272 580 1.41
2001 Dec 1	2	+20.32 752 − 0.00 045	29.56 580 15.04 487	−0.153 259 +0.574 301 +0.000 000	+0.466 539 +0.129 093 −0.000 008	0.272 581 1.44
2001 Dec 28	8	+20.08 772 − 0.00 033	148.66 976 15.04 425	+0.013 252 +0.571 409 +0.000 013	+0.231 671 +0.137 015 +0.000 003	0.272 580 1.42

Table III   (cont.)

Date	T_0	D0 D1	H0 H1	X0 X1 X2	Y0 Y1 Y2	k F

Saturn

2002 Jan 24	16	+19.98 300 − 0.00 004	297.03 611 15.04 244	+0.190 642 +0.561 098 +0.000 009	+0.128 735 +0.139 107 +0.000 002	0.272 577 1.37
2002 Feb 21	0	+20.09 270 + 0.00 030	84.03 471 15.04 019	−0.239 358 +0.549 962 +0.000 015	+0.128 267 +0.136 084 −0.000 008	0.272 575 1.31
2002 Mar 20	10	+20.40 419 + 0.00 056	259.68 346 15.03 815	+0.097 389 +0.544 476 −0.000 017	+0.513 587 +0.130 267 −0.000 034	0.272 571 1.24
2002 Apr 16	20	+20.83 491 + 0.00 066	74.15 357 15.03 662	−0.081 779 +0.546 339 −0.000 016	+0.829 753 +0.122 701 −0.000 053	0.272 569 1.19
2002 May 14	8	+21.28 516 + 0.00 062	277.89 592 15.03 571	−0.015 440 +0.553 181 −0.000 010	+1.165 475 +0.112 761 −0.000 071	0.272 566 1.16
2006 Dec 10	12	+14.23 699 + 0.00 031	111.35 230 15.04 149	+0.500 923 +0.507 252 −0.000 036	+1.043 267 −0.227 545 −0.000 038	0.272 578 1.33
2007 Jan 6	19	+14.61 839 + 0.00 094	244.08 099 15.04 339	+0.311 618 +0.514 466 −0.000 032	+0.840 971 −0.229 879 −0.000 024	0.272 581 1.39
2007 Feb 2	24	+15.30 948 + 0.00 123	347.72 868 15.04 446	+0.514 220 +0.518 562 −0.000 024	+0.760 552 −0.228 053 −0.000 023	0.272 583 1.42
2007 Mar 2	2	+16.03 323 + 0.00 105	46.52 179 15.04 422	+0.234 852 +0.517 864 +0.000 009	+1.077 085 −0.222 087 −0.000 044	0.272 583 1.41
2007 Mar 29	5	+16.52 148 + 0.00 053	119.79 145 15.04 279	+0.409 508 +0.515 516 +0.000 006	+1.133 416 −0.215 987 −0.000 051	0.272 580 1.36
2007 Apr 25	10	+16.62 283 − 0.00 011	222.00 338 15.04 082	+0.213 563 +0.515 442 −0.000 001	+1.105 728 −0.214 138 −0.000 051	0.272 576 1.30
2007 May 22	20	+16.31 152 − 0.00 071	38.12 378 15.03 891	+0.527 378 +0.517 407 −0.000 037	+0.633 843 −0.218 189 −0.000 030	0.272 571 1.24

Table III (cont.)

Date	T_0	D0 D1	H0 H1	X0 X1 X2	Y0 Y1 Y2	k F

Saturn

2007 Jun 19	8	+15.63 712 − 0.00 120	243.19 734 15.03 743	+0.082 004 +0.518 382 −0.000 031	+0.372 094 −0.226 199 −0.000 018	0.272 567 1.19
2007 Jul 16	23	+14.68 109 − 0.00 155	132.58 151 15.03 648	+0.213 706 +0.515 904 −0.000 036	−0.141 594 −0.235 222 +0.000 003	0.272 565 1.15
2007 Aug 13	13	+13.55 021 − 0.00 173	6.48 795 15.03 608	−0.283 785 +0.510 120 −0.000 000	−0.328 434 −0.242 621 +0.000 006	0.272 565 1.14
2007 Sep 10	3	+12.36 212 − 0.00 171	240.31 014 15.03 622	−0.278 132 +0.502 758 +0.000 017	−0.747 656 −0.247 249 +0.000 018	0.272 565 1.14
2007 Oct 7	15	+11.25 845 − 0.00 147	84.32 757 15.03 689	−0.514 221 +0.496 568 +0.000 039	−1.115 298 −0.249 741 +0.000 031	0.272 567 1.16
2013 Dec 1	10	−14.92 373 − 0.00 127	355.08 941 15.03 647	+0.041 155 +0.588 325 +0.000 018	−1.279 608 −0.114 368 +0.000 073	0.272 557 1.08
2013 Dec 29	1	−15.67 529 − 0.00 092	244.41 119 15.03 728	−0.238 628 +0.584 634 +0.000 040	−0.888 528 −0.106 895 +0.000 049	0.272 559 1.10
2014 Jan 25	14	−16.15 950 − 0.00 048	104.39 539 15.03 869	+0.011 493 +0.576 856 +0.000 016	−0.583 400 −0.100 695 +0.000 032	0.272 563 1.15
2014 Feb 21	22	−16.33 072 + 0.00 001	250.27 249 15.04 054	−0.179 674 +0.572 157 +0.000 003	−0.287 266 −0.097 347 +0.000 017	0.272 567 1.20
2014 Mar 21	3	−16.19 051 + 0.00 046	352.28 743 15.04 247	−0.184 267 +0.575 009 −0.000 013	−0.224 537 −0.098 028 +0.000 012	0.272 570 1.25
2014 Apr 17	7	−15.79 610 + 0.00 077	80.38 927 15.04 392	−0.201 779 +0.582 217 −0.000 009	−0.360 061 −0.102 741 +0.000 013	0.272 572 1.29
2014 May 14	12	−15.27 759 + 0.00 082	184.16 957 15.04 436	−0.188 037 +0.586 144 +0.000 008	−0.553 910 −0.109 528 +0.000 017	0.272 572 1.31

Table III (cont.)

Date	T_0	D0 D1	H0 H1	X0 X1 X2	Y0 Y1 Y2	k F

Saturn

Date	T_0	D0 / D1	H0 / H1	X0 / X1 / X2	Y0 / Y1 / Y2	k / F
2014 Jun 10	19	−14.82 402 / + 0.00 057	317.94 419 / 15.04 365	+0.102 429 / +0.582 089 / +0.000 009	−0.665 404 / −0.115 286 / +0.000 020	0.272 571 / 1.29
2014 Jul 8	2	−14.61 473 / + 0.00 010	90.96 007 / 15.04 212	−0.248 582 / +0.571 708 / +0.000 028	−0.410 256 / −0.117 725 / +0.000 012	0.272 569 / 1.24
2014 Aug 4	11	−14.74 052 / − 0.00 042	252.93 473 / 15.04 029	+0.260 534 / +0.560 908 / −0.000 009	−0.129 498 / −0.116 283 / +0.000 007	0.272 567 / 1.19
2014 Aug 31	19	−15.18 747 / − 0.00 086	38.69 901 / 15.03 859	+0.000 065 / +0.555 230 / −0.000 013	+0.380 609 / −0.111 982 / −0.000 007	0.272 564 / 1.14
2014 Sep 28	5	−15.87 176 / − 0.00 114	213.54 109 / 15.03 726	+0.319 545 / +0.556 882 / −0.000 034	+0.716 217 / −0.106 175 / −0.000 017	0.272 562 / 1.09
2014 Oct 25	16	−16.67 590 / − 0.00 123	42.70 313 / 15.03 642	+0.146 797 / +0.563 946 / −0.000 019	+1.044 139 / −0.099 817 / −0.000 033	0.272 560 / 1.07
2018 Dec 9	5	−22.63 647 / + 0.00 025	233.42 937 / 15.03 612	−0.171 049 / +0.551 364 / −0.000 012	+1.239 474 / −0.011 183 / −0.000 034	0.272 561 / 1.06
2019 Jan 5	19	−22.42 346 / + 0.00 039	107.14 395 / 15.03 589	+0.157 170 / +0.549 354 / −0.000 004	+0.957 961 / −0.000 250 / −0.000 020	0.272 561 / 1.05
2019 Feb 2	7	−22.13 345 / + 0.00 047	310.85 350 / 15.03 632	−0.059 546 / +0.546 157 / +0.000 023	+0.685 032 / +0.011 648 / −0.000 008	0.272 562 / 1.06
2019 Mar 1	18	−21.82 562 / + 0.00 044	140.01 837 / 15.03 738	−0.264 162 / +0.544 478 / +0.000 034	+0.333 169 / +0.022 644 / +0.000 003	0.272 564 / 1.09
2019 Mar 29	5	−21.58 187 / + 0.00 027	330.05 800 / 15.03 896	−0.001 240 / +0.546 683 / +0.000 010	−0.058 507 / +0.030 697 / +0.000 011	0.272 567 / 1.14
2019 Apr 25	15	−21.48 111 / + 0.00 001	146.22 582 / 15.04 089	+0.296 667 / +0.552 620 / −0.000 022	−0.388 604 / +0.034 123 / +0.000 015	0.272 569 / 1.19

Table III (cont.)

Date	T_0	$D0$ $D1$	$H0$ $H1$	$X0$ $X1$ $X2$	$Y0$ $Y1$ $Y2$	k F

Saturn

Date	T_0	$D0$/$D1$	$H0$/$H1$	$X0$/$X1$/$X2$	$Y0$/$Y1$/$Y2$	k/F
2019 May 22	22	−21.56 122 − 0.000 26	278.547 48 15.042 79	−0.137 699 +0.559 742 −0.000 014	−0.571 755 +0.032 231 +0.000 015	0.272 572 1.24
2019 Jun 19	4	−21.79 198 − 0.000 43	36.920 77 15.044 15	+0.116 922 +0.564 297 −0.000 018	−0.472 945 +0.026 075 +0.000 007	0.272 574 1.28
2019 Jul 16	7	−22.08 237 − 0.000 44	110.729 49 15.044 48	−0.150 369 +0.564 113 +0.000 013	−0.246 357 +0.018 182 −0.000 002	0.272 575 1.29
2019 Aug 12	10	−22.33 407 − 0.000 32	184.360 82 15.043 62	+0.057 426 +0.560 485 +0.000 019	−0.041 070 +0.011 665 −0.000 006	0.272 574 1.26
2019 Sep 8	14	−22.48 523 − 0.000 14	272.178 71 15.041 92	+0.160 111 +0.557 583 +0.000 012	−0.041 337 +0.008 618 +0.000 000	0.272 571 1.22
2019 Oct 5	21	−22.51 218 + 0.000 06	43.885 42 15.039 92	+0.209 215 +0.558 650 −0.000 009	−0.271 412 +0.009 626 +0.000 016	0.272 567 1.17
2019 Nov 2	7	−22.40 398 + 0.000 27	219.441 93 15.038 11	−0.206 079 +0.562 953 −0.000 012	−0.638 114 +0.014 485 +0.000 037	0.272 563 1.12
2019 Nov 29	21	−22.15 149 + 0.000 49	94.104 26 15.036 78	−0.038 266 +0.566 541 −0.000 026	−0.987 515 +0.023 078 +0.000 057	0.272 561 1.08

Uranus

Date	T_0	$D0$/$D1$	$H0$/$H1$	$X0$/$X1$/$X2$	$Y0$/$Y1$/$Y2$	k/F
1991 Feb 11	5	−23.22 206 + 0.000 16	292.634 05 15.038 91	+0.327 613 +0.539 891 +0.000 011	−1.187 119 +0.098 029 +0.000 058	0.272 531 0.58
1991 Mar 10	15	−23.12 511 + 0.000 10	108.432 44 15.039 69	+0.127 021 +0.537 576 +0.000 013	−0.948 140 +0.101 493 +0.000 045	0.272 532 0.59
1991 Apr 7	0	−23.07 838 + 0.000 01	269.786 93 15.040 70	−0.096 823 +0.536 246 +0.000 002	−0.647 109 +0.103 163 +0.000 027	0.272 533 0.60

Table III   (cont.)

Date	T_0	D0 D1	H0 H1	X0 X1 X2	Y0 Y1 F Y2	k
Uranus						
1991 May 4	8	−23.09531 − 0.00009	56.78060 15.04173	+0.132223 +0.536965 −0.000024	−0.314824 +0.103594 +0.000008	0.272534 0.61
1991 May 31	13	−23.16767 − 0.00015	159.26869 15.04256	−0.148831 +0.539274 −0.000014	−0.249970 +0.102552 +0.000002	0.272535 0.62
1991 Jun 27	17	−23.26850 − 0.00017	247.12306 15.04296	−0.210960 +0.541375 +0.000004	−0.317335 +0.099655 +0.000005	0.272535 0.63
1991 Jul 24	21	−23.36451 − 0.00014	335.06566 15.04282	−0.163282 +0.541778 +0.000022	−0.424229 +0.095218 +0.000012	0.272535 0.63
1991 Aug 21	2	−23.42991 − 0.00008	77.79220 15.04218	−0.138501 +0.540488 +0.000029	−0.438967 +0.090717 +0.000015	0.272534 0.62
1991 Sep 17	9	−23.45184 − 0.00001	210.06853 15.04122	−0.017552 +0.538796 +0.000013	−0.252844 +0.088221 +0.000010	0.272533 0.61
1991 Oct 14	18	−23.42622 + 0.00006	11.75555 15.04017	+0.183813 +0.537895 −0.000018	+0.096034 +0.089242 −0.000001	0.272532 0.59
1991 Nov 11	3	−23.35259 + 0.00014	172.78990 15.03926	−0.263683 +0.537939 −0.000016	+0.360152 +0.093854 −0.000010	0.272532 0.58
1991 Dec 8	13	−23.23332 + 0.00020	348.35908 15.03866	−0.342217 +0.537807 −0.000009	+0.599635 +0.100529 −0.000017	0.272531 0.57
1992 Jan 4	23	−23.07828 + 0.00024	163.65701 15.03845	−0.249814 +0.536332 +0.000008	+0.786726 +0.107080 −0.000021	0.272531 0.57
1992 Feb 1	8	−22.90882 + 0.00024	323.91877 15.03866	−0.462279 +0.533687 +0.000037	+0.925791 +0.112124 −0.000025	0.272531 0.57
1999 Apr 11	7	−16.62537 + 0.00036	345.27074 15.03980	−0.089120 +0.560710 +0.000022	+1.071785 +0.121696 −0.000040	0.272529 0.57

Table III (cont.)

Date	T_0	D0 D1	H0 H1	X0 X1 X2	Y0 Y1 Y2	k F
Uranus						
1999 May 8	16	−16.46190 + 0.00010	146.65387 15.04068	−0.062486 +0.553679 +0.000003	+0.778743 +0.122592 −0.000034	0.272531 0.59
1999 Jun 4	22	−16.48005 − 0.00018	263.51315 15.04159	−0.139535 +0.551236 −0.000011	+0.523595 +0.121833 −0.000030	0.272532 0.60
1999 Jul 2	3	−16.66428 − 0.00040	5.89049 15.04236	+0.176879 +0.554281 −0.000028	+0.511410 +0.120002 −0.000032	0.272532 0.61
1999 Jul 29	7	−16.95661 − 0.00050	93.61794 15.04277	−0.071385 +0.559544 −0.000003	+0.529499 +0.117713 −0.000030	0.272533 0.61
1999 Aug 25	13	−17.27038 − 0.00046	211.54705 15.04270	+0.054235 +0.562004 +0.000010	+0.665043 +0.115230 −0.000030	0.272532 0.61
1999 Sep 21	20	−17.51221 − 0.00029	344.31475 15.04216	−0.157770 +0.558721 +0.000029	+0.620316 +0.112987 −0.000024	0.272532 0.61
1999 Oct 19	4	−17.61137 − 0.00003	131.65087 15.04129	−0.315716 +0.550979 +0.000026	+0.395498 +0.111298 −0.000014	0.272532 0.59
1999 Nov 15	13	−17.53419 + 0.00024	293.41242 15.04032	+0.102989 +0.543493 −0.000013	+0.144292 +0.110642 −0.000006	0.272531 0.58
1999 Dec 12	21	−17.28561 + 0.00048	79.52925 15.03948	+0.169388 +0.540560 −0.000031	−0.180544 +0.111747 +0.000005	0.272531 0.57
2000 Jan 9	5	−16.90099 + 0.00065	225.18065 15.03892	+0.079832 +0.542385 −0.000024	−0.432039 +0.115048 +0.000018	0.272530 0.56
2000 Feb 5	14	−16.43732 + 0.00072	25.62184 15.03872	−0.090120 +0.545249 +0.000002	−0.615975 +0.120195 +0.000033	0.272530 0.56
2000 Mar 4	1	−15.96563 + 0.00067	216.13875 15.03891	+0.140642 +0.545521 +0.000012	−0.733394 +0.126020 +0.000044	0.272530 0.56

T a b l e   III   (cont.)

Date	T_0	D0 D1	H0 H1	X0 X1 X2	Y0 Y1 Y2	k F
U r a n u s						
2000 Mar 31	12	− 15.56 368 + 0.00 050	46.89 377 15.03 944	+0.135 722 +0.542 390 +0.000 016	−1.003 047 +0.130 805 +0.000 055	0.272 530 0.57
2006 Mar 27	16	− 7.67 103 + 0.00 081	81.08 336 15.03 905	+0.417 618 +0.546 363 −0.000 009	−1.179 933 +0.283 204 +0.000 076	0.272 526 0.55
2006 Apr 24	3	− 7.17 557 + 0.00 061	271.91 736 15.03 952	+0.355 290 +0.536 981 +0.000 000	−1.046 325 +0.279 972 +0.000 067	0.272 527 0.56
2006 May 21	11	− 6.84 710 + 0.00 031	58.02 625 15.04 023	+0.294 935 +0.530 258 −0.000 006	−0.808 715 +0.276 822 +0.000 047	0.272 528 0.57
2006 Jun 17	17	− 6.73 437 − 0.00 004	174.56 848 15.04 107	+0.228 969 +0.530 701 −0.000 019	−0.521 412 +0.277 317 +0.000 024	0.272 529 0.59
2006 Jul 14	23	− 6.84 860 − 0.00 037	291.65 899 15.04 189	+0.224 161 +0.538 082 −0.000 030	−0.273 557 +0.281 865 +0.000 007	0.272 529 0.60
2006 Aug 11	6	− 7.15 286 − 0.00 061	64.26 197 15.04 248	+0.035 641 +0.547 399 −0.000 016	−0.291 419 +0.287 365 +0.000 009	0.272 529 0.61
2006 Sep 7	15	− 7.55 953 − 0.00 067	227.22 015 15.04 269	+0.119 401 +0.551 937 −0.000 005	−0.323 758 +0.289 409 +0.000 018	0.272 529 0.61
2006 Oct 5	0	− 7.94 534 − 0.00 054	30.15 837 15.04 242	−0.025 463 +0.547 944 +0.000 019	−0.507 355 +0.285 655 +0.000 034	0.272 529 0.60
2006 Nov 1	9	− 8.19 169 − 0.00 026	192.78 066 15.04 175	+0.408 359 +0.537 791 −0.000 004	−0.263 867 +0.278 302 +0.000 021	0.272 529 0.59
2006 Nov 28	15	− 8.22 042 + 0.00 011	309.75 719 15.04 087	+0.080 498 +0.528 909 −0.000 001	−0.216 775 +0.272 707 +0.000 013	0.272 529 0.58
2006 Dec 25	21	− 8.01 222 + 0.00 046	66.15 118 15.04 000	−0.131 114 +0.527 841 −0.000 010	+0.010 221 +0.273 406 −0.000 006	0.272 528 0.57

Table III (cont.)

Date	T_0	$D0$ $D1$	$H0$ $H1$	$X0$ $X1$ $X2$	$Y0$ $Y1$ $Y2$	k F

Uranus

2007 Jan 22	5	− 7.59 703	212.11 395	−0.399 109	+0.174 285	0.272 527
		+ 0.00 073	15.03 933	+0.534 016	+0.279 741	0.56
				−0.000 002	−0.000 016	
2007 Feb 18	17	− 7.04 056	57.88 982	−0.189 535	+0.484 255	0.272 526
		+ 0.00 087	15.03 896	+0.540 903	+0.286 653	0.55
				−0.000 001	−0.000 028	
2007 Mar 18	6	− 6.43 410	278.57 908	−0.502 656	+0.471 822	0.272 526
		+ 0.00 087	15.03 895	+0.542 449	+0.289 388	0.55
				+0.000 036	−0.000 020	
2007 Apr 14	19	− 5.87 364	139.37 108	−0.645 700	+0.613 068	0.272 527
		+ 0.00 073	15.03 927	+0.537 352	+0.286 787	0.56
				+0.000 055	−0.000 023	
2007 May 12	6	− 5.44 721	330.39 166	−0.614 071	+0.960 049	0.272 528
		+ 0.00 048	15.03 987	+0.529 063	+0.281 621	0.57
				+0.000 049	−0.000 042	
2014 Aug 14	16	+ 5.76 645	187.72 411	−0.571 351	+0.994 744	0.272 530
		− 0.00 033	15.04 184	+0.570 205	+0.180 470	0.60
				+0.000 012	−0.000 058	
2014 Sep 11	1	+ 5.47 354	350.39 059	−0.398 602	+0.982 378	0.272 530
		− 0.00 058	15.04 247	+0.579 843	+0.182 850	0.61
				+0.000 006	−0.000 057	
2014 Oct 8	10	+ 5.06 528	153.34 872	−0.372 856	+1.072 912	0.272 530
		− 0.00 067	15.04 271	+0.581 889	+0.183 700	0.61
				+0.000 020	−0.000 055	
2014 Nov 4	18	+ 4.66 638	301.25 944	−0.193 726	+1.229 426	0.272 530
		− 0.00 056	15.04 245	+0.575 023	+0.182 857	0.61
				+0.000 027	−0.000 055	
2014 Dec 1	23	+ 4.40 412	43.73 582	−0.586 451	+1.055 894	0.272 530
		− 0.00 027	15.04 179	+0.565 162	+0.181 640	0.60
				+0.000 049	−0.000 044	
2014 Dec 29	4	+ 4.36 274	145.69 633	−0.557 975	+0.808 030	0.272 529
		+ 0.00 010	15.04 091	+0.561 691	+0.181 452	0.58
				+0.000 031	−0.000 039	
2015 Jan 25	12	+ 4.56 730	292.19 594	+0.070 230	+0.647 207	0.272 528
		+ 0.00 047	15.04 004	+0.568 282	+0.182 531	0.57
				−0.000 019	−0.000 041	

T a b l e   III   (cont.)

Date	T_0	$D0$ $D1$	$H0$ $H1$	$X0$ $X1$ $X2$	$Y0$ $Y1$ $Y2$	k F
U r a n u s						
2015 Feb 21	22	+ 4.98 444 + 0.00 074	108.26 076 15.03 936	−0.165 042 +0.579 423 −0.000 013	+0.259 334 +0.183 656 −0.000 025	0.272 527 0.56
2015 Mar 21	11	+ 5.54 139 + 0.00 088	329.09 360 15.03 899	−0.184 756 +0.586 402 −0.000 000	+0.053 137 +0.183 300 −0.000 009	0.272 526 0.55
2015 Apr 18	1	+ 6.14 488 + 0.00 088	204.82 788 15.03 895	+0.240 484 +0.584 971 −0.000 006	+0.038 542 +0.181 363 −0.000 001	0.272 526 0.55
2015 May 15	12	+ 6.69 941 + 0.00 074	35.52 935 15.03 924	+0.036 589 +0.577 318 +0.000 018	−0.212 440 +0.178 674 +0.000 017	0.272 527 0.56
2015 Jun 11	21	+ 7.12 627 + 0.00 050	196.43 820 15.03 981	+0.313 528 +0.569 608 +0.000 002	−0.396 659 +0.176 453 +0.000 026	0.272 528 0.57
2015 Jul 9	3	+ 7.36 527 + 0.00 018	312.67 400 15.04 059	+0.116 431 +0.567 975 −0.000 003	−0.759 096 +0.175 778 +0.000 040	0.272 529 0.58
2015 Aug 5	9	+ 7.38 467 − 0.00 016	69.45 154 15.04 144	+0.147 117 +0.574 716 −0.000 021	−0.971 422 +0.177 310 +0.000 047	0.272 529 0.59
2015 Sep 1	16	+ 7.18 870 − 0.00 047	201.81 051 15.04 221	+0.013 108 +0.586 163 −0.000 020	−1.071 517 +0.180 535 +0.000 054	0.272 529 0.60
2015 Sep 29	1	+ 6.83 085 − 0.00 064	4.66 317 15.04 266	+0.052 511 +0.594 414 −0.000 012	−0.977 170 +0.184 002 +0.000 055	0.272 529 0.61
2015 Oct 26	11	+ 6.41 612 − 0.00 063	182.71 047 15.04 265	+0.341 998 +0.592 896 −0.000 011	−0.791 306 +0.185 908 +0.000 051	0.272 529 0.61
2015 Nov 22	19	+ 6.07 639 − 0.00 042	330.50 634 15.04 217	+0.178 020 +0.582 152 +0.000 010	−0.883 162 +0.185 292 +0.000 058	0.272 529 0.60
2015 Dec 20	1	+ 5.92 122 − 0.00 008	87.78 259 15.04 135	+0.080 732 +0.570 675 +0.000 008	−1.141 080 +0.183 343 +0.000 065	0.272 529 0.59

Table III (cont.)

Date	T_0	$D0$ $D1$	$H0$ $H1$	$X0$ $X1$ $X2$	$Y0$ $Y1$ $Y2$	k F

N e p t u n e

Date	T_0	$D0$ $D1$	$H0$ $H1$	$X0$ $X1$ $X2$	$Y0$ $Y1$ $Y2$	k F
1991 May 31	20	−21.63858 − 0.00011	261.24445 15.04202	+0.223577 +0.536133 −0.000030	−1.187001 +0.114738 +0.000040	0.272520 0.40
1991 Jun 28	1	−21.71480 − 0.00014	3.75884 15.04230	+0.321808 +0.538050 −0.000019	−1.191479 +0.113395 +0.000041	0.272520 0.40
1991 Sep 17	18	−21.92583 − 0.00005	341.04694 15.04132	+0.356020 +0.536164 +0.000001	−1.068115 +0.104652 +0.000043	0.272520 0.39
1991 Oct 15	2	−21.93392 + 0.00001	127.90203 15.04065	+0.226189 +0.535169 −0.000016	−0.771850 +0.104542 +0.000032	0.272519 0.38
1991 Nov 11	11	−21.89796 + 0.00008	289.36714 15.04003	+0.212864 +0.535336 −0.000033	−0.443631 +0.107270 +0.000019	0.272519 0.38
1991 Dec 8	20	−21.82016 + 0.00014	90.48158 15.03960	+0.186391 +0.535842 −0.000030	−0.223759 +0.111676 +0.000011	0.272519 0.37
1992 Jan 5	4	−21.71057 + 0.00017	236.35523 15.03943	−0.138766 +0.535248 +0.000002	−0.176664 +0.115836 +0.000012	0.272519 0.37
1992 Feb 1	12	−21.58693 + 0.00018	22.21341 15.03955	−0.248311 +0.533217 +0.000028	−0.085193 +0.118532 +0.000011	0.272519 0.37
1992 Feb 28	21	−21.47271 + 0.00014	183.28278 15.03994	+0.182571 +0.531528 +0.000017	+0.233070 +0.119996 −0.000002	0.272519 0.38
1992 Mar 27	5	−21.39247 + 0.00008	329.63811 15.04053	−0.053046 +0.532089 +0.000012	+0.496677 +0.121195 −0.000017	0.272519 0.38
1992 Apr 23	13	−21.36360 − 0.00001	116.41046 15.04119	−0.295786 +0.535189 +0.000001	+0.735242 +0.122527 −0.000032	0.272519 0.39
1992 May 20	21	−21.39088 − 0.00009	263.60714 15.04180	−0.184132 +0.539091 −0.000014	+0.913715 +0.123192 −0.000043	0.272520 0.39

Table III (cont.)

Date	T_0	D0 D1	H0 H1	X0 X1 X2	Y0 Y1 Y2	k F
Neptune						
1992 Jun 17	3	−21.46374 − 0.00015	21.05821 15.04220	−0.467337 +0.541654 +0.000005	+0.839988 +0.122203 −0.000041	0.272520 0.40
1992 Jul 14	9	−21.56038 − 0.00016	138.68148 15.04231	−0.037850 +0.541787 +0.000007	+0.855036 +0.119073 −0.000039	0.272520 0.40
1992 Aug 10	13	−21.65440 − 0.00014	226.18414 15.04210	−0.343573 +0.540166 +0.000035	+0.777244 +0.115003 −0.000033	0.272520 0.40
1992 Sep 6	18	−21.72333 − 0.00009	328.49314 15.04160	−0.398429 +0.539042 +0.000037	+0.913944 +0.111780 −0.000037	0.272520 0.39
1992 Oct 4	1	−21.75126 − 0.00002	100.50558 15.04096	−0.280084 +0.540458 +0.000015	+1.217943 +0.111146 −0.000049	0.272519 0.39
1999 Apr 10	9	−18.93225 + 0.00013	26.72228 15.04050	+0.084718 +0.559299 +0.000008	+1.221682 +0.085354 −0.000050	0.272518 0.38
1999 May 7	16	−18.88686 − 0.00001	158.42306 15.04113	−0.301128 +0.553914 +0.000004	+0.879719 +0.085497 −0.000040	0.272519 0.39
1999 Jun 3	22	−18.93610 − 0.00015	275.49152 15.04172	−0.029847 +0.553457 −0.000020	+0.701327 +0.084049 −0.000037	0.272519 0.39
1999 Jul 1	2	−19.06258 − 0.00024	2.81217 15.04214	−0.362841 +0.557104 −0.000001	+0.581447 +0.081733 −0.000032	0.272520 0.40
1999 Jul 28	8	−19.23117 − 0.00027	120.40272 15.04228	+0.103459 +0.561009 −0.000003	+0.707929 +0.079089 −0.000034	0.272520 0.40
1999 Aug 24	14	−19.39583 − 0.00023	237.98025 15.04210	−0.223101 +0.561371 +0.000028	+0.735867 +0.076953 −0.000031	0.272519 0.40
1999 Sep 20	22	−19.51373 − 0.00013	25.42816 15.04164	−0.159149 +0.556993 +0.000028	+0.701853 +0.075516 −0.000027	0.272519 0.39

Table III (cont.)

Date	T_0	D0 D1	H0 H1	X0 X1 X2	Y0 Y1 Y2	k F

N e p t u n e

Date	T_0	D0 D1	H0 H1	X0 X1 X2	Y0 Y1 Y2	k F
1999 Oct 18	7	−19.55 344 + 0.00 000	187.55 302 15.04 101	+0.199 342 +0.550 121 −0.000 003	+0.538 468 +0.074 786 −0.000 021	0.272 519 0.39
1999 Nov 14	15	−19.50 147 + 0.00 014	334.20 996 15.04 036	+0.225 657 +0.545 006 −0.000 025	+0.229 211 +0.074 797 −0.000 011	0.272 519 0.38
1999 Dec 11	22	−19.36 287 + 0.00 026	105.43 535 15.03 984	−0.010 855 +0.544 465 −0.000 025	−0.079 918 +0.075 806 +0.000 001	0.272 519 0.38
2000 Jan 8	6	−19.15 809 + 0.00 034	251.42 511 15.03 953	+0.187 998 +0.547 056 −0.000 021	−0.209 430 +0.078 185 +0.000 010	0.272 519 0.37
2000 Feb 4	14	−18.92 108 + 0.00 036	37.30 591 15.03 951	−0.134 409 +0.549 044 +0.000 014	−0.345 183 +0.081 807 +0.000 021	0.272 518 0.37
2000 Mar 2	24	−18.69 278 + 0.00 031	213.34 033 15.03 976	+0.108 568 +0.548 057 +0.000 019	−0.449 323 +0.085 884 +0.000 029	0.272 519 0.38
2000 Mar 30	10	−18.51 601 + 0.00 020	29.61 645 15.04 023	+0.247 026 +0.544 736 +0.000 009	−0.685 514 +0.089 083 +0.000 036	0.272 519 0.38
2000 Apr 26	19	−18.42 421 + 0.00 006	191.21 425 15.04 084	+0.325 278 +0.541 531 −0.000 013	−0.996 011 +0.090 405 +0.000 041	0.272 519 0.39
2000 Aug 13	17	−18.87 853 − 0.00 027	270.41 098 15.04 221	+0.142 749 +0.545 478 +0.000 007	−1.243 881 +0.083 764 +0.000 049	0.272 520 0.40
2007 Jul 3	20	−14.54 912 − 0.00 029	257.36 573 15.04 189	+0.334 369 +0.538 052 −0.000 036	−1.216 846 +0.236 274 +0.000 050	0.272 519 0.40
2007 Jul 31	2	−14.75 881 − 0.00 038	14.84 157 15.04 219	+0.303 378 +0.544 692 −0.000 024	−1.189 002 +0.238 877 +0.000 054	0.272 519 0.40
2007 Aug 27	10	−14.99 806 − 0.00 037	162.50 321 15.04 219	+0.390 935 +0.548 001 −0.000 010	−1.214 950 +0.239 081 +0.000 062	0.272 518 0.40

Table III   (cont.)

Date	T_0	D0 D1	H0 H1	X0 X1 X2	Y0 Y1 Y2	k F

Neptune

2007 Oct 21	4	− 15.31229 − 0.00010	127.43445 15.04136	+ 0.294529 + 0.535527 + 0.000001	− 1.190907 + 0.229640 + 0.000062	0.272518 0.39
2007 Nov 17	12	− 15.29813 + 0.00010	274.32928 15.04072	+ 0.398892 + 0.526610 − 0.000023	− 0.871183 + 0.225501 + 0.000040	0.272518 0.39
2007 Dec 14	18	− 15.15520 + 0.00029	30.73792 15.04014	+ 0.004915 + 0.523488 − 0.000023	− 0.703482 + 0.226100 + 0.000027	0.272518 0.38
2008 Jan 11	2	− 14.90474 + 0.00043	176.89575 15.03973	+ 0.379970 + 0.526456 − 0.000039	− 0.281877 + 0.231176 + 0.000010	0.272518 0.38
2008 Feb 7	11	− 14.58907 + 0.00048	337.90011 15.03956	+ 0.249295 + 0.530870 − 0.000016	− 0.197788 + 0.236997 + 0.000012	0.272517 0.37
2008 Mar 5	22	− 14.26228 + 0.00045	168.96135 15.03966	+ 0.174986 + 0.531609 + 0.000007	− 0.117765 + 0.239942 + 0.000015	0.272517 0.38
2008 Apr 2	9	− 13.98258 + 0.00034	0.17041 15.04000	− 0.130256 + 0.527224 + 0.000028	− 0.056589 + 0.239013 + 0.000013	0.272518 0.38
2008 Apr 29	19	− 13.79940 + 0.00017	176.62664 15.04052	− 0.238217 + 0.520241 + 0.000021	+ 0.207369 + 0.236082 − 0.000004	0.272518 0.38
2008 May 27	3	− 13.74347 − 0.00004	323.37389 15.04112	− 0.114026 + 0.514940 − 0.000003	+ 0.603231 + 0.233893 − 0.000030	0.272519 0.39
2008 Jun 23	8	− 13.81890 − 0.00023	65.38297 15.04168	− 0.397886 + 0.514325 − 0.000003	+ 0.705585 + 0.233977 − 0.000039	0.272519 0.40
2008 Jul 20	13	− 14.00148 − 0.00036	167.71186 15.04208	− 0.243748 + 0.517943 − 0.000005	+ 0.826509 + 0.235372 − 0.000044	0.272520 0.40
2008 Aug 16	18	− 14.24132 − 0.00040	270.22166 15.04222	− 0.470595 + 0.522281 + 0.000023	+ 0.648993 + 0.235909 − 0.000031	0.272519 0.40

Table III (cont.)

Date	T_0	D0 D1	H0 H1	X0 X1 X2	Y0 Y1 Y2	k F

N e p t u n e

Date	T_0	D0 D1	H0 H1	X0 X1 X2	Y0 Y1 Y2	k F
2008 Sep 13	1	−14.47 417 − 0.00 034	42.80 779 15.04 205	−0.495 251 +0.523 300 +0.000 041	+0.586 555 +0.233 750 −0.000 022	0.272 519 0.40
2008 Oct 10	10	−14.63 604 − 0.00 018	205.27 519 15.04 161	−0.191 099 +0.519 311 +0.000 029	+0.828 480 +0.229 237 −0.000 031	0.272 519 0.39
2008 Nov 6	18	−14.68 034 + 0.00 002	352.35 172 15.04 101	−0.564 941 +0.512 193 +0.000 029	+0.952 365 +0.224 913 −0.000 039	0.272 519 0.39
2016 Jun 25	24	− 7.84 332 − 0.00 014	290.82 228 15.04 136	−0.326 946 +0.561 202 +0.000 002	+1.119 186 +0.175 356 −0.000 061	0.272 518 0.39
2016 Jul 23	5	− 7.99 249 − 0.00 034	32.96 473 15.04 185	−0.348 448 +0.567 072 −0.000 002	+0.977 218 +0.175 796 −0.000 058	0.272 518 0.40
2016 Aug 19	12	− 8.24 398 − 0.00 044	165.45 299 15.04 214	−0.037 945 +0.575 187 −0.000 007	+1.077 576 +0.176 321 −0.000 061	0.272 518 0.40
2016 Sep 15	20	− 8.52 994 − 0.00 044	313.08 654 15.04 215	−0.292 761 +0.578 626 +0.000 024	+1.073 937 +0.176 104 −0.000 054	0.272 518 0.40
2016 Oct 13	5	− 8.77 108 − 0.00 031	115.67 090 15.04 187	−0.587 801 +0.573 411 +0.000 050	+0.995 596 +0.174 449 −0.000 044	0.272 518 0.40
2016 Nov 9	14	− 8.89 804 − 0.00 010	277.98 712 15.04 135	−0.514 382 +0.562 103 +0.000 040	+0.866 164 +0.171 692 −0.000 037	0.272 518 0.39
2016 Dec 6	22	− 8.87 123 + 0.00 015	64.88 596 15.04 074	−0.038 195 +0.552 409 −0.000 003	+0.714 930 +0.169 285 −0.000 035	0.272 518 0.39
2017 Jan 3	4	− 8.68 907 + 0.00 037	181.31 424 15.04 019	−0.125 710 +0.550 809 −0.000 015	+0.375 836 +0.168 887 −0.000 024	0.272 518 0.38
2017 Jan 30	11	− 8.38 197 + 0.00 052	312.46 700 15.03 979	−0.245 854 +0.556 464 −0.000 008	+0.131 914 +0.170 707 −0.000 012	0.272 518 0.38

Table III   (cont.)

Date	T_0	D0 D1	H0 H1	X0 X1 X2	Y0 Y1 Y2	k F

N e p t u n e

2017 Feb 26	21	− 8.00 261 + 0.000 58	128.55 596 15.039 63	−0.006 786 +0.562 647 −0.000 002	+0.099 952 +0.173 655 −0.000 003	0.272 517 0.38
2017 Mar 26	8	− 7.61 771 + 0.000 54	319.66 313 15.039 73	−0.225 213 +0.563 430 +0.000 027	−0.075 573 +0.176 232 +0.000 013	0.272 517 0.38
2017 Apr 22	20	− 7.29 299 + 0.000 40	165.94 878 15.040 05	+0.091 433 +0.557 807 +0.000 017	−0.176 039 +0.177 335 +0.000 019	0.272 517 0.38
2017 May 20	6	− 7.08 394 + 0.000 20	342.42 486 15.040 54	+0.257 867 +0.549 344 −0.000 002	−0.411 101 +0.176 762 +0.000 025	0.272 518 0.38
2017 Jun 16	13	− 7.02 307 − 0.000 04	114.13 133 15.041 10	+0.180 322 +0.543 438 −0.000 016	−0.715 597 +0.175 373 +0.000 029	0.272 519 0.39
2017 Jul 13	18	− 7.11 405 − 0.000 26	216.12 382 15.041 64	+0.086 603 +0.543 459 −0.000 024	−0.898 442 +0.174 468 +0.000 032	0.272 519 0.40
2017 Aug 9	23	− 7.32 953 − 0.000 41	318.42 674 15.042 04	+0.198 842 +0.548 271 −0.000 025	−0.848 822 +0.174 704 +0.000 031	0.272 519 0.40
2017 Sep 6	5	− 7.61 230 − 0.000 46	75.95 123 15.042 18	+0.224 536 +0.553 107 −0.000 010	−0.742 603 +0.175 565 +0.000 034	0.272 519 0.40
2017 Oct 3	13	− 7.88 568 − 0.000 38	223.55 639 15.042 02	+0.412 091 +0.552 952 −0.000 002	−0.651 228 +0.175 920 +0.000 037	0.272 519 0.40
2017 Oct 30	21	− 8.07 173 − 0.000 20	10.96 729 15.041 60	+0.035 466 +0.546 370 +0.000 018	−0.916 459 +0.174 755 +0.000 049	0.272 519 0.40
2017 Nov 27	6	− 8.11 555 + 0.000 04	173.08 199 15.041 02	+0.362 733 +0.537 118 −0.000 012	−1.114 455 +0.172 572 +0.000 051	0.272 519 0.39

6

Sunspot Activity

1749 − 1994

Sunspot Activity

In 1848 R. Wolf introduced the quantity $(10g + f)$ as a measure of the sunspot activity, where g is the number of groups of sunspots on the solar hemisphere visible from Earth, and f is the number of individual spots. However, an empirical constant k has been introduced in order to remove systematic differences between values determined by different observers. Using the usual notation, therefore, the *Zürich relative sunspot number* is defined by

$$R = k(10g + f)$$

These sunspot numbers have been found useful as a broad indication, particularly for statistical studies. Until December 1980, they were regularly calculated and published by the Swiss Federal Observatory at Zürich, Switzerland. They are now issued by the Sunspot Index Data Center (SIDC), Uccle Observatory, near Brussels, Belgium.

Table A provides the yearly means of the definitive Zürich sunspot numbers, from 1749 to 1993. Maxima and minima are indicated by M and m, respectively. The diagrams on pages 352 and 353 illustrate the variation of these yearly means.

Table B contains the monthly means of the definitive Zürich sunspot numbers, from January 1749 to June 1994.

In Table C we give, for the same period, the smoothed monthly means, based on the definitive Zürich sunspot numbers, but calculated by means of the formula advocated by J. Meeus (*Ciel et Terre*, Vol. 74, 445-449; November-December 1958). In this formula, a greater weight is given to the central months. If $A, B, C, D, E, F, G, H, J, K, L, M, N$ are the definitive monthly means of 13 consecutive months, then the smoothed mean for the central month is given by

$$\frac{A + 3B + 5C + 7D + 9E + 10F + 11G + 10H + 9J + 7K + 5L + 3M + N}{81}$$

The diagrams on pages 367 − 369 illustrate the smoothed monthly means from 1901 to 1993. Secondary variations are well shown. For instance, we see that the sunspot maxima of 1905-1907 and 1968-1970 were much broader than the maximum of 1957-1958.

In Table D we give the following data:

— the epochs of maxima and minima of the sunspot activity, to the nearest tenth of a year, as calculated by the method used at the Swiss Federal Observatory;

— the same epochs, but deduced from the data of Table C, considering the month having the highest (or lowest) smoothed mean as the epoch of the sunspot maximum (or minimum). The corresponding smoothed means are also given;

— the month(s) with the highest or lowest monthly mean.

Table E gives the monthly and yearly numbers of spotless days, that is, days with $R = 0$, in the period 1850 to June 1994. For the years not mentioned in the table, there were no spotless days.

The synodic rotations of the Sun are numbered in continuation of Carrington's Greenwich photoheliographic series, of which No. 1 commenced on 1853 November 9. The *mean* synodic period is 27.2752 days; the true value varies between 27.20 and 27.34 days, due to the non-uniform speed of the Earth in its orbit. See also Chapter 28 in our *Astronomical Algorithms* (Willmann-Bell, Inc., 1991).

Table F gives the dates of the beginning of the Sun's synodic rotations, as seen from the Earth, from A.D. 1952 to 2020, to the nearest 0.01 day (Universal Time).

The definitive Zürich sunspot numbers (Tables A and B) are reproduced by kind permission of Dr. M. Waldmeier. The values from 1981 onwards are taken from SIDC publications.

TABLE A
YEARLY MEANS OF SUNSPOT NUMBERS

Year	Value		Year	Value		Year	Value		Year	Value	
1749	80.9		1792	60.0		1835	56.9		1878	3.4	*m*
1750	83.4	*M*	1793	46.9		1836	121.5		1879	6.0	
1751	47.7		1794	41.0		1837	138.3	*M*	1880	32.3	
1752	47.8		1795	21.3		1838	103.2		1881	54.3	
1753	30.7		1796	16.0		1839	85.7		1882	59.7	
1754	12.2		1797	6.4		1840	64.6		1883	63.7	*M*
1755	9.6	*m*	1798	4.1	*m*	1841	36.7		1884	63.5	
1756	10.2		1799	6.8		1842	24.2		1885	52.2	
1757	32.4		1800	14.5		1843	10.7	*m*	1886	25.4	
1758	47.6		1801	34.0		1844	15.0		1887	13.1	
1759	54.0		1802	45.0		1845	40.1		1888	6.8	
1760	62.9		1803	43.1		1846	61.5		1889	6.3	*m*
1761	85.9	*M*	1804	47.5	*M*	1847	98.5		1890	7.1	
1762	61.2		1805	42.2		1848	124.7	*M*	1891	35.6	
1763	45.1		1806	28.1		1849	96.3		1892	73.0	
1764	36.4		1807	10.1		1850	66.6		1893	85.1	*M*
1765	20.9		1808	8.1		1851	64.5		1894	78.0	
1766	11.4	*m*	1809	2.5		1852	54.1		1895	64.0	
1767	37.8		1810	0.0	*m*	1853	39.0		1896	41.8	
1768	69.8		1811	1.4		1854	20.6		1897	26.2	
1769	106.1	*M*	1812	5.0		1855	6.7		1898	26.7	
1770	100.8		1813	12.2		1856	4.3	*m*	1899	12.1	
1771	81.6		1814	13.9		1857	22.7		1900	9.5	
1772	66.5		1815	35.4		1858	54.8		1901	2.7	*m*
1773	34.8		1816	45.8	*M*	1859	93.8		1902	5.0	
1774	30.6		1817	41.1		1860	95.8	*M*	1903	24.4	
1775	7.0	*m*	1818	30.1		1861	77.2		1904	42.0	
1776	19.8		1819	23.9		1862	59.1		1905	63.5	*M*
1777	92.5		1820	15.6		1863	44.0		1906	53.8	
1778	154.4	*M*	1821	6.6		1864	47.0		1907	62.0	
1779	125.9		1822	4.0		1865	30.5		1908	48.5	
1780	84.8		1823	1.8	*m*	1866	16.3		1909	43.9	
1781	68.1		1824	8.5		1867	7.3	*m*	1910	18.6	
1782	38.5		1825	16.6		1868	37.6		1911	5.7	
1783	22.8		1826	36.3		1869	74.0		1912	3.6	
1784	10.2	*m*	1827	49.6		1870	139.0	*M*	1913	1.4	*m*
1785	24.1		1828	64.2		1871	111.2		1914	9.6	
1786	82.9		1829	67.0		1872	101.6		1915	47.4	
1787	132.0	*M*	1830	70.9	*M*	1873	66.2		1916	57.1	
1788	130.9		1831	47.8		1874	44.7		1917	103.9	*M*
1789	118.1		1832	27.5		1875	17.0		1918	80.6	
1790	89.9		1833	8.5	*m*	1876	11.3		1919	63.6	
1791	66.6		1834	13.2		1877	12.4		1920	37.6	

Table A : yearly means (cont.)

1921	26.1	1940	67.8	1959	159.0	1978	92.5
1922	14.2	1941	47.5	1960	112.3	1979	155.4 *M*
1923	5.8 *m*	1942	30.6	1961	53.9	1980	154.6
1924	16.7	1943	16.3	1962	37.5	1981	140.5
1925	44.3	1944	9.6 *m*	1963	27.9	1982	115.9
1926	63.9	1945	33.2	1964	10.2 *m*	1983	66.6
1927	69.0	1946	92.6	1965	15.1	1984	45.9
1928	77.8 *M*	1947	151.6 *M*	1966	47.0	1985	17.9
1929	64.9	1948	136.3	1967	93.8	1986	13.4 *m*
1930	35.7	1949	134.7	1968	105.9 *M*	1987	29.2
1931	21.2	1950	83.9	1969	105.5	1988	100.2
1932	11.1	1951	69.4	1970	104.5	1989	157.6 *M*
1933	5.7 *m*	1952	31.5	1971	66.6	1990	142.6
1934	8.7	1953	13.9	1972	68.9	1991	145.7
1935	36.1	1954	4.4 *m*	1973	38.0	1992	94.3
1936	79.7	1955	38.0	1974	34.5	1993	54.6
1937	114.4 *M*	1956	141.7	1975	15.5		
1938	109.6	1957	190.2 *M*	1976	12.6 *m*		
1939	88.8	1958	184.8	1977	27.5		

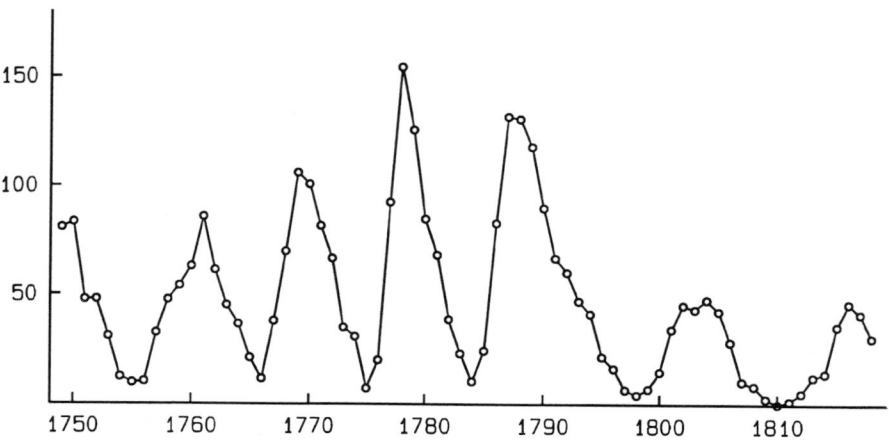

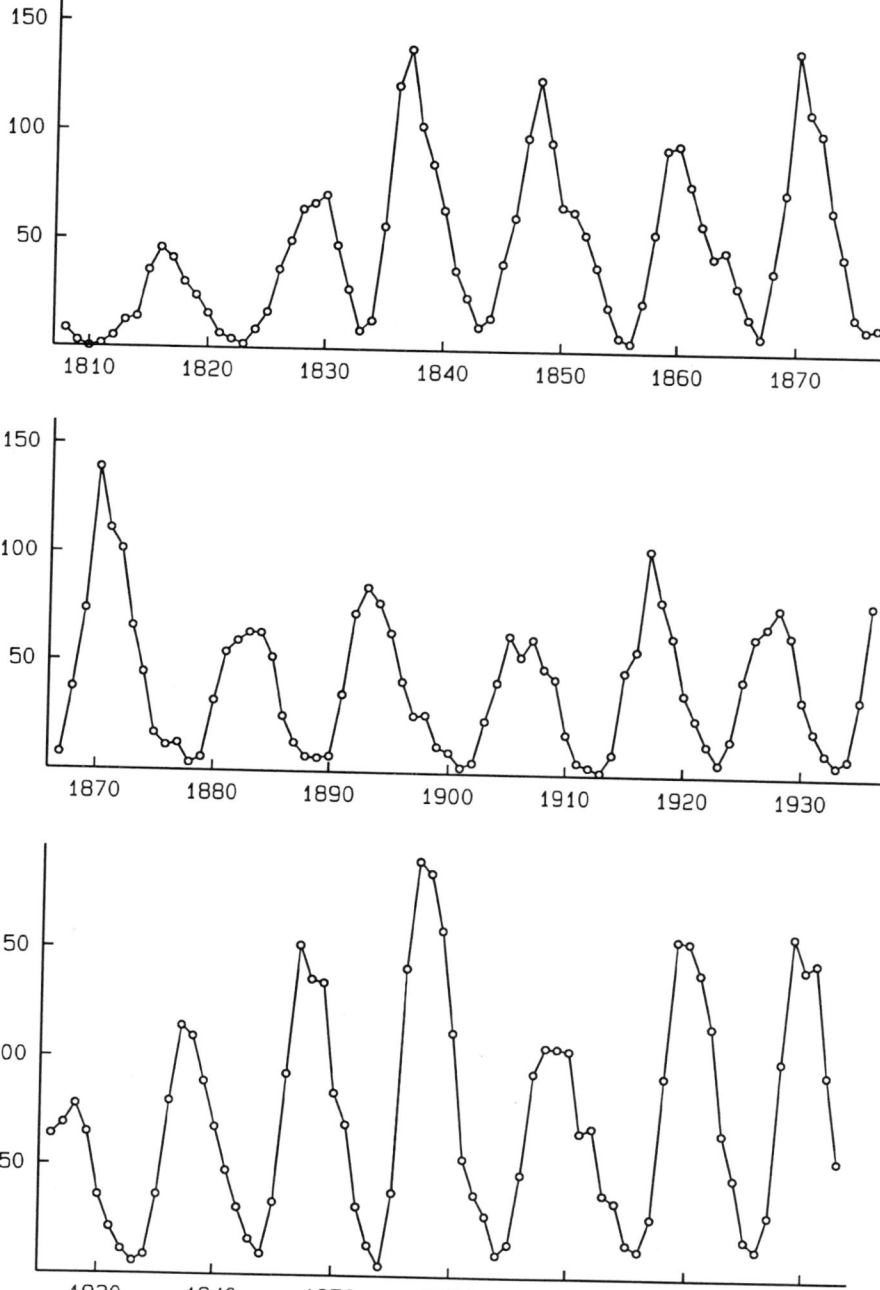

TABLE B
MONTHLY MEANS OF SUNSPOT NUMBERS

Year	Jan	Feb	Mar	Apr	May	Jun	Jul	Aug	Sep	Oct	Nov	Dec
1749	58.0	62.6	70.0	55.7	85.0	83.5	94.8	66.3	75.9	75.5	158.6	85.2
1750	73.3	75.9	89.2	88.3	90.0	100.0	85.4	103.0	91.2	65.7	63.3	75.4
1751	70.0	43.5	45.3	56.4	60.7	50.7	66.3	59.8	23.5	23.2	28.5	44.0
1752	35.0	50.0	71.0	59.3	59.7	39.6	78.4	29.3	27.1	46.6	37.6	40.0
1753	44.0	32.0	45.7	38.0	36.0	31.7	22.0	39.0	28.0	25.0	20.0	6.7
1754	0.0	3.0	1.7	13.7	20.7	26.7	18.8	12.3	8.2	24.1	13.2	4.2
1755	10.2	11.2	6.8	6.5	0.0	0.0	8.6	3.2	17.8	23.7	6.8	20.0
1756	12.5	7.1	5.4	9.4	12.5	12.9	3.6	6.4	11.8	14.3	17.0	9.4
1757	14.1	21.2	26.2	30.0	38.1	12.8	25.0	51.3	39.7	32.5	64.7	33.5
1758	37.6	52.0	49.0	72.3	46.4	45.0	44.0	38.7	62.5	37.7	43.0	43.0
1759	48.3	44.0	46.8	47.0	49.0	50.0	51.0	71.3	77.2	59.7	46.3	57.0
1760	67.3	59.5	74.7	58.3	72.0	48.3	66.0	75.6	61.3	50.6	59.7	61.0
1761	70.0	91.0	80.7	71.7	107.2	99.3	94.1	91.1	100.7	88.7	89.7	46.0
1762	43.8	72.8	45.7	60.2	39.9	77.1	33.8	67.7	68.5	69.3	77.8	77.2
1763	56.5	31.9	34.2	32.9	32.7	35.8	54.2	26.5	68.1	46.3	60.9	61.4
1764	59.7	59.7	40.2	34.4	44.3	30.0	30.0	30.0	28.2	28.0	26.0	25.7
1765	24.0	26.0	25.0	22.0	20.2	20.0	27.0	29.7	16.0	14.0	14.0	13.0
1766	12.0	11.0	36.6	6.0	26.8	3.0	3.3	4.0	4.3	5.0	5.7	19.2
1767	27.4	30.0	43.0	32.9	29.8	33.3	21.9	40.8	42.7	44.1	54.7	53.3
1768	53.5	66.1	46.3	42.7	77.7	77.4	52.6	66.8	74.8	77.8	90.6	111.8
1769	73.9	64.2	64.3	96.7	73.6	94.4	118.6	120.3	148.8	158.2	148.1	112.0
1770	104.0	142.5	80.1	51.0	70.1	83.3	109.8	126.3	104.4	103.6	132.2	102.3
1771	36.0	46.2	46.7	64.9	152.7	119.5	67.7	58.5	101.4	90.0	99.7	95.7
1772	100.9	90.8	31.1	92.2	38.0	57.0	77.3	56.2	50.5	78.6	61.3	64.0
1773	54.6	29.0	51.2	32.9	41.1	28.4	27.7	12.7	29.3	26.3	40.9	43.2
1774	46.8	65.4	55.7	43.8	51.3	28.5	17.5	6.6	7.9	14.0	17.7	12.2
1775	4.4	0.0	11.6	11.2	3.9	12.3	1.0	7.9	3.2	5.6	15.1	7.9
1776	21.7	11.6	6.3	21.8	11.2	19.0	1.0	24.2	16.0	30.0	35.0	40.0
1777	45.0	36.5	39.0	95.5	80.3	80.7	95.0	112.0	116.2	106.5	146.0	157.3
1778	177.3	109.3	134.0	145.0	238.9	171.6	153.0	140.0	171.7	156.3	150.3	105.0
1779	114.7	165.7	118.0	145.0	140.0	113.7	143.0	112.0	111.0	124.0	114.0	110.0
1780	70.0	98.0	98.0	95.0	107.2	88.0	86.0	86.0	93.7	77.0	60.0	58.7
1781	98.7	74.7	53.0	68.3	104.7	97.7	73.5	66.0	51.0	27.3	67.0	35.2
1782	54.0	37.5	37.0	41.0	54.3	38.0	37.0	44.0	34.0	23.2	31.5	30.0
1783	28.0	38.7	26.7	28.3	23.0	25.2	32.2	20.0	18.0	8.0	15.0	10.5
1784	13.0	8.0	11.0	10.0	6.0	9.0	6.0	10.0	10.0	8.0	17.0	14.0
1785	6.5	8.0	9.0	15.7	20.7	26.3	36.3	20.0	32.0	47.2	40.2	27.3

Table B :   Monthly means   (cont.)

Year	Jan	Feb	Mar	Apr	May	Jun	Jul	Aug	Sep	Oct	Nov	Dec
1786	37.2	47.6	47.7	85.4	92.3	59.0	83.0	89.7	111.5	112.3	116.0	112.7
1787	134.7	106.0	87.4	127.2	134.8	99.2	128.0	137.2	157.3	157.0	141.5	174.0
1788	138.0	129.2	143.3	108.5	113.0	154.2	141.5	136.0	141.0	142.0	94.7	129.5
1789	114.0	125.3	120.0	123.3	123.5	120.0	117.0	103.0	112.0	89.7	134.0	135.5
1790	103.0	127.5	96.3	94.0	93.0	91.0	69.3	87.0	77.3	84.3	82.0	74.0
1791	72.7	62.0	74.0	77.2	73.7	64.2	71.0	43.0	66.5	61.7	67.0	66.0
1792	58.0	64.0	63.0	75.7	62.0	61.0	45.8	60.0	59.0	59.0	57.0	56.0
1793	56.0	55.0	55.5	53.0	52.3	51.0	50.0	29.3	24.0	47.0	44.0	45.7
1794	45.0	44.0	38.0	28.4	55.7	41.5	41.0	40.0	11.1	28.5	67.4	51.4
1795	21.4	39.9	12.6	18.6	31.0	17.1	12.9	25.7	13.5	19.5	25.0	18.0
1796	22.0	23.8	15.7	31.7	21.0	6.7	26.9	1.5	18.4	11.0	8.4	5.1
1797	14.4	4.2	4.0	4.0	7.3	11.1	4.3	6.0	5.7	6.9	5.8	3.0
1798	2.0	4.0	12.4	1.1	0.0	0.0	0.0	3.0	2.4	1.5	12.5	9.9
1799	1.6	12.6	21.7	8.4	8.2	10.6	2.1	0.0	0.0	4.6	2.7	8.6
1800	6.9	9.3	13.9	0.0	5.0	23.7	21.0	19.5	11.5	12.3	10.5	40.1
1801	27.0	29.0	30.0	31.0	32.0	31.2	35.0	38.7	33.5	32.6	39.8	48.2
1802	47.8	47.0	40.8	42.0	44.0	46.0	48.0	50.0	51.8	38.5	34.5	50.0
1803	50.0	50.8	29.5	25.0	44.3	36.0	48.3	34.1	45.3	54.3	51.0	48.0
1804	45.3	48.3	48.0	50.6	33.4	34.8	29.8	43.1	53.0	62.3	61.0	60.0
1805	61.0	44.1	51.4	37.5	39.0	40.5	37.6	42.7	44.4	29.4	41.0	38.3
1806	39.0	29.6	32.7	27.7	26.4	25.6	30.0	26.3	24.0	27.0	25.0	24.0
1807	12.0	12.2	9.6	23.8	10.0	12.0	12.7	12.0	5.7	8.0	2.6	0.0
1808	0.0	4.5	0.0	12.3	13.5	13.5	6.7	8.0	11.7	4.7	10.5	12.3
1809	7.2	9.2	0.9	2.5	2.0	7.7	0.3	0.2	0.4	0.0	0.0	0.0
1810	0.0	0.0	0.0	0.0	0.0	0.0	0.0	0.0	0.0	0.0	0.0	0.0
1811	0.0	0.0	0.0	0.0	0.0	0.0	6.6	0.0	2.4	6.1	0.8	1.1
1812	11.3	1.9	0.7	0.0	1.0	1.3	0.5	15.6	5.2	3.9	7.9	10.1
1813	0.0	10.3	1.9	16.6	5.5	11.2	18.3	8.4	15.3	27.8	16.7	14.3
1814	22.2	12.0	5.7	23.8	5.8	14.9	18.5	2.3	8.1	19.3	14.5	20.1
1815	19.2	32.2	26.2	31.6	9.8	55.9	35.5	47.2	31.5	33.5	37.2	65.0
1816	26.3	68.8	73.7	58.8	44.3	43.6	38.8	23.2	47.8	56.4	38.1	29.9
1817	36.4	57.9	96.2	26.4	21.2	40.0	50.0	45.0	36.7	25.6	28.9	28.4
1818	34.9	22.4	25.4	34.5	53.1	36.4	28.0	31.5	26.1	31.6	10.9	25.8
1819	32.8	20.7	3.7	20.2	19.6	35.0	31.4	26.1	14.9	27.5	25.1	30.6
1820	19.2	26.6	4.5	19.4	29.3	10.8	20.6	25.9	5.2	8.9	7.9	9.1
1821	21.5	4.2	5.7	9.2	1.7	1.8	2.5	4.8	4.4	18.8	4.4	0.2
1822	0.0	0.9	16.1	13.5	1.5	5.6	7.9	2.1	0.0	0.4	0.0	0.0
1823	0.0	0.0	0.6	0.0	0.0	0.0	0.5	0.0	0.0	0.0	0.0	20.4
1824	21.7	10.8	0.0	19.4	2.8	0.0	0.0	1.4	20.5	25.2	0.0	0.8
1825	5.0	15.5	22.4	3.8	15.5	15.4	30.9	25.7	15.7	15.6	11.7	22.0

Table B :   Monthly means   (cont.)

Year	Jan	Feb	Mar	Apr	May	Jun	Jul	Aug	Sep	Oct	Nov	Dec
1826	17.7	18.2	36.7	24.0	32.4	37.1	52.5	39.6	18.9	50.6	39.5	68.1
1827	34.6	47.4	57.8	46.0	56.3	56.7	42.3	53.7	49.6	56.1	48.2	46.1
1828	52.8	64.4	65.0	61.1	89.1	98.0	54.2	76.4	50.4	54.7	57.0	46.9
1829	43.0	49.4	72.3	95.0	67.4	73.9	90.8	77.6	52.8	57.2	67.6	56.5
1830	52.2	72.1	84.6	106.3	66.3	65.1	43.9	50.7	62.1	84.4	81.2	82.1
1831	47.5	50.1	93.4	54.5	38.1	33.4	45.2	55.0	37.9	46.3	43.5	28.9
1832	30.9	55.6	55.1	26.9	41.3	26.7	14.0	8.9	8.2	21.1	14.3	27.5
1833	11.3	14.9	11.8	2.8	12.9	1.0	7.0	5.7	11.6	7.5	5.9	9.9
1834	4.9	18.1	3.9	1.4	8.8	7.8	8.7	4.0	11.5	24.8	30.5	34.5
1835	7.5	24.5	19.7	61.5	43.6	33.2	59.8	59.0	100.8	95.2	100.0	77.5
1836	88.6	107.6	98.2	142.9	111.4	124.7	116.7	107.8	95.1	137.4	120.9	206.2
1837	188.0	175.6	134.6	138.2	111.7	158.0	162.8	134.0	96.3	123.7	107.0	129.8
1838	144.9	84.8	140.8	126.6	137.6	94.5	108.2	78.8	73.6	90.8	77.4	79.8
1839	105.6	102.5	77.7	61.8	53.8	54.6	84.8	131.2	132.7	90.9	68.8	63.7
1840	81.2	87.7	67.8	65.9	69.2	48.5	60.7	57.8	74.0	55.0	54.3	53.7
1841	24.1	29.9	29.7	40.2	67.5	55.7	30.8	39.3	36.5	28.5	19.8	38.8
1842	20.4	22.1	21.7	26.9	24.9	20.5	12.6	26.6	18.4	38.1	40.5	17.6
1843	13.3	3.5	8.3	9.5	21.1	10.5	9.5	11.8	4.2	5.3	19.1	12.7
1844	9.4	14.7	13.6	20.8	11.6	3.7	21.2	23.9	7.0	21.5	10.7	21.6
1845	25.7	43.6	43.3	57.0	47.8	31.1	30.6	32.3	29.6	40.7	39.4	59.7
1846	38.7	51.0	63.9	69.3	59.9	65.1	46.5	54.8	107.1	55.9	60.4	65.5
1847	62.6	44.9	85.7	44.7	75.4	85.3	52.2	140.6	160.9	180.4	138.9	109.6
1848	159.1	111.8	108.6	107.1	102.2	129.0	139.2	132.6	100.3	132.4	114.6	159.5
1849	157.0	131.7	96.2	102.5	80.6	81.1	78.0	67.7	93.7	71.5	99.0	97.0
1850	78.0	89.4	82.6	44.1	61.6	70.0	39.1	61.6	86.2	71.0	54.8	61.0
1851	75.5	105.4	64.6	56.5	62.6	63.2	36.1	57.4	67.9	62.5	51.0	71.4
1852	68.4	66.4	61.2	65.4	54.9	46.9	42.1	39.7	37.5	67.3	54.3	45.4
1853	41.1	42.9	37.7	47.6	34.7	40.0	45.9	50.4	33.5	42.3	28.8	23.4
1854	15.4	20.0	20.7	26.5	24.0	21.1	18.7	15.8	22.4	12.6	28.2	21.6
1855	12.3	11.4	17.4	4.4	9.1	5.3	0.4	3.1	0.0	9.6	4.2	3.1
1856	0.5	4.9	0.4	6.5	0.0	5.2	4.6	5.9	4.4	4.5	7.7	7.2
1857	13.7	7.4	5.2	11.1	28.6	16.0	22.2	16.9	42.4	40.6	31.4	37.2
1858	39.0	34.9	57.5	38.3	41.4	44.5	56.7	55.3	80.1	91.2	51.9	66.9
1859	83.7	87.6	90.3	85.7	91.0	87.1	95.2	106.8	105.8	114.6	97.2	81.0
1860	82.4	88.3	98.9	71.4	107.1	108.6	116.7	100.3	92.2	90.1	97.9	95.6
1861	62.3	77.7	101.0	98.5	56.8	88.1	78.0	82.5	79.9	67.2	53.7	80.5
1862	63.1	64.5	43.6	53.7	64.4	84.0	73.4	62.5	66.6	41.9	50.6	40.9
1863	48.3	56.7	66.4	40.6	53.8	40.8	32.7	48.1	22.0	39.9	37.7	41.2
1864	57.7	47.1	66.3	35.8	40.6	57.8	54.7	54.8	28.5	33.9	57.6	28.6
1865	48.7	39.3	39.5	29.4	34.5	33.6	26.8	37.8	21.6	17.1	24.6	12.8

Table B : Monthly means (cont.)

Year	Jan	Feb	Mar	Apr	May	Jun	Jul	Aug	Sep	Oct	Nov	Dec
1866	31.6	38.4	24.6	17.6	12.9	16.5	9.3	12.7	7.3	14.1	9.0	1.5
1867	0.0	0.7	9.2	5.1	2.9	1.5	5.0	4.8	9.8	13.5	9.6	25.2
1868	15.6	15.7	26.5	36.6	26.7	31.1	29.0	34.4	47.2	61.6	59.1	67.6
1869	60.9	59.9	52.7	41.0	103.9	108.4	59.2	79.6	80.6	59.3	78.1	104.3
1870	77.3	114.9	157.6	160.0	176.0	135.6	132.4	153.8	136.0	146.4	147.5	130.0
1871	88.3	125.3	143.2	162.4	145.5	91.7	103.0	110.1	80.3	89.0	105.4	90.4
1872	79.5	120.1	88.4	102.1	107.6	109.9	105.5	92.9	114.6	102.6	112.0	83.9
1873	86.7	107.0	98.3	76.2	47.9	44.8	66.9	68.2	47.1	47.1	55.4	49.2
1874	60.8	64.2	46.4	32.0	44.6	38.2	67.8	61.3	28.0	34.3	28.9	29.3
1875	14.6	21.5	33.8	29.1	11.5	23.9	12.5	14.6	2.4	12.7	17.7	9.9
1876	14.3	15.0	30.6	2.3	5.1	1.6	15.2	8.8	9.9	14.3	9.9	8.2
1877	24.4	8.7	11.9	15.8	21.6	14.2	6.0	6.3	16.9	6.7	14.2	2.2
1878	3.3	6.6	7.8	0.1	5.9	6.4	0.1	0.0	5.3	1.1	4.1	0.5
1879	1.0	0.6	0.0	6.2	2.4	4.8	7.5	10.7	6.1	12.3	13.1	7.3
1880	24.0	27.2	19.3	19.5	23.5	34.1	21.9	48.1	66.0	43.0	30.7	29.6
1881	36.4	53.2	51.5	51.6	43.5	60.5	76.9	58.4	53.2	64.4	54.8	47.3
1882	45.0	69.5	66.8	95.8	64.1	45.2	45.4	40.4	57.7	59.2	84.4	41.8
1883	60.6	46.9	42.8	82.1	31.5	76.3	80.6	46.0	52.6	83.8	84.5	75.9
1884	91.5	86.9	87.5	76.1	66.5	51.2	53.1	55.8	61.9	47.8	36.6	47.2
1885	42.8	71.8	49.8	55.0	73.0	83.7	66.5	50.0	39.6	38.7	30.9	21.7
1886	29.9	25.9	57.3	43.7	30.7	27.1	30.3	16.9	21.4	8.6	0.3	13.0
1887	10.3	13.2	4.2	6.9	20.0	15.7	23.3	21.4	7.4	6.6	6.9	20.7
1888	12.7	7.1	7.8	5.1	7.0	7.1	3.1	2.8	8.8	2.1	10.7	6.7
1889	0.8	8.5	6.7	4.3	2.4	6.4	9.4	20.6	6.5	2.1	0.2	6.7
1890	5.3	0.6	5.1	1.6	4.8	1.3	11.6	8.5	17.2	11.2	9.6	7.8
1891	13.5	22.2	10.4	20.5	41.1	48.3	58.8	33.0	53.8	51.5	41.9	32.5
1892	69.1	75.6	49.9	69.6	79.6	76.3	76.5	101.4	62.8	70.5	65.4	78.6
1893	75.0	73.0	65.7	88.1	84.7	89.9	88.6	129.2	77.9	80.0	75.1	93.8
1894	83.2	84.6	52.3	81.6	101.2	98.9	106.0	70.3	65.9	75.5	56.6	60.0
1895	63.3	67.2	61.0	76.9	67.5	71.5	47.8	68.9	57.7	67.9	47.2	70.7
1896	29.0	57.4	52.0	43.8	27.7	49.0	45.0	27.2	61.3	28.7	38.0	42.6
1897	40.6	29.4	29.1	31.0	20.0	11.3	27.6	21.8	48.1	14.3	8.4	33.3
1898	30.2	36.4	38.3	14.5	25.8	22.3	9.0	31.4	34.8	34.4	30.9	12.6
1899	19.5	9.2	18.1	14.2	7.7	20.5	13.5	2.9	8.4	13.0	7.8	10.5
1900	9.4	13.6	8.6	16.0	15.2	12.1	8.3	4.3	8.3	12.9	4.5	0.3
1901	0.2	2.4	4.5	0.0	10.2	5.8	0.7	1.0	0.6	3.7	3.8	0.0
1902	5.5	0.0	12.4	0.0	2.8	1.4	0.9	2.3	7.6	16.3	10.3	1.1
1903	8.3	17.0	13.5	26.1	14.6	16.3	27.9	28.8	11.1	38.9	44.5	45.6
1904	31.6	24.5	37.2	43.0	39.5	41.9	50.6	58.2	30.1	54.2	38.0	54.6
1905	54.8	85.8	56.5	39.3	48.0	49.0	73.0	58.8	55.0	78.7	107.2	55.5

Table B :   Monthly means   (cont.)

Year	Jan	Feb	Mar	Apr	May	Jun	Jul	Aug	Sep	Oct	Nov	Dec
1906	45.5	31.3	64.5	55.3	57.7	63.2	103.6	47.7	56.1	17.8	38.9	64.7
1907	76.4	108.2	60.7	52.6	42.9	40.4	49.7	54.3	85.0	65.4	61.5	47.3
1908	39.2	33.9	28.7	57.6	40.8	48.1	39.5	90.5	86.9	32.3	45.5	39.5
1909	56.7	46.6	66.3	32.3	36.0	22.6	35.8	23.1	38.8	58.4	55.8	54.2
1910	26.4	31.5	21.4	8.4	22.2	12.3	14.1	11.5	26.2	38.3	4.9	5.8
1911	3.4	9.0	7.8	16.5	9.0	2.2	3.5	4.0	4.0	2.6	4.2	2.2
1912	0.3	0.0	4.9	4.5	4.4	4.1	3.0	0.3	9.5	4.6	1.1	6.4
1913	2.3	2.9	0.5	0.9	0.0	0.0	1.7	0.2	1.2	3.1	0.7	3.8
1914	2.8	2.6	3.1	17.3	5.2	11.4	5.4	7.7	12.7	8.2	16.4	22.3
1915	23.0	42.3	38.8	41.3	33.0	68.8	71.6	69.6	49.5	53.5	42.5	34.5
1916	45.3	55.4	67.0	71.8	74.5	67.7	53.5	35.2	45.1	50.7	65.6	53.0
1917	74.7	71.9	94.8	74.7	114.1	114.9	119.8	154.5	129.4	72.2	96.4	129.3
1918	96.0	65.3	72.2	80.5	76.7	59.4	107.6	101.7	79.9	85.0	83.4	59.2
1919	48.1	79.5	66.5	51.8	88.1	111.2	64.7	69.0	54.7	52.8	42.0	34.9
1920	51.1	53.9	70.2	14.8	33.3	38.7	27.5	19.2	36.3	49.6	27.2	29.9
1921	31.5	28.3	26.7	32.4	22.2	33.7	41.9	22.8	17.8	18.2	17.8	20.3
1922	11.8	26.4	54.7	11.0	8.0	5.8	10.9	6.5	4.7	6.2	7.4	17.5
1923	4.5	1.5	3.3	6.1	3.2	9.1	3.5	0.5	13.2	11.6	10.0	2.8
1924	0.5	5.1	1.8	11.3	20.8	24.0	28.1	19.3	25.1	25.6	22.5	16.5
1925	5.5	23.2	18.0	31.7	42.8	47.5	38.5	37.9	60.2	69.2	58.6	98.6
1926	71.8	69.9	62.5	38.5	64.3	73.5	52.3	61.6	60.8	71.5	60.5	79.4
1927	81.6	93.0	69.6	93.5	79.1	59.1	54.9	53.8	68.4	63.1	67.2	45.2
1928	83.5	73.5	85.4	80.6	77.0	91.4	98.0	83.8	89.7	61.4	50.3	59.0
1929	68.9	62.8	50.2	52.8	58.2	71.9	70.2	65.8	34.4	54.0	81.1	108.0
1930	65.3	49.9	35.0	38.2	36.8	28.8	21.9	24.9	32.1	34.4	35.6	25.8
1931	14.6	43.1	30.0	31.2	24.6	15.3	17.4	13.0	19.0	10.0	18.7	17.8
1932	12.1	10.6	11.2	11.2	17.9	22.2	9.6	6.8	4.0	8.9	8.2	11.0
1933	12.3	22.2	10.1	2.9	3.2	5.2	2.8	0.2	5.1	3.0	0.6	0.3
1934	3.4	7.8	4.3	11.3	19.7	6.7	9.3	8.3	4.0	5.7	8.7	15.4
1935	18.6	20.5	23.1	12.2	27.3	45.7	33.9	30.1	42.1	53.2	64.2	61.5
1936	62.8	74.3	77.1	74.9	54.6	70.0	52.3	87.0	76.0	89.0	115.4	123.4
1937	132.5	128.5	83.9	109.3	116.7	130.3	145.1	137.7	100.7	124.9	74.4	88.8
1938	98.4	119.2	86.5	101.0	127.4	97.5	165.3	115.7	89.6	99.1	122.2	92.7
1939	80.3	77.4	64.6	109.1	118.3	101.0	97.6	105.8	112.6	88.1	68.1	42.1
1940	50.5	59.4	83.3	60.7	54.4	83.9	67.5	105.5	66.5	55.0	58.4	68.3
1941	45.6	44.5	46.4	32.8	29.5	59.8	66.9	60.0	65.9	46.3	38.4	33.7
1942	35.6	52.8	54.2	60.7	25.0	11.4	17.7	20.2	17.2	19.2	30.7	22.5
1943	12.4	28.9	27.4	26.1	14.1	7.6	13.2	19.4	10.0	7.8	10.2	18.8
1944	3.7	0.5	11.0	0.3	2.5	5.0	5.0	16.7	14.3	16.9	10.8	28.4
1945	18.5	12.7	21.5	32.0	30.6	36.2	42.6	25.9	34.9	68.8	46.0	27.4

Table B : Monthly means (cont.)

Year	Jan	Feb	Mar	Apr	May	Jun	Jul	Aug	Sep	Oct	Nov	Dec
1946	47.6	86.2	76.6	75.7	84.9	73.5	116.2	107.2	94.4	102.3	123.8	121.7
1947	115.7	133.4	129.8	149.8	201.3	163.9	157.9	188.8	169.4	163.6	128.0	116.5
1948	108.5	86.1	94.8	189.7	174.0	167.8	142.2	157.9	143.3	136.3	95.8	138.0
1949	119.1	182.3	157.5	147.0	106.2	121.7	125.8	123.8	145.3	131.6	143.5	117.6
1950	101.6	94.8	109.7	113.4	106.2	83.6	91.0	85.2	51.3	61.4	54.8	54.1
1951	59.9	59.9	55.9	92.9	108.5	100.6	61.5	61.0	83.1	51.6	52.4	45.8
1952	40.7	22.7	22.0	29.1	23.4	36.4	39.3	54.9	28.2	23.8	22.1	34.3
1953	26.5	3.9	10.0	27.8	12.5	21.8	8.6	23.5	19.3	8.2	1.6	2.5
1954	0.2	0.5	10.9	1.8	0.8	0.2	4.8	8.4	1.5	7.0	9.2	7.6
1955	23.1	20.8	4.9	11.3	28.9	31.7	26.7	40.7	42.7	58.5	89.2	76.9
1956	73.6	124.0	118.4	110.7	136.6	116.6	129.1	169.6	173.2	155.3	201.3	192.1
1957	165.0	130.2	157.4	175.2	164.6	200.7	187.2	158.0	235.8	253.8	210.9	239.4
1958	202.5	164.9	190.7	196.0	175.3	171.5	191.4	200.2	201.2	181.5	152.3	187.6
1959	217.4	143.1	185.7	163.3	172.0	168.7	149.6	199.6	145.2	111.4	124.0	125.0
1960	146.3	106.0	102.2	122.0	119.6	110.2	121.7	134.1	127.2	82.8	89.6	85.6
1961	57.9	46.1	53.0	61.4	51.0	77.4	70.2	55.9	63.6	37.7	32.6	40.0
1962	38.7	50.3	45.6	46.4	43.7	42.0	21.8	21.8	51.3	39.5	26.9	23.2
1963	19.8	24.4	17.1	29.3	43.0	35.9	19.6	33.2	38.8	35.3	23.4	14.9
1964	15.3	17.7	16.5	8.6	9.5	9.1	3.1	9.3	4.7	6.1	7.4	15.1
1965	17.5	14.2	11.7	6.8	24.1	15.9	11.9	8.9	16.8	20.1	15.8	17.0
1966	28.2	24.4	25.3	48.7	45.3	47.7	56.7	51.2	50.2	57.2	57.2	70.4
1967	110.9	93.6	111.8	69.5	86.5	67.3	91.5	107.2	76.8	88.2	94.3	126.4
1968	121.8	111.9	92.2	81.2	127.2	110.3	96.1	109.3	117.2	107.7	86.0	109.8
1969	104.4	120.5	135.8	106.8	120.0	106.0	96.8	98.0	91.3	95.7	93.5	97.9
1970	111.5	127.8	102.9	109.5	127.5	106.8	112.5	93.0	99.5	86.6	95.2	83.5
1971	91.3	79.0	60.7	71.8	57.5	49.8	81.0	61.4	50.2	51.7	63.2	82.2
1972	61.5	88.4	80.1	63.2	80.5	88.0	76.5	76.8	64.0	61.3	41.6	45.3
1973	43.4	42.9	46.0	57.7	42.4	39.5	23.1	25.6	59.3	30.7	23.9	23.3
1974	27.6	26.0	21.3	40.3	39.5	36.0	55.8	33.6	40.2	47.1	25.0	20.5
1975	18.9	11.5	11.5	5.1	9.0	11.4	28.2	39.7	13.9	9.1	19.4	7.8
1976	8.1	4.3	21.9	18.8	12.4	12.2	1.9	16.4	13.5	20.6	5.2	15.3
1977	16.4	23.1	8.7	12.9	18.6	38.5	21.4	30.1	44.0	43.8	29.1	43.2
1978	51.9	93.6	76.5	99.7	82.7	95.1	70.4	58.1	138.2	125.1	97.9	122.7
1979	166.6	137.5	138.0	101.5	134.4	149.5	159.4	142.2	188.4	186.2	183.3	176.3
1980	159.6	155.0	126.2	164.1	179.9	157.3	136.3	135.4	155.0	164.7	147.9	174.4
1981	114.0	141.3	135.5	156.4	127.5	90.9	143.8	158.7	167.3	162.4	137.5	150.1
1982	111.2	163.6	153.8	122.0	82.2	110.4	106.1	107.6	118.8	94.7	98.1	127.0
1983	84.3	51.0	66.5	80.7	99.2	91.1	82.2	71.8	50.3	55.8	33.3	33.4
1984	57.0	85.4	83.5	69.7	76.4	46.1	37.4	25.5	15.7	12.0	22.8	18.7
1985	16.5	15.9	17.2	16.2	27.5	24.2	30.7	11.1	3.9	18.6	16.2	17.3

Table B :   Monthly means   (cont.)

Year	Jan	Feb	Mar	Apr	May	Jun	Jul	Aug	Sep	Oct	Nov	Dec
1986	2.5	23.2	15.1	18.5	13.7	1.1	18.1	7.4	3.8	35.4	15.2	6.8
1987	10.4	2.4	14.7	39.6	33.0	17.4	33.0	38.7	33.9	60.6	39.9	27.1
1988	59.0	40.0	76.2	88.0	60.1	101.8	113.8	111.6	120.1	125.1	125.1	179.2
1989	161.3	165.1	131.4	130.6	138.5	196.2	126.9	168.9	176.7	159.4	173.0	165.5
1990	177.3	130.5	140.3	140.3	132.2	105.4	149.4	200.3	125.2	145.5	131.4	129.7
1991	136.9	167.5	141.9	140.0	121.3	169.7	173.7	176.3	125.3	144.1	108.2	144.4
1992	150.0	161.1	106.7	99.8	73.8	65.2	85.7	64.5	63.9	88.7	91.8	82.6
1993	59.3	91.0	69.8	62.2	61.3	49.8	57.9	42.2	22.4	56.4	35.6	48.9
1994	57.8	35.5	31.7	16.1	17.8	28.0						

TABLE C
SMOOTHED MONTHLY MEANS OF SUNSPOT NUMBERS

Year	Jan	Feb	Mar	Apr	May	Jun	Jul	Aug	Sep	Oct	Nov	Dec
1749							81.8	84.5	87.2	88.6	90.1	89.9
1750	90.2	90.0	90.3	90.0	88.8	87.9	86.9	84.9	81.3	76.9	72.3	67.7
1751	63.7	60.5	58.0	55.6	53.3	50.7	48.0	45.5	43.4	42.4	42.2	43.2
1752	45.1	47.7	50.1	51.4	51.8	50.4	49.0	46.4	44.0	42.1	40.7	39.8
1753	38.8	37.9	37.5	36.2	34.7	32.9	30.3	27.5	24.0	20.5	17.2	14.6
1754	12.9	12.0	11.7	12.4	13.5	14.5	15.2	15.3	14.8	13.9	12.4	10.9
1755	9.5	8.3	7.3	6.8	6.8	7.3	8.4	9.5	10.6	11.5	11.9	12.2
1756	11.9	11.3	10.5	9.9	9.7	9.8	9.9	10.6	11.5	12.8	14.6	16.5
1757	18.5	20.7	23.3	25.6	28.3	31.0	33.6	36.1	38.3	40.6	43.2	45.2
1758	47.1	48.1	48.7	49.5	49.1	48.7	48.2	47.2	46.5	45.6	45.5	45.5
1759	45.9	46.6	48.0	49.9	52.1	54.1	56.1	57.9	59.3	60.5	61.3	61.9
1760	62.2	62.4	62.9	63.1	63.5	63.2	62.9	62.9	63.5	64.6	66.6	69.6
1761	73.3	77.2	81.3	85.3	88.8	90.4	90.2	88.2	85.1	80.6	75.4	69.8
1762	65.0	61.1	58.1	56.9	56.6	58.4	60.1	61.8	62.3	62.0	60.3	57.3
1763	53.6	49.5	45.7	43.0	41.8	41.9	43.9	46.3	49.5	51.2	52.6	52.6
1764	51.3	49.2	46.2	42.7	39.3	36.0	33.2	30.9	29.3	28.1	26.9	25.9
1765	25.2	24.6	24.1	23.6	23.0	22.2	21.2	20.1	18.9	18.1	17.3	16.8
1766	16.2	15.3	14.7	13.5	12.4	10.8	9.9	9.6	10.3	12.2	15.1	18.6
1767	22.3	25.5	28.5	30.9	33.0	35.0	36.9	39.5	42.2	44.9	47.7	50.7
1768	53.5	55.8	57.8	59.9	62.4	65.5	68.9	72.2	75.0	77.5	79.1	80.2
1769	81.3	83.2	85.8	90.5	96.5	103.8	111.2	118.2	123.7	125.3	123.7	119.3
1770	112.9	106.6	100.6	96.3	94.7	95.8	97.4	97.9	97.2	95.0	91.8	88.4
1771	84.8	82.1	81.1	81.8	83.7	85.8	88.8	91.0	91.7	90.5	88.1	84.7
1772	81.7	78.5	73.7	69.9	66.1	64.0	62.4	61.4	60.6	59.4	57.2	54.7
1773	51.1	47.0	43.1	38.7	35.3	32.6	31.4	31.4	33.0	35.2	38.3	41.3
1774	43.6	44.5	43.4	40.5	36.5	31.4	26.3	21.2	16.9	13.8	11.5	10.0
1775	9.2	8.6	8.1	7.5	7.0	7.0	7.1	7.8	8.3	9.0	10.1	11.0
1776	12.0	12.7	13.3	14.1	14.8	16.4	18.3	21.1	24.2	28.2	33.3	39.1
1777	45.6	53.0	60.9	69.5	78.0	87.4	97.3	107.1	115.7	122.9	130.6	138.2
1778	145.3	150.4	155.4	158.8	161.8	161.5	160.0	157.2	153.9	149.0	144.0	139.8
1779	137.1	135.5	133.1	131.4	129.6	127.7	125.2	120.9	116.7	112.5	108.3	104.9
1780	101.4	98.8	96.6	94.7	92.7	90.2	88.0	85.6	82.6	79.5	77.2	76.0
1781	76.0	76.1	76.3	75.9	74.9	72.8	69.1	64.8	60.4	55.5	51.3	47.5
1782	45.5	43.9	43.2	42.3	41.2	39.7	38.2	36.7	35.2	33.7	32.2	31.0
1783	30.1	29.4	28.5	27.6	26.2	24.4	22.5	20.3	18.3	16.2	14.5	12.9
1784	11.5	10.5	9.8	9.4	9.1	9.1	9.3	9.5	9.8	10.1	10.6	11.2
1785	12.3	13.8	15.5	17.9	20.8	23.8	26.8	29.6	32.4	35.3	38.9	43.0

Table C :　Smoothed monthly means　(cont.)

Year	Jan	Feb	Mar	Apr	May	Jun	Jul	Aug	Sep	Oct	Nov	Dec
1786	47.6	53.1	59.2	65.9	73.0	80.1	87.5	94.1	99.8	104.1	108.0	111.1
1787	113.4	114.9	116.5	119.1	122.4	126.4	131.5	136.0	140.4	143.1	143.4	143.1
1788	141.6	139.3	137.1	135.3	134.2	133.2	132.3	131.6	130.2	128.7	126.6	124.8
1789	122.7	121.5	120.2	118.9	117.8	116.9	116.0	115.2	114.9	114.2	113.6	112.3
1790	110.1	107.3	103.2	99.0	94.3	90.0	86.4	83.5	80.8	79.0	77.4	75.8
1791	74.6	73.1	71.4	69.7	68.1	66.6	65.5	64.2	63.6	63.0	63.1	63.4
1792	63.5	63.5	63.1	62.5	61.5	60.6	59.5	58.7	57.8	57.1	56.5	56.1
1793	55.7	54.7	53.1	51.0	48.9	46.8	44.9	43.4	42.2	41.5	41.1	41.3
1794	41.5	41.8	41.7	40.6	39.8	39.3	38.8	38.4	38.0	37.2	36.1	34.6
1795	32.5	30.4	27.9	25.6	23.0	21.2	20.3	20.0	19.8	20.1	20.5	20.7
1796	20.9	20.8	20.3	19.6	18.5	17.0	15.7	14.0	12.7	11.1	9.9	9.0
1797	8.2	7.5	7.1	6.7	6.5	6.4	6.2	5.9	5.8	5.6	5.3	4.9
1798	4.5	4.0	3.6	3.1	2.8	2.8	3.0	3.4	4.2	5.3	6.6	7.9
1799	8.8	9.4	9.4	8.9	8.0	7.0	6.0	5.3	4.8	4.8	5.1	5.8
1800	6.9	8.2	9.6	10.9	12.0	13.4	14.8	16.5	18.1	20.1	22.0	24.0
1801	25.7	27.7	29.5	31.0	32.1	33.1	34.2	35.7	37.2	38.7	40.1	41.5
1802	42.6	43.6	44.4	45.1	45.4	45.4	45.4	45.5	45.5	45.0	44.1	43.2
1803	42.1	41.2	40.3	39.9	40.3	40.8	41.8	43.0	44.6	46.0	46.9	47.0
1804	46.5	45.6	44.4	43.6	43.2	43.8	45.1	47.0	49.1	50.9	52.2	52.5
1805	51.9	50.1	48.0	45.5	43.4	41.7	40.4	39.5	38.8	37.9	37.1	35.8
1806	34.5	33.0	31.6	30.2	29.1	28.0	27.0	25.9	24.6	23.2	21.7	20.1
1807	18.4	16.9	15.4	14.3	13.0	11.8	10.5	9.2	7.8	6.4	5.4	5.0
1808	5.0	5.4	6.1	7.1	7.9	8.7	9.2	9.4	9.4	9.0	8.4	7.7
1809	7.0	6.3	5.3	4.5	3.5	2.7	2.0	1.4	1.0	0.7	0.4	0.2
1810	0.0	0.0	0.0	0.0	0.0	0.0	0.0	0.0	0.0	0.0	0.0	0.0
1811	0.1	0.2	0.4	0.7	1.1	1.4	1.9	2.4	2.8	3.0	3.1	3.1
1812	2.9	2.8	2.9	3.0	3.3	3.8	4.3	5.0	5.5	6.1	6.5	6.9
1813	7.3	7.9	8.4	9.5	10.7	12.0	13.4	14.6	15.5	16.1	16.3	16.2
1814	15.8	15.3	14.4	13.7	13.0	12.8	12.9	13.3	14.5	15.9	17.5	19.6
1815	21.9	24.7	27.3	29.8	31.6	34.1	35.8	37.8	39.9	42.7	45.3	47.9
1816	49.3	50.7	50.8	50.4	49.1	47.1	45.0	43.2	42.6	43.2	43.9	44.7
1817	45.6	46.0	46.1	44.8	43.6	41.9	40.0	37.7	35.1	33.4	32.7	32.1
1818	31.8	31.7	32.3	32.8	33.0	32.4	31.6	30.5	28.7	26.8	24.6	23.1
1819	22.3	22.0	21.9	22.2	22.8	23.6	24.1	24.6	24.8	24.5	23.8	22.9
1820	21.7	21.1	20.3	19.5	18.4	17.2	16.2	15.2	14.0	12.9	11.6	10.4
1821	9.4	8.1	7.2	6.6	6.1	5.7	5.3	5.2	5.2	5.5	5.7	5.9
1822	6.1	6.2	6.2	6.0	5.6	5.1	4.4	3.5	2.5	1.6	1.0	0.6
1823	0.3	0.1	0.1	0.1	0.1	0.4	1.1	2.3	3.6	5.1	6.6	8.0
1824	8.7	8.8	8.5	8.5	8.1	7.7	7.4	7.5	8.2	8.8	9.3	10.0
1825	10.9	12.1	13.3	14.5	16.0	17.2	18.2	18.5	18.8	19.4	20.0	21.0

Table C : Smoothed monthly means (cont.)

Year	Jan	Feb	Mar	Apr	May	Jun	Jul	Aug	Sep	Oct	Nov	Dec
1826	22.5	24.7	27.3	29.8	32.4	34.9	37.3	39.3	41.0	43.0	44.5	46.4
1827	47.5	48.8	50.0	50.7	51.1	51.1	51.0	51.4	51.6	52.3	53.5	55.9
1828	58.9	62.0	65.0	67.2	68.6	68.8	67.2	65.0	61.8	59.8	58.4	58.1
1829	59.7	62.5	65.6	68.5	70.4	71.5	71.3	70.0	68.2	67.0	67.0	67.6
1830	68.3	69.1	69.5	69.8	69.2	68.9	68.2	67.4	66.7	66.8	66.7	66.2
1831	64.6	62.2	59.3	55.4	52.1	49.0	46.6	44.7	43.1	42.2	41.8	41.5
1832	40.6	38.9	36.7	33.7	30.5	27.0	23.8	20.8	18.3	16.8	15.5	14.7
1833	13.7	12.7	11.5	10.1	9.0	8.0	7.6	7.4	7.6	7.8	7.9	8.0
1834	7.9	7.9	7.6	7.7	8.4	9.7	11.4	13.2	15.4	18.1	21.0	24.0
1835	27.0	30.8	34.9	40.5	46.8	53.8	61.4	68.7	75.9	82.2	88.5	94.3
1836	99.6	104.1	107.3	110.4	112.7	116.0	120.3	125.6	131.3	137.7	142.8	147.9
1837	150.9	152.8	151.7	148.8	144.1	139.1	134.2	130.5	127.2	125.4	123.9	122.7
1838	121.6	120.1	118.9	115.6	111.7	106.2	101.0	96.5	92.8	89.5	86.8	84.5
1839	82.5	81.1	80.9	81.6	82.8	84.3	85.9	87.4	88.4	88.1	86.5	83.5
1840	79.4	74.7	70.6	67.8	65.9	63.9	61.8	58.9	56.1	52.8	49.8	47.5
1841	45.2	43.6	42.4	41.8	41.4	40.6	39.6	37.9	35.5	32.9	30.1	28.0
1842	26.2	25.0	23.9	23.3	23.2	23.5	23.7	24.0	23.5	22.8	21.5	19.8
1843	18.0	16.0	14.2	12.5	11.3	10.8	10.7	10.8	10.9	11.2	11.6	12.0
1844	12.5	13.1	13.7	14.2	14.4	14.7	15.3	16.2	17.7	20.1	23.2	27.0
1845	30.4	33.6	36.0	37.7	38.5	38.7	38.8	38.7	39.3	41.0	43.4	46.8
1846	50.0	52.9	55.8	58.6	60.7	62.4	63.7	64.4	65.0	64.7	64.6	64.5
1847	64.7	65.4	68.6	74.2	83.0	92.8	103.0	114.0	122.2	128.4	131.3	131.1
1848	129.9	126.7	123.4	120.8	119.8	120.5	122.1	124.7	126.8	128.6	128.4	126.8
1849	122.8	117.4	110.9	104.0	96.9	91.1	86.8	84.4	84.1	83.7	83.2	82.1
1850	79.9	76.8	73.3	70.0	67.3	65.0	63.9	64.3	65.6	67.3	68.8	69.7
1851	69.8	69.1	67.1	65.3	63.3	61.6	60.0	59.4	59.5	60.6	61.5	62.2
1852	62.2	61.4	59.3	57.1	54.9	52.8	50.8	49.3	48.1	47.5	46.7	46.0
1853	45.1	44.4	43.5	42.7	41.8	41.1	40.0	38.4	36.1	33.6	30.8	28.3
1854	25.9	23.9	22.5	21.6	21.0	20.7	20.5	20.1	19.6	18.7	17.8	16.6
1855	15.2	13.5	11.6	9.7	8.0	6.4	5.3	4.6	4.0	3.7	3.5	3.4
1856	3.3	3.4	3.4	3.5	3.7	4.1	4.6	5.2	5.7	6.3	7.2	8.3
1857	9.6	11.0	12.9	15.3	18.1	20.9	24.0	26.9	30.1	32.9	35.2	37.3
1858	39.3	41.1	43.3	46.0	49.5	52.9	56.8	60.8	64.9	69.0	72.6	76.2
1859	79.2	82.2	85.1	88.0	91.4	94.2	95.9	96.9	97.1	96.6	95.4	94.3
1860	93.7	93.5	94.2	94.9	96.5	97.8	98.5	97.5	96.0	94.2	91.9	89.2
1861	86.8	85.2	84.0	82.9	81.2	79.8	78.1	76.4	73.8	70.9	68.4	66.5
1862	64.7	63.9	63.4	63.6	63.5	63.1	62.0	60.4	58.9	56.9	54.9	52.8
1863	51.3	50.1	49.0	47.4	45.8	43.9	42.2	41.1	40.5	41.3	42.3	43.8
1864	45.7	47.2	48.6	48.7	48.6	48.0	46.8	45.7	44.4	43.2	42.2	40.6
1865	39.4	38.0	36.9	35.4	33.5	31.6	29.6	28.0	26.8	25.9	25.2	24.3

Table C : Smoothed monthly means (cont.)

Year	Jan	Feb	Mar	Apr	May	Jun	Jul	Aug	Sep	Oct	Nov	Dec
1866	23.7	22.6	21.3	19.8	18.0	16.0	13.8	11.6	9.8	8.5	7.3	6.2
1867	5.3	4.6	4.3	4.3	4.7	5.5	6.8	8.2	9.9	12.0	14.4	17.0
1868	19.5	22.0	24.5	27.3	30.4	34.0	38.0	42.4	46.5	50.1	53.5	57.4
1869	61.0	64.2	67.2	69.7	72.2	74.4	76.3	78.4	81.3	85.8	91.9	99.8
1870	108.9	118.7	127.8	135.3	141.1	144.1	144.8	143.3	140.2	137.4	135.2	133.7
1871	131.9	130.2	127.7	124.5	120.5	115.8	111.1	106.2	101.6	98.1	96.2	96.3
1872	97.2	98.5	99.7	101.4	102.6	103.4	103.7	103.1	102.7	101.4	99.3	95.6
1873	91.4	86.8	81.4	75.7	70.1	65.3	61.3	58.2	56.0	54.7	53.8	52.8
1874	51.7	50.7	50.0	49.0	47.9	46.2	44.3	41.7	38.8	36.2	33.1	30.2
1875	27.3	24.9	23.1	21.4	19.5	18.0	16.3	15.1	14.0	13.6	13.4	13.2
1876	13.1	12.7	12.4	11.6	11.1	10.4	10.3	10.3	10.6	11.2	12.1	12.9
1877	13.5	13.6	13.8	13.6	13.3	12.8	11.9	11.0	10.2	9.1	8.1	7.2
1878	6.5	5.8	5.1	4.4	3.9	3.5	3.2	2.8	2.5	2.2	2.1	2.0
1879	2.1	2.5	3.0	3.7	4.6	5.7	6.9	8.4	10.1	11.9	13.7	15.6
1880	17.7	19.8	22.5	25.7	28.9	31.9	34.4	36.7	38.8	40.5	41.8	42.9
1881	44.5	46.5	48.5	51.0	53.7	55.9	57.2	57.6	57.7	58.3	59.2	60.0
1882	60.5	61.1	61.0	60.8	60.1	59.4	58.4	57.3	56.4	55.7	56.0	56.0
1883	56.9	57.5	57.6	58.5	59.3	61.6	64.0	66.9	70.1	73.2	75.9	77.7
1884	78.1	77.3	75.6	72.4	68.2	63.9	59.7	56.1	53.6	51.9	51.4	52.3
1885	54.0	56.2	57.8	59.2	59.6	58.4	55.5	51.4	46.9	43.1	39.6	36.8
1886	35.2	34.3	33.9	32.8	31.0	28.4	25.3	21.8	18.3	15.1	12.8	11.6
1887	10.9	11.2	11.8	12.6	13.5	13.9	14.2	14.1	13.6	13.0	12.0	11.1
1888	10.1	9.1	8.3	7.6	6.9	6.3	5.8	5.6	5.7	5.7	5.7	5.7
1889	5.7	5.9	6.2	6.6	6.9	7.1	7.2	7.2	6.7	6.2	5.5	4.9
1890	4.3	4.0	4.2	4.9	5.8	6.7	7.8	8.9	10.1	11.1	12.4	14.4
1891	17.1	20.5	24.3	28.9	33.5	37.5	41.1	44.1	47.0	49.3	51.7	54.2
1892	57.3	61.2	65.0	68.5	71.4	73.7	74.9	75.4	75.2	75.0	74.6	74.8
1893	75.4	77.1	79.8	82.8	85.1	87.3	88.5	89.5	88.7	87.3	85.4	84.2
1894	83.4	83.3	83.3	84.0	84.1	83.1	81.3	78.7	75.8	72.7	69.6	67.4
1895	65.9	65.4	65.3	65.3	64.8	64.4	63.2	61.8	59.9	58.2	55.7	53.7
1896	51.2	49.2	46.9	45.3	43.6	42.5	41.8	41.0	40.3	39.2	38.3	36.5
1897	34.4	32.2	30.1	28.4	26.8	25.4	24.8	24.6	25.3	25.7	26.3	27.0
1898	27.0	26.7	26.3	26.0	26.2	25.9	25.5	25.2	24.7	24.1	23.0	21.3
1899	19.8	17.8	15.9	14.3	13.0	12.3	11.5	11.0	10.6	10.3	10.4	10.6
1900	10.9	11.3	11.3	11.3	11.1	10.5	9.7	8.6	7.5	6.4	5.3	4.6
1901	4.1	3.8	3.5	3.3	3.3	3.2	3.1	3.0	2.8	2.9	2.9	3.0
1902	3.2	3.2	3.3	3.4	3.8	4.2	4.8	5.5	6.4	7.7	9.1	10.5
1903	12.1	13.9	15.5	17.2	19.1	21.7	24.2	26.6	28.6	30.8	32.7	34.5
1904	35.8	37.3	38.8	40.1	41.3	42.5	44.0	46.0	47.9	49.9	51.2	52.7
1905	53.7	54.9	55.6	56.3	57.8	59.6	61.4	62.5	63.4	63.9	63.7	62.2

Table C : Smoothed monthly means (cont.)

Year	Jan	Feb	Mar	Apr	May	Jun	Jul	Aug	Sep	Oct	Nov	Dec
1906	61.2	60.3	60.2	59.6	58.8	58.0	58.2	57.9	58.5	58.5	59.3	60.0
1907	60.1	60.5	60.3	60.3	59.6	58.6	57.6	56.4	55.6	54.5	53.2	51.1
1908	48.7	46.9	46.5	47.5	49.0	51.2	52.9	54.4	55.0	54.4	53.3	51.2
1909	49.0	45.9	43.0	40.8	39.8	39.0	38.9	39.0	39.4	39.5	38.8	37.2
1910	34.2	30.6	26.7	23.2	20.7	18.7	17.6	16.8	16.0	15.2	14.0	12.8
1911	11.3	10.1	8.7	7.6	6.8	6.2	5.5	4.7	3.9	3.3	2.9	2.8
1912	2.8	2.9	3.0	3.3	3.5	3.7	3.9	4.0	4.0	3.8	3.6	3.2
1913	2.8	2.4	1.8	1.4	1.2	1.1	1.1	1.3	1.5	2.0	2.6	3.4
1914	4.3	5.2	6.0	7.0	7.7	8.7	9.7	11.3	13.5	16.2	19.4	23.4
1915	28.1	33.5	38.8	43.9	48.0	51.3	52.9	53.5	53.2	53.1	53.1	53.5
1916	54.5	55.9	57.3	58.3	58.6	58.3	57.5	56.7	56.9	58.1	61.0	65.5
1917	71.7	79.1	87.7	95.0	101.5	106.8	110.6	112.1	110.9	107.8	103.5	97.9
1918	92.4	88.0	84.9	83.5	83.1	82.5	82.2	81.7	80.7	78.5	75.9	74.0
1919	72.5	71.7	71.6	71.8	71.5	70.5	68.1	64.9	61.1	57.5	53.4	49.5
1920	46.8	44.1	41.8	39.4	37.7	35.8	34.3	33.1	32.2	32.1	31.8	31.3
1921	31.0	30.8	30.2	29.4	28.6	27.8	26.6	25.0	24.1	23.4	22.7	22.1
1922	21.3	20.4	19.5	17.6	15.6	13.4	11.6	9.8	8.2	7.4	7.1	6.9
1923	6.5	6.0	5.7	5.6	5.6	5.8	6.1	6.3	6.5	6.4	6.5	6.9
1924	7.7	9.2	10.9	13.2	15.6	18.0	19.8	20.8	21.2	21.1	21.0	21.5
1925	22.6	24.6	27.3	31.1	35.4	40.5	46.1	51.6	56.6	60.8	63.4	65.6
1926	66.2	66.1	65.0	63.6	62.6	62.0	62.2	63.9	65.9	68.8	71.5	74.0
1927	75.5	76.3	75.7	74.4	72.2	69.5	66.9	65.0	64.4	64.7	66.1	68.3
1928	71.6	74.7	77.9	80.3	81.8	82.0	80.9	78.5	75.6	71.8	68.1	64.9
1929	62.4	61.0	60.2	59.8	60.0	60.9	62.1	63.5	64.3	64.7	63.8	61.9
1930	58.3	53.9	48.9	43.8	38.8	34.6	31.7	30.3	29.8	29.6	29.5	29.4
1931	28.9	28.3	26.8	25.2	23.3	21.5	19.9	18.1	16.6	15.4	14.7	14.4
1932	14.2	14.0	13.7	13.2	12.7	12.0	11.4	11.0	10.7	10.5	10.3	10.1
1933	9.9	9.5	8.7	7.7	6.5	5.3	4.2	3.3	2.9	3.0	3.5	4.2
1934	5.2	6.3	7.3	8.0	8.6	8.9	9.2	9.4	10.1	10.9	12.1	14.0
1935	16.5	19.2	22.1	25.2	28.7	32.7	36.8	41.3	46.2	51.2	55.9	59.7
1936	62.8	65.3	67.1	68.5	70.0	72.9	76.8	82.5	88.5	94.7	100.8	106.2
1937	110.8	114.6	117.0	119.1	119.8	119.3	117.9	116.0	113.3	110.3	106.7	104.3
1938	103.1	104.3	106.2	108.5	111.1	112.6	113.1	111.2	108.4	104.5	100.7	97.4
1939	94.8	93.6	94.3	95.9	96.8	96.9	95.7	92.7	88.5	83.0	77.0	71.9
1940	68.4	66.5	66.4	67.2	68.6	70.3	70.6	70.1	67.7	64.8	61.0	56.8
1941	53.1	50.3	48.9	48.9	49.4	50.0	50.4	50.5	50.3	49.6	48.7	46.8
1942	44.3	41.9	39.0	36.0	32.7	29.6	26.5	23.9	22.1	21.4	21.5	21.8
1943	21.7	21.4	20.5	19.4	17.8	16.3	15.0	13.6	12.2	11.0	9.9	9.0
1944	7.8	6.9	6.5	6.5	7.0	7.9	9.4	11.3	12.9	14.8	16.7	18.7
1945	20.7	22.9	25.0	27.6	30.6	33.4	35.9	38.8	42.2	45.9	49.9	54.5

Table C :   Smoothed monthly means   (cont.)

Year	Jan	Feb	Mar	Apr	May	Jun	Jul	Aug	Sep	Oct	Nov	Dec
1946	59.6	65.8	72.0	77.7	83.8	89.7	95.5	100.1	104.9	109.8	115.6	122.2
1947	128.8	136.1	143.9	151.2	157.2	160.4	161.2	158.9	153.1	146.5	139.9	135.3
1948	133.1	133.2	135.3	139.5	142.9	145.5	145.8	145.3	143.6	141.5	139.0	138.3
1949	137.4	137.4	136.4	135.9	134.8	133.7	131.8	129.4	126.9	124.6	122.4	119.2
1950	115.2	110.8	106.0	100.6	95.1	89.2	83.5	77.7	71.9	67.4	64.9	64.9
1951	66.5	68.9	71.9	74.8	76.2	76.0	73.9	70.1	64.9	58.4	51.7	45.4
1952	40.7	37.1	34.6	33.2	32.5	32.5	32.6	32.4	31.2	29.5	27.2	24.8
1953	22.3	20.1	18.7	17.9	16.9	16.0	14.7	13.5	11.9	10.0	8.0	6.3
1954	4.9	3.9	3.4	3.2	3.5	3.9	4.6	5.7	6.8	8.1	9.7	11.5
1955	13.5	15.7	18.3	21.6	25.8	31.2	37.2	44.5	53.2	62.5	72.3	82.0
1956	91.3	100.7	109.7	118.3	126.9	135.7	144.7	152.1	158.0	162.6	165.9	167.7
1957	169.1	169.3	170.3	173.6	178.3	184.6	192.2	199.2	204.7	207.8	208.4	206.8
1958	203.2	199.2	195.0	191.0	188.2	186.2	185.8	185.8	185.1	183.9	182.4	180.7
1959	178.6	175.8	174.1	171.2	168.1	163.7	158.3	153.4	147.1	140.6	134.7	129.1
1960	124.9	121.3	119.5	118.8	117.8	116.0	113.3	109.1	103.5	96.2	88.3	80.5
1961	73.6	68.0	64.0	61.9	60.1	58.7	57.0	55.0	52.7	49.9	47.5	45.5
1962	43.7	42.5	41.4	40.7	39.8	38.5	36.7	34.8	32.9	30.9	29.4	28.4
1963	27.9	27.7	27.7	28.3	29.3	30.1	30.3	29.8	28.6	26.8	24.3	21.7
1964	19.2	16.6	14.1	12.0	10.4	9.3	8.6	8.5	8.7	9.2	10.0	11.1
1965	12.1	12.9	13.5	13.9	14.4	14.6	15.0	15.6	16.6	18.1	20.1	22.8
1966	26.2	30.1	34.2	38.3	42.2	46.0	50.0	54.5	59.7	65.0	70.3	75.2
1967	79.7	83.1	86.0	87.0	87.9	88.3	89.5	91.5	94.0	97.0	99.8	102.6
1968	104.3	105.2	105.9	106.4	106.6	106.0	105.6	105.8	106.5	107.7	108.4	109.7
1969	110.8	111.6	111.6	110.5	109.2	106.8	104.2	102.3	101.0	100.9	102.0	104.0
1970	106.2	108.4	109.6	109.9	109.2	107.3	104.7	101.2	97.7	93.7	89.6	85.0
1971	80.8	77.0	73.1	69.6	66.4	64.1	62.9	62.3	63.2	64.6	66.3	68.8
1972	70.8	73.1	74.9	75.9	75.6	74.0	71.3	67.8	63.5	59.5	55.6	52.0
1973	48.9	46.3	44.3	42.9	41.3	39.7	37.8	35.8	33.9	31.9	30.7	30.1
1974	30.3	31.2	32.2	33.9	35.9	37.2	37.7	36.9	35.2	32.3	28.5	24.6
1975	20.8	18.1	16.6	15.7	15.8	16.4	16.9	17.0	16.7	16.2	15.4	14.4
1976	13.5	12.6	12.4	12.6	12.6	12.7	12.8	13.2	13.3	13.6	13.8	14.7
1977	15.7	16.9	18.4	20.7	23.2	26.0	28.9	32.6	37.1	42.5	48.5	55.1
1978	61.7	68.0	73.3	79.3	84.2	89.0	93.9	99.3	105.7	111.7	117.5	122.8
1979	127.9	131.7	135.0	138.3	143.3	148.5	153.9	158.8	163.3	165.9	167.1	166.6
1980	164.7	162.2	159.0	156.4	154.5	153.4	152.8	151.9	151.0	149.9	147.9	146.0
1981	143.0	141.0	138.9	138.1	138.2	139.1	141.1	142.8	144.6	145.6	144.9	143.2
1982	139.2	134.4	128.7	123.5	118.3	114.4	111.1	107.6	103.7	100.1	96.9	93.7
1983	90.0	86.8	83.9	81.2	78.5	74.7	70.7	67.2	63.8	61.1	59.4	59.2
1984	59.9	60.6	60.4	58.3	54.6	49.1	42.5	35.5	29.2	24.3	21.0	19.2
1985	18.7	18.9	19.3	19.5	19.5	19.1	18.5	17.6	16.7	16.0	15.2	14.7

Table C : Smoothed monthly means (cont.)

Year	Jan	Feb	Mar	Apr	May	Jun	Jul	Aug	Sep	Oct	Nov	Dec
1986	14.2	14.1	13.8	13.5	13.3	13.2	13.3	13.1	12.8	13.2	13.8	14.9
1987	16.2	17.9	20.3	23.0	26.3	29.6	32.8	35.9	38.6	41.8	45.2	49.3
1988	54.9	60.9	68.3	76.2	84.5	94.0	103.5	113.1	122.0	129.6	135.9	141.6
1989	145.3	148.2	150.2	152.3	153.8	156.0	157.6	160.2	161.6	162.1	161.1	157.7
1990	154.0	150.3	147.2	144.5	143.0	142.0	141.6	142.4	142.7	143.3	142.9	143.0
1991	143.2	144.1	145.6	147.8	148.8	149.4	148.8	148.0	146.0	143.8	139.8	134.6
1992	127.5	120.1	111.7	103.5	95.6	88.7	83.1	79.2	77.5	77.1	76.7	76.1
1993	74.2	72.0	68.4	64.1	59.6	55.3	51.9	49.0	46.4	44.6	42.4	40.4

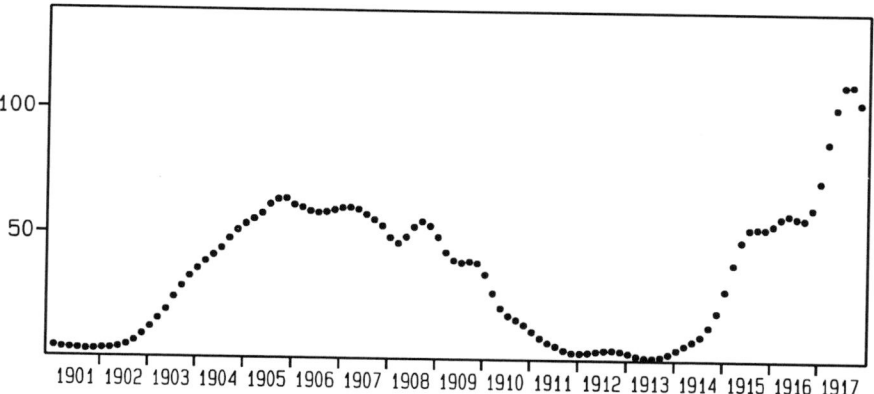

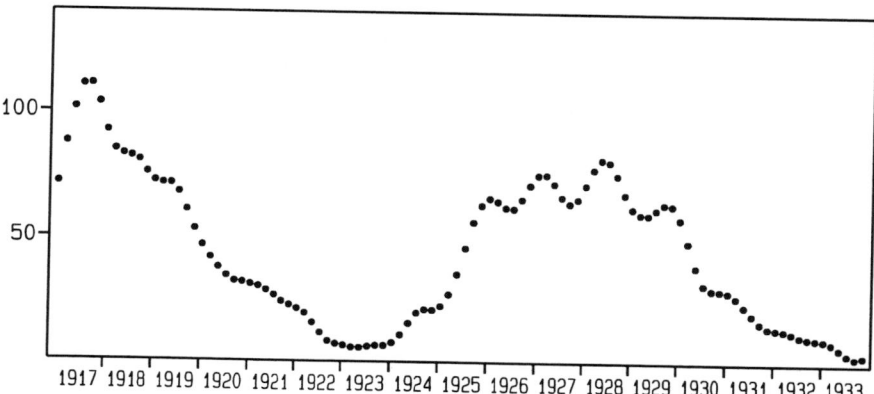

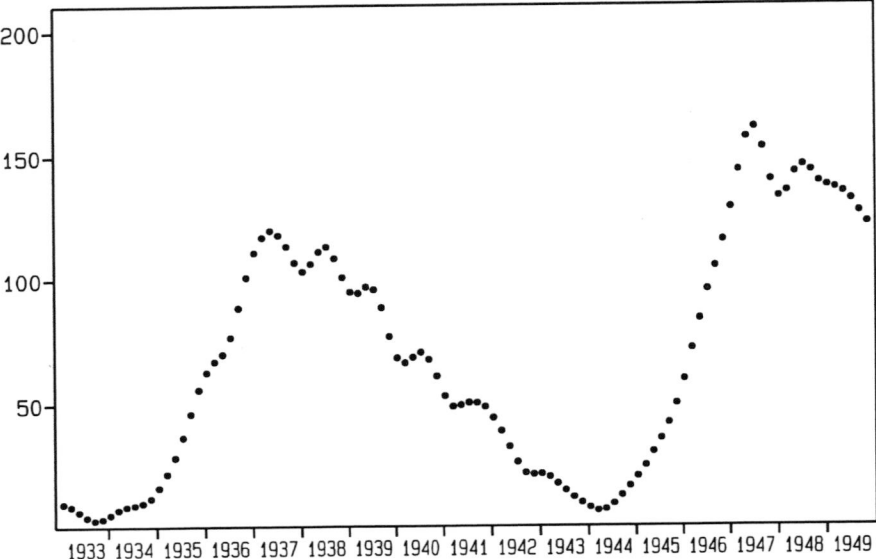

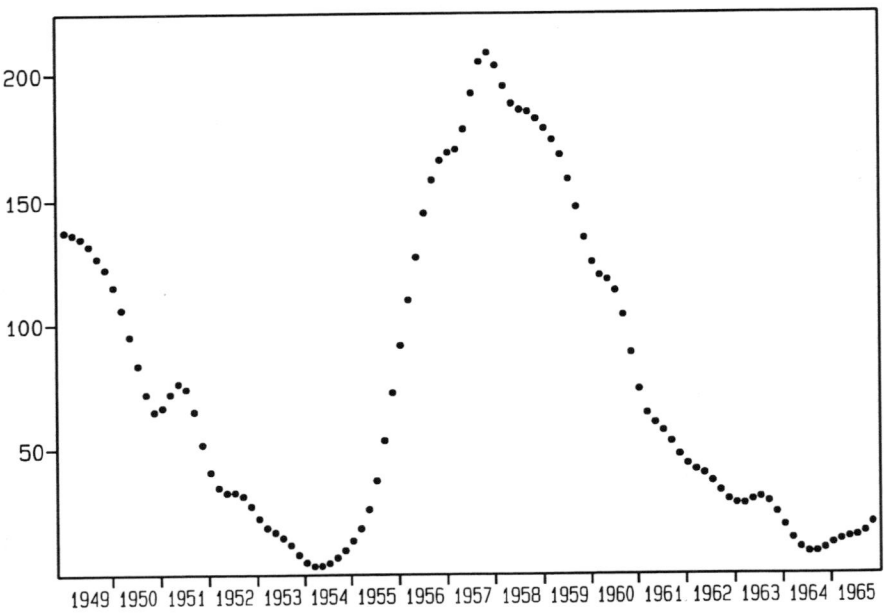

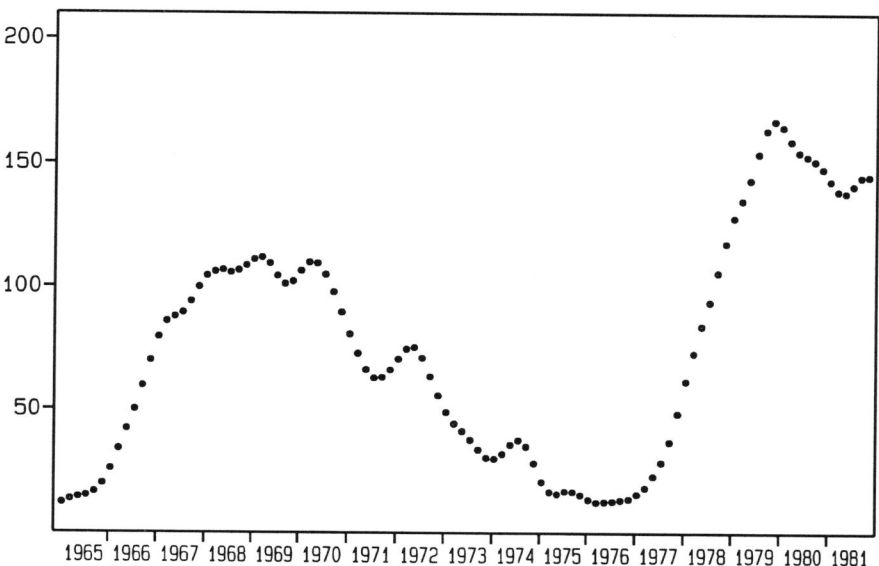

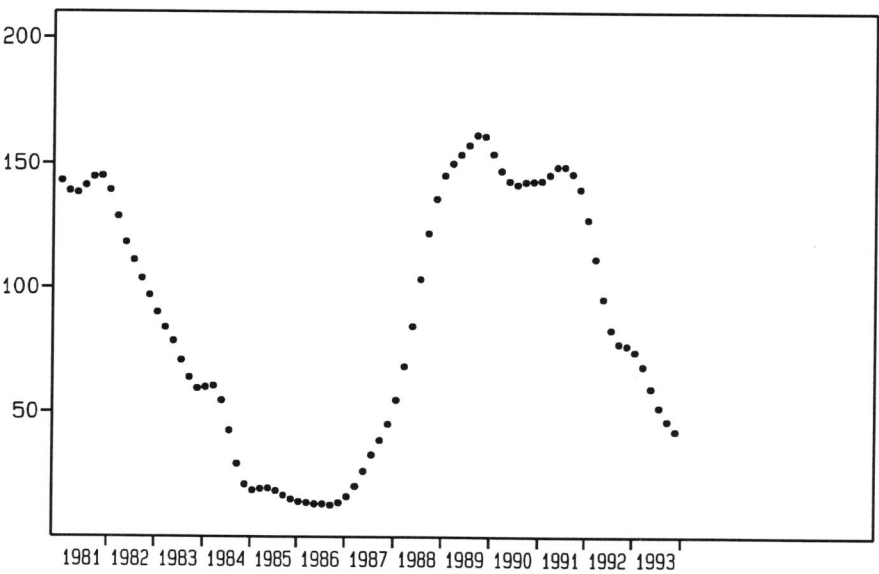

TABLE D
EPOCHS OF MAXIMA AND MINIMA

	Epoch (Zürich)	Epoch and Value (Meeus' formula)		Highest or Lowest Monthly Mean	
Max.	1750.3	1750 March	90.3	1749 November	158.6
min.	1755.2	1755 May	6.8	1755 May – June	0.0
Max.	1761.5	1761 June	90.4	1761 May	107.2
min.	1766.5	1766 August	9.6	1766 June	3.0
Max.	1769.7	1769 October	125.3	1769 October	158.2
min.	1775.5	1775 June	7.0	1775 February	0.0
Max.	1778.4	1778 May	161.8	1778 May	238.9
min.	1784.7	1784 May	9.1	1784 May, July	6.0
Max.	1788.1	1787 November	143.4	1787 December	174.0
min.	1798.3	1798 June	2.8	1798 May – July	0.0
Max.	1805.2	1804 December *(a)*	52.5	1804 October	62.3
min.	1810.6	1810 July ?	0.0	1809 Oct. − 1811 June	0.0
Max.	1816.4	1816 March *(b)*	50.8	1817 March	96.2
min.	1823.3	1823 April	0.1	1822 Nov. − 1823 Feb. 1823 April – June 1823 August – November	0.0
Max.	1829.9	1829 June *(c)*	71.5	1830 April	106.3
min.	1833.9	1833 August	7.4	1833 June	1.0
Max.	1837.2	1837 February	152.8	1836 December	206.2
min.	1843.5	1843 July	10.7	1843 February	3.5
Max.	1848.1	1847 November *(d)*	131.3	1847 October	180.4
min.	1856.0	1856 January	3.3	1855 Sept., 1856 May	0.0
Max.	1860.1	1860 July *(e)*	98.5	1860 July	116.7
min.	1867.2	1867 April	4.3	1867 January	0.0
Max.	1870.6	1870 July	144.8	1870 May	176.0
min.	1878.9	1878 December	2.0	1878 Aug., 1879 March	0.0
Max.	1883.9	1884 January	78.1	1882 April	95.8
min.	1889.6	1890 February	4.0	1889 November	0.2
Max.	1894.1	1893 August	89.5	1893 August	129.2
min.	1901.7	1901 September	2.8	1901 April, December 1902 February, April	0.0
Max.	1907.0	1905 October *(f)*	63.9	1907 February	108.2
min.	1913.6	1913 June	1.1	1913 May – June	0.0

Table D :   Epochs of maxima and minima   (cont.)

	Epoch (Zürich)	Epoch and Value (Meeus' formula)		Highest or Lowest Monthly Mean	
Max.	1917.6	1917 August	112.1	1917 August	154.5
min.	1923.6	1923 April	5.6	1923 Aug., 1924 Jan.	0.5
Max.	1928.4	1928 June *(g)*	82.0	1929 December	108.0
min.	1933.8	1933 September	2.9	1933 August	0.2
Max.	1937.4	1937 May *(h)*	119.8	1938 July	165.3
min.	1944.2	1944 April	6.5	1944 April	0.3
Max.	1947.5	1947 July	161.2	1947 May	201.3
min.	1954.3	1954 April	3.2	1954 January, June	0.2
Max.	1957.9	1957 November	208.4	1957 October	253.8
min.	1964.7	1964 August	8.5	1964 July	3.1
Max.	1968.9	1969 February *(i)*	111.6	1969 March	135.8
min.		1976 March	12.4	1976 July	1.9
Max.		1979 November	167.1	1979 September	188.4
min.		1986 September	12.8	1986 June	1.1
Max.		1989 October *(j)*	162.1	1990 August	200.3

Notes

(a)   Secondary maxima :   1802 August (45.5) and 1803 December (47.0).

(b)   Secondary maximum : 1817 March (46.1).

(c)   Secondary maxima :   1828 June (68.8) and 1830 April (69.8).

(d)   Secondary maximum : 1848 October (128.6).

(e)   Secondary maximum : 1859 September (97.1).

(f)   Secondary maximum : 1907 February (60.5).

(g)   Secondary maximum : 1927 February (76.3).

(h)   Secondary maximum : 1938 July (113.1).

(i)   Secondary maximum : 1970 April (109.9).

(j)   Secondary maximum : 1991 June (149.4).

TABLE E
NUMBER OF SPOTLESS DAYS

Year	Jan	Feb	Mar	Apr	May	Jun	Jul	Aug	Sep	Oct	Nov	Dec	Total
1850	0	0	0	1	0	0	5	0	0	0	1	0	7
1852	0	0	0	0	0	1	2	0	1	0	0	0	4
1853	0	0	2	1	0	1	1	0	1	0	0	0	6
1854	6	8	7	4	8	1	5	4	6	16	0	5	70
1855	17	9	6	21	19	21	30	25	30	9	21	26	234
1856	30	19	30	17	31	22	20	17	21	22	15	17	261
1857	4	14	18	13	0	6	6	9	0	0	0	0	70
1858	1	0	0	1	0	0	0	0	0	0	0	0	2
1861	0	0	0	0	0	0	0	0	0	2	0	0	2
1862	0	0	1	0	0	0	0	0	0	0	0	3	4
1863	0	0	0	0	0	0	0	0	2	0	0	0	2
1864	0	0	0	3	0	0	0	2	1	0	0	1	7
1865	0	0	2	2	2	2	2	1	7	10	4	12	44
1866	0	1	0	2	6	3	9	5	13	5	15	27	86
1867	31	26	12	20	24	26	18	20	16	13	9	7	222
1868	13	5	3	0	3	2	10	0	1	0	0	0	37
1869	0	0	0	1	0	0	1	0	0	0	0	0	2
1873	0	0	0	0	4	7	0	0	0	1	1	1	14
1874	0	0	0	3	0	0	0	0	2	2	0	5	12
1875	11	6	0	1	16	3	10	15	26	11	17	15	131
1876	17	9	8	26	21	26	11	17	14	7	15	19	190
1877	7	8	15	11	2	4	16	17	4	23	10	23	140
1878	22	20	15	29	22	18	30	31	18	28	18	29	280
1879	26	26	31	17	24	20	12	15	12	12	10	12	217
1880	7	6	5	2	4	0	5	0	0	0	3	1	33
1881	1	0	0	0	0	0	0	3	0	0	0	1	5
1882	0	0	0	0	0	0	0	0	0	0	0	2	2
1883	0	1	1	0	1	0	0	0	1	0	0	0	4

Table E : Spotless days (cont.)

Year	Jan	Feb	Mar	Apr	May	Jun	Jul	Aug	Sep	Oct	Nov	Dec	Total
1885	1	0	1	0	0	0	0	0	0	3	3	5	13
1886	7	0	0	1	2	2	4	2	4	6	27	7	62
1887	8	10	15	11	5	0	4	11	12	11	13	4	104
1888	7	12	12	11	17	10	16	20	6	21	7	11	150
1889	28	15	15	17	24	17	11	8	13	20	27	17	212
1890	16	25	17	20	16	23	10	14	2	9	12	7	171
1891	11	3	5	2	0	0	0	1	0	0	0	2	24
1895	0	0	0	0	0	0	0	0	0	0	1	0	1
1896	0	0	0	3	1	0	0	2	0	1	0	0	7
1897	0	0	1	3	6	6	0	0	0	7	8	1	32
1898	1	3	5	9	0	1	11	2	0	0	0	7	39
1899	2	10	4	1	7	2	9	24	11	16	10	8	104
1900	10	13	14	3	8	9	14	21	13	8	16	29	158
1901	30	20	23	30	18	15	28	27	28	21	16	31	287
1902	20	28	17	30	20	25	28	22	18	4	17	28	257
1903	9	0	11	2	6	6	0	0	10	0	0	1	45
1904	1	0	0	0	0	0	0	0	0	0	0	0	1
1905	1	0	0	0	1	0	1	0	0	0	0	0	3
1906	0	0	0	0	0	0	0	0	0	4	0	0	4
1908	0	0	1	0	0	0	0	0	0	3	0	0	4
1909	0	0	0	0	0	1	2	1	0	0	0	2	6
1910	0	6	0	9	0	5	5	10	4	1	19	16	75
1911	19	13	14	5	11	22	19	17	20	22	15	23	200
1912	30	29	17	21	16	16	22	30	8	21	27	17	254
1913	22	19	29	27	31	30	25	30	27	21	27	23	311
1914	24	21	21	1	19	15	14	15	9	12	2	0	153
1915	0	1	0	2	9	0	0	0	0	0	0	0	12
1916	0	0	0	0	0	0	0	3	0	1	0	0	4
1920	0	0	0	2	0	0	0	1	4	0	0	0	7
1921	0	0	3	1	6	1	0	5	4	7	9	10	46
1922	10	4	3	13	14	14	13	17	12	14	9	11	134
1923	21	23	21	13	19	14	20	29	4	3	9	24	200
1924	29	22	25	17	5	0	1	0	0	0	8	9	116
1925	17	2	6	0	0	2	1	1	0	0	0	0	29

Table E : Spotless days (cont.)

Year	Jan	Feb	Mar	Apr	May	Jun	Jul	Aug	Sep	Oct	Nov	Dec	Total
1926	0	0	0	0	0	0	2	0	0	0	0	0	2
1930	0	0	0	0	0	0	0	0	1	0	1	1	3
1931	4	1	0	0	0	11	3	10	0	5	7	2	43
1932	7	11	8	12	5	1	10	14	16	6	12	6	108
1933	12	12	9	22	21	14	24	30	15	23	28	30	240
1934	21	8	18	10	4	17	12	13	18	11	11	11	154
1935	0	0	4	9	6	0	1	0	0	0	0	0	20
1941	0	0	0	0	0	0	0	0	0	0	3	2	5
1942	1	0	1	0	2	6	5	2	1	3	3	0	24
1943	4	2	1	0	0	9	7	3	13	10	10	6	65
1944	20	27	12	29	26	10	13	5	5	2	6	4	159
1945	3	3	1	2	1	1	4	1	0	0	0	0	16
1950	0	0	0	0	0	0	0	0	0	0	0	3	3
1952	0	6	9	0	0	0	0	0	2	2	2	2	23
1953	7	17	11	8	8	0	14	9	1	9	25	22	131
1954	30	26	14	24	28	29	14	10	24	12	15	15	241
1955	0	3	18	11	4	4	3	1	1	3	0	0	48
1961	0	0	0	0	0	0	0	0	0	0	3	3	6
1962	1	0	0	0	0	0	1	3	0	0	1	4	10
1963	0	1	2	6	0	0	0	1	3	2	1	5	21
1964	1	8	2	7	4	10	20	11	18	15	10	6	112
1965	2	4	4	13	8	4	10	9	3	3	7	3	70
1966	4	0	2	0	1	0	0	1	0	0	0	0	8
1973	0	0	0	0	0	0	1	6	2	5	6	7	27
1974	7	2	1	0	5	0	0	0	1	0	0	4	20
1975	2	9	9	18	13	9	0	0	7	9	6	13	95
1976	16	18	5	2	6	7	24	0	3	3	15	6	105
1977	6	2	7	4	2	0	4	0	0	0	0	0	25
1983	0	0	0	0	0	0	0	0	0	0	4	0	4
1984	0	0	0	0	0	0	0	0	8	4	1	0	13
1985	11	0	5	7	0	1	0	7	16	14	11	11	83
1986	24	4	6	5	9	26	5	8	18	2	5	17	129
1987	7	19	2	0	0	7	9	0	0	0	0	0	44
1994	0	0	0	5	6	5							

TABLE F
BEGINNING OF THE SUN'S SYNODIC ROTATIONS

Rotation No.	Date of beginning			Rotation No.	Date of beginning			Rotation No.	Date of beginning		
1316	**1952**	Jan	22.96	1356	**1955**	Jan	17.96	1396	**1958**	Jan	12.96
1317		Feb	19.30	1357		Feb	14.30	1397		Feb	9.30
1318		Mar	17.63	1358		Mar	13.63	1398		Mar	8.63
1319		Apr	13.92	1359		Apr	9.93	1399		Apr	4.94
1320		May	11.16	1360		May	7.18	1400		May	2.19
1321		Jun	7.37	1361		Jun	3.39	1401		May	29.41
1322		Jul	4.57	1362		Jun	30.59	1402		Jun	25.61
1323		Jul	31.78	1363		Jul	27.80	1403		Jul	22.82
1324		Aug	28.01	1364		Aug	24.03	1404		Aug	19.04
1325		Sep	24.27	1365		Sep	20.28	1405		Sep	15.29
1326		Oct	21.56	1366		Oct	17.57	1406		Oct	12.57
1327		Nov	17.86	1367		Nov	13.87	1407		Nov	8.87
1328		Dec	15.18	1368		Dec	11.18	1408		Dec	6.18
1329	**1953**	Jan	11.51	1369	**1956**	Jan	7.51	1409	**1959**	Jan	2.51
1330		Feb	7.86	1370		Feb	3.85	1410		Jan	29.85
1331		Mar	7.19	1371		Mar	2.19	1411		Feb	26.19
1332		Apr	3.49	1372		Mar	29.50	1412		Mar	25.51
1333		Apr	30.76	1373		Apr	25.77	1413		Apr	21.78
1334		May	27.98	1374		May	23.00	1414		May	19.02
1335		Jun	24.18	1375		Jun	19.20	1415		Jun	15.22
1336		Jul	21.38	1376		Jul	16.40	1416		Jul	12.42
1337		Aug	17.60	1377		Aug	12.62	1417		Aug	8.63
1338		Sep	13.85	1378		Sep	8.87	1418		Sep	4.88
1339		Oct	11.13	1379		Oct	6.14	1419		Oct	2.15
1340		Nov	7.43	1380		Nov	2.43	1420		Oct	29.44
1341		Dec	4.74	1381		Nov	29.74	1421		Nov	25.75
				1382		Dec	27.07	1422		Dec	23.07
1342	**1954**	Jan	1.07								
1343		Jan	28.41	1383	**1957**	Jan	23.41	1423	**1960**	Jan	19.40
1344		Feb	24.75	1384		Feb	19.75	1424		Feb	15.75
1345		Mar	24.07	1385		Mar	19.07	1425		Mar	14.07
1346		Apr	20.34	1386		Apr	15.36	1426		Apr	10.37
1347		May	17.58	1387		May	12.60	1427		May	7.62
1348		Jun	13.78	1388		Jun	8.81	1428		Jun	3.83
1349		Jul	10.98	1389		Jul	6.00	1429		Jul	1.03
1350		Aug	7.20	1390		Aug	2.21	1430		Jul	28.23
1351		Sep	3.44	1391		Aug	29.45	1431		Aug	24.46
1352		Sep	30.71	1392		Sep	25.72	1432		Sep	20.72
1353		Oct	28.00	1393		Oct	23.00	1433		Oct	18.01
1354		Nov	24.30	1394		Nov	19.31	1434		Nov	14.31
1355		Dec	21.62	1395		Dec	16.63	1435		Dec	11.63

Table F : Beginning of rotations   (cont.)

Rotation No.	Date of beginning			Rotation No.	Date of beginning			Rotation No.	Date of beginning		
1436	**1961**	Jan	7.96	1476	**1964**	Jan	3.95	1517	**1967**	Jan	26.29
1437		Feb	4.30	1477		Jan	31.29	1518		Feb	22.63
1438		Mar	3.63	1478		Feb	27.63	1519		Mar	21.95
1439		Mar	30.94	1479		Mar	25.95	1520		Apr	18.24
1440		Apr	27.21	1480		Apr	22.22	1521		May	15.47
1441		May	24.44	1481		May	19.46	1522		Jun	11.68
1442		Jun	20.64	1482		Jun	15.66	1523		Jul	8.88
1443		Jul	17.84	1483		Jul	12.86	1524		Aug	5.09
1444		Aug	14.06	1484		Aug	9.07	1525		Sep	1.33
1445		Sep	10.31	1485		Sep	5.32	1526		Sep	28.60
1446		Oct	7.58	1486		Oct	2.59	1527		Oct	25.89
1447		Nov	3.88	1487		Oct	29.88	1528		Nov	22.19
1448		Dec	1.19	1488		Nov	26.19	1529		Dec	19.51
1449		Dec	28.51	1489		Dec	23.51				
1450	**1962**	Jan	24.85	1490	**1965**	Jan	19.85	1530	**1968**	Jan	15.84
1451		Feb	21.19	1491		Feb	16.19	1531		Feb	12.19
1452		Mar	20.51	1492		Mar	15.52	1532		Mar	10.52
1453		Apr	16.80	1493		Apr	11.81	1533		Apr	6.82
1454		May	14.04	1494		May	9.05	1534		May	4.07
1455		Jun	10.24	1495		Jun	5.27	1535		May	31.29
1456		Jul	7.44	1496		Jul	2.46	1536		Jun	27.49
1457		Aug	3.65	1497		Jul	29.67	1537		Jul	24.69
1458		Aug	30.89	1498		Aug	25.90	1538		Aug	20.92
1459		Sep	27.16	1499		Sep	22.16	1539		Sep	17.17
1460		Oct	24.44	1500		Oct	19.45	1540		Oct	14.46
1461		Nov	20.75	1501		Nov	15.75	1541		Nov	10.76
1462		Dec	18.07	1502		Dec	13.07	1542		Dec	8.07
1463	**1963**	Jan	14.40	1503	**1966**	Jan	9.40	1543	**1969**	Jan	4.40
1464		Feb	10.74	1504		Feb	5.74	1544		Jan	31.74
1465		Mar	10.08	1505		Mar	5.08	1545		Feb	28.08
1466		Apr	6.38	1506		Apr	1.38	1546		Mar	27.39
1467		May	3.63	1507		Apr	28.65	1547		Apr	23.66
1468		May	30.85	1508		May	25.87	1548		May	20.89
1469		Jun	27.05	1509		Jun	22.07	1549		Jun	17.10
1470		Jul	24.25	1510		Jul	19.27	1550		Jul	14.29
1471		Aug	20.48	1511		Aug	15.50	1551		Aug	10.51
1472		Sep	16.73	1512		Sep	11.75	1552		Sep	6.76
1473		Oct	14.02	1513		Oct	9.02	1553		Oct	4.03
1474		Nov	10.31	1514		Nov	5.32	1554		Oct	31.32
1475		Dec	7.63	1515		Dec	2.63	1555		Nov	27.63
				1516		Dec	29.95	1556		Dec	24.95

Table F : Beginning of rotations (cont.)

Rotation No.	Date of beginning			Rotation No.	Date of beginning			Rotation No.	Date of beginning		
1557	**1970**	Jan	21.29	1597	**1973**	Jan	16.29	1637	**1976**	Jan	12.29
1558		Feb	17.63	1598		Feb	12.63	1638		Feb	8.63
1559		Mar	16.96	1599		Mar	11.96	1639		Mar	6.96
1560		Apr	13.25	1600		Apr	8.26	1640		Apr	3.27
1561		May	10.49	1601		May	5.51	1641		Apr	30.53
1562		Jun	6.70	1602		Jun	1.73	1642		May	27.75
1563		Jul	3.90	1603		Jun	28.92	1643		Jun	23.95
1564		Jul	31.11	1604		Jul	26.13	1644		Jul	21.15
1565		Aug	27.34	1605		Aug	22.36	1645		Aug	17.37
1566		Sep	23.61	1606		Sep	18.61	1646		Sep	13.63
1567		Oct	20.89	1607		Oct	15.90	1647		Oct	10.90
1568		Nov	17.19	1608		Nov	12.20	1648		Nov	7.20
1569		Dec	14.51	1609		Dec	9.51	1649		Dec	4.51
								1650		Dec	31.84
1570	**1971**	Jan	10.84	1610	**1974**	Jan	5.84	1651	**1977**	Jan	28.18
1571		Feb	7.18	1611		Feb	2.18	1652		Feb	24.52
1572		Mar	6.52	1612		Mar	1.52	1653		Mar	23.84
1573		Apr	2.83	1613		Mar	28.83	1654		Apr	20.12
1574		Apr	30.09	1614		Apr	25.10	1655		May	17.35
1575		May	27.31	1615		May	22.33	1656		Jun	13.56
1576		Jun	23.51	1616		Jun	18.53	1657		Jul	10.75
1577		Jul	20.71	1617		Jul	15.73	1658		Aug	6.97
1578		Aug	16.93	1618		Aug	11.95	1659		Sep	3.21
1579		Sep	13.18	1619		Sep	8.20	1660		Sep	30.48
1580		Oct	10.46	1620		Oct	5.47	1661		Oct	27.77
1581		Nov	6.76	1621		Nov	1.76	1662		Nov	24.08
1582		Dec	4.07	1622		Nov	29.07	1663		Dec	21.40
1583		Dec	31.40	1623		Dec	26.40				
1584	**1972**	Jan	27.74	1624	**1975**	Jan	22.73	1664	**1978**	Jan	17.73
1585		Feb	24.08	1625		Feb	19.07	1665		Feb	14.07
1586		Mar	22.40	1626		Mar	18.40	1666		Mar	13.40
1587		Apr	18.68	1627		Apr	14.69	1667		Apr	9.70
1588		May	15.91	1628		May	11.93	1668		May	6.95
1589		Jun	12.12	1629		Jun	8.14	1669		Jun	3.16
1590		Jul	9.32	1630		Jul	5.34	1670		Jun	30.36
1591		Aug	5.53	1631		Aug	1.55	1671		Jul	27.57
1592		Sep	1.77	1632		Aug	28.78	1672		Aug	23.80
1593		Sep	29.04	1633		Sep	25.05	1673		Sep	20.06
1594		Oct	26.33	1634		Oct	22.33	1674		Oct	17.34
1595		Nov	22.63	1635		Nov	18.64	1675		Nov	13.64
1596		Dec	19.95	1636		Dec	15.95	1676		Dec	10.95

Table F :   Beginning of rotations   (cont.)

Rotation No.	Date of beginning			Rotation No.	Date of beginning			Rotation No.	Date of beginning		
1677	**1979**	Jan	7.28	1717	**1982**	Jan	2.28	1758	**1985**	Jan	24.62
1678		Feb	3.62	1718		Jan	29.62	1759		Feb	20.96
1679		Mar	2.96	1719		Feb	25.96	1760		Mar	20.28
1680		Mar	30.27	1720		Mar	25.28	1761		Apr	16.57
1681		Apr	26.54	1721		Apr	21.56	1762		May	13.81
1682		May	23.77	1722		May	18.79	1763		Jun	10.02
1683		Jun	19.97	1723		Jun	14.99	1764		Jul	7.21
1684		Jul	17.17	1724		Jul	12.19	1765		Aug	3.42
1685		Aug	13.39	1725		Aug	8.41	1766		Aug	30.66
1686		Sep	9.64	1726		Sep	4.65	1767		Sep	26.93
1687		Oct	6.91	1727		Oct	1.92	1768		Oct	24.22
1688		Nov	3.21	1728		Oct	29.21	1769		Nov	20.52
1689		Nov	30.51	1729		Nov	25.52	1770		Dec	17.84
1690		Dec	27.84	1730		Dec	22.84				
1691	**1980**	Jan	24.18	1731	**1983**	Jan	19.17	1771	**1986**	Jan	14.17
1692		Feb	20.52	1732		Feb	15.52	1772		Feb	10.51
1693		Mar	18.84	1733		Mar	14.84	1773		Mar	9.85
1694		Apr	15.13	1734		Apr	11.14	1774		Apr	6.15
1695		May	12.37	1735		May	8.39	1775		May	3.40
1696		Jun	8.58	1736		Jun	4.60	1776		May	30.62
1697		Jul	5.78	1737		Jul	1.80	1777		Jun	26.82
1698		Aug	1.99	1738		Jul	29.00	1778		Jul	24.02
1699		Aug	29.22	1739		Aug	25.24	1779		Aug	20.25
1700		Sep	25.49	1740		Sep	21.50	1780		Sep	16.51
1701		Oct	22.77	1741		Oct	18.78	1781		Oct	13.79
1702		Nov	19.08	1742		Nov	15.08	1782		Nov	10.08
1703		Dec	16.40	1743		Dec	12.40	1783		Dec	7.40
1704	**1981**	Jan	12.73	1744	**1984**	Jan	8.73	1784	**1987**	Jan	3.73
1705		Feb	9.07	1745		Feb	5.07	1785		Jan	31.07
1706		Mar	8.40	1746		Mar	3.40	1786		Feb	27.40
1707		Apr	4.71	1747		Mar	30.71	1787		Mar	26.72
1708		May	1.97	1748		Apr	26.98	1788		Apr	23.00
1709		May	29.19	1749		May	24.21	1789		May	20.23
1710		Jun	25.38	1750		Jun	20.41	1790		Jun	16.43
1711		Jul	22.59	1751		Jul	17.61	1791		Jul	13.63
1712		Aug	18.81	1752		Aug	13.83	1792		Aug	9.84
1713		Sep	15.07	1753		Sep	10.08	1793		Sep	6.09
1714		Oct	12.35	1754		Oct	7.35	1794		Oct	3.36
1715		Nov	8.64	1755		Nov	3.65	1795		Oct	30.65
1716		Dec	5.96	1756		Nov	30.96	1796		Nov	26.96
				1757		Dec	28.28	1797		Dec	24.28

Table F : Beginning of rotations (cont.)

Rotation No.	Date of beginning			Rotation No.	Date of beginning			Rotation No.	Date of beginning		
1798	**1988**	Jan	20.62	1838	**1991**	Jan	15.61	1878	**1994**	Jan	10.61
1799		Feb	16.96	1839		Feb	11.96	1879		Feb	6.95
1800		Mar	15.29	1840		Mar	11.29	1880		Mar	6.29
1801		Apr	11.58	1841		Apr	7.59	1881		Apr	2.60
1802		May	8.83	1842		May	4.84	1882		Apr	29.86
1803		Jun	5.04	1843		Jun	1.06	1883		May	27.08
1804		Jul	2.24	1844		Jun	28.26	1884		Jun	23.28
1805		Jul	29.44	1845		Jul	25.46	1885		Jul	20.48
1806		Aug	25.67	1846		Aug	21.69	1886		Aug	16.71
1807		Sep	21.94	1847		Sep	17.95	1887		Sep	12.96
1808		Oct	19.22	1848		Oct	15.23	1888		Oct	10.23
1809		Nov	15.52	1849		Nov	11.53	1889		Nov	6.53
1810		Dec	12.84	1850		Dec	8.84	1890		Dec	3.84
								1891		Dec	31.17
1811	**1989**	Jan	9.17	1851	**1992**	Jan	5.17	1892	**1995**	Jan	27.51
1812		Feb	5.51	1852		Feb	1.51	1893		Feb	23.85
1813		Mar	4.85	1853		Feb	28.85	1894		Mar	23.17
1814		Apr	1.15	1854		Mar	27.16	1895		Apr	19.45
1815		Apr	28.42	1855		Apr	23.43	1896		May	16.68
1816		May	25.64	1856		May	20.66	1897		Jun	12.89
1817		Jun	21.84	1857		Jun	16.87	1898		Jul	10.09
1818		Jul	19.05	1858		Jul	14.07	1899		Aug	6.30
1819		Aug	15.27	1859		Aug	10.28	1900		Sep	2.54
1820		Sep	11.52	1860		Sep	6.53	1901		Sep	29.81
1821		Oct	8.79	1861		Oct	3.80	1902		Oct	27.10
1822		Nov	5.09	1862		Oct	31.09	1903		Nov	23.40
1823		Dec	2.40	1863		Nov	27.40	1904		Dec	20.72
1824		Dec	29.72	1864		Dec	24.72				
1825	**1990**	Jan	26.06	1865	**1993**	Jan	21.06	1905	**1996**	Jan	17.06
1826		Feb	22.40	1866		Feb	17.40	1906		Feb	13.40
1827		Mar	21.72	1867		Mar	16.73	1907		Mar	11.73
1828		Apr	18.01	1868		Apr	13.02	1908		Apr	8.03
1829		May	15.25	1869		May	10.26	1909		May	5.28
1830		Jun	11.45	1870		Jun	6.48	1910		Jun	1.50
1831		Jul	8.65	1871		Jul	3.67	1911		Jun	28.70
1832		Aug	4.86	1872		Jul	30.88	1912		Jul	25.90
1833		Sep	1.10	1873		Aug	27.11	1913		Aug	22.13
1834		Sep	28.37	1874		Sep	23.38	1914		Sep	18.39
1835		Oct	25.66	1875		Oct	20.66	1915		Oct	15.67
1836		Nov	21.96	1876		Nov	16.97	1916		Nov	11.97
1837		Dec	19.28	1877		Dec	14.28	1917		Dec	9.28

Table F : Beginning of rotations (cont.)

Rotation No.	Date of beginning		Rotation No.	Date of beginning		Rotation No.	Date of beginning	
1918	**1997** Jan	5.61	1958	**2000** Jan	1.61	1999	**2003** Jan	23.95
1919	Feb	1.95	1959	Jan	28.95	2000	Feb	20.29
1920	Mar	1.29	1960	Feb	25.29	2001	Mar	19.61
1921	Mar	28.60	1961	Mar	23.61	2002	Apr	15.90
1922	Apr	24.87	1962	Apr	19.89	2003	May	13.14
1923	May	22.10	1963	May	17.12	2004	Jun	9.35
1924	Jun	18.30	1964	Jun	13.33	2005	Jul	6.55
1925	Jul	15.50	1965	Jul	10.53	2006	Aug	2.76
1926	Aug	11.72	1966	Aug	6.74	2007	Aug	29.99
1927	Sep	7.97	1967	Sep	2.98	2008	Sep	26.26
1928	Oct	5.24	1968	Sep	30.25	2009	Oct	23.55
1929	Nov	1.53	1969	Oct	27.54	2010	Nov	19.85
1930	Nov	28.84	1970	Nov	23.85	2011	Dec	17.17
1931	Dec	26.17	1971	Dec	21.17			
1932	**1998** Jan	22.50	1972	**2001** Jan	17.50	2012	**2004** Jan	13.50
1933	Feb	18.84	1973	Feb	13.84	2013	Feb	9.84
1934	Mar	18.17	1974	Mar	13.17	2014	Mar	8.17
1935	Apr	14.46	1975	Apr	9.47	2015	Apr	4.48
1936	May	11.70	1976	May	6.72	2016	May	1.74
1937	Jun	7.91	1977	Jun	2.94	2017	May	28.96
1938	Jul	5.11	1978	Jun	30.13	2018	Jun	25.16
1939	Aug	1.32	1979	Jul	27.34	2019	Jul	22.36
1940	Aug	28.55	1980	Aug	23.57	2020	Aug	18.58
1941	Sep	24.82	1981	Sep	19.83	2021	Sep	14.84
1942	Oct	22.10	1982	Oct	17.11	2022	Oct	12.12
1943	Nov	18.41	1983	Nov	13.41	2023	Nov	8.41
1944	Dec	15.72	1984	Dec	10.72	2024	Dec	5.73
1945	**1999** Jan	12.06	1985	**2002** Jan	7.05	2025	**2005** Jan	2.05
1946	Feb	8.40	1986	Feb	3.39	2026	Jan	29.39
1947	Mar	7.73	1987	Mar	2.73	2027	Feb	25.73
1948	Apr	4.04	1988	Mar	30.04	2028	Mar	25.05
1949	May	1.30	1989	Apr	26.31	2029	Apr	21.33
1950	May	28.52	1990	May	23.54	2030	May	18.56
1951	Jun	24.72	1991	Jun	19.74	2031	Jun	14.76
1952	Jul	21.92	1992	Jul	16.94	2032	Jul	11.96
1953	Aug	18.14	1993	Aug	13.16	2033	Aug	8.18
1954	Sep	14.40	1994	Sep	9.41	2034	Sep	4.42
1955	Oct	11.68	1995	Oct	6.68	2035	Oct	1.69
1956	Nov	7.97	1996	Nov	2.98	2036	Oct	28.98
1957	Dec	5.28	1997	Nov	30.29	2037	Nov	25.29
			1998	Dec	27.61	2038	Dec	22.61

Table C : Beginning of rotations (cont.)

Rotation No.	Date of beginning		Rotation No.	Date of beginning		Rotation No.	Date of beginning	
2039	2006 Jan	18.94	2079	2009 Jan	13.94	2119	2012 Jan	9.94
2040	Feb	15.29	2080	Feb	10.28	2120	Feb	6.28
2041	Mar	14.61	2081	Mar	9.62	2121	Mar	4.62
2042	Apr	10.91	2082	Apr	5.92	2122	Mar	31.93
2043	May	8.16	2083	May	3.18	2123	Apr	28.19
2044	Jun	4.37	2084	May	30.39	2124	May	25.42
2045	Jul	1.57	2085	Jun	26.59	2125	Jun	21.62
2046	Jul	28.78	2086	Jul	23.80	2126	Jul	18.82
2047	Aug	25.01	2087	Aug	20.02	2127	Aug	15.04
2048	Sep	21.27	2088	Sep	16.28	2128	Sep	11.29
2049	Oct	18.55	2089	Oct	13.56	2129	Oct	8.56
2050	Nov	14.85	2090	Nov	9.86	2130	Nov	4.86
2051	Dec	12.17	2091	Dec	7.17	2131	Dec	2.17
						2132	Dec	29.50
2052	2007 Jan	8.50	2092	2010 Jan	3.50	2133	2013 Jan	25.83
2053	Feb	4.84	2093	Jan	30.84	2134	Feb	22.17
2054	Mar	4.18	2094	Feb	27.17	2135	Mar	21.50
2055	Mar	31.48	2095	Mar	26.49	2136	Apr	17.78
2056	Apr	27.75	2096	Apr	22.77	2137	May	15.02
2057	May	24.98	2097	May	20.00	2138	Jun	11.22
2058	Jun	21.18	2098	Jun	16.20	2139	Jul	8.42
2059	Jul	18.38	2099	Jul	13.40	2140	Aug	4.63
2060	Aug	14.60	2100	Aug	9.62	2141	Aug	31.87
2061	Sep	10.85	2101	Sep	5.86	2142	Sep	28.14
2062	Oct	8.12	2102	Oct	3.13	2143	Oct	25.43
2063	Nov	4.42	2103	Oct	30.42	2144	Nov	21.73
2064	Dec	1.73	2104	Nov	26.73	2145	Dec	19.05
2065	Dec	29.05	2105	Dec	24.05			
2066	2008 Jan	25.39	2106	2011 Jan	20.39	2146	2014 Jan	15.39
2067	Feb	21.73	2107	Feb	16.73	2147	Feb	11.73
2068	Mar	20.05	2108	Mar	16.06	2148	Mar	11.06
2069	Apr	16.34	2109	Apr	12.35	2149	Apr	7.36
2070	May	13.58	2110	May	9.60	2150	May	4.61
2071	Jun	9.79	2111	Jun	5.81	2151	May	31.83
2072	Jul	6.99	2112	Jul	3.01	2152	Jun	28.03
2073	Aug	3.20	2113	Jul	30.21	2153	Jul	25.23
2074	Aug	30.43	2114	Aug	26.45	2154	Aug	21.46
2075	Sep	26.70	2115	Sep	22.71	2155	Sep	17.72
2076	Oct	23.99	2116	Oct	19.99	2156	Oct	15.00
2077	Nov	20.29	2117	Nov	16.29	2157	Nov	11.30
2078	Dec	17.61	2118	Dec	13.61	2158	Dec	8.61

Table F :   Beginning of rotations   (cont.)

Rotation No.	Date of beginning			Rotation No.	Date of beginning			Rotation No.	Date of beginning		
2159	2015	Jan	4.94	2186	2017	Jan	10.38	2213	2019	Jan	16.83
2160		Feb	1.28	2187		Feb	6.73	2214		Feb	13.17
2161		Feb	28.62	2188		Mar	6.06	2215		Mar	12.50
2162		Mar	27.93	2189		Apr	2.37	2216		Apr	8.80
2163		Apr	24.21	2190		Apr	29.63	2217		May	6.05
2164		May	21.44	2191		May	26.85	2218		Jun	2.27
2165		Jun	17.64	2192		Jun	23.05	2219		Jun	29.47
2166		Jul	14.84	2193		Jul	20.25	2220		Jul	26.67
2167		Aug	11.05	2194		Aug	16.48	2221		Aug	22.90
2168		Sep	7.30	2195		Sep	12.73	2222		Sep	19.16
2169		Oct	4.57	2196		Oct	10.01	2223		Oct	16.44
2170		Oct	31.86	2197		Nov	6.30	2224		Nov	12.74
2171		Nov	28.17	2198		Dec	3.61	2225		Dec	10.05
2172		Dec	25.49	2199		Dec	30.94				
2173	2016	Jan	21.83	2200	2018	Jan	27.28	2226	2020	Jan	6.38
2174		Feb	18.17	2201		Feb	23.62	2227		Feb	2.72
2175		Mar	16.50	2202		Mar	22.94	2228		Mar	1.06
2176		Apr	12.79	2203		Apr	19.22	2229		Mar	28.37
2177		May	10.04	2204		May	16.46	2230		Apr	24.65
2178		Jun	6.25	2205		Jun	12.66	2231		May	21.87
2179		Jul	3.44	2206		Jul	9.86	2232		Jun	18.08
2180		Jul	30.65	2207		Aug	6.07	2233		Jul	15.28
2181		Aug	26.89	2208		Sep	2.31	2234		Aug	11.49
2182		Sep	23.15	2209		Sep	29.58	2235		Sep	7.74
2183		Oct	20.43	2210		Oct	26.87	2236		Oct	5.01
2184		Nov	16.74	2211		Nov	23.17	2237		Nov	1.31
2185		Dec	14.05	2212		Dec	20.49	2238		Nov	28.61
								2239		Dec	25.94

7

Other Tables

Contents

I

TABLE FOR CALCULATING THE JULIAN DAY

This table (in three parts) can be used for the calculation of the Julian Day Number for any date between the years -1900 and $+2999$. Simply add the three numbers corresponding to the century, the year, and the month (from tables a to c), and then the day of the month. The result is the Julian Day Number at $0h$ (Dynamical Time *or* Universal Time) of the date.

Example. — Find the JD corresponding to 1981 February 15, at $0h$.

Table a	1900	....	2415 019.5
Table b	81	....	29 585
Table c	Feb	....	31
Day of month		....	15
	Sum :		2444 650.5

Note that the Julian Days begin at Greenwich mean *noon* (12 h UT). If they are measured from 12 h Dynamical Time, they are called Julian Ephemeris Days (JDE).

The number characterizing *negative* years should be split in such a way as to have the last two figures positive. For example, the year -328 should be split up as -400 and $+72$. Then, table a should be consulted for -400, and table b for $+72$.

In table a, two values are given for the year 1500, one for the Julian calendar (J), the other for the Gregorian calendar (G).

In table c, the months marked with B should be used in the case of a bissextile (leap) year. Remember that, in the Gregorian calendar, the century years are bissextile only when their numerical designation is divisible by 400. These are the years 1600, 2000, 2400, etc. The years 1700, 1800, 1900, 2100, 2200, 2300, 2500, etc., are common years in the Gregorian calendar.

Table a : Century Year

Year	JD	Year	JD
-1900	1027 082.5	600	1940 207.5
-1800	1063 607.5	700	1976 732.5
-1700	1100 132.5	800	2013 257.5
-1600	1136 657.5	900	2049 782.5
-1500	1173 182.5	1000	2086 307.5
-1400	1209 707.5	1100	2122 832.5
-1300	1246 232.5	1200	2159 357.5
-1200	1282 757.5	1300	2195 882.5
-1100	1319 282.5	1400	2232 407.5
-1000	1355 807.5	1500 J	2268 932.5
$-$ 900	1392 332.5	1500 G	2268 922.5
$-$ 800	1428 857.5	1600	2305 447.5
$-$ 700	1465 382.5	1700	2341 971.5
$-$ 600	1501 907.5	1800	2378 495.5
$-$ 500	1538 432.5	1900	2415 019.5
$-$ 400	1574 957.5	2000	2451 544.5
$-$ 300	1611 482.5	2100	2488 068.5
$-$ 200	1648 007.5	2200	2524 592.5
$-$ 100	1684 532.5	2300	2561 116.5
0	1721 057.5	2400	2597 641.5
$+$ 100	1757 582.5	2500	2634 165.5
200	1794 107.5	2600	2670 689.5
300	1830 632.5	2700	2707 213.5
400	1867 157.5	2800	2743 738.5
500	1903 682.5	2900	2780 262.5

Table b : Additional Year

Year	JD	Year	JD	Year	JD
0	0	40	14 610	80	29 220
1	365	41	14 975	81	29 585
2	730	42	15 340	82	29 950
3	1 095	43	15 705	83	30 315
4	1 461	44	16 071	84	30 681
5	1 826	45	16 436	85	31 046
6	2 191	46	16 801	86	31 411
7	2 556	47	17 166	87	31 776
8	2 922	48	17 532	88	32 142
9	3 287	49	17 897	89	32 507
10	3 652	50	18 262	90	32 872
11	4 017	51	18 627	91	33 237
12	4 383	52	18 993	92	33 603
13	4 748	53	19 358	93	33 968
14	5 113	54	19 723	94	34 333
15	5 478	55	20 088	95	34 698
16	5 844	56	20 454	96	35 064
17	6 209	57	20 819	97	35 429
18	6 574	58	21 184	98	35 794
19	6 939	59	21 549	99	36 159
20	7 305	60	21 915		
21	7 670	61	22 280		
22	8 035	62	22 645		
23	8 400	63	23 010		
24	8 766	64	23 376		
25	9 131	65	23 741		
26	9 496	66	24 106		
27	9 861	67	24 471		
28	10 227	68	24 837		
29	10 592	69	25 202		
30	10 957	70	25 567		
31	11 322	71	25 932		
32	11 688	72	26 298		
33	12 053	73	26 663		
34	12 418	74	27 028		
35	12 783	75	27 393		
36	13 149	76	27 759		
37	13 514	77	28 124		
38	13 879	78	28 489		
39	14 244	79	28 854		

Table c : Month

Month	JD
Jan.	0
Jan. B	−1
Feb.	31
Feb. B	30
March	59
April	90
May	120
June	151
July	181
Aug.	212
Sept.	243
Oct.	273
Nov.	304
Dec.	334

II

Perpetual Calendar
(Table for determining the Weekday of a given date)

In the first table, take the number at the intersection of the secular part (left) and the last two figures (upper part) of the given year.

With that number, enter the second table, and take the number at the intersection with the given month. Note the special columns for January and February in the case of a bissextile year.

With that new number, enter the third table. The weekday is found at the intersection with the given day of the month.

00	01	02	03	—	04	05
06	07	—	08	09	10	11
—	12	13	14	15	—	16
17	18	19	—	20	21	22
23	—	24	25	26	27	—
28	29	30	31	—	32	33
34	35	—	36	37	38	39
—	40	41	42	43	—	44
45	46	47	—	48	49	50
51	—	52	53	54	55	—
56	57	58	59	—	60	61
62	63	—	64	65	66	67
—	68	69	70	71	—	72
73	74	75	—	76	77	78
79	—	80	81	82	83	—
84	85	86	87	—	88	89
90	91	—	92	93	94	95
—	96	97	98	99		

Year

0	7	14	17	21	6	0	1	2	3	4	5
1	8	15 J			5	6	0	1	2	3	4
2	9		18	22	4	5	6	0	1	2	3
3	10				3	4	5	6	0	1	2
4	11	15 G	19	23	2	3	4	5	6	0	1
5	12	16	20	24	1	2	3	4	5	6	0
6	13				0	1	2	3	4	5	6

J :   until 1582 October 4 inclusively (Julian Calendar)

G :   from 1582 October 15 onwards (Gregorian Calendar)

Month	May	Feb. *(B)* Aug.	Feb. March Nov.	June	Sept. Dec.	Jan. *(B)* April July	Jan. Oct.
1	2	3	4	5	6	0	1
2	3	4	5	6	0	1	2
3	4	5	6	0	1	2	3
4	5	6	0	1	2	3	4
5	6	0	1	2	3	4	5
6	0	1	2	3	4	5	6
0	1	2	3	4	5	6	0

(B) = Bissextile (lear) year

Day of Month	1 8 15 22 29	2 9 16 23 30	3 10 17 24 31	4 11 18 25	5 12 19 26	6 13 20 27	7 14 21 28
1	Sun.	Mon.	Tue.	Wed.	Thu.	Fri.	Sat.
2	Mon.	Tue.	Wed.	Thu.	Fri.	Sat.	Sun.
3	Tue.	Wed.	Thu.	Fri.	Sat.	Sun.	Mon.
4	Wed.	Thu.	Fri.	Sat.	Sun.	Mon.	Tue.
5	Thu.	Fri.	Sat.	Sun.	Mon.	Tue.	Wed.
6	Fri.	Sat.	Sun.	Mon.	Tue.	Wed.	Thu.
0	Sat.	Sun.	Mon.	Tue.	Wed.	Thu.	Fri.

Example: 1970 March 7.

In the first table, we find **5** at the intersection of 19 and 70.
In the second table, we find **1** at the intersection of 5 and March.
In the third table, we find *Saturday* at the intersection of 1 and 7.

III

Date of Easter Sunday

In the Gregorian calendar, the earliest possible date for the Christian Easter Day is March 22, and the latest is April 25. During the period 1583 – 3000, these extreme Easter dates occur in the following years:

March 22 :   in 1598, 1693, 1761, 1818, 2285, 2353, 2437, 2505, 2972 ;
April 25 :   in 1666, 1734, 1886, 1943, 2038, 2190, 2258, 2326, 2410, 2573, 2630, 2782, 2877, 2945.

The table which follows gives the dates of Easter Sunday for the years 1583 to 2119 in the Gregorian calendar. The dates indicated by M correspond to March, and all other dates to April. For example, in A.D. 1982 Easter Sunday was April 11; this date is found at the intersection of line 1980 and column 2.

Year	0	1	2	3	4	5	6	7	8	9
1580				10	1	21	6	29 M	17	2
1590	22	14	29 M	18	10	26 M	14	6	22 M	11
1600	2	22	7	30 M	18	10	26 M	15	6	19
1610	11	3	22	7	30 M	19	3	26 M	15	31 M
1620	19	11	27 M	16	7	30 M	12	4	23	15
1630	31 M	20	11	27 M	16	8	23 M	12	4	24
1640	8	31 M	20	5	27 M	16	1	21	12	4
1650	17	9	31 M	13	5	28 M	16	·1	21	13
1660	28 M	17	9	25 M	13	5	25	10	1	21
1670	6	29 M	17	2	25 M	14	5	18	10	2
1680	21	6	29 M	18	2	22	14	30 M	18	10
1690	26 M	15	6	22 M	11	3	22	7	30 M	19
1700	11	27 M	16	8	23 M	12	4	24	8	31 M
1710	20	5	27 M	16	1	21	12	28 M	17	9
1720	31 M	13	5	28 M	16	1	21	13	28 M	17
1730	9	25 M	13	5	25	10	1	21	6	29 M
1740	17	2	25 M	14	5	18	10	2	14	6
1750	29 M	11	2	22	14	30 M	18	10	26 M	15
1760	6	22 M	11	3	22	7	30 M	19	3	26 M
1770	15	31 M	19	11	3	16	7	30 M	19	4
1780	26 M	15	31 M	20	11	27 M	16	8	23 M	12

Year	0	1	2	3	4	5	6	7	8	9
1790	4	24	8	31M	20	5	27M	16	8	24M
1800	13	5	18	10	1	14	6	29M	17	2
1810	22	14	29M	18	10	26M	14	6	22M	11
1820	2	22	7	30M	18	3	26M	15	6	19
1830	11	3	22	7	30M	19	3	26M	15	31M
1840	19	11	27M	16	7	23M	12	4	23	8
1850	31M	20	11	27M	16	8	23M	12	4	24
1860	8	31M	20	5	27M	16	1	21	12	28M
1870	17	9	31M	13	5	28M	16	1	21	13
1880	28M	17	9	25M	13	5	25	10	1	21
1890	6	29M	17	2	25M	14	5	18	10	2
1900	15	7	30M	12	3	23	15	31M	19	11
1910	27M	16	7	23M	12	4	23	8	31M	20
1920	4	27M	16	1	20	12	4	17	8	31M
1930	20	5	27M	16	1	21	12	28M	17	9
1940	24M	13	5	25	9	1	21	6	28M	17
1950	9	25M	13	5	18	10	1	21	6	29M
1960	17	2	22	14	29M	18	10	26M	14	6
1970	29M	11	2	22	14	30M	18	10	26M	15
1980	6	19	11	3	22	7	30M	19	3	26M
1990	15	31M	19	11	3	16	7	30M	12	4
2000	23	15	31M	20	11	27M	16	8	23M	12
2010	4	24	8	31M	20	5	27M	16	1	21
2020	12	4	17	9	31M	20	5	28M	16	1
2030	21	13	28M	17	9	25M	13	5	25	10
2040	1	21	6	29M	17	9	25M	14	5	18
2050	10	2	21	6	29M	18	2	22	14	30M
2060	18	10	26M	15	6	29M	11	3	22	14
2070	30M	19	10	26M	15	7	19	11	3	23
2080	7	30M	19	4	26M	15	31M	20	11	3
2090	16	8	30M	12	4	24	15	31M	20	12
2100	28M	17	9	25M	13	5	18	10	1	21
2110	6	29M	17	2	22	14	29M	18	10	26M

IV

JEWISH CALENDAR

In the Jewish (or Hebrew) calendar, a common year may contain 353, 354, or 355 days, and an embolismic or leap year 383, 384, or 385 days. For the Jewish years 5750 to 5781, the table gives the Gregorian date corresponding to New Year (1 Tishri), the length of the year, and the Gregorian date corresponding to 15 Nisan (Pesach).

Jewish Year	Beginning in Gregorian cal.	Length (days)	15 Nisan in Gregorian cal.
5750	1989 Sep 30	355	1990 Apr 10
5751	1990 Sep 20	354	1991 Mar 30
5752	1991 Sep 9	385	1992 Apr 18
5753	1992 Sep 28	353	1993 Apr 6
5754	1993 Sep 16	355	1994 Mar 27
5755	1994 Sep 6	384	1995 Apr 15
5756	1995 Sep 25	355	1996 Apr 4
5757	1996 Sep 14	383	1997 Apr 22
5758	1997 Oct 2	354	1998 Apr 11
5759	1998 Sep 21	355	1999 Apr 1
5760	1999 Sep 11	385	2000 Apr 20
5761	2000 Sep 30	353	2001 Apr 8
5762	2001 Sep 18	354	2002 Mar 28
5763	2002 Sep 7	385	2003 Apr 17
5764	2003 Sep 27	355	2004 Apr 6
5765	2004 Sep 16	383	2005 Apr 24
5766	2005 Oct 4	354	2006 Apr 13
5767	2006 Sep 23	355	2007 Apr 3
5768	2007 Sep 13	383	2008 Apr 20
5769	2008 Sep 30	354	2009 Apr 9
5770	2009 Sep 19	355	2010 Mar 30
5771	2010 Sep 9	385	2011 Apr 19
5772	2011 Sep 29	354	2012 Apr 7
5773	2012 Sep 17	353	2013 Mar 26
5774	2013 Sep 5	385	2014 Apr 15
5775	2014 Sep 25	354	2015 Apr 4
5776	2015 Sep 14	385	2016 Apr 23
5777	2016 Oct 3	353	2017 Apr 11
5778	2017 Sep 21	354	2018 Mar 31
5779	2018 Sep 10	385	2019 Apr 20
5780	2019 Sep 30	355	2020 Apr 9
5781	2020 Sep 19	353	2021 Mar 28

V

MOSLEM CALENDAR

In the Moslem (or Islamic) calendar, a year contains 354 or 355 days. The Moslem New Year consequently regresses through the seasons over a period of about 33 years. For the Moslem years 1410 to 1442, the following table gives the Gregorian date corresponding to New Year (1 Muharram) and the length of the year.

Moslem Year	Beginning in Gregorian cal.	Length (days)	Moslem Year	Beginning in Gregorian cal.	Length (days)
1410	1989 Aug 4	354	1427	2006 Jan 31	354
1411	1990 Jul 24	354	1428	2007 Jan 20	355
1412	1991 Jul 13	355	1429	2008 Jan 10	354
1413	1992 Jul 2	354	1430	2008 Dec 29	354
1414	1993 Jun 21	354	1431	2009 Dec 18	355
1415	1994 Jun 10	355	1432	2010 Dec 8	354
1416	1995 May 31	354	1433	2011 Nov 27	354
1417	1996 May 19	355	1434	2012 Nov 15	355
1418	1997 May 9	354	1435	2013 Nov 5	354
1419	1998 Apr 28	354	1436	2014 Oct 25	355
1420	1999 Apr 17	355	1437	2015 Oct 15	354
1421	2000 Apr 6	354	1438	2016 Oct 3	354
1422	2001 Mar 26	354	1439	2017 Sep 22	355
1423	2002 Mar 15	355	1440	2018 Sep 12	354
1424	2003 Mar 5	354	1441	2019 Sep 1	354
1425	2004 Feb 22	354	1442	2020 Aug 20	355
1426	2005 Feb 10	355			

The data on pages 392 and 393 were calculated by means of the algorithms given by Denis Savoie in *Observations et Travaux* (Société Astronomique de France), No. 22/23 (1990), pp. 32-38 (with corrections in No. 25, p. 43); No. 26 (1991), pp. 12-19; and No. 28 (1991), pp. 40-41.

VI

PERIGEE AND APOGEE OF THE MOON

The table gives, for the years 1990 to 2020, the times of perigee and apogee of the Moon, rounded to the nearest integer hour *Universal Time*, together with the distance d (in kilometers) between the *centers* of Earth and Moon at these instants.

The corresponding value of the Moon's equatorial horizontal parallax π can be found from the formula $\sin \pi = 6378.14 / d$, while the Moon's geocentric semidiameter, expressed in seconds of arc, is equal to $358\,473\,400 / d$.

	Perigee				Apogee		
1990	Jan 7	19 *h*	366976 *km*	Jan 19	16 *h*	404494 *km*	
	Feb 2	3	370179	Feb 16	13	404508	
	Feb 28	8	365735	Mar 16	8	405215	
	Mar 28	8	360558	Apr 12	20	406040	
	Apr 25	17	357502	May 10	0	406431	
	May 24	3	357339	Jun 6	4	406184	
	Jun 21	11	359942	Jul 3	16	405355	
	Jul 19	11	364493	Jul 31	8	404479	
	Aug 15	10	369075	Aug 28	3	404219	
	Sep 9	11	368381	Sep 24	22	404805	
	Oct 6	18	363245	Oct 22	16	405824	
	Nov 3	23	358625	Nov 19	3	406529	
	Dec 2	11	356525	Dec 16	4	406584	
	Dec 30	24	357751				
1991				Jan 12	11 *h*	406037 *km*	
	Jan 28	9 *h*	361980 *km*	Feb 9	4	405042	
	Feb 25	1	367580	Mar 9	1	404280	
	Mar 22	5	369899	Apr 5	21	404305	
	Apr 17	17	365440	May 3	15	405034	
	May 15	17	360627	May 31	3	405871	
	Jun 13	0	357779	Jun 27	7	406244	
	Jul 11	10	357599	Jul 24	11	405966	
	Aug 8	18	360100	Aug 20	23	405155	
	Sep 5	19	364662	Sep 17	15	404342	
	Oct 2	18	369418	Oct 15	11	404162	
	Oct 27	16	368398	Nov 12	8	404810	
	Nov 24	2	362941	Dec 10	2	405823	
	Dec 22	9	358354				

Perigee

Apogee

1992

Jan 19	22 h	356550 km
Feb 17	11	358096
Mar 16	18	362411
Apr 13	7	367731
May 8	12	369586
Jun 4	2	365329
Jul 2	1	360619
Jul 30	8	357674
Aug 27	18	357371
Sep 25	3	359928
Oct 23	5	364778
Nov 19	0	369741
Dec 13	21	367972

Jan 6	12 h	406471 km
Feb 2	12	406509
Feb 29	21	405916
Mar 28	14	404904
Apr 25	10	404202
May 23	5	404320
Jun 19	22	405130
Jul 17	11	406012
Aug 13	16	406373
Sep 9	19	406080
Oct 7	6	405299
Nov 3	23	404538
Dec 1	20	404411
Dec 29	17	405070

1993

Jan 10	12 h	362264 km
Feb 7	20	357905
Mar 8	9	356528
Apr 5	19	358381
May 4	0	362698
May 31	11	367759
Jun 25	17	369357
Jul 22	8	365146
Aug 19	7	360388
Sep 16	15	357407
Oct 15	2	357243
Nov 12	12	360143
Dec 10	14	365357

Jan 26	10 h	406028 km
Feb 22	18	406627
Mar 21	19	406632
Apr 18	5	405951
May 15	22	404908
Jun 12	16	404242
Jul 10	11	404412
Aug 7	4	405257
Sep 3	17	406125
Sep 30	21	406426
Oct 27	24	406103
Nov 24	13	405302
Dec 22	8	404515

1994

Jan 6	1 h	370142 km
Jan 31	4	367408
Feb 27	22	361845
Mar 28	6	357959
Apr 25	17	356929
May 24	3	358816
Jun 21	7	362954
Jul 18	18	367865
Aug 12	23	369464
Sep 8	14	365148
Oct 6	14	360244
Nov 3	24	357240
Dec 2	12	357264
Dec 30	23	360488

Jan 19	5 h	404362 km
Feb 16	2	404980
Mar 15	17	405892
Apr 11	24	406468
May 9	2	406423
Jun 5	13	405693
Jul 3	5	404677
Jul 30	23	404086
Aug 27	18	404343
Sep 24	12	405245
Oct 22	2	406098
Nov 18	5	406347
Dec 15	8	406019

Perigee   Apogee

				Jan	11	22 h	405204 km
1995				Feb	8	18	404419
Jan	27	23 h	365887 km	Mar	8	15	404306
Feb	23	2	370181	Apr	5	10	404974
Mar	20	13	366972	May	3	1	405926
Apr	17	8	361702	May	30	8	406515
May	15	15	358040	Jun	26	11	406441
Jun	13	1	357009	Jul	23	20	405717
Jul	11	10	358775	Aug	20	12	404763
Aug	8	14	362863	Sep	17	6	404262
Sep	5	1	367912	Oct	15	2	404602
Sep	30	4	369505	Nov	11	21	405531
Oct	26	21	364791	Dec	9	10	406324
Nov	23	23	359670				
Dec	22	10	356798				

				Jan	5	12 h	406525 km
1996				Feb	1	16	406163
Jan	19	23 h	357251 km	Feb	29	7	405274
Feb	17	9	360884	Mar	28	3	404463
Mar	16	6	366290	Apr	24	22	404374
Apr	11	3	369910	May	22	16	405072
May	6	22	366531	Jun	19	6	406033
Jun	3	16	361491	Jul	16	14	406598
Jul	1	22	357949	Aug	12	16	406477
Jul	30	8	356948	Sep	9	2	405735
Aug	27	17	358777	Oct	6	18	404791
Sep	24	22	363053	Nov	3	14	404312
Oct	22	9	368350	Dec	1	11	404654
Nov	16	5	369460	Dec	29	5	405522
Dec	13	4	364237				

1997	Jan	10	9 h	359233 km	Jan	25	17 h	406224 km
Feb	7	21	356847	Feb	21	17	406395
Mar	8	9	357758	Mar	20	24	405958
Apr	5	17	361498	Apr	17	15	405003
May	3	11	366626	May	15	10	404211
May	29	7	369788	Jun	12	5	404185
Jun	24	5	366494	Jul	9	23	404945
Jul	21	23	361581	Aug	6	14	405936
Aug	19	5	358020	Sep	2	21	406479
Sep	16	15	356966	Sep	29	23	406330
Oct	15	2	358860	Oct	27	9	405603
Nov	12	8	363380	Nov	24	2	404695
Dec	9	17	368874	Dec	21	23	404262

	Perigee				Apogee		

1998

Jan	3	9 h	369240 km	Jan	18	21 h	404635 km
Jan	30	14	363801	Feb	15	15	405496
Feb	27	20	359087	Mar	15	1	406189
Mar	28	7	357025	Apr	11	2	406347
Apr	25	18	358042	May	8	9	405861
May	24	0	361660	Jun	4	24	404925
Jun	20	17	366593	Jul	2	17	404220
Jul	16	14	369700	Jul	30	12	404301
Aug	11	12	366447	Aug	27	6	405148
Sep	8	6	361374	Sep	23	22	406169
Oct	6	13	357633	Oct	21	5	406669
Nov	4	1	356614	Nov	17	6	406493
Dec	2	12	358842	Dec	14	17	405758
Dec	30	18	363783				

1999

				Jan	11	12 h	404830 km
Jan	26	21 h	369258 km	Feb	8	9	404387
Feb	20	15	368654	Mar	8	5	404751
Mar	20	0	363267	Apr	4	22	405596
Apr	17	5	358894	May	2	6	406277
May	15	15	357093	May	29	8	406398
Jun	13	1	358191	Jun	25	16	405859
Jul	11	6	361777	Jul	23	6	404925
Aug	7	24	366705	Aug	19	23	404262
Sep	2	18	369820	Sep	16	19	404392
Sep	28	17	366251	Oct	14	14	405256
Oct	26	13	360949	Nov	11	6	406216
Nov	23	22	357278	Dec	8	11	406624
Dec	22	11	356654				

2000

				Jan	4	12 h	406418 km
Jan	19	23 h	359361 km	Feb	1	1	405608
Feb	17	3	364494	Feb	28	21	404615
Mar	14	24	369533	Mar	27	17	404167
Apr	8	22	368258	Apr	24	12	404559
May	6	9	363169	May	22	4	405428
Jun	3	13	359089	Jun	18	13	406112
Jul	1	22	357362	Jul	15	16	406200
Jul	30	8	358375	Aug	11	22	405652
Aug	27	14	361906	Sep	8	13	404761
Sep	24	8	366961	Oct	6	7	404170
Oct	19	22	370116	Nov	3	3	404377
Nov	14	23	366047	Nov	30	24	405273
Dec	12	22	360602	Dec	28	15	406193

	Perigee			Apogee		
2001	Jan 10	9 *h*	357 130 *km*	Jan 24	19 *h*	406 562 *km*
	Feb 7	22	356 852	Feb 20	22	406 332
	Mar 8	9	359 775	Mar 20	11	405 474
	Apr 5	10	364 810	Apr 17	6	404 506
	May 2	4	369 420	May 15	1	404 144
	May 27	7	368 033	Jun 11	20	404 629
	Jun 23	17	363 132	Jul 9	11	405 567
	Jul 21	21	359 027	Aug 5	21	406 269
	Aug 19	6	357 157	Sep 1	23	406 330
	Sep 16	16	358 128	Sep 29	6	405 787
	Oct 14	23	361 860	Oct 26	20	404 935
	Nov 11	17	367 256	Nov 23	16	404 394
	Dec 6	23	370 117	Dec 21	13	404 633
2002	Jan 2	7 *h*	365 406 *km*	Jan 18	9 *h*	405 505 *km*
	Jan 30	9	359 996	Feb 14	22	406 363
	Feb 27	20	356 897	Mar 14	1	406 707
	Mar 28	8	357 010	Apr 10	5	406 408
	Apr 25	16	360 085	May 7	19	405 483
	May 23	16	364 984	Jun 4	13	404 522
	Jun 19	7	369 309	Jul 2	8	404 210
	Jul 14	13	367 847	Jul 30	2	404 743
	Aug 10	23	362 927	Aug 26	18	405 695
	Sep 8	3	358 746	Sep 23	3	406 352
	Oct 6	13	356 918	Oct 20	5	406 360
	Nov 4	1	358 154	Nov 16	11	405 796
	Dec 2	9	362 289	Dec 14	4	404 914
	Dec 30	1	367 902			
2003				Jan 11	1 *h*	404 343 *km*
	Jan 23	22 *h*	369 898 *km*	Feb 7	22	404 552
	Feb 19	16	364 845	Mar 7	17	405 382
	Mar 19	19	359 816	Apr 4	4	406 209
	Apr 17	5	357 157	May 1	8	406 529
	May 15	16	357 449	May 28	13	406 168
	Jun 12	23	360 425	Jun 25	2	405 232
	Jul 10	22	365 145	Jul 22	20	404 328
	Aug 6	14	369 433	Aug 19	14	404 102
	Aug 31	19	367 926	Sep 16	9	404 714
	Sep 28	6	362 835	Oct 14	2	405 692
	Oct 26	12	358 547	Nov 10	12	406 301
	Nov 23	23	356 811	Dec 7	12	406 279
	Dec 22	12	358 338			

Perigee   Apogee

2004							
				Jan	3	20 h	405 707 km
Jan	19	19 h	362 770 km	Jan	31	14	404 807
Feb	16	8	368 322	Feb	28	11	404 258
Mar	12	4	369 506	Mar	27	7	404 521
Apr	8	2	364 547	Apr	24	0	405 403
May	6	5	359 811	May	21	12	406 264
Jun	3	13	357 247	Jun	17	16	406 575
Jul	1	23	357 448	Jul	14	21	406 192
Jul	30	6	360 324	Aug	11	10	405 292
Aug	27	6	365 105	Sep	8	3	404 464
Sep	22	21	369 589	Oct	5	22	404 326
Oct	17	24	367 758	Nov	2	18	404 998
Nov	14	14	362 311	Nov	30	11	405 953
Dec	12	21	357 983	Dec	27	19	406 489

2005							
Jan	10	10 h	356 570 km	Jan	23	19 h	406 445 km
Feb	7	22	358 565	Feb	20	5	405 805
Mar	8	4	363 233	Mar	19	23	404 847
Apr	4	11	368 492	Apr	16	19	404 304
Apr	29	10	369 029	May	14	14	404 600
May	26	11	364 241	Jun	11	6	405 506
Jun	23	12	359 672	Jul	8	18	406 363
Jul	21	20	357 158	Aug	4	22	406 632
Aug	19	6	357 393	Sep	1	3	406 213
Sep	16	14	360 405	Sep	28	15	405 308
Oct	14	14	365 449	Oct	26	10	404 494
Nov	10	0	370 010	Nov	23	6	404 370
Dec	5	5	367 365	Dec	21	3	405 014

2006							
Jan	1	23 h	361 750 km	Jan	17	19 h	405 885 km
Jan	30	8	357 778	Feb	14	1	406 359
Feb	27	20	356 884	Mar	13	2	406 278
Mar	28	7	359 168	Apr	9	13	405 551
Apr	25	11	363 732	May	7	7	404 572
May	22	15	368 608	Jun	4	2	404 081
Jun	16	17	368 919	Jul	1	20	404 448
Jul	13	18	364 288	Jul	29	13	405 406
Aug	10	18	359 749	Aug	26	1	406 269
Sep	8	3	357 175	Sep	22	5	406 500
Oct	6	14	357 410	Oct	19	10	406 074
Nov	3	24	360 596	Nov	15	23	405 194
Dec	2	0	365 923	Dec	13	19	404 418
Dec	28	2	370 323				

Perigee   Apogee

				Jan	10	16 h	404 335 km
2007				Feb	7	13	404 992
Jan	22	13 h	366 926 km	Mar	7	4	405 853
Feb	19	10	361 436	Apr	3	9	406 329
Mar	19	19	357 814	Apr	30	11	406 209
Apr	17	6	357 136	May	27	22	405 460
May	15	15	359 390	Jun	24	14	404 540
Jun	12	17	363 779	Jul	22	9	404 150
Jul	9	22	368 528	Aug	19	3	404 618
Aug	3	24	368 891	Sep	15	21	405 642
Aug	31	0	364 171	Oct	13	10	406 492
Sep	28	2	359 419	Nov	9	13	406 671
Oct	26	12	356 753	Dec	6	17	406 235
Nov	24	0	357 194				
Dec	22	10	360 815				
				Jan	3	8 h	405 331 km
2008				Jan	31	4	404 533
Jan	19	9 h	366 430 km	Feb	28	1	404 443
Feb	14	1	370 219	Mar	26	20	405 092
Mar	10	22	366 298	Apr	23	10	405 943
Apr	7	19	361 080	May	20	14	406 403
May	6	3	357 771	Jun	16	18	406 228
Jun	3	13	357 251	Jul	14	4	405 452
Jul	1	21	359 513	Aug	10	20	404 556
Jul	29	23	363 883	Sep	7	15	404 214
Aug	26	4	368 696	Oct	5	11	404 721
Sep	20	3	368 886	Nov	2	5	405 724
Oct	17	6	363 823	Nov	29	17	406 480
Nov	14	10	358 971	Dec	26	18	406 601
Dec	12	22	356 566				
2009				Jan	23	0 h	406 118 km
Jan	10	11 h	357 497 km	Feb	19	17	405 129
Feb	7	20	361 488	Mar	19	13	404 299
Mar	7	15	367 017	Apr	16	9	404 232
Apr	2	2	370 013	May	14	3	404 915
Apr	28	6	366 040	Jun	10	16	405 787
May	26	4	361 153	Jul	7	22	406 232
Jun	23	11	358 014	Aug	4	1	406 028
Jul	21	20	357 463	Aug	31	11	405 269
Aug	19	5	359 639	Sep	28	4	404 432
Sep	16	8	364 053	Oct	25	23	404 166
Oct	13	12	369 067	Nov	22	20	404 733
Nov	7	7	368 903	Dec	20	15	405 731
Dec	4	14	363 479				

	Perigee			Apogee		
2010	Jan 1	21 h	358682 km	Jan 17	2 h	406435 km
	Jan 30	9	356593	Feb 13	2	406540
	Feb 27	22	357829	Mar 12	10	406008
	Mar 28	5	361876	Apr 9	3	405002
	Apr 24	21	367141	May 6	22	404236
	May 20	9	369733	Jun 3	17	404266
	Jun 15	15	365932	Jul 1	10	405036
	Jul 13	11	361115	Jul 28	24	405955
	Aug 10	18	357857	Aug 25	6	406389
	Sep 8	4	357190	Sep 21	8	406165
	Oct 6	14	359455	Oct 18	18	405428
	Nov 3	17	364191	Nov 15	12	404631
	Nov 30	19	369430	Dec 13	9	404406
	Dec 25	12	368465			
2011				Jan 10	6 h	404977 km
	Jan 22	0 h	362792 km	Feb 6	23	405924
	Feb 19	7	358247	Mar 6	8	406583
	Mar 19	19	356575	Apr 2	9	406656
	Apr 17	6	358090	Apr 29	18	406039
	May 15	11	362135	May 27	10	405003
	Jun 12	2	367189	Jun 24	4	404271
	Jul 7	14	369570	Jul 21	23	404355
	Aug 2	21	365761	Aug 18	16	405161
	Aug 30	18	360858	Sep 15	6	406065
	Sep 28	1	357557	Oct 12	12	406434
	Oct 26	12	357052	Nov 8	13	406177
	Nov 23	23	359691	Dec 6	1	405414
	Dec 22	3	364800			
2012				Jan 2	20 h	404579 km
	Jan 17	21 h	369886 km	Jan 30	18	404323
	Feb 11	19	367922	Feb 27	14	404863
	Mar 10	10	362400	Mar 26	6	405777
	Apr 7	17	358315	Apr 22	14	406419
	May 6	4	356955	May 19	16	406448
	Jun 3	13	358485	Jun 16	1	405787
	Jul 1	18	362366	Jul 13	17	404779
	Jul 29	8	367315	Aug 10	11	404123
	Aug 23	19	369728	Sep 7	6	404294
	Sep 19	3	365752	Oct 5	1	405160
	Oct 17	1	360672	Nov 1	15	406050
	Nov 14	10	357361	Nov 28	20	406362
	Dec 12	23	357075	Dec 25	21	406098

Perigee   Apogee

2013	Jan 10	10 *h*	360048 *km*	Jan 22	11 *h*	405311 *km*
	Feb 7	12	365318	Feb 19	6	404472
	Mar 5	23	369957	Mar 19	3	404261
	Mar 31	4	367504	Apr 15	22	404862
	Apr 27	20	362268	May 13	14	405825
	May 26	2	358377	Jun 9	22	406486
	Jun 23	11	356991	Jul 7	1	406490
	Jul 21	20	358401	Aug 3	9	405833
	Aug 19	1	362264	Aug 30	24	404881
	Sep 15	17	367391	Sep 27	18	404308
	Oct 10	23	369814	Oct 25	14	404557
	Nov 6	9	365361	Nov 22	10	405443
	Dec 4	10	360067	Dec 19	24	406269
2014	Jan 1	21 *h*	356923 *km*	Jan 16	2 *h*	406532 *km*
	Jan 30	10	357080	Feb 12	5	406231
	Feb 27	20	360440	Mar 11	20	405364
	Mar 27	19	365703	Apr 8	15	404500
	Apr 23	0	369765	May 6	10	404318
	May 18	12	367102	Jun 3	4	404954
	Jun 15	3	362065	Jun 30	19	405930
	Jul 13	8	358260	Jul 28	3	406567
	Aug 10	18	356896	Aug 24	6	406523
	Sep 8	4	358389	Sep 20	14	405845
	Oct 6	10	362476	Oct 18	6	404897
	Nov 3	0	367879	Nov 15	2	404336
	Nov 27	23	369827	Dec 12	23	404581
	Dec 24	17	364797			
2015				Jan 9	18 *h*	405408 *km*
	Jan 21	20 *h*	359645 *km*	Feb 6	6	406150
	Feb 19	7	356995	Mar 5	8	406385
	Mar 19	20	357584	Apr 1	13	406012
	Apr 17	4	361023	Apr 29	4	405083
	May 15	0	366024	May 26	22	404244
	Jun 10	5	369711	Jun 23	17	404132
	Jul 5	19	367093	Jul 21	11	404836
	Aug 2	10	362139	Aug 18	3	405848
	Aug 30	15	358290	Sep 14	11	406464
	Sep 28	2	356877	Oct 11	13	406388
	Oct 26	13	358463	Nov 7	22	405721
	Nov 23	20	362817	Dec 5	15	404800
	Dec 21	9	368417			

<table>
<tr><td colspan="4">Perigee</td><td colspan="3">Apogee</td></tr>
</table>

	Perigee			Apogee		
2016				Jan 2	12 h	404277 km
	Jan 15	2 h	369619 km	Jan 30	9	404553
	Feb 11	3	364360	Feb 27	3	405383
	Mar 10	7	359510	Mar 25	14	406125
	Apr 7	18	357163	Apr 21	16	406351
	May 6	4	357827	May 18	22	405933
	Jun 3	11	361140	Jun 15	12	405024
	Jul 1	7	365983	Jul 13	5	404269
	Jul 27	12	369662	Aug 10	0	404262
	Aug 22	1	367050	Sep 6	19	405055
	Sep 18	17	361896	Oct 4	11	406096
	Oct 16	24	357861	Oct 31	19	406662
	Nov 14	11	356509	Nov 27	20	406554
	Dec 12	23	358461	Dec 25	6	405870
2017	Jan 10	6 h	363238 km	Jan 22	0 h	404914 km
	Feb 6	14	368816	Feb 18	21	404376
	Mar 3	8	369062	Mar 18	17	404650
	Mar 30	13	363854	Apr 15	10	405475
	Apr 27	16	359327	May 12	20	406210
	May 26	1	357207	Jun 8	22	406401
	Jun 23	11	357937	Jul 6	4	405934
	Jul 21	17	361236	Aug 2	18	405025
	Aug 18	13	366121	Aug 30	11	404308
	Sep 13	16	369860	Sep 27	7	404348
	Oct 9	6	366855	Oct 25	2	405154
	Nov 6	0	361438	Nov 21	19	406132
	Dec 4	9	357492	Dec 19	1	406603
2018	Jan 1	22 h	356565 km	Jan 15	2 h	406464 km
	Jan 30	10	358994	Feb 11	14	405700
	Feb 27	15	363933	Mar 11	9	404678
	Mar 26	17	369106	Apr 8	6	404144
	Apr 20	15	368714	May 6	1	404457
	May 17	21	363776	Jun 2	17	405317
	Jun 14	24	359503	Jun 30	3	406061
	Jul 13	8	357431	Jul 27	6	406223
	Aug 10	18	358078	Aug 23	11	405746
	Sep 8	1	361351	Sep 20	1	404876
	Oct 5	22	366392	Oct 17	19	404227
	Oct 31	20	370204	Nov 14	16	404339
	Nov 26	12	366620	Dec 12	12	405177
	Dec 24	10	361062			

P e r i g e e   A p o g e e

2019								
	Jan	21	20 *h*	357342 *km*	Jan	9	4 *h*	406117 *km*
	Feb	19	9	356761	Feb	5	9	406555
	Mar	19	20	359377	Mar	4	11	406391
	Apr	16	22	364205	Apr	1	0	405577
	May	13	22	369009	Apr	28	18	404582
	Jun	7	23	368504	May	26	13	404138
	Jul	5	5	363726	Jun	23	8	404548
	Aug	2	7	359398	Jul	20	24	405481
	Aug	30	16	357176	Aug	17	11	406245
	Sep	28	2	357802	Sep	13	14	406377
	Oct	26	11	361311	Oct	10	18	405899
	Nov	23	8	366716	Nov	7	9	405058
	Dec	18	20	370265	Dec	5	4	404446

2020								
	Jan	13	20 *h*	365958 *km*	Jan	2	1 *h*	404580 *km*
	Feb	10	20	360461	Jan	29	21	405393
	Mar	10	6	357122	Feb	26	12	406278
	Apr	7	18	356907	Mar	24	15	406692
	May	6	3	359654	Apr	20	19	406462
	Jun	3	4	364366	May	18	8	405583
	Jun	30	2	368958	Jun	15	1	404595
	Jul	25	5	368361	Jul	12	19	404199
	Aug	21	11	363513	Aug	9	14	404659
	Sep	18	14	359082	Sep	6	6	405607
	Oct	16	24	356912	Oct	3	17	406321
	Nov	14	12	357837	Oct	30	19	406394
	Dec	12	21	361773	Nov	27	0	405894
					Dec	24	17	405012

During the period 1990 − 2020 considered here, the smallest and the largest distances between the centers of Earth and Moon are: 356509 km on 2016 November 14, and 406707 km on 2002 March 14.

During the longer period from A.D. 1500 to 2500, the extreme distances are 356371 km on 2257 January 1, and 406720 km on 2125 February 3 and on 2266 January 7.

During the 20th century, the extremes are 356375 km on 1912 January 4, and 406712 km on 1984 March 2.

VII

TABLE FOR CALCULATING THE ILLUMINATED FRACTION OF THE MOON'S DISK

The illuminated fraction of the Moon's disk, as seen from the Earth, can be calculated by means of the tables on pages 405 to 409 for any instant between the years −900 and +2999.

Firstly, express the instant in Dynamical Time (see pages 5 − 8). Between the years +1000 and +3000, the difference between Universal Time and Dynamical Time can be ignored for our purpose.

From the tables a to e, take the values of the quantities A, B, C, and D corresponding to the century year, the additional year, the month, the day of the month, and the hour.

In table a, two values are given for the year 1500, the first one for the Julian calendar (J), the second one for the Gregorian calendar (G).

In the case of a negative year number, split it up in such a way as to have the last two figures positive. For instance, the year number −328 should be split up as −400 and +72. Then, enter table a with −400, and table b with +72.

Add up the numbers in order to obtain the values of the arguments A, B, C, and D for the given instant. If an argument is larger than 1000, subtract 1000 or a convenient multiple of 1000.

With the arguments A, B, C, and D, take from tables f to i the corrections I, II, III, and IV, respectively. Then calculate

$$d = D + \text{corrections I to IV.}$$

If d is larger than 1000, subtract 1000. If d is negative, add 1000.

(text continues at bottom of next page)

Table a : Century Year

Year	A	B	C	D
−900	110	829	186	507
−800	107	380	341	361
−700	104	931	497	214
−600	102	482	652	67
−500	99	33	808	921
−400	96	584	963	774
−300	94	136	119	627
−200	91	687	274	480
−100	89	239	429	334
0	86	790	584	187
+100	83	341	739	40
200	81	893	894	894
300	78	445	49	747
400	75	996	204	600
500	73	548	359	453
600	70	100	513	307
700	68	651	668	160
800	65	203	823	13
900	62	755	977	866
1000	60	307	132	719
1100	57	859	286	572
1200	54	411	440	426
1300	52	963	595	279
1400	49	515	749	132
1500 J	47	67	903	985
1500 G	19	704	589	646
1600	17	256	743	500
1700	11	772	865	319
1800	6	288	988	138
1900	0	804	110	957
2000	998	357	264	810
2100	992	873	386	630
2200	987	389	509	449
2300	982	905	631	268
2400	979	458	784	121
2500	974	974	906	940
2600	968	491	28	759
2700	963	7	150	579
2800	960	560	303	432
2900	955	76	425	251

ASTRONOMICAL TABLES

Table b : Additional Year

Year	A	B	C	D	Year	A	B	C	D	Year	A	B	C	D
0	5	0	0	0	35	2	916	830	873	70	2	869	692	780
1	4	246	474	360	36	4	199	335	267	71	1	115	166	140
2	4	493	947	720	37	3	445	809	627	72	3	398	671	534
3	3	739	421	80	38	3	692	283	987	73	2	644	144	894
4	5	22	926	474	39	2	938	756	347	74	2	891	618	254
5	4	269	400	834	40	4	221	262	741	75	1	137	92	614
6	3	515	874	194	41	3	467	735	101	76	3	420	597	8
7	3	761	347	554	42	3	714	209	461	77	2	666	71	368
8	5	44	852	948	43	2	960	683	821	78	2	913	544	728
9	4	291	326	308	44	4	243	188	215	79	1	159	18	89
10	3	537	800	668	45	3	489	661	575	80	3	442	523	482
11	3	784	273	28	46	2	736	135	935	81	2	688	997	843
12	5	66	778	422	47	2	982	609	296	82	1	935	470	203
13	4	313	252	782	48	4	265	114	689	83	1	181	944	563
14	3	559	726	143	49	3	512	588	50	84	3	464	449	957
15	3	806	199	503	50	2	758	61	410	85	2	710	923	317
16	5	88	705	896	51	2	4	535	770	86	1	957	397	677
17	4	335	178	257	52	4	287	40	164	87	1	203	870	37
18	3	581	652	617	53	3	534	514	524	88	3	486	375	431
19	2	828	126	977	54	2	780	987	884	89	2	733	849	791
20	4	110	631	371	55	1	27	461	244	90	1	979	323	151
21	4	357	104	731	56	4	309	966	638	91	1	225	796	511
22	3	603	578	91	57	3	556	440	998	92	3	508	302	905
23	2	850	52	451	58	2	802	914	358	93	2	755	775	265
24	4	133	557	845	59	1	49	387	718	94	1	1	249	625
25	4	379	31	205	60	3	331	892	112	95	0	248	723	985
26	3	625	504	565	61	3	578	366	472	96	2	530	228	379
27	2	872	978	925	62	2	824	840	832	97	2	777	701	739
28	4	155	483	319	63	1	71	313	192	98	1	23	175	99
29	4	401	957	679	64	3	354	818	586	99	0	270	649	459
30	3	648	430	39	65	3	600	292	946					
31	2	894	904	399	66	2	846	766	306					
32	4	177	409	793	67	1	93	239	666					
33	3	423	883	153	68	3	376	745	60					
34	3	670	357	513	69	2	622	218	420					

The critical table k on page 409 then gives the illuminated fraction k of the lunar disk as a function of d. For instance, if $d = 134$ then $k = 0.17$, because this value is situated between the tabular values $d = 132$ and $d = 137$.

When d equals one of the tabular or 'critical' values, the appropriate value of k is the upper of the two values. For example, if $d = 171$, then k is to be taken as 0.26. This rule is given as 'In critical cases ascend', mentioned under the table.

The resulting value of k will never be more than 0.01 in error.

Table c : Month

Month	A	B	C	D
Jan.	989	0	0	0
Jan. B	986	964	969	966
Feb.	74	125	974	50
Feb. B	71	89	943	16
March	151	141	855	998
April	235	266	829	48
May	318	355	772	64
June	402	480	747	113
July	485	569	690	129
Aug.	569	694	664	179
Sept.	654	819	639	229
Oct.	736	908	582	245
Nov.	821	33	556	294
Dec.	903	121	499	310

The months marked with B should be used in the case of a bissextile (leap) year.

Table d : Day of month

Day	A	B	C	D
1	3	36	31	34
2	5	73	63	68
3	8	109	94	102
4	11	145	126	135
5	14	181	157	169
6	16	218	189	203
7	19	254	220	237
8	22	290	251	271
9	25	327	283	305
10	27	363	314	339
11	30	399	346	372
12	33	435	377	406
13	36	472	409	440
14	38	508	440	474
15	41	544	472	508
16	44	581	503	542
17	47	617	534	576
18	49	653	566	610
19	52	690	597	643
20	55	726	629	677
21	57	762	660	711
22	60	798	692	745
23	63	835	723	779
24	66	871	754	813
25	68	907	786	847
26	71	944	817	880
27	74	980	849	914
28	77	16	880	948
29	79	52	912	982
30	82	89	943	16
31	85	125	974	50

Table e : Hour

Hour	A	B	C	D
0	0	0	0	0
1	0	2	1	1
2	0	3	3	3
3	0	5	4	4
4	0	6	5	6
5	1	8	7	7
6	1	9	8	8
7	1	11	9	10
8	1	12	10	11
9	1	14	12	13
10	1	15	13	14
11	1	17	14	16
12	1	18	16	17
13	1	20	17	18
14	2	21	18	20
15	2	23	20	21
16	2	24	21	23
17	2	26	22	24
18	2	27	24	25
19	2	29	25	27
20	2	30	26	28
21	2	32	28	30
22	3	33	29	31
23	3	35	30	32
24	3	36	31	34

EXAMPLE. — Illuminated fraction of the Moon's disk on 1982 November 7 at 20h UT.

		A	B	C	D
Table a	1900	0	804	110	957
Table b	82	1	935	470	203
Table c	Nov.	821	33	556	294
Table d	7	19	254	220	237
Table e	20h	2	30	26	28
Sum		843	2056	1382	1719
			= 56	= 382	= 719

Table f	A = 834	I = +5
	D	719
Table g	B = 56	II = +7
Table h	C = 382	III = +2
Table i	D = 719	IV = +1
Sum		d = 734

whence k = 0.55, waning Moon

Table f	
A	I
0	0
20	−1
40	−1
60	−2
80	−3
100	−3
120	−4
140	−5
160	−5
180	−5
200	−6
220	−6
240	−6
260	−6
280	−6
300	−6
320	−5
340	−5
360	−4
380	−4
400	−3
420	−3
440	−2
460	−1
480	−1
500	0
520	+1
540	+1
560	+2
580	+3
600	+3
620	+4
640	+4
660	+5
680	+5
700	+6
720	+6
740	+6
760	+6
780	+6
800	+6
820	+5
840	+5
860	+5
880	+4
900	+3
920	+3
940	+2
960	+1
980	+1
1000	0

Table g	
B	II
0	0
20	+ 2
40	+ 5
60	+ 7
80	+ 9
100	+11
120	+13
140	+14
160	+15
180	+16
200	+17
220	+17
240	+17
260	+17
280	+17
300	+16
320	+15
340	+14
360	+13
380	+11
400	+10
420	+ 8
440	+ 6
460	+ 4
480	+ 2
500	0
520	− 2
540	− 4
560	− 6
580	− 8
600	−10
620	−11
640	−13
660	−14
680	−15
700	−16
720	−17
740	−17
760	−17
780	−17
800	−17
820	−16
840	−15
860	−14
880	−13
900	−11
920	− 9
940	− 7
960	− 5
980	− 2
1000	0

Table h	
C	III
0	0
20	+0
40	+1
60	+1
80	+2
100	+2
120	+2
140	+3
160	+3
180	+3
200	+3
220	+3
240	+4
260	+4
280	+3
300	+3
320	+3
340	+3
360	+3
380	+2
400	+2
420	+2
440	+1
460	+1
480	+0
500	0
520	−0
540	−1
560	−1
580	−2
600	−2
620	−2
640	−3
660	−3
680	−3
700	−3
720	−3
740	−4
760	−4
780	−3
800	−3
820	−3
840	−3
860	−3
880	−2
900	−2
920	−2
940	−1
960	−1
980	−0
1000	0

Table i	
D	IV
0	0
20	+0
40	+1
60	+1
80	+2
100	+2
120	+2
140	+2
160	+2
180	+1
200	+1
220	+1
240	+0
260	−0
280	−1
300	−1
320	−1
340	−2
360	−2
380	−2
400	−2
420	−2
440	−1
460	−1
480	−0
500	0
520	+0
540	+1
560	+1
580	+2
600	+2
620	+2
640	+2
660	+2
680	+1
700	+1
720	+1
740	+0
760	−0
780	−1
800	−1
820	−1
840	−2
860	−2
880	−2
900	−2
920	−2
940	−1
960	−1
980	−0
1000	0

Table *k*

d	k	d	d	k	d	d	k	d	d	k	d
0	0.00	1000	164	0.25	836	248	0.50	752	331	0.75	669
22	0.01	978	168	0.26	832	251	0.51	749	334	0.76	666
38	0.02	962	171	0.27	829	254	0.52	746	338	0.77	662
50	0.03	950	175	0.28	825	257	0.53	743	342	0.78	658
59	0.04	941	178	0.29	822	260	0.54	740	346	0.79	654
67	0.05	933	182	0.30	818	263	0.55	737	350	0.80	650
75	0.06	925	185	0.31	815	267	0.56	733	354	0.81	646
81	0.07	919	189	0.32	811	270	0.57	730	358	0.82	642
88	0.08	912	192	0.33	808	273	0.58	727	362	0.83	638
93	0.09	907	196	0.34	804	276	0.59	724	366	0.84	634
99	0.10	901	199	0.35	801	280	0.60	720	370	0.85	630
104	0.11	896	202	0.36	798	283	0.61	717	375	0.86	625
109	0.12	891	206	0.37	794	286	0.62	714	379	0.87	621
114	0.13	886	209	0.38	791	289	0.63	711	384	0.88	616
119	0.14	881	212	0.39	788	293	0.64	707	389	0.89	611
124	0.15	876	215	0.40	785	296	0.65	704	394	0.90	606
128	0.16	872	219	0.41	781	299	0.66	701	400	0.91	600
132	0.17	868	222	0.42	778	303	0.67	697	405	0.92	595
137	0.18	863	225	0.43	775	306	0.68	694	411	0.93	589
141	0.19	859	228	0.44	772	309	0.69	691	417	0.94	583
145	0.20	855	232	0.45	768	313	0.70	687	424	0.95	576
149	0.21	851	235	0.46	765	316	0.71	684	431	0.96	569
153	0.22	847	238	0.47	762	320	0.72	680	439	0.97	561
156	0.23	844	241	0.48	759	323	0.73	677	449	0.98	551
160	0.24	840	244	0.49	756	327	0.74	673	460	0.99	540
164		836	248		752	331		669	477	1.00	523
									500		500

In critical cases ascend

If *d* is between 0 and 500, then the Moon is waxing (phase increasing).
If *d* is between 500 and 1000, the Moon is waning (phase decreasing).

VIII
TABLE FOR CALCULATING
THE SELENOGRAPHIC COLONGITUDE OF THE SUN

Table *a* : Century Year

Year	Colong.	A
1600	89.86°	17
1700	24.80	11
1800	319.72	6
1900	254.64	0
2000	201.76	998
2100	136.67	992
2200	71.58	987
2300	6.50	982

Table *b* : Additional Year

Year	Colong.	A	Year	Colong.	A	Year	Colong.	A
0	0.00°	5	35	314.35°	2	70	280.88°	2
1	129.62	4	36	96.16	4	71	50.51	1
2	259.25	4	37	225.78	3	72	192.32	3
3	28.87	3	38	355.41	3	73	321.95	2
4	170.68	5	39	125.03	2	74	91.57	2
5	300.31	4	40	266.85	4	75	221.19	1
6	69.93	3	41	36.47	3	76	3.01	3
7	199.55	3	42	166.09	3	77	132.63	2
8	341.37	5	43	295.72	2	78	262.25	2
9	110.99	4	44	77.53	4	79	31.88	1
10	240.62	3	45	207.15	3	80	173.69	3
11	10.24	3	46	336.78	2	81	303.31	2
12	152.05	5	47	106.40	2	82	72.94	1
13	281.68	4	48	248.21	4	83	202.56	1
14	51.30	3	49	17.84	3	84	344.38	3
15	180.92	3	50	147.46	2	85	114.00	2
16	322.74	5	51	277.09	2	86	243.62	1
17	92.36	4	52	58.90	4	87	13.25	1
18	221.99	3	53	188.52	3	88	155.06	3
19	351.61	2	54	318.15	2	89	284.68	2
20	133.42	4	55	87.77	1	90	54.31	1
21	263.05	4	56	229.58	4	91	183.93	1
22	32.67	3	57	359.21	3	92	325.75	3
23	162.29	2	58	128.83	2	93	95.37	2
24	304.11	4	59	258.45	1	94	224.99	1
25	73.73	4	60	40.27	3	95	354.62	0
26	203.35	3	61	169.89	3	96	136.43	2
27	332.98	2	62	299.52	2	97	266.05	2
28	114.79	4	63	69.14	1	98	35.68	1
29	244.42	4	64	210.95	3	99	165.30	0
30	14.04	3	65	340.58	3			
31	143.66	2	66	110.20	2			
32	285.48	4	67	239.82	1			
33	55.10	3	68	21.64	3			
34	184.72	3	69	151.26	2			

The selenographic longitude and latitude of the Sun specify the point on the lunar surface where the Sun is in the zenith. The selenographic *colongitude* of the Sun is equal to 90° minus the selenographic longitude of the Sun. Its value and that of the selenographic latitude of the Sun are given each year in the great astronomical almanacs.

The selenographic latitude of the Sun never exceeds 1°.6 north or south, and may be neglected in a first approximation when the position of the terminator on the Moon is required. The selenographic colongitude of the Sun can be calculated with a good approximation with the aid of the tables on pages 410 to 412. The error will rarely exceed 0.05 degree, and never exceed 0.20 degree.

Table c : Month

Month	Colong.	A
Jan.	0.00°	989
Jan. (B)	347.81	986
Feb.	17.91	74
Feb. (B)	5.72	71
March	359.25	151
April	17.17	235
May	22.89	318
June	40.80	402
July	46.53	485
Aug.	64.44	569
Sept.	82.35	654
Oct.	88.07	736
Nov.	105.99	821
Dec.	111.71	903

Use the months marked *(B)* in the case of a bissextile (leap) year. All years whose millesime is divisible by 4 are bissextile years, except 1700, 1800, 1900, 2100, 2200, and 2300.

Table d : Day

Day	Colong.	A
1	12.19°	3
2	24.38	5
3	36.57	8
4	48.76	11
5	60.95	14
6	73.14	16
7	85.34	19
8	97.53	22
9	109.72	25
10	121.91	27
11	134.10	30
12	146.29	33
13	158.48	36
14	170.67	38
15	182.86	41
16	195.05	44
17	207.24	47
18	219.43	49
19	231.62	52
20	243.81	55
21	256.01	57
22	268.20	60
23	280.39	63
24	292.58	66
25	304.77	68
26	316.96	71
27	329.15	74
28	341.34	77
29	353.53	79
30	5.72	82
31	17.91	85

Table e : Hour

Hour	Colong.	A
0	0.00°	0
1	0.51	0
2	1.02	0
3	1.52	0
4	2.03	0
5	2.54	1
6	3.05	1
7	3.56	1
8	4.06	1
9	4.57	1
10	5.08	1
11	5.59	1
12	6.10	1
13	6.60	1
14	7.11	2
15	7.62	2
16	8.13	2
17	8.64	2
18	9.14	2
19	9.65	2
20	10.16	2
21	10.67	2
22	11.17	3
23	11.68	3
24	12.19	3

Table f

A	Corr.
0	0.00°
20	−0.25
40	−0.49
60	−0.72
80	−0.94
100	−1.15
120	−1.33
140	−1.50
160	−1.64
180	−1.75
200	−1.83
220	−1.89
240	−1.91
260	−1.91
280	−1.87
300	−1.81
320	−1.72
340	−1.60
360	−1.46
380	−1.29
400	−1.11
420	−0.91
440	−0.69
460	−0.47
480	−0.24
500	0.00
520	+0.24
540	+0.47
560	+0.69
580	+0.91
600	+1.11
620	+1.29
640	+1.46
660	+1.60
680	+1.72
700	+1.81
720	+1.87
740	+1.91
760	+1.91
780	+1.89
800	+1.83
820	+1.75
840	+1.64
860	+1.50
880	+1.33
900	+1.15
920	+0.94
940	+0.72
960	+0.49
980	+0.25
1000	0.00

From tables *a* to *e*, take the values of the Colong. and of the quantity *A* for the century year, the additional year, the month, the day of the month, and the hour (*Universal Time!*). Add the values thus found, in order to obtain the Colong. and the value of the argument *A* for the given instant. If *A* exceeds 1000, subtract 1000 or a convenient multiple. If Colong. exceeds 360 degrees, subtract 360 degrees.

With *A* as argument, take the correction by interpolation from table *f*, and add it to the Colong. With the new value of the colongitude as argument, take the second correction from table *g*, and add it to the colongitude.

No interpolation is required for tables *a* to *d*. Table *g* is a so-called 'critical table' which gives the value of the correction without interpolation.

Table *g* : Second Correction

Col.	Corr.	Col.	Col.	Corr.	Col.
0.0°		360.0°	90.0°		270.0°
	+0.15			0.00	
9.5		350.5	92.0		268.0
	+0.14			−0.01	
23.3		336.7	95.9		264.1
	+0.13			−0.02	
31.8		328.2	99.8		260.2
	+0.12			−0.03	
38.5		321.5	103.8		256.2
	+0.11			−0.04	
44.4		315.6	107.8		252.2
	+0.10			−0.05	
49.7		310.3	112.0		248.0
	+0.09			−0.06	
54.7		305.3	116.2		243.8
	+0.08			−0.07	
59.3		300.7	120.7		239.3
	+0.07			−0.08	
63.8		296.2	125.3		234.7
	+0.06			−0.09	
68.0		292.0	130.3		229.7
	+0.05			−0.10	
72.2		287.8	135.6		224.4
	+0.04			−0.11	
76.2		283.8	141.5		218.5
	+0.03			−0.12	
80.2		279.8	148.2		211.8
	+0.02			−0.13	
84.1		275.9	156.7		203.3
	+0.01			−0.14	
88.0		272.0	170.5		189.5
	0.00			−0.15	
90.0		270.0	180.0		180.0

As an example, let us calculate the Sun's selenographic colongitude on 1963 November 22, at 17h UT.
We find :

		Colong.	A
Table *a*	1900	254.64°	0
Table *b*	63	69.14	1
Table *c*	Nov.	105.99	821
Table *d*	22	268.20	60
Table *e*	17h	8.64	2
Sum		706.61°	884
		= 346.61	

With $A = 884$ as argument, we find in table *f* that the first correction is +1°.29. Hence, the new value of the colongitude is

$$346.61 + 1.29 = 347.90$$

With 347°.90 as argument, we find in table *g* that the second correction is +0°.14. Our final result is then

colongitude = 347°.90 + 0°.14
            = 348°.04.

The Sun's selenographic colongitude is *approximately* 270° at New Moon, 0° at First Quarter, 90° at Full Moon, and 180° at Last Quarter.

If the selenographic longitudes λ are counted positively from the mean center of the lunar disk towards Mare Crisium, that is, westward on the sky, then we have the following rules, *c* being the Sun's selenographic colongitude, and its latitude being neglected :

— for a place on the Moon whose selenographic longitude is λ, we have

$$\text{sunrise when} \quad c = 360° − λ$$
$$\text{noon when} \quad c = 90° − λ$$
$$\text{sunset when} \quad c = 180° − λ$$

— if, at a given instant, the Sun's selenographic colongitude is *c*, then the morning (sunrise) terminator lies in longitude λ = 360° − *c*, while the evening (sunset) terminator is at longitude λ = 180° − *c*.

Remarks about the Tables IX to XII

Tables IX to XII list all inferior conjunctions of Venus, superior conjunctions of Venus, oppositions of Jupiter, and oppositions of Saturn, respectively, taking place during the period A.D. 0 to 2500.

The times are rounded to the nearest tenth of a day, *Universal Time.*

For instance, 1973 Jul 30.5, in Table XI, means that the opposition of Jupiter with the Sun took place on 1973 July 30 at approximately 12 h UT.

Note that a date such as March 16.9 should *not* be rounded to March 17. Indeed, March 16.9 denotes an instant belonging to March 16, namely 0.9 day after March 16, 0 h UT; consequently, the event takes place on March 16, not on March 17.

The tables contain some 'odd' dates. For instance, we read that in A.D. 491 Saturn was in opposition with the Sun on July 32.0, and that there will be an inferior conjunction of Venus on 2210 August 32.0. In such cases, the '32.0' means that the event occurs between 31.9500 and 31.9999, that is, late in the evening (Universal Time) of the 31th.

IX
INFERIOR CONJUNCTIONS OF VENUS, 0 − 2500

1 Mar 30.8	39 Aug 9.4	77 Dec 25.3	116 May 7.0	154 Sep 17.0
2 Nov 3.0	41 Mar 19.2	79 Jul 28.5	117 Dec 12.9	156 Apr 25.4
4 Jun 8.3	42 Oct 21.5	81 Mar 7.5	119 Jul 16.8	157 Nov 30.5
6 Jan 16.3	44 May 27.7	82 Oct 9.1	121 Feb 23.7	159 Jul 5.1
7 Aug 18.9	46 Jan 4.1	84 May 16.2	122 Sep 26.8	161 Feb 11.8
9 Mar 28.5	47 Aug 7.0	85 Dec 22.8	124 May 4.7	162 Sep 14.6
10 Oct 31.5	49 Mar 16.8	87 Jul 26.2	125 Dec 10.4	164 Apr 23.1
12 Jun 6.0	50 Oct 19.1	89 Mar 5.1	127 Jul 14.4	165 Nov 28.0
14 Jan 13.9	52 May 25.4	90 Oct 6.7	129 Feb 21.3	167 Jul 2.7
15 Aug 16.6	54 Jan 1.7	92 May 13.9	130 Sep 24.4	169 Feb 9.4
17 Mar 26.2	55 Aug 4.6	93 Dec 20.3	132 May 2.3	170 Sep 12.1
18 Oct 29.0	57 Mar 14.5	95 Jul 23.8	133 Dec 7.9	172 Apr 20.8
20 Jun 3.7	58 Oct 16.6	97 Mar 2.8	135 Jul 12.1	173 Nov 25.5
22 Jan 11.4	60 May 23.1	98 Oct 4.2	137 Feb 19.0	175 Jun 30.4
23 Aug 14.2	61 Dec 30.2	100 May 11.6	138 Sep 21.9	177 Feb 7.0
25 Mar 23.8	63 Aug 2.3	101 Dec 17.9	140 Apr 30.0	178 Sep 9.7
26 Oct 26.5	65 Mar 12.2	103 Jul 21.5	141 Dec 5.4	180 Apr 18.5
28 Jun 1.4	66 Oct 14.1	105 Feb 28.4	143 Jul 9.7	181 Nov 23.0
30 Jan 9.0	68 May 20.8	106 Oct 1.7	145 Feb 16.6	183 Jun 28.1
31 Aug 11.8	69 Dec 27.7	108 May 9.3	146 Sep 19.5	185 Feb 4.6
33 Mar 21.5	71 Jul 30.9	109 Dec 15.4	148 Apr 27.7	186 Sep 7.3
34 Oct 24.0	73 Mar 9.8	111 Jul 19.1	149 Dec 3.0	188 Apr 16.2
36 May 30.1	74 Oct 11.6	113 Feb 26.0	151 Jul 7.4	189 Nov 20.5
38 Jan 6.6	76 May 18.5	114 Sep 29.3	153 Feb 14.2	191 Jun 25.7

193 Feb 2.3	274 Aug 11.9	356 Feb 27.1	437 Sep 4.9	519 Mar 22.7
194 Sep 4.9	276 Mar 21.6	357 Sep 29.3	439 Apr 15.0	520 Oct 24.0
196 Apr 13.9	277 Oct 24.0	359 May 8.1	440 Nov 18.0	522 May 31.4
197 Nov 18.0	279 May 31.2	360 Dec 12.9	442 Jun 23.6	524 Jan 7.6
199 Jun 23.4	281 Jan 6.6	362 Jul 16.9	444 Jan 31.9	525 Aug 9.6
201 Jan 30.9	282 Aug 9.5	364 Feb 24.8	445 Sep 2.5	527 Mar 20.4
202 Sep 2.5	284 Mar 19.3	365 Sep 26.8	447 Apr 12.7	528 Oct 21.5
204 Apr 11.5	285 Oct 21.5	367 May 5.8	448 Nov 15.5	530 May 29.1
205 Nov 15.5	287 May 28.9	368 Dec 10.4	450 Jun 21.2	532 Jan 5.1
207 Jun 21.1	289 Jan 4.1	370 Jul 14.6	452 Jan 29.5	533 Aug 7.2
209 Jan 28.4	290 Aug 7.1	372 Feb 22.4	453 Aug 31.1	535 Mar 18.1
210 Aug 31.0	292 Mar 17.0	373 Sep 24.4	455 Apr 10.4	536 Oct 19.0
212 Apr 9.2	293 Oct 19.1	375 May 3.5	456 Nov 13.0	538 May 26.8
213 Nov 13.0	295 May 26.6	376 Dec 7.9	458 Jun 18.9	540 Jan 2.7
215 Jun 18.8	297 Jan 1.7	378 Jul 12.2	460 Jan 27.1	541 Aug 4.9
217 Jan 26.0	298 Aug 4.8	380 Feb 20.0	461 Aug 28.7	543 Mar 15.7
218 Aug 28.6	300 Mar 14.6	381 Sep 21.9	463 Apr 8.1	544 Oct 16.6
220 Apr 6.9	301 Oct 16.6	383 May 1.2	464 Nov 10.5	546 May 24.5
221 Nov 10.5	303 May 24.3	384 Dec 5.4	466 Jun 16.6	547 Dec 31.2
223 Jun 16.4	304 Dec 30.2	386 Jul 9.9	468 Jan 24.6	549 Aug 2.5
225 Jan 23.6	306 Aug 2.4	388 Feb 17.6	469 Aug 26.3	551 Mar 13.4
226 Aug 26.2	308 Mar 12.3	389 Sep 19.5	471 Apr 5.7	552 Oct 14.1
228 Apr 4.6	309 Oct 14.1	391 Apr 28.9	472 Nov 8.0	554 May 22.2
229 Nov 8.0	311 May 22.0	392 Dec 2.9	474 Jun 14.3	555 Dec 28.7
231 Jun 14.1	312 Dec 27.7	394 Jul 7.5	476 Jan 22.2	557 Jul 31.1
233 Jan 21.2	314 Jul 31.0	396 Feb 15.3	477 Aug 23.9	559 Mar 11.0
234 Aug 23.8	316 Mar 9.9	397 Sep 17.1	479 Apr 3.4	560 Oct 11.6
236 Apr 2.3	317 Oct 11.6	399 Apr 26.6	480 Nov 5.5	562 May 19.8
237 Nov 5.5	319 May 19.7	400 Nov 30.4	482 Jun 12.0	563 Dec 26.3
239 Jun 11.8	320 Dec 25.3	402 Jul 5.2	484 Jan 19.8	565 Jul 28.8
241 Jan 18.8	322 Jul 28.7	404 Feb 12.9	485 Aug 21.5	567 Mar 8.7
242 Aug 21.4	324 Mar 7.6	405 Sep 14.6	487 Apr 1.1	568 Oct 9.2
244 Mar 30.9	325 Oct 9.1	407 Apr 24.3	488 Nov 3.0	570 May 17.5
245 Nov 3.0	327 May 17.4	408 Nov 27.9	490 Jun 9.7	571 Dec 23.8
247 Jun 9.5	328 Dec 22.8	410 Jul 2.9	492 Jan 17.4	573 Jul 26.4
249 Jan 16.3	330 Jul 26.3	412 Feb 10.5	493 Aug 19.1	575 Mar 6.3
250 Aug 19.0	332 Mar 5.2	413 Sep 12.2	495 Mar 29.7	576 Oct 6.7
252 Mar 28.6	333 Oct 6.7	415 Apr 22.0	496 Oct 31.5	578 May 15.2
253 Oct 31.5	335 May 15.1	416 Nov 25.4	498 Jun 7.3	579 Dec 21.3
255 Jun 7.2	336 Dec 20.3	418 Jun 30.5	500 Jan 14.9	581 Jul 24.1
257 Jan 13.9	338 Jul 23.9	420 Feb 8.1	501 Aug 16.7	583 Mar 4.0
258 Aug 16.6	340 Mar 2.9	421 Sep 9.8	503 Mar 27.4	584 Oct 4.2
260 Mar 26.3	341 Oct 4.2	423 Apr 19.6	504 Oct 29.0	586 May 12.9
261 Oct 29.0	343 May 12.8	424 Nov 22.9	506 Jun 5.0	587 Dec 18.8
263 Jun 4.9	344 Dec 17.8	426 Jun 28.2	508 Jan 12.5	589 Jul 21.7
265 Jan 11.5	346 Jul 21.6	428 Feb 5.7	509 Aug 14.4	591 Mar 1.6
266 Aug 14.3	348 Feb 29.5	429 Sep 7.3	511 Mar 25.1	592 Oct 1.8
268 Mar 24.0	349 Oct 1.7	431 Apr 17.3	512 Oct 26.5	594 May 10.6
269 Oct 26.5	351 May 10.4	432 Nov 20.4	514 Jun 2.7	595 Dec 16.3
271 Jun 2.5	352 Dec 15.4	434 Jun 25.9	516 Jan 10.0	597 Jul 19.4
273 Jan 9.0	354 Jul 19.2	436 Feb 3.3	517 Aug 12.0	599 Feb 27.2

600	Sep	29.3	682	Apr	15.2	763	Oct	25.0	845	May	8.5	926	Nov	18.9
602	May	8.3	683	Nov	18.9	765	May	31.6	846	Dec	13.8	928	Jun	23.9
603	Dec	13.9	685	Jun	23.7	767	Jan	7.6	848	Jul	17.2	930	Jan	31.9
605	Jul	17.0	687	Jan	31.9	768	Aug	9.7	850	Feb	24.9	931	Sep	3.6
607	Feb	24.8	688	Sep	2.6	770	Mar	20.5	851	Sep	27.9	933	Apr	13.0
608	Sep	26.9	690	Apr	12.8	771	Oct	22.5	853	May	6.1	934	Nov	16.4
610	May	6.0	691	Nov	16.4	773	May	29.3	854	Dec	11.3	936	Jun	21.6
611	Dec	11.4	693	Jun	21.4	775	Jan	5.1	856	Jul	14.8	938	Jan	29.5
613	Jul	14.7	695	Jan	29.5	776	Aug	7.3	858	Feb	22.5	939	Sep	1.2
615	Feb	22.5	696	Aug	31.2	778	Mar	18.2	859	Sep	25.4	941	Apr	10.6
616	Sep	24.4	698	Apr	10.5	779	Oct	20.0	861	May	3.8	942	Nov	13.9
618	May	3.7	699	Nov	13.9	781	May	26.9	862	Dec	8.8	944	Jun	19.2
619	Dec	8.9	701	Jun	19.1	783	Jan	2.6	864	Jul	12.5	946	Jan	27.1
621	Jul	12.4	703	Jan	27.1	784	Aug	5.0	866	Feb	20.2	947	Aug	29.9
623	Feb	20.1	704	Aug	28.8	786	Mar	15.8	867	Sep	23.0	949	Apr	8.3
624	Sep	22.0	706	Apr	8.2	787	Oct	17.6	869	May	1.5	950	Nov	11.4
626	May	1.4	707	Nov	11.4	789	May	24.6	870	Dec	6.4	952	Jun	16.9
627	Dec	6.4	709	Jun	16.8	790	Dec	31.2	872	Jul	10.2	954	Jan	24.7
629	Jul	10.0	711	Jan	24.7	792	Aug	2.6	874	Feb	17.8	955	Aug	27.5
631	Feb	17.7	712	Aug	26.4	794	Mar	13.5	875	Sep	20.6	957	Apr	6.0
632	Sep	19.5	714	Apr	5.9	795	Oct	15.1	877	Apr	29.2	958	Nov	8.9
634	Apr	29.1	715	Nov	8.9	797	May	22.3	878	Dec	3.9	960	Jun	14.6
635	Dec	3.9	717	Jun	14.5	798	Dec	28.7	880	Jul	7.8	962	Jan	22.2
637	Jul	7.7	719	Jan	22.2	800	Jul	31.3	882	Feb	15.4	963	Aug	25.1
639	Feb	15.3	720	Aug	24.0	802	Mar	11.1	883	Sep	18.1	965	Apr	3.7
640	Sep	17.1	722	Apr	3.5	803	Oct	12.6	885	Apr	26.9	966	Nov	6.4
642	Apr	26.7	723	Nov	6.4	805	May	20.0	886	Dec	1.4	968	Jun	12.3
643	Dec	1.4	725	Jun	12.1	806	Dec	26.2	888	Jul	5.5	970	Jan	19.8
645	Jul	5.4	727	Jan	19.8	808	Jul	28.9	890	Feb	13.0	971	Aug	22.7
647	Feb	12.9	728	Aug	21.6	810	Mar	8.8	891	Sep	15.7	973	Apr	1.3
648	Sep	14.7	730	Apr	1.2	811	Oct	10.1	893	Apr	24.6	974	Nov	3.9
650	Apr	24.4	731	Nov	3.9	813	May	17.7	894	Nov	28.9	976	Jun	10.0
651	Nov	28.9	733	Jun	9.8	814	Dec	23.8	896	Jul	3.2	978	Jan	17.4
653	Jul	3.0	735	Jan	17.4	816	Jul	26.5	898	Feb	10.6	979	Aug	20.3
655	Feb	10.5	736	Aug	19.2	818	Mar	6.4	899	Sep	13.3	981	Mar	30.0
656	Sep	12.2	738	Mar	29.9	819	Oct	7.7	901	Apr	22.3	982	Nov	1.4
658	Apr	22.1	739	Nov	1.4	821	May	15.4	902	Nov	26.4	984	Jun	7.7
659	Nov	26.4	741	Jun	7.5	822	Dec	21.3	904	Jun	30.9	986	Jan	14.9
661	Jun	30.7	743	Jan	14.9	824	Jul	24.2	906	Feb	8.2	987	Aug	17.9
663	Feb	8.1	744	Aug	16.8	826	Mar	4.0	907	Sep	10.9	989	Mar	27.6
664	Sep	9.8	746	Mar	27.5	827	Oct	5.2	909	Apr	19.9	990	Oct	29.9
666	Apr	19.8	747	Oct	30.0	829	May	13.1	910	Nov	23.9	992	Jun	5.4
667	Nov	23.9	749	Jun	5.2	830	Dec	18.8	912	Jun	28.5	994	Jan	12.5
669	Jun	28.4	751	Jan	12.5	832	Jul	21.8	914	Feb	5.8	995	Aug	15.5
671	Feb	5.7	752	Aug	14.5	834	Mar	1.7	915	Sep	8.5	997	Mar	25.3
672	Sep	7.4	754	Mar	25.2	835	Oct	2.8	917	Apr	17.6	998	Oct	27.5
674	Apr	17.5	755	Oct	27.5	837	May	10.8	918	Nov	21.4	1000	Jun	3.0
675	Nov	21.4	757	Jun	2.9	838	Dec	16.3	920	Jun	26.2	1002	Jan	10.0
677	Jun	26.0	759	Jan	10.0	840	Jul	19.5	922	Feb	3.4	1003	Aug	13.2
679	Feb	3.3	760	Aug	12.1	842	Feb	27.3	923	Sep	6.0	1005	Mar	23.0
680	Sep	5.0	762	Mar	22.9	843	Sep	30.3	925	Apr	15.3	1006	Oct	25.0

1008	May	31.7	1089	Dec	13.8	1171	Jun	25.0	1253	Jan	7.5	1334	Jul	18.4
1010	Jan	7.6	1091	Jul	18.3	1173	Jan	32.0	1254	Aug	10.9	1336	Feb	26.0
1011	Aug	10.8	1093	Feb	25.0	1174	Sep	3.7	1256	Mar	20.7	1337	Sep	27.9
1013	Mar	20.6	1094	Sep	27.9	1176	Apr	13.1	1257	Oct	22.5	1339	May	7.4
1014	Oct	22.5	1096	May	6.3	1177	Nov	16.3	1259	May	30.6	1340	Dec	11.3
1016	May	29.4	1097	Dec	11.3	1179	Jun	22.7	1261	Jan	5.1	1342	Jul	16.1
1018	Jan	5.1	1099	Jul	16.0	1181	Jan	29.5	1262	Aug	8.5	1344	Feb	23.6
1019	Aug	8.4	1101	Feb	22.6	1182	Sep	1.3	1264	Mar	18.3	1345	Sep	25.5
1021	Mar	18.3	1102	Sep	25.5	1184	Apr	10.8	1265	Oct	20.0	1347	May	5.1
1022	Oct	20.0	1104	May	4.0	1185	Nov	13.9	1267	May	28.3	1348	Dec	8.8
1024	May	27.1	1105	Dec	8.8	1187	Jun	20.4	1269	Jan	2.6	1350	Jul	13.8
1026	Jan	2.6	1107	Jul	13.6	1189	Jan	27.1	1270	Aug	6.2	1352	Feb	21.2
1027	Aug	6.1	1109	Feb	20.2	1190	Aug	29.9	1272	Mar	16.0	1353	Sep	23.0
1029	Mar	15.9	1110	Sep	23.0	1192	Apr	8.4	1273	Oct	17.5	1355	May	2.8
1030	Oct	17.6	1112	May	1.7	1193	Nov	11.4	1275	May	25.9	1356	Dec	6.3
1032	May	24.8	1113	Dec	6.3	1195	Jun	18.1	1276	Dec	31.1	1358	Jul	11.4
1033	Dec	31.2	1115	Jul	11.3	1197	Jan	24.7	1278	Aug	3.8	1360	Feb	18.8
1035	Aug	3.7	1117	Feb	17.8	1198	Aug	27.5	1280	Mar	13.6	1361	Sep	20.6
1037	Mar	13.6	1118	Sep	20.6	1200	Apr	6.1	1281	Oct	15.1	1363	Apr	30.5
1038	Oct	15.1	1120	Apr	29.3	1201	Nov	8.9	1283	May	23.6	1364	Dec	3.8
1040	May	22.5	1121	Dec	3.8	1203	Jun	15.8	1284	Dec	28.7	1366	Jul	9.1
1041	Dec	28.7	1123	Jul	9.0	1205	Jan	22.2	1286	Aug	1.5	1368	Feb	16.4
1043	Aug	1.4	1125	Feb	15.4	1206	Aug	25.1	1288	Mar	11.3	1369	Sep	18.2
1045	Mar	11.2	1126	Sep	18.2	1208	Apr	3.8	1289	Oct	12.6	1371	Apr	28.2
1046	Oct	12.6	1128	Apr	27.0	1209	Nov	6.4	1291	May	21.3	1372	Dec	1.3
1048	May	20.2	1129	Dec	1.3	1211	Jun	13.5	1292	Dec	26.2	1374	Jul	6.8
1049	Dec	26.2	1131	Jul	6.6	1213	Jan	19.8	1294	Jul	30.1	1376	Feb	14.0
1051	Jul	30.0	1133	Feb	13.0	1214	Aug	22.8	1296	Mar	8.9	1377	Sep	15.8
1053	Mar	8.8	1134	Sep	15.7	1216	Apr	1.4	1297	Oct	10.1	1379	Apr	25.8
1054	Oct	10.2	1136	Apr	24.7	1217	Nov	3.9	1299	May	19.0	1380	Nov	28.8
1056	May	17.9	1137	Nov	28.8	1219	Jun	11.1	1300	Dec	23.7	1382	Jul	4.5
1057	Dec	23.7	1139	Jul	4.3	1221	Jan	17.3	1302	Jul	27.8	1384	Feb	11.6
1059	Jul	27.7	1141	Feb	10.6	1222	Aug	20.4	1304	Mar	6.5	1385	Sep	13.4
1061	Mar	6.5	1142	Sep	13.3	1224	Mar	30.1	1305	Oct	7.7	1387	Apr	23.5
1062	Oct	7.7	1144	Apr	22.4	1225	Nov	1.4	1307	May	16.7	1388	Nov	26.3
1064	May	15.5	1145	Nov	26.3	1227	Jun	8.8	1308	Dec	21.2	1390	Jul	2.1
1065	Dec	21.2	1147	Jul	2.0	1229	Jan	14.9	1310	Jul	25.4	1392	Feb	9.2
1067	Jul	25.3	1149	Feb	8.2	1230	Aug	18.0	1312	Mar	4.2	1393	Sep	11.0
1069	Mar	4.1	1150	Sep	10.9	1232	Mar	27.7	1313	Oct	5.2	1395	Apr	21.2
1070	Oct	5.2	1152	Apr	20.1	1233	Oct	29.9	1315	May	14.4	1396	Nov	23.8
1072	May	13.2	1153	Nov	23.8	1235	Jun	6.5	1316	Dec	18.7	1398	Jun	29.8
1073	Dec	18.8	1155	Jun	29.7	1237	Jan	12.5	1318	Jul	23.1	1400	Feb	6.8
1075	Jul	23.0	1157	Feb	5.8	1238	Aug	15.6	1320	Mar	1.8	1401	Sep	8.6
1077	Mar	1.7	1158	Sep	8.5	1240	Mar	25.4	1321	Oct	2.8	1403	Apr	18.9
1078	Oct	2.8	1160	Apr	17.7	1241	Oct	27.4	1323	May	12.1	1404	Nov	21.3
1080	May	10.9	1161	Nov	21.3	1243	Jun	4.2	1324	Dec	16.2	1406	Jun	27.5
1081	Dec	16.3	1163	Jun	27.3	1245	Jan	10.0	1326	Jul	20.8	1408	Feb	4.4
1083	Jul	20.6	1165	Feb	3.4	1246	Aug	13.3	1328	Feb	28.4	1409	Sep	6.2
1085	Feb	27.4	1166	Sep	6.1	1248	Mar	23.1	1329	Sep	30.4	1411	Apr	16.5
1086	Sep	30.3	1168	Apr	15.4	1249	Oct	25.0	1331	May	9.8	1412	Nov	18.8
1088	May	8.6	1169	Nov	18.8	1251	Jun	1.9	1332	Dec	13.7	1414	Jun	25.2

1416 Feb 2.0	1497 Aug 11.0	1579 Feb 26.1	1660 Sep 13.8	1742 Apr 1.8
1417 Sep 3.8	1499 Mar 21.8	1580 Sep 27.9	1662 Apr 24.3	1743 Nov 3.4
1419 Apr 14.2	1500 Oct 22.5	1582 May 7.6	1663 Nov 27.3	1745 Jun 10.9
1420 Nov 16.3	1502 May 30.7	1583 Dec 22.2	1665 Jul 3.0	1747 Jan 17.0
1422 Jun 22.9	1504 Jan 6.0	1585 Jul 26.2	1667 Feb 9.5	1748 Aug 19.7
1424 Jan 30.5	1505 Aug 8.6	1587 Mar 5.7	1668 Sep 11.4	1750 Mar 30.5
1425 Sep 1.4	1507 Mar 19.4	1588 Oct 5.5	1670 Apr 22.0	1751 Oct 32.0
1427 Apr 11.9	1508 Oct 20.0	1590 May 15.2	1671 Nov 24.8	1753 Jun 8.5
1428 Nov 13.8	1510 May 28.4	1591 Dec 19.7	1673 Jun 30.7	1755 Jan 14.5
1430 Jun 20.5	1512 Jan 3.6	1593 Jul 23.9	1675 Feb 7.1	1756 Aug 17.4
1432 Jan 28.1	1513 Aug 6.3	1595 Mar 3.3	1676 Sep 9.1	1758 Mar 28.1
1433 Aug 30.0	1515 Mar 17.1	1596 Oct 3.1	1678 Apr 19.6	1759 Oct 29.5
1435 Apr 9.5	1516 Oct 17.5	1598 May 12.9	1679 Nov 22.3	1761 Jun 6.2
1436 Nov 11.3	1518 May 26.1	1599 Dec 17.2	1681 Jun 28.4	1763 Jan 12.0
1438 Jun 18.2	1520 Jan 1.1	1601 Jul 21.6	1683 Feb 4.7	1764 Aug 15.0
1440 Jan 25.7	1521 Aug 3.9	1603 Feb 28.9	1684 Sep 6.7	1766 Mar 25.8
1441 Aug 27.6	1523 Mar 14.7	1604 Sep 30.7	1686 Apr 17.3	1767 Oct 27.1
1443 Apr 7.2	1524 Oct 15.1	1606 May 10.6	1687 Nov 19.8	1769 Jun 3.9
1444 Nov 8.8	1526 May 23.8	1607 Dec 14.7	1689 Jun 26.1	1771 Jan 9.6
1446 Jun 15.9	1527 Dec 29.6	1609 Jul 19.2	1691 Feb 2.2	1772 Aug 12.7
1448 Jan 23.2	1529 Aug 1.6	1611 Feb 26.5	1692 Sep 4.3	1774 Mar 23.4
1449 Aug 25.2	1531 Mar 12.3	1612 Sep 28.2	1694 Apr 15.0	1775 Oct 24.6
1451 Apr 4.9	1532 Oct 12.6	1614 May 8.3	1695 Nov 17.3	1777 Jun 1.6
1452 Nov 6.3	1534 May 21.5	1615 Dec 12.2	1697 Jun 23.7	1779 Jan 7.1
1454 Jun 13.6	1535 Dec 27.1	1617 Jul 16.9	1699 Jan 30.8	1780 Aug 10.4
1456 Jan 20.8	1537 Jul 30.2	1619 Feb 24.1	1700 Sep 2.9	1782 Mar 21.0
1457 Aug 22.8	1539 Mar 10.0	1620 Sep 25.8	1702 Apr 13.6	1783 Oct 22.1
1459 Apr 2.5	1540 Oct 10.2	1622 May 6.0	1703 Nov 15.8	1785 May 30.3
1460 Nov 3.9	1542 May 19.1	1623 Dec 9.7	1705 Jun 22.4	1787 Jan 4.6
1462 Jun 11.3	1543 Dec 24.7	1625 Jul 14.6	1707 Jan 29.3	1788 Aug 8.0
1464 Jan 18.3	1545 Jul 27.9	1627 Feb 21.7	1708 Aug 31.5	1790 Mar 18.6
1465 Aug 20.5	1547 Mar 7.6	1628 Sep 23.4	1710 Apr 11.3	1791 Oct 19.7
1467 Mar 31.2	1548 Oct 7.7	1630 May 3.6	1711 Nov 13.3	1793 May 28.0
1468 Nov 1.4	1550 May 16.8	1631 Dec 7.2	1713 Jun 20.1	1795 Jan 2.1
1470 Jun 9.0	1551 Dec 22.2	1633 Jul 12.3	1715 Jan 26.8	1796 Aug 5.7
1472 Jan 15.9	1553 Jul 25.6	1635 Feb 19.2	1716 Aug 29.2	1798 Mar 16.2
1473 Aug 18.1	1555 Mar 5.2	1636 Sep 21.0	1718 Apr 8.9	1799 Oct 17.3
1475 Mar 28.8	1556 Oct 5.3	1638 May 1.3	1719 Nov 10.9	1801 May 26.6
1476 Oct 29.9	1558 May 14.5	1639 Dec 4.7	1721 Jun 17.8	1802 Dec 31.6
1478 Jun 6.7	1559 Dec 19.7	1641 Jul 10.0	1723 Jan 24.4	1804 Aug 4.3
1480 Jan 13.4	1561 Jul 23.2	1643 Feb 16.8	1724 Aug 26.8	1806 Mar 14.9
1481 Aug 15.7	1563 Mar 2.8	1644 Sep 18.6	1726 Apr 6.6	1807 Oct 15.8
1483 Mar 26.5	1564 Oct 2.8	1646 Apr 29.0	1727 Nov 8.4	1809 May 24.3
1484 Oct 27.4	1566 May 12.2	1647 Dec 2.3	1729 Jun 15.5	1810 Dec 29.1
1486 Jun 4.3	1567 Dec 17.2	1649 Jul 7.6	1731 Jan 21.9	1812 Aug 2.0
1488 Jan 11.0	1569 Jul 20.9	1651 Feb 14.4	1732 Aug 24.5	1814 Mar 12.5
1489 Aug 13.4	1571 Feb 28.4	1652 Sep 16.2	1734 Apr 4.2	1815 Oct 13.4
1491 Mar 24.1	1572 Sep 30.4	1654 Apr 26.7	1735 Nov 5.9	1817 May 22.0
1492 Oct 24.9	1574 May 9.9	1655 Nov 29.8	1737 Jun 13.2	1818 Dec 26.6
1494 Jun 2.0	1575 Dec 14.7	1657 Jul 5.3	1739 Jan 19.5	1820 Jul 30.7
1496 Jan 8.5	1577 Jul 18.6	1659 Feb 12.0	1740 Aug 22.1	1822 Mar 10.1

Year			Year			Year			Year			Year		
1823	Oct	11.0	1905	Apr	27.4	1986	Nov	5.4	2068	May	20.8	2149	Dec	1.2
1825	May	19.7	1906	Nov	30.2	1988	Jun	13.0	2069	Dec	25.1	2151	Jul	8.3
1826	Dec	24.2	1908	Jul	6.1	1990	Jan	18.9	2071	Jul	30.5	2153	Feb	13.5
1828	Jul	28.3	1910	Feb	12.5	1991	Aug	22.8	2073	Mar	8.7	2154	Sep	16.6
1830	Mar	7.7	1911	Sep	15.5	1993	Apr	1.5	2074	Oct	9.5	2156	Apr	26.2
1831	Oct	8.5	1913	Apr	25.1	1994	Nov	3.0	2076	May	18.5	2157	Nov	28.7
1833	May	17.4	1914	Nov	27.7	1996	Jun	10.7	2077	Dec	22.6	2159	Jul	6.0
1834	Dec	21.7	1916	Jul	3.8	1998	Jan	16.5	2079	Jul	28.1	2161	Feb	11.0
1836	Jul	26.0	1918	Feb	10.1	1999	Aug	20.5	2081	Mar	6.3	2162	Sep	14.2
1838	Mar	5.3	1919	Sep	13.1	2001	Mar	30.2	2082	Oct	7.1	2164	Apr	23.8
1839	Oct	6.1	1921	Apr	22.7	2002	Oct	31.5	2084	May	16.2	2165	Nov	26.2
1841	May	15.0	1922	Nov	25.2	2004	Jun	8.4	2085	Dec	20.1	2167	Jul	3.6
1842	Dec	19.2	1924	Jul	1.5	2006	Jan	14.0	2087	Jul	25.8	2169	Feb	8.6
1844	Jul	23.7	1926	Feb	7.6	2007	Aug	18.1	2089	Mar	3.9	2170	Sep	11.8
1846	Mar	2.9	1927	Sep	10.7	2009	Mar	27.8	2090	Oct	4.7	2172	Apr	21.5
1847	Oct	3.7	1929	Apr	20.4	2010	Oct	29.0	2092	May	13.8	2173	Nov	23.7
1849	May	12.7	1930	Nov	22.8	2012	Jun	6.0	2093	Dec	17.6	2175	Jul	1.3
1850	Dec	16.7	1932	Jun	29.2	2014	Jan	11.5	2095	Jul	23.5	2177	Feb	6.1
1852	Jul	21.4	1934	Feb	5.2	2015	Aug	15.8	2097	Mar	1.5	2178	Sep	9.4
1854	Feb	28.5	1935	Sep	8.4	2017	Mar	25.4	2098	Oct	2.3	2180	Apr	19.1
1855	Oct	1.3	1937	Apr	18.0	2018	Oct	26.6	2100	May	12.5	2181	Nov	21.2
1857	May	10.4	1938	Nov	20.3	2020	Jun	3.7	2101	Dec	16.1	2183	Jun	29.0
1858	Dec	14.2	1940	Jun	26.9	2022	Jan	9.0	2103	Jul	22.2	2185	Feb	3.7
1860	Jul	19.0	1942	Feb	2.7	2023	Aug	13.5	2105	Feb	28.1	2186	Sep	7.1
1862	Feb	26.1	1943	Sep	6.0	2025	Mar	23.0	2106	Sep	30.9	2188	Apr	16.8
1863	Sep	28.9	1945	Apr	15.7	2026	Oct	24.2	2108	May	10.2	2189	Nov	18.8
1865	May	8.1	1946	Nov	17.8	2028	Jun	1.4	2109	Dec	13.6	2191	Jun	26.7
1866	Dec	11.7	1948	Jun	24.6	2030	Jan	6.5	2111	Jul	19.9	2193	Feb	1.2
1868	Jul	16.7	1950	Jan	31.3	2031	Aug	11.1	2113	Feb	25.6	2194	Sep	4.7
1870	Feb	23.6	1951	Sep	3.6	2033	Mar	20.7	2114	Sep	28.5	2196	Apr	14.4
1871	Sep	26.5	1953	Apr	13.3	2034	Oct	21.7	2116	May	7.8	2197	Nov	16.3
1873	May	5.7	1954	Nov	15.3	2036	May	30.1	2117	Dec	11.1	2199	Jun	24.4
1874	Dec	9.2	1956	Jun	22.3	2038	Jan	4.1	2119	Jul	17.5	2201	Jan	30.8
1876	Jul	14.4	1958	Jan	28.8	2039	Aug	8.8	2121	Feb	23.2	2202	Sep	3.3
1878	Feb	21.2	1959	Sep	1.3	2041	Mar	18.3	2122	Sep	26.1	2204	Apr	13.1
1879	Sep	24.1	1961	Apr	11.0	2042	Oct	19.3	2124	May	5.5	2205	Nov	14.8
1881	May	3.4	1962	Nov	12.8	2044	May	27.8	2125	Dec	8.6	2207	Jun	23.1
1882	Dec	6.7	1964	Jun	19.9	2046	Jan	1.6	2127	Jul	15.2	2209	Jan	28.3
1884	Jul	12.1	1966	Jan	26.4	2047	Aug	6.5	2129	Feb	20.8	2210	Aug	32.0
1886	Feb	18.8	1967	Aug	29.9	2049	Mar	15.9	2130	Sep	23.7	2212	Apr	10.7
1887	Sep	21.7	1969	Apr	8.6	2050	Oct	16.8	2132	May	3.2	2213	Nov	12.3
1889	May	1.1	1970	Nov	10.4	2052	May	25.5	2133	Dec	6.2	2215	Jun	20.8
1890	Dec	4.2	1972	Jun	17.6	2053	Dec	30.1	2135	Jul	12.9	2217	Jan	25.8
1892	Jul	9.8	1974	Jan	23.9	2055	Aug	4.1	2137	Feb	18.4	2218	Aug	29.6
1894	Feb	16.4	1975	Aug	27.5	2057	Mar	13.5	2138	Sep	21.3	2220	Apr	8.3
1895	Sep	19.3	1977	Apr	6.3	2058	Oct	14.4	2140	Apr	30.8	2221	Nov	9.9
1897	Apr	28.7	1978	Nov	7.9	2060	May	23.1	2141	Dec	3.7	2223	Jun	18.4
1898	Dec	1.7	1980	Jun	15.3	2061	Dec	27.6	2143	Jul	10.6	2225	Jan	23.4
1900	Jul	8.5	1982	Jan	21.4	2063	Aug	1.8	2145	Feb	15.9	2226	Aug	27.3
1902	Feb	15.0	1983	Aug	25.2	2065	Mar	11.1	2146	Sep	18.9	2228	Apr	6.0
1903	Sep	17.9	1985	Apr	3.9	2066	Oct	12.0	2148	Apr	28.5	2229	Nov	7.4

2231	Jun	16.1	2285	Oct	21.3	2340	Mar	4.5	2394	Jul	10.4	2448	Nov	15.8

2231	Jun	16.1	2285	Oct	21.3	2340	Mar	4.5	2394	Jul	10.4	2448	Nov	15.8
2233	Jan	20.9	2287	May	30.9	2341	Oct	5.3	2396	Feb	16.5	2450	Jun	24.2
2234	Aug	24.9	2289	Jan	3.5	2343	May	15.6	2397	Sep	18.6	2452	Jan	30.3
2236	Apr	3.6	2290	Aug	8.6	2344	Dec	18.1	2399	Apr	29.2	2453	Sep	2.1
2237	Nov	4.9	2292	Mar	17.9	2346	Jul	24.3	2400	Nov	30.6	2455	Apr	12.7
2239	Jun	13.8	2293	Oct	18.8	2348	Mar	2.1	2402	Jul	8.1	2456	Nov	13.3
2241	Jan	18.4	2295	May	28.6	2349	Oct	2.9	2404	Feb	14.0	2458	Jun	21.9
2242	Aug	22.6	2297	Jan	1.0	2351	May	13.3	2405	Sep	16.2	2460	Jan	27.8
2244	Apr	1.2	2298	Aug	6.2	2352	Dec	15.6	2407	Apr	26.9	2461	Aug	30.7
2245	Nov	2.5	2300	Mar	16.5	2354	Jul	22.0	2408	Nov	28.2	2463	Apr	10.4
2247	Jun	11.5	2301	Oct	17.4	2356	Feb	28.6	2410	Jul	5.8	2464	Nov	10.9
2249	Jan	15.9	2303	May	27.3	2357	Sep	30.5	2412	Feb	11.6	2466	Jun	19.6
2250	Aug	20.2	2304	Dec	30.5	2359	May	10.9	2413	Sep	13.9	2468	Jan	25.3
2252	Mar	29.8	2306	Aug	4.9	2360	Dec	13.1	2415	Apr	24.5	2469	Aug	28.4
2253	Oct	31.0	2308	Mar	14.1	2362	Jul	19.7	2416	Nov	25.7	2471	Apr	8.0
2255	Jun	9.2	2309	Oct	15.0	2364	Feb	26.2	2418	Jul	3.5	2472	Nov	8.4
2257	Jan	13.5	2311	May	24.9	2365	Sep	28.2	2420	Feb	9.1	2474	Jun	17.3
2258	Aug	17.9	2312	Dec	28.0	2367	May	8.6	2421	Sep	11.5	2476	Jan	22.8
2260	Mar	27.5	2314	Aug	2.6	2368	Dec	10.6	2423	Apr	22.2	2477	Aug	26.0
2261	Oct	28.6	2316	Mar	11.7	2370	Jul	17.3	2424	Nov	23.2	2479	Apr	5.6
2263	Jun	6.9	2317	Oct	12.6	2372	Feb	23.8	2426	Jul	1.1	2480	Nov	5.9
2265	Jan	11.0	2319	May	22.6	2373	Sep	25.8	2428	Feb	6.6	2482	Jun	14.9
2266	Aug	15.6	2320	Dec	25.5	2375	May	6.3	2429	Sep	9.2	2484	Jan	20.4
2268	Mar	25.1	2322	Jul	31.3	2376	Dec	8.1	2431	Apr	19.8	2485	Aug	23.7
2269	Oct	26.1	2324	Mar	9.3	2378	Jul	15.0	2432	Nov	20.7	2487	Apr	3.3
2271	Jun	4.5	2325	Oct	10.2	2380	Feb	21.3	2434	Jun	28.8	2488	Nov	3.5
2273	Jan	8.5	2327	May	20.3	2381	Sep	23.4	2436	Feb	4.2	2490	Jun	12.6
2274	Aug	13.2	2328	Dec	23.0	2383	May	3.9	2437	Sep	6.8	2492	Jan	17.9
2276	Mar	22.7	2330	Jul	28.9	2384	Dec	5.6	2439	Apr	17.5	2493	Aug	21.4
2277	Oct	23.7	2332	Mar	6.9	2386	Jul	12.7	2440	Nov	18.2	2495	Mar	31.9
2279	Jun	2.2	2333	Oct	7.7	2388	Feb	18.9	2442	Jun	26.5	2496	Nov	1.0
2281	Jan	6.0	2335	May	17.9	2389	Sep	21.0	2444	Feb	1.7	2498	Jun	10.3
2282	Aug	10.9	2336	Dec	20.6	2391	May	1.6	2445	Sep	4.4	2500	Jan	15.4
2284	Mar	20.3	2338	Jul	26.6	2392	Dec	3.1	2447	Apr	15.1			

The greatest eastern elongation (evening apparition) of Venus occurs approximately 71 days before inferior conjunction.

The greatest western elongation (morning apparition) of Venus occurs approximately 71 days after inferior conjunction.

X
SUPERIOR CONJUNCTIONS OF VENUS, 0 − 2500

0 Jun 10.7	70 Oct 9.9	141 Feb 19.8	211 Jun 20.8	281 Oct 19.9
2 Jan 18.2	72 May 21.7	142 Sep 18.0	213 Jan 28.3	283 May 31.8
3 Aug 18.3	73 Dec 26.0	144 May 1.4	214 Aug 27.7	285 Jan 5.3
5 Apr 1.4	75 Jul 28.8	145 Dec 2.5	216 Apr 10.8	286 Aug 7.1
6 Oct 30.0	77 Mar 11.2	147 Jul 8.6	217 Nov 9.2	288 Mar 20.8
8 Jun 8.5	78 Oct 7.4	149 Feb 17.3	219 Jun 18.6	289 Oct 17.4
10 Jan 15.7	80 May 19.4	150 Sep 15.6	221 Jan 25.8	291 May 29.6
11 Aug 16.0	81 Dec 23.4	152 Apr 29.1	222 Aug 25.4	293 Jan 2.7
13 Mar 30.1	83 Jul 26.6	153 Nov 29.9	224 Apr 8.5	294 Aug 4.8
14 Oct 27.4	85 Mar 8.8	155 Jul 6.4	225 Nov 6.6	296 Mar 18.4
16 Jun 6.3	86 Oct 4.9	157 Feb 14.9	227 Jun 16.4	297 Oct 14.9
18 Jan 13.1	88 May 17.2	158 Sep 13.2	229 Jan 23.3	299 May 27.3
19 Aug 13.7	89 Dec 20.8	160 Apr 26.9	230 Aug 23.1	300 Dec 31.1
21 Mar 27.8	91 Jul 24.3	161 Nov 27.3	232 Apr 6.2	302 Aug 2.5
22 Oct 24.9	93 Mar 6.4	163 Jul 4.2	233 Nov 4.1	304 Mar 16.0
24 Jun 4.0	94 Oct 2.5	165 Feb 12.4	235 Jun 14.1	305 Oct 12.4
26 Jan 10.5	96 May 14.9	166 Sep 10.8	237 Jan 20.7	307 May 25.1
27 Aug 11.4	97 Dec 18.2	168 Apr 24.6	238 Aug 20.8	308 Dec 28.5
29 Mar 25.4	99 Jul 22.1	169 Nov 24.7	240 Apr 3.9	310 Jul 31.3
30 Oct 22.4	101 Mar 4.0	171 Jul 1.9	241 Nov 1.5	312 Mar 13.6
32 Jun 1.8	102 Sep 30.1	173 Feb 9.9	243 Jun 11.9	313 Oct 9.9
34 Jan 7.9	104 May 12.7	174 Sep 8.5	245 Jan 18.1	315 May 22.8
35 Aug 9.2	105 Dec 15.5	176 Apr 22.3	246 Aug 18.5	316 Dec 25.9
37 Mar 23.1	107 Jul 19.8	177 Nov 22.1	248 Apr 1.5	318 Jul 29.0
38 Oct 19.9	109 Mar 1.6	179 Jun 29.7	249 Oct 30.0	320 Mar 11.2
40 May 30.6	110 Sep 27.6	181 Feb 7.4	251 Jun 9.7	321 Oct 7.5
42 Jan 5.3	112 May 10.4	182 Sep 6.1	253 Jan 15.6	323 May 20.6
43 Aug 6.9	113 Dec 12.9	184 Apr 20.0	254 Aug 16.2	324 Dec 23.3
45 Mar 20.7	115 Jul 17.6	185 Nov 19.5	256 Mar 30.2	326 Jul 26.8
46 Oct 17.4	117 Feb 27.1	187 Jun 27.5	257 Oct 27.4	328 Mar 8.8
48 May 28.4	118 Sep 25.2	189 Feb 4.9	259 Jun 7.5	329 Oct 5.0
50 Jan 2.8	120 May 8.2	190 Sep 3.8	261 Jan 13.0	331 May 18.3
51 Aug 4.6	121 Dec 10.3	192 Apr 17.7	262 Aug 13.9	332 Dec 20.7
53 Mar 18.3	123 Jul 15.3	193 Nov 16.9	264 Mar 27.9	334 Jul 24.5
54 Oct 14.9	125 Feb 24.7	195 Jun 25.3	265 Oct 24.9	336 Mar 6.4
56 May 26.1	126 Sep 22.8	197 Feb 2.4	267 Jun 5.2	337 Oct 2.6
57 Dec 31.2	128 May 5.9	198 Sep 1.4	269 Jan 10.4	339 May 16.1
59 Aug 2.3	129 Dec 7.7	200 Apr 15.4	270 Aug 11.6	340 Dec 18.1
61 Mar 16.0	131 Jul 13.1	201 Nov 14.3	272 Mar 25.5	342 Jul 22.3
62 Oct 12.4	133 Feb 22.2	203 Jun 23.0	273 Oct 22.4	344 Mar 4.0
64 May 23.9	134 Sep 20.4	205 Jan 30.9	275 Jun 3.0	345 Sep 30.1
65 Dec 28.6	136 May 3.7	206 Aug 30.1	277 Jan 7.9	347 May 13.9
67 Jul 31.1	137 Dec 5.1	208 Apr 13.1	278 Aug 9.3	348 Dec 15.5
69 Mar 13.6	139 Jul 10.9	209 Nov 11.7	280 Mar 23.1	350 Jul 20.0

352 Mar 1.6	433 Sep 3.9	515 Mar 26.6	596 Sep 27.8	678 Apr 19.0
353 Sep 27.7	435 Apr 18.8	516 Oct 22.4	598 May 11.8	679 Nov 17.8
355 May 11.6	436 Nov 16.9	518 Jun 3.2	599 Dec 13.8	681 Jun 25.7
356 Dec 12.9	438 Jun 25.5	520 Jan 8.8	601 Jul 18.0	683 Feb 3.3
358 Jul 17.8	440 Feb 3.3	521 Aug 9.5	603 Feb 28.1	684 Sep 1.7
360 Feb 28.1	441 Sep 1.5	523 Mar 24.2	604 Sep 25.4	686 Apr 16.7
361 Sep 25.3	443 Apr 16.6	524 Oct 19.9	606 May 9.5	687 Nov 15.3
363 May 9.4	444 Nov 14.3	526 May 32.0	607 Dec 11.2	689 Jun 23.4
364 Dec 10.2	446 Jun 23.2	528 Jan 6.2	609 Jul 15.7	691 Jan 31.8
366 Jul 15.5	448 Jan 31.8	529 Aug 7.2	611 Feb 25.7	692 Aug 30.4
368 Feb 25.7	449 Aug 30.2	531 Mar 21.8	612 Sep 23.0	694 Apr 14.4
369 Sep 22.9	451 Apr 14.2	532 Oct 17.4	614 May 7.2	695 Nov 12.7
371 May 7.1	452 Nov 11.7	534 May 29.7	615 Dec 8.6	697 Jun 21.2
372 Dec 7.6	454 Jun 21.0	536 Jan 3.6	617 Jul 13.5	699 Jan 29.2
374 Jul 13.3	456 Jan 29.3	537 Aug 5.0	619 Feb 23.2	700 Aug 28.0
376 Feb 23.2	457 Aug 27.9	539 Mar 19.4	620 Sep 20.6	702 Apr 12.0
377 Sep 20.5	459 Apr 11.9	540 Oct 14.9	622 May 5.0	703 Nov 10.1
379 May 4.8	460 Nov 9.2	542 May 27.5	623 Dec 6.0	705 Jun 19.0
380 Dec 5.0	462 Jun 18.8	544 Jan 1.0	625 Jul 11.3	707 Jan 26.7
382 Jul 11.1	464 Jan 26.7	545 Aug 2.7	627 Feb 20.8	708 Aug 25.7
384 Feb 20.8	465 Aug 25.6	547 Mar 17.1	628 Sep 18.2	710 Apr 9.7
385 Sep 18.1	467 Apr 9.6	548 Oct 12.4	630 May 2.7	711 Nov 7.6
387 May 2.6	468 Nov 6.6	550 May 25.3	631 Dec 3.4	713 Jun 16.8
388 Dec 2.4	470 Jun 16.6	551 Dec 29.4	633 Jul 9.0	715 Jan 24.1
390 Jul 8.8	472 Jan 24.2	553 Jul 31.5	635 Feb 18.3	716 Aug 23.4
392 Feb 18.3	473 Aug 23.3	555 Mar 14.7	636 Sep 15.8	718 Apr 7.4
393 Sep 15.7	475 Apr 7.3	556 Oct 10.0	638 Apr 30.4	· 719 Nov 5.0
395 Apr 30.3	476 Nov 4.1	558 May 23.0	639 Nov 30.8	721 Jun 14.5
396 Nov 29.8	478 Jun 14.3	559 Dec 26.8	641 Jul 6.8	723 Jan 21.6
398 Jul 6.6	480 Jan 21.6	561 Jul 29.2	643 Feb 15.8	724 Aug 21.1
400 Feb 15.8	481 Aug 21.0	563 Mar 12.3	644 Sep 13.4	726 Apr 5.0
401 Sep 13.3	483 Apr 4.9	564 Oct 7.5	646 Apr 28.1	727 Nov 2.5
403 Apr 28.0	484 Nov 1.5	566 May 20.8	647 Nov 28.2	729 Jun 12.3
404 Nov 27.2	486 Jun 12.1	567 Dec 24.2	649 Jul 4.6	731 Jan 19.0
406 Jul 4.4	488 Jan 19.1	569 Jul 26.9	651 Feb 13.3	732 Aug 18.8
408 Feb 13.4	489 Aug 18.7	571 Mar 9.9	652 Sep 11.1	734 Apr 2.7
409 Sep 10.9	491 Apr 2.6	572 Oct 5.1	654 Apr 25.9	735 Oct 31.0
411 Apr 25.7	492 Oct 30.0	574 May 18.5	655 Nov 25.6	737 Jun 10.1
412 Nov 24.6	494 Jun 9.9	575 Dec 21.6	657 Jul 2.3	739 Jan 16.5
414 Jul 2.1	496 Jan 16.5	577 Jul 24.7	659 Feb 10.8	740 Aug 16.5
416 Feb 10.9	497 Aug 16.4	579 Mar 7.4	660 Sep 8.7	742 Mar 31.3
417 Sep 8.6	499 Mar 31.3	580 Oct 2.6	662 Apr 23.6	743 Oct 28.5
419 Apr 23.4	500 Oct 27.4	582 May 16.3	663 Nov 23.0	745 Jun 7.8
420 Nov 22.0	502 Jun 7.7	583 Dec 19.0	665 Jun 30.1	747 Jan 13.9
422 Jun 29.9	504 Jan 13.9	585 Jul 22.4	667 Feb 8.3	748 Aug 14.2
424 Feb 8.4	505 Aug 14.1	587 Mar 5.0	668 Sep 6.4	750 Mar 29.0
425 Sep 6.2	507 Mar 28.9	588 Sep 30.2	670 Apr 21.3	751 Oct 25.9
427 Apr 21.1	508 Oct 24.9	590 May 14.0	671 Nov 20.4	753 Jun 5.6
428 Nov 19.5	510 Jun 5.4	591 Dec 16.4	673 Jun 27.9	755 Jan 11.3
430 Jun 27.7	512 Jan 11.4	593 Jul 20.2	675 Feb 5.8	756 Aug 12.0
432 Feb 5.9	513 Aug 11.8	595 Mar 2.6	676 Sep 4.0	758 Mar 26.6

759 Oct 23.4	841 May 11.9	922 Nov 17.8	1004 Jun 3.6	1085 Dec 13.6
761 Jun 3.4	842 Dec 13.7	924 Jun 25.8	1006 Jan 8.6	1087 Jul 19.3
763 Jan 8.7	844 Jul 18.2	926 Feb 3.2	1007 Aug 10.9	1089 Feb 28.1
764 Aug 9.7	846 Feb 28.1	927 Sep 2.8	1009 Mar 24.3	1090 Sep 26.5
766 Mar 24.2	847 Sep 26.5	929 Apr 16.8	1010 Oct 21.0	1092 May 9.8
767 Oct 20.9	849 May 9.6	930 Nov 15.2	1012 Jun 1.3	1093 Dec 11.0
769 Jun 1.1	850 Dec 11.1	932 Jun 23.6	1014 Jan 6.0	1095 Jul 17.1
771 Jan 6.1	852 Jul 15.9	934 Jan 31.7	1015 Aug 8.6	1097 Feb 25.7
772 Aug 7.4	854 Feb 25.7	935 Aug 31.5	1017 Mar 21.9	1098 Sep 24.1
774 Mar 21.9	855 Sep 24.1	937 Apr 14.4	1018 Oct 18.5	1100 May 7.5
775 Oct 18.4	857 May 7.4	938 Nov 12.7	1020 May 30.1	1101 Dec 8.4
777 May 29.9	858 Dec 8.5	940 Jun 21.4	1022 Jan 3.4	1103 Jul 14.9
779 Jan 3.5	860 Jul 13.7	942 Jan 29.2	1023 Aug 6.3	1105 Feb 23.2
780 Aug 5.2	862 Feb 23.2	943 Aug 29.2	1025 Mar 19.5	1106 Sep 21.8
782 Mar 19.5	863 Sep 21.7	945 Apr 12.1	1026 Oct 16.0	1108 May 5.2
783 Oct 16.0	865 May 5.1	946 Nov 10.1	1028 May 27.8	1109 Dec 5.8
785 May 27.7	866 Dec 5.9	948 Jun 19.2	1029 Dec 31.8	1111 Jul 12.6
786 Dec 31.9	868 Jul 11.4	950 Jan 26.6	1031 Aug 4.1	1113 Feb 20.7
788 Aug 2.9	870 Feb 20.7	951 Aug 26.9	1033 Mar 17.1	1114 Sep 19.4
790 Mar 17.1	871 Sep 19.3	953 Apr 9.8	1034 Oct 13.5	1116 May 2.9
791 Oct 13.5	873 May 2.8	954 Nov 7.6	1036 May 25.6	1117 Dec 3.2
793 May 25.4	874 Dec 3.3	956 Jun 17.0	1037 Dec 29.2	1119 Jul 10.4
794 Dec 29.3	876 Jul 9.2	958 Jan 24.1	1039 Aug 1.8	1121 Feb 18.2
796 Jul 31.6	878 Feb 18.3	959 Aug 24.6	1041 Mar 14.7	1122 Sep 17.0
798 Mar 14.7	879 Sep 16.9	961 Apr 7.4	1042 Oct 11.1	1124 Apr 30.7
799 Oct 11.0	881 Apr 30.5	962 Nov 5.0	1044 May 23.3	1125 Nov 30.6
801 May 23.2	882 Nov 30.7	964 Jun 14.7	1045 Dec 26.7	1127 Jul 8.2
802 Dec 26.7	884 Jul 7.0	966 Jan 21.5	1047 Jul 30.6	1129 Feb 15.7
804 Jul 29.4	886 Feb 15.8	967 Aug 22.3	1049 Mar 12.3	1130 Sep 14.7
806 Mar 12.3	887 Sep 14.5	969 Apr 5.1	1050 Oct 8.6	1132 Apr 28.4
807 Oct 8.6	889 Apr 28.3	970 Nov 2.5	1052 May 21.1	1133 Nov 28.1
809 May 20.9	890 Nov 28.1	972 Jun 12.5	1053 Dec 24.0	1135 Jul 6.0
810 Dec 24.1	892 Jul 4.8	974 Jan 18.9	1055 Jul 28.3	1137 Feb 13.2
812 Jul 27.1	894 Feb 13.3	975 Aug 20.0	1057 Mar 9.9	1138 Sep 12.3
814 Mar 9.9	895 Sep 12.2	977 Apr 2.8	1058 Oct 6.2	1140 Apr 26.1
815 Oct 6.1	897 Apr 26.0	978 Oct 31.0	1060 May 18.8	1141 Nov 25.5
817 May 18.7	898 Nov 25.5	980 Jun 10.3	1061 Dec 21.4	1143 Jul 3.7
818 Dec 21.5	900 Jul 2.5	982 Jan 16.4	1063 Jul 26.1	1145 Feb 10.7
820 Jul 24.9	902 Feb 10.8	983 Aug 17.7	1065 Mar 7.4	1146 Sep 10.0
822 Mar 7.5	903 Sep 9.8	985 Mar 31.4	1066 Oct 3.8	1148 Apr 23.8
823 Oct 3.7	905 Apr 23.7	986 Oct 28.5	1068 May 16.6	1149 Nov 22.9
825 May 16.4	906 Nov 22.9	988 Jun 8.0	1069 Dec 18.8	1151 Jul 1.5
826 Dec 18.9	908 Jun 30.3	990 Jan 13.8	1071 Jul 23.8	1153 Feb 8.2
828 Jul 22.6	910 Feb 8.3	991 Aug 15.4	1073 Mar 5.0	1154 Sep 7.6
830 Mar 5.0	911 Sep 7.5	993 Mar 29.0	1074 Oct 1.4	1156 Apr 21.5
831 Oct 1.3	913 Apr 21.4	994 Oct 25.9	1076 May 14.3	1157 Nov 20.3
833 May 14.2	914 Nov 20.4	996 Jun 5.8	1077 Dec 16.2	1159 Jun 29.3
834 Dec 16.3	916 Jun 28.1	998 Jan 11.2	1079 Jul 21.6	1161 Feb 5.7
836 Jul 20.4	918 Feb 5.8	999 Aug 13.1	1081 Mar 2.6	1162 Sep 5.3
838 Mar 2.6	919 Sep 5.2	1001 Mar 26.7	1082 Sep 28.9	1164 Apr 19.2
839 Sep 28.9	921 Apr 19.1	1002 Oct 23.4	1084 May 12.1	1165 Nov 17.8

1167 Jun 27.0	1249 Jan 8.5	1330 Jul 19.5	1412 Feb 4.1	1493 Aug 11.2
1169 Feb 3.2	1250 Aug 11.0	1332 Feb 29.1	1413 Sep 3.1	1495 Mar 25.3
1170 Sep 3.0	1252 Mar 24.3	1333 Sep 26.6	1415 Apr 17.9	1496 Oct 21.0
1172 Apr 16.8	1253 Oct 21.0	1335 May 10.9	1416 Nov 15.1	1498 Jun 2.6
1173 Nov 15.2	1255 Jun 2.5	1336 Dec 11.0	1418 Jun 25.0	1500 Jan 6.8
1175 Jun 24.8	1257 Jan 5.9	1338 Jul 17.3	1420 Feb 1.6	1501 Aug 8.9
1177 Jan 31.6	1258 Aug 8.8	1340 Feb 26.6	1421 Aug 31.8	1503 Mar 22.9
1178 Aug 31.6	1260 Mar 21.9	1341 Sep 24.2	1423 Apr 15.6	1504 Oct 18.5
1180 Apr 14.5	1261 Oct 18.5	1343 May 8.6	1424 Nov 12.6	1506 May 31.4
1181 Nov 12.6	1263 May 31.2	1344 Dec 8.3	1426 Jun 22.8	1508 Jan 4.3
1183 Jun 22.6	1265 Jan 3.3	1346 Jul 15.1	1428 Jan 30.0	1509 Aug 6.7
1185 Jan 29.1	1266 Aug 6.5	1348 Feb 24.1	1429 Aug 29.5	1511 Mar 20.6
1186 Aug 29.3	1268 Mar 19.5	1349 Sep 21.9	1431 Apr 13.3	1512 Oct 16.1
1188 Apr 12.2	1269 Oct 16.0	1351 May 6.4	1432 Nov 10.1	1514 May 29.1
1189 Nov 10.1	1271 May 29.0	1352 Dec 5.8	1434 Jun 20.5	1516 Jan 1.7
1191 Jun 20.4	1272 Dec 31.8	1354 Jul 12.8	1436 Jan 27.5	1517 Aug 4.4
1193 Jan 26.5	1274 Aug 4.2	1356 Feb 21.7	1437 Aug 27.2	1519 Mar 18.1
1194 Aug 27.0	1276 Mar 17.1	1357 Sep 19.5	1439 Apr 10.9	1520 Oct 13.6
1196 Apr 9.9	1277 Oct 13.6	1359 May 4.1	1440 Nov 7.5	1522 May 26.9
1197 Nov 7.6	1279 May 26.7	1360 Dec 3.2	1442 Jun 18.3	1523 Dec 30.1
1199 Jun 18.1	1280 Dec 29.2	1362 Jul 10.6	1444 Jan 24.9	1525 Aug 2.2
1201 Jan 24.0	1282 Aug 2.0	1364 Feb 19.2	1445 Aug 24.9	1527 Mar 15.7
1202 Aug 24.7	1284 Mar 14.7	1365 Sep 17.1	1447 Apr 8.6	1528 Oct 11.2
1204 Apr 7.5	1285 Oct 11.1	1367 May 1.8	1448 Nov 5.0	1530 May 24.6
1205 Nov 5.0	1287 May 24.5	1368 Nov 30.6	1450 Jun 16.1	1531 Dec 27.5
1207 Jun 15.9	1288 Dec 26.6	1370 Jul 8.4	1452 Jan 22.3	1533 Jul 30.9
1209 Jan 21.4	1290 Jul 30.7	1372 Feb 16.7	1453 Aug 22.6	1535 Mar 13.3
1210 Aug 22.4	1292 Mar 12.3	1373 Sep 14.8	1455 Apr 6.2	1536 Oct 8.7
1212 Apr 5.2	1293 Oct 8.7	1375 Apr 29.5	1456 Nov 2.5	1538 May 22.4
1213 Nov 2.5	1295 May 22.2	1376 Nov 28.0	1458 Jun 13.8	1539 Dec 24.9
1215 Jun 13.7	1296 Dec 24.0	1378 Jul 6.1	1460 Jan 19.8	1541 Jul 28.7
1217 Jan 18.9	1298 Jul 28.5	1380 Feb 14.2	1461 Aug 20.3	1543 Mar 10.9
1218 Aug 20.1	1300 Mar 9.9	1381 Sep 12.4	1463 Apr 3.9	1544 Oct 6.3
1220 Apr 2.8	1301 Oct 6.3	1383 Apr 27.2	1464 Oct 31.0	1546 May 20.1
1221 Oct 31.0	1303 May 20.0	1384 Nov 25.4	1466 Jun 11.6	1547 Dec 22.3
1223 Jun 11.4	1304 Dec 21.4	1386 Jul 3.9	1468 Jan 17.2	1549 Jul 26.4
1225 Jan 16.3	1306 Jul 26.2	1388 Feb 11.7	1469 Aug 18.0	1551 Mar 8.4
1226 Aug 17.8	1308 Mar 7.4	1389 Sep 10.1	1471 Apr 1.5	1552 Oct 3.9
1228 Mar 31.5	1309 Oct 3.8	1391 Apr 24.9	1472 Oct 28.5	1554 May 17.8
1229 Oct 28.5	1311 May 17.7	1392 Nov 22.8	1474 Jun 9.4	1555 Dec 19.7
1231 Jun 9.2	1312 Dec 18.7	1394 Jul 1.7	1476 Jan 14.6	1557 Jul 24.2
1233 Jan 13.7	1314 Jul 24.0	1396 Feb 9.1	1477 Aug 15.7	1559 Mar 6.0
1234 Aug 15.6	1316 Mar 5.0	1397 Sep 7.7	1479 Mar 30.1	1560 Oct 1.5
1236 Mar 29.1	1317 Oct 1.4	1399 Apr 22.6	1480 Oct 26.0	1562 May 15.6
1237 Oct 26.0	1319 May 15.4	1400 Nov 20.3	1482 Jun 7.1	1563 Dec 17.1
1239 Jun 7.0	1320 Dec 16.1	1402 Jun 29.4	1484 Jan 12.0	1565 Jul 21.9
1241 Jan 11.1	1322 Jul 21.8	1404 Feb 6.6	1485 Aug 13.5	1567 Mar 3.5
1242 Aug 13.3	1324 Mar 2.5	1405 Sep 5.4	1487 Mar 27.7	1568 Sep 29.1
1244 Mar 26.7	1325 Sep 29.0	1407 Apr 20.2	1488 Oct 23.5	1570 May 13.3
1245 Oct 23.5	1327 May 13.2	1408 Nov 17.7	1490 Jun 4.9	1571 Dec 14.5
1247 Jun 4.7	1328 Dec 13.6	1410 Jun 27.2	1492 Jan 9.4	1573 Jul 19.7

1575 Mar 1.1	1656 Sep 13.2	1738 Apr 5.4	1819 Oct 9.8	1901 May 1.1
1576 Sep 26.7	1658 Apr 28.0	1739 Nov 2.0	1821 May 23.1	1902 Nov 29.1
1578 May 11.0	1659 Nov 26.1	1741 Jun 13.8	1822 Dec 23.8	1904 Jul 8.3
1579 Dec 11.9	1661 Jul 5.2	1743 Jan 17.7	1824 Jul 29.6	1906 Feb 14.4
1581 Jul 17.5	1663 Feb 11.5	1744 Aug 20.1	1826 Mar 10.5	1907 Sep 15.0
1583 Mar 8.6	1664 Sep 10.9	1746 Apr 3.0	1827 Oct 7.4	1909 Apr 28.7
1584 Oct 4.3	1666 Apr 25.7	1747 Oct 30.6	1829 May 20.9	1910 Nov 26.6
1586 May 18.7	1667 Nov 23.6	1749 Jun 11.5	1830 Dec 21.2	1912 Jul 6.1
1587 Dec 19.3	1669 Jul 2.9	1751 Jan 15.1	1832 Jul 27.4	1914 Feb 11.8
1589 Jul 25.2	1671 Feb 8.9	1752 Aug 17.8	1834 Mar 8.0	1915 Sep 12.7
1591 Mar 6.1	1672 Sep 8.6	1754 Mar 31.6	1835 Oct 5.0	1917 Apr 26.4
1592 Oct 1.9	1674 Apr 23.3	1755 Oct 28.1	1837 May 18.6	1918 Nov 24.0
1594 May 16.5	1675 Nov 21.0	1757 Jun 9.3	1838 Dec 18.6	1920 Jul 3.8
1595 Dec 16.7	1677 Jun 30.7	1759 Jan 12.6	1840 Jul 25.2	1922 Feb 9.3
1597 Jul 23.0	1679 Feb 6.4	1760 Aug 15.6	1842 Mar 5.6	1923 Sep 10.4
1599 Mar 3.6	1680 Sep 6.3	1762 Mar 29.1	1843 Oct 2.7	1925 Apr 24.0
1600 Sep 29.6	1682 Apr 21.0	1763 Oct 25.7	1845 May 16.3	1926 Nov 21.5
1602 May 14.2	1683 Nov 18.5	1765 Jun 7.0	1846 Dec 16.0	1928 Jul 1.6
1603 Dec 14.1	1685 Jun 28.5	1767 Jan 10.0	1848 Jul 22.9	1930 Feb 6.7
1605 Jul 20.8	1687 Feb 3.8	1768 Aug 13.3	1850 Mar 3.1	1931 Sep 8.1
1607 Mar 1.1	1688 Sep 4.0	1770 Mar 26.7	1851 Sep 30.3	1933 Apr 21.7
1608 Sep 27.2	1690 Apr 18.6	1771 Oct 23.2	1853 May 14.0	1934 Nov 19.0
1610 May 11.9	1691 Nov 16.0	1773 Jun 4.8	1854 Dec 13.4	1936 Jun 29.4
1611 Dec 11.5	1693 Jun 26.2	1775 Jan 7.4	1856 Jul 20.7	1938 Feb 4.1
1613 Jul 18.5	1695 Feb 1.2	1776 Aug 11.1	1858 Feb 28.6	1939 Sep 5.9
1615 Feb 26.6	1696 Sep 1.7	1778 Mar 24.3	1859 Sep 28.0	1941 Apr 19.3
1616 Sep 24.9	1698 Apr 16.3	1779 Oct 20.8	1861 May 11.7	1942 Nov 16.5
1618 May 9.6	1699 Nov 13.5	1781 Jun 2.5	1862 Dec 10.9	1944 Jun 27.1
1619 Dec 8.9	1701 Jun 25.0	1783 Jan 4.8	1864 Jul 18.5	1946 Feb 1.6
1621 Jul 16.3	1703 Jan 30.7	1784 Aug 8.8	1866 Feb 26.1	1947 Sep 3.6
1623 Feb 24.1	1704 Aug 31.4	1786 Mar 21.8	1867 Sep 25.6	1949 Apr 16.9
1624 Sep 22.5	1706 Apr 14.9	1787 Oct 18.4	1869 May 9.4	1950 Nov 14.0
1626 May 7.3	1707 Nov 12.0	1789 May 31.2	1870 Dec 8.3	1952 Jun 24.9
1627 Dec 6.4	1709 Jun 22.7	1791 Jan 2.2	1872 Jul 16.2	1954 Jan 30.0
1629 Jul 14.1	1711 Jan 28.1	1792 Aug 6.6	1874 Feb 23.5	1955 Sep 1.3
1631 Feb 21.6	1712 Aug 29.2	1794 Mar 19.4	1875 Sep 23.3	1957 Apr 14.5
1632 Sep 20.2	1714 Apr 12.5	1795 Oct 16.0	1877 May 7.0	1958 Nov 11.5
1634 May 5.0	1715 Nov 9.5	1797 May 29.0	1878 Dec 5.7	1960 Jun 22.7
1635 Dec 3.8	1717 Jun 20.5	1798 Dec 30.6	1880 Jul 14.0	1962 Jan 27.4
1637 Jul 11.8	1719 Jan 25.5	1800 Aug 5.3	1882 Feb 21.0	1963 Aug 30.0
1639 Feb 19.1	1720 Aug 26.9	1802 Mar 17.9	1883 Sep 21.0	1965 Apr 12.2
1640 Sep 17.9	1722 Apr 10.1	1803 Oct 14.6	1885 May 4.7	1966 Nov 9.0
1642 May 2.7	1723 Nov 7.0	1805 May 27.7	1886 Dec 3.2	1968 Jun 20.4
1643 Dec 1.2	1725 Jun 18.3	1806 Dec 29.0	1888 Jul 11.8	1970 Jan 24.8
1645 Jul 9.6	1727 Jan 22.9	1808 Aug 3.1	1890 Feb 18.5	1971 Aug 27.8
1647 Feb 16.6	1728 Aug 24.6	1810 Mar 15.5	1891 Sep 18.7	1973 Apr 9.8
1648 Sep 15.5	1730 Apr 7.7	1811 Oct 12.2	1893 May 2.4	1974 Nov 6.5
1650 Apr 30.3	1731 Nov 4.5	1813 May 25.4	1894 Nov 30.6	1976 Jun 18.2
1651 Nov 28.7	1733 Jun 16.0	1814 Dec 26.4	1896 Jul 9.5	1978 Jan 22.2
1653 Jul 7.4	1735 Jan 20.3	1816 Jul 31.9	1898 Feb 15.9	1979 Aug 25.5
1655 Feb 14.0	1736 Aug 22.3	1818 Mar 13.0	1899 Sep 16.3	1981 Apr 7.4

1982 Nov 4.1	2064 May 24.2	2145 Nov 30.1	2227 Jun 19.1	2308 Dec 27.6
1984 Jun 15.9	2065 Dec 24.7	2147 Jul 10.5	2229 Jan 21.5	2310 Aug 4.0
1986 Jan 19.6	2067 Jul 31.8	2149 Feb 15.3	2230 Aug 25.4	2312 Mar 14.4
1987 Aug 23.2	2069 Mar 11.5	2150 Sep 16.2	2232 Apr 7.0	2313 Oct 11.6
1989 Apr 5.0	2070 Oct 8.5	2152 Apr 29.8	2233 Nov 3.6	2315 May 26.0
1990 Nov 1.6	2072 May 21.9	2153 Nov 27.5	2235 Jun 16.8	2316 Dec 25.0
1992 Jun 13.7	2073 Dec 22.1	2155 Jul 8.2	2237 Jan 18.9	2318 Aug 1.7
1994 Jan 17.1	2075 Jul 29.6	2157 Feb 12.7	2238 Aug 23.1	2320 Mar 11.9
1995 Aug 21.0	2077 Mar 9.0	2158 Sep 13.9	2240 Apr 4.5	2321 Oct 9.2
1997 Apr 2.5	2078 Oct 6.1	2160 Apr 27.4	2241 Nov 1.2	2323 May 23.7
1998 Oct 30.2	2080 May 19.7	2161 Nov 25.0	2243 Jun 14.5	2324 Dec 22.5
2000 Jun 11.4	2081 Dec 19.5	2163 Jul 6.0	2245 Jan 16.4	2326 Jul 30.5
2002 Jan 14.5	2083 Jul 27.3	2165 Feb 10.2	2246 Aug 20.9	2328 Mar 9.4
2003 Aug 18.7	2085 Mar 6.5	2166 Sep 11.6	2248 Apr 2.1	2329 Oct 6.9
2005 Mar 31.1	2086 Oct 3.8	2168 Apr 25.0	2249 Oct 29.8	2331 May 21.4
2006 Oct 27.7	2088 May 17.3	2169 Nov 22.5	2251 Jun 12.3	2332 Dec 19.9
2008 Jun 9.2	2089 Dec 17.0	2171 Jul 3.8	2253 Jan 13.8	2334 Jul 28.3
2010 Jan 11.9	2091 Jul 25.1	2173 Feb 7.6	2254 Aug 18.6	2336 Mar 6.9
2011 Aug 16.5	2093 Mar 4.0	2174 Sep 9.3	2256 Mar 30.7	2337 Oct 4.5
2013 Mar 28.7	2094 Oct 1.4	2176 Apr 22.7	2257 Oct 27.3	2339 May 19.1
2014 Oct 25.3	2096 May 15.0	2177 Nov 20.0	2259 Jun 10.0	2340 Dec 17.3
2016 Jun 6.9	2097 Dec 14.4	2179 Jul 1.5	2261 Jan 11.2	2342 Jul 26.0
2018 Jan 9.3	2099 Jul 22.9	2181 Feb 5.0	2262 Aug 16.4	2344 Mar 4.4
2019 Aug 14.2	2101 Mar 2.5	2182 Sep 7.0	2264 Mar 28.2	2345 Oct 2.2
2021 Mar 26.3	2102 Sep 30.1	2184 Apr 20.3	2265 Oct 24.9	2347 May 16.8
2022 Oct 22.9	2104 May 13.7	2185 Nov 17.5	2267 Jun 7.7	2348 Dec 14.8
2024 Jun 4.6	2105 Dec 12.8	2187 Jun 29.3	2269 Jan 8.6	2350 Jul 23.8
2026 Jan 6.7	2107 Jul 21.6	2189 Feb 2.5	2270 Aug 14.1	2352 Mar 1.9
2027 Aug 12.0	2109 Feb 28.0	2190 Sep 4.7	2272 Mar 25.8	2353 Sep 29.9
2029 Mar 23.8	2110 Sep 27.7	2192 Apr 17.9	2273 Oct 22.5	2355 May 14.5
2030 Oct 20.4	2112 May 11.4	2193 Nov 15.0	2275 Jun 5.5	2356 Dec 12.2
2032 Jun 2.4	2113 Dec 10.3	2195 Jun 27.0	2277 Jan 6.0	2358 Jul 21.6
2034 Jan 4.1	2115 Jul 19.4	2197 Jan 30.9	2278 Aug 11.9	2360 Feb 28.4
2035 Aug 9.8	2117 Feb 25.4	2198 Sep 2.4	2280 Mar 23.3	2361 Sep 27.5
2037 Mar 21.4	2118 Sep 25.4	2200 Apr 16.6	2281 Oct 20.1	2363 May 12.2
2038 Oct 18.0	2120 May 9.1	2201 Nov 13.5	2283 Jun 3.2	2364 Dec 9.6
2040 May 31.1	2121 Dec 7.7	2203 Jun 25.8	2285 Jan 3.4	2366 Jul 19.3
2042 Jan 1.5	2123 Jul 17.2	2205 Jan 29.3	2286 Aug 9.7	2368 Feb 25.8
2043 Aug 7.5	2125 Feb 22.9	2206 Sep 1.2	2288 Mar 20.9	2369 Sep 25.2
2045 Mar 18.9	2126 Sep 23.1	2208 Apr 14.2	2289 Oct 17.7	2371 May 9.8
2046 Oct 15.6	2128 May 6.8	2209 Nov 11.0	2291 May 31.9	2372 Dec 7.1
2048 May 28.8	2129 Dec 5.1	2211 Jun 23.6	2292 Dec 31.8	2374 Jul 17.1
2049 Dec 29.9	2131 Jul 14.9	2213 Jan 26.7	2294 Aug 7.4	2376 Feb 23.3
2051 Aug 5.3	2133 Feb 20.4	2214 Aug 29.9	2296 Mar 18.4	2377 Sep 22.9
2053 Mar 16.4	2134 Sep 20.8	2216 Apr 11.8	2297 Oct 15.3	2379 May 7.5
2054 Oct 13.3	2136 May 4.4	2217 Nov 8.6	2299 May 29.6	2380 Dec 4.6
2056 May 26.5	2137 Dec 2.6	2219 Jun 21.3	2300 Dec 30.2	2382 Jul 14.9
2057 Dec 27.3	2139 Jul 12.7	2221 Jan 24.1	2302 Aug 6.2	2384 Feb 20.7
2059 Aug 3.0	2141 Feb 17.8	2222 Aug 27.6	2304 Mar 16.9	2385 Sep 20.6
2061 Mar 14.0	2142 Sep 18.5	2224 Apr 9.4	2305 Oct 14.0	2387 May 5.2
2062 Oct 10.9	2144 May 2.1	2225 Nov 6.1	2307 May 28.3	2388 Dec 2.0

2390	Jul	12.6	2412	Nov	24.5	2435	Apr	21.0	2457	Aug	31.1	2480	Jan	20.8
2392	Feb	18.2	2414	Jul	5.9	2436	Nov	17.0	2459	Apr	13.8	2481	Aug	24.3
2393	Sep	18.3	2416	Feb	10.5	2438	Jun	29.2	2460	Nov	9.6	2483	Apr	6.5
2395	May	2.8	2417	Sep	11.4	2440	Feb	2.8	2462	Jun	22.4	2484	Nov	2.2
2396	Nov	29.5	2419	Apr	25.7	2441	Sep	4.6	2464	Jan	26.0	2486	Jun	15.7
2398	Jul	10.4	2420	Nov	22.0	2443	Apr	18.6	2465	Aug	28.8	2488	Jan	18.2
2400	Feb	15.6	2422	Jul	3.7	2444	Nov	14.5	2467	Apr	11.4	2489	Aug	22.0
2401	Sep	16.0	2424	Feb	7.9	2446	Jun	26.9	2468	Nov	7.1	2491	Apr	4.1
2403	Apr	30.4	2425	Sep	9.1	2448	Jan	31.2	2470	Jun	20.2	2492	Oct	30.8
2404	Nov	27.0	2427	Apr	23.3	2449	Sep	2.3	2472	Jan	23.4	2494	Jun	13.4
2406	Jul	8.1	2428	Nov	19.5	2451	Apr	16.2	2473	Aug	26.5	2496	Jan	15.7
2408	Feb	13.1	2430	Jul	1.4	2452	Nov	12.0	2475	Apr	8.9	2497	Aug	19.8
2409	Sep	13.7	2432	Feb	5.4	2454	Jun	24.7	2476	Nov	4.7	2499	Apr	1.6
2411	Apr	28.1	2433	Sep	6.9	2456	Jan	28.6	2478	Jun	17.9	2500	Oct	29.4

XI
OPPOSITIONS OF JUPITER, 0 – 2500

0	Feb	26.3	48	Mar	14.5	96	Mar	31.9	144	Apr	18.7	192	May	6.8
1	Mar	28.3	49	Apr	15.1	97	May	3.3	145	May	22.0	193	Jun	10.3
2	Apr	29.5	50	May	18.2	98	Jun	6.4	146	Jun	26.3	194	Jul	16.6
3	Jun	2.6	51	Jun	22.3	99	Jul	12.5	147	Aug	2.3	195	Aug	23.1
4	Jul	7.6	52	Jul	28.1	100	Aug	17.8	148	Sep	7.8	196	Sep	28.3
5	Aug	13.9	53	Sep	3.6	101	Sep	24.2	149	Oct	14.5	197	Nov	3.1
6	Sep	20.4	54	Oct	10.5	102	Oct	30.2	150	Nov	18.4	198	Dec	6.8
7	Oct	26.5	55	Nov	14.7	103	Dec	3.2	151	Dec	21.3	200	Jan	7.8
8	Nov	28.7	56	Dec	16.8	105	Jan	3.4	153	Jan	20.7	201	Feb	6.7
9	Dec	30.9	58	Jan	17.3	106	Feb	3.4	154	Feb	20.4	202	Mar	9.4
11	Jan	30.9	59	Feb	17.0	107	Mar	6.1	155	Mar	23.3	203	Apr	9.8
12	Mar	1.6	60	Mar	18.8	108	Apr	5.3	156	Apr	23.2	204	May	11.6
13	Apr	1.7	61	Apr	19.5	109	May	7.9	157	May	26.8	205	Jun	15.3
14	May	4.2	62	May	22.9	110	Jun	11.3	158	Jul	1.2	206	Jul	21.8
15	Jun	7.5	63	Jun	27.3	111	Jul	17.6	159	Aug	7.3	207	Aug	28.2
16	Jul	12.8	64	Aug	2.4	112	Aug	23.0	160	Sep	12.7	208	Oct	3.3
17	Aug	19.2	65	Sep	8.9	113	Sep	29.3	161	Oct	19.3	209	Nov	7.9
18	Sep	25.5	66	Oct	15.6	114	Nov	4.1	162	Nov	22.9	210	Dec	11.3
19	Oct	31.3	67	Nov	19.4	115	Dec	7.8	163	Dec	25.7	212	Jan	12.2
20	Dec	3.2	68	Dec	21.2	117	Jan	7.7	165	Jan	25.0	213	Feb	11.0
22	Jan	4.2	70	Jan	21.6	118	Feb	7.6	166	Feb	24.7	214	Mar	13.8
23	Feb	4.2	71	Feb	21.2	119	Mar	10.3	167	Mar	27.7	215	Apr	14.3
24	Mar	5.8	72	Mar	23.1	120	Apr	9.6	168	Apr	27.7	216	May	16.3
25	Apr	6.2	73	Apr	24.1	121	May	12.5	169	May	31.5	217	Jun	20.2
26	May	8.9	74	May	27.8	122	Jun	16.2	170	Jul	6.3	218	Jul	26.8
27	Jun	12.5	75	Jul	2.5	123	Jul	22.8	171	Aug	12.5	219	Sep	2.3
28	Jul	17.9	76	Aug	7.6	124	Aug	28.4	172	Sep	17.9	220	Oct	8.2
29	Aug	24.4	77	Sep	14.1	125	Oct	4.5	173	Oct	24.3	221	Nov	12.5
30	Sep	30.5	78	Oct	20.5	126	Nov	9.0	174	Nov	27.6	222	Dec	15.8
31	Nov	5.1	79	Nov	24.0	127	Dec	12.4	175	Dec	30.1	224	Jan	16.5
32	Dec	7.7	80	Dec	25.6	129	Jan	12.1	177	Jan	29.2	225	Feb	15.3
34	Jan	8.6	82	Jan	25.8	130	Feb	11.9	178	Feb	28.8	226	Mar	18.1
35	Feb	8.5	83	Feb	25.5	131	Mar	14.6	179	Mar	31.9	227	Apr	18.7
36	Mar	10.2	84	Mar	27.5	132	Apr	14.1	180	May	2.2	228	May	21.0
37	Apr	10.6	85	Apr	28.7	133	May	17.2	181	Jun	5.3	229	Jun	25.1
38	May	13.5	86	Jun	1.7	134	Jun	21.3	182	Jul	11.4	230	Jul	31.9
39	Jun	17.3	87	Jul	7.6	135	Jul	28.1	183	Aug	17.8	231	Sep	7.5
40	Jul	22.9	88	Aug	12.8	136	Sep	2.6	184	Sep	23.2	232	Oct	13.3
41	Aug	29.4	89	Sep	19.2	137	Oct	9.5	185	Oct	29.3	233	Nov	17.3
42	Oct	5.4	90	Oct	25.4	138	Nov	13.7	186	Dec	2.3	234	Dec	20.3
43	Nov	9.9	91	Nov	28.7	139	Dec	16.8	188	Jan	3.5	236	Jan	20.8
44	Dec	12.3	92	Dec	30.0	141	Jan	16.3	189	Feb	2.5	237	Feb	19.5
46	Jan	13.0	94	Jan	30.1	142	Feb	16.1	190	Mar	5.1	238	Mar	22.3
47	Feb	12.8	95	Mar	1.8	143	Mar	18.9	191	Apr	5.3	239	Apr	23.1

240	May	25.6	296	Feb	11.1	351	Oct	28.2	407	Jun	29.2
241	Jun	30.1	297	Mar	12.8	352	Nov	30.4	408	Aug	4.2
242	Aug	6.2	298	Apr	13.2	354	Jan	1.8	409	Sep	10.7
243	Sep	12.8	299	May	16.1	355	Feb	1.8	410	Oct	17.4
244	Oct	18.4	300	Jun	19.0	356	Mar	3.4	411	Nov	21.3
245	Nov	22.1	301	Jul	25.7	357	Apr	3.5	412	Dec	23.2
246	Dec	24.8	302	Sep	1.3	358	May	5.9	414	Jan	23.5
248	Jan	25.1	303	Oct	8.3	359	Jun	9.2	415	Feb	23.2
249	Feb	23.7	304	Nov	11.7	360	Jul	14.3	416	Mar	25.1
250	Mar	26.6	305	Dec	15.0	361	Aug	20.8	417	Apr	25.9
251	Apr	27.7	307	Jan	15.7	362	Sep	27.2	418	May	29.5
252	May	30.5	308	Feb	15.4	363	Nov	2.2	419	Jul	4.2
253	Jul	5.2	309	Mar	17.1	364	Dec	5.1	420	Aug	9.3
254	Aug	11.5	310	Apr	17.7	366	Jan	6.2	421	Sep	16.0
255	Sep	18.0	311	May	20.9	367	Feb	6.1	422	Oct	22.5
256	Oct	23.3	312	Jun	24.1	368	Mar	7.8	423	Nov	26.1
257	Nov	26.7	313	Jul	31.0	369	Apr	8.0	424	Dec	27.7
258	Dec	29.2	314	Sep	6.6	370	May	10.6	426	Jan	27.9
260	Jan	29.3	315	Oct	13.4	371	Jun	14.2	427	Feb	27.5
261	Feb	28.0	316	Nov	16.5	372	Jul	19.6	428	Mar	29.5
262	Mar	31.0	317	Dec	19.5	373	Aug	26.1	429	Apr	30.6
263	May	2.3	319	Jan	19.9	374	Oct	2.4	430	Jun	3.5
264	Jun	4.4	320	Feb	19.7	375	Nov	7.1	431	Jul	9.4
265	Jul	10.4	321	Mar	21.5	376	Dec	9.7	432	Aug	14.7
266	Aug	16.7	322	Apr	22.3	378	Jan	10.6	433	Sep	21.2
267	Sep	23.1	323	May	25.8	379	Feb	10.4	434	Oct	27.5
268	Oct	28.2	324	Jun	29.2	380	Mar	12.1	435	Nov	30.8
269	Dec	1.3	325	Aug	5.2	381	Apr	12.5	437	Jan	1.1
271	Jan	2.6	326	Sep	11.8	382	May	15.4	438	Feb	1.2
272	Feb	2.6	327	Oct	18.4	383	Jun	19.2	439	Mar	3.8
273	Mar	4.3	328	Nov	21.1	384	Jul	24.9	440	Apr	2.9
274	Apr	4.5	329	Dec	23.9	385	Aug	31.4	441	May	5.2
275	May	7.0	331	Jan	24.3	386	Oct	7.5	442	Jun	8.4
276	Jun	9.3	332	Feb	24.0	387	Nov	11.9	443	Jul	14.6
277	Jul	15.4	333	Mar	25.9	388	Dec	14.2	444	Aug	20.0
278	Aug	21.8	334	Apr	26.9	390	Jan	14.9	445	Sep	26.4
279	Sep	28.0	335	May	30.6	391	Feb	14.7	446	Nov	1.4
280	Nov	1.9	336	Jul	4.2	392	Mar	16.4	447	Dec	5.4
281	Dec	5.8	337	Aug	10.3	393	Apr	17.0	449	Jan	5.5
283	Jan	6.9	338	Sep	16.8	394	May	20.2	450	Feb	5.5
284	Feb	6.9	339	Oct	23.2	395	Jun	24.2	451	Mar	8.1
285	Mar	8.6	340	Nov	25.7	396	Jul	30.0	452	Apr	7.4
286	Apr	8.9	341	Dec	28.3	397	Sep	5.5	453	May	10.0
287	May	11.6	343	Jan	28.6	398	Oct	12.4	454	Jun	13.4
288	Jun	14.2	344	Feb	28.3	399	Nov	16.5	455	Jul	19.8
289	Jul	20.5	345	Mar	30.3	400	Dec	18.7	456	Aug	25.2
290	Aug	27.0	346	May	1.5	402	Jan	19.2	457	Oct	1.4
291	Oct	3.2	347	Jun	4.4	403	Feb	19.0	458	Nov	6.2
292	Nov	6.8	348	Jul	9.3	404	Mar	20.8	459	Dec	9.9
293	Dec	10.4	349	Aug	15.5	405	Apr	21.6	461	Jan	9.9
295	Jan	11.3	350	Sep	21.9	406	May	24.9	462	Feb	9.8

463	Mar	12.5
464	Apr	11.9
465	May	14.7
466	Jun	18.4
467	Jul	24.9
468	Aug	30.4
469	Oct	6.5
470	Nov	11.1
471	Dec	14.5
473	Jan	14.3
474	Feb	14.0
475	Mar	16.7
476	Apr	16.2
477	May	19.2
478	Jun	23.2
479	Jul	30.0
480	Sep	4.7
481	Oct	11.7
482	Nov	16.0
483	Dec	19.2
485	Jan	18.7
486	Feb	18.4
487	Mar	21.1
488	Apr	20.7
489	May	24.0
490	Jun	28.3
491	Aug	4.4
492	Sep	10.0
493	Oct	16.8
494	Nov	20.8
495	Dec	23.7
497	Jan	23.0
498	Feb	22.6
499	Mar	25.5
500	Apr	25.3
501	May	28.9
502	Jul	3.5
503	Aug	9.7
504	Sep	15.3
505	Oct	21.8
506	Nov	25.5
507	Dec	28.1
509	Jan	27.3
510	Feb	26.9
511	Mar	29.9
512	Apr	30.0
513	Jun	2.9
514	Jul	8.7
515	Aug	14.9
516	Sep	20.4
517	Oct	26.7

518	Nov	30.1	574	Aug	3.7	630	Apr	11.7
520	Jan	1.5	575	Sep	10.2	631	May	14.3
521	Jan	31.6	576	Oct	16.1	632	Jun	17.0
522	Mar	3.3	577	Nov	20.1	633	Jul	23.5
523	Apr	3.3	578	Dec	23.0	634	Aug	30.0
524	May	4.6	580	Jan	23.5	635	Oct	6.2
525	Jun	7.7	581	Feb	22.1	636	Nov	9.9
526	Jul	13.8	582	Mar	25.0	637	Dec	13.4
527	Aug	20.1	583	Apr	25.8	639	Jan	14.3
528	Sep	25.6	584	May	28.3	640	Feb	14.1
529	Oct	31.7	585	Jul	2.8	641	Mar	15.7
530	Dec	4.8	586	Aug	8.9	642	Apr	16.2
532	Jan	6.0	587	Sep	15.5	643	May	19.1
533	Feb	4.9	588	Oct	21.1	644	Jun	22.0
534	Mar	7.5	589	Nov	24.8	645	Jul	28.7
535	Apr	7.7	590	Dec	27.6	646	Sep	4.3
536	May	9.1	592	Jan	27.8	647	Oct	11.3
537	Jun	12.5	593	Feb	26.4	648	Nov	14.7
538	Jul	18.8	594	Mar	29.3	649	Dec	18.0
539	Aug	25.3	595	Apr	30.2	651	Jan	18.7
540	Sep	30.7	596	Jun	2.0	652	Feb	18.4
541	Nov	5.7	597	Jul	7.7	653	Mar	20.1
542	Dec	9.5	598	Aug	14.0	654	Apr	20.6
544	Jan	10.5	599	Sep	20.6	655	May	23.7
545	Feb	9.3	600	Oct	26.1	656	Jun	26.8
546	Mar	11.9	601	Nov	29.6	657	Aug	2.7
547	Apr	12.1	603	Jan	1.1	658	Sep	9.4
548	May	13.8	604	Feb	1.2	659	Oct	16.4
549	Jun	17.5	605	Mar	2.8	660	Nov	19.6
550	Jul	24.1	606	Apr	2.7	661	Dec	22.7
551	Aug	30.7	607	May	4.9	663	Jan	23.1
552	Oct	6.0	608	Jun	6.9	664	Feb	22.7
553	Nov	10.6	609	Jul	12.9	665	Mar	24.4
554	Dec	14.1	610	Aug	19.4	666	Apr	25.1
556	Jan	14.8	611	Sep	26.0	667	May	28.5
557	Feb	13.6	612	Oct	31.2	668	Jul	1.9
558	Mar	16.2	613	Dec	4.3	669	Aug	8.1
559	Apr	16.7	615	Jan	5.5	670	Sep	14.8
560	May	18.7	616	Feb	5.5	671	Oct	21.6
561	Jun	22.6	617	Mar	7.1	672	Nov	24.4
562	Jul	29.4	618	Apr	7.1	673	Dec	27.2
563	Sep	5.1	619	May	9.6	675	Jan	27.4
564	Oct	11.1	620	Jun	11.9	676	Feb	27.0
565	Nov	15.4	621	Jul	18.2	677	Mar	28.8
566	Dec	18.6	622	Aug	24.8	678	Apr	29.7
568	Jan	19.1	623	Oct	1.2	679	Jun	2.4
569	Feb	17.8	624	Nov	5.1	680	Jul	7.1
570	Mar	20.6	625	Dec	8.9	681	Aug	13.5
571	Apr	21.2	627	Jan	9.9	682	Sep	20.1
572	May	23.5	628	Feb	9.8	683	Oct	26.6
573	Jun	27.7	629	Mar	11.4	684	Nov	29.1

685	Dec	31.6	741	Sep	8.8
687	Jan	31.7	742	Oct	15.8
688	Mar	2.3	743	Nov	20.0
689	Apr	2.2	744	Dec	22.1
690	May	4.4	746	Jan	22.5
691	Jun	7.4	747	Feb	22.2
692	Jul	12.4	748	Mar	23.9
693	Aug	18.7	749	Apr	24.6
694	Sep	25.2	750	May	28.0
695	Oct	31.5	751	Jul	2.3
696	Dec	3.7	752	Aug	7.4
698	Jan	5.0	753	Sep	14.1
699	Feb	5.0	754	Oct	20.8
700	Mar	6.6	755	Nov	24.7
701	Apr	6.7	756	Dec	26.5
702	May	9.0	758	Jan	26.8
703	Jun	12.3	759	Feb	26.5
704	Jul	17.5	760	Mar	28.3
705	Aug	24.0	761	Apr	29.2
706	Sep	30.4	762	Jun	1.8
707	Nov	5.4	763	Jul	7.4
708	Dec	8.3	764	Aug	12.6
710	Jan	9.4	765	Sep	19.2
711	Feb	9.3	766	Oct	25.8
712	Mar	10.9	767	Nov	29.4
713	Apr	11.0	768	Dec	31.0
714	May	13.6	770	Jan	31.2
715	Jun	17.0	771	Mar	2.8
716	Jul	22.5	772	Apr	1.6
717	Aug	29.1	773	May	3.7
718	Oct	5.5	774	Jun	6.5
719	Nov	10.3	775	Jul	12.3
720	Dec	13.0	776	Aug	17.7
722	Jan	13.9	777	Sep	24.3
723	Feb	13.7	778	Oct	30.7
724	Mar	15.2	779	Dec	4.1
725	Apr	15.5	781	Jan	4.5
726	May	18.3	782	Feb	4.6
727	Jun	22.1	783	Mar	7.1
728	Jul	27.7	784	Apr	6.1
729	Sep	3.5	785	May	8.3
730	Oct	10.7	786	Jun	11.4
731	Nov	15.2	787	Jul	17.5
732	Dec	17.6	788	Aug	23.0
734	Jan	18.2	789	Sep	29.6
735	Feb	17.9	790	Nov	4.8
736	Mar	19.5	791	Dec	8.8
737	Apr	20.0	793	Jan	8.9
738	May	23.1	794	Feb	8.8
739	Jun	27.2	795	Mar	11.3
740	Aug	2.1	796	Apr	10.4

Year			Year			Year			Year			Year		
797	May	12.9	853	Jan	30.6	908	Oct	14.0	964	Jun	14.5	1020	Feb	29.7
798	Jun	16.4	854	Mar	2.2	909	Nov	18.4	965	Jul	20.7	1021	Mar	31.5
799	Jul	22.8	855	Apr	2.0	910	Dec	21.7	966	Aug	27.2	1022	May	2.3
800	Aug	28.4	856	May	2.9	912	Jan	22.3	967	Oct	3.7	1023	Jun	4.8
801	Oct	4.8	857	Jun	5.7	913	Feb	20.9	968	Nov	7.8	1024	Jul	9.4
802	Nov	9.7	858	Jul	11.6	914	Mar	23.6	969	Dec	11.8	1025	Aug	15.6
803	Dec	13.4	859	Aug	17.9	915	Apr	24.1	971	Jan	12.8	1026	Sep	22.3
805	Jan	13.3	860	Sep	23.6	916	May	26.2	972	Feb	12.7	1027	Oct	28.9
806	Feb	13.1	861	Oct	30.1	917	Jun	30.4	973	Mar	14.2	1028	Dec	1.6
807	Mar	15.6	862	Dec	3.5	918	Aug	6.4	974	Apr	14.4	1030	Jan	3.2
808	Apr	14.9	864	Jan	4.9	919	Sep	13.2	975	May	16.9	1031	Feb	3.3
809	May	17.7	865	Feb	3.9	920	Oct	19.1	976	Jun	19.4	1032	Mar	4.9
810	Jun	21.4	866	Mar	6.4	921	Nov	23.2	977	Jul	25.9	1033	Apr	4.7
811	Jul	28.1	867	Apr	6.4	922	Dec	26.2	978	Sep	1.6	1034	May	6.7
812	Sep	2.7	868	May	7.6	924	Jan	26.6	979	Oct	9.0	1035	Jun	9.6
813	Oct	9.9	869	Jun	10.7	925	Feb	25.2	980	Nov	12.7	1036	Jul	14.5
814	Nov	14.4	870	Jul	16.8	926	Mar	27.9	981	Dec	16.4	1037	Aug	20.9
815	Dec	17.9	871	Aug	23.2	927	Apr	28.6	983	Jan	17.2	1038	Sep	27.6
817	Jan	17.6	872	Sep	28.8	928	May	31.1	984	Feb	16.9	1039	Nov	3.0
818	Feb	17.3	873	Nov	3.9	929	Jul	5.6	985	Mar	18.5	1040	Dec	6.3
819	Mar	20.0	874	Dec	8.0	930	Aug	11.7	986	Apr	18.8	1042	Jan	7.6
820	Apr	19.4	876	Jan	9.2	931	Sep	18.4	987	May	21.6	1043	Feb	7.6
821	May	22.4	877	Feb	8.1	932	Oct	24.1	988	Jun	24.5	1044	Mar	9.1
822	Jun	26.4	878	Mar	10.7	933	Nov	27.9	989	Jul	31.2	1045	Apr	9.1
823	Aug	2.3	879	Apr	10.8	934	Dec	30.6	990	Sep	6.9	1046	May	11.4
824	Sep	7.9	880	May	12.3	936	Jan	30.8	991	Oct	14.0	1047	Jun	14.5
825	Oct	15.0	881	Jun	15.6	937	Mar	1.4	992	Nov	17.5	1048	Jul	19.7
826	Nov	19.2	882	Jul	21.9	938	Apr	1.3	993	Dec	20.8	1049	Aug	26.2
827	Dec	22.4	883	Aug	28.5	939	May	3.2	995	Jan	21.4	1050	Oct	2.7
829	Jan	22.0	884	Oct	3.9	940	Jun	4.9	996	Feb	21.1	1051	Nov	7.8
830	Feb	21.6	885	Nov	8.8	941	Jul	10.6	997	Mar	22.8	1052	Dec	10.8
831	Mar	24.3	886	Dec	12.6	942	Aug	16.9	998	Apr	23.3	1054	Jan	11.9
832	Apr	23.9	888	Jan	13.6	943	Sep	23.5	999	May	26.4	1055	Feb	11.8
833	May	27.1	889	Feb	12.4	944	Oct	29.0	1000	Jun	29.5	1056	Mar	13.4
834	Jul	1.3	890	Mar	15.0	945	Dec	2.5	1001	Aug	5.4	1057	Apr	13.5
835	Aug	7.3	891	Apr	15.2	947	Jan	4.0	1002	Sep	12.0	1058	May	16.0
836	Sep	13.0	892	May	16.8	948	Feb	4.2	1003	Oct	19.0	1059	Jun	19.5
837	Oct	19.9	893	Jun	20.4	949	Mar	5.7	1004	Nov	22.2	1060	Jul	24.9
838	Nov	24.0	894	Jul	26.9	950	Apr	5.7	1005	Dec	25.3	1061	Aug	31.4
839	Dec	27.0	895	Sep	2.5	951	May	7.7	1007	Jan	25.8	1062	Oct	7.8
841	Jan	26.4	896	Oct	8.8	952	Jun	9.6	1008	Feb	25.4	1063	Nov	12.6
842	Feb	26.0	897	Nov	13.6	953	Jul	15.6	1009	Mar	27.2	1064	Dec	15.3
843	Mar	28.7	898	Dec	17.2	954	Aug	21.9	1010	Apr	27.8	1066	Jan	16.3
844	Apr	28.4	900	Jan	18.0	955	Sep	28.5	1011	May	31.1	1067	Feb	16.1
845	May	31.9	901	Feb	16.8	956	Nov	2.8	1012	Jul	4.4	1068	Mar	17.7
846	Jul	6.4	902	Mar	19.3	957	Dec	7.1	1013	Aug	10.4	1069	Apr	18.0
847	Aug	12.6	903	Apr	19.7	959	Jan	8.5	1014	Sep	17.0	1070	May	20.7
848	Sep	18.3	904	May	21.5	960	Feb	8.5	1015	Oct	23.9	1071	Jun	24.3
849	Oct	25.0	905	Jun	25.4	961	Mar	10.0	1016	Nov	26.8	1072	Jul	29.9
850	Nov	28.8	906	Aug	1.1	962	Apr	10.0	1017	Dec	29.7	1073	Sep	5.5
851	Dec	31.4	907	Sep	7.8	963	May	12.3	1019	Jan	30.1	1074	Oct	12.7

1075	Nov	17.3	1131	Jul	19.5	1187	Mar	30.7	1242	Dec	19.0	1298	Aug	23.9
1076	Dec	19.8	1132	Aug	24.9	1188	Apr	30.4	1244	Jan	19.8	1299	Sep	30.5
1078	Jan	20.6	1133	Oct	1.4	1189	Jun	2.8	1245	Feb	18.6	1300	Nov	4.8
1079	Feb	20.4	1134	Nov	6.6	1190	Jul	8.2	1246	Mar	21.2	1301	Dec	9.1
1080	Mar	22.0	1135	Dec	10.7	1191	Aug	14.3	1247	Apr	21.6	1303	Jan	10.4
1081	Apr	22.4	1137	Jan	11.0	1192	Sep	19.9	1248	May	23.4	1304	Feb	10.3
1082	May	25.3	1138	Feb	11.0	1193	Oct	26.7	1249	Jun	27.2	1305	Mar	11.9
1083	Jun	29.3	1139	Mar	13.6	1194	Nov	30.5	1250	Aug	2.9	1306	Apr	11.9
1084	Aug	4.0	1140	Apr	12.6	1196	Jan	2.3	1251	Sep	9.5	1307	May	14.2
1085	Sep	10.7	1141	May	15.0	1197	Feb	1.6	1252	Oct	15.6	1308	Jun	16.4
1086	Oct	17.8	1142	Jun	18.3	1198	Mar	4.2	1253	Nov	20.1	1309	Jul	22.6
1087	Nov	22.1	1143	Jul	24.6	1199	Apr	4.0	1254	Dec	23.5	1310	Aug	29.0
1088	Dec	24.3	1144	Aug	30.1	1200	May	4.9	1256	Jan	24.1	1311	Oct	5.5
1090	Jan	24.9	1145	Oct	6.5	1201	Jun	7.6	1257	Feb	22.9	1312	Nov	9.6
1091	Feb	24.5	1146	Nov	11.5	1202	Jul	13.3	1258	Mar	25.5	1313	Dec	13.6
1092	Mar	26.2	1147	Dec	15.4	1203	Aug	19.5	1259	Apr	26.0	1315	Jan	14.7
1093	Apr	26.7	1149	Jan	15.3	1204	Sep	25.1	1260	May	28.1	1316	Feb	14.6
1094	May	30.0	1150	Feb	15.2	1205	Oct	31.7	1261	Jul	2.2	1317	Mar	16.2
1095	Jul	4.2	1151	Mar	17.7	1206	Dec	5.2	1262	Aug	8.0	1318	Apr	16.3
1096	Aug	9.3	1152	Apr	16.9	1208	Jan	6.7	1263	Sep	14.7	1319	May	18.9
1097	Sep	16.0	1153	May	19.5	1209	Feb	5.8	1264	Oct	20.7	1320	Jun	21.3
1098	Oct	22.9	1154	Jun	23.2	1210	Mar	8.3	1265	Nov	24.9	1321	Jul	27.7
1099	Nov	27.0	1155	Jul	29.7	1211	Apr	8.2	1266	Dec	28.0	1322	Sep	3.3
1100	Dec	28.9	1156	Sep	4.4	1212	May	9.3	1268	Jan	28.5	1323	Oct	10.6
1102	Jan	29.2	1157	Oct	11.7	1213	Jun	12.3	1269	Feb	27.1	1324	Nov	14.5
1103	Feb	28.8	1158	Nov	16.4	1214	Jul	18.3	1270	Mar	29.7	1325	Dec	18.2
1104	Mar	30.5	1159	Dec	20.0	1215	Aug	24.8	1271	Apr	30.4	1327	Jan	19.1
1105	May	1.3	1161	Jan	19.7	1216	Sep	30.4	1272	Jun	1.7	1328	Feb	18.9
1106	Jun	3.8	1162	Feb	19.4	1217	Nov	5.7	1273	Jul	7.1	1329	Mar	20.4
1107	Jul	9.4	1163	Mar	22.0	1218	Dec	9.9	1274	Aug	13.2	1330	Apr	20.6
1108	Aug	14.6	1164	Apr	21.4	1220	Jan	11.2	1275	Sep	20.0	1331	May	23.4
1109	Sep	21.3	1165	May	24.3	1221	Feb	10.1	1276	Oct	25.8	1332	Jun	26.1
1110	Oct	27.9	1166	Jun	28.2	1222	Mar	12.6	1277	Nov	29.8	1333	Aug	1.8
1111	Dec	1.6	1167	Aug	4.0	1223	Apr	12.6	1279	Jan	1.6	1334	Sep	8.5
1113	Jan	2.2	1168	Sep	9.7	1224	May	14.0	1280	Feb	1.8	1335	Oct	15.8
1114	Feb	2.4	1169	Oct	16.8	1225	Jun	17.3	1281	Mar	3.4	1336	Nov	19.4
1115	Mar	5.0	1170	Nov	21.2	1226	Jul	23.5	1282	Apr	3.1	1337	Dec	22.9
1116	Apr	3.8	1171	Dec	24.4	1227	Aug	30.1	1283	May	4.9	1339	Jan	23.5
1117	May	5.9	1173	Jan	24.0	1228	Oct	5.6	1284	Jun	6.6	1340	Feb	23.2
1118	Jun	8.7	1174	Feb	23.6	1229	Nov	10.6	1285	Jul	12.3	1341	Mar	24.7
1119	Jul	14.5	1175	Mar	26.3	1230	Dec	14.5	1286	Aug	18.6	1342	Apr	25.1
1120	Aug	19.8	1176	Apr	25.9	1232	Jan	15.5	1287	Sep	25.3	1343	May	28.1
1121	Sep	26.4	1177	May	29.1	1233	Feb	14.3	1288	Oct	30.9	1344	Jul	1.2
1122	Nov	1.8	1178	Jul	3.3	1234	Mar	16.9	1289	Dec	4.4	1345	Aug	7.1
1123	Dec	6.2	1179	Aug	9.2	1235	Apr	17.1	1291	Jan	6.0	1346	Sep	13.9
1125	Jan	6.6	1180	Sep	14.9	1236	May	18.7	1292	Feb	6.1	1347	Oct	21.0
1126	Feb	6.7	1181	Oct	21.8	1237	Jun	22.3	1293	Mar	7.6	1348	Nov	24.2
1127	Mar	9.3	1182	Nov	25.9	1238	Jul	28.8	1294	Apr	7.5	1349	Dec	27.4
1128	Apr	8.3	1183	Dec	28.9	1239	Sep	4.4	1295	May	9.6	1351	Jan	27.8
1129	May	10.5	1185	Jan	28.3	1240	Oct	10.7	1296	Jun	11.5	1352	Feb	27.4
1130	Jun	13.5	1186	Feb	27.9	1241	Nov	15.4	1297	Jul	17.5	1353	Mar	29.1

1354 Apr 29.7	1410 Jan 18.5	1465 Sep 28.8	1521 May 31.4	1577 Feb 17.3
1355 Jun 2.0	1411 Feb 18.3	1466 Nov 4.3	1522 Jul 5.6	1578 Mar 19.8
1356 Jul 6.3	1412 Mar 19.8	1467 Dec 8.8	1523 Aug 11.7	1579 Apr 19.8
1357 Aug 12.4	1413 Apr 20.0	1469 Jan 9.2	1524 Sep 17.5	1580 May 21.3
1358 Sep 19.2	1414 May 22.7	1470 Feb 9.2	1525 Oct 24.6	1581 Jun 24.8
1359 Oct 26.0	1415 Jun 26.4	1471 Mar 11.7	1526 Nov 28.8	1582 Jul 31.4
1360 Nov 29.0	1416 Aug 1.1	1472 Apr 10.5	1527 Dec 31.7	1583 Sep 17.2
1361 Dec 31.8	1417 Sep 7.8	1473 May 12.7	1529 Jan 31.1	1584 Oct 23.7
1363 Feb 1.1	1418 Oct 15.1	1474 Jun 15.8	1530 Mar 2.6	1585 Nov 28.5
1364 Mar 2.7	1419 Nov 19.7	1475 Jul 21.9	1531 Apr 2.2	1587 Jan 1.1
1365 Apr 2.4	1420 Dec 22.2	1476 Aug 27.4	1532 May 2.9	1588 Feb 1.9
1366 May 4.3	1422 Jan 22.9	1477 Oct 4.1	1533 Jun 5.3	1589 Mar 3.6
1367 Jun 6.9	1423 Feb 22.5	1478 Nov 9.3	1534 Jul 10.8	1590 Apr 3.1
1368 Jul 11.4	1424 Mar 24.1	1479 Dec 13.4	1535 Aug 17.1	1591 May 4.3
1369 Aug 17.6	1425 Apr 24.5	1481 Jan 13.6	1536 Sep 22.9	1592 Jun 5.1
1370 Sep 24.3	1426 May 27.5	1482 Feb 13.5	1537 Oct 29.7	1593 Jul 10.0
1371 Oct 30.9	1427 Jul 1.5	1483 Mar 16.0	1538 Dec 3.5	1594 Aug 15.8
1372 Dec 3.6	1428 Aug 6.3	1484 Apr 15.0	1540 Jan 5.2	1595 Sep 22.6
1374 Jan 5.2	1429 Sep 13.0	1485 May 17.4	1541 Feb 4.4	1596 Oct 28.9
1375 Feb 5.4	1430 Oct 20.1	1486 Jun 20.8	1542 Mar 6.9	1597 Dec 3.4
1376 Mar 7.0	1431 Nov 24.4	1487 Jul 27.1	1543 Apr 6.6	1599 Jan 5.7
1377 Apr 6.8	1432 Dec 26.6	1488 Sep 1.7	1544 May 7.5	1600 Feb 6.2
1378 May 8.9	1434 Jan 27.2	1489 Oct 9.1	1545 Jun 10.2	1601 Mar 7.8
1379 Jun 11.7	1435 Feb 26.8	1490 Nov 14.1	1546 Jul 16.0	1602 Apr 7.4
1380 Jul 16.5	1436 Mar 28.5	1491 Dec 17.9	1547 Aug 22.4	1603 May 8.8
1381 Aug 22.9	1437 Apr 29.1	1493 Jan 17.9	1548 Sep 28.1	1604 Jun 9.9
1382 Sep 29.5	1438 Jun 1.3	1494 Feb 17.7	1549 Nov 3.6	1605 Jul 15.1
1383 Nov 4.9	1439 Jul 6.5	1495 Mar 20.3	1550 Dec 8.1	1606 Aug 21.1
1384 Dec 8.3	1440 Aug 11.5	1496 Apr 19.5	1552 Jan 9.6	1607 Sep 27.8
1386 Jan 9.7	1441 Sep 18.2	1497 May 22.1	1553 Feb 8.6	1608 Nov 2.9
1387 Feb 9.7	1442 Oct 25.2	1498 Jun 25.7	1554 Mar 11.2	1609 Dec 8.1
1388 Mar 11.2	1443 Nov 29.2	1499 Aug 1.3	1555 Apr 11.1	1611 Jan 10.1
1389 Apr 11.1	1444 Dec 31.2	1500 Sep 6.9	1556 May 12.2	1612 Feb 10.5
1390 May 13.3	1446 Jan 31.5	1501 Oct 14.3	1557 Jun 15.1	1613 Mar 12.1
1391 Jun 16.4	1447 Mar 3.1	1502 Nov 19.0	1558 Jul 21.1	1614 Apr 11.8
1392 Jul 21.5	1448 Apr 1.8	1503 Dec 22.6	1559 Aug 27.6	1615 May 13.4
1393 Aug 28.1	1449 May 3.4	1505 Jan 22.3	1560 Oct 3.2	1616 Jun 14.7
1394 Oct 4.7	1450 Jun 5.9	1506 Feb 22.0	1561 Nov 8.6	1617 Jul 20.1
1395 Nov 9.9	1451 Jul 11.4	1507 Mar 24.6	1562 Dec 12.8	1618 Aug 26.3
1396 Dec 13.0	1452 Aug 16.7	1508 Apr 23.8	1564 Jan 14.1	1619 Oct 3.1
1398 Jan 14.2	1453 Sep 23.5	1509 May 26.6	1565 Feb 13.0	1620 Nov 7.9
1399 Feb 14.0	1454 Oct 30.3	1510 Jun 30.5	1566 Mar 15.5	1621 Dec 12.8
1400 Mar 15.5	1455 Dec 4.1	1511 Aug 6.3	1567 Apr 15.4	1623 Jan 14.6
1401 Apr 15.5	1457 Jan 4.8	1512 Sep 12.1	1568 May 16.6	1624 Feb 14.9
1402 May 18.0	1458 Feb 4.9	1513 Oct 19.4	1569 Jun 19.8	1625 Mar 16.4
1403 Jun 21.4	1459 Mar 7.4	1514 Nov 23.9	1570 Jul 26.1	1626 Apr 16.1
1404 Jul 26.8	1460 Apr 6.2	1515 Dec 27.2	1571 Sep 1.8	1627 May 17.8
1405 Sep 2.5	1461 May 8.0	1517 Jan 26.8	1572 Oct 8.4	1628 Jun 19.4
1406 Oct 10.0	1462 Jun 10.8	1518 Feb 26.4	1573 Nov 13.5	1629 Jul 25.0
1407 Nov 14.9	1463 Jul 16.6	1519 Mar 28.9	1574 Dec 17.5	1630 Aug 31.4
1408 Dec 17.6	1464 Aug 22.1	1520 Apr 28.3	1576 Jan 18.5	1631 Oct 8.2

1632 Nov 12.9	1688 Jul 14.1	1744 Mar 29.7	1799 Dec 18.2	1855 Aug 21.3
1633 Dec 17.6	1689 Aug 20.0	1745 Apr 29.7	1801 Jan 19.9	1856 Sep 27.1
1635 Jan 19.2	1690 Sep 26.8	1746 Jun 1.0	1802 Feb 20.1	1857 Nov 3.4
1636 Feb 19.3	1691 Nov 3.1	1747 Jul 5.3	1803 Mar 22.6	1858 Dec 8.8
1637 Mar 20.7	1692 Dec 7.4	1748 Aug 9.7	1804 Apr 21.3	1860 Jan 11.2
1638 Apr 20.5	1694 Jan 9.7	1749 Sep 16.4	1805 May 23.1	1861 Feb 10.7
1639 May 22.4	1695 Feb 10.1	1750 Oct 24.0	1806 Jun 25.7	1862 Mar 13.3
1640 Jun 24.3	1696 Mar 11.7	1751 Nov 29.0	1807 Jul 31.5	1863 Apr 12.9
1641 Jul 30.2	1697 Apr 11.2	1752 Dec 31.9	1808 Sep 5.8	1864 May 13.3
1642 Sep 5.8	1698 May 12.7	1754 Feb 1.8	1809 Oct 13.6	1865 Jun 15.3
1643 Oct 13.5	1699 Jun 14.9	1755 Mar 4.6	1810 Nov 19.3	1866 Jul 20.4
1644 Nov 18.0	1700 Jul 20.2	1756 Apr 3.0	1811 Dec 23.9	1867 Aug 26.3
1645 Dec 22.3	1701 Aug 26.3	1757 May 4.1	1813 Jan 24.4	1868 Oct 2.1
1647 Jan 23.6	1702 Oct 3.2	1758 Jun 5.7	1814 Feb 24.4	1869 Nov 8.3
1648 Feb 23.5	1703 Nov 9.2	1759 Jul 10.3	1815 Mar 26.8	1870 Dec 13.5
1649 Mar 24.9	1704 Dec 13.3	1760 Aug 14.9	1816 Apr 25.6	1872 Jan 15.7
1650 Apr 24.9	1706 Jan 15.2	1761 Sep 21.7	1817 May 27.7	1873 Feb 15.1
1651 May 27.1	1707 Feb 15.4	1762 Oct 29.2	1818 Jun 30.6	1874 Mar 17.6
1652 Jun 29.3	1708 Mar 16.9	1763 Dec 3.9	1819 Aug 5.6	1875 Apr 17.2
1653 Aug 4.5	1709 Apr 16.5	1765 Jan 5.5	1820 Sep 11.1	1876 May 17.8
1654 Sep 11.2	1710 May 18.2	1766 Feb 6.2	1821 Oct 18.9	1877 Jun 20.0
1655 Oct 18.8	1711 Jun 20.7	1767 Mar 8.8	1822 Nov 24.3	1878 Jul 25.4
1656 Nov 23.0	1712 Jul 25.4	1768 Apr 7.2	1823 Dec 28.5	1879 Aug 31.6
1657 Dec 27.0	1713 Aug 31.7	1769 May 8.5	1825 Jan 28.7	1880 Oct 7.4
1659 Jan 28.0	1714 Oct 8.6	1770 Jun 10.4	1826 Feb 28.6	1881 Nov 13.4
1660 Feb 27.8	1715 Nov 14.3	1771 Jul 15.4	1827 Mar 31.0	1882 Dec 18.3
1661 Mar 29.2	1716 Dec 18.0	1772 Aug 20.3	1828 Apr 29.9	1884 Jan 20.1
1662 Apr 29.3	1718 Jan 19.6	1773 Sep 27.1	1829 Jun 1.2	1885 Feb 19.3
1663 May 31.8	1719 Feb 19.7	1774 Nov 3.4	1830 Jul 5.5	1886 Mar 21.8
1664 Jul 4.3	1720 Mar 21.1	1775 Dec 8.8	1831 Aug 10.8	1887 Apr 21.4
1665 Aug 9.8	1721 Apr 20.9	1777 Jan 10.0	1832 Sep 16.6	1888 May 22.2
1666 Sep 16.5	1722 May 22.8	1778 Feb 10.5	1833 Oct 24.2	1889 Jun 24.8
1667 Oct 23.9	1723 Jun 25.7	1779 Mar 13.0	1834 Nov 29.3	1890 Jul 30.5
1668 Nov 27.8	1724 Jul 30.6	1780 Apr 11.6	1836 Jan 2.2	1891 Sep 5.9
1669 Dec 31.5	1725 Sep 6.1	1781 May 13.0	1837 Feb 2.1	1892 Oct 12.8
1671 Feb 1.3	1726 Oct 13.8	1782 Jun 15.2	1838 Mar 4.8	1893 Nov 18.5
1672 Mar 3.0	1727 Nov 19.2	1783 Jul 20.5	1839 Apr 4.3	1894 Dec 23.1
1673 Apr 2.5	1728 Dec 22.6	1784 Aug 25.6	1840 May 4.4	1896 Jan 24.6
1674 May 3.8	1730 Jan 23.9	1785 Oct 2.4	1841 Jun 5.9	1897 Feb 23.6
1675 Jun 5.5	1731 Feb 23.9	1786 Nov 8.4	1842 Jul 10.5	1898 Mar 26.0
1676 Jul 9.3	1732 Mar 25.4	1787 Dec 13.5	1843 Aug 16.1	1899 Apr 25.8
1677 Aug 15.0	1733 Apr 25.3	1789 Jan 14.4	1844 Sep 21.9	1900 May 27.8
1678 Sep 21.7	1734 May 27.5	1790 Feb 14.7	1845 Oct 29.3	1901 Jun 30.7
1679 Oct 29.0	1735 Jun 30.6	1791 Mar 17.2	1846 Dec 4.1	1902 Aug 5.7
1680 Dec 2.6	1736 Aug 4.7	1792 Apr 15.9	1848 Jan 6.6	1903 Sep 12.3
1682 Jan 5.1	1737 Sep 11.3	1793 May 17.6	1849 Feb 6.4	1904 Oct 19.0
1683 Feb 5.7	1738 Oct 18.9	1794 Jun 20.0	1850 Mar 9.1	1905 Nov 24.4
1684 Mar 7.4	1739 Nov 24.1	1795 Jul 25.6	1851 Apr 8.6	1906 Dec 28.6
1685 Apr 6.9	1740 Dec 27.3	1796 Aug 30.8	1852 May 8.8	1908 Jan 29.9
1686 May 8.2	1742 Jan 28.4	1797 Oct 7.6	1853 Jun 10.6	1909 Feb 28.8
1687 Jun 10.1	1743 Feb 28.3	1798 Nov 13.4	1854 Jul 15.5	1910 Mar 31.3

1911	May	1.2	1967	Jan	20.2	2022	Sep	26.8	
1912	Jun	1.4	1968	Feb	20.5	2023	Nov	3.2	
1913	Jul	5.6	1969	Mar	21.9	2024	Dec	7.9	
1914	Aug	10.9	1970	Apr	21.6	2026	Jan	10.4	
1915	Sep	17.5	1971	May	23.4	2027	Feb	11.0	
1916	Oct	24.1	1972	Jun	24.9	2028	Mar	12.6	
1917	Nov	29.2	1973	Jul	30.5	2029	Apr	12.2	
1919	Jan	2.2	1974	Sep	5.8	2030	May	13.5	
1920	Feb	3.3	1975	Oct	13.6	2031	Jun	15.4	
1921	Mar	5.1	1976	Nov	18.3	2032	Jul	19.3	
1922	Apr	4.6	1977	Dec	23.0	2033	Aug	25.2	
1923	May	5.6	1979	Jan	24.6	2034	Oct	2.0	
1924	Jun	6.0	1980	Feb	24.7	2035	Nov	8.2	
1925	Jul	10.4	1981	Mar	26.2	2036	Dec	12.6	
1926	Aug	15.8	1982	Apr	26.0	2038	Jan	14.8	
1927	Sep	22.5	1983	May	27.9	2039	Feb	15.3	
1928	Oct	29.0	1984	Jun	29.7	2040	Mar	16.9	
1929	Dec	4.0	1985	Aug	4.5	2041	Apr	16.5	
1931	Jan	6.7	1986	Sep	10.9	2042	May	18.0	
1932	Feb	7.6	1987	Oct	18.6	2043	Jun	20.1	
1933	Mar	9.4	1988	Nov	23.1	2044	Jul	24.3	
1934	Apr	8.8	1989	Dec	27.6	2045	Aug	30.3	
1935	May	10.0	1991	Jan	29.0	2046	Oct	7.1	
1936	Jun	10.7	1992	Feb	29.0	2047	Nov	13.1	
1937	Jul	15.3	1993	Mar	30.5	2048	Dec	17.2	
1938	Aug	21.0	1994	Apr	30.4	2050	Jan	19.2	
1939	Sep	27.8	1995	Jun	1.5	2051	Feb	19.6	
1940	Nov	3.2	1996	Jul	4.5	2052	Mar	21.2	
1941	Dec	8.8	1997	Aug	9.6	2053	Apr	20.8	
1943	Jan	11.3	1998	Sep	16.1	2054	May	22.5	
1944	Feb	11.9	1999	Oct	23.8	2055	Jun	24.8	
1945	Mar	13.5	2000	Nov	28.1	2056	Jul	29.3	
1946	Apr	13.0	2002	Jan	1.2	2057	Sep	4.5	
1947	May	14.3	2003	Feb	2.4	2058	Oct	12.3	
1948	Jun	15.3	2004	Mar	4.2	2059	Nov	18.1	
1949	Jul	20.3	2005	Apr	3.6	2060	Dec	22.0	
1950	Aug	26.3	2006	May	4.6	2062	Jan	23.7	
1951	Oct	3.2	2007	Jun	6.0	2063	Feb	23.8	
1952	Nov	8.4	2008	Jul	9.3	2064	Mar	25.3	
1953	Dec	13.7	2009	Aug	14.7	2065	Apr	25.0	
1955	Jan	15.8	2010	Sep	21.5	2066	May	26.8	
1956	Feb	16.2	2011	Oct	29.1	2067	Jun	29.5	
1957	Mar	17.7	2012	Dec	3.1	2068	Aug	3.3	
1958	Apr	17.3	2014	Jan	5.9	2069	Sep	9.8	
1959	May	18.8	2015	Feb	6.8	2070	Oct	17.6	
1960	Jun	20.1	2016	Mar	8.4	2071	Nov	23.2	
1961	Jul	25.4	2017	Apr	7.9	2072	Dec	26.7	
1962	Aug	31.6	2018	May	9.0	2074	Jan	28.2	
1963	Oct	8.4	2019	Jun	10.6	2075	Feb	28.1	
1964	Nov	13.4	2020	Jul	14.3	2076	Mar	29.5	
1965	Dec	18.4	2021	Aug	20.0	2077	Apr	29.4	

2078	May	31.4	2134	Feb	19.8
2079	Jul	4.4	2135	Mar	22.3
2080	Aug	8.5	2136	Apr	20.8
2081	Sep	15.1	2137	May	22.4
2082	Oct	22.8	2138	Jun	24.7
2083	Nov	28.1	2139	Jul	30.2
2084	Dec	31.3	2140	Sep	4.5
2086	Feb	1.5	2141	Oct	12.3
2087	Mar	4.3	2142	Nov	18.2
2088	Apr	2.8	2143	Dec	23.1
2089	May	3.7	2145	Jan	23.8
2090	Jun	5.1	2146	Feb	24.0
2091	Jul	9.4	2147	Mar	26.5
2092	Aug	13.7	2148	Apr	25.2
2093	Sep	20.4	2149	May	27.0
2094	Oct	28.0	2150	Jun	29.6
2095	Dec	3.0	2151	Aug	4.4
2097	Jan	4.9	2152	Sep	9.8
2098	Feb	5.8	2153	Oct	17.5
2099	Mar	8.6	2154	Nov	23.2
2100	Apr	8.1	2155	Dec	27.7
2101	May	9.2	2157	Jan	28.2
2102	Jun	10.7	2158	Feb	28.3
2103	Jul	15.3	2159	Mar	30.7
2104	Aug	19.8	2160	Apr	29.6
2105	Sep	26.5	2161	May	31.6
2106	Nov	2.9	2162	Jul	4.5
2107	Dec	8.7	2163	Aug	9.4
2109	Jan	10.3	2164	Sep	14.9
2110	Feb	11.1	2165	Oct	22.6
2111	Mar	13.8	2166	Nov	27.9
2112	Apr	12.4	2168	Jan	1.3
2113	May	13.6	2169	Feb	1.6
2114	Jun	15.4	2170	Mar	4.5
2115	Jul	20.2	2171	Apr	4.0
2116	Aug	24.9	2172	May	3.9
2117	Oct	1.7	2173	Jun	5.2
2118	Nov	8.0	2174	Jul	9.3
2119	Dec	13.5	2175	Aug	14.5
2121	Jan	14.9	2176	Sep	20.1
2122	Feb	15.4	2177	Oct	27.7
2123	Mar	18.0	2178	Dec	2.9
2124	Apr	16.5	2180	Jan	5.9
2125	May	17.9	2181	Feb	6.0
2126	Jun	19.9	2182	Mar	8.7
2127	Jul	25.1	2183	Apr	8.2
2128	Aug	30.1	2184	May	8.2
2129	Oct	7.0	2185	Jun	9.6
2130	Nov	13.2	2186	Jul	14.1
2131	Dec	18.4	2187	Aug	19.6
2133	Jan	19.4	2188	Sep	25.4

2189	Nov	2.0	2245	Jul	4.3	2301	Mar	22.8
2190	Dec	7.9	2246	Aug	9.3	2302	Apr	22.3
2192	Jan	10.6	2247	Sep	15.8	2303	May	23.7
2193	Feb	10.4	2248	Oct	22.6	2304	Jun	24.9
2194	Mar	13.0	2249	Nov	28.2	2305	Jul	30.1
2195	Apr	12.5	2251	Jan	1.6	2306	Sep	5.3
2196	May	12.6	2252	Feb	2.9	2307	Oct	13.2
2197	Jun	14.3	2253	Mar	4.8	2308	Nov	18.3
2198	Jul	19.1	2254	Apr	4.2	2309	Dec	23.5
2199	Aug	24.9	2255	May	5.0	2311	Jan	25.4
2200	Oct	1.8	2256	Jun	5.2	2312	Feb	25.6
2201	Nov	8.1	2257	Jul	9.3	2313	Mar	27.1
2202	Dec	13.7	2258	Aug	14.5	2314	Apr	26.6
2204	Jan	16.1	2259	Sep	21.2	2315	May	28.3
2205	Feb	15.7	2260	Oct	27.9	2316	Jun	29.7
2206	Mar	18.2	2261	Dec	3.2	2317	Aug	4.3
2207	Apr	17.7	2263	Jan	6.2	2318	Sep	10.7
2208	May	18.1	2264	Feb	7.3	2319	Oct	18.6
2209	Jun	20.1	2265	Mar	9.0	2320	Nov	23.5
2210	Jul	25.2	2266	Apr	8.5	2321	Dec	28.2
2211	Aug	31.2	2267	May	9.5	2323	Jan	29.8
2212	Oct	7.1	2268	Jun	9.9	2324	Feb	29.9
2213	Nov	13.2	2269	Jul	14.3	2325	Mar	31.3
2214	Dec	18.5	2270	Aug	19.8	2326	Apr	31.0
2216	Jan	20.5	2271	Sep	26.6	2327	Jun	1.9
2217	Feb	19.9	2272	Nov	2.1	2328	Jul	4.7
2218	Mar	22.5	2273	Dec	8.0	2329	Aug	9.6
2219	Apr	22.1	2275	Jan	10.8	2330	Sep	16.1
2220	May	22.7	2276	Feb	11.6	2331	Oct	23.9
2221	Jun	24.9	2277	Mar	13.3	2332	Nov	28.4
2222	Jul	30.3	2278	Apr	12.8	2334	Jan	1.9
2223	Sep	5.4	2279	May	13.9	2335	Feb	3.2
2224	Oct	12.2	2280	Jun	14.6	2336	Mar	5.2
2225	Nov	18.1	2281	Jul	19.3	2337	Apr	4.6
2226	Dec	23.1	2282	Aug	25.0	2338	May	5.4
2228	Jan	24.9	2283	Oct	1.8	2339	Jun	6.6
2229	Feb	24.2	2284	Nov	7.1	2340	Jul	9.6
2230	Mar	26.7	2285	Dec	12.7	2341	Aug	14.7
2231	Apr	26.4	2287	Jan	15.2	2342	Sep	21.3
2232	May	27.2	2288	Feb	15.9	2343	Oct	29.0
2233	Jun	29.7	2289	Mar	17.5	2344	Dec	3.3
2234	Aug	4.3	2290	Apr	17.1	2346	Jan	6.4
2235	Sep	10.6	2291	May	18.4	2347	Feb	7.5
2236	Oct	17.4	2292	Jun	19.3	2348	Mar	9.4
2237	Nov	23.1	2293	Jul	24.3	2349	Apr	8.9
2238	Dec	27.8	2294	Aug	30.2	2350	May	9.8
2240	Jan	29.4	2295	Oct	7.0	2351	Jun	11.2
2241	Feb	28.5	2296	Nov	12.2	2352	Jul	14.5
2242	Mar	30.9	2297	Dec	17.6	2353	Aug	19.9
2243	Apr	30.6	2299	Jan	19.8	2354	Sep	26.6
2244	May	31.6	2300	Feb	20.3	2355	Nov	3.1

2356	Dec	8.2	2412	Aug	8.7
2358	Jan	11.0	2413	Sep	15.2
2359	Feb	12.0	2414	Oct	23.0
2360	Mar	13.7	2415	Nov	28.6
2361	Apr	13.1	2417	Jan	1.2
2362	May	14.1	2418	Feb	2.6
2363	Jun	15.7	2419	Mar	5.6
2364	Jul	19.3	2420	Apr	4.0
2365	Aug	24.9	2421	May	4.7
2366	Oct	1.8	2422	Jun	5.7
2367	Nov	8.3	2423	Jul	9.6
2368	Dec	13.1	2424	Aug	13.7
2370	Jan	15.7	2425	Sep	20.3
2371	Feb	16.4	2426	Oct	28.2
2372	Mar	18.0	2427	Dec	3.6
2373	Apr	17.4	2429	Jan	5.9
2374	May	18.6	2430	Feb	7.1
2375	Jun	20.4	2431	Mar	10.0
2376	Jul	24.3	2432	Apr	8.3
2377	Aug	30.3	2433	May	9.2
2378	Oct	7.2	2434	Jun	10.4
2379	Nov	13.6	2435	Jul	14.6
2380	Dec	18.0	2436	Aug	19.0
2382	Jan	20.3	2437	Sep	25.8
2383	Feb	20.7	2438	Nov	2.5
2384	Mar	22.2	2439	Dec	8.7
2385	Apr	21.7	2441	Jan	10.6
2386	May	23.1	2442	Feb	11.5
2387	Jun	25.3	2443	Mar	14.2
2388	Jul	29.5	2444	Apr	12.6
2389	Sep	4.7	2445	May	13.6
2390	Oct	12.6	2446	Jun	15.1
2391	Nov	18.7	2447	Jul	19.7
2392	Dec	22.8	2448	Aug	24.3
2394	Jan	24.8	2449	Oct	1.2
2395	Feb	25.0	2450	Nov	7.7
2396	Mar	26.5	2451	Dec	13.6
2397	Apr	26.1	2453	Jan	15.2
2398	May	27.7	2454	Feb	15.8
2399	Jun	30.1	2455	Mar	18.4
2400	Aug	3.7	2456	Apr	16.9
2401	Sep	9.9	2457	May	18.1
2402	Oct	17.8	2458	Jun	19.9
2403	Nov	23.6	2459	Jul	24.8
2404	Dec	27.4	2460	Aug	29.6
2406	Jan	29.1	2461	Oct	6.5
2407	Mar	1.3	2462	Nov	12.8
2408	Mar	30.7	2463	Dec	18.3
2409	Apr	30.4	2465	Jan	19.6
2410	Jun	1.3	2466	Feb	20.1
2411	Jul	5.0	2467	Mar	22.7

2468	Apr	21.2	2474	Nov	17.9	2481	May	27.0	2487	Dec	27.9	2494	Jul	4.1

2468	Apr	21.2	2474	Nov	17.9	2481	May	27.0	2487	Dec	27.9	2494	Jul	4.1
2469	May	22.6	2475	Dec	23.1	2482	Jun	29.2	2489	Jan	28.7	2495	Aug	8.8
2470	Jun	24.6	2477	Jan	24.2	2483	Aug	3.7	2490	Feb	28.9	2496	Sep	14.3
2471	Jul	29.8	2478	Feb	24.5	2484	Sep	8.9	2491	Mar	31.3	2497	Oct	22.3
2472	Sep	3.8	2479	Mar	27.0	2485	Oct	16.9	2492	Apr	29.9	2498	Nov	28.1
2473	Oct	11.7	2480	Apr	25.5	2486	Nov	22.9	2493	May	31.6	2500	Jan	1.7

XII
OPPOSITIONS OF SATURN, 0 – 2500

0 Dec 6.4	46 Jun 19.5	92 Jan 13.3	137 Jul 22.1	183 Feb 17.8
1 Dec 20.5	47 Jul 1.7	93 Jan 25.9	138 Aug 3.9	184 Mar 1.6
3 Jan 3.5	48 Jul 13.2	94 Feb 8.2	139 Aug 17.0	185 Mar 14.3
4 Jan 17.2	49 Jul 25.8	95 Feb 21.3	140 Aug 29.3	186 Mar 26.7
5 Jan 29.7	50 Aug 7.7	96 Mar 5.1	141 Sep 11.9	187 Apr 8.0
6 Feb 12.0	51 Aug 20.9	97 Mar 17.7	142 Sep 25.7	188 Apr 19.1
7 Feb 24.9	52 Sep 2.3	98 Mar 30.1	143 Oct 9.8	189 May 1.2
8 Mar 8.7	53 Sep 15.9	99 Apr 11.4	144 Oct 23.0	190 May 13.2
9 Mar 21.2	54 Sep 29.8	100 Apr 22.5	145 Nov 6.3	191 May 25.2
10 Apr 2.6	55 Oct 13.9	101 May 4.6	146 Nov 20.7	192 Jun 5.2
11 Apr 14.8	56 Oct 27.1	102 May 16.6	147 Dec 5.0	193 Jun 17.4
12 Apr 25.9	57 Nov 10.5	103 May 28.7	148 Dec 18.2	194 Jun 29.6
13 May 7.9	58 Nov 24.8	104 Jun 8.7	150 Jan 1.2	195 Jul 12.0
14 May 19.9	59 Dec 9.1	105 Jun 20.9	151 Jan 15.0	196 Jul 23.6
15 May 31.9	60 Dec 22.2	106 Jul 3.2	152 Jan 28.5	197 Aug 5.5
16 Jun 12.0	62 Jan 5.1	107 Jul 15.7	153 Feb 9.8	198 Aug 18.6
17 Jun 24.2	63 Jan 18.8	108 Jul 27.4	154 Feb 22.9	199 Aug 32.0
18 Jul 6.6	64 Feb 1.3	109 Aug 9.3	155 Mar 7.6	200 Sep 13.6
19 Jul 19.1	65 Feb 13.5	110 Aug 22.5	156 Mar 19.2	201 Sep 27.4
20 Jul 30.8	66 Feb 26.4	111 Sep 5.0	157 Mar 31.6	202 Oct 11.5
21 Aug 12.8	67 Mar 11.1	112 Sep 17.7	158 Apr 12.9	203 Oct 25.8
22 Aug 26.1	68 Mar 22.6	113 Oct 1.6	159 Apr 25.0	204 Nov 8.1
23 Sep 8.6	69 Apr 4.0	114 Oct 15.7	160 May 6.1	205 Nov 22.4
24 Sep 21.3	70 Apr 16.2	115 Oct 30.0	161 May 18.1	206 Dec 6.7
25 Oct 5.3	71 Apr 28.3	116 Nov 12.3	162 May 30.1	207 Dec 20.9
26 Oct 19.5	72 May 9.3	117 Nov 26.6	163 Jun 11.2	209 Jan 2.9
27 Nov 2.8	73 May 21.3	118 Dec 10.8	164 Jun 22.4	210 Jan 16.7
28 Nov 16.2	74 Jun 2.3	119 Dec 24.9	165 Jul 4.7	211 Jan 30.2
29 Nov 30.5	75 Jun 14.4	121 Jan 6.8	166 Jul 17.2	212 Feb 12.5
30 Dec 14.7	76 Jun 25.6	122 Jan 20.5	167 Jul 29.9	213 Feb 24.5
31 Dec 28.8	77 Jul 8.0	123 Feb 2.9	168 Aug 10.9	214 Mar 9.2
33 Jan 10.7	78 Jul 20.6	124 Feb 16.1	169 Aug 24.1	215 Mar 21.8
34 Jan 24.3	79 Aug 2.3	125 Feb 28.0	170 Sep 6.6	216 Apr 2.1
35 Feb 6.7	80 Aug 14.4	126 Mar 12.7	171 Sep 20.4	217 Apr 14.3
36 Feb 19.8	81 Aug 27.7	127 Mar 25.2	172 Oct 3.3	218 Apr 26.5
37 Mar 3.6	82 Sep 10.2	128 Apr 5.5	173 Oct 17.5	219 May 8.5
38 Mar 16.3	83 Sep 24.0	129 Apr 17.6	174 Oct 31.8	220 May 19.5
39 Mar 28.7	84 Oct 7.0	130 Apr 29.7	175 Nov 15.1	221 May 31.5
40 Apr 9.0	85 Oct 21.2	131 May 11.7	176 Nov 28.4	222 Jun 12.6
41 Apr 21.2	86 Nov 4.5	132 May 22.7	177 Dec 12.6	223 Jun 24.8
42 May 3.2	87 Nov 18.9	133 Jun 3.8	178 Dec 26.7	224 Jul 6.2
43 May 15.2	88 Dec 2.2	134 Jun 15.9	180 Jan 9.6	225 Jul 18.7
44 May 26.3	89 Dec 16.4	135 Jun 28.1	181 Jan 22.3	226 Jul 31.5
45 Jun 7.3	90 Dec 30.5	136 Jul 9.5	182 Feb 4.6	227 Aug 13.5

228 Aug 25.7	281 Jun 14.0	334 Apr 6.0	387 Jan 22.1	439 Nov 1.4
229 Sep 8.2	282 Jun 26.2	335 Apr 18.1	388 Feb 4.6	440 Nov 14.8
230 Sep 22.0	283 Jul 8.6	336 Apr 29.2	389 Feb 16.8	441 Nov 29.1
231 Oct 6.0	284 Jul 20.1	337 May 11.2	390 Mar 1.7	442 Dec 13.4
232 Oct 19.2	285 Aug 1.9	338 May 23.2	391 Mar 14.5	443 Dec 27.6
233 Nov 2.5	286 Aug 14.9	339 Jun 4.2	392 Mar 25.9	445 Jan 9.5
234 Nov 16.8	287 Aug 28.2	340 Jun 15.3	393 Apr 7.2	446 Jan 23.3
235 Dec 1.2	288 Sep 9.8	341 Jun 27.5	394 Apr 19.4	447 Feb 5.7
236 Dec 14.4	289 Sep 23.6	342 Jul 9.9	395 May 1.4	448 Feb 18.9
237 Dec 28.5	290 Oct 7.6	343 Jul 22.4	396 May 12.4	449 Mar 2.9
239 Jan 11.3	291 Oct 21.8	344 Aug 3.2	397 May 24.4	450 Mar 15.6
240 Jan 24.9	292 Nov 4.1	345 Aug 16.2	398 Jun 5.4	451 Mar 28.1
241 Feb 6.3	293 Nov 18.5	346 Aug 29.5	399 Jun 17.5	452 Apr 8.4
242 Feb 19.4	294 Dec 2.8	347 Sep 12.1	400 Jun 28.7	453 Apr 20.5
243 Mar 4.2	295 Dec 17.0	348 Sep 24.9	401 Jul 11.0	454 May 2.6
244 Mar 15.9	296 Dec 30.1	349 Oct 9.0	402 Jul 23.6	455 May 14.5
245 Mar 28.3	298 Jan 12.9	350 Oct 23.2	403 Aug 5.4	456 May 25.5
246 Apr 9.5	299 Jan 26.5	351 Nov 6.5	404 Aug 17.4	457 Jun 6.5
247 Apr 21.6	300 Feb 8.9	352 Nov 19.9	405 Aug 30.7	458 Jun 18.6
248 May 2.7	301 Feb 21.0	353 Dec 4.2	406 Sep 13.3	459 Jun 30.8
249 May 14.7	302 Mar 5.8	354 Dec 18.5	407 Sep 27.1	460 Jul 12.1
250 May 26.7	303 Mar 18.4	356 Jan 1.5	408 Oct 10.2	461 Jul 24.6
251 Jun 7.7	304 Mar 29.8	357 Jan 14.4	409 Oct 24.5	462 Aug 6.4
252 Jun 18.8	305 Apr 11.0	358 Jan 28.0	410 Nov 7.8	463 Aug 19.4
253 Jul 1.1	306 Apr 23.1	359 Feb 10.3	411 Nov 22.2	464 Aug 31.8
254 Jul 13.5	307 May 5.1	360 Feb 23.4	412 Dec 5.5	465 Sep 14.3
255 Jul 26.1	308 May 16.1	361 Mar 7.2	413 Dec 19.7	466 Sep 28.2
256 Aug 7.0	309 May 28.1	362 Mar 19.8	415 Jan 2.8	467 Oct 12.3
257 Aug 20.1	310 Jun 9.1	363 Apr 1.1	416 Jan 16.6	468 Oct 25.5
258 Sep 2.5	311 Jun 21.2	364 Apr 12.3	417 Jan 29.2	469 Nov 8.9
259 Sep 16.1	312 Jul 2.4	365 Apr 24.4	418 Feb 11.6	470 Nov 23.3
260 Sep 29.0	313 Jul 14.9	366 May 6.4	419 Feb 24.6	471 Dec 7.6
261 Oct 13.1	314 Jul 27.5	367 May 18.4	420 Mar 8.4	472 Dec 20.8
262 Oct 27.4	315 Aug 9.4	368 May 29.3	421 Mar 21.0	474 Jan 3.9
263 Nov 10.7	316 Aug 21.5	369 Jun 10.4	422 Apr 2.4	475 Jan 17.7
264 Nov 24.1	317 Sep 3.9	370 Jun 22.5	423 Apr 14.6	476 Jan 31.3
265 Dec 8.4	318 Sep 17.6	371 Jul 4.7	424 Apr 25.6	477 Feb 12.6
266 Dec 22.6	319 Oct 1.5	372 Jul 16.1	425 May 7.6	478 Feb 25.7
268 Jan 5.5	320 Oct 14.6	373 Jul 28.8	426 May 19.6	479 Mar 10.5
269 Jan 18.3	321 Oct 28.9	374 Aug 10.7	427 May 31.5	480 Mar 22.1
270 Jan 31.8	322 Nov 12.3	375 Aug 23.8	428 Jun 11.5	481 Apr 3.5
271 Feb 14.0	323 Nov 26.6	376 Sep 5.2	429 Jun 23.6	482 Apr 15.7
272 Feb 27.0	324 Dec 9.9	377 Sep 18.9	430 Jul 5.9	483 Apr 27.7
273 Mar 10.7	325 Dec 24.1	378 Oct 2.8	431 Jul 18.3	484 May 8.7
274 Mar 23.3	327 Jan 7.0	379 Oct 17.0	432 Jul 29.9	485 May 20.7
275 Apr 4.6	328 Jan 20.8	380 Oct 30.3	433 Aug 11.8	486 Jun 1.6
276 Apr 15.8	329 Feb 2.3	381 Nov 13.6	434 Aug 24.9	487 Jun 13.6
277 Apr 27.9	330 Feb 15.5	382 Nov 28.0	435 Sep 7.4	488 Jun 24.7
278 May 9.9	331 Feb 28.4	383 Dec 12.3	436 Sep 20.1	489 Jul 6.9
279 May 21.9	332 Mar 12.2	384 Dec 25.4	437 Oct 4.0	490 Jul 19.4
280 Jun 1.9	333 Mar 24.7	386 Jan 8.4	438 Oct 18.1	491 Jul 32.0

492 Aug 12.8	545 Jun 2.6	598 Mar 24.9	651 Jan 7.5	703 Oct 16.9
493 Aug 26.0	546 Jun 14.6	599 Apr 6.3	652 Jan 21.3	704 Oct 30.1
494 Sep 8.4	547 Jun 26.7	600 Apr 17.6	653 Feb 2.9	705 Nov 13.4
495 Sep 22.1	548 Jul 7.9	601 Apr 29.6	654 Feb 16.3	706 Nov 27.8
496 Oct 5.0	549 Jul 20.3	602 May 11.6	655 Mar 1.3	707 Dec 12.1
497 Oct 19.2	550 Aug 2.0	603 May 23.6	656 Mar 13.2	708 Dec 25.3
498 Nov 2.4	551 Aug 14.8	604 Jun 3.6	657 Mar 25.8	710 Jan 8.3
499 Nov 16.8	552 Aug 27.0	605 Jun 15.6	658 Apr 7.2	711 Jan 22.1
500 Nov 30.2	553 Sep 9.4	606 Jun 27.7	659 Apr 19.4	712 Feb 4.7
501 Dec 14.4	554 Sep 23.0	607 Jul 9.9	660 Apr 30.5	713 Feb 17.1
502 Dec 28.6	555 Oct 7.0	608 Jul 21.3	661 May 12.6	714 Mar 2.2
504 Jan 11.5	556 Oct 20.1	609 Aug 2.9	662 May 24.5	715 Mar 15.0
505 Jan 24.3	557 Nov 3.4	610 Aug 15.8	663 Jun 5.5	716 Mar 26.6
506 Feb 6.7	558 Nov 17.7	611 Aug 28.9	664 Jun 16.5	717 Apr 8.1
507 Feb 19.9	559 Dec 2.1	612 Sep 10.3	665 Jun 28.6	718 Apr 20.3
508 Mar 3.9	560 Dec 15.4	613 Sep 24.0	666 Jul 10.9	719 May 2.5
509 Mar 16.6	561 Dec 29.5	614 Oct 7.9	667 Jul 23.3	720 May 13.5
510 Mar 29.1	563 Jan 12.4	615 Oct 22.0	668 Aug 3.9	721 May 25.5
511 Apr 10.4	564 Jan 26.2	616 Nov 4.3	669 Aug 16.8	722 Jun 6.5
512 Apr 21.5	565 Feb 7.6	617 Nov 18.6	670 Aug 29.9	723 Jun 18.5
513 May 3.6	566 Feb 20.9	618 Dec 2.9	671 Sep 12.3	724 Jun 29.6
514 May 15.6	567 Mar 5.8	619 Dec 17.2	672 Sep 24.9	725 Jul 11.9
515 May 27.5	568 Mar 17.5	620 Dec 30.3	673 Oct 8.8	726 Jul 24.3
516 Jun 7.5	569 Mar 30.0	622 Jan 13.3	674 Oct 22.9	727 Aug 6.0
517 Jun 19.6	570 Apr 11.3	623 Jan 27.0	675 Nov 6.2	728 Aug 17.8
518 Jul 1.8	571 Apr 23.5	624 Feb 9.5	676 Nov 19.5	729 Aug 31.0
519 Jul 14.1	572 May 4.5	625 Feb 21.7	677 Dec 3.8	730 Sep 13.3
520 Jul 25.6	573 May 16.5	626 Mar 6.6	678 Dec 18.1	731 Sep 27.0
521 Aug 7.4	574 May 28.5	627 Mar 19.4	680 Jan 1.2	732 Oct 9.9
522 Aug 20.4	575 Jun 9.5	628 Mar 30.9	681 Jan 14.1	733 Oct 23.9
523 Sep 2.7	576 Jun 20.5	629 Apr 12.2	682 Jan 27.8	734 Nov 7.2
524 Sep 15.3	577 Jul 2.7	630 Apr 24.3	683 Feb 10.3	735 Nov 21.5
525 Sep 29.1	578 Jul 15.0	631 May 6.4	684 Feb 23.5	736 Dec 4.8
526 Oct 13.2	579 Jul 27.6	632 May 17.4	685 Mar 7.5	737 Dec 19.0
527 Oct 27.4	580 Aug 8.3	633 May 29.4	686 Mar 20.2	739 Jan 2.1
528 Nov 9.8	581 Aug 21.3	634 Jun 10.4	687 Apr 1.7	740 Jan 16.0
529 Nov 24.2	582 Sep 3.6	635 Jun 22.5	688 Apr 13.1	741 Jan 28.7
530 Dec 8.5	583 Sep 17.2	636 Jul 3.6	689 Apr 25.2	742 Feb 11.2
531 Dec 22.7	584 Sep 30.0	637 Jul 16.0	690 May 7.3	743 Feb 24.4
533 Jan 4.8	585 Oct 14.1	638 Jul 28.5	691 May 19.3	744 Mar 8.4
534 Jan 18.6	586 Oct 28.3	639 Aug 10.2	692 May 30.3	745 Mar 21.1
535 Feb 1.2	587 Nov 11.7	640 Aug 22.3	693 Jun 11.3	746 Apr 2.6
536 Feb 14.6	588 Nov 25.0	641 Sep 4.5	694 Jun 23.4	747 Apr 15.0
537 Feb 26.7	589 Dec 9.4	642 Sep 18.1	695 Jul 5.6	748 Apr 26.2
538 Mar 11.5	590 Dec 23.6	643 Oct 1.9	696 Jul 16.9	749 May 8.3
539 Mar 24.1	592 Jan 6.6	644 Oct 14.9	697 Jul 29.5	750 May 20.3
540 Apr 4.4	593 Jan 19.5	645 Oct 29.2	698 Aug 11.2	751 Jun 1.3
541 Apr 16.6	594 Feb 2.1	646 Nov 12.5	699 Aug 24.2	752 Jun 12.3
542 Apr 28.7	595 Feb 15.4	647 Nov 26.9	700 Sep 5.5	753 Jun 24.4
543 May 10.7	596 Feb 28.5	648 Dec 10.2	701 Sep 19.1	754 Jul 6.6
544 May 21.7	597 Mar 12.3	649 Dec 24.4	702 Oct 2.9	755 Jul 19.0

756 Jul 30.5	809 May 21.3	862 Mar 11.4	914 Dec 23.8	967 Oct 3.6
757 Aug 12.3	810 Jun 2.3	863 Mar 24.2	916 Jan 6.8	968 Oct 16.5
758 Aug 25.3	811 Jun 14.4	864 Apr 4.7	917 Jan 19.6	969 Oct 30.6
759 Sep 7.6	812 Jun 25.5	865 Apr 17.0	918 Feb 2.2	970 Nov 13.7
760 Sep 20.1	813 Jul 7.7	866 Apr 29.2	919 Feb 15.6	971 Nov 28.0
761 Oct 3.9	814 Jul 20.1	867 May 11.4	920 Feb 28.8	972 Dec 11.2
762 Oct 17.9	815 Aug 1.7	868 May 22.4	921 Mar 12.7	973 Dec 25.3
763 Nov 1.1	816 Aug 13.5	869 Jun 3.5	922 Mar 25.4	975 Jan 8.3
764 Nov 14.4	817 Aug 26.5	870 Jun 15.5	923 Apr 6.9	976 Jan 22.2
765 Nov 28.7	818 Sep 8.8	871 Jun 27.7	924 Apr 18.2	977 Feb 3.7
766 Dec 13.0	819 Sep 22.3	872 Jul 8.9	925 Apr 30.5	978 Feb 17.1
767 Dec 27.2	820 Oct 5.1	873 Jul 21.4	926 May 12.6	979 Mar 2.2
769 Jan 9.3	821 Oct 19.1	874 Aug 3.0	927 May 24.7	980 Mar 14.1
770 Jan 23.1	822 Nov 2.3	875 Aug 15.8	928 Jun 4.7	981 Mar 26.7
771 Feb 5.6	823 Nov 16.6	876 Aug 27.8	929 Jun 16.8	982 Apr 8.2
772 Feb 19.0	824 Nov 29.9	877 Sep 10.1	930 Jun 29.0	983 Apr 20.6
773 Mar 3.1	825 Dec 14.1	878 Sep 23.7	931 Jul 11.3	984 May 1.8
774 Mar 15.9	826 Dec 28.3	879 Oct 7.5	932 Jul 22.7	985 May 13.9
775 Mar 28.6	828 Jan 11.3	880 Oct 20.5	933 Aug 4.3	986 May 26.0
776 Apr 9.0	829 Jan 24.1	881 Nov 3.7	934 Aug 17.2	987 Jun 7.0
777 Apr 21.3	830 Feb 6.7	882 Nov 17.9	935 Aug 30.3	988 Jun 18.2
778 May 3.4	831 Feb 20.0	883 Dec 2.2	936 Sep 11.6	989 Jun 30.4
779 May 15.5	832 Mar 4.1	884 Dec 15.5	937 Sep 25.2	990 Jul 12.7
780 May 26.5	833 Mar 16.9	885 Dec 29.6	938 Oct 9.0	991 Jul 25.1
781 Jun 7.5	834 Mar 29.6	887 Jan 12.6	939 Oct 23.0	992 Aug 5.8
782 Jun 19.6	835 Apr 11.0	888 Jan 26.3	940 Nov 5.2	993 Aug 18.7
783 Jul 1.7	836 Apr 22.3	889 Feb 7.9	941 Nov 19.4	994 Aug 31.8
784 Jul 13.0	837 May 4.5	890 Feb 21.2	942 Dec 3.7	995 Sep 14.2
785 Jul 25.5	838 May 16.5	891 Mar 6.2	943 Dec 17.9	996 Sep 26.8
786 Aug 7.1	839 May 28.6	892 Mar 18.0	944 Dec 31.0	997 Oct 10.6
787 Aug 20.0	840 Jun 8.6	893 Mar 30.7	946 Jan 14.0	998 Oct 24.7
788 Sep 1.1	841 Jun 20.7	894 Apr 12.1	947 Jan 27.7	999 Nov 7.8
789 Sep 14.5	842 Jul 2.9	895 Apr 24.4	948 Feb 10.2	1000 Nov 21.1
790 Sep 28.2	843 Jul 15.2	896 May 5.6	949 Feb 22.5	1001 Dec 5.3
791 Oct 12.0	844 Jul 26.7	897 May 17.7	950 Mar 7.5	1002 Dec 19.5
792 Oct 25.1	845 Aug 8.4	898 May 29.7	951 Mar 20.3	1004 Jan 2.6
793 Nov 8.3	846 Aug 21.3	899 Jun 10.8	952 Mar 31.9	1005 Jan 15.5
794 Nov 22.6	847 Sep 3.4	900 Jun 21.9	953 Apr 13.4	1006 Jan 29.2
795 Dec 6.9	848 Sep 15.8	901 Jul 4.1	954 Apr 25.6	1007 Feb 11.7
796 Dec 20.1	849 Sep 29.5	902 Jul 16.5	955 May 7.8	1008 Feb 24.9
798 Jan 3.2	850 Oct 13.4	903 Jul 29.0	956 May 18.9	1009 Mar 8.9
799 Jan 17.1	851 Oct 27.4	904 Aug 9.7	957 May 31.0	1010 Mar 21.7
800 Jan 30.8	852 Nov 9.6	905 Aug 22.7	958 Jun 12.1	1011 Apr 3.3
801 Feb 12.2	853 Nov 23.9	906 Sep 4.9	959 Jun 24.2	1012 Apr 14.7
802 Feb 25.4	854 Dec 8.2	907 Sep 18.3	960 Jul 5.5	1013 Apr 27.0
803 Mar 10.3	855 Dec 22.3	908 Sep 31.0	961 Jul 17.9	1014 May 9.2
804 Mar 22.1	857 Jan 4.4	909 Oct 14.8	962 Jul 30.4	1015 May 21.3
805 Apr 3.6	858 Jan 18.3	910 Oct 28.9	963 Aug 12.2	1016 Jun 1.4
806 Apr 15.9	859 Jan 31.9	911 Nov 12.1	964 Aug 24.2	1017 Jun 13.5
807 Apr 28.2	860 Feb 14.3	912 Nov 25.4	965 Sep 6.4	1018 Jun 25.6
808 May 9.3	861 Feb 26.5	913 Dec 9.6	966 Sep 19.9	1019 Jul 7.9

1020	Jul	19.3	1073	May	10.6	1126	Feb	28.1	1178	Dec	11.5	1231	Sep	21.5
1021	Jul	31.9	1074	May	22.7	1127	Mar	13.0	1179	Dec	25.6	1232	Oct	4.2
1022	Aug	13.7	1075	Jun	3.7	1128	Mar	24.7	1181	Jan	7.7	1233	Oct	18.2
1023	Aug	26.7	1076	Jun	14.9	1129	Apr	6.2	1182	Jan	21.5	1234	Nov	1.3
1024	Sep	8.0	1077	Jun	27.0	1130	Apr	18.6	1183	Feb	4.1	1235	Nov	15.5
1025	Sep	21.5	1078	Jul	9.3	1131	Apr	30.8	1184	Feb	17.5	1236	Nov	28.8
1026	Oct	5.3	1079	Jul	21.8	1132	May	12.0	1185	Mar	1.7	1237	Dec	13.0
1027	Oct	19.2	1080	Aug	2.4	1133	May	24.1	1186	Mar	14.6	1238	Dec	27.2
1028	Nov	1.3	1081	Aug	15.3	1134	Jun	5.2	1187	Mar	27.2	1240	Jan	10.2
1029	Nov	15.5	1082	Aug	28.3	1135	Jun	17.3	1188	Apr	7.7	1241	Jan	23.1
1030	Nov	29.7	1083	Sep	10.7	1136	Jun	28.5	1189	Apr	20.1	1242	Feb	5.7
1031	Dec	13.9	1084	Sep	23.2	1137	Jul	10.8	1190	May	2.3	1243	Feb	19.0
1032	Dec	27.0	1085	Oct	7.0	1138	Jul	23.2	1191	May	14.4	1244	Mar	3.2
1034	Jan	10.0	1086	Oct	20.9	1139	Aug	4.9	1192	May	25.5	1245	Mar	16.0
1035	Jan	23.8	1087	Nov	4.1	1140	Aug	16.8	1193	Jun	6.6	1246	Mar	28.7
1036	Feb	6.3	1088	Nov	17.3	1141	Aug	29.9	1194	Jun	18.7	1247	Apr	10.2
1037	Feb	18.7	1089	Dec	1.5	1142	Sep	12.2	1195	Jun	30.9	1248	Apr	21.5
1038	Mar	3.7	1090	Dec	15.7	1143	Sep	25.8	1196	Jul	12.2	1249	May	3.7
1039	Mar	16.6	1091	Dec	29.8	1144	Oct	8.6	1197	Jul	24.7	1250	May	15.8
1040	Mar	28.2	1093	Jan	11.8	1145	Oct	22.7	1198	Aug	6.3	1251	May	27.8
1041	Apr	9.7	1094	Jan	25.5	1146	Nov	5.8	1199	Aug	19.2	1252	Jun	7.9
1042	Apr	22.0	1095	Feb	8.0	1147	Nov	20.0	1200	Aug	31.4	1253	Jun	20.0
1043	May	4.2	1096	Feb	21.3	1148	Dec	3.3	1201	Sep	13.7	1254	Jul	2.2
1044	May	15.3	1097	Mar	5.3	1149	Dec	17.5	1202	Sep	27.4	1255	Jul	14.5
1045	May	27.4	1098	Mar	18.2	1150	Dec	31.6	1203	Oct	11.2	1256	Jul	26.0
1046	Jun	8.5	1099	Mar	30.8	1152	Jan	14.5	1204	Oct	24.3	1257	Aug	7.7
1047	Jun	20.6	1100	Apr	11.2	1153	Jan	27.2	1205	Nov	7.4	1258	Aug	20.6
1048	Jul	1.8	1101	Apr	23.5	1154	Feb	9.7	1206	Nov	21.7	1259	Sep	2.7
1049	Jul	14.1	1102	May	5.6	1155	Feb	23.0	1207	Dec	5.9	1260	Sep	15.1
1050	Jul	26.6	1103	May	17.7	1156	Mar	7.0	1208	Dec	19.1	1261	Sep	28.8
1051	Aug	8.3	1104	May	28.8	1157	Mar	19.8	1210	Jan	2.2	1262	Oct	12.7
1052	Aug	20.3	1105	Jun	9.9	1158	Apr	1.3	1211	Jan	16.1	1263	Oct	26.7
1053	Sep	2.4	1106	Jun	22.0	1159	Apr	13.7	1212	Jan	29.8	1264	Nov	8.9
1054	Sep	15.8	1107	Jul	4.3	1160	Apr	25.0	1213	Feb	11.3	1265	Nov	23.2
1055	Sep	29.5	1108	Jul	15.6	1161	May	7.1	1214	Feb	24.6	1266	Dec	7.4
1056	Oct	12.3	1109	Jul	28.1	1162	May	19.2	1215	Mar	9.5	1267	Dec	21.6
1057	Oct	26.4	1110	Aug	9.9	1163	May	31.3	1216	Mar	21.3	1269	Jan	3.7
1058	Nov	9.6	1111	Aug	22.8	1164	Jun	11.4	1217	Apr	2.9	1270	Jan	17.7
1059	Nov	23.8	1112	Sep	4.0	1165	Jun	23.5	1218	Apr	15.2	1271	Jan	31.4
1060	Dec	7.0	1113	Sep	17.5	1166	Jul	5.7	1219	Apr	27.5	1272	Feb	13.8
1061	Dec	21.2	1114	Oct	1.1	1167	Jul	18.1	1220	May	8.6	1273	Feb	26.0
1063	Jan	4.3	1115	Oct	15.0	1168	Jul	29.6	1221	May	20.7	1274	Mar	11.0
1064	Jan	18.2	1116	Oct	28.1	1169	Aug	11.4	1222	Jun	1.7	1275	Mar	23.8
1065	Jan	30.9	1117	Nov	11.3	1170	Aug	24.3	1223	Jun	13.8	1276	Apr	4.3
1066	Feb	13.3	1118	Nov	25.5	1171	Sep	6.6	1224	Jun	24.9	1277	Apr	16.6
1067	Feb	26.5	1119	Dec	9.8	1172	Sep	19.0	1225	Jul	7.1	1278	Apr	28.9
1068	Mar	10.4	1120	Dec	22.9	1173	Oct	2.8	1226	Jul	19.5	1279	May	11.0
1069	Mar	23.2	1122	Jan	6.0	1174	Oct	16.7	1227	Aug	1.0	1280	May	22.0
1070	Apr	4.8	1123	Jan	19.9	1175	Oct	30.8	1228	Aug	12.8	1281	Jun	3.0
1071	Apr	17.1	1124	Feb	2.5	1176	Nov	13.0	1229	Aug	25.8	1282	Jun	15.1
1072	Apr	28.4	1125	Feb	14.9	1177	Nov	27.2	1230	Sep	8.0	1283	Jun	27.2

1284 Jul 8.4	1337 Apr 30.2	1390 Feb 16.3	1442 Nov 27.6	1495 Sep 8.0
1285 Jul 20.8	1338 May 12.2	1391 Mar 1.5	1443 Dec 11.9	1496 Sep 20.4
1286 Aug 2.3	1339 May 24.3	1392 Mar 13.5	1444 Dec 25.1	1497 Oct 4.1
1287 Aug 15.1	1340 Jun 4.3	1393 Mar 26.2	1446 Jan 8.2	1498 Oct 18.0
1288 Aug 27.1	1341 Jun 16.3	1394 Apr 7.8	1447 Jan 22.1	1499 Nov 1.0
1289 Sep 9.4	1342 Jun 28.4	1395 Apr 20.1	1448 Feb 4.8	1500 Nov 14.2
1290 Sep 22.9	1343 Jul 10.7	1396 May 1.3	1449 Feb 17.3	1501 Nov 28.5
1291 Oct 6.6	1344 Jul 22.0	1397 May 13.4	1450 Mar 2.5	1502 Dec 12.8
1292 Oct 19.6	1345 Aug 3.5	1398 May 25.4	1451 Mar 15.5	1503 Dec 27.0
1293 Nov 2.7	1346 Aug 16.3	1399 Jun 6.4	1452 Mar 27.3	1505 Jan 9.1
1294 Nov 16.9	1347 Aug 29.3	1400 Jun 17.5	1453 Apr 8.8	1506 Jan 23.0
1295 Dec 1.2	1348 Sep 10.6	1401 Jun 29.6	1454 Apr 21.2	1507 Feb 5.7
1296 Dec 14.5	1349 Sep 24.1	1402 Jul 11.8	1455 May 3.4	1508 Feb 19.2
1297 Dec 28.6	1350 Oct 7.8	1403 Jul 24.1	1456 May 14.5	1509 Mar 3.4
1299 Jan 11.6	1351 Oct 21.8	1404 Aug 4.7	1457 May 26.5	1510 Mar 16.4
1300 Jan 25.5	1352 Nov 3.9	1405 Aug 17.4	1458 Jun 7.5	1511 Mar 29.2
1301 Feb 7.1	1353 Nov 18.2	1406 Aug 30.4	1459 Jun 19.5	1512 Apr 9.7
1302 Feb 20.4	1354 Dec 2.5	1407 Sep 12.7	1460 Jun 30.6	1513 Apr 22.1
1303 Mar 5.5	1355 Dec 16.7	1408 Sep 25.2	1461 Jul 12.8	1514 May 4.3
1304 Mar 17.4	1356 Dec 29.9	1409 Oct 8.9	1462 Jul 25.2	1515 May 16.4
1305 Mar 30.1	1358 Jan 12.9	1410 Oct 22.9	1463 Aug 6.7	1516 May 27.5
1306 Apr 11.5	1359 Jan 26.7	1411 Nov 6.0	1464 Aug 18.5	1517 Jun 8.5
1307 Apr 23.8	1360 Feb 9.3	1412 Nov 19.2	1465 Aug 31.4	1518 Jun 20.5
1308 May 5.0	1361 Feb 21.6	1413 Dec 3.5	1466 Sep 13.7	1519 Jul 2.6
1309 May 17.1	1362 Mar 6.8	1414 Dec 17.8	1467 Sep 27.2	1520 Jul 13.8
1310 May 29.1	1363 Mar 19.6	1415 Dec 31.9	1468 Oct 9.9	1521 Jul 26.2
1311 Jun 10.2	1364 Mar 31.3	1417 Jan 13.9	1469 Oct 23.9	1522 Aug 7.7
1312 Jun 21.3	1365 Apr 12.7	1418 Jan 27.8	1470 Nov 7.0	1523 Aug 20.5
1313 Jul 3.4	1366 Apr 25.0	1419 Feb 10.4	1471 Nov 21.2	1524 Sep 1.4
1314 Jul 15.7	1367 May 7.2	1420 Feb 23.7	1472 Dec 4.5	1525 Sep 14.7
1315 Jul 28.2	1368 May 18.2	1421 Mar 7.8	1473 Dec 18.7	1526 Sep 28.2
1316 Aug 8.9	1369 May 30.3	1422 Mar 20.7	1475 Jan 1.9	1527 Oct 11.9
1317 Aug 21.8	1370 Jun 11.3	1423 Apr 2.3	1476 Jan 15.9	1528 Oct 24.8
1318 Sep 3.9	1371 Jun 23.4	1424 Apr 13.8	1477 Jan 28.7	1529 Nov 7.9
1319 Sep 17.4	1372 Jul 4.6	1425 Apr 26.1	1478 Feb 11.3	1530 Nov 22.1
1320 Sep 30.0	1373 Jul 16.9	1426 May 8.2	1479 Feb 24.7	1531 Dec 6.4
1321 Oct 13.9	1374 Jul 29.3	1427 May 20.3	1480 Mar 8.8	1532 Dec 19.6
1322 Oct 28.0	1375 Aug 11.0	1428 May 31.3	1481 Mar 21.7	1534 Jan 2.8
1323 Nov 11.2	1376 Aug 22.9	1429 Jun 12.4	1482 Apr 3.3	1535 Jan 16.8
1324 Nov 24.5	1377 Sep 5.1	1430 Jun 24.5	1483 Apr 15.8	1536 Jan 30.6
1325 Dec 8.8	1378 Sep 18.5	1431 Jul 6.6	1484 Apr 27.1	1537 Feb 12.2
1326 Dec 23.0	1379 Oct 2.2	1432 Jul 17.9	1485 May 9.2	1538 Feb 25.5
1328 Jan 6.1	1380 Oct 15.1	1433 Jul 30.4	1486 May 21.3	1539 Mar 10.6
1329 Jan 19.0	1381 Oct 29.1	1434 Aug 12.0	1487 Jun 2.3	1540 Mar 22.5
1330 Feb 1.7	1382 Nov 12.4	1435 Aug 24.9	1488 Jun 13.4	1541 Apr 4.2
1331 Feb 15.2	1383 Nov 26.6	1436 Sep 6.1	1489 Jun 25.5	1542 Apr 16.7
1332 Feb 28.4	1384 Dec 9.9	1437 Sep 19.5	1490 Jul 7.6	1543 Apr 29.0
1333 Mar 12.4	1385 Dec 24.1	1438 Oct 3.2	1491 Jul 19.9	1544 May 10.1
1334 Mar 25.1	1387 Jan 7.2	1439 Oct 17.0	1492 Jul 31.4	1545 May 22.2
1335 Apr 6.6	1388 Jan 21.1	1440 Oct 30.1	1493 Aug 13.0	1546 Jun 3.3
1336 Apr 18.0	1389 Feb 2.8	1441 Nov 13.3	1494 Aug 25.9	1547 Jun 15.3

1548 Jun 26.4	1601 Apr 27.5	1654 Feb 11.3	1706 Nov 22.9	1759 Sep 5.8
1549 Jul 8.6	1602 May 9.8	1655 Feb 24.9	1707 Dec 7.1	1760 Sep 17.8
1550 Jul 20.9	1603 May 22.0	1656 Mar 9.2	1708 Dec 20.3	1761 Oct 1.1
1551 Aug 2.3	1604 Jun 2.1	1657 Mar 22.3	1710 Jan 3.5	1762 Oct 14.6
1552 Aug 14.0	1605 Jun 14.2	1658 Apr 4.2	1711 Jan 17.6	1763 Oct 28.3
1553 Aug 26.9	1606 Jun 26.2	1659 Apr 16.9	1712 Jan 31.5	1764 Nov 10.1
1554 Sep 9.0	1607 Jul 8.3	1660 Apr 28.4	1713 Feb 13.3	1765 Nov 24.2
1555 Sep 22.4	1608 Jul 19.5	1661 May 10.7	1714 Feb 26.8	1766 Dec 8.3
1556 Oct 5.0	1609 Jul 31.8	1662 May 22.9	1715 Mar 12.2	1767 Dec 22.5
1557 Oct 18.9	1610 Aug 13.3	1663 Jun 4.1	1716 Mar 24.3	1769 Jan 4.7
1558 Nov 1.9	1611 Aug 25.9	1664 Jun 15.1	1717 Apr 6.2	1770 Jan 18.7
1559 Nov 16.1	1612 Sep 6.8	1665 Jun 27.2	1718 Apr 18.8	1771 Feb 1.7
1560 Nov 29.4	1613 Sep 19.9	1666 Jul 9.3	1719 May 1.3	1772 Feb 15.4
1561 Dec 13.6	1614 Oct 3.3	1667 Jul 21.5	1720 May 12.7	1773 Feb 27.9
1562 Dec 27.8	1615 Oct 16.9	1668 Aug 1.8	1721 May 24.9	1774 Mar 13.3
1564 Jan 10.9	1616 Oct 29.8	1669 Aug 14.3	1722 Jun 6.0	1775 Mar 26.4
1565 Jan 23.8	1617 Nov 12.8	1670 Aug 27.0	1723 Jun 18.1	1776 Apr 7.2
1566 Feb 6.5	1618 Nov 27.0	1671 Sep 8.9	1724 Jun 29.2	1777 Apr 19.9
1567 Feb 20.0	1619 Dec 11.2	1672 Sep 21.0	1725 Jul 11.3	1778 May 2.4
1568 Mar 4.3	1620 Dec 24.5	1673 Oct 4.4	1726 Jul 23.6	1779 May 14.7
1569 Mar 17.3	1622 Jan 7.7	1674 Oct 18.0	1727 Aug 4.9	1780 May 26.0
1570 Mar 30.0	1623 Jan 21.8	1675 Oct 31.8	1728 Aug 16.4	1781 Jun 7.1
1571 Apr 11.6	1624 Feb 4.7	1676 Nov 13.8	1729 Aug 29.1	1782 Jun 19.2
1572 Apr 23.0	1625 Feb 17.4	1677 Nov 28.0	1730 Sep 11.0	1783 Jul 1.3
1573 May 5.2	1626 Mar 2.8	1678 Dec 12.2	1731 Sep 24.1	1784 Jul 12.5
1574 May 17.3	1627 Mar 16.1	1679 Dec 26.4	1732 Oct 6.5	1785 Jul 24.7
1575 May 29.4	1628 Mar 28.1	1681 Jan 8.6	1733 Oct 20.1	1786 Aug 6.1
1576 Jun 9.4	1629 Apr 9.9	1682 Jan 22.7	1734 Nov 2.9	1787 Aug 18.6
1577 Jun 21.5	1630 Apr 22.5	1683 Feb 5.6	1735 Nov 16.9	1788 Aug 30.3
1578 Jul 3.6	1631 May 4.9	1684 Feb 19.2	1736 Nov 30.1	1789 Sep 12.3
1579 Jul 15.8	1632 May 16.1	1685 Mar 3.7	1737 Dec 14.3	1790 Sep 25.4
1580 Jul 27.2	1633 May 28.3	1686 Mar 17.0	1738 Dec 28.5	1791 Oct 8.8
1581 Aug 8.7	1634 Jun 9.3	1687 Mar 30.0	1740 Jan 11.7	1792 Oct 21.4
1582 Aug 21.5	1635 Jun 21.4	1688 Apr 10.8	1741 Jan 24.7	1793 Nov 4.2
1583 Sep 13.4	1636 Jul 2.4	1689 Apr 23.3	1742 Feb 7.6	1794 Nov 18.2
1584 Sep 25.7	1637 Jul 14.6	1690 May 5.8	1743 Feb 21.3	1795 Dec 2.4
1585 Oct 9.1	1638 Jul 26.8	1691 May 18.0	1744 Mar 5.7	1796 Dec 15.6
1586 Oct 22.8	1639 Aug 8.2	1692 May 29.2	1745 Mar 18.9	1797 Dec 29.7
1587 Nov 5.8	1640 Aug 19.7	1693 Jun 10.3	1746 Mar 31.9	1799 Jan 12.9
1588 Nov 18.8	1641 Sep 1.5	1694 Jun 22.4	1747 Apr 13.7	1800 Jan 26.9
1589 Dec 3.0	1642 Sep 14.5	1695 Jul 4.5	1748 Apr 25.3	1801 Feb 9.8
1590 Dec 17.3	1643 Sep 27.7	1696 Jul 15.6	1749 May 7.7	1802 Feb 23.4
1591 Dec 31.5	1644 Oct 10.1	1697 Jul 27.9	1750 May 20.0	1803 Mar 8.8
1593 Jan 13.6	1645 Oct 23.9	1698 Aug 9.3	1751 Jun 1.2	1804 Mar 21.0
1594 Jan 27.6	1646 Nov 6.8	1699 Aug 21.8	1752 Jun 12.3	1805 Apr 3.0
1595 Feb 10.4	1647 Nov 20.8	1700 Sep 3.6	1753 Jun 24.4	1806 Apr 15.8
1596 Feb 24.0	1648 Dec 4.0	1701 Sep 16.6	1754 Jul 6.5	1807 Apr 28.4
1597 Mar 8.4	1649 Dec 18.2	1702 Sep 29.8	1755 Jul 18.7	1808 May 9.8
1598 Mar 21.5	1651 Jan 1.4	1703 Oct 13.3	1756 Jul 30.0	1809 May 22.1
1599 Apr 3.4	1652 Jan 15.5	1704 Oct 26.0	1757 Aug 11.4	1810 Jun 3.3
1600 Apr 15.0	1653 Jan 28.5	1705 Nov 8.9	1758 Aug 24.0	1811 Jun 15.4

1812	Jun	26.6	1865	Apr	17.0	1918	Jan	31.8	1970	Nov	11.9	2023	Aug	27.3
1813	Jul	8.7	1866	Apr	29.6	1919	Feb	14.6	1971	Nov	26.0	2024	Sep	8.2
1814	Jul	20.9	1867	May	12.0	1920	Feb	28.2	1972	Dec	9.1	2025	Sep	21.2
1815	Aug	2.2	1868	May	23.3	1921	Mar	12.5	1973	Dec	23.2	2026	Oct	4.5
1816	Aug	13.7	1869	Jun	4.5	1922	Mar	25.7	1975	Jan	6.4	2027	Oct	18.0
1817	Aug	26.3	1870	Jun	16.7	1923	Apr	7.6	1976	Jan	20.4	2028	Oct	30.7
1818	Sep	8.1	1871	Jun	28.8	1924	Apr	19.4	1977	Feb	2.4	2029	Nov	13.6
1819	Sep	21.2	1872	Jul	10.0	1925	May	1.9	1978	Feb	16.2	2030	Nov	27.7
1820	Oct	3.4	1873	Jul	22.2	1926	May	14.3	1979	Mar	1.7	2031	Dec	11.8
1821	Oct	16.9	1874	Aug	3.6	1927	May	26.6	1980	Mar	14.1	2032	Dec	24.9
1822	Oct	30.6	1875	Aug	16.0	1928	Jun	6.8	1981	Mar	27.2	2034	Jan	8.1
1823	Nov	13.5	1876	Aug	27.7	1929	Jun	19.0	1982	Apr	9.1	2035	Jan	22.1
1824	Nov	26.6	1877	Sep	9.6	1930	Jul	1.1	1983	Apr	21.8	2036	Feb	5.1
1825	Dec	10.7	1878	Sep	22.6	1931	Jul	13.3	1984	May	3.3	2037	Feb	17.8
1826	Dec	24.9	1879	Oct	6.0	1932	Jul	24.6	1985	May	15.7	2038	Mar	3.3
1828	Jan	8.0	1880	Oct	18.5	1933	Aug	6.0	1986	May	28.0	2039	Mar	16.7
1829	Jan	21.1	1881	Nov	1.2	1934	Aug	18.5	1987	Jun	9.2	2040	Mar	28.8
1830	Feb	4.0	1882	Nov	15.1	1935	Aug	31.2	1988	Jun	20.4	2041	Apr	10.6
1831	Feb	17.7	1883	Nov	29.2	1936	Sep	12.1	1989	Jul	2.5	2042	Apr	23.3
1832	Mar	2.2	1884	Dec	12.3	1937	Sep	25.2	1990	Jul	14.7	2043	May	5.8
1833	Mar	15.5	1885	Dec	26.5	1938	Oct	8.5	1991	Jul	27.0	2044	May	17.2
1834	Mar	28.6	1887	Jan	9.6	1939	Oct	22.1	1992	Aug	7.4	2045	May	29.4
1835	Apr	10.4	1888	Jan	23.6	1940	Nov	3.9	1993	Aug	19.9	2046	Jun	10.6
1836	Apr	22.1	1889	Feb	5.5	1941	Nov	17.8	1994	Sep	1.7	2047	Jun	22.8
1837	May	4.6	1890	Feb	19.2	1942	Dec	1.9	1995	Sep	14.6	2048	Jul	4.0
1838	May	16.9	1891	Mar	4.6	1943	Dec	16.0	1996	Sep	26.8	2049	Jul	16.2
1839	May	29.1	1892	Mar	16.9	1944	Dec	29.1	1997	Oct	10.2	2050	Jul	28.4
1840	Jun	9.3	1893	Mar	29.9	1946	Jan	12.2	1998	Oct	23.8	2051	Aug	9.9
1841	Jun	21.4	1894	Apr	11.8	1947	Jan	26.2	1999	Nov	6.6	2052	Aug	21.4
1842	Jul	3.6	1895	Apr	24.4	1948	Feb	9.1	2000	Nov	19.5	2053	Sep	3.2
1843	Jul	15.7	1896	May	5.9	1949	Feb	21.7	2001	Dec	3.6	2054	Sep	16.2
1844	Jul	27.0	1897	May	18.2	1950	Mar	7.2	2002	Dec	17.7	2055	Sep	29.4
1845	Aug	8.4	1898	May	30.4	1951	Mar	20.4	2003	Dec	31.9	2056	Oct	11.8
1846	Aug	20.9	1899	Jun	11.6	1952	Apr	1.4	2005	Jan	14.0	2057	Oct	25.4
1847	Sep	2.7	1900	Jun	23.7	1953	Apr	14.2	2006	Jan	27.9	2058	Nov	8.3
1848	Sep	14.6	1901	Jul	5.9	1954	Apr	26.8	2007	Feb	10.8	2059	Nov	22.2
1849	Sep	27.8	1902	Jul	18.1	1955	May	9.3	2008	Feb	24.4	2060	Dec	5.3
1850	Oct	11.2	1903	Jul	30.3	1956	May	20.6	2009	Mar	8.8	2061	Dec	19.5
1851	Oct	24.8	1904	Aug	10.8	1957	Jun	1.8	2010	Mar	22.0	2063	Jan	2.6
1852	Nov	6.7	1905	Aug	23.3	1958	Jun	14.0	2011	Apr	4.0	2064	Jan	16.7
1853	Nov	20.7	1906	Sep	5.1	1959	Jun	26.1	2012	Apr	15.8	2065	Jan	29.7
1854	Dec	4.8	1907	Sep	18.1	1960	Jul	7.3	2013	Apr	28.3	2066	Feb	12.5
1855	Dec	19.0	1908	Sep	30.3	1961	Jul	19.5	2014	May	10.8	2067	Feb	26.1
1857	Jan	1.2	1909	Oct	13.7	1962	Jul	31.8	2015	May	23.1	2068	Mar	10.5
1858	Jan	15.3	1910	Oct	27.4	1963	Aug	13.2	2016	Jun	3.3	2069	Mar	23.6
1859	Jan	29.3	1911	Nov	10.2	1964	Aug	24.8	2017	Jun	15.4	2070	Apr	5.6
1860	Feb	12.1	1912	Nov	23.3	1965	Sep	6.6	2018	Jun	27.6	2071	Apr	18.3
1861	Feb	24.7	1913	Dec	7.4	1966	Sep	19.7	2019	Jul	9.7	2072	Apr	29.9
1862	Mar	10.1	1914	Dec	21.5	1967	Oct	2.9	2020	Jul	20.9	2073	May	12.3
1863	Mar	23.3	1916	Jan	4.7	1968	Oct	15.4	2021	Aug	2.3	2074	May	24.5
1864	Apr	4.3	1917	Jan	17.8	1969	Oct	29.1	2022	Aug	14.7	2075	Jun	5.7

2076 Jun 16.9	2129 Apr 8.2	2182 Jan 20.9	2234 Nov 1.8	2287 Aug 18.0
2077 Jun 29.0	2130 Apr 20.9	2183 Feb 3.9	2235 Nov 15.7	2288 Aug 29.6
2078 Jul 11.2	2131 May 3.4	2184 Feb 17.6	2236 Nov 28.7	2289 Sep 11.4
2079 Jul 23.4	2132 May 14.8	2185 Mar 2.2	2237 Dec 12.9	2290 Sep 24.4
2080 Aug 3.7	2133 May 27.0	2186 Mar 15.6	2238 Dec 27.0	2291 Oct 7.7
2081 Aug 16.2	2134 Jun 8.2	2187 Mar 28.7	2240 Jan 10.2	2292 Oct 20.2
2082 Aug 28.8	2135 Jun 20.3	2188 Apr 9.6	2241 Jan 23.3	2293 Nov 2.9
2083 Sep 10.7	2136 Jul 1.4	2189 Apr 22.3	2242 Feb 6.2	2294 Nov 16.8
2084 Sep 22.8	2137 Jul 13.6	2190 May 4.8	2243 Feb 20.0	2295 Nov 30.9
2085 Oct 6.1	2138 Jul 25.8	2191 May 17.2	2244 Mar 4.6	2296 Dec 14.1
2086 Oct 19.6	2139 Aug 7.1	2192 May 28.4	2245 Mar 18.0	2297 Dec 28.2
2087 Nov 2.4	2140 Aug 18.6	2193 Jun 9.6	2246 Mar 31.1	2299 Jan 11.4
2088 Nov 15.3	2141 Aug 31.3	2194 Jun 21.7	2247 Apr 13.0	2300 Jan 25.5
2089 Nov 29.3	2142 Sep 13.1	2195 Jul 3.8	2248 Apr 24.7	2301 Feb 8.4
2090 Dec 13.5	2143 Sep 26.2	2196 Jul 14.9	2249 May 7.2	2302 Feb 22.2
2091 Dec 27.6	2144 Oct 8.6	2197 Jul 27.1	2250 May 19.5	2303 Mar 7.8
2093 Jan 9.8	2145 Oct 22.1	2198 Aug 8.5	2251 May 31.7	2304 Mar 20.2
2094 Jan 23.8	2146 Nov 4.9	2199 Aug 21.0	2252 Jun 11.9	2305 Apr 2.3
2095 Feb 6.7	2147 Nov 18.8	2200 Sep 2.6	2253 Jun 24.0	2306 Apr 15.2
2096 Feb 20.4	2148 Dec 1.9	2201 Sep 15.5	2254 Jul 6.0	2307 Apr 27.9
2097 Mar 4.9	2149 Dec 16.1	2202 Sep 28.6	2255 Jul 18.2	2308 May 9.4
2098 Mar 18.2	2150 Dec 30.2	2203 Oct 12.0	2256 Jul 29.4	2309 May 21.7
2099 Mar 31.3	2152 Jan 13.4	2204 Oct 24.5	2257 Aug 10.7	2310 Jun 2.9
2100 Apr 13.1	2153 Jan 26.4	2205 Nov 7.3	2258 Aug 23.2	2311 Jun 15.0
2101 Apr 25.8	2154 Feb 9.3	2206 Nov 21.3	2259 Sep 4.9	2312 Jun 26.1
2102 May 8.3	2155 Feb 23.0	2207 Dec 5.3	2260 Sep 16.7	2313 Jul 8.2
2103 May 20.6	2156 Mar 7.5	2208 Dec 18.5	2261 Sep 29.9	2314 Jul 20.3
2104 May 31.9	2157 Mar 20.8	2210 Jan 1.7	2262 Oct 13.2	2315 Aug 1.5
2105 Jun 13.0	2158 Apr 2.8	2211 Jan 15.8	2263 Oct 26.8	2316 Aug 12.9
2106 Jun 25.2	2159 Apr 15.6	2212 Jan 29.8	2264 Nov 8.6	2317 Aug 25.3
2107 Jul 7.3	2160 Apr 27.3	2213 Feb 11.7	2265 Nov 22.5	2318 Sep 7.0
2108 Jul 18.6	2161 May 9.7	2214 Feb 25.4	2266 Dec 6.6	2319 Sep 19.9
2109 Jul 30.9	2162 May 22.0	2215 Mar 10.9	2267 Dec 20.8	2320 Oct 2.0
2110 Aug 12.3	2163 Jun 3.2	2216 Mar 23.2	2269 Jan 3.0	2321 Oct 15.3
2111 Aug 24.9	2164 Jun 14.4	2217 Apr 5.2	2270 Jan 17.1	2322 Oct 28.9
2112 Sep 5.6	2165 Jun 26.5	2218 Apr 18.0	2271 Jan 31.1	2323 Nov 11.7
2113 Sep 18.6	2166 Jul 8.7	2219 Apr 30.6	2272 Feb 14.0	2324 Nov 24.7
2114 Oct 1.9	2167 Jul 20.9	2220 May 12.0	2273 Feb 26.7	2325 Dec 8.8
2115 Oct 15.3	2168 Aug 1.2	2221 May 24.3	2274 Mar 12.2	2326 Dec 22.9
2116 Oct 28.0	2169 Aug 13.6	2222 Jun 5.6	2275 Mar 25.4	2328 Jan 6.1
2117 Nov 10.9	2170 Aug 26.2	2223 Jun 17.7	2276 Apr 6.4	2329 Jan 19.2
2118 Nov 24.9	2171 Sep 8.0	2224 Jun 28.8	2277 Apr 19.3	2330 Feb 2.3
2119 Dec 9.0	2172 Sep 20.0	2225 Jul 10.9	2278 May 1.8	2331 Feb 16.1
2120 Dec 22.1	2173 Oct 3.3	2226 Jul 23.1	2279 May 14.3	2332 Feb 29.8
2122 Jan 5.3	2174 Oct 16.8	2227 Aug 4.4	2280 May 25.6	2333 Mar 14.3
2123 Jan 19.4	2175 Oct 30.5	2228 Aug 15.9	2281 Jun 6.8	2334 Mar 27.6
2124 Feb 2.3	2176 Nov 12.3	2229 Aug 28.5	2282 Jun 18.9	2335 Apr 9.6
2125 Feb 15.1	2177 Nov 26.4	2230 Sep 10.3	2283 Jul 1.0	2336 Apr 21.4
2126 Feb 28.7	2178 Dec 10.5	2231 Sep 23.3	2284 Jul 12.1	2337 May 4.0
2127 Mar 14.1	2179 Dec 24.7	2232 Oct 5.5	2285 Jul 24.3	2338 May 16.4
2128 Mar 26.2	2181 Jan 6.8	2233 Oct 19.0	2286 Aug 5.6	2339 May 28.7

2340 Jun 8.9	2373 Jul 21.4	2406 Sep 2.7	2439 Oct 18.4	2472 Dec 4.8
2341 Jun 21.0	2374 Aug 2.6	2407 Sep 15.5	2440 Oct 30.9	2473 Dec 18.9
2342 Jul 3.1	2375 Aug 15.0	2408 Sep 27.5	2441 Nov 13.7	2475 Jan 2.1
2343 Jul 15.2	2376 Aug 26.4	2409 Oct 10.7	2442 Nov 27.6	2476 Jan 16.3
2344 Jul 26.4	2377 Sep 8.1	2410 Oct 24.2	2443 Dec 11.7	2477 Jan 29.3
2345 Aug 7.7	2378 Sep 21.0	2411 Nov 6.9	2444 Dec 24.9	2478 Feb 12.3
2346 Aug 20.1	2379 Oct 4.1	2412 Nov 19.8	2446 Jan 8.0	2479 Feb 26.1
2347 Sep 1.7	2380 Oct 16.4	2413 Dec 3.9	2447 Jan 22.2	2480 Mar 10.7
2348 Sep 13.5	2381 Oct 30.0	2414 Dec 18.0	2448 Feb 5.2	2481 Mar 24.0
2349 Sep 26.5	2382 Nov 12.7	2416 Jan 1.2	2449 Feb 18.1	2482 Apr 6.2
2350 Oct 9.7	2383 Nov 26.7	2417 Jan 14.4	2450 Mar 3.7	2483 Apr 19.1
2351 Oct 23.2	2384 Dec 9.8	2418 Jan 28.5	2451 Mar 17.2	2484 Apr 30.8
2352 Nov 5.0	2385 Dec 23.9	2419 Feb 11.4	2452 Mar 29.5	2485 May 13.4
2353 Nov 18.9	2387 Jan 7.1	2420 Feb 25.2	2453 Apr 11.5	2486 May 25.7
2354 Dec 2.9	2388 Jan 21.2	2421 Mar 9.8	2454 Apr 24.3	2487 Jun 7.0
2355 Dec 17.1	2389 Feb 3.3	2422 Mar 23.2	2455 May 6.9	2488 Jun 18.1
2356 Dec 30.3	2390 Feb 17.1	2423 Apr 5.3	2456 May 18.4	2489 Jun 30.2
2358 Jan 13.5	2391 Mar 2.8	2424 Apr 17.2	2457 May 30.7	2490 Jul 12.3
2359 Jan 27.5	2392 Mar 15.3	2425 Apr 29.9	2458 Jun 11.9	2491 Jul 24.5
2360 Feb 10.5	2393 Mar 28.5	2426 May 12.5	2459 Jun 24.0	2492 Aug 4.7
2361 Feb 23.3	2394 Apr 10.6	2427 May 24.8	2460 Jul 5.1	2493 Aug 17.0
2362 Mar 8.9	2395 Apr 23.4	2428 Jun 5.0	2461 Jul 17.2	2494 Aug 29.5
2363 Mar 22.2	2396 May 5.0	2429 Jun 17.2	2462 Jul 29.4	2495 Sep 11.1
2364 Apr 3.4	2397 May 17.4	2430 Jun 29.2	2463 Aug 10.7	2496 Sep 23.0
2365 Apr 16.3	2398 May 29.7	2431 Jul 11.3	2464 Aug 22.1	2497 Oct 6.0
2366 Apr 29.0	2399 Jun 10.9	2432 Jul 22.5	2465 Sep 3.7	2498 Oct 19.4
2367 May 11.5	2400 Jun 22.0	2433 Aug 3.7	2466 Sep 16.4	2499 Nov 1.9
2368 May 22.8	2401 Jul 4.1	2434 Aug 16.0	2467 Sep 29.4	2500 Nov 15.7
2369 Jun 4.0	2402 Jul 16.2	2435 Aug 28.4	2468 Oct 11.7	
2370 Jun 16.1	2403 Jul 28.4	2436 Sep 9.1	2469 Oct 25.1	
2371 Jun 28.2	2404 Aug 8.7	2437 Sep 22.0	2470 Nov 7.8	
2372 Jul 9.3	2405 Aug 21.1	2438 Oct 5.1	2471 Nov 21.7	

During the period A.D. 0 − 2500, the three bright superior planets (Mars, Jupiter, Saturn) all reach opposition within an interval less than 30.0 days in the following years :

214, 451, 453, 551, 690, 788, 829, 927, 929, 1027, 1068, 1166, 1264, 1266, 1503, 1740-1741, 1743, 1980, 1982, 2219, and 2458.

The shortest interval was 4.4 days, in December 1503.

During the same period 0-2500, in the following years neither Jupiter nor Saturn are in opposition with the Sun : 591, 650, 709, 1445, 1504, 1563, 2239, 2298, and 2357.

In the years 591, 650, 1445, 1504, 2239, and 2357, there is no opposition of Mars, Jupiter, and Saturn.

XIII

TRANSITS OF MERCURY, 1600 – 2300

The table lists all transits of Mercury over the Sun's disk taking place in the period A.D. 1600 to 2300. The calculations are based on Bretagnon's VSOP 87 theory. More information can be found in the author's *Transits* (Wilmann-Bell, ed.; 1989).

All instants are given in *Dynamical Time*. To convert them to Universal Time, subtract the quantity ΔT, as explained on pages 5 to 8.

The data are *geocentric*, that is, as seen from the center of the Earth.

Columns 2 to 4 give the time t_1 of exterior contact at ingress, the time t_m of the least distance between the centers of Mercury and the Sun, and the time t_4 of exterior contact at egress, respectively. The date mentioned in the first column refers to the instant of least distance; for instance, in 1924, the transit began at $21^h 44^m$ TD on May 7, not on May 8.

The fifth column provides the least geocentric distance d_m between the centers of Mercury and the Sun, in seconds of arc. This quantity is positive when Mercury passes to the north of the center of the solar disk, negative if it passes to the south.

The last two columns give the position angles P_1 and P_4 of the center of Mercury at the times of exterior contacts at ingress and egress, respectively. These angles are measured from the North Point of the solar disk in the usual way, that is, eastward from the North.

Transits of Mercury occur either in May (when Mercury, at inferior conjunction with the Sun, is near the descending node of its orbit) or in November (when Mercury is near the ascending node). At the May transits, the semidiameter of Mercury is 6.0 arcseconds, that of the Sun is 950″ or 951″. At the November transits, the semidiameter of Mercury is equal to 5.0 arcseconds, that of the Sun is 968″ to 971″. These values apply to the period A.D. 1600 to 2300; they slowly vary in the course of the centuries.

Date			t_1	t_m	t_4	d_m	P_1	P_4
			h m	h m	h m	″	°	°
1605	Nov	1	18 46	20 05	21 24	−856	179	235
1615	May	3	6 43	10 11	13 38	+468	33	271
1618	Nov	4	11 10	13 44	16 18	−353	138	275
1628	May	5	14 21	17 33	20 46	−571	99	206
1631	Nov	7	4 39	7 22	10 05	+146	107	304
1644	Nov	9	22 54	0 58	3 01	+641	74	336
1651	Nov	3	23 08	0 53	2 39	−751	167	246
1661	May	3	13 05	16 55	20 44	+263	46	258
1664	Nov	4	15 53	18 33	21 12	−250	131	281
1674	May	7	21 56	0 17	2 37	−775	117	189
1677	Nov	7	9 32	12 11	14 50	+249	100	310
1690	Nov	10	3 57	5 43	7 30	+742	65	344

Date			t_1	t_m	t_4	d_m	P_1	P_4
			h m	h m	h m	"	°	°
1697	Nov	3	3 38	5 42	7 43	−647	158	255
1707	May	5	19 34	23 32	3 30	+ 65	59	246
1710	Nov	6	20 39	23 22	2 05	−145	124	287
1723	Nov	9	14 25	16 59	19 32	+351	94	316
1736	Nov	11	9 08	10 30	11 53	+843	54	354
1740	May	2	21 34	23 02	0 30	+889	354	310
1743	Nov	5	8 13	10 30	12 47	−542	150	262
1753	May	6	2 17	6 13	10 09	−139	71	234
1756	Nov	7	1 26	4 11	6 55	− 43	118	293
1769	Nov	9	19 21	21 47	0 12	+454	87	322
1776	Nov	2	20 56	21 36	22 17	−944	192	221
1782	Nov	12	14 35	15 16	15 57	+945	38	9
1786	May	4	2 56	5 41	8 26	+689	16	288
1789	Nov	5	12 52	15 19	17 46	−440	143	269
1799	May	7	9 07	12 51	16 34	−340	84	222
1802	Nov	9	6 14	8 59	11 43	+ 61	112	299
1815	Nov	12	0 19	2 33	4 48	+556	80	329
1822	Nov	5	1 01	2 25	3 49	−839	176	237
1832	May	5	9 00	12 25	15 50	+485	32	273
1835	Nov	7	17 33	20 08	22 43	−336	136	275
1845	May	8	16 20	19 37	22 53	−547	98	208
1848	Nov	9	11 05	13 48	16 30	+163	105	304
1861	Nov	12	5 18	7 20	9 21	+658	72	336
1868	Nov	5	5 26	7 14	9 03	−735	165	247
1878	May	6	15 13	19 00	22 47	+287	45	260
1881	Nov	8	22 17	0 57	3 38	−232	129	281
1891	May	10	23 54	2 22	4 49	−754	116	192
1894	Nov	10	15 56	18 35	21 13	+266	99	310
1907	Nov	14	10 24	12 07	13 50	+759	63	345
1914	Nov	7	9 57	12 03	14 10	−631	156	255
1924	May	8	21 44	1 41	5 39	+ 85	58	248
1927	Nov	10	3 02	5 46	8 30	−129	123	287
1937	May	11	8 53	9 00	9 06	−956	152	156
1940	Nov	11	20 49	23 22	1 54	+368	92	316
1953	Nov	14	15 38	16 54	18 11	+862	51	355
1957	May	6	23 59	1 15	2 30	+907	351	314
1960	Nov	7	14 34	16 53	19 13	−528	148	263
1970	May	9	4 20	8 17	12 14	−114	70	237
1973	Nov	10	7 48	10 33	13 18	− 26	116	293
1986	Nov	13	1 44	4 08	6 32	+471	85	323

Date	t_1	t_m	t_4	d_m	P_1	P_4
	h m	h m	h m	"	°	°
1993 Nov 6	3 07	3 58	4 48	−927	188	224
1999 Nov 15	21 16	21 42	22 08	+963	32	14
2003 May 7	5 14	7 53	10 33	+708	15	291
2006 Nov 8	19 13	21 42	0 11	−423	141	269
2016 May 9	11 13	14 59	18 44	−319	83	224
2019 Nov 11	12 37	15 21	18 05	+ 76	110	299
2032 Nov 13	6 42	8 55	11 09	+572	78	329
2039 Nov 7	7 19	8 48	10 17	−822	173	238
2049 May 7	11 05	14 26	17 46	+512	31	276
2052 Nov 9	23 55	2 32	5 08	−319	134	276
2062 May 10	18 18	21 39	0 59	−521	97	211
2065 Nov 11	17 26	20 08	22 50	+181	103	305
2078 Nov 14	11 45	13 44	15 42	+674	69	337
2085 Nov 7	11 45	13 37	15 29	−719	163	248
2095 May 8	17 24	21 09	0 53	+310	45	263
2098 Nov 10	4 39	7 20	10 01	−215	127	282
2108 May 12	1 43	4 20	6 56	−725	114	195
2111 Nov 14	22 19	0 57	3 35	+283	97	310
2124 Nov 15	16 54	18 33	20 11	+779	60	346
2131 Nov 9	16 19	18 27	20 36	−614	154	256
2141 May 10	23 51	3 48	7 44	+108	57	250
2144 Nov 11	9 23	12 07	14 51	−113	121	287
2154 May 13	10 08	11 03	11 58	−931	141	168
2157 Nov 14	3 14	5 45	8 16	+387	90	317
2170 Nov 16	22 11	23 21	0 32	+880	48	357
2174 May 8	2 30	3 32	4 33	+924	349	318
2177 Nov 9	20 54	23 15	1 36	−510	146	263
2187 May 11	6 34	10 31	14 27	− 96	70	239
2190 Nov 12	14 10	16 55	19 40	− 9	114	293
2203 Nov 16	8 12	10 34	12 56	+489	83	323
2210 Nov 9	9 22	10 20	11 19	−911	185	226
2220 May 9	7 31	10 04	12 38	+729	14	293
2223 Nov 12	1 33	4 03	6 34	−406	139	270
2233 May 12	13 21	17 08	20 54	−296	83	227
2236 Nov 13	18 59	21 43	0 28	+ 95	108	299
2249 Nov 16	13 11	15 22	17 32	+592	75	330
2256 Nov 9	13 36	15 09	16 41	−807	171	239
2266 May 10	13 26	16 44	20 01	+530	31	278
2269 Nov 12	6 14	8 52	11 29	−302	132	276
2279 May 13	20 25	23 48	3 12	−499	96	213
2282 Nov 15	23 51	2 32	5 14	+198	101	305
2295 Nov 17	18 15	20 10	22 05	+695	67	338

XIV

TRANSITS OF VENUS, 0 – 4000

The table lists all transits of Venus over the Sun's disk taking place in the period A.D. 0 to 4000. The calculations are based on Bretagnon's VSOP 87 theory. More information can be found in the author's *Transits* (Willmann-Bell, ed.; 1989).

All instants are given in *Dynamical Time*. To convert them to Universal Time, subtract the quantity ΔT, as explained on pages 5 to 8. Note that the Julian calendar is used before A.D. 1583. The data are *geocentric*, that is, as seen from the center of the Earth. The data in the columns are similar to those for the transits of Mercury; see the explanation on page 447.

On 2854 December 14 and on 3705 June 24, there will be no transit as seen from the center of the Earth, but there will be a partial transit of Venus at the Sun's southern limb for some places in the southern hemisphere.

During the period A.D. 1600 to 3000, the values of the semidiameters are: at the June transits, Sun 945″ to 947″, Venus 29″; at the December transits, Sun 973″ to 975″, Venus 32″.

Date	t_1	t_m	t_4	d_m	P_1	P_4
	h m	h m	h m	″	°	°
60 May 23	1 51	5 54	9 57	− 87	74	244
181 Nov 23	21 44	1 47	5 50	+114	105	298
303 May 24	5 14	9 14	13 14	−158	79	241
424 Nov 23	20 28	0 32	4 37	+ 10	110	291
546 May 24	8 45	12 41	16 38	−232	85	237
554 May 22	4 51	6 00	7 09	+934	356	324
667 Nov 23	19 16	23 19	3 23	− 99	115	284
789 May 24	12 16	16 07	19 58	−308	90	233
797 May 22	7 23	9 14	11 05	+867	8	314
910 Nov 23	18 02	22 01	2 00	−208	121	277
1032 May 24	15 43	19 28	23 13	−373	95	230
1040 May 22	10 26	12 48	15 10	+792	17	306
1153 Nov 23	17 03	20 56	0 49	−308	126	270
1275 May 25	19 13	22 49	2 26	−445	101	226
1283 May 23	13 16	15 56	18 37	+734	24	302
1396 Nov 23	15 50	19 32	23 14	−424	132	262

Date	t_1	t_m	t_4	d_m	P_1	P_4
	h m	h m	h m	"	°	°
1518 May 26	22 32	2 00	5 28	−505	106	223
1526 May 23	16 17	19 15	22 12	+667	30	297
1631 Dec 7	3 53	5 21	6 49	+939	37	354
1639 Dec 4	14 58	18 27	21 56	−524	138	255
1761 Jun 6	2 02	5 19	8 37	−570	111	220
1769 Jun 3	19 16	22 26	1 36	+609	36	293
1874 Dec 9	1 49	4 07	6 26	+830	49	340
1882 Dec 6	13 56	17 06	20 15	−637	145	246
2004 Jun 8	5 15	8 21	11 27	−627	116	216
2012 Jun 6	22 11	1 31	4 51	+554	41	290
2117 Dec 11	0 02	2 52	5 42	+724	58	330
2125 Dec 8	13 19	16 06	18 52	−736	152	238
2247 Jun 11	8 51	11 42	14 34	−691	122	212
2255 Jun 9	1 17	4 48	8 18	+492	46	287
2360 Dec 13	22 47	1 59	5 10	+626	65	321
2368 Dec 10	12 45	15 00	17 16	−836	160	228
2490 Jun 12	12 02	14 40	17 18	−741	128	209
2498 Jun 10	4 12	7 49	11 25	+443	50	284
2603 Dec 16	21 15	0 44	4 14	+517	71	313
2611 Dec 13	12 36	14 06	15 36	−935	172	215
2733 Jun 15	15 45	18 01	20 17	−808	135	203
2741 Jun 13	7 17	11 00	14 44	+386	55	282
2846 Dec 17	20 24	0 05	3 46	+432	76	307
2854 Dec 14	—	13 14	—	−1027	—	—
2976 Jun 16	18 54	20 53	22 52	−850	141	200
2984 Jun 14	10 10	13 59	17 47	+336	59	280
3089 Dec 18	19 02	22 54	2 46	+321	82	299
3219 Jun 20	22 31	0 00	1 29	−908	150	193
3227 Jun 17	13 03	16 55	20 47	+293	63	278
3332 Dec 20	18 14	22 12	2 10	+235	86	293
3462 Jun 22	1 48	2 47	3 45	−948	158	186
3470 Jun 19	15 51	19 47	23 42	+248	66	276
3575 Dec 23	17 08	21 11	1 13	+132	91	286
3705 Jun 24	—	5 35	—	−989	—	—
3713 Jun 21	18 30	22 27	2 24	+215	69	275
3818 Dec 25	16 23	20 27	0 32	+ 41	95	280
3956 Jun 24	21 17	1 17	5 16	+175	73	274

XV
SOLAR ECLIPSES, 1951 – 2050

On the next pages the list is given of all solar eclipses taking place during the period A.D. 1951 to 2050, in two parts :

(A) The first part lists the total and annular eclipses. The successive columns give :
— the calendar date ;
— the instant of the least distance of the axis of the lunar shadow cone to the center of the Earth, rounded to the nearest integer hour of Universal Time ;
— the type of the eclipse, for which the following symbols have been used :

> t = central total eclipse ;
> a = central annular eclipse ;
> at = annular-total eclipse. This is a central eclipse which is total for a part of the path, and annular for the rest ;
> (t) = non-central total eclipse ;
> (a) = non-central annular eclipse.

At a (t)- or (a)-eclipse, only a part of the umbral cone passes over the surface of the Earth (within the polar regions), and the axis of the cone does not intersect it. Such eclipses therefore have no central line ;
— the maximum duration of the total or annular phase on the central line, in minutes and seconds. In the case of an eclipse of type at, the maximum duration is that of the total phase ;
— the maximum width of the path of total or annular eclipse, in kilometers. Again, in the case of an at-eclipse, the maximum width refers to the part of the path where the eclipse is total. When only a part of the shadow cone passes over the Earth, no value is given for the width ;
— the maximum altitude of the center of the solar disk at central eclipse, in degrees ;
— a short description of the region where the total or annular phase is visible. The letters N, E, S, W mean 'northern part of ', etc.

(B) In the second part, all *partial* solar eclipses in the period 1951 – 2050 are given. These eclipses are nowhere total or annular. The successive columns give :
— the date ;
— the instant of the least distance of the axis of the lunar shadow cone to the center of the Earth, rounded to the nearest integer hour of Universal Time ;
— the greatest magnitude on the Earth's surface (that is, the greatest possible fraction of the solar diameter obscured by the Moon) ;
— whether the eclipse is visible from the northern (N) or southern (S) hemisphere of the Earth.

Accurate Besselian elements for these eclipses can be found in the author's *Elements of Solar Eclipses 1951 – 2200* (Willmann-Bell, ed.; 1989).

(A) TOTAL AND ANNULAR ECLIPSES

Date	UT	Type	Max. Dur.	Max. Width	Max. Alt.	Area of Visibility
	h		m s	km	°	
1951 Mar 7	21	a	1 39	101	76	New Zealand, Pacific Ocean, Central America
1951 Sep 1	13	a	2 44	155	81	E of U.S.A., Atlantic Ocean, Africa, Madagascar
1952 Feb 25	9	t	3 09	142	62	Atlantic, Africa, Arabia, Asia
1952 Aug 20	15	a	6 41	309	52	South America, S Atlantic Oc.
1954 Jan 5	3	a	1 49	402	21	Antarctica
1954 Jun 30	13	t	2 35	153	52	North America, S Greenland, Scandinavia, Russia, SW Asia
1954 Dec 25	8	a	7 40	347	75	Indian Ocean, Indonesia
1955 Jun 20	4	t	7 08	254	81	Indian Ocean, Ceylon, SE Asia, Philippines, Pacific Ocean
1955 Dec 14	7	a	12 10	378	65	NE Africa, Indian Ocean, SE Asia
1956 Jun 8	21	t	4 44	430	26	S Pacific Ocean
1957 Apr 30	0	(a)	—	—	—	Arctic Ocean, Novaya Zemlya
1957 Oct 23	5	(t)	—	—	—	Very small part of Antarctica
1958 Apr 19	3	a	7 07	297	74	Indian Ocean, SE Asia, Pacific
1958 Oct 12	21	t	5 11	210	73	Pacific Ocean, Chile
1959 Apr 8	3	a	7 26	301	63	Australia
1959 Oct 2	12	t	3 02	122	65	Atlantic, Canary Islands, Africa
1961 Feb 15	8	t	2 45	264	28	Europe, NW Asia
1961 Aug 11	11	a	6 35	522	27	S Atlantic Ocean
1962 Feb 5	0	t	4 08	147	78	Borneo, Celebes, New Guinea, Pacific Ocean
1962 Jul 31	12	a	3 33	162	84	N of South America, Atlantic Ocean, Africa, Madagascar
1963 Jan 25	14	a	1 08	89	60	S of South America, S Atlantic Ocean, S Africa, Madagascar
1963 Jul 20	21	t	1 40	101	49	N Pacific Ocean, North America
1965 May 30	21	t	5 16	199	65	New Zealand, Pacific Ocean
1965 Nov 23	4	a	4 05	187	67	Asia, Borneo, Celebes, New Guinea, Pacific Ocean

Date	UT	Type	Max. Dur.	Max. Width	Max. Alt.	Area of Visibility
	h		m s	km	°	
1966 May 20	10	a	0 59	69	70	Africa, Greece, Asia
1966 Nov 12	14	t	1 57	85	71	South America, S Atlantic Ocean
1967 Nov 2	6	(t)	—	—	—	S Atlantic Ocean
1968 Sep 22	11	t	0 40	110	19	Siberia, China
1969 Mar 18	5	a	1 18	79	74	Indian Ocean, Indonesia, Pacific
1969 Sep 11	20	a	3 15	177	77	Pacific Ocean, South America
1970 Mar 7	18	t	3 28	158	63	Pacific Ocean, North America, N Atlantic Ocean
1970 Aug 31	22	a	6 48	310	57	Pacific Ocean
1972 Jan 16	11	a	1 58	451	20	Antarctica
1972 Jul 10	20	t	2 36	178	46	E Siberia, Alaska, Canada, Atlantic Ocean
1973 Jan 4	16	a	7 50	356	74	S Pacific Ocean, South America, Atlantic Ocean
1973 Jun 30	12	t	7 04	256	86	N South America, Atlantic Ocean, Africa, Indian Ocean
1973 Dec 24	15	a	12 03	378	65	Central America, N of South America, Atlantic Ocean, NW Africa
1974 Jun 20	5	t	5 08	346	34	Indian Ocean, extr. SW Australia
1976 Apr 29	10	a	6 41	302	70	Atlantic Ocean, N Africa, Turkey, Asia
1976 Oct 23	5	t	4 47	200	71	E Africa, Indian Ocean, S Australia
1977 Apr 18	11	a	7 05	272	66	S Atlantic, S Africa, Indian Oc.
1977 Oct 12	20	t	2 37	99	67	Pacific Oc., N of South America
1979 Feb 26	17	t	2 49	307	26	U.S.A., Canada, Greenland
1979 Aug 22	17	a	6 03	1010	15	S Pacific Ocean, Antarctica
1980 Feb 16	9	t	4 08	149	77	Africa, Indian Ocean, S Asia
1980 Aug 10	19	a	3 24	158	79	Pacific Ocean, South America
1981 Feb 4	22	a	1 13	94	61	Tasmania, Pacific Ocean
1981 Jul 31	4	t	2 03	108	54	Siberia, Pacific Ocean
1983 Jun 11	5	t	5 11	200	60	Indian Ocean, Java, Celebes, New Guinea

Date	UT	Type	Max. Dur.		Max. Width	Max. Alt.	Area of Visibility
	h		m	s	km	°	
1983 Dec 4	13	a	4	03	183	66	Atlantic Ocean, Africa
1984 May 30	17	a	1	05	73	74	E Pacific Oc., Mexico, U.S.A., Atlantic Ocean, NW Africa
1984 Nov 22	23	t	2	00	85	72	New Guinea, S Pacific Ocean
1985 Nov 12	14	t	1	59	772	11	S Pacific Ocean
1986 Oct 3	19	at	0	00	1	5	N Atlantic Ocean
1987 Mar 29	13	at	0	08	5	72	S South America, Atlantic Ocean, Africa
1987 Sep 23	3	a	3	50	200	74	Asia, Pacific Ocean
1988 Mar 18	2	t	3	46	174	65	Sumatra, Borneo, Philippines, Pacific Ocean
1988 Sep 11	5	a	6	57	315	62	Indian Ocean
1990 Jan 26	20	a	2	06	505	18	S Atlantic Ocean, Antarctica
1990 Jul 22	3	t	2	33	209	40	Finland, Novaya Zemlya, N and NE Siberia, Pacific Ocean
1991 Jan 15	24	a	7	55	361	74	SW Australia, Tasmania, New Zealand, Pacific Ocean
1991 Jul 11	19	t	6	54	258	90	Pacific Ocean, Hawaii, Mexico, Central and South America
1992 Jan 4	23	a	11	43	377	66	Pacific Ocean
1992 Jun 30	12	t	5	20	297	41	Atlantic Ocean
1994 May 10	17	a	6	14	311	66	E Pacific Ocean, North America, Atlantic, extreme NW Africa
1994 Nov 3	14	t	4	23	190	69	South America, S Atlantic Ocean
1995 Apr 29	18	a	6	38	247	70	Pacific Ocean, South America
1995 Oct 24	5	t	2	10	78	69	Asia, Borneo, Pacific Ocean
1997 Mar 9	1	t	2	50	371	23	Mongolia, Siberia
1998 Feb 26	17	t	4	08	152	76	Pacific Ocean, N of South America, Atlantic Ocean
1998 Aug 22	2	a	3	14	156	75	Sumatra, Borneo, Pacific Ocean
1999 Feb 16	7	a	1	19	96	62	Indian Ocean, Australia
1999 Aug 11	11	t	2	23	112	59	Atlantic Ocean, Europe (France to Black Sea), SE and S Asia
2001 Jun 21	12	t	4	56	201	55	Atlantic Oc., Africa, Madagascar

Date	UT	Type	Max. Dur.	Max. Width	Max. Alt.	Area of Visibility
	h		m s	km	°	
2001 Dec 14	21	a	3 54	177	66	Pacific Ocean, Central America
2002 Jun 10	24	a	1 13	78	78	Pacific Ocean
2002 Dec 4	8	t	2 04	87	72	S Africa, Indian Oc., Australia
2003 May 31	4	a	3 37	—	3	Scotland, Iceland, Greenland
2003 Nov 23	23	t	1 57	544	15	Antarctica
2005 Apr 8	21	at	0 42	27	70	Pacific Ocean, Central America, N of South America
2005 Oct 3	11	a	4 32	223	71	Atlantic Ocean, Spain, Africa, Indian Ocean
2006 Mar 29	10	t	4 07	189	67	Atlantic Ocean, Africa, Turkey, Asia
2006 Sep 22	12	a	7 09	324	66	NE of South America, Atlantic
2008 Feb 7	4	a	2 14	581	16	S Pacific Ocean, Antarctica
2008 Aug 1	10	t	2 27	251	34	N Canada, N Greenland, Novaya Zemlya, Siberia, China
2009 Jan 26	8	a	7 56	363	73	S Atlantic Ocean, Indian Ocean, Sumatra, Borneo
2009 Jul 22	3	t	6 40	258	86	India, China, Pacific Ocean
2010 Jan 15	7	a	11 11	373	66	Africa, Indian Ocean, S and SE Asia
2010 Jul 11	20	t	5 20	262	47	Pacific, extreme S South America
2012 May 20	24	a	5 47	325	61	China, Japan, N Pacific, U.S.A.
2012 Nov 13	22	t	4 02	179	68	N Australia, Pacific Ocean
2013 May 10	0	a	6 04	226	74	Australia, Pacific Ocean
2013 Nov 3	13	at	1 40	58	71	Atlantic Ocean, Africa
2014 Apr 29	6	(a)	—	—	—	Antarctica
2015 Mar 20	10	t	2 47	486	198	N Atlantic Ocean, Svalbard, Arctic Ocean
2016 Mar 9	2	t	4 09	155	75	Sumatra, Borneo, Celebes, Pacific
2016 Sep 1	9	a	3 06	158	70	Atlantic Ocean, Africa, Madagascar, Indian Ocean
2017 Feb 26	15	a	1 23	97	63	S Pacific Ocean, South America, Atlantic Ocean, Africa
2017 Aug 21	18	t	2 40	115	64	Pacific, U.S.A., Atlantic Ocean

Date	UT	Type	Max. Dur.	Max. Width	Max. Alt.	Area of Visibility
	h		m s	km	°	
2019 Jul 2	19	t	4 32	201	50	Pacific Ocean, South America
2019 Dec 26	5	a	3 40	169	66	Arabia, India, Sumatra, Borneo, Pacific Ocean
2020 Jun 21	7	a	1 23	85	83	Africa, Arabia, Asia, Pacific
2020 Dec 14	16	t	2 10	90	73	Pacific, South America, Atlantic
2021 Jun 10	11	a	3 51	692	23	Canada, Arctic Oc., NE Siberia
2021 Dec 4	8	t	1 55	451	17	Antarctica, S Atlantic Ocean
2023 Apr 20	4	at	1 16	49	67	Indian Ocean, extr. W Australia, New Guinea, Pacific Ocean
2023 Oct 14	18	a	5 18	245	68	U.S.A., Central & South America
2024 Apr 8	18	t	4 28	203	70	Pacific Ocean, Mexico, U.S.A., E Canada, Atlantic Ocean
2024 Oct 2	19	a	7 25	333	69	Pacific Oc., S of South America
2026 Feb 17	12	a	2 21	766	12	Antarctica
2026 Aug 12	18	t	2 18	318	26	Arctic Ocean, Greenland, Iceland, N Atlantic Ocean, Spain
2027 Feb 6	16	a	7 54	362	73	S Pacific Ocean, South America, Atlantic Ocean
2027 Aug 2	10	t	6 23	259	82	Atlantic Ocean, N Africa, Arabia, Indian Ocean
2028 Jan 26	15	a	10 31	366	67	E Pacific Ocean, South America, Atlantic Ocean, Spain
2028 Jul 22	3	t	5 09	234	53	Indian Ocean, Australia, New Zealand
2030 Jun 1	6	a	5 21	345	56	N Africa, Greece, Asia, Japan
2030 Nov 25	7	t	3 44	169	67	S Africa, Indian Oc., Australia
2031 May 21	7	a	5 26	208	79	Africa, Indian Ocean, S India, SE Asia, Borneo, Celebes
2031 Nov 14	21	at	1 08	38	72	Pacific Ocean
2032 May 9	13	a	0 43	91	20	S Atlantic Ocean
2033 Mar 30	18	t	2 37	829	11	Alaska, Arctic Ocean
2034 Mar 20	10	t	4 09	160	73	Atlantic, Africa, Arabia, Asia
2034 Sep 12	16	a	2 58	162	67	Pacific, South America, S Atlantic
2035 Mar 9	23	a	1 26	96	64	New Zealand, Pacific Ocean

Date			UT	Type	Max. Dur.		Max. Width	Max. Alt.	Area of Visibility
			h		m	s	km	°	
2035 Sep	2		2	t	2	54	117	68	Asia, Japan, Pacific Ocean
2037 Jul	13		3	t	3	58	201	43	Indian Ocean, Australia, New Zealand
2038 Jan	5		14	a	3	19	159	65	Cuba, Haiti, Atlantic, Africa
2038 Jul	2		14	a	1	35	94	88	N of South America, Atlantic Ocean, Africa
2038 Dec	26		1	t	2	18	95	73	Australia, New Zealand, Pacific
2039 Jun	21		17	a	4	05	498	33	Alaska, Greenland, Scandinavia, Russia
2039 Dec	15		16	t	1	52	401	18	Antarctica
2041 Apr	30		12	t	1	50	72	63	S Atlantic, Africa, Indian Ocean
2041 Oct	25		2	a	6	08	267	66	E Asia, Japan, Pacific Ocean
2042 Apr	20		2	t	4	51	215	73	Indian Ocean, Sumatra, Borneo, Philippines, Pacific Ocean
2042 Oct	14		2	a	7	45	343	72	SE Asia, Borneo, Celebes, Australia, N. Zealand, S Pacific
2043 Apr	9		19	(t)	—		—	—	E Siberia
2043 Oct	3		3	(a)	—		—	—	Very small part SW Indian Ocean
2044 Feb	28		20	a	2	27	—	4	S Atlantic Ocean
2044 Aug	23		1	t	2	04	501	15	Greenland, Canada, U.S.A.
2045 Feb	16		24	a	7	49	358	72	New Zealand, Pacific Ocean
2045 Aug	12		18	t	6	06	259	78	E Pacific Ocean, U.S.A., Haiti, N of South America
2046 Feb	5		23	a	9	46	358	68	New Guinea, Pacific Ocean, Hawaii, W of U.S.A.
2046 Aug	2		10	t	4	51	210	58	Atlantic Ocean, S Africa, S Indian Ocean
2048 Jun	11		13	a	4	59	377	49	North America, Greenland, Iceland, Scandinavia, Russia
2048 Dec	5		16	t	3	28	160	66	Pacific Ocean, South America, S Atlantic Ocean, SW Africa
2049 May	31		14	a	4	46	194	83	South America, Atlantic, Africa
2049 Nov	25		6	at	0	37	21	73	Arabia, Indian Ocean, Sumatra, Borneo, Celebes, Pacific Oc.
2050 May	20		21	at	0	21	26	29	S Pacific Ocean

(B) PARTIAL ECLIPSES

Date			UT	Max. Magn.	
1953	Feb	14	1 h	0.760	N
1953	Jul	11	3	0.202	N
1953	Aug	9	16	0.373	S
1956	Dec	2	8	0.805	N
1960	Mar	27	7	0.705	S
1960	Sep	20	23	0.614	N
1964	Jan	14	20	0.559	S
1964	Jun	10	5	0.754	S
1964	Jul	9	11	0.322	N
1964	Dec	4	2	0.752	N
1967	May	9	15	0.720	N
1968	Mar	28	23	0.899	S
1971	Feb	25	10	0.787	N
1971	Jul	22	10	0.069	N
1971	Aug	20	23	0.508	S
1974	Dec	13	16	0.827	N
1975	May	11	7	0.864	N
1975	Nov	3	13	0.958	S
1978	Apr	7	15	0.788	S
1978	Oct	2	6	0.691	N
1982	Jan	25	5	0.566	S
1982	Jun	21	12	0.616	S
1982	Jul	20	19	0.465	N
1982	Dec	15	10	0.735	N
1985	May	19	21	0.841	N
1986	Apr	9	6	0.823	S
1989	Mar	7	18	0.827	N
1989	Aug	31	6	0.634	S
1992	Dec	24	1	0.842	N
1993	May	21	14	0.735	N
1993	Nov	13	22	0.928	S
1996	Apr	17	23	0.880	S
1996	Oct	12	14	0.758	N
1997	Sep	2	0	0.898	S
2000	Feb	5	13	0.579	S
2000	Jul	1	20	0.476	S
2000	Jul	31	2	0.604	N
2000	Dec	25	18	0.723	N

Date			UT	Max. Magn.	
2004	Apr	19	14 h	0.736	S
2004	Oct	14	3	0.929	N
2007	Mar	19	3	0.876	N
2007	Sep	11	13	0.750	S
2011	Jan	4	9	0.858	N
2011	Jun	1	21	0.601	N
2011	Jul	1	9	0.097	S
2011	Nov	25	6	0.904	S
2014	Oct	23	22	0.812	N
2015	Sep	13	7	0.787	S
2018	Feb	15	21	0.599	S
2018	Jul	13	3	0.336	S
2018	Aug	11	10	0.737	N
2019	Jan	6	2	0.715	N
2022	Apr	30	21	0.639	S
2022	Oct	25	11	0.862	N
2025	Mar	29	11	0.938	N
2025	Sep	21	20	0.855	S
2029	Jan	14	17	0.872	N
2029	Jun	12	4	0.458	N
2029	Jul	11	16	0.230	S
2029	Dec	5	15	0.891	S
2032	Nov	3	6	0.856	N
2033	Sep	23	14	0.689	S
2036	Feb	27	5	0.628	S
2036	Jul	23	11	0.199	S
2036	Aug	21	17	0.863	N
2037	Jan	16	10	0.705	N
2040	May	11	4	0.530	S
2040	Nov	4	19	0.808	N
2047	Jan	26	2	0.891	N
2047	Jun	23	11	0.313	N
2047	Jul	22	23	0.360	S
2047	Dec	16	24	0.881	S
2050	Nov	14	13	0.888	N

XVI
LUNAR ECLIPSES, 1951 – 2050

The table on the next pages mentions all lunar eclipses taking place from A.D. 1951 to 2050, including the penumbral eclipses. More details about these eclipses can be found in works mentioned in the Bibliography at the end of this book.

The first two columns give the date and the instant of the maximum of the eclipse, in *Universal Time*.

The next column gives the (greatest) magnitude of the eclipse, the Moon's diameter being taken as unity. The eclipse is total if the magnitude is greater than 1. In the case of a penumbral eclipse, the magnitude in the penumbra is given, and this value is placed between parentheses. In all other cases, the magnitude in the umbra is given.

It should be noted that penumbral eclipses are not observable unless their magnitude is larger than approximately 0.7. Small partial penumbral eclipses, such as that of 1998 August 8, are undistinguishible and are given only for reason of completeness ; the statistics of eclipses would be incomplete without them.

Observation has shown that the atmosphere of the Earth has the effect of increasing the apparent radius of the Earth's shadow by about one fiftieth. Following an old tradition, most astronomical almanacs increase by 1/50 the geometric radii of both the umbra and the penumbra.

However, since 1951 the French almanac *Connaissance des Temps* uses another theory for the calculation of the radii. A. Danjon correctly pointed out that the only reasonable way to take into account the presence of an opaque atmospheric layer around the Earth is to increase the Earth's radius, which can be performed by increasing proportionally the Moon's *parallax*. In that case, the radii of umbra and penumbra will undergo the same *absolute* correction, not the same relative correction as the traditional rule states. Thus, even if the rule of 1/50 were true for the umbra, it cannot be correct for the penumbra. As a consequence, the magnitudes of lunar eclipses, as calculated using the traditional rule, are too large (as compared to the values obtained by using the 'French' rule) by about 0.005 for umbral eclipses, but by about 0.026 for penumbral ones.

In the present list, the magnitudes of the eclipses have been calculated in accordance with the French rule. This should be taken into account when our results are compared with the data of other sources. In the calculation, we neglected the flattening of the Earth, and a mean radius for the Earth has been used, a convention adopted in the astronomical almanacs. For this reason, and because the edge of the Earth's shadow is rather poorly defined, we rounded the magnitudes of the eclipses to two decimal places.

Small penumbral 'eclipses', found by using the classical rule of 1/50 for the enlargement of the penumbra, do in fact not exist. This was the case for the 'eclipse' of 1951 February 21. It will occur again at the Full Moons of 2016 August 18 and 2042 October 28.

Finally, the last column of the table indicates whether the center of the Moon passes to the North (N) or to the South (S) of the axis of the Earth's shadow.

Date	UT (h m)	Magn.		Date	UT (h m)	Magn.	
1951 Mar 23	10 37	(0.64)	S	1968 Oct 6	11 42	1.17	N
1951 Aug 17	3 14	(0.12)	S	1969 Apr 2	18 32	(0.70)	S
1951 Sep 15	12 26	(0.80)	N	1969 Aug 27	10 48	(0.01)	S
1952 Feb 11	0 39	0.08	N	1969 Sep 25	20 10	(0.90)	N
1952 Aug 5	19 47	0.53	S	1970 Feb 21	8 30	0.05	N
1953 Jan 29	23 47	1.33	N	1970 Aug 17	3 23	0.41	S
1953 Jul 26	12 21	1.86	S	1971 Feb 10	7 45	1.31	N
1954 Jan 19	2 32	1.03	S	1971 Aug 6	19 43	1.73	S
1954 Jul 16	0 20	0.41	N	1972 Jan 30	10 53	1.05	S
1955 Jan 8	12 33	(0.85)	S	1972 Jul 26	7 16	0.54	N
1955 Jun 5	14 23	(0.62)	S	1973 Jan 18	21 17	(0.87)	S
1955 Nov 29	16 59	0.12	N	1973 Jun 15	20 50	(0.47)	S
1956 May 24	15 31	0.96	S	1973 Jul 15	11 39	(0.10)	N
1956 Nov 18	6 48	1.32	N	1973 Dec 10	1 44	0.10	N
1957 May 13	22 31	1.30	N	1974 Jun 4	22 16	0.83	S
1957 Nov 7	14 27	1.03	S	1974 Nov 29	15 13	1.29	N
1958 Apr 4	4 00	(0.02)	S	1975 May 25	5 48	1.43	N
1958 May 3	12 13	0.01	N	1975 Nov 18	22 23	1.06	S
1958 Oct 27	15 27	(0.78)	S	1976 May 13	19 54	0.12	N
1959 Mar 24	20 11	0.26	S	1976 Nov 6	23 01	(0.84)	S
1959 Sep 17	1 03	(0.99)	N	1977 Apr 4	4 18	0.19	S
1960 Mar 13	8 28	1.51	S	1977 Sep 27	8 29	(0.90)	N
1960 Sep 5	11 21	1.42	N	1978 Mar 24	16 22	1.45	S
1961 Mar 2	13 28	0.80	N	1978 Sep 16	19 04	1.33	N
1961 Aug 26	3 08	0.99	S	1979 Mar 13	21 08	0.85	N
1962 Feb 19	13 03	(0.61)	N	1979 Sep 6	10 54	1.09	S
1962 Jul 17	11 54	(0.39)	N	1980 Mar 1	20 45	(0.65)	N
1962 Aug 15	19 57	(0.60)	S	1980 Jul 27	19 08	(0.25)	N
1963 Jan 9	23 19	(1.02)	S	1980 Aug 26	3 30	(0.71)	S
1963 Jul 6	22 02	0.71	N	1981 Jan 20	7 50	(1.01)	S
1963 Dec 30	11 07	1.34	S	1981 Jul 17	4 47	0.55	N
1964 Jun 25	1 06	1.56	S	1982 Jan 9	19 56	1.33	S
1964 Dec 19	2 37	1.18	N	1982 Jul 6	7 31	1.72	S
1965 Jun 14	1 49	0.18	S	1982 Dec 30	11 29	1.18	N
1965 Dec 8	17 10	(0.88)	N	1983 Jun 25	8 22	0.33	S
1966 May 4	21 12	(0.92)	N	1983 Dec 20	1 49	(0.89)	N
1966 Oct 29	10 12	(0.95)	S	1984 May 15	4 40	(0.81)	N
1967 Apr 24	12 06	1.34	N	1984 Jun 13	14 26	(0.06)	S
1967 Oct 18	10 15	1.14	S	1984 Nov 8	17 55	(0.90)	S
1968 Apr 13	4 47	1.11	S	1985 May 4	19 56	1.24	N

Date	UT	Magn.		Date	UT	Magn.	
	h m				h m		
1985 Oct 28	17 42	1.07	S	2003 May 16	3 40	1.13	N
1986 Apr 24	12 43	1.20	S	−2003 Nov 9	1 19	1.02	S
1986 Oct 17	19 18	1.24	N	2004 May 4	20 30	1.30	S
1987 Apr 14	2 19	(0.78)	S	⌐ 2004 Oct 28	3 04	1.31	N
1987 Oct 7	4 01	(0.99)	N	2005 Apr 24	9 55	(0.87)	S
1988 Mar 3	16 13	(1.09)	N	⪤ 2005 Oct 17	12 03	0.06	N
1988 Aug 27	11 05	0.29	S	2006 Mar 14	23 47	(1.03)	N
1989 Feb 20	15 35	1.27	N	− 2006 Sep 7	18 51	0.18	S
1989 Aug 17	3 08	1.60	S	2007 Mar 3	23 21	1.23	N
1990 Feb 9	19 11	1.07	S	− 2007 Aug 28	10 37	1.48	S
1990 Aug 6	14 12	0.68	N	2008 Feb 21	3 26	1.11	S
1991 Jan 30	5 59	(0.88)	S	⁻ 2008 Aug 16	21 10	0.81	N
1991 Jun 27	3 15	(0.31)	S	2009 Feb 9	14 38	(0.90)	S
1991 Jul 26	18 08	(0.25)	N	2009 Jul 7	9 39	(0.16)	S
1991 Dec 21	10 33	0.09	N	⪤ 2009 Aug 6	0 39	(0.40)	N
1992 Jun 15	4 57	0.68	S	2009 Dec 31	19 23	0.08	N
1992 Dec 9	23 44	1.27	N	− 2010 Jun 26	11 38	0.54	S
1993 Jun 4	13 00	1.56	N	2010 Dec 21	8 17	1.26	N
1993 Nov 29	6 26	1.09	S	⪤ 2011 Jun 15	20 13	1.70	N
1994 May 25	3 30	0.24	N	2011 Dec 10	14 32	1.11	S
1994 Nov 18	6 44	(0.88)	S	⪤ 2012 Jun 4	11 03	0.37	N
1995 Apr 15	12 18	0.11	S	2012 Nov 28	14 33	(0.92)	S
1995 Oct 8	16 04	(0.83)	N	2013 Apr 25	20 07	0.01	S
1996 Apr 4	0 10	1.38	S	⪤ 2013 May 25	4 10	(0.02)	N
1996 Sep 27	2 54	1.24	N	2013 Oct 18	23 50	(0.76)	N
1997 Mar 24	4 39	0.92	N	⪤ 2014 Apr 15	7 46	1.29	S
1997 Sep 16	18 47	1.19	S	2014 Oct 8	10 54	1.17	N
1998 Mar 13	4 20	(0.71)	N	− 2015 Apr 4	12 00	1.00	N
1998 Aug 8	2 25	(0.12)	N	2015 Sep 28	2 47	1.28	S
1998 Sep 6	11 10	(0.81)	S.	⌐2016 Mar 23	11 47	(0.77)	N
1999 Jan 31	16 18	(1.00)	S	2016 Sep 16	18 54	(0.91)	S
1999 Jul 28	11 34	0.40	N	− 2017 Feb 11	0 44	(0.99)	S
− 2000 Jan 21	4 44	1.32	S	2017 Aug 7	18 20	0.25	N
2000 Jul 16	13 56	1.77	N	⌐ 2018 Jan 31	13 30	1.32	S
⌐2001 Jan 9	20 21	1.19	N	2018 Jul 27	20 22	1.61	N
2001 Jul 5	14 55	0.49	S	⌐ 2019 Jan 21	5 12	1.20	N
− 2001 Dec 30	10 29	(0.89)	N	2019 Jul 16	21 31	0.65	S
2002 May 26	12 03	(0.69)	N	2020 Jan 10	19 10	(0.90)	N
2002 Jun 24	21 27	(0.21)	S	2020 Jun 5	19 25	(0.57)	N
⪤ 2002 Nov 20	1 47	(0.86)	S	2020 Jul 5	4 30	(0.35)	S

Date	UT	Magn.		Date	UT	Magn.	
	h m				h m		
2020 Nov 30	9 43	(0.83)	S	2036 Aug 7	2 51	1.45	N
2021 May 26	11 19	1.01	N	2037 Jan 31	14 00	1.21	N
2021 Nov 19	9 03	0.97	S	2037 Jul 27	4 08	0.81	S
2022 May 16	4 11	1.41	S	2038 Jan 21	3 48	(0.90)	N
2022 Nov 8	10 59	1.36	N	2038 Jun 17	2 43	(0.44)	N
2023 May 5	17 23	(0.96)	S	2038 Jul 16	11 34	(0.50)	S
2023 Oct 28	20 14	0.12	N	2038 Dec 11	17 43	(0.80)	S
2024 Mar 25	7 13	(0.96)	N	2039 Jun 6	18 53	0.88	N
2024 Sep 18	2 44	0.08	S	2039 Nov 30	16 55	0.94	S
2025 Mar 14	6 59	1.18	N	2040 May 26	11 45	1.53	S
2025 Sep 7	18 12	1.36	S	2040 Nov 18	19 03	1.40	N
2026 Mar 3	11 33	1.15	S	2041 May 16	0 41	0.06	S
2026 Aug 28	4 13	0.93	N	2041 Nov 8	4 33	0.17	N
2027 Feb 20	23 13	(0.93)	S	2042 Apr 5	14 28	(0.87)	N
2027 Jul 18	16 03	(0.00)	S	2042 Sep 29	10 44	(0.95)	S
2027 Aug 17	7 13	(0.55)	N	2043 Mar 25	14 30	1.11	N
2028 Jan 12	4 13	0.07	N	2043 Sep 19	1 50	1.26	S
2028 Jul 6	18 19	0.39	S	2044 Mar 13	19 37	1.20	S
2028 Dec 31	16 52	1.25	N	2044 Sep 7	11 19	1.04	N
2029 Jun 26	3 22	1.84	N	2045 Mar 3	7 42	(0.96)	S
2029 Dec 20	22 42	1.12	S	2045 Aug 27	13 53	(0.68)	N
2030 Jun 15	18 33	0.50	N	2046 Jan 22	13 01	0.05	N
2030 Dec 9	22 27	(0.94)	S	2046 Jul 18	1 04	0.25	S
2031 May 7	3 51	(0.88)	S	2047 Jan 12	1 24	1.23	N
2031 Jun 5	11 44	(0.13)	N	2047 Jul 7	10 34	1.75	S
2031 Oct 30	7 45	(0.72)	N	2048 Jan 1	6 52	1.13	S
2032 Apr 25	15 13	1.19	S	2048 Jun 26	2 01	0.64	N
2032 Oct 18	19 02	1.10	N	2048 Dec 20	6 26	(0.96)	S
2033 Apr 14	19 12	1.09	N	2049 May 17	11 25	(0.77)	S
2033 Oct 8	10 55	1.35	S	2049 Jun 15	19 12	(0.25)	N
2034 Apr 3	19 05	(0.85)	N	2049 Nov 9	15 50	(0.68)	N
2034 Sep 28	2 46	0.01	S	2050 May 6	22 30	1.08	S
2035 Feb 22	9 05	(0.97)	S	2050 Oct 30	3 20	1.05	N
2035 Aug 19	1 11	0.10	N				
2036 Feb 11	22 12	1.30	S				

XVII
EQUINOXES AND SOLSTICES ON MARS

This table gives the times of the equinoxes and solstices on the planet Mars for the years 1646 to 2060. On each line one finds, from the left to the right, respectively: the time of the spring equinox for the *northern* hemisphere of Mars (corresponding to the March equinox on the Earth), that of the summer solstice, that of the autumn equinox, and finally the instant of the winter solstice. Each instant has been rounded to the nearest hundredth of a day, *Universal Time*.

For the calculation of the heliocentric positions of Mars, use was made of the VSOP87 theory by P. Bretagnon. We adopted the following expressions for the longitude and latitude of the North Pole of Mars, referred to the ecliptic and mean equinox of the date, as given on page 272 of our *Astronomical Algorithms* (Willmann-Bell, ed.; 1991):

$$\lambda_0 = 352°.9065 + 1°.17330\,T$$
$$\beta_0 = +63°.2818 - 0°.00394\,T$$

where T is the time in Julian centuries (36525 days) from the epoch 2000.0.

The following equinoxes and solstices (for Mars' northern hemisphere) occur less than ten days from an opposition of Mars:

Spring Equinox :	1691 Dec 14, 1770 Dec 13, 1849 Dec 12, 1896 Dec 19, 1975 Dec 19, 2054 Dec 16
Summer Solstice :	1713 Mar 10, 1760 Mar 17, 1792 Mar 7, 1839 Mar 17, 1918 Mar 15, 1997 Mar 13, 2044 Mar 20
Autumn Equinox :	1717 Jun 13, 1796 Jun 11, 1875 Jun 11, 1922 Jun 19, 2001 Jun 17
Winter Solstice :	1672 Sep 15, 1751 Sep 15, 1830 Sep 14, 2035 Sep 20

Finally the durations (in *terrestrial* days) of the astronomical seasons, for the northern hemisphere of Mars, are as follows:

	Epoch 1650	*Epoch 1850*	*Epoch 2050*
Spring	199.07 days	198.80 days	198.53 days
Summer	182.34	183.00	183.65
Autumn	146.34	146.51	146.71
Winter	159.22	158.66	158.09

Spring Equinox	*Summer Solstice*	*Autumn Equinox*	*Winter Solstice*
1646 Oct 24.25	1647 May 11.33	1647 Nov 9.67	1648 Apr 3.98
1648 Sep 10.19	1649 Mar 28.28	1649 Sep 26.62	1650 Feb 19.98
1650 Jul 29.18	1651 Feb 13.24	1651 Aug 14.57	1652 Jan 7.93
1652 Jun 15.17	1652 Dec 31.21	1653 Jul 1.56	1653 Nov 24.89
1654 May 3.13	1654 Nov 18.19	1655 May 19.53	1655 Oct 12.89
1656 Mar 20.10	1656 Oct 5.18	1657 Apr 5.51	1657 Aug 29.85
1658 Feb 5.08	1658 Aug 23.15	1659 Feb 21.52	1659 Jul 17.84
1659 Dec 24.03	1660 Jul 10.09	1661 Jan 8.48	1661 Jun 3.84
1661 Nov 10.00	1662 May 28.05	1662 Nov 26.42	1663 Apr 21.80
1663 Sep 27.99	1664 Apr 14.02	1664 Oct 13.40	1665 Mar 8.75
1665 Aug 14.96	1666 Mar 1.98	1666 Aug 31.37	1667 Jan 24.74
1667 Jul 2.92	1668 Jan 17.96	1668 Jul 18.32	1668 Dec 11.69
1669 May 19.90	1669 Dec 4.95	1670 Jun 5.33	1670 Oct 29.65
1671 Apr 6.84	1671 Oct 22.92	1672 Apr 22.35	1672 Sep 15.69
1673 Feb 21.82	1673 Sep 8.86	1674 Mar 10.29	1674 Aug 3.68
1675 Jan 9.82	1675 Jul 27.83	1676 Jan 26.25	1676 Jun 20.62
1676 Nov 26.80	1677 Jun 13.81	1677 Dec 13.24	1678 May 8.60
1678 Oct 14.75	1679 May 1.78	1679 Oct 31.18	1680 Mar 25.58
1680 Aug 31.73	1681 Mar 18.77	1681 Sep 17.18	1682 Feb 10.52
1682 Jul 19.69	1683 Feb 3.72	1683 Aug 5.18	1683 Dec 29.52
1684 Jun 5.64	1684 Dec 21.68	1685 Jun 22.13	1685 Nov 15.52
1686 Apr 23.63	1686 Nov 8.65	1687 May 10.10	1687 Oct 3.48
1688 Mar 10.61	1688 Sep 25.62	1689 Mar 27.10	1689 Aug 20.46
1690 Jan 26.58	1690 Aug 13.58	1691 Feb 12.05	1691 Jul 8.46
1691 Dec 14.56	1692 Jun 30.57	1692 Dec 30.01	1693 May 25.39
1693 Oct 31.52	1694 May 18.54	1694 Nov 17.02	1695 Apr 12.37
1695 Sep 18.46	1696 Apr 4.48	1696 Oct 3.99	1697 Feb 27.38
1697 Aug 5.44	1698 Feb 20.44	1698 Aug 21.93	1699 Jan 15.34
1699 Jun 23.43	1700 Jan 8.41	1700 Jul 9.91	1700 Dec 3.28
1701 May 11.39	1701 Nov 26.39	1702 May 27.89	1702 Oct 21.29
1703 Mar 29.36	1703 Oct 14.38	1704 Apr 13.86	1704 Sep 7.26
1705 Feb 13.35	1705 Aug 31.36	1706 Mar 1.87	1706 Jul 26.23
1707 Jan 1.30	1707 Jul 19.31	1708 Jan 17.85	1708 Jun 12.24
1708 Nov 18.27	1709 Jun 5.27	1709 Dec 4.80	1710 Apr 30.21
1710 Oct 6.26	1711 Apr 23.23	1711 Oct 22.77	1712 Mar 17.16
1712 Aug 23.23	1713 Mar 10.20	1713 Sep 8.75	1714 Feb 2.15
1714 Jul 11.19	1715 Jan 26.17	1715 Jul 27.69	1715 Dec 21.12
1716 May 28.18	1716 Dec 13.17	1717 Jun 13.69	1717 Nov 7.06
1718 Apr 15.13	1718 Oct 31.13	1719 May 1.71	1719 Sep 25.09
1720 Mar 2.08	1720 Sep 17.08	1721 Mar 18.67	1721 Aug 12.09

Spring Equinox	Summer Solstice	Autumn Equinox	Winter Solstice
1722 Jan 18.09	1722 Aug 5.04	1723 Feb 3.62	1723 Jun 30.04
1723 Dec 6.07	1724 Jun 22.01	1724 Dec 21.60	1725 May 17.00
1725 Oct 23.02	1726 May 9.98	1726 Nov 8.55	1727 Apr 3.99
1727 Sep 10.00	1728 Mar 26.97	1728 Sep 25.52	1729 Feb 18.92
1729 Jul 27.96	1730 Feb 11.93	1730 Aug 13.53	1731 Jan 6.91
1731 Jun 14.90	1731 Dec 30.88	1732 Jun 30.49	1732 Nov 23.92
1733 May 1.88	1733 Nov 16.85	1734 May 18.45	1734 Oct 11.88
1735 Mar 19.87	1735 Oct 4.82	1736 Apr 4.45	1736 Aug 28.86
1737 Feb 3.85	1737 Aug 21.79	1738 Feb 20.42	1738 Jul 16.86
1738 Dec 22.83	1739 Jul 9.78	1740 Jan 8.37	1740 Jun 2.80
1740 Nov 8.80	1741 May 26.76	1741 Nov 25.38	1742 Apr 20.77
1742 Sep 26.74	1743 Apr 13.70	1743 Oct 13.37	1744 Mar 7.78
1744 Aug 13.71	1745 Feb 28.66	1745 Aug 30.31	1746 Jan 23.76
1746 Jul 1.71	1747 Jan 16.63	1747 Jul 18.28	1747 Dec 11.70
1748 May 18.68	1748 Dec 3.60	1749 Jun 4.26	1749 Oct 28.69
1750 Apr 5.63	1750 Oct 21.59	1751 Apr 22.23	1751 Sep 15.68
1752 Feb 21.63	1752 Sep 7.57	1753 Mar 9.23	1753 Aug 2.63
1754 Jan 8.59	1754 Jul 26.53	1755 Jan 25.22	1755 Jun 20.64
1755 Nov 26.54	1756 Jun 12.47	1756 Dec 12.17	1757 May 7.62
1757 Oct 13.53	1758 Apr 30.44	1758 Oct 30.12	1759 Mar 25.57
1759 Aug 31.50	1760 Mar 17.40	1760 Sep 16.11	1761 Feb 9.54
1761 Jul 18.46	1762 Feb 2.37	1762 Aug 4.05	1762 Dec 28.52
1763 Jun 5.44	1763 Dec 21.36	1764 Jun 21.03	1764 Nov 14.46
1765 Apr 22.40	1765 Nov 7.34	1766 May 9.06	1766 Oct 2.47
1767 Mar 10.34	1767 Sep 25.29	1768 Mar 26.04	1768 Aug 19.49
1769 Jan 25.35	1769 Aug 12.25	1770 Feb 10.98	1770 Jul 7.44
1770 Dec 13.34	1771 Jun 30.22	1771 Dec 29.96	1772 May 24.40
1772 Oct 30.29	1773 May 17.19	1773 Nov 15.93	1774 Apr 11.40
1774 Sep 17.26	1775 Apr 4.17	1775 Oct 3.88	1776 Feb 27.34
1776 Aug 4.24	1777 Feb 19.15	1777 Aug 20.90	1778 Jan 14.31
1778 Jun 22.18	1779 Jan 7.10	1779 Jul 8.87	1779 Dec 2.32
1780 May 9.16	1780 Nov 24.06	1781 May 25.82	1781 Oct 19.30
1782 Mar 27.15	1782 Oct 12.04	1783 Apr 12.82	1783 Sep 6.27
1784 Feb 12.13	1784 Aug 29.01	1785 Feb 27.80	1785 Jul 24.26
1785 Dec 30.10	1786 Jul 16.98	1787 Jan 15.74	1787 Jun 11.22
1787 Nov 17.08	1788 Jun 2.97	1788 Dec 2.74	1789 Apr 28.17
1789 Oct 4.02	1790 Apr 20.91	1790 Oct 20.73	1791 Mar 16.18
1791 Aug 21.98	1792 Mar 7.86	1792 Sep 6.67	1793 Jan 31.16
1793 Jul 8.97	1794 Jan 23.83	1794 Jul 25.63	1794 Dec 19.10
1795 May 26.95	1795 Dec 11.80	1796 Jun 11.62	1796 Nov 5.08

Spring Equinox	*Summer Solstice*	*Autumn Equinox*	*Winter Solstice*
1797 Apr 12.90	1797 Oct 28.79	1798 Apr 29.59	1798 Sep 23.08
1799 Feb 28.89	1799 Sep 15.77	1800 Mar 17.58	1800 Aug 11.03
1801 Jan 16.86	1801 Aug 3.73	1802 Feb 2.59	1802 Jun 29.03
1802 Dec 4.80	1803 Jun 21.68	1803 Dec 21.54	1804 May 16.03
1804 Oct 21.79	1805 May 8.64	1805 Nov 7.49	1806 Apr 2.98
1806 Sep 8.78	1807 Mar 26.61	1807 Sep 25.47	1808 Feb 18.94
1808 Jul 26.74	1809 Feb 10.58	1809 Aug 12.43	1810 Jan 5.94
1810 Jun 13.71	1810 Dec 29.57	1811 Jun 30.39	1811 Nov 23.87
1812 Apr 30.68	1812 Nov 15.56	1813 May 17.42	1813 Oct 10.86
1814 Mar 18.62	1814 Oct 3.51	1815 Apr 4.41	1815 Aug 28.89
1816 Feb 3.62	1816 Aug 20.46	1817 Feb 19.35	1817 Jul 15.86
1817 Dec 21.61	1818 Jul 8.43	1819 Jan 7.32	1819 Jun 2.81
1819 Nov 8.57	1820 May 25.40	1820 Nov 24.29	1821 Apr 19.80
1821 Sep 25.53	1822 Apr 12.37	1822 Oct 12.24	1823 Mar 7.75
1823 Aug 13.51	1824 Feb 28.35	1824 Aug 29.25	1825 Jan 22.71
1825 Jun 30.46	1826 Jan 15.31	1826 Jul 17.23	1826 Dec 10.72
1827 May 18.42	1827 Dec 3.27	1828 Jun 3.18	1828 Oct 27.70
1829 Apr 4.41	1829 Oct 20.24	1830 Apr 21.17	1830 Sep 14.67
1831 Feb 20.39	1831 Sep 7.21	1832 Mar 8.16	1832 Aug 1.66
1833 Jan 7.36	1833 Jul 25.18	1834 Jan 24.10	1834 Jun 19.63
1834 Nov 25.34	1835 Jun 12.17	1835 Dec 12.09	1836 May 6.57
1836 Oct 12.29	1837 Apr 29.13	1837 Oct 29.09	1838 Mar 24.57
1838 Aug 30.24	1839 Mar 17.08	1839 Sep 16.04	1840 Feb 9.57
1840 Jul 17.24	1841 Feb 1.04	1841 Aug 2.99	1841 Dec 27.52
1842 Jun 4.22	1842 Dec 20.02	1843 Jun 20.98	1843 Nov 14.48
1844 Apr 21.18	1844 Nov 6.00	1845 May 7.96	1845 Oct 1.49
1846 Mar 9.16	1846 Sep 23.98	1847 Mar 25.94	1847 Aug 19.45
1848 Jan 25.14	1848 Aug 10.95	1849 Feb 9.95	1849 Jul 6.43
1849 Dec 12.08	1850 Jun 28.90	1850 Dec 28.91	1851 May 24.44
1851 Oct 30.06	1852 May 15.85	1852 Nov 14.85	1853 Apr 10.39
1853 Sep 16.05	1854 Apr 2.82	1854 Oct 2.83	1855 Feb 26.34
1855 Aug 4.01	1856 Feb 18.79	1856 Aug 19.80	1857 Jan 13.34
1857 Jun 20.98	1858 Jan 5.77	1858 Jul 7.75	1858 Dec 1.28
1859 May 8.95	1859 Nov 23.76	1860 May 24.77	1860 Oct 18.25
1861 Mar 25.89	1861 Oct 10.71	1862 Apr 11.78	1862 Sep 5.29
1863 Feb 10.87	1863 Aug 28.66	1864 Feb 27.72	1864 Jul 23.27
1864 Dec 28.88	1865 Jul 15.63	1866 Jan 14.68	1866 Jun 10.21
1866 Nov 15.84	1867 Jun 2.60	1867 Dec 2.66	1868 Apr 27.20
1868 Oct 2.80	1869 Apr 19.58	1869 Oct 19.61	1870 Mar 15.17
1870 Aug 20.78	1871 Mar 7.56	1871 Sep 6.60	1872 Jan 31.11

Spring Equinox	*Summer Solstice*	*Autumn Equinox*	*Winter Solstice*
1872 Jul 7.74	1873 Jan 22.52	1873 Jul 24.60	1873 Dec 18.12
1874 May 25.69	1874 Dec 10.48	1875 Jun 11.56	1875 Nov 5.11
1876 Apr 11.68	1876 Oct 27.45	1877 Apr 28.53	1877 Sep 22.08
1878 Feb 27.67	1878 Sep 14.42	1879 Mar 16.53	1879 Aug 10.06
1880 Jan 15.64	1880 Aug 1.39	1881 Jan 31.47	1881 Jun 27.05
1881 Dec 2.62	1882 Jun 19.38	1882 Dec 19.44	1883 May 14.98
1883 Oct 20.58	1884 May 6.35	1884 Nov 5.45	1885 Mar 31.97
1885 Sep 6.52	1886 Mar 24.29	1886 Sep 23.42	1887 Feb 16.98
1887 Jul 25.50	1888 Feb 9.25	1888 Aug 10.36	1889 Jan 3.93
1889 Jun 11.49	1889 Dec 27.22	1890 Jun 28.34	1890 Nov 21.88
1891 Apr 29.45	1891 Nov 14.19	1892 May 15.32	1892 Oct 8.89
1893 Mar 16.42	1893 Oct 1.18	1894 Apr 2.29	1894 Aug 26.85
1895 Feb 1.41	1895 Aug 19.15	1896 Feb 18.30	1896 Jul 13.83
1896 Dec 19.35	1897 Jul 6.10	1898 Jan 5.27	1898 May 31.83
1898 Nov 6.32	1899 May 24.06	1899 Nov 23.22	1900 Apr 18.80
1900 Sep 24.32	1901 Apr 11.03	1901 Oct 11.19	1902 Mar 6.75
1902 Aug 12.28	1903 Feb 27.00	1903 Aug 29.17	1904 Jan 22.74
1904 Jun 29.25	1905 Jan 13.97	1905 Jul 16.11	1905 Dec 9.70
1906 May 17.23	1906 Dec 1.97	1907 Jun 3.12	1907 Oct 27.66
1908 Apr 3.17	1908 Oct 18.93	1909 Apr 20.14	1909 Sep 13.69
1910 Feb 19.15	1910 Sep 5.88	1911 Mar 8.09	1911 Aug 1.69
1912 Jan 7.15	1912 Jul 23.85	1913 Jan 23.04	1913 Jun 18.63
1913 Nov 24.13	1914 Jun 10.82	1914 Dec 11.03	1915 May 6.60
1915 Oct 12.08	1916 Apr 27.79	1916 Oct 27.98	1917 Mar 23.58
1917 Aug 29.06	1918 Mar 15.77	1918 Sep 14.96	1919 Feb 8.52
1919 Jul 17.02	1920 Jan 31.73	1920 Aug 1.97	1920 Dec 26.51
1921 Jun 2.96	1921 Dec 18.68	1922 Jun 19.92	1922 Nov 13.52
1923 Apr 20.95	1923 Nov 5.65	1924 May 6.89	1924 Sep 30.49
1925 Mar 7.94	1925 Sep 22.62	1926 Mar 24.88	1926 Aug 18.46
1927 Jan 23.91	1927 Aug 10.59	1928 Feb 9.84	1928 Jul 5.45
1928 Dec 10.88	1929 Jun 27.57	1929 Dec 27.79	1930 May 23.39
1930 Oct 28.85	1931 May 15.55	1931 Nov 14.81	1932 Apr 9.36
1932 Sep 14.79	1933 Apr 1.50	1933 Oct 1.78	1934 Feb 25.38
1934 Aug 2.77	1935 Feb 17.45	1935 Aug 19.72	1936 Jan 13.34
1936 Jun 19.76	1937 Jan 4.42	1937 Jul 6.70	1937 Nov 30.29
1938 May 7.73	1938 Nov 22.40	1939 May 24.69	1939 Oct 18.29
1940 Mar 24.69	1940 Oct 9.38	1941 Apr 10.65	1941 Sep 4.27
1942 Feb 9.69	1942 Aug 27.37	1943 Feb 26.66	1943 Jul 23.23
1943 Dec 28.64	1944 Jul 14.33	1945 Jan 13.65	1945 Jun 9.24
1945 Nov 14.60	1946 Jun 1.28	1946 Dec 1.59	1947 Apr 27.22

Spring Equinox	Summer Solstice	Autumn Equinox	Winter Solstice
1947 Oct 2.59	1948 Apr 18.24	1948 Oct 18.55	1949 Mar 14.17
1949 Aug 19.57	1950 Mar 6.21	1950 Sep 5.54	1951 Jan 30.14
1951 Jul 7.52	1952 Jan 22.18	1952 Jul 23.48	1952 Dec 17.12
1953 May 24.50	1953 Dec 9.17	1954 Jun 10.47	1954 Nov 4.06
1955 Apr 11.46	1955 Oct 27.14	1956 Apr 27.50	1956 Sep 21.08
1957 Feb 26.41	1957 Sep 13.08	1958 Mar 15.46	1958 Aug 9.09
1959 Jan 14.41	1959 Aug 1.05	1960 Jan 31.40	1960 Jun 26.04
1960 Dec 1.40	1961 Jun 18.02	1961 Dec 18.38	1962 May 13.99
1962 Oct 19.35	1963 May 5.98	1963 Nov 5.34	1964 Mar 30.99
1964 Sep 5.32	1965 Mar 22.97	1965 Sep 22.31	1966 Feb 15.93
1966 Jul 24.29	1967 Feb 7.94	1967 Aug 10.32	1968 Jan 3.91
1968 Jun 10.24	1968 Dec 25.89	1969 Jun 27.29	1969 Nov 20.92
1970 Apr 28.21	1970 Nov 12.86	1971 May 15.25	1971 Oct 8.89
1972 Mar 15.21	1972 Sep 29.83	1973 Apr 1.24	1973 Aug 25.86
1974 Jan 31.19	1974 Aug 17.80	1975 Feb 17.21	1975 Jul 13.86
1975 Dec 19.16	1976 Jul 4.78	1977 Jan 4.16	1977 May 30.81
1977 Nov 5.13	1978 May 22.77	1978 Nov 22.17	1979 Apr 17.77
1979 Sep 23.07	1980 Apr 8.72	1980 Oct 9.16	1981 Mar 4.78
1981 Aug 10.04	1982 Feb 24.66	1982 Aug 27.10	1983 Jan 20.76
1983 Jun 28.04	1984 Jan 12.63	1984 Jul 14.06	1984 Dec 7.70
1985 May 15.01	1985 Nov 29.61	1986 Jun 1.05	1986 Oct 25.69
1987 Apr 1.96	1987 Oct 17.59	1988 Apr 18.02	1988 Sep 11.68
1989 Feb 16.96	1989 Sep 3.57	1990 Mar 6.01	1990 Jul 30.63
1991 Jan 4.92	1991 Jul 22.53	1992 Jan 22.01	1992 Jun 16.63
1992 Nov 21.86	1993 Jun 8.48	1993 Dec 8.95	1994 May 4.62
1994 Oct 9.85	1995 Apr 26.44	1995 Oct 26.91	1996 Mar 21.57
1996 Aug 26.84	1997 Mar 13.41	1997 Sep 12.89	1998 Feb 6.54
1998 Jul 14.79	1999 Jan 29.38	1999 Jul 31.85	1999 Dec 25.53
2000 May 31.77	2000 Dec 16.37	2001 Jun 17.82	2001 Nov 11.47
2002 Apr 18.74	2002 Nov 3.35	2003 May 5.85	2003 Sep 29.47
2004 Mar 5.68	2004 Sep 20.30	2005 Mar 22.83	2005 Aug 16.49
2006 Jan 21.68	2006 Aug 8.26	2007 Feb 7.77	2007 Jul 4.46
2007 Dec 9.67	2008 Jun 25.23	2008 Dec 25.75	2009 May 21.40
2009 Oct 26.63	2010 May 13.20	2010 Nov 12.72	2011 Apr 8.40
2011 Sep 13.60	2012 Mar 30.18	2012 Sep 29.67	2013 Feb 23.35
2013 Jul 31.58	2014 Feb 15.16	2014 Aug 17.68	2015 Jan 11.31
2015 Jun 18.52	2016 Jan 3.11	2016 Jul 4.66	2016 Nov 28.32
2017 May 5.49	2017 Nov 20.07	2018 May 22.61	2018 Oct 16.30
2019 Mar 23.48	2019 Oct 8.04	2020 Apr 8.60	2020 Sep 2.27
2021 Feb 7.46	2021 Aug 25.01	2022 Feb 24.58	2022 Jul 21.26

Spring Equinox	*Summer Solstice*	*Autumn Equinox*	*Winter Solstice*
2022 Dec 26.43	2023 Jul 12.98	2024 Jan 12.52	2024 Jun 7.22
2024 Nov 12.40	2025 May 29.97	2025 Nov 29.51	2026 Apr 25.17
2026 Sep 30.35	2027 Apr 16.92	2027 Oct 17.51	2028 Mar 12.17
2028 Aug 17.30	2029 Mar 3.87	2029 Sep 3.46	2030 Jan 28.17
2030 Jul 5.30	2031 Jan 19.83	2031 Jul 22.41	2031 Dec 16.11
2032 May 22.28	2032 Dec 6.81	2033 Jun 8.41	2033 Nov 2.08
2034 Apr 9.23	2034 Oct 24.79	2035 Apr 26.38	2035 Sep 20.09
2036 Feb 25.22	2036 Sep 10.78	2037 Mar 13.36	2037 Aug 7.04
2038 Jan 12.20	2038 Jul 29.75	2039 Jan 29.37	2039 Jun 25.03
2039 Nov 30.14	2040 Jun 15.69	2040 Dec 16.33	2041 May 12.04
2041 Oct 17.13	2042 May 3.65	2042 Nov 3.28	2043 Mar 29.99
2043 Sep 4.12	2044 Mar 20.62	2044 Sep 20.26	2045 Feb 13.95
2045 Jul 22.08	2046 Feb 5.59	2046 Aug 8.22	2047 Jan 1.94
2047 Jun 9.05	2047 Dec 24.58	2048 Jun 25.18	2048 Nov 18.88
2049 Apr 26.02	2049 Nov 10.56	2050 May 13.20	2050 Oct 6.87
2051 Mar 13.96	2051 Sep 28.51	2052 Mar 30.20	2052 Aug 23.90
2053 Jan 28.95	2053 Aug 15.46	2054 Feb 15.13	2054 Jul 11.87
2054 Dec 16.94	2055 Jul 3.43	2056 Jan 3.10	2056 May 28.81
2056 Nov 2.91	2057 May 20.40	2057 Nov 20.08	2058 Apr 15.80
2058 Sep 20.86	2059 Apr 7.37	2059 Oct 8.02	2060 Mar 2.76

XVIII

Positions of Bright Zodiacal Stars

For 51 bright zodiacal stars, the ecliptical longitude and latitude are given from the year −2000 to +2800. These coordinates, referred to the ecliptic and mean equinox of the date, are expressed in degrees and decimals. In the calculation, the precession (including the rotation of the ecliptic) and the proper motions of the stars have been taken into account.

The tabular interval is 400 years. The positions are given for the beginning of the year mentioned in the column at left. Strictly speaking, the 'years' are measured as Julian years (of 365.25 days) from the epoch J2000.0 = 2000 January 1.5 = JDE 2451545.0. For example, the year +1600 in the table corresponds to Julian Ephemeris Day

$$2451545 - (400 \times 365.25) = 2305445.0$$

which corresponds to 1599 December 29.5, Gregorian calendar.

The year +1200 corresponds to Julian Ephemeris Day

$$2451545 - (800 \times 365.25) = 2159345.0$$

which corresponds to December 19.5 of the year 1199, Julian calendar.

The tabulated star positions can be linearly interpolated, and then rounded to the nearest hundredth of a degree. The celestial longitude of each star increases by approximately five-and-a-half degrees in four centuries, while the latitude varies very slowly. The latitudes are not exactly constant, due to the proper motion of the stars and to the slow rotation of the plane of the ecliptic.

From the table we see, for instance, that δ Cancri was exactly on the ecliptic near the year +80; presently, this star is 5′ to the north of the ecliptic. The star ρ Leonis was on the ecliptic near the year −720. On the other hand, the latitude of η Virginis reached a maximum value (+1°25′) about the year −1000, while that of β Virginis will be a maximum (+0°42′) about the year +2700.

YEAR	ETA PSC		ALPHA ARI		ETA TAU	
−2000	331.450	+5.191	342.213	+9.899	4.619	+3.662
−1600	336.942	+5.203	347.713	+9.900	10.112	+3.698
−1200	342.444	+5.216	353.222	+9.902	15.614	+3.734
− 800	347.955	+5.231	358.742	+9.905	21.126	+3.771
− 400	353.477	+5.247	4.271	+9.910	26.649	+3.809
0	359.009	+5.265	9.811	+9.916	32.181	+3.848
+ 400	4.550	+5.285	15.361	+9.923	37.723	+3.887
800	10.102	+5.306	20.921	+9.932	43.276	+3.927
1200	15.663	+5.329	26.492	+9.942	48.838	+3.968
1600	21.235	+5.352	32.072	+9.953	54.410	+4.009
2000	26.816	+5.378	37.662	+9.965	59.992	+4.051
2400	32.407	+5.404	43.263	+9.978	65.585	+4.093
2800	38.009	+5.432	48.874	+9.993	71.187	+4.136

YEAR	GAMMA TAU		EPSILON TAU		ALPHA TAU	
−2000	10.272	−6.154	12.958	−2.990	14.343	−5.737
−1600	15.781	−6.115	18.464	−2.950	19.844	−5.712
−1200	21.301	−6.074	23.981	−2.909	25.355	−5.687
− 800	26.830	−6.033	29.507	−2.868	30.875	−5.661
− 400	32.369	−5.992	35.043	−2.826	36.405	−5.634
0	37.918	−5.950	40.589	−2.784	41.945	−5.607
+ 400	43.476	−5.907	46.145	−2.741	47.494	−5.580
800	49.044	−5.864	51.710	−2.698	53.053	−5.552
1200	54.621	−5.821	57.285	−2.655	58.622	−5.524
1600	60.209	−5.777	62.870	−2.611	64.201	−5.496
2000	65.806	−5.732	68.465	−2.567	69.789	−5.467
2400	71.412	−5.688	74.070	−2.523	75.387	−5.439
2800	77.029	−5.643	79.684	−2.479	80.995	−5.410

YEAR	BETA TAU		ZETA TAU		ETA GEM	
−2000	27.187	+5.054	29.396	−2.703	38.116	−1.412
−1600	32.681	+5.087	34.891	−2.652	43.603	−1.358
−1200	38.184	+5.120	40.396	−2.601	49.101	−1.305
− 800	43.698	+5.154	45.910	−2.550	54.608	−1.251
− 400	49.222	+5.187	51.434	−2.499	60.126	−1.198
0	54.755	+5.220	56.968	−2.448	65.653	−1.146
+ 400	60.299	+5.253	62.512	−2.397	71.190	−1.093
800	65.853	+5.286	68.065	−2.346	76.737	−1.041
1200	71.417	+5.319	73.629	−2.296	82.293	−0.990
1600	76.991	+5.352	79.202	−2.246	87.860	−0.939
2000	82.575	+5.385	84.785	−2.196	93.436	−0.888
2400	88.169	+5.418	90.377	−2.146	99.023	−0.838
2800	93.773	+5.450	95.980	−2.096	104.619	−0.789

YEAR	MU GEM		GAMMA GEM		EPSILON GEM	
−2000	39.852	−1.237	43.675	−7.236	44.557	+1.550
−1600	45.353	−1.194	49.174	−7.184	50.051	+1.604
−1200	50.863	−1.151	54.683	−7.133	55.554	+1.658
− 800	56.384	−1.108	60.202	−7.082	61.067	+1.711
− 400	61.914	−1.065	65.731	−7.032	66.591	+1.764
0	67.454	−1.023	71.269	−6.982	72.124	+1.817
+ 400	73.004	−0.982	76.817	−6.933	77.667	+1.869
800	78.563	−0.941	82.374	−6.885	83.220	+1.920
1200	84.133	−0.900	87.941	−6.837	88.783	+1.971
1600	89.713	−0.860	93.518	−6.789	94.356	+2.021
2000	95.302	−0.820	99.105	−6.743	99.939	+2.070
2400	100.901	−0.781	104.701	−6.697	105.532	+2.118
2800	106.510	−0.743	110.307	−6.651	111.134	+2.166

YEAR	DELTA	GEM	LAMBDA	GEM	ALPHA	GEM
−2000	53.162	−0.683	53.455	−6.109	55.003	+9.718
−1600	58.653	−0.629	58.943	−6.058	60.481	+9.759
−1200	64.155	−0.576	64.442	−6.008	65.969	+9.800
− 800	69.666	−0.523	69.950	−5.959	71.467	+9.840
− 400	75.186	−0.471	75.468	−5.910	76.976	+9.879
0	80.717	−0.420	80.995	−5.862	82.495	+9.918
+ 400	86.258	−0.370	86.532	−5.815	88.023	+9.955
800	91.808	−0.321	92.079	−5.769	93.563	+9.992
1200	97.369	−0.273	97.636	−5.723	99.112	+10.027
1600	102.939	−0.225	103.203	−5.679	104.671	+10.062
2000	108.519	−0.178	108.779	−5.635	110.241	+10.096
2400	114.109	−0.133	114.365	−5.593	115.820	+10.129
2800	119.709	−0.088	119.960	−5.551	121.410	+10.161

YEAR	BETA	GEM	KAPPA	GEM	DELTA	CNC
−2000	58.491	+6.352	58.298	+2.634	73.289	−0.118
−1600	63.919	+6.389	63.790	+2.683	78.788	−0.093
−1200	69.356	+6.426	69.292	+2.730	84.297	−0.069
− 800	74.803	+6.461	74.804	+2.777	89.815	−0.046
− 400	80.261	+6.496	80.326	+2.823	95.344	−0.024
0	85.728	+6.529	85.857	+2.868	100.882	−0.004
+ 400	91.205	+6.562	91.399	+2.912	106.430	+0.015
800	96.693	+6.594	96.951	+2.955	111.988	+0.033
1200	102.190	+6.625	102.512	+2.998	117.556	+0.049
1600	107.698	+6.655	108.084	+3.039	123.134	+0.064
2000	113.216	+6.684	113.666	+3.079	128.722	+0.077
2400	118.743	+6.712	119.257	+3.117	134.320	+0.089
2800	124.281	+6.739	124.859	+3.155	139.927	+0.100

YEAR	OMIKRON	LEO	ETA	LEO	GAMMA	LEO
−2000	89.026	−4.014	92.486	+4.545	93.792	+8.544
−1600	94.504	−3.981	97.983	+4.585	99.329	+8.579
−1200	99.991	−3.950	103.490	+4.623	104.876	+8.612
− 800	105.489	−3.920	109.007	+4.659	110.433	+8.643
− 400	110.996	−3.892	114.534	+4.694	116.001	+8.673
0	116.513	−3.866	120.071	+4.727	121.578	+8.701
+ 400	122.040	−3.841	125.618	+4.758	127.166	+8.727
800	127.577	−3.817	131.175	+4.787	132.763	+8.752
1200	133.124	−3.796	136.742	+4.815	138.371	+8.775
1600	138.681	−3.776	142.318	+4.841	143.988	+8.796
2000	144.247	−3.757	147.905	+4.866	149.615	+8.815
2400	149.823	−3.740	153.502	+4.888	155.253	+8.832
2800	155.409	−3.725	159.108	+4.909	160.900	+8.848

YEAR	ALPHA LEO		RHO LEO		BETA VIR	
−2000	94.702	+0.251	101.009	−0.099	120.897	+0.565
−1600	100.170	+0.280	106.503	−0.066	126.479	+0.587
−1200	105.648	+0.307	112.006	−0.035	132.071	+0.607
− 800	111.136	+0.332	117.519	−0.006	137.673	+0.625
− 400	116.634	+0.356	123.042	+0.022	143.285	+0.641
0	122.142	+0.379	128.575	+0.048	148.907	+0.655
+ 400	127.660	+0.399	134.118	+0.072	154.538	+0.666
800	133.187	+0.418	139.671	+0.094	160.180	+0.676
1200	138.725	+0.435	145.234	+0.114	165.832	+0.684
1600	144.272	+0.451	150.806	+0.133	171.493	+0.690
2000	149.829	+0.465	156.389	+0.150	177.164	+0.694
2400	155.396	+0.477	161.981	+0.165	182.845	+0.696
2800	160.973	+0.488	167.584	+0.178	188.536	+0.697

YEAR	ETA VIR		GAMMA VIR		ALPHA VIR	
−2000	129.490	+1.414	135.316	+3.085	148.504	−1.830
−1600	134.980	+1.418	140.754	+3.063	153.993	−1.846
−1200	140.480	+1.419	146.202	+3.040	159.492	−1.863
− 800	145.989	+1.419	151.660	+3.014	165.001	−1.881
− 400	151.509	+1.416	157.128	+2.987	170.520	−1.901
0	157.038	+1.412	162.606	+2.958	176.049	−1.923
+ 400	162.577	+1.406	168.093	+2.928	181.588	−1.946
800	168.126	+1.399	173.591	+2.896	187.136	−1.971
1200	173.685	+1.389	179.098	+2.862	192.695	−1.998
1600	179.254	+1.378	184.615	+2.827	198.263	−2.025
2000	184.832	+1.365	190.141	+2.790	203.841	−2.054
2400	190.421	+1.351	195.678	+2.752	209.430	−2.085
2800	196.019	+1.335	201.224	+2.712	215.028	−2.117

YEAR	ALPHA LIB		DELTA SCO		PI SCO	
−2000	169.784	+0.789	187.203	−1.503	187.587	−4.987
−1600	175.270	+0.748	192.695	−1.548	193.078	−5.033
−1200	180.765	+0.707	198.197	−1.594	198.578	−5.079
− 800	186.271	+0.664	203.709	−1.641	204.088	−5.127
− 400	191.786	+0.620	209.231	−1.689	209.608	−5.175
0	197.311	+0.574	214.763	−1.737	215.138	−5.224
+ 400	202.845	+0.528	220.305	−1.786	220.679	−5.273
800	208.390	+0.481	225.857	−1.835	226.229	−5.323
1200	213.944	+0.433	231.418	−1.885	231.789	−5.373
1600	219.509	+0.383	236.990	−1.935	237.360	−5.424
2000	225.083	+0.333	242.571	−1.986	242.940	−5.475
2400	230.667	+0.282	248.163	−2.037	248.530	−5.527
2800	236.260	+0.230	253.764	−2.089	254.131	−5.579

YEAR	BETA SCO		SIGMA SCO		ALPHA SCO	
−2000	187.801	+1.489	192.438	−3.532	194.401	−4.058
−1600	193.296	+1.444	197.929	−3.581	199.892	−4.107
−1200	198.800	+1.398	203.431	−3.629	205.394	−4.157
− 800	204.315	+1.351	208.942	−3.679	210.905	−4.207
− 400	209.839	+1.304	214.463	−3.729	216.426	−4.258
0	215.373	+1.256	219.994	−3.779	221.957	−4.309
+ 400	220.917	+1.207	225.535	−3.830	227.498	−4.361
800	226.470	+1.158	231.086	−3.881	233.049	−4.412
1200	232.034	+1.108	236.647	−3.933	238.610	−4.465
1600	237.607	+1.058	242.219	−3.985	244.181	−4.517
2000	243.190	+1.008	247.800	−4.037	249.762	−4.570
2400	248.783	+0.957	253.391	−4.090	255.353	−4.623
2800	254.386	+0.906	258.992	−4.143	260.954	−4.676

YEAR	TAU SCO		ETA OPH		THETA OPH	
−2000	196.099	−5.600	202.521	+7.638	206.016	−1.297
−1600	201.590	−5.650	208.022	+7.595	211.509	−1.352
−1200	207.091	−5.701	213.534	+7.551	217.012	−1.406
− 800	212.601	−5.752	219.054	+7.508	222.525	−1.461
− 400	218.122	−5.804	224.585	+7.464	228.048	−1.515
0	223.653	−5.856	230.125	+7.419	233.581	−1.570
+ 400	229.193	−5.908	235.675	+7.375	239.124	−1.625
800	234.744	−5.961	241.234	+7.331	244.677	−1.680
1200	240.305	−6.014	246.803	+7.286	250.240	−1.734
1600	245.876	−6.067	252.381	+7.242	255.812	−1.789
2000	251.457	−6.120	257.970	+7.198	261.395	−1.843
2400	257.048	−6.174	263.567	+7.153	266.987	−1.898
2800	262.649	−6.227	269.175	+7.109	272.590	−1.952

YEAR	GAMMA SGR		DELTA SGR		LAMBDA SGR	
−2000	215.942	−6.249	219.158	−5.902	220.991	−1.395
−1600	221.428	−6.324	224.655	−5.960	226.478	−1.471
−1200	226.925	−6.399	230.162	−6.019	231.976	−1.546
− 800	232.432	−6.474	235.679	−6.077	237.484	−1.622
− 400	237.949	−6.549	241.206	−6.135	243.002	−1.696
0	243.476	−6.624	246.743	−6.192	248.530	−1.771
+ 400	249.013	−6.698	252.291	−6.249	254.067	−1.845
800	254.560	−6.772	257.848	−6.306	259.615	−1.918
1200	260.117	−6.845	263.416	−6.362	265.172	−1.991
1600	265.684	−6.918	268.993	−6.417	270.740	−2.064
2000	271.261	−6.991	274.581	−6.472	276.317	−2.136
2400	276.849	−7.063	280.179	−6.527	281.904	−2.207
2800	282.446	−7.135	285.786	−6.581	287.502	−2.277

YEAR	PHI SGR		SIGMA SGR		KSI SGR	
−2000	224.734	−3.414	226.987	−2.855	228.034	+2.209
−1600	230.234	−3.471	232.482	−2.917	233.531	+2.152
−1200	235.743	−3.527	237.987	−2.979	239.039	+2.095
− 800	241.263	−3.582	243.502	−3.040	244.556	+2.038
− 400	246.793	−3.637	249.027	−3.100	250.083	+1.982
0	252.333	−3.691	254.562	−3.160	255.620	+1.927
+ 400	257.882	−3.745	260.106	−3.220	261.166	+1.872
800	263.442	−3.798	265.661	−3.278	266.723	+1.818
1200	269.012	−3.851	271.226	−3.336	272.289	+1.765
1600	274.592	−3.903	276.801	−3.393	277.865	+1.713
2000	280.181	−3.954	282.385	−3.450	283.451	+1.661
2400	285.781	−4.004	287.980	−3.505	289.047	+1.611
2800	291.391	−4.054	293.585	−3.560	294.653	+1.561

YEAR	ZETA SGR		TAU SGR		PI SGR	
−2000	228.264	−6.647	229.533	−4.290	230.875	+2.003
−1600	233.756	−6.703	235.018	−4.373	236.368	+1.943
−1200	239.258	−6.758	240.513	−4.455	241.872	+1.884
− 800	244.770	−6.813	246.018	−4.537	247.385	+1.825
− 400	250.292	−6.868	251.533	−4.618	252.908	+1.768
0	255.825	−6.921	257.058	−4.698	258.440	+1.711
+ 400	261.367	−6.974	262.593	−4.778	263.983	+1.654
800	266.920	−7.027	268.138	−4.857	269.535	+1.599
1200	272.483	−7.078	273.694	−4.935	275.098	+1.544
1600	278.055	−7.129	279.259	−5.012	280.670	+1.490
2000	283.638	−7.179	284.834	−5.089	286.252	+1.437
2400	289.231	−7.228	290.420	−5.164	291.844	+1.385
2800	294.834	−7.276	296.015	−5.239	297.445	+1.334

YEAR	ALPHA 2 CAP		BETA CAP		THETA CAP	
−2000	248.436	+7.412	248.639	+5.066	258.388	−0.079
−1600	253.935	+7.358	254.135	+5.013	263.889	−0.136
−1200	259.443	+7.306	259.642	+4.961	269.400	−0.191
− 800	264.961	+7.254	265.158	+4.910	274.921	−0.246
− 400	270.488	+7.204	270.685	+4.860	280.452	−0.299
0	276.026	+7.155	276.221	+4.812	285.992	−0.350
+ 400	281.573	+7.108	281.766	+4.765	291.543	−0.400
800	287.130	+7.062	287.322	+4.719	297.103	−0.449
1200	292.696	+7.016	292.887	+4.674	302.673	−0.496
1600	298.273	+6.973	298.462	+4.631	308.254	−0.542
2000	303.859	+6.930	304.047	+4.589	313.844	−0.586
2400	309.454	+6.889	309.642	+4.548	319.444	−0.628
2800	315.060	+6.849	315.247	+4.509	325.054	−0.670

YEAR	GAMMA CAP		DELTA CAP		LAMBDA AQR	
−2000	266.199	−2.097	267.977	−1.839	286.161	−0.210
−1600	271.713	−2.150	273.488	−1.923	291.658	−0.235
−1200	277.238	−2.202	279.010	−2.005	297.165	−0.259
− 800	282.772	−2.252	284.542	−2.085	302.682	−0.282
− 400	288.316	−2.300	290.083	−2.164	308.208	−0.302
0	293.870	−2.347	295.635	−2.241	313.745	−0.321
+ 400	299.435	−2.393	301.196	−2.316	319.291	−0.337
800	305.009	−2.436	306.768	−2.390	324.848	−0.352
1200	310.593	−2.478	312.349	−2.462	330.414	−0.366
1600	316.187	−2.519	317.941	−2.533	335.990	−0.377
2000	321.791	−2.557	323.543	−2.602	341.576	−0.387
2400	327.405	−2.594	329.154	−2.669	347.172	−0.394
2800	333.029	−2.630	334.776	−2.734	352.778	−0.400

Alternatively, the celestial (ecliptical) longitudes λ and latitudes β of the 51 zodiacal stars can be calculated, for any epoch between the years −2000 and +2800, by means of the coefficients A to F listed below. The formulae are

$$\lambda = A + BT + CT^2$$
$$\beta = D + ET + FT^2$$

where T is measured in Julian *millennia* from 2000.0, that is, $T = $ (year − 2000)/1000. The results are expressed in degrees and decimals. The values A, B, and C are always positive.

This method should be used only for the years −2000 to +2800. Large values of T would give meaningless results. For the year +150000, for instance, that is, $T = +148$, we would obtain for the latitude of Regulus (α Leonis) the absurd value $\beta = −109°$.

Star	A	B	C	D	E	F
η Psc	26.816	13.9660	0.03115	+5.378	+0.0647	+0.00419
α Ari	37.662	13.9889	0.03168	+9.965	+0.0318	+0.00347
η Tau	59.992	13.9681	0.03120	+4.051	+0.1047	+0.00149
γ Tau	65.806	14.0046	0.03027	−5.732	+0.1110	+0.00096
ε Tau	68.465	13.9992	0.03059	−2.567	+0.1102	+0.00070
α Tau	69.789	13.9828	0.03033	−5.467	+0.0713	+0.00056
β Tau	82.575	13.9726	0.03139	+5.385	+0.0814	−0.00069
ζ Tau	84.785	13.9696	0.03065	−2.196	+0.1246	−0.00091
η Gem	93.436	13.9533	0.03080	−0.888	+0.1255	−0.00172
μ Gem	95.302	13.9856	0.03081	−0.820	+0.0981	−0.00190

Star	A	B	C	D	E	F
γ Gem	99.105	13.9785	0.03029	−6.743	+0.1157	−0.00225
ε Gem	99.939	13.9697	0.03108	+2.070	+0.1222	−0.00232
δ Gem	108.519	13.9628	0.03090	−0.178	+0.1152	−0.00304
λ Gem	108.779	13.9527	0.03045	−5.635	+0.1075	−0.00305
α Gem	110.241	13.9364	0.03168	+10.096	+0.0831	−0.00314
β Gem	113.216	13.8065	0.03133	+6.684	+0.0713	−0.00313
κ Gem	113.666	13.9665	0.03114	+3.079	+0.0984	−0.00345
δ Cnc	128.722	13.9816	0.03084	+0.077	+0.0317	−0.00449
o Leo	144.247	13.9283	0.03075	−3.757	+0.0439	−0.00513
η Leo	147.905	13.9792	0.03107	+4.866	+0.0589	−0.00533
γ Leo	149.615	14.0810	0.03116	+8.815	+0.0458	−0.00556
α Leo	149.829	13.9054	0.03088	+0.465	+0.0327	−0.00523
ρ Leo	156.389	13.9686	0.03091	+0.150	+0.0400	−0.00551
β Vir	177.164	14.1906	0.03088	+0.694	+0.0077	−0.00596
η Vir	184.832	13.9587	0.03077	+1.365	−0.0341	−0.00524
γ Vir	190.141	13.8289	0.03060	+2.790	−0.0938	−0.00477
α Vir	203.841	13.9580	0.03090	−2.054	−0.0747	−0.00435
α Lib	225.083	13.9475	0.03073	+0.333	−0.1266	−0.00279
δ Sco	242.571	13.9660	0.03100	−1.986	−0.1274	−0.00124
π Sco	242.940	13.9634	0.03133	−5.475	−0.1286	−0.00120
β Sco	243.190	13.9701	0.03071	+1.008	−0.1267	−0.00118
σ Sco	247.800	13.9651	0.03121	−4.037	−0.1308	−0.00074
α Sco	249.762	13.9652	0.03127	−4.570	−0.1319	−0.00055
τ Sco	251.457	13.9650	0.03142	−6.120	−0.1333	−0.00038
η Oph	257.970	13.9826	0.03015	+7.198	−0.1107	+0.00024
θ Oph	261.395	13.9688	0.03103	−1.843	−0.1358	+0.00058
γ Sgr	271.261	13.9561	0.03155	−6.991	−0.1810	+0.00152
δ Sgr	274.581	13.9818	0.03149	−6.472	−0.1367	+0.00183
λ Sgr	276.317	13.9560	0.03111	−2.136	−0.1787	+0.00198
φ Sgr	280.181	13.9869	0.03126	−3.954	−0.1269	+0.00235
σ Sgr	282.385	13.9743	0.03122	−3.450	−0.1398	+0.00253
ξ Sgr	283.451	13.9773	0.03077	+1.661	−0.1278	+0.00263
ζ Sgr	283.638	13.9698	0.03153	−7.179	−0.1238	+0.00264
τ Sgr	284.834	13.9509	0.03140	−5.089	−0.1901	+0.00272
π Sgr	286.252	13.9673	0.03080	+1.437	−0.1313	+0.00286
α₂ Cap	303.859	13.9774	0.03049	+6.930	−0.1044	+0.00419
β Cap	304.047	13.9746	0.03064	+4.589	−0.1034	+0.00420
θ Cap	313.844	13.9879	0.03100	−0.586	−0.1083	+0.00478
γ Cap	321.791	14.0224	0.03108	−2.557	−0.0946	+0.00520
δ Cap	323.543	14.0163	0.03120	−2.602	−0.1699	+0.00526
λ Aqr	341.576	13.9774	0.03088	−0.387	−0.0218	+0.00558

XIX
CONJUNCTIONS OF THE SUN WITH BRIGHT STARS

The table on the next two pages gives the UT times of conjunction in celestial (ecliptical) *longitude* of the Sun with ten bright zodiacal stars, to the nearest tenth of a day.

These times are given for century years only. For other years, add the correction (in days) given in the table at the right; this correction is always positive. The error in the final result will not exceed 0.1 day.

For example, calculate the time of the conjunction of the Sun with α Tauri in 1975. We find :

1900	May 30.3
75	1.2
S u m	May 31.5

Numbers characterizing negative years should be split up in such a way as to have the last two figures positive. For example, the year −328 should be split up as −400 + 72 ; then, take −400 from the main table, and add the correction for +72 years.

For each star, two values are given for the year 1500, the first one for the Julian calendar (*J*), and the second one for the Gregorian calendar (*G*).

For σ Sagittarii, the last five dates are given as later than December 31, in order to allow the use of the correction table at the right. Remember that, for example, December 33 is the same as January 2 of the following year.

Correction for additional years

Year	Corr.	Year	Corr.	Year	Corr.
	d		d		d
0	0.0	35	1.0	70	0.9
1	0.3	36	0.2	71	1.2
2	0.5	37	0.5	72	0.5
3	0.8	38	0.7	73	0.7
4	0.0	39	1.0	74	1.0
5	0.3	40	0.3	75	1.2
6	0.5	41	0.5	76	0.5
7	0.8	42	0.8	77	0.7
8	0.1	43	1.0	78	1.0
9	0.3	44	0.3	79	1.3
10	0.6	45	0.5	80	0.5
11	0.8	46	0.8	81	0.8
12	0.1	47	1.0	82	1.0
13	0.3	48	0.3	83	1.3
14	0.6	49	0.6	84	0.5
15	0.8	50	0.8	85	0.8
16	0.1	51	1.1	86	1.0
17	0.4	52	0.3	87	1.3
18	0.6	53	0.6	88	0.6
19	0.9	54	0.8	89	0.8
20	0.1	55	1.1	90	1.1
21	0.4	56	0.4	91	1.3
22	0.6	57	0.6	92	0.6
23	0.9	58	0.9	93	0.8
24	0.2	59	1.1	94	1.1
25	0.4	60	0.4	95	1.4
26	0.7	61	0.6	96	0.6
27	0.9	62	0.9	97	0.9
28	0.2	63	1.2	98	1.1
29	0.4	64	0.4	99	1.4
30	0.7	65	0.7		
31	0.9	66	0.9		
32	0.2	67	1.2		
33	0.5	68	0.4		
34	0.7	69	0.7		

Year	α Ari Hamal	η Tau Alcyone	α Tau Aldebaran	β Tau Elnath	β Gem Pollux
− 1500	Mar 23.0	Apr 15.3	Apr 25.5	May 9.0	Jun 10.8
− 1400	23.6	15.9	26.1	9.6	11.4
− 1300	24.3	16.6	26.8	10.3	12.0
− 1200	24.9	17.2	27.4	10.9	12.7
− 1100	25.6	17.9	28.1	11.5	13.3
− 1000	Mar 26.2	Apr 18.5	Apr 28.7	May 12.2	Jun 13.9
− 900	26.9	19.2	29.4	12.8	14.5
− 800	27.6	19.8	30.0	13.5	15.2
− 700	28.2	20.5	30.7	14.1	15.8
− 600	28.9	21.1	May 1.3	14.8	16.4
− 500	Mar 29.5	Apr 21.8	May 2.0	May 15.4	Jun 17.0
− 400	30.2	22.4	2.6	16.1	17.7
− 300	30.8	23.1	3.3	16.7	18.3
− 200	31.5	23.7	3.9	17.3	18.9
− 100	Apr 1.1	24.4	4.6	18.0	19.5
0	Apr 1.8	Apr 25.0	May 5.2	May 18.6	Jun 20.2
+ 100	2.4	25.6	5.8	19.3	20.8
200	3.1	26.3	6.5	19.9	21.4
300	3.7	26.9	7.1	20.5	22.0
400	4.4	27.6	7.8	21.2	22.6
500	Apr 5.0	Apr 28.2	May 8.4	May 21.8	Jun 23.2
600	5.7	28.8	9.0	22.5	23.9
700	6.3	29.5	9.7	23.1	24.5
800	6.9	30.1	10.3	23.7	25.1
900	7.6	30.8	11.0	24.4	25.7
1000	Apr 8.2	May 1.4	May 11.6	May 25.0	Jun 26.3
1100	8.9	2.0	12.2	25.6	26.9
1200	9.5	2.7	12.9	26.3	27.5
1300	10.2	3.3	13.5	26.9	28.2
1400	10.8	3.9	14.1	27.5	28.8
1500 J	Apr 11.4	May 4.6	May 14.8	May 28.1	Jun 29.4
1500 G	21.4	14.6	24.8	Jun 7.1	Jul 9.4
1600	22.1	15.2	25.4	7.8	10.0
1700	23.7	16.8	27.0	9.4	11.6
1800	25.4	18.5	28.7	11.0	13.2
1900	Apr 27.0	May 20.1	May 30.3	Jun 12.7	Jul 14.8
2000	27.6	20.7	30.9	13.3	15.4
2100	29.3	22.4	Jun 1.6	14.9	17.0
2200	30.9	24.0	3.2	16.5	18.6
2300	May 2.5	25.6	4.8	18.2	20.2

Year	α Leo Regulus	α Vir Spica	α Lib Zubenelgenubi	α Sco Antares	σ Sgr Nunki
−1500	Jul 18.6	Sep 11.3	Oct 2.5	Oct 26.7	Nov 27.6
−1400	19.2	12.0	3.1	27.4	28.3
−1300	19.8	12.6	3.8	28.1	29.0
−1200	20.5	13.3	4.5	28.7	29.7
−1100	21.1	13.9	5.1	29.4	30.3
−1000	Jul 21.8	Sep 14.6	Oct 5.8	Oct 30.1	Dec 1.0
− 900	22.4	15.3	6.4	30.7	1.7
− 800	23.1	15.9	7.1	31.4	2.3
− 700	23.7	16.6	7.8	Nov 1.1	3.0
− 600	24.3	17.2	8.4	1.7	3.7
− 500	Jul 25.0	Sep 17.9	Oct 9.1	Nov 2.4	Dec 4.3
− 400	25.6	18.5	9.7	3.0	5.0
− 300	26.2	19.2	10.4	3.7	5.7
− 200	26.9	19.8	11.0	4.4	6.3
− 100	27.5	20.5	11.7	5.0	7.0
0	Jul 28.1	Sep 21.1	Oct 12.3	Nov 5.7	Dec 7.6
+ 100	28.8	21.8	13.0	6.3	8.3
200	29.4	22.4	13.6	7.0	9.0
300	30.0	23.1	14.3	7.6	9.6
400	30.7	23.7	14.9	8.3	10.3
500	Jul 31.3	Sep 24.4	Oct 15.6	Nov 8.9	Dec 10.9
600	31.9	25.0	16.2	9.6	11.6
700	Aug 1.6	25.6	16.9	10.3	12.3
800	2.2	26.3	17.5	10.9	12.9
900	2.8	26.9	18.2	11.6	13.6
1000	Aug 3.4	Sep 27.6	Oct 18.8	Nov 12.2	Dec 14.2
1100	4.1	28.2	19.5	12.9	14.9
1200	4.7	28.8	20.1	13.5	15.5
1300	5.3	29.5	20.7	14.1	16.2
1400	5.9	30.1	21.4	14.8	16.8
1500 J	Aug 6.5	Sep 30.7	Oct 22.0	Nov 15.4	Dec 17.5
1500 G	16.5	Oct 10.7	Nov 1.0	25.4	27.5
1600	17.2	11.4	1.7	26.1	28.1
1700	18.8	13.0	3.3	27.7	29.8
1800	20.4	14.6	4.9	29.4	31.4
1900	Aug 22.0	Oct 16.3	Nov 6.6	Dec 1.0	Dec 33.1
2000	22.6	16.9	7.2	1.7	33.7
2100	24.3	18.5	8.8	3.3	35.4
2200	25.9	20.2	10.5	4.9	37.0
2300	27.5	21.8	12.1	6.6	38.7

Bibliography

General

Explanatory Supplement to the Astronomical Ephemeris; H.M. Nautical Almanac Office, London (1961).

Seidelmann, P.K., ed., *Explanatory Supplement to the Astronomical Almanac*; University Science Books, Mill Valley, CA (1992).

Meeus, J., *Astronomical Algorithms*; Willmann-Bell, Inc., Richmond, Virginia (1991).

Sun, Moon, Planets

Chapront-Touzé, M., and Chapront, J., *Lunar Tables and Programs from 4000 B.C. to A.D. 8000*; Willmann-Bell, Inc., Richmond, Virginia (1991).
>   Tables for the calculation of accurate geocentric positions and osculating orbital elements of the Moon.

Bretagnon, P., and Simon, J.-L., *Planetary Programs and Tables from −4000 to +2800*; Willman-Bell, Inc., Richmond, Virginia (1986).
>   Tables for the calculation of geocentric positions of the Sun and the planets Mercury to Neptune with an accuracy of 0.01 degree or better.

Goldstine, H.H., *New and Full Moons, 1001 B.C. to A.D. 1651*; Mem. American Phil. Soc., Vol. 94; Philadelphia, PA (1973).

Tuckerman, B., *Planetary, Lunar, and Solar Positions, 601 B.C. to A.D. 1*; Mem. American Phil. Soc., Vol. 56; Philadelphia, PA (1962).

Tuckerman, B., *Planetary, Lunar, and Solar Positions, A.D. 2 to A.D. 1649*; Mem. American Phil. Soc., Vol. 59; Philadelphia, PA (1964).

Stahlman, W.D., and Gingerich, O., *Solar and Planetary Longitudes for Years −2500 to +2000*; University of Wisconsin Press, Madison, Wisconsin (1963).
>   Gives the celestial longitudes (to the nearest degree) of the Sun and the five naked-eye planets at intervals of 10 days.

Minor Planets, Comets

Ephemerides of Minor Planets, issued annually by the Institute of Theoretical Astronomy, Academy of Sciences, Leningrad / St. Petersburg.
>   Contains the orbital elements and ephemerides of all numbered minor planets.

Marsden, B.G., *Catalogue of Cometary Orbits 1994*; ninth edition. Minor Planet Center, 60 Garden Street, Cambridge, MA.

Eclipses, Transits

Meeus, J., and Mucke, H., *Canon of Lunar Eclipses, −2002 to +2526*; Astronomisches Büro, Wien (1979; 2nd ed. 1983).

Mucke, H., and Meeus, J., *Canon of Solar Eclipses, −2003 to +2526*; Astronomisches Büro, Wien (1983).

Liu, B.-L, and Fiala, A.D., *Canon of Lunar Eclipses, 1500 B.C. — A.D. 3000*; Willmann-Bell, Inc., Richmond, Virginia (1992).

Meeus, J., *Elements of Solar Eclipses 1951 − 2200*; Willmann-Bell, Inc., Richmond, Virginia (1989).

> Gives accurate Besselian elements for all solar eclipses from A.D. 1951 to 2200, with formulae for their use and numerical examples.

Meeus, J., *Transits*; Willmann-Bell, Inc., Richmond, Virginia (1989).

> Gives accurate data and elements for all transits of Mercury over the Sun's disk from A.D. 1600 to 2300. and for all transits of Venus from 2000 B.C. to A.D. 4000.

Espenak, F., *Fifty Year Canon of Solar Eclipses : 1986 − 2035* ; NASA Reference Publication 1178; Washington, D.C. (1987).

Espenak, F., *Fifty Year Canon of Lunar Eclipses : 1986 − 2035* ; NASA Reference Publication 1216; Washington, D.C. (1989).

Stars

Smithsonian Astrophysical Observatory: Star Catalog; Smithsonian Institution, Washington, D.C. (1966).

> This set of 4 volumes contains the positions and proper motions of 258997 stars for the epoch and equinox of 1950.0.

Fifth Fundamental Catalogue (FK5). Part I: the basic fundamental stars; Veröffentlichungen, Astronomisches Rechen-Institut Heidelberg, No. 32 (1988). *Part II: the FK5 extension — new fundamental stars*; ibid., No. 33 (1991).

> These publications contain accurate positions and proper motions of 1535 and 3117 stars, respectively, for the equinox and epoch J2000.0 and B1950.0.

Astronomisches Rechen-Institut Heidelberg, *PPM Star Catalogue*; published by Spektrum Akademischer Verlag, Heidelberg (1991 and 1993).

> This set of 4 volumes contains the positions and proper motions of 181731 + 197179 stars for the equinox and epoch J2000.0. Compared to the SAO Star Catalog, the mean star density is higher and the positions and proper motions are more accurate.

Robertson, J., *Catalog of 3539 Zodiacal Stars*; Astronomical Papers, Vol. X, Part II, Washington, D.C. (1940).

> Also known as the 'Zodiacal Catalog' (ZC), this is the reference work for occultations of stars by the Moon. This catalogue is now outdated, however, as much more accurate star positions can be found in the PPM.

Hoffleit, D., *Bright Star Catalogue*; Yale University Observatory; fourth revised edition; New Haven, Connecticut (1982).

Hirshfeld, A., and Sinnott, R.W., *Sky Catalogue 2000.0*; *Vol. 1: Stars to Magnitude 8.0* (1982); *Vol. 2: Double Stars, Variable Stars and Nonstellar Objects* (1985); Sky Publishing Corporation, Cambridge, MA.

Wepner, W., *291 Doppelstern-Ephemeriden für die Jahre 1975 – 2000*; Treugesell Verlag, Düsseldorf (1976).

Annual Publications

Nautical Almanac Office, U.S. Naval Observatory, *The Astronomical Almanac for the Year 19--*. Published jointly with the Royal Greenwich Observatory. Government Printing Office, Washington, D.C. and Her Majesty's Stationery Office, London.

Nautical Almanac Office, U.S. Naval Observatory, *Astronomical Phenomena for the Year 19--*; Government Printing Office, Washington, D.C.
> Pamphlet which is, in part, a reprint of selected pages from The Astronomical Almanac.

Royal Astronomical Society of Canada, *Observer's Handbook 19--*; Toronto, Ontario, Canada.
> This annual volume is intended for the serious observer who wants the yearly observing opportunities readily at hand. While the emphasis is primarily on specific events occurring during the year, much other information is included.

British Astronomical Association, *Handbook 19--*; London, England.
> Similar to Observer's Handbook above, but more easily obtained in Europe.

Ottewell, G., *Astronomical Calendar 19--*; Astronomical Workshop, Furman University, Greenville, SC.
> Large format, profusely illustrated with sketches. Appeals to both the beginner and experienced observer.

Victor, R.C., and Pon, J.L., *Sky Calendar*; Abrams Planetarium, Michigan State University, East Lansing, MI 48824.
> Each month has his own page divided into daily boxes showing horizon at dawn and dusk. Positions of the Moon, visible planets and bright stars are shown. Available for $7.50 per year.

Index

The references to tables are to their first page only